Lecture Notes in Computer Science   11282

Commenced Publication in 1973
Founding and Former Series Editors:
Gerhard Goos, Juris Hartmanis, and Jan van Leeuwen

More information about this series at http://www.springer.com/series/7407

Therese Biedl · Andreas Kerren (Eds.)

Graph Drawing and Network Visualization

26th International Symposium, GD 2018
Barcelona, Spain, September 26–28, 2018
Proceedings

Editors
Therese Biedl ⓘD
University of Waterloo
Waterloo, ON, Canada

Andreas Kerren ⓘD
Linnaeus University
Växjö, Sweden

ISSN 0302-9743   ISSN 1611-3349   (electronic)
Lecture Notes in Computer Science
ISBN 978-3-030-04413-8   ISBN 978-3-030-04414-5   (eBook)
https://doi.org/10.1007/978-3-030-04414-5

Library of Congress Control Number: 2018961593

LNCS Sublibrary: SL1 – Theoretical Computer Science and General Issues

This Springer imprint is published by the registered company Springer Nature Switzerland AG
The registered company address is: Gewerbestrasse 11, 6330 Cham, Switzerland

Preface

This volume contains the papers presented at GD 2018: 26th International Symposium on Graph Drawing and Network Visualization held during September 26–28, 2018, in Barcelona. Graph drawing is concerned with the geometric representation of graphs and constitutes the algorithmic core of network visualization. Graph drawing and network visualization are motivated by applications where it is crucial to visually analyze and interact with relational datasets. Information about the conference series and past symposia is maintained at http://www.graphdrawing.org. The 2018 edition of the conference was hosted by Universitat Politècnica de Catalunya, with Vera Sacristán and Rodrigo Silveira as co-chairs of the Organizing Committee. A total of 99 participants attended the conference.

Regular papers could be submitted to one of two distinct tracks: Track 1 for papers on combinatorial and algorithmic aspects of graph drawing and Track 2 for papers on experimental, applied, and network visualization aspects. Short papers were given a separate category, which welcomed both theoretical and applied contributions. An additional track was devoted to poster submissions. All the tracks were handled by a single Program Committee. In response to the call for papers, the Program Committee received a total of 102 submissions, consisting of 85 papers (45 in Track 1, 23 in Track 2, and 17 in the short paper category; two papers that were withdrawn by the authors are not included in these statistics) and 17 posters. More than 330 expert single-blind reviews were provided, roughly a third of which were contributed by external sub-reviewers. After extensive electronic discussions via EasyChair, the Program Committee selected 41 papers and 14 posters for inclusion in the scientific program of GD 2018. This resulted in an overall paper acceptance rate of 48% (58% in Track 1, 43% in Track 2, and 29% in the short paper category). Authors published an electronic version of their accepted papers on an arXiv repository; a conference index with links to these contributions was made available before the conference.

There were two keynote talks at GD 2018. Alexandru Telea, from University of Groningen, The Netherlands, talked about methods for "Image-Based Graph Visualization: Advances and Challenge." Bojan Mohar, from Simon Fraser University, Canada, spoke about the "Beauty and Challenges of Crossing Numbers." The abstracts of both talks are included in the proceedings.

The conference gave out best paper awards in Track 1 and Track 2, plus a best presentation award and a best poster award. As decided by a subcommittee of the Program Committee, the award for the best paper in Track 1 was assigned to "Pole Dancing: 3D Morphs for Tree Drawings" by Elena Arseneva, Prosenjit Bose, Pilar Cano, Anthony D'Angelo, Vida Dujmović, Fabrizio Frati, Stefan Langerman, and Alessandra Tappini, and the award for the best paper in Track 2 was assigned to "Aesthetic Discrimination of Graph Layouts" by Moritz Klammler, Tamara Mchedlidze, and Alexey Pak. The participants of the conference voted to determine as the best presentation the one given jointly by Elena Arseneva and Pilar Cano for the paper

"Pole Dancing: 3D Morphs for Tree Drawings" and as the best poster the one by Charles Camacho, Silvia Fernández-Merchant, Marija Jelic, Rachel Kirsch, Linda Kleist, Elizabeth Bailey Matson, and Jennifer White entitled "Bounding the Tripartite-Circle Crossing Number of Complete Tripartite Graphs." Congratulations to all the award winners for their excellent contributions, and many thanks to Springer and MDPI whose sponsorship funded the prize money for these awards.

Following the tradition, the 25th Annual Graph Drawing Contest was held during the conference. The contest was divided into two parts, creative topics and the live challenge. The creative topics featured two graphs, one about Games of Thrones and one about the Mathematics Genealogy Project. The live challenge focused on drawings that maximize the crossing-angles, and had two categories: manual and automatic. Awards were given in each of the four categories. We thank the Contest Committee, chaired by Maarten Löffler, for preparing interesting and challenging contest problems. A report about the contest is included in these proceedings.

Many people and organizations contributed to the success of GD 2018. We would like to thank the Program Committee members and the external reviewers for carefully reviewing and discussing the submitted papers and posters; this was crucial for putting together a strong and interesting program. Thanks to all the authors who chose GD 2018 as the publication venue for their research. We are indebted to the gold sponsors Tom Sawyer Software and yWorks, the silver sponsor Microsoft, and the bronze sponsor Springer. Their generous support helps to ensure the continued success of this conference. Last but not least, the organizing co-chairs, Vera Sacristán and Rodrigo Silveira, did a terrific job; they in turn would like to express their thanks to other local organizers and volunteers, including Therese Biedl, Pilar Cano, Karla García, Carmen Hernando, Clemens Huemer, Maarten Löffler, Mercè Mora, Carlos Seara, and Roger Solí.

The 27th International Symposium on Graph Drawing and Network Visualization (GD 2019) will take place September 17–20, 2019 in Průhonice (near Prague), Czech Republic. Daniel Archambault and Csaba Tóth will co-chair the Program Committee. Jiří Fiala and Pavel Valtr will co-chair the Organizing Committee.

October 2018

Therese Biedl
Andreas Kerren

Organization

Steering Committee

Daniel Archambault	Swansea University, UK
Therese Biedl	University of Waterloo, Canada
Giuseppe Di Battista	Università Roma Tre, Italy
Fabrizio Frati	Università Roma Tre, Italy
Andreas Kerren	Linnaeus University, Sweden
Stephen G. Kobourov (Incoming Chair)	University of Arizona, USA
Giuseppe Liotta (Outgoing Chair)	Università di Perugia, Italy
Kwan-Liu Ma	University of California at Davis, USA
Martin Nöllenburg	Technische Universität Wien, Austria
Roberto Tamassia	Brown University, USA
Ioannis G. Tollis	University of Crete, Greece and Tom Sawyer Software, USA
Csaba D. Tóth	California State University Northridge, USA

Program Committee

Patrizio Angelini	Tübingen University, Germany
Daniel Archambault	Swansea University, UK
David Auber	LaBRI, Université Bordeaux, France
Therese Biedl (Co-chair)	University of Waterloo, Canada
Carla Binucci	University of Perugia, Italy
Erin Chambers	Saint Louis University, USA
Steven Chaplick	Universität Würzburg, Germany
Giuseppe Di Battista	Università Roma Tre, Italy
Tim Dwyer	Monash University, Australia
Radoslav Fulek	IST Austria, Austria
Christophe Hurter	ENAC – École Nationale de l'Aviation Civile, France
Andreas Kerren (Co-chair)	Linnaeus University, Sweden
Karsten Klein	University of Konstanz, Germany
Debajyoti Mondal	University of Saskatchewan, Canada
Petra Mutzel	TU Dortmund University, Germany
Yoshio Okamoto	The University of Electro-Communications and RIKEN Center for Advanced Intelligence Project (AIP), Japan
Sergey Pupyrev	University of Arizona, USA
Helen Purchase	University of Glasgow, UK
Marcus Schaefer	DePaul University, USA

Gerik Scheuermann   Leipzig University, Germany
Darren Strash   Hamilton College, USA
Shigeo Takahashi   University of Aizu, Japan
Tatiana von Landesberger   Technische Universität Darmstadt, Germany
Sue Whitesides   University of Victoria, Canada
David R. Wood   Monash University, Australia
Hsu-Chun Yen   National Taiwan University, Taiwan

Organizing Committee

Vera Sacristán   Universitat Politècnica de Catalunya, Spain
Rodrigo Silveira   Universitat Politècnica de Catalunya, Spain

Contest Committee

Will Devanny   University of California at Irvine, USA
Philipp Kindermann   University of Waterloo, Canada
Maarten Löffler (Chair)   Utrecht University, The Netherlands
Ignaz Rutter   Universität Passau, Germany

External Referees

Ackerman, Eyal	Didimo, Walter	Kaufmann, Michael
Aichem, Michael	Dujmović, Vida	Keil, Mark
Akitaya, Hugo	Döring, Hanna	Keszegh, Balázs
Alam, Jawaherul	Evans, William	Kieffer, Steve
Arroyo, Alan	Fink, Martin	Kindermann, Philipp
Balko, Martin	Firman, Oksana	Klemz, Boris
Ballweg, Kathrin	Frati, Fabrizio	Klimenta, Mirza
Barth, Lukas	Förster, Henry	Knauer, Kolja
Bekos, Michael	Grilli, Luca	Kryven, Myroslav
Bereg, Sergey	Gronemann, Martin	Kurz, Denis
Biniaz, Ahmad	Gupta, Siddharth	Kynčl, Jan
Blaszczyszyn, Bartek	Hassoumi, Almoctar	Lau, Lap Chi
Bläsius, Thomas	Healy, Patrick	Lin, Chun-Cheng
Brandenburg, Franz	Hossain, Md. Iqbal	Lu, Hsueh-I
Chang, Hsien-Chih	Hu, Yifan	Lubiw, Anna
Chimani, Markus	Irvine, Veronika	Löffler, Andre
Cornelsen, Sabine	Jaeger, Sabrina	Mehrabi, Saeed
Da Lozzo, Giordano	Jesus Lobo, Maria	Michael, Traoré
Di Giacomo, Emilio	Joret, Gwenaël	Montecchiani, Fabrizio

Nöllenburg, Martin
Otachi, Yota
Patakova, Zuzana
Pálvölgyi, Dömötör
Rahman, Md. Saidur
Roselli, Vincenzo
Rutter, Ignaz
Rzążewski, Paweł

Rüegg, Ulf
Schulz, André
Sommer, Björn
Tappini, Alessandra
Uehara, Ryuhei
van Renssen, André
Wallner, Günter
Wang, Hung-Lung

Wang, Yunhai
Wiechert, Veit
Wild, Pascal
Wolff, Alexander
Wunderlich, Marcel
Wybrow, Michael
Yu, Tian-Li
Zink, Johannes

Sponsors

Gold Sponsors

Silver Sponsor

Bronze Sponsor

Springer

Beauty and Challenges of Crossing Numbers (Keynote Presentation)

Bojan Mohar

Simon Fraser University, Burnaby and IMFM, Ljubljana
mohar@sfu.ca

Abstract. One of the initial goals of the graph drawing community was trying to understand what it means for a drawing of a graph to be nice or even beautiful. These attempts failed due to lack of a formal description how to measure how beautiful a drawing of a graph is. However, there is a lot of beauty of the results and methods in this area.

In this talk, the speaker will outline some of his favorite results in crossing number theory that demonstrate extreme beauty and elegance. Yet, there are some very basic problems that elude our proper understanding of this area. The speaker will touch upon some of these as well.

Contents

Best Paper Track 2

Orders

Crossings

Crossing Angles

Experiments

Orthogonal Drawings

Realizability

Miscellaneous

Graph Drawing Contest Report

Poster Abstracts

Invited Talk

Image-Based Graph Visualization: Advances and Challenges

Alexandru Telea[✉]

Bernoulli Institute, University of Groningen, Groningen, The Netherlands
a.c.telea@rug.nl

Abstract. Visualizing large, multiply-attributed, and time-dependent graphs is one of the grand challenges of information visualization. In recent years, image-based techniques have emerged as a strong competitor in the arena of solutions for this task. While many papers on this topic have been published, the precise advantages and limitations of such techniques, and also how they relate to similar techniques in the more traditional fields of scientific visualization (scivis) and image processing, have not been sufficiently outlined. In this paper, we aim to provide such an overview and comparison. We highlight the main advantages of image-based graph visualization and propose a simple taxonomy for such techniques. Next, we highlight the differences between graph and scivis/image datasets that lead to limitations of current image-based graph visualization techniques. Finally, we consider these limitations to propose a number of future work directions for extending the effectiveness and range of image-based graph visualization.

Keywords: Large graph visualization
Image-based information visualization · Multiscale visualization

1 Introduction

Relational data, also called networks or graphs, is a central and ubiquitous element of many types of data collections generated by multiple application domains such as traffic analysis and planning, social media, business intelligence, biology, software engineering, and the internet. Since the first moments when such data was collected, visualization has been a key tool for its exploration and analysis, leading to the emergence and development of the research domains of *graph drawing* and *graph visualization* [15,23]. Last-decade developments in processing power, data-acquisition tools, and techniques, have led to what is today globally called *big data* – collections of tens of millions of samples having hundreds of measurement values (attributes), all which can evolve over thousands of time steps. A particular case hereof, big-data graphs, pose fundamental problems for visual exploration.

On the other hand, several solutions, techniques, and tools have been developed for the scalable visual exploration of other types of big data collections,

© Springer Nature Switzerland AG 2018
T. Biedl and A. Kerren (Eds.): GD 2018, LNCS 11282, pp. 3–19, 2018.
https://doi.org/10.1007/978-3-030-04414-5_1

such as 2D images, 3D scalar or vector field volumes, or more generally multidimensional fields, in the domains of *scientific visualization* and *imaging sciences* [52]. Recent developments have tried to approach the two traditionally separately evolving fields of graph visualization and scientific visualization, thereby aiming at leveraging the (visual) scalability of the latter methods to address big graph related challenges from the former [20]. This has led to interesting parallels and links between concepts, methods, and applications between the two fields, and the development of hybrid visualization methods that inherit strengths from both graph visualization and scientific visualization. However, large graph visualization still has many unsolved challenges [23].

In this paper (and related talk) we aim to provide an overview of the research at the crossroads of large graph visualization and scientific visualization. We start highlighting the main challenges in large graph visualization (Sect. 2). Next, we outline the high-level directions proposed by current research towards addressing these (Sect. 3). We focus next on one type of technique that aims to solve these challenges by adapting methods from scientific visualization and imaging to the particularities of graph visualization – image-based graph visualization (Sect. 4). Based on the structure of graph data outlined in Sect. 2, we discuss here various types of image-based methods for graph visualization and highlight parallels to simplification methods for multivariate field and image data. In the light of these methods, we next highlight open challenges for image-based graph visualization (Sect. 5) and attempt to clarify some of the more subtle points related to this new emerging visualization field which, we believe, have not been sufficiently discussed in current literature. Section 6 concludes the paper outlining promising directions for future research in image-based graph visualization.

2   Problem Definition

2.1   Preliminaries

To better outline the large graph visualization challenges, we first introduce some notations. Let $G = (V, E \subset V \times V)$ be a graph with vertices, or nodes, $V = \{\mathbf{v}_i\}$ and edges $E = \{\mathbf{e}_i\}$. Both nodes and edges typically have one or multiple attributes (also called features, dimensions, or variables). We denote by v_i^j, $1 \leq j \leq N_V$, the individual attributes of node $\mathbf{v}_i$, and by e_i^j, $1 \leq j \leq N_E$, the individual attributes of edge $\mathbf{e}_i$, respectively. As a shorthand, let $\mathbf{v}^j$ denote all values of the jth attribute of all nodes V; let $\mathbf{e}^j$ denote all values of the jth attribute of all edges E; let $\mathbf{V} = (\mathbf{v}^1, \ldots, \mathbf{v}^{N_V})$ denote all values of all node attributes; and let $\mathbf{E} = (\mathbf{e}^1, \ldots, \mathbf{e}^{N_E})$ denote all values of all edge attributes, respectively. Attributes can be of all types, *e.g.*, quantitative (values in $\mathbb{R}$), integral (values in $\mathbb{N}$), ordinal, categorical, text, hyperlinks, but also more complex data types such as images or video. In this sense, the ordered collections $\mathbf{V}$ and $\mathbf{E}$ are very similar to so-called multidimensional datasets as well known in information visualization [20,34,52]. That is, every node $\mathbf{v}_i$ or edge $\mathbf{e}_i$ can be seen as a sample, or observation, of a respectively N_V and N_E dimensional dataset. Finally, as graphs can evolve over time, all their ingredients (sets V, E, $\mathbf{V}$, and $\mathbf{E}$) can be seen as functions of (continuous or discrete) time [2].

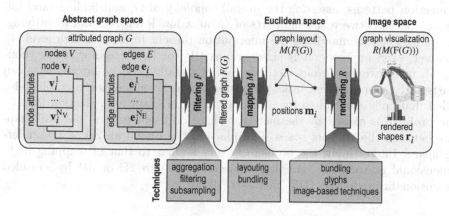

Fig. 1. Graph visualization pipeline.

With these notations, visualizing a graph can be defined in terms of the traditional data visualization pipeline [52] in terms of *filtering, mapping* and *rendering* operations (see also Fig. 1). Filtering F reads the input graph G and produces another graph $F(G)$ which is (more) suitable for subsequent visualization, *e.g.*, by removing nodes, edges, and/or attributes that are not of interest, or aggregating such elements into fewer and/or semantically richer ones. Mapping M is a function that takes as input $F(G)$ and outputs a set of shapes $M(F(G)) = \{\mathbf{m}_i\}$ embedded in $\mathbb{R}^2$ or, less frequently, $\mathbb{R}^3$. Typically, nodes are mapped to individual points, and edges are mapped to straight lines or, less commonly, curves. Other layout methods, such as adjacency matrices [1], exist but are less intuitive, less common, and thus not discussed here. Most often, M takes into account only the graph topology (V, E), and computes only positions $\mathbf{m}_i$ for nodes. This is the case of so-called *graph layout* techniques [51,55]. Rendering R takes as input the layout $M(F(G))$ and creates actual visible shapes $R(M(F(G))) = \{\mathbf{r}_i\}$, where each $\mathbf{r}_i$ is placed at the corresponding layout positions $\mathbf{m}_i$. Visual variables [57] of $\mathbf{r}_i$ such as size, color, texture, transparency, orientation, texture, and annotation are used to encode the attributes $\mathbf{v}_i$ and $\mathbf{e}_i$ of the respective node or edge. Interactive exploration techniques such as zooming, panning, brushing, and lensing can be subsumed to the rendering operator R as they are essentially customized ways to perform rendering; hence, we do not discuss them separately.

2.2 Scalability Challenge

With the above notations, we can decompose the challenge of visualizing big-data graphs into the following three elements:

Layout: A good layout should arguably allow end users to detect structures of interest present in G by examining the rendering $R(M(G))$. These include, but are not limited to, finding groups of strongly-connected nodes; finding specific connection patterns; assessing the overall topology of G; and finding (and following) paths between specific parts of G, at a low level [24]; and identifying, comparing, and summarizing the information present in G, at a high level [4]. However, even for moderately-sized graphs ($|V|$ or $|E|$ exceeding a few thousands), most existing layout methods cannot usually produce layouts that can consistently support these tasks [23]. Suboptimal layouts of large graphs, also called 'hairballs', are all to frequent a problem in graph visualization [38,47]. The problem is caused by the fact that there does not exist a 'natural' mapping between the abstract space of graphs and the Euclidean 2D or 3D rendering space. Interestingly, the problem is very similar to that of mapping high-dimensional scatterplots (sampled datasets in $\mathbb{R}^n$) to 2D or 3D by so-called dimensionality reduction (DR) methods [29,50].

Dimensionality: An effective graph visualization should allow users to answer questions on all elements of interest of the original graph. Apart from the topology (V, E) which should be captured by the layout $M(G)$, this includes the node and edge attributes $\mathbf{V}$ and $\mathbf{E}$. The problem is that, when N_V and N_E are large, nodes and edges essentially become points in high-dimensional spaces. Since, as explained, each node and/or edge is typically mapped to a separate location $\mathbf{m}_i$, the challenge is how to depict a high-dimensional data sample, consisting of potentially different attribute types, to the space at or around $\mathbf{m}_i$. A similar problem exists in scientific visualization when using glyphs to depict high-dimensional fields [3,44]: The higher-dimensional our data points are, the more space one needs to show all dimensions, so the fewer such points (in our case, nodes and/or edges) can one show on a given screen size. At one extreme, we can display (tens of) thousands of nodes on a typical computer screen if we only show 2 or 3 attributes per node (encoded *e.g.* in hue, luminance, and size); at the other extreme, we can display tens of attributes per node, like in UML diagrams, but for only a few tens up to hundreds of nodes [5]. The problem is well known also in multidimensional information visualization.

Clutter and Overdraw: Finally, a scalable graph visualization should accommodate (very) large graphs consisting of millions of nodes and/or edges. Even if we abstract from the aforementioned layout and dimensionality challenges, a fundamental difficulty here resides in the fact that a node-link visualization cannot exceed a given *density*: If nodes and/or edges are drawn too close to each other, they will form a compact *cluttered* mass where they cannot be distinguished from each other. Additionally, an edge (in the node-link visual model) is drawn as a line (or curve) so in the limit it needs to use at least a few (tens of) pixels of screen space to be visible as such (if the edge is too show, we cannot *e.g.* see its direction); in Tufte's terms, there is an upper bound to the data-ink ratio [57] when drawing a graph edge. Moreover, when attributes must be rendered atop of the edge, the amount of surrounding whitespace needs to be increased [17]. This leads in turn to inherent *overdraw, i.e.* edges that partially occlude each other, even for moderately-sized graphs of thousands of nodes. A detailed overview of clutter reduction techniques in information visualization is given by Ellis and Dix [8]. In large graph visualization, clutter and overdraw are hard to jointly optimize for: Spatial distortion, *e.g.* via edge bundling (discussed next in Sect. 4.2), creates more white space, thus reduces clutter, but increases overdraw; space-filling techniques are of limited effect since, as noted, edges must be surrounded by white space to be visible as such; apart from these, reducing clutter and overdraw is not fully possible in the rendering phase only, as this phase works within the constraints of the layout fed to it by the M operator.

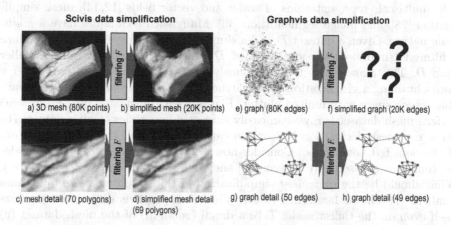

Fig. 2. Data-space simplification of scivis data (a) *vs* graph data (b). (Color figure online)

3 Simplification: Ways Towards a Solution

For a given screen resolution for the target image R, how can we approach large graph visualization? Given the scalability challenges outlined in Sect. 2.2, two types of approaches exist, as follows.

Data-Space Simplification: First, we can simplify the graph G in the filtering stage F in the visualization pipeline (Fig. 1). This reduces the number of nodes ($|V|$), edges ($|E|$), and/or attributes (N_V, N_E) to be next passed to the mapping operator M. Following the clutter reduction taxonomy of Ellis and Dix [8], this includes subsampling, filtering, and clustering (aggregation) [45], all applicable to V, E, and ($\mathbf{V}, \mathbf{E}$) respectively. While effective in tackling clutter, overdraw, and dimensionality issues, such approaches have two limitations. First, they require *a priori* knowledge on which data items (samples or dimensions) can be filtered or clustered together. Secondly, performing such operations on graphs can easily affect the *semantics* of the underlying data.

At this point, it is instructive to compare graph visualization (graphvis) with image and field visualization as done in classical scientific visualization (scivis). Consider a multidimensional dataset $D : \mathbb{R}^m \to \mathbb{R}^n$; for each point of the Euclidean m-dimensional domain, n quantitative values are measured. Scivis provides many methods for visualizing such datasets, *e.g.* for 2D and 3D vector fields ($m \in \{2,3\}, n \in \{2,3\}$) or 2D and 3D scalar fields ($m \in \{2,3\}, n = 1$) [52]. Many techniques exist in scivis (and, by extension, in imaging and signal processing) for simplifying large fields – we mention here just a few, *e.g.*, perceptually-based image downscaling [39], feature extraction from vector fields [43], multiscale representations of scalar and vector fields [12,14], mesh simplification [28], and image segmentation [40]. Many such techniques have a *multiscale* nature: Given a dataset D and a simplification level $\tau \in \mathbb{R}^+$, they produce a filtered (simplified) version $F(D)$ of D which is (roughly) τ times smaller than D. This allows users to continuously vary the level-of-detail parameter τ until obtaining a visualization that matches their goals, as well as fits the available screen space with limited clutter. Figure 2(left) illustrates this: From a 3D surface-mesh dataset (a), we can easily extract a four times smaller dataset (b) using *e.g.* mesh decimation [46], which captures very well the overall structure of the depicted bone shape. Consider now a graph of similar size, whose nodes are functions in a software system [54] and edges function calls respectively (e). What should be the *equivalent* simplification of this graph to a size four times smaller? (f) This is far from evident. The scivis-graphvis difference manifests itself even on the tiniest scale: Take a detail (zoom-in) of the mesh dataset (c) from which we decimate a *single* polygon (data point). The result (d) is visually identical. Consider now the analogous zoom-in on a small portion of our call graph (g) from which we remove a *single* edge. The result (h) may be visually similar to the input (g), but can have a completely different *semantics* – just imagine that the removed function call edge is vital to the understanding of the operation of the underlying software.

All in all, most scivis data-space simplification methods succeed in keeping the overall semantics of their data. In contrast, even tiny changes to graph data can massively affect the underlying semantics. More formally put, scivis data-space simplification methods appear (in general) to be Cauchy or Lipschitz continuous (small data changes imply small semantic changes). This clearly does not hold in general for graph data. We believe the difference is due to two factors:

1. **Scivis** data is defined over *Euclidean* domains ($\mathbb{R}^m$). This allows simplification operators to readily use *continuous* Euclidean distances to *e.g.* aggregate and cluster data. An entire machinery is available for this, including basis functions and interpolation methods [12,14].
2. All data samples have (roughly) the same importance, and the phenomenon (signal) sampled by the scivis dataset D is of bounded frequency. Hence, discarding a few samples does not affect data semantics.

In contrast:

1. **Graph** data is defined over an *abstract* graph space, whose dimensions, and even dimensionality, are not known or even properly defined. It is not always evident how to define 'proximity' between graph nodes and/or edges. There is no comparable (continuous) interpolation theory for graph data. Graph-theoretic distances are not continuous. Simply put: There is *nothing* (no information) between two nodes connected by an edge;
2. Nodes and edges can have widely different importances. There is, as we know, no similar notion of 'maximal frequency' of a graph dataset as in scivis. Hence, discarding a few samples can massively affect graph data semantics.

Image-Level Simplification: A second way to handle large graph visualizations is to simplify them in the image domain. That is, given the limitations of data-space graph simplification listed earlier, rather than designing simplification operators F that act on the graph datasets, we embed the simplification into the graph rendering operator R. The key advantage here is that R acts, by definition, upon an Euclidean space (the 2D target image), where all samples (pixels) are equally important. Hence, the main proposal of image-based graph visualization is to *delay* simplification to the moment where we can reuse/adapt known scivis techniques for data simplification. Rather than first simplifying the graph data (F) and then mapping (M) and rendering (R) it, image-level techniques first map the data, and then simplify it during rendering[1]. We detail the advantages and challenges of image-based graph visualization next.

4   Image-Based Graph Visualization

Image-based graph visualization is a subfield of the larger field of image-based information visualization [20]. The name of this field can be traced back to 2002, when image-based flow visualization (IBFV) was proposed to depict large, complex, and time-dependent 2D vector fields using animated textures [60]. Key to IBFV (and its sequels) was the manipulation of the *image-space* pixels to produce the final visualization. Several advantages followed from this approach:

- *Dense* visualizations: Every target image pixel encodes a certain amount of information, thus maximizing the data-ink ratio [57];
- Clutter is avoided by construction: Rather than *scattering* dataset samples over the image space (which can lead to clutter when several such samples inadvertently overlap), samples are *gathered* and explicitly aggregated for each pixel. The aggregation function is fully controlled by the algorithm;
- Implicit *multiscale* visualizations: By simply changing the resolution of the target image (zooming in or out), users can continuously control the amount of information displayed per screen area unit;

---

[1] The underlying assumption here is that mapping and simplification are conceptually commutative. As discussed next, this is not always the case.

- Exploitation of existing knowledge about image *perception* when synthesizing and/or simplifying a graph visualization;
- *Accclcrated* implementations: Image-based techniques parallelize naturally over the target image pixels (much as raycasting does), so they optimally fit to modern GPU architectures [25, 62];
- Simpler *implementations vs* data-space graph simplification techniques.

A more subtle (but present) advantage of image-based graph visualizations is their ability to reuse principles and techniques grounded in the theory and practice of image and signal processing, thereby allowing a more principled reasoning about, and control of, the resulting visualization. We next outline the main advances of this field, along with the challenges that we still see open. Given the structure of a graph in terms of nodes, edges, and attributes thereof (Sect. 2.1), we structure our discussion along the same concepts.

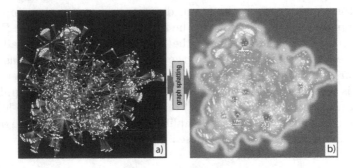

Fig. 3. (a) Node-link graph drawing (dataset from Fig. 2e) and (b) its graph splatting. (Color figure online)

4.1   Node-Centric Techniques

The first image-based graph visualization, to our knowledge, is *graph splatting*, proposed in 2003 by De Leeuw and Van Liere [27]. Its intuition is simple: Given a graph drawing (layout) $M(G)$, its visualization $R(M(G))$ is the convolution of $M(G)$ with an isotropic 2D Gaussian kernel in image space. This is simply a low-pass filter that emphasizes high-density node and/or edge areas in the layout. The visualization's level-of-detail, or multiscale nature, is controlled by the filter's radius. The samples' (nodes or edges) weights can be set to reflect their importance. Figure 3(b) shows the splatting of the graph in Fig. 3(a), for the same call graph as in Fig. 2e, where the nodes' weights are set to their number of outgoing edges (fan-out factor). The resulting density map, visualized with a rainbow colormap, thus emphasizes nodes (functions) that call many other functions as red spots. This allows easily detecting such suitably-called 'hot spots' in the software system's architecture.

Graph splatting is extremely simple to implement, fast to execute (linear in the number of splatted nodes and/or edges), and easy to control by users via its kernel-radius parameter. It also forms the basis of more advanced techniques such as graph bundling (Sect. 4.2). Formally, it is a variant of the more general kernel density estimation (KDE) set of techniques used in multidimensional data analysis [49]. Its key limitation is that it assumes a good layout $M(G)$: Density hot spots appear when nodes and/or edges show up *closely* in a graph layout. So, when layout methods M place unrelated nodes close to each other, 'false positive' hot spots appear (and analogously for false negatives).

4.2   Edge-Centric Techniques

Graph bundling is the foremost image-based technique focusing on graph edges. Bundling has a long history (see Fig. 4 for an overview of its most important moments). We distinguish five phases, as follows (for a comprehensive recent survey, we refer to [26,61]):

Early Phase: Minard hand-drew a so-called 'flow map' (a single-root directed acyclic graph) showing the French wine exports in 1864 [33]. While not properly a bundled graph, as no edges are grouped together, the visual style featuring curved edges whose thickness maps edge weights, suggests later bundling techniques. The design was refined in 1898 to create the so-called Sankey diagrams, which can display more complex (multiple source, cyclic) graphs;

First Computer Methods: One of the first computer-computed bundling-like visualizations was proposed by Newbery in 1989 [35]. The key novelty *vs* earlier methods is grouping edges sharing the same end nodes (so, this technique can be seen as a particular case of graph simplification by aggregation). Dickerson *et al.* coined the term 'edge bundling' in 2003 for their method that optimizes node placement and groups same-endpoints edges (via splines) to simplify graph drawings. All these techniques could handle only small graphs of tens up to hundreds of nodes and edges.

Establishment Phase: Subsequent methods focused on larger-size graphs (thousands of nodes and edges). Flow map layouts [42] generalized in 2005 the computation of Sankey-like diagrams, also first featuring the 'organic' branch-like structure to be encountered in many later techniques [7,9,53]. At roughly the same time (2006), two key bundling techniques emerged: Gansner *et al.* presented improved circular layouts [11], which grouped edges based on their spatial proximity in $M(G)$; Holten proposed hierarchical edge bundling [16] which grouped edges based on the graph-theoretic distance of their start and end nodes in a hierarchy of the graph's nodes. Holten also pioneered several advanced blending techniques to cope with edge overdraw (see also Sect. 4.3).

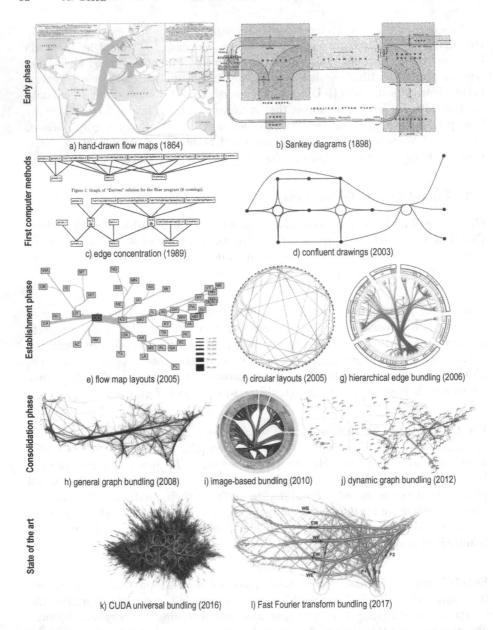

Fig. 4. Key moments in edge bundling history.

Consolidation Phase: The next phase focused on treating general graphs [18], time-dependent graphs [36], and, most importantly for our context, *image-based methods*. The latter include image-based edge bundles (IBEB [53], following the name-giving of IBFV [60]) which introduced clustering and grouped rendering of spatially close edges in the form of shaded cushions [58] to both simplify the rendered graph and emphasize distinct/crossing, bundles. IBEB reused several image-processing operators such as KDE [49], distance transforms [10], and medial axes [48] for computational speed. Next, skeleton-based edge bundles (SBEB) [9] used medial axes to actually *perform* graph bundlings, by following the simple but effective intuition that bundling a set of (close) curves means moving them towards the centerline of their hull.

State of the Art: Most recent methods focus mainly on scalability, using image-based techniques. Kernel density edge bundling (KDEEB) [21] showed that bundling a graph drawing is identical to applying mean shift, well known in data clustering [6], on the KDE edge-density field. CUDA Universal Bundling (CUBu) [62] next accelerated KDEEB to bundle 1 million-edge graphs in subsecond time by parallelizing KDE on the GPU. Fast Fourier Transform Edge Bundling (FFTEB) [25] further accelerated CUBu by computing the KDE convolution in frequency space, thus bundling graphs of tens of millions of edges at interactive rates. As such, scalability seems to have been addressed successfully.

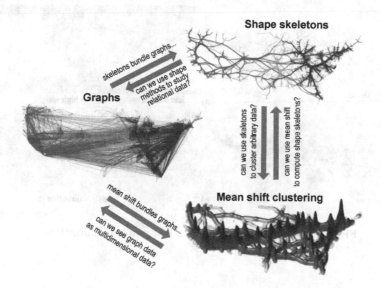

Fig. 5. Puzzling connections between graph visualization, shape analysis, and multidimensional data analysis.

Several points can be made about edge bundling. First, bundling is an *image-space* simplification technique of the graph *drawing* $R(M(G))$ that reduces clutter by creating whitespace between bundles, but increases overdraw (of same-bundle edges); a recent bundling formal definition as an image-processing operator is given in [26]. Image-based bundling is a *multiscale* technique, where the KDE kernel radius controls the extent over which close edges get bundled, thereby allowing users to easily and continuously specify how much they want to simplify (bundle) their graphs. Image-based methods are clearly the fastest, most scalable, bundling methods, due to the high GPU parallelization of their underlying image processing operations. Edge similarity, the bundling driving factor, can be easily defined in terms of a mix of spatial (Euclidean) and attribute-based distances [41]. More interestingly from a theoretical point, bundling exposes some puzzling connections between domains as different as data clustering [6], shape simplification [48], and graph visualization itself (Fig. 5). Briefly put:

- If skeletons can be used to bundle graphs [9], how can we further use the wealth of shape analysis methods to analyze/visualize graphs?
- If skeletons and mean shift bundle graphs [9,21], can we use skeletons to cluster multidimensional data, or mean shift to compute shape skeletons?
- If mean shift simplifies graphs [21], could we see graphs as yet another form of multidimensional data?

These questions open, we think, a wealth of new vistas on data visualization.

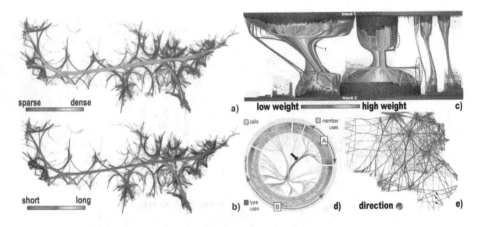

Fig. 6. Attribute encoding in bundled graph visualizations.

4.3 Attribute-Centric Techniques

Graph visualization scalability also means handing high-dimensional node and/or edge attributes (Sect. 2.2). Visualizing these is hard, since the method of choice for handling geometric scalability – bundling – massively increases edge overdraw. Several image-based techniques address attribute visualization, as follows.

One can directly visualize the edge-density (KDE) map *e.g.* by alpha blending [16], which is a simple form of graph splatting using a one-pixel-wide kernel. Additionally, hue mapping can encode edge attributes, such as density [18,62] (Fig. 6a), length [16,62] (Fig. 6b), quantitative weights [56] (Fig. 6c), categorical edge types [53] (Fig. 6d), and edge directions [41] (Fig. 6e). Two main challenges exist here. First, at edge overlap locations, attribute values e_i^j of multiple edges $\mathbf{e}_j$ have to be *aggregated* together prior to color coding. While this is straightforward to do for *e.g.* edge density, it becomes problematic for other attributes such as edge categorical types or edge directions. This issue parallels known challenges in scivis (interpolation of vector fields) and infovis (aggregation of categorical data). Secondly, there is currently no scalable method that can render *at the same time* more than roughly two attributes per edge in high-density graph visualizations. Visualizing graphs having tens of attributes per edge ($N_E > 2$) is an open problem. Separately, animation has been used to encode edge directions by using particle-based techniques [19]. Interestingly, this approach resembles a form of IBFV [60] applied to the vector field defined by the edges' tangent vectors. However, in a typical vector field, the number of singularities (where IBFV would have problems rendering a smooth, informative, animation) is quite limited; in a dense graph, this number is very high, equalling the amount of edge crossings or, in the bundled case, overlaps of different-direction edges [9]. Hence, IBFV cannot be directly used to visualize large/complex graphs.

5   Open Challenges

Image-based techniques have shown high potential for the efficient and effective visualization of large graphs. Yet, we also see a number of key challenges that they would need to tackle to become (more) effective in practice, as follows.

Layouts: Current image-based techniques address the rendering (R) phase, but assume a suitable node layout to be given as input. As explained, computing such a layout (for large graphs) is challenging. A promising direction is to further explore analogies between dimensionality reduction (DR, used to efficiently and effectively visualize high-dimensional sample sets embedded in $\mathbb{R}^n$) and graph drawing [22]. An additional advantage of doing this is that DR can easily accommodate a wide range of similarity functions, *e.g.*, accounting for both graph structure and attributes [32]. This could open new ways to visualizing graphs having many node and/or edge attributes. Separately, it would be interesting to consider image-based bundling approaches for the *layout* of a graph's nodes.

Aggregation: Graph splatting and bundling are the techniques of choice for generating images of large graphs. However, the way in which the multiple node and/or edge attribute values that cover a given pixels are to be aggregated is currently limited to simple operations (sum, average, minimum, or maximum) [16,62]. Such operations cannot aggregate attributes such as categorical types or edge directions. For edge directions, it is interesting to consider analogies with scivis techniques for dense tensor field interpolation [59] which address related problems. Separately, image processing has proposed a wealth of operators for detecting and emphasizing specific features present in images such as edges, lines, textures, or even more complex shapes [13]. Such operators could be readily adapted to highlight patterns of interest in image-based graph visualizations.

Quality: Measuring the quality of an (image-based) graph visualization is an open topic [37], much due to the fact that there is typically no ground truth to compare against. Still, image processing techniques can be helpful in this area, *e.g.* by providing quantitative measures for the amount of edge intersections, bends, preservation of graph-theoretic distances, or edge-angle spatial distributions, in the final image. Such image-based metrics have been successful in assessing the quality of DR scatterplot projections [31], bringing added value beyond simple aggregate metrics. Exploring their extension to graph visualizations is potentially effective. Also, such metrics could be easily used to locally *constrain* the mapping and/or rendering phases, *e.g.* to limit the amount of undesired deformations that bundling produces.

Applications: An interesting and potentially rich field for graph visualization is the exploration of deep neural networks (DNNs), currently the favored technique in machine learning. DNNs are large (millions of nodes and/or edges), attributed by several values (*e.g.* activations and weights), and time-dependent (*e.g.* during the network training). Understanding how DNNs work, and why/where they do not work, is a major challenge in deep learning [30]. Visualizing DNNs is also very difficult, as their tightly-connected structure yields significant edge crossings and overdraw, and it is not evident how *e.g.* bundling would help for these topologies. Exploring image-based techniques for this use-case is promising.

6   Conclusion

In this paper, we surveyed current developments of image-based techniques for the visualization of large, high-dimensional, and time-dependent graphs. These techniques have major advantages – the ability of creating dense visualizations with high data-ink ratios, treatment of clutter by construction, an implicit multiscale nature able to handle large and dense graphs, and scalable implementations. We highlighted analogies and differences between image-based techniques and related techniques for the visualization of densely-sampled fields in scientific visualization. While graph data has several important differences as compared to field data, the existing similarities make us believe that existing scivis and image-processing techniques can be further adapted to further assist graph visualization. From a practical perspective, this would lead to the creation of novel efficient and effective tools for graph visual exploration. Equally important, from a theoretical perspective, this could lead to further unification of the currently still separated disciplines of scientific and information visualization.

References

1. Abello, J., van Ham, F.: Matrix zoom: a visual interface to semi-external graphs. In: Ward, M., Munzner, T. (eds.) Proceedings of IEEE InfoVis, pp. 127–135 (2005)
2. Archambault, D., et al.: Temporal multivariate networks. In: Kerren, A., Purchase, H.C., Ward, M.O. (eds.) Multivariate Network Visualization. LNCS, vol. 8380, pp. 151–174. Springer, Cham (2014). https://doi.org/10.1007/978-3-319-06793-3_8
3. Borgo, R., et al.: Glyph-based visualization: foundations, design guidelines, techniques and applications. In: Sbert, M., Szirmay-Kalos, L. (eds.) Eurographics - State of the Art Reports. The Eurographics Association (2013)
4. Brehmer, M., Munzner, T.: A multi-level typology of abstract visualization tasks. IEEE TVCG **19**(12), 2376–2385 (2013)
5. Byelas, H., Telea, A.: Visualizing multivariate attributes on software diagrams. In: Proceedings of IEEE CSMR, pp. 335–338 (2009)
6. Comaniciu, D., Meer, P.: Mean shift: a robust approach toward feature space analysis. IEEE TPAMI **24**(5), 603–619 (2002)
7. Cui, W., Zhou, H., Qu, H., Wong, P.C., Li, X.: Geometry-based edge clustering for graph visualization. IEEE TVCG **14**(6), 1277–1284 (2008)
8. Ellis, G., Dix, A.: A taxonomy of clutter reduction for information visualisation. IEEE TVCG **13**(6), 1216–1223 (2007)
9. Ersoy, O., Hurter, C., Paulovich, F., Cantareiro, G., Telea, A.: Skeleton-based edge bundles for graph visualization. IEEE TVCG **17**(2), 2364–2373 (2011)
10. Fabbri, R., da F. Costa, L., Torelli, J., Bruno, O.: 2D Euclidean distance transform algorithms: a comparative survey. ACM Comput. Surv. **40**(1), 1–44 (2008)
11. Gansner, E.R., Koren, Y.: Improved circular layouts. In: Kaufmann, M., Wagner, D. (eds.) GD 2006. LNCS, vol. 4372, pp. 386–398. Springer, Heidelberg (2007). https://doi.org/10.1007/978-3-540-70904-6_37
12. Garcke, H., Preusser, T., Rumpf, M., Telea, A., Weikard, U., van Wijk, J.J.: A continuous clustering method for vector fields. In: Moorhead, R. (ed.) Proceedings of IEEE Visualization, pp. 351–358 (2000)
13. Gonzalez, R.C., Woods, R.E.: Digital Image Processing. Pearson, London (2011)
14. Griebel, M., Preusser, T., Rumpf, M., Schweitzer, M.A., Telea, A.: Flow field clustering via algebraic multigrid. In: Proceedings of IEEE Visualization, pp. 35–42 (2004)
15. Herman, I., Melancon, G., Marshall, M.S.: Graph visualization and navigation in information visualization: a survey. IEEE TVCG **6**(1), 24–43 (2000)
16. Holten, D.: Hierarchical edge bundles: visualization of adjacency relations in hierarchical data. IEEE TVCG **12**(5), 741–748 (2006)
17. Holten, D., Isenberg, P., Van Wijk, J.J., Fekete, J.D.: An extended evaluation of the readability of tapered, animated, and textured directed-edge representations in node-link graphs. In: Battista, G.D., Fekete, J.D., Qu, H. (eds.) Proceedings of IEEE PacificVis, pp. 195–202 (2011)
18. Holten, D., Van Wijk, J.J.: Force-directed edge bundling for graph visualization. Comput. Graph. Forum **28**(3), 983–990 (2009)
19. Hurter, C., Ersoy, O., Fabrikant, S.I., Klein, T.R., Telea, A.C.: Bundled visualization of dynamic graph and trail data. IEEE TVCG **20**(8), 1141–1157 (2014)
20. Hurter, C.: Image-Based Visualization: Interactive Multidimensional Data Exploration. Morgan & Claypool Publishers, San Rafael (2015)
21. Hurter, C., Ersoy, O., Telea, A.: Graph bundling by kernel density estimation. Comput. Graph. Forum **31**(3), 865–874 (2012)

22. Kruiger, J.F., Rauber, P.E., Martins, R.M., Kerren, A., Kobourov, S., Telea, A.C.: Graph layouts by t-SNE. Comput. Graph. Forum **36**(3), 283–294 (2017)
23. Landesberger, T.V., et al.: Visual analysis of large graphs: state-of-the-art and future research challenges. Comput. Graph. Forum **30**(6), 1719–1749 (2011)
24. Lee, B., Plaisant, C., Parr, C.S., Fekete, J.D., Henry, N.: Task taxonomy for graph visualization. In: Bertini, E., Plaisant, C., Santucci, G. (eds.) Proceedings of AVI BELIV, pp. 1–5. ACM (2006)
25. Lhuillier, A., Hurter, C., Telea, A.: FFTEB: edge bundling of huge graphs by the Fast Fourier Transform. In: Seo, J., Lee, B. (eds.) Proceedings of IEEE PacificVis (2017)
26. Lhuillier, A., Hurter, C., Telea, A.: State of the art in edge and trail bundling techniques. Comput. Graph. Forum **36**(3), 619–645 (2017)
27. van Liere, R., de Leeuw, W.: GraphSplatting: visualizing graphs as continuous fields. IEEE TVCG **9**(2), 206–212 (2003)
28. Luebke, D.P.: A developer's survey of polygonal simplification algorithms. IEEE CG&A **21**(3), 24–35 (2001)
29. van der Maaten, L., Postma, E.: Dimensionality reduction: a comparative review. Technicla report TiCC TR 2009-005, Tilburg University, Netherlands (2009). http://www.uvt.nl/ticc
30. Marcus, G.: Deep learning: a critical appraisal (2018). arXiv:1801.00631[cs.AI]
31. Martins, R., Coimbra, D., Minghim, R., Telea, A.: Visual analysis of dimensionality reduction quality for parameterized projections. Comput. Graph. **41**, 26–42 (2014)
32. Martins, R.M., Kruiger, J.F., Minghim, R., Telea, A.C., Kerren, A.: MVN-Reduce: dimensionality reduction for the visual analysis of multivariate networks. In: Kozlikova, B., Schreck, T., Wischgoll, T. (eds.) Proceedings of Eurographics - Short Papers (2017)
33. Minard, C.J.: Carte figurative et approximative des quantités de vin français exportés par mer en 1864 (1865)
34. Munzner, T.: Visualization Analysis and Design. CRC Press, Boca Raton (2014)
35. Newbery, F.: Edge concentration: a method for clustering directed graphs. ACM SIGSOFT Softw. Eng. Notes **14**(7), 76–85 (1989)
36. Nguyen, Q., Eades, P., Hong, S.-H.: StreamEB: stream edge bundling. In: Didimo, W., Patrignani, M. (eds.) GD 2012. LNCS, vol. 7704, pp. 400–413. Springer, Heidelberg (2013). https://doi.org/10.1007/978-3-642-36763-2_36
37. Nguyen, Q., Eades, P., Hong, S.H.: On the faithfulness of graph visualizations. In: Carpendale, S., Chen, W., Hong, S. (eds.) Proceedings of IEEE PacificVis (2013)
38. Nocaj, A., Ortmann, M., Brandes, U.: Untangling hairballs. In: Duncan, C., Symvonis, A. (eds.) GD 2014. LNCS, vol. 8871, pp. 101–112. Springer, Heidelberg (2014). https://doi.org/10.1007/978-3-662-45803-7_9
39. Oztireli, A.C., Gross, M.: Perceptually based downscaling of images. ACM TOG **34**(4), 77 (2015)
40. Pal, N., Pal, S.K.: A review on image segmentation techniques. Pattern Recogn. **26**(9), 1277–1294 (1993)
41. Peysakhovich, V., Hurter, C., Telea, A.: Attribute-driven edge bundling for general graphs with applications in trail analysis. In: Liu, S., Scheuermann, G., Takahashi, S. (eds.) Proceedings of IEEE PacificVis, pp. 39–46 (2015)
42. Phan, D., Xiao, L., Yeh, R., Hanrahan, P., Winograd, T.: Flow map layout. In: Stasko, J., Ward, M. (eds.) Proceedings of InfoVis, pp. 219–224 (2005)
43. Post, F.H., Vrolijk, B., Hauser, H., Laramee, R., Doleisch, H.: The state of the art in flow visualisation: feature extraction and tracking. Comput. Graph. Forum **22**(4), 775–792 (2003)

44. Ropinski, T., Oeltze, S., Preim, B.: Survey of glyph-based visualization techniques for spatial multivariate medical data. Comput. Graph. **35**(2), 392–401 (2011)
45. Schaeffer, S.: Graph clustering. Comput. Sci. Rev. **1**(1), 27–64 (2007)
46. Schroeder, W., Zarge, J., Lorensen, W.: Decimation of triangle meshes. In: Thomas, J.J. (ed.) Proceedings of ACM SIGGRAPH, pp. 65–70 (1992)
47. Schulz, H.J., Hurter, C.: Grooming the hairball-how to tidy up network visualizations? In: Proceedings of IEEE InfoVis (Tutorials) (2013)
48. Siddiqi, K., Pizer, S.: Medial Representations: Mathematics, Algorithms and Applications. Springer, Dordrecht (2009). https://doi.org/10.1007/978-1-4020-8658-8
49. Silverman, B.: Density Estimation for Statistics and Data Analysis. Monographs on Statistics and Applied Probability, vol. 26 (1992)
50. Sorzano, C., Vargas, J., Pascual-Montano, A.: A survey of dimensionality reduction techniques (2014). arxiv.org/pdf/1403.2877
51. Tamassia, R.: Handbook of Graph Drawing and Visualization. CRC Press, Boca Raton (2013)
52. Telea, A.: Data Visualization: Principles and Practice, 2nd edn. CRC Press, Boca Raton (2015)
53. Telea, A., Ersoy, O.: Image-based edge bundles: simplified visualization of large graphs. Comput. Graph. Forum **29**(3), 543–551 (2010)
54. Telea, A., Maccari, A., Riva, C.: An open toolkit for prototyping reverse engineering visualizations. In: Ebert, D., Brunet, P., Navazzo, I. (eds.) Proceedings of Data Visualization (IEEE VisSym), pp. 67–75 (2002)
55. Tollis, I., Battista, G.D., Eades, P., Tamassia, R.: Graph Drawing: Algorithms for the Visualization of Graphs. Prentice Hall, Upper Saddle River (1999)
56. Trümper, J., Döllner, J., Telea, A.: Multiscale visual comparison of execution traces. In: Kagdi, H., Poshyvanyk, D., Penta, M.D. (eds.) Proceedings of IEEE ICPC (2013)
57. Tufte, E.R.: The Visual Display of Quantitative Information. Graphics Press, Cheshire (1992)
58. Van Wijk, J.J., van de Wetering, H.: Cushion treemaps: visualization of hierarchical information. In: Wills, G., Keim, D. (eds.) Proceedings of IEEE InfoVis, pp. 73–82 (1999)
59. Weickert, J., Hagen, H.: Visualization and Processing of Tensor Fields. Springer, Heidelberg (2007). https://doi.org/10.1007/3-540-31272-2
60. van Wijk, J.J.: Image based flow visualization. Proc. ACM TOG (SIGGRAPH) **21**(3), 745–754 (2002)
61. Zhou, H., Xu, P., Yuan, X., Qu, H.: Edge bundling in information visualization. Tsinghua Sci. Technol. **18**(2), 145–156 (2013)
62. van der Zwan, M., Codreanu, V., Telea, A.: CUBu: universal real-time bundling for large graphs. IEEE TVCG **22**(12), 2550–2563 (2016)

Planarity Variants

Clustered Planarity = Flat Clustered Planarity

Pier Francesco Cortese and Maurizio Patrignani[(✉)]

Roma Tre University, Rome, Italy
pierfrancesco@pfcortese.it, maurizio.patrignani@uniroma3.it

Abstract. The complexity of deciding whether a clustered graph admits a clustered planar drawing is a long-standing open problem in the graph drawing research area. Several research efforts focus on a restricted version of this problem where the hierarchy of the clusters is 'flat', i.e., no cluster different from the root contains other clusters. We prove that this restricted problem, that we call FLAT CLUSTERED PLANARITY, retains the same complexity of the general CLUSTERED PLANARITY problem, where the clusters are allowed to form arbitrary hierarchies. We strengthen this result by showing that FLAT CLUSTERED PLANARITY is polynomial-time equivalent to INDEPENDENT FLAT CLUSTERED PLANARITY, where each cluster induces an independent set. We discuss the consequences of these results.

1 Introduction

A clustered graph (c-graph) is a planar graph with a recursive hierarchy defined on its vertices. A clustered planar (c-planar) drawing of a c-graph is a planar drawing of the underlying graph where: (i) each cluster is represented by a simple closed region of the plane containing only the vertices of the corresponding cluster, (ii) cluster borders never intersect, and (iii) any edge and any cluster border intersect at most once (more formal definitions are given in Sect. 2). The complexity of deciding whether a c-graph admits a c-planar drawing is still an open problem after more than 20 years of intense research [12,14,17–19,26,32,34–36,38,42–45,49,50,53].

If we had an efficient c-planarity testing and embedding algorithm we could produce straight-line drawings of clustered trees [28] and straight-line drawings [11,33] and orthogonal drawings [27] of c-planar c-graphs with rectangular regions for the clusters.

In order to shed light on the complexity of CLUSTERED PLANARITY, this problem has been compared with other problems whose complexity is likewise challenging. This line of investigation was opened by Marcus Schaefer's

---

This research was partially supported by MIUR project "MODE – MOrphing graph Drawings Efficiently", prot. 20157EFM5C_001.

T. Biedl and A. Kerren (Eds.): GD 2018, LNCS 11282, pp. 23–38, 2018.
https://doi.org/10.1007/978-3-030-04414-5_2

polynomial-time reduction of CLUSTERED PLANARITY to SEFE [53]. SIMUL-
TANEOUS EMBEDDING WITH FIXED EDGES (SEFE) takes as input two planar graphs $G_1 = (V, E_1)$ and $G_2 = (V, E_2)$ and asks whether a planar drawing $\Gamma_1(G_1)$ and a planar drawing $\Gamma_2(G_2)$ exist such that: (i) each vertex $v \in V$
is mapped to the same point in Γ_1 and in Γ_2 and (ii) every edge $e \in E_1 \cap E_2$ is
mapped to the same Jordan curve in Γ_1 and in Γ_2.

However, the polynomial-time equivalence of the two problems is open and
the reverse reduction of SEFE to CLUSTERED PLANARITY is known only for the
case when the intersection graph $G_\cap = (V, E_1 \cap E_2)$ of the instance of SEFE is
connected [4]. Also in this special case, the complexity of the problem is unknown,
with the exception of the case when $G_\cap$ is a star, which produces a c-graph with
only two clusters, a known polynomial case for CLUSTERED PLANARITY [10,47].

Since the general CLUSTERED PLANARITY problem appears to be elusive, several authors focused on a restricted version of it where the hierarchy of the clusters is 'flat', i.e., only the root cluster contains other clusters and it does not directly contain vertices of the underlying graph [2,3,5–
7,9,16,20,21,25,29,37,39–41,47,51]. This restricted problem, that we call FLAT
CLUSTERED PLANARITY, is expressive enough to be useful in several applica-
tive domains, as for example in computer networks where routers are grouped
into Autonomous Systems [15], or social networks where people are grouped
into communities [13,30], or software diagrams where classes are grouped into
packages [52]. Also, several hybrid representations have been proposed for the
visual analysis of (not necessarily planar) flat clustered graphs, such as mixed
matrix and node-link representations [13,23,24,31,46], mixed intersection and
node-link representations [8], and mixed space-filling and node-link representations [1,48,54].

Unfortunately, the complexity of FLAT CLUSTERED PLANARITY is open as
the complexity of the general problem. The authors of [14], after recasting FLAT
CLUSTERED PLANARITY as an embedding problem on planar multi-graphs, conclude that we are still far away from solving it. The authors of [4] wonder whether
FLAT CLUSTERED PLANARITY retains the same complexity of CLUSTERED PLA-
NARITY. In this paper we answer this question in the affirmative. Obviously, a
reduction of FLAT CLUSTERED PLANARITY to CLUSTERED PLANARITY is trivial, since the instances of FLAT CLUSTERED PLANARITY are simply a subset of
those of CLUSTERED PLANARITY. The reverse reduction is the subject of Sect. 3,
that proves the following theorem.

Theorem 1. *There exists a quadratic-time transformation that maps an
instance of* CLUSTERED PLANARITY *to an equivalent instance of* FLAT CLUS-
TERED PLANARITY.

With very similar techniques we are able to prove also a stronger result.

Theorem 2. *There exists a linear-time transformation that maps an instance
of* FLAT CLUSTERED PLANARITY *to an equivalent instance of* INDEPENDENT
FLAT CLUSTERED PLANARITY.

Here, by INDEPENDENT FLAT CLUSTERED PLANARITY we mean the restriction of FLAT CLUSTERED PLANARITY to instances where each non-root cluster induces an independent set.

The paper is structured as follows. Section 2 contains basic definitions. Section 3 contains the proof of Theorem 1 under some simplifying hypotheses (which are removed in [22]). Some immediate consequences of Theorem 1 are discussed in Sect. 4. The proof of Theorem 2 and some remarks about it are in Sects. 5 and 6, respectively. Conclusions and open problems are in Sect. 7. For space reasons some proofs are sketched or, when trivial, omitted.

2   Preliminaries

Let T be a rooted tree. We denote by $r(T)$ the root of T and by $T[\mu]$ the subtree of T rooted at one of its nodes μ. The *depth* of a node μ of T is the length (number of edges) of the path from $r(T)$ to μ. The *height* $h(T)$ of a tree T is the maximum depth of its nodes.

The nodes of a tree can be partitioned into *leaves*, that do not have children, and *internal nodes*. In turn, the internal nodes can be partitioned into two sets: *lower nodes*, whose children are all leaves, and *higher nodes*, that have at least one internal-node child. We say that a node is *homogeneous* if its children are either all leaves or all internal nodes. A tree is *homogeneous* if all its nodes are homogeneous. We say that a tree is *flat* if all its leaves have depth 2. A flat tree is homogeneous. Figure 1 shows a non-homogeneous tree (Fig. 1(a)), a homogeneous tree (Fig. 1(b)), and a flat tree (Fig. 1(c)).

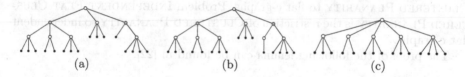

(a)   (b)   (c)

Fig. 1. (a) A tree that is not homogeneous. (b) A homogeneous tree. (c) A flat tree.

We also need a special notion of size: the *size* of a tree T, denoted by $\mathcal{S}(T)$, is the number of higher nodes of T different from the root of T. Observe that a homogeneous tree T is flat if and only if $\mathcal{S}(T) = 0$. For example, the sizes of the trees represented in Figs. 1(a), (b), and (c) are 2, 2, and 0, respectively (filled gray nodes in Fig. 1). The proof of the following lemma is trivial.

Lemma 1. *A homogeneous tree T of height $h(T) \geq 2$ and size $\mathcal{S}(T) > 0$ contains at least one node $\mu^* \neq r(T)$ such that $T[\mu^*]$ is flat.*

A *graph* $G = (V, E)$ is a set V of *vertices* and a set E of *edges*, where each edge is an unordered pair of vertices. A *drawing* $\Gamma(G)$ of G is a mapping of its vertices to distinct points on the plane and of its edges to Jordan curves joining the incident vertices. Drawing $\Gamma(G)$ is *planar* if no two edges intersect except at common end-vertices. A graph is *planar* if it admits a planar drawing.

A *clustered graph* (or *c-graph*) C is a pair (G, T) where $G = (V, E)$ is a planar graph, called the *underlying graph* of C, and T, called the *inclusion tree* of C, is a rooted tree such that the set of leaves of T coincides with V. A *cluster* μ is an internal node of T. When it is not ambiguous we also identify a cluster with the respective subset of the vertex set. An *inter-cluster edge* of a cluster μ of T is an edge of G that has one end-vertex inside μ and the other end-vertex outside μ. An *independent set* of vertices is a set of pairwise non-adjacent vertices. A cluster μ of T is *independent* if its vertices form an independent set. A c-graph is *independent* if all its clusters, with the exception of the root, are independent clusters. A cluster μ of T is a *lower cluster* (*higher cluster*) of C if μ is a lower node (higher node) of T.

A c-graph is *flat* if its inclusion tree is flat. The clusters of a flat c-graph are all lower clusters with the exception of the root cluster. A cluster is called *singleton* if it contains a single cluster or a single vertex.

A *drawing* $\Gamma(C)$ of a c-graph $C(G, T)$ is a mapping of vertices and edges of G to points and to Jordan curves joining their incident vertices, respectively, and of each internal node μ of T to a simple closed region $R(\mu)$ containing exactly the vertices of μ. Drawing $\Gamma(C)$ is *c-planar* if: (i) curves representing edges of G do not intersect except at common end-points; (ii) the boundaries of the regions representing clusters do not intersect; and (iii) each edge intersects the boundary of a region at most one time. A c-graph is *c-planar* if it admits a c-planar drawing.

Problem CLUSTERED PLANARITY is the problem of deciding whether a c-graph is c-planar. Problem FLAT CLUSTERED PLANARITY is the restriction of CLUSTERED PLANARITY to flat c-graphs. Problem INDEPENDENT FLAT CLUS-TERED PLANARITY is the restriction of CLUSTERED PLANARITY to independent flat c-graphs.

The proof of the following lemma can be found in [22].

Lemma 2. *An instance $C(G, T)$ of* CLUSTERED PLANARITY *with n vertices and c clusters can be reduced in time $O(n+c)$ to an equivalent instance such that: (1) T is homogeneous, (2) $r(T)$ has at least two children, and (3) $h(T) \leq n-1$.*

3   Proof of Theorem 1

We describe a polynomial-time reduction of CLUSTERED PLANARITY to FLAT CLUSTERED PLANARITY. Let $C(G, T)$ be a clustered graph, let n be the number of vertices of G, and let c be the number of clusters of C. Due to Lemma 2 we can achieve in $O(n+c)$ time that T is homogeneous and $\mathcal{S}(T) \in O(n)$. We reduce C to an equivalent instance $C_f(G_f, T_f)$ where T_f is flat. The reduction consists of a sequence of transformations of $C = C_0$ into $C_1, C_2, \ldots, C_{\mathcal{S}(T)} = C_f$, where each $C_i(G_i, T_i)$, $i = 0, 1, \ldots, \mathcal{S}(T)$, has an homogeneous inclusion tree T_i and each transformation takes $O(n)$ time.

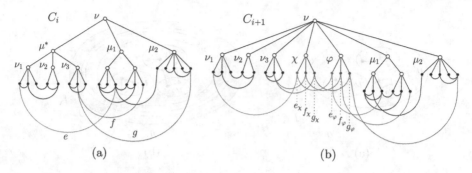

Fig. 2. (a) A c-graph C_i. Inter-cluster edges of μ^* are colored red, green, and blue. (b) The construction of C_{i+1}. (Color figure online)

Consider any $C_i(G_i, T_i)$, with $i = 0, \ldots, \mathcal{S}(T) - 1$, where T_i is a homogeneous, non-flat tree of height $h(T_i) \geq 2$ (refer to Fig. 2(a)). By Lemma 1, T_i has at least one node $\mu^* \neq r(T_i)$ such that $T_i[\mu^*]$ is flat. Since $\mu^* \neq r(T_i)$, node μ^* has a parent ν. Also, denote by $\nu_1, \nu_2, \ldots, \nu_h$ the children of μ^* and by $\mu_1, \mu_2, \ldots, \mu_k$ the siblings of μ^* in T_i. We construct $C_{i+1}(G_{i+1}, T_{i+1})$ as follows (refer to Fig. 2(b)). Graph G_{i+1} is obtained from G_i by introducing, for each inter-cluster edge $e = (u, v)$ of μ^*, two new vertices e_χ and e_φ and by replacing e with a path $(u, e_\chi)(e_\chi, e_\varphi)(e_\varphi, v)$. Tree T_{i+1} is obtained from T_i by removing node μ^*, attaching its children $\nu_1, \nu_2, \ldots, \nu_h$ directly to ν and adding to ν two new children χ and φ, where cluster χ (cluster φ, respectively) contains all vertices e_χ (e_φ, respectively) introduced when replacing each inter-cluster edge e of μ^* with a path. The following lemmas are trivial.

Lemma 3. *If T_i is homogeneous then T_{i+1} is homogeneous.*

Lemma 4. *We have that $\mathcal{S}(T_{i+1}) = \mathcal{S}(T_i) - 1$.*

Lemma 5. *The c-graph $C_f = C_{\mathcal{S}(T)}$ is flat.*

The proof of the following lemma is given here under two simplifying hypotheses (the proof of the general case can be found in [22]):

$\mathcal{H}$-*conn*: The underlying graph G_i is connected
$\mathcal{H}$-*not-root*: Cluster ν is not the root of T

Observe that Hypothesis $\mathcal{H}$-*conn* implies that also G_{i+1} is connected. Observe, also, that Hypothesis $\mathcal{H}$-*not-root* and Property 2 of Lemma 2 imply that there is at least one vertex of G_i that is not part of ν (this hypothesis is not satisfied, for example, by the c-graph depicted in Fig. 2(a)).

Lemma 6. *$C_i(G_i, T_i)$ is c-planar if and only if $C_{i+1}(G_{i+1}, T_{i+1})$ is c-planar.*

Proof sketch. The first direction of the proof is straightforward. Let $\Gamma(C_i)$ be a c-planar drawing of C_i (refer to Fig. 3(a)). We show how to construct a c-planar

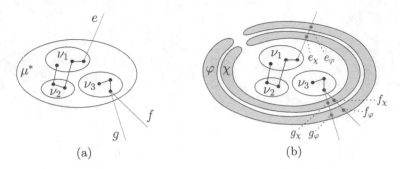

Fig. 3. (a) A c-planar drawing $\Gamma(C_i)$ of c-graph C_i. (b) The construction of a c-planar drawing $\Gamma(C_{i+1})$.

drawing of C_{i+1} (refer to Fig. 3(b)). Consider the region $R(\mu^*)$ that contains $R(\nu_i)$, with $i = 1, \dots, h$. The boundary of $R(\mu^*)$ is crossed exactly once by each inter-cluster edge of μ^*. Identify outside the boundary of $R(\mu^*)$ two arbitrarily thin regions $R(\chi)$ and $R(\varphi)$ that turn around $R(\mu^*)$ and that intersect exactly once all and only the inter-cluster edges of μ^*. Insert into each inter-cluster edge e of μ^* two vertices e_χ and e_φ, placing e_χ inside $R(\chi)$ and e_φ inside $R(\varphi)$. By ignoring $R(\mu^*)$ you have a c-planar drawing $\Gamma(C_{i+1})$ of C_{i+1}.

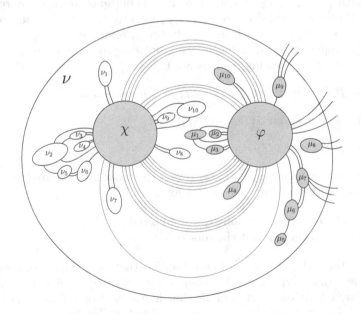

Fig. 4. A c-planar drawing of clusters ν, χ, and φ in $\Gamma(C_{i+1})$. (Color figure online)

Suppose now to have a c-planar drawing $\Gamma(C_{i+1})$ of C_{i+1}. We show how to construct a c-planar drawing $\Gamma(C_i)$ of C_i under the Hypotheses $\mathcal{H}$-*conn* and

$\mathcal{H}$-*not-root*. Consider the regions $R(\chi)$ and $R(\varphi)$ inside $R(\nu)$ (refer to Fig. 4). Regions $R(\chi)$ and $R(\varphi)$ are joined by the p inter-cluster edges introduced when replacing each inter-cluster edge e_i of μ^*, where $i = 1, \ldots, p$, with a path (red edges of Fig. 4). Such inter-cluster edges of χ and φ partition $R(\nu)$ into p regions that have to host the remaining children of ν and the inter-cluster edges among them. In particular, $p-1$ of these regions are bounded by two inter-cluster edges and two portions of the boundaries of $R(\chi)$ and $R(\varphi)$. One of such regions, instead, is also externally bounded by the boundary of $R(\nu)$.

Now consider the regions $R(\nu_i)$ corresponding to the children ν_i of ν, with $i = 1, \ldots, h$, that were originally children of μ^*. These regions (filled white in Fig. 4) may have inter-cluster edges among them and may be connected to χ, but by construction cannot have inter-cluster edges connecting them to φ, or connecting them to the original children $\mu_i \neq \mu^*$ of ν, or exiting the border of $R(\nu)$. In particular, due to Hypothesis $\mathcal{H}$-*conn*, these regions must be directly or indirectly connected to χ. Finally, consider the regions $R(\mu_i)$ corresponding to the original children $\mu_i \neq \mu^*$ of ν (filled gray in Fig. 4). These regions may have inter-cluster edges among them, connecting them to φ, or connecting them to the rest of the graph outside ν. In particular, due to Hypotheses $\mathcal{H}$-*conn* and $\mathcal{H}$-*not-root*, each μ_i (and also φ) must be directly or indirectly connected to the border of $R(\nu)$. It follows that the drawing in $\Gamma(C_{i+1})$ of the subgraph G_{μ^*} composed by the regions of $\chi, \nu_1, \nu_2, \ldots, \nu_h$ and their inter-cluster edges cannot contain in one of its internal faces any other cluster of ν. Hence, the sub-region $R(\mu^*)$ of $R(\nu)$ that is the union of $R(\chi)$ and the region enclosed by G_{μ^*} is a closed and simple region that only contains the regions $R(\nu_1), \ldots R(\nu_h)$ plus the region $R(\chi)$ and all the inter-cluster edges among them (see Fig. 5). By ignoring $R(\chi)$ and $R(\varphi)$ and by removing vertices e_χ and e_φ and joining their incident edges we obtain a c-planar drawing $\Gamma(C_i)$.   □

The proof of Theorem 1 descends from Lemmas 5 and 6 and from the consideration that each construction of C_{i+1} from C_i takes at most $O(n)$ time and, hence, the time needed to construct C_f is $O(n^2)$. Due to the $O(n+c)$-time preprocessing (Lemma 2), the overall time complexity of the reduction is $O(n^2+c)$.

4   Remarks About Theorem 1

In this section we discuss some consequences of Theorem 1 that descend from the properties of the reduction described in Sect. 3. Such properties are summarized in the following lemma.

Lemma 7. *Let $C(G,T)$ be an n-vertex clustered graph with c clusters. The flat clustered graph $C_f(G_f, T_f)$ equivalent to C built as described in the proof of Theorem 1 has the following properties:*

1. *Graph G_f is a subdivision of G*
2. *Each edge of G is replaced by a path of length at most $4h(T) - 8$*
3. *The number of vertices of G_f is $n_f \in O(n \cdot h(T))$*

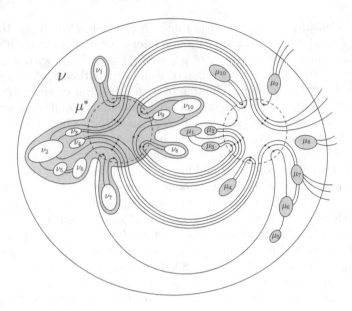

Fig. 5. The drawing of cluster μ in $\Gamma(C_i)$ corresponding to the drawing $\Gamma(C_{i+1})$ of Fig. 4.

4. The number of clusters of C_f is $c_f = c + \mathcal{S}(T)$

Proof. Regarding Property 1, observe that, for $i = 1, \ldots, \mathcal{S}(T)$, each G_i is obtained from G_{i-1} by replacing edges with paths. Hence $G_{\mathcal{S}(T)} = G_f$ is a subdivision of $G_0 = G$. To prove Property 2 observe that each time an edge e is subdivided, a pair of vertices e_χ and e_φ is inserted and that edges are subdivided when the boundary of a higher cluster is removed. Edges that traverse more boundaries are those that link two vertices whose lowest common ancestor is the root of T. These edges traverse $2h(T) - 4$ higher-cluster boundaries in C. Hence, the number of vertices inserted into these edges is $4h(T) - 8$. Property 3 can be proved by considering that G has $O(n)$ edges and each edge, by Property 2, is replaced by a path of length at most $O(h(T))$. Finally, Property 4 descends from the fact that at each step C_{i+1} has exactly one cluster more than C_i, since new clusters χ and φ are inserted but cluster μ^* is removed. $\square$

An immediate consequence of Property 1 of Lemma 7 is that the number of faces of G_f is equal to the number of faces of G. Also, if G is connected, biconnected, or a subdivision of a triconnected graph, G_f is also connected, biconnected, or a subdivision of a triconnected graph, respectively. If G is a cycle or a tree, G_f is also a cycle or a tree, respectively. Hence, the complexity of CLUSTERED PLANARITY restricted to these kinds of graphs can be related to the complexity of FLAT CLUSTERED PLANARITY restricted to the same kinds of graphs. Further, since a subdivision preserves the embedding of the original graph, the problem of deciding whether a c-graph $C(G, T)$ admits a c-planar drawing where G has a fixed embedding is polynomially equivalent to deciding whether a flat c-graph $C_f(G_f, T_f)$ admits a c-planar drawing where G_f has a fixed embedding.

By the above observations some results on flat clustered graphs can be immediately exported to general c-graphs. Consider for example the following.

Theorem 3 *([16, Theorem 1]). There exists an $O(n^3)$-time algorithm to test the c-planarity of an n-vertex embedded flat c-graph C with at most two vertices per cluster on each face.*

We generalize Theorem 3 to non-flat c-graphs.

Theorem 4. *Let $C(G, T)$ be an n-vertex c-graph where G has a fixed embedding. There exists an $O(n^3 \cdot h(T)^3)$-time algorithm to test the c-planarity of C if each lower cluster has at most two vertices on the same face of G and each higher cluster has at most two inter-cluster edges on the same face of G.*

Proof sketch. The proof is based on showing that, starting from a c-graph $C(G, T)$ that satisfies the hypotheses of the statement, the equivalent flat c-graph $C_f(G_f, T_f)$ built as described in the proof of Theorem 1 satisfies the hypotheses of Theorem 3. Hence, we first transform $C(G, T)$ into $C_f(G_f, T_f)$ in $O(n^2)$ time and then apply Theorem 3 to $C_f(G_f, T_f)$, which gives an answer to the c-planarity test in $O(n_f^3)$ time, which is, by Property 3 of Lemma 7, $O(n^3 \cdot h(T)^3)$ time. □

In [25] it has been proven that FLAT CLUSTERED PLANARITY admits a subexponential-time algorithm when the underlying graph has a fixed embedding and its maximum face size ℓ belongs to $o(n)$.

Theorem 5 *([25, Theorem 3]). FLAT CLUSTERED PLANARITY can be solved in $2^{O(\sqrt{\ell n} \cdot \log n)}$ time for n-vertex embedded flat c-graphs with maximum face size ℓ.*

The authors of [25] ask whether their results can be generalized to non-flat c-graphs. We give an affirmative answer with the following theorem.

Theorem 6. CLUSTERED PLANARITY *can be solved in $2^{O(h(T) \cdot \sqrt{\ell n} \cdot \log(n \cdot h(T))}$ time for n-vertex embedded c-graphs with maximum face size ℓ and height $h(T)$ of the inclusion tree.*

Proof sketch. The proof is based on applying Theorem 5 to the equivalent flat c-graph $C_f(G_f, T_f)$ built as described in the proof of Theorem 1. □

Observe that Theorem 6 gives a subexponential-time upper bound for CLUSTERED PLANARITY whenever $\ell \cdot h(T)^2 \in o(n)$. Also observe that Theorems 4 and 6 are actual generalizations of the corresponding Theorems 3 and 5, respectively, as they yield the same bounds when applied to flat clustered graphs.

5   Proof of Theorem 2

In this section we reduce FLAT CLUSTERED PLANARITY to INDEPENDENT FLAT CLUSTERED PLANARITY by applying a transformation very similar to the one described in Sect. 3 to each non-independent cluster.

Let $C(G,T)$ be a flat c-graph. Let k be the number of lower clusters of C that are not independent. The reduction consists of a sequence of transformations of $C = C_0$ into $C_1, C_2, \ldots, C_k$ where each C_i, $i = 0, \ldots, k$, is a flat c-graph with $k - i$ non-independent lower clusters.

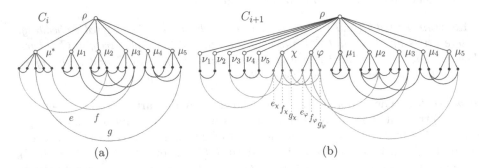

(a)   (b)

Fig. 6. (a) A flat c-graph C_i with a non-independent cluster μ^*. (b) The construction of C_{i+1} where μ^* is replaced by independent clusters $\nu_1, \ldots, \nu_5, \chi$, and φ.

Consider a flat c-graph $C_i(G_i, T_i)$, with $i = 0, \ldots, k - 1$, such that C_i has $k - i$ non-independent clusters and let μ^* be a non-independent cluster of C. We show how to construct an flat c-graph $C_{i+1}(G_{i+1}, T_{i+1})$ equivalent to C_i and such that C_{i+1} has $k - i - 1$ non-independent clusters (refer to Fig. 6). Denote by μ_j, with $j = 1, 2, \ldots, l$, those children of $r(T_i)$ such that $\mu_j \neq \mu^*$. Suppose that μ^* has children $v_1, v_2, \ldots, v_h$, which are vertices of G_i.

The underlying graph G_{i+1} of C_{i+1} is obtained from G_i by introducing, for each inter-cluster edge $e = (u, v)$ of μ^*, two new vertices e_χ and e_φ and replacing e with a path $(u, e_\chi)(e_\chi, e_\varphi)(e_\varphi, v)$. The inclusion tree T_{i+1} of C_{i+1} is obtained from T_i by removing cluster μ^* and introducing, for each $j = 1, 2, \ldots, h$, a lower cluster ν_j child of $r(T_{i+1})$ containing only v_j. We also introduce two lower clusters χ and φ as children of $r(T_{i+1})$ that contain all the vertices e_χ and e_φ, respectively, introduced when replacing each inter-cluster edge e of μ^* with a path. It is easy to see that C_{i+1} is a flat clustered graph and that it has one non-independent cluster less than C_i.

We prove the following lemma assuming that Hypothesis $\mathcal{H}$-*conn* holds. The complete proof is in [22].

Lemma 8. $C_i(G_i, T_i)$ *is c-planar if and only if* $C_{i+1}(G_{i+1}, T_{i+1})$ *is c-planar.*

Proof sketch. The proof is very similar to the proof of Lemma 6. First, we show that, given a c-planar drawing $\Gamma(C_i)$ of the flat c-graph C_i it is easy to construct a

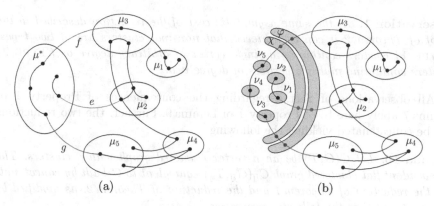

Fig. 7. (a) A c-planar drawing of the flat c-graph of Fig. 6(a). (b) The corresponding c-planar drawing the flat c-graph of Fig. 6(b) where the non-independent cluster μ^* is replaced by independent clusters $\nu_1, \ldots, \nu_5, \chi$, and φ.

c-planar drawing $\Gamma(C_{i+1})$ of C_{i+1} (see, as an example, Fig. 7). Second we show that, given a c-planar drawing $\Gamma(C_{i+1})$ of the flat c-graph C_{i+1} it is possible to construct a c-planar drawing $\Gamma(C_i)$ of C_i. This second part of the proof is complicated by the fact that, since in this case Hypothesis $\mathcal{H}$-*not-root* does not apply, we may have that in $\Gamma(C_{i+1})$ the region $R(\varphi)$ is embraced by inter-cluster edges and region boundaries of $R(\nu_1)$, $R(\nu_2)$, ... $R(\nu_l)$, and $R(\chi)$. Hence, before identifying the region $R(\mu^*)$ the drawing $\Gamma(C_{i+1})$ needs to be modified so that the external face touches $R(\varphi)$. This can be easily done by rerouting edges. $\square$

The proof of Theorem 2 is concluded by showing that each G_{i+1} can be obtained from G_i in time proportional to the number of vertices and inter-cluster edges of μ^*, which gives an overall $O(n)$ time for the reduction.

6   Remarks About Theorem 2

Starting from a flat c-graph, the reduction described in Sect. 5 allows us to find an equivalent independent flat c-graph with the properties stated in the following lemma (the proof can be found in [22]).

Lemma 9. *Let $C_f(G_f, T_f)$ be an n_f-vertex flat clustered graph with c_f clusters. The independent flat clustered graph $C_{if}(G_{if}, T_{if})$ equivalent to C_f built as described in the proof of Theorem 2 has the following properties:*

1. *Graph G_{if} is a subdivision of G_f*
2. *Each inter-cluster edge of G_f is replaced by a path of length at most 4.*
3. *The number of vertices of G_{if} is $O(n_f)$*
4. *The number of clusters of C_{if} (including the root) is $c_{if} \leq 2c_f + n_f - 1$*

Also, a further property can be pursued.

Observation 1. *At the same asymptotic cost of the reduction described in the proof of Theorem 2 it can be achieved that non-root clusters are of two types: (*TYPE 1*) clusters containing a single vertex of arbitrary degree or (*TYPE 2*) clusters containing multiple vertices of degree two.*

All observations of Sect. 4 regarding the consequences of Property 1 of Lemma 7 apply here to of Property 1 of Lemma 9. Further, the two reductions can be concatenated yielding the following.

Lemma 10. *Let $C(G,T)$ be an n-vertex clustered graph with c clusters. The independent flat clustered graph $C_{if}(G_{if}, T_{if})$ equivalent to C built by concatenating the reduction of Theorem 1 and the reduction of Theorem 2, as modified by Observation 1, has the following properties:*

1. *Graph G_{if} is a subdivision of G*
2. *Each inter-cluster edge of G_f is replaced by a path of length at most $4h(T) - 4$*
3. *The number of vertices of G_{if} is $O(n^2)$*
4. *The number of clusters of C_{if} is $O(n \cdot h(T))$*
5. *Non-root clusters are of two types: (*TYPE 1*) clusters containing a single vertex of arbitrary degree or (*TYPE 2*) clusters containing multiple vertices of degree two*

Lemma 10 describes the most constrained version of CLUSTERED PLANARITY that is known to be polynomially equivalent to the general problem. Observe that if all non-root clusters of a c-graph $C(G,T)$ are of TYPE 1 then INDEPENDENT FLAT CLUSTERED PLANARITY is linear, since C is c-planar if and only if G is planar. Conversely, if all clusters are of TYPE 2 then the underlying graph is a collection of cycles, and the problem has unknown complexity [20, 21].

7   Conclusions and Open Problems

We showed that CLUSTERED PLANARITY can be reduced to FLAT CLUSTERED PLANARITY and that this problem, in turn, can be reduced to INDEPENDENT FLAT CLUSTERED PLANARITY. The consequences of these results are twofold: on one side the investigations about the complexity of CLUSTERED PLANARITY could legitimately be restricted to (independent) flat clustered graphs, neglecting more complex hierarchies of the inclusion tree; on the other side some polynomial-time results on flat clustered graphs could be easily exported to general c-graphs (we gave some examples in Sect. 4).

We remark that while Theorems 1 and 2 are formulated in terms of decision problems, their proofs offer a solution of the corresponding search problems, meaning that they actually describe a polynomial-time algorithm to compute a c-planar drawing of a c-graph, provided to have a c-planar drawing of the corresponding flat c-graph or a c-planar drawing of the corresponding independent flat c-graph.

Several interesting questions are left open:

- Can the reduction presented in this paper be used to generalize some other polynomial-time testing algorithm for FLAT CLUSTERED PLANARITY to plain CLUSTERED PLANARITY?
- What is the complexity of INDEPENDENT FLAT CLUSTERED PLANARITY when the underlying graph is a cycle? We know that this problem is polynomial only for constrained drawings of the inter-cluster edges [20,21].
- What is the complexity of INDEPENDENT FLAT CLUSTERED PLANARITY when the number of TYPE 2 clusters is bounded?

References

1. Abello, J., Kobourov, S.G., Yusufov, R.: Visualizing large graphs with compound-fisheye views and treemaps. In: Pach, J. (ed.) GD 2004. LNCS, vol. 3383, pp. 431–441. Springer, Heidelberg (2005). https://doi.org/10.1007/978-3-540-31843-9_44

2. Akitaya, H.A., Fulek, R., Tóth, C.D.: Recognizing weak embeddings of graphs. In: Czumaj, A. (ed.) SODA 2018, pp. 274–292 (2018). https://doi.org/10.1137/1.9781611975031.20

3. Angelini, P., Da Lozzo, G.: Clustered planarity with pipes. In: Hong, S.H., Eades, P., Meidiana, A. (eds.) ISAAC 2016. LIPIcs, vol. 64, pp. 13:1–13:13 (2016). https://doi.org/10.4230/LIPIcs.ISAAC.2016.13

4. Angelini, P., Da Lozzo, G.: SEFE = C-planarity? Comput. J. **59**(12), 1831–1838 (2016). https://doi.org/10.1093/comjnl/bxw035

5. Angelini, P., Da Lozzo, G., Di Battista, G., Frati, F.: Strip planarity testing. In: Wismath, S., Wolff, A. (eds.) GD 2013. LNCS, vol. 8242, pp. 37–48. Springer, Cham (2013). https://doi.org/10.1007/978-3-319-03841-4_4

6. Angelini, P., Da Lozzo, G., Di Battista, G., Frati, F.: Strip planarity testing for embedded planar graphs. Algorithmica **77**(4), 1022–1059 (2017). https://doi.org/10.1007/s00453-016-0218-9

7. Angelini, P., Da Lozzo, G., Di Battista, G., Frati, F., Patrignani, M., Roselli, V.: Relaxing the constraints of clustered planarity. Comput. Geom. Theory Appl. **48**(2), 42–75 (2015). https://doi.org/10.1016/j.comgeo.2014.08.001

8. Angelini, P., Da Lozzo, G., Di Battista, G., Frati, F., Patrignani, M., Rutter, I.: Intersection-link representations of graphs. J. Graph Algorithms Appl. **21**(4), 731–755 (2017). https://doi.org/10.7155/jgaa.00437

9. Angelini, P., Da Lozzo, G., Di Battista, G., Frati, F., Roselli, V.: The importance of being proper (in clustered-level planarity and T-level planarity). Theor. Comput. Sci. **571**, 1–9 (2015). https://doi.org/10.1016/j.tcs.2014.12.019

10. Angelini, P., Di Battista, G., Frati, F., Patrignani, M., Rutter, I.: Testing the simultaneous embeddability of two graphs whose intersection is a biconnected or a connected graph. J. Discret. Algorithms **14**, 150–172 (2012). https://doi.org/10.1016/j.jda.2011.12.015

11. Angelini, P., Frati, F., Kaufmann, M.: Straight-line rectangular drawings of clustered graphs. Discret. Comput. Geom. **45**(1), 88–140 (2011). https://doi.org/10.1007/s00454-010-9302-z

12. Angelini, P., Frati, F., Patrignani, M.: Splitting clusters to get C-planarity. In: Eppstein, D., Gansner, E.R. (eds.) GD 2009. LNCS, vol. 5849, pp. 57–68. Springer, Heidelberg (2010). https://doi.org/10.1007/978-3-642-11805-0_8

13. Batagelj, V., Brandenburg, F.J., Didimo, W., Liotta, G., Palladino, P., Patrignani, M.: Visual analysis of large graphs using (X, Y)-clustering and hybrid visualizations. IEEE Trans. Visual. Comput. Graph. **17**(11), 1587–1598 (2011). https://doi.org/10.1109/TVCG.2010.265

14. Bläsius, T., Rutter, I.: A new perspective on clustered planarity as a combinatorial embedding problem. Theor. Comput. Sci. **609**, 306–315 (2016). https://doi.org/10.1016/j.tcs.2015.10.011

15. Candela, M., Di Bartolomeo, M., Di Battista, G., Squarcella, C.: Radian: visual exploration of traceroutes. IEEE Trans. Visual. Comput. Graph. **24**, 2194–2208 (2018). https://doi.org/10.1109/TVCG.2017.2716937

16. Chimani, M., Di Battista, G., Frati, F., Klein, K.: Advances on testing C-planarity of embedded flat clustered graphs. In: Duncan, C., Symvonis, A. (eds.) GD 2014. LNCS, vol. 8871, pp. 416–427. Springer, Heidelberg (2014). https://doi.org/10.1007/978-3-662-45803-7_35

17. Cornelsen, S., Wagner, D.: Completely connected clustered graphs. J. Discret. Algorithms **4**(2), 313–323 (2006). https://doi.org/10.1016/j.jda.2005.06.002

18. Cortese, P.F., Di Battista, G.: Clustered planarity (invited lecture). In: SoCG 2005, pp. 30–32. ACM (2005)

19. Cortese, P.F., Di Battista, G., Frati, F., Patrignani, M., Pizzonia, M.: C-planarity of c-connected clustered graphs. J. Graph Algorithms Appl. **12**(2), 225–262 (2008)

20. Cortese, P.F., Di Battista, G., Patrignani, M., Pizzonia, M.: Clustering cycles into cycles of clusters. J. Graph Algorithms Appl. **9**(3), 391–413 (2005)

21. Cortese, P.F., Di Battista, G., Patrignani, M., Pizzonia, M.: On embedding a cycle in a plane graph. Discret. Math. **309**(7), 1856–1869 (2009). https://doi.org/10.1016/j.disc.2007.12.090

22. Cortese, P., Patrignani, M.: Clustered planarity = Flat clustered planarity. Technical report arXiv:1808.07437v1, Cornell University (2018). http://arxiv.org/abs/1808.07437v2

23. Da Lozzo, G., Di Battista, G., Frati, F., Patrignani, M.: Computing NodeTrix representations of clustered graphs. In: Hu, Y., Nöllenburg, M. (eds.) GD 2016. LNCS, vol. 9801, pp. 107–120. Springer, Cham (2016). https://doi.org/10.1007/978-3-319-50106-2_9

24. Da Lozzo, G., Di Battista, G., Frati, F., Patrignani, M.: Computing NodeTrix representations of clustered graphs. J. Graph Algorithms Appl. **22**(2), 139–176 (2018)

25. Da Lozzo, G., Eppstein, D., Goodrich, M.T., Gupta, S.: Subexponential-time and FPT algorithms for embedded flat clustered planarity. In: WG 2018 (2018, to appear)

26. Dahlhaus, E.: A linear time algorithm to recognize clustered planar graphs and its parallelization. In: Lucchesi, C.L., Moura, A.V. (eds.) LATIN 1998. LNCS, vol. 1380, pp. 239–248. Springer, Heidelberg (1998). https://doi.org/10.1007/BFb0054325

27. Di Battista, G., Didimo, W., Marcandalli, A.: Planarization of clustered graphs. In: Mutzel, P., Jünger, M., Leipert, S. (eds.) GD 2001. LNCS, vol. 2265, pp. 60–74. Springer, Heidelberg (2002). https://doi.org/10.1007/3-540-45848-4_5

28. Di Battista, G., Drovandi, G., Frati, F.: How to draw a clustered tree. J. Discret. Algorithms **7**(4), 479–499 (2009). https://doi.org/10.1016/j.jda.2008.09.015

29. Di Battista, G., Frati, F.: Efficient c-planarity testing for embedded flat clustered graphs with small faces. J. Graph Algorithms Appl. **13**(3), 349–378 (2009)
30. Di Giacomo, E., Didimo, W., Liotta, G., Palladino, P.: Visual analysis of one-to-many matched graphs. J. Graph Algorithms Appl. **14**(1), 97–119 (2010)
31. Di Giacomo, E., Liotta, G., Patrignani, M., Tappini, A.: NodeTrix planarity testing with small clusters. In: Frati, F., Ma, K.-L. (eds.) GD 2017. LNCS, vol. 10692, pp. 479–491. Springer, Cham (2018). https://doi.org/10.1007/978-3-319-73915-1_37
32. Eades, P., Feng, Q.-W.: Multilevel visualization of clustered graphs. In: North, S. (ed.) GD 1996. LNCS, vol. 1190, pp. 101–112. Springer, Heidelberg (1997). https://doi.org/10.1007/3-540-62495-3_41
33. Eades, P., Feng, Q., Lin, X., Nagamochi, H.: Straight-line drawing algorithms for hierarchical graphs and clustered graphs. Algorithmica **44**(1), 1–32 (2006). https://doi.org/10.1007/s00453-004-1144-8
34. Eades, P., Feng, Q., Nagamochi, H.: Drawing clustered graphs on an orthogonal grid. J. Graph Algorithms Appl. **3**(4), 3–29 (1999). https://doi.org/10.7155/jgaa.00016
35. Eades, P., Huang, M.L.: Navigating clustered graphs using force-directed methods. J. Graph Algorithms Appl. **4**(3), 157–181 (2000). https://doi.org/10.7155/jgaa.00029
36. Feng, Q.-W., Cohen, R.F., Eades, P.: Planarity for clustered graphs. In: Spirakis, P. (ed.) ESA 1995. LNCS, vol. 979, pp. 213–226. Springer, Heidelberg (1995). https://doi.org/10.1007/3-540-60313-1_145
37. Frati, F.: Multilayer drawings of clustered graphs. J. Graph Algorithms Appl. **18**(5), 633–675 (2014). https://doi.org/10.7155/jgaa.00340
38. Frati, F.: Clustered graph drawing. In: Kao, M.Y. (ed.) Encyclopedia of Algorithms, 2nd edn, pp. 1–6. Springer, New York (2015). https://doi.org/10.1007/978-3-642-27848-8_655-1
39. Fulek, R.: C-planarity of embedded cyclic c-graphs. Comput. Geom. **66**, 1–13 (2017). https://doi.org/10.1016/j.comgeo.2017.06.016
40. Fulek, R.: Embedding graphs into embedded graphs. In: Okamoto, Y., Tokuyama, T. (eds.) ISAAC 2017. LIPIcs, vol. 92, pp. 34:1–34:12 (2017). https://doi.org/10.4230/LIPIcs.ISAAC.2017.34
41. Fulek, R., Kynčl, J.: Hanani-Tutte for approximating maps of graphs. In: Speckmann, B., Tóth, C.D. (eds.) SoCG 18. LIPIcs, pp. 39:1–39:15 (2018). https://doi.org/10.4230/LIPIcs.SoCG.2018.39
42. Fulek, R., Kynčl, J., Malinović, I., Pálvölgyi, D.: Clustered planarity testing revisited. Electr. J. Comb. **22**(4) (2015). http://www.combinatorics.org/ojs/index.php/eljc/article/view/v22i4p24, Paper P4.24
43. Goodrich, M.T., Lueker, G.S., Sun, J.Z.: C-planarity of extrovert clustered graphs. In: Healy, P., Nikolov, N.S. (eds.) GD 2005. LNCS, vol. 3843, pp. 211–222. Springer, Heidelberg (2006). https://doi.org/10.1007/11618058_20
44. Gutwenger, C., Jünger, M., Leipert, S., Mutzel, P., Percan, M., Weiskircher, R.: Advances in C-planarity testing of clustered graphs. In: Goodrich, M.T., Kobourov, S.G. (eds.) GD 2002. LNCS, vol. 2528, pp. 220–236. Springer, Heidelberg (2002). https://doi.org/10.1007/3-540-36151-0_21
45. Gutwenger, C., Mutzel, P., Schaefer, M.: Practical experience with Hanani-Tutte for testing c-planarity. In: McGeoch, C.C., Meyer, U. (eds.) ALENEX 2014, pp. 86–97. SIAM (2014). https://doi.org/10.1137/1.9781611973198.9
46. Henry, N., Fekete, J., McGuffin, M.J.: NodeTrix: a hybrid visualization of social networks. IEEE Trans. Vis. Comput. Graph. **13**(6), 1302–1309 (2007)

47. Hong, S.H., Nagamochi, H.: Two-page book embedding and clustered graph planarity. Technical report 2009–004, Department of Applied Mathematics & Physics, University of Kyoto, Japan (2009)
48. Itoh, T., Muelder, C., Ma, K., Sese, J.: A hybrid space-filling and force-directed layout method for visualizing multiple-category graphs. In: Eades, P., Ertl, T., Shen, H. (eds.) IEEE PacificVis 2009, pp. 121–128 (2009)
49. Jelínek, V., Jelínková, E., Kratochvíl, J., Lidický, B.: Clustered planarity: embedded clustered graphs with two-component clusters. In: Tollis, I.G., Patrignani, M. (eds.) GD 2008. LNCS, vol. 5417, pp. 121–132. Springer, Heidelberg (2009). https://doi.org/10.1007/978-3-642-00219-9_13
50. Jelínek, V., Suchý, O., Tesař, M., Vyskočil, T.: Clustered planarity: clusters with few outgoing edges. In: Tollis, I.G., Patrignani, M. (eds.) GD 2008. LNCS, vol. 5417, pp. 102–113. Springer, Heidelberg (2009). https://doi.org/10.1007/978-3-642-00219-9_11
51. Jelínková, E., Kára, J., Kratochvíl, J., Pergel, M., Suchý, O., Vyskocil, T.: Clustered planarity: small clusters in cycles and eulerian graphs. J. Graph Algorithms Appl. **13**(3), 379–422 (2009). https://doi.org/10.7155/jgaa.00192
52. Sander, G.: Visualisierungstechniken für den compilerbau. Ph.D. thesis, Universität Saarbrücken, Germany (1996)
53. Schaefer, M.: Toward a theory of planarity: Hanani-Tutte and planarity variants. J. Graph Algorithms Appl. **17**(4), 367–440 (2013). https://doi.org/10.7155/jgaa.00298
54. Zhao, S., McGuffin, M.J., Chignell, M.H.: Elastic hierarchies: combining treemaps and node-link diagrams. In: Stasko, J.T., Ward, M.O. (eds.) IEEE InfoVis 2005, p. 8 (2005). https://doi.org/10.1109/INFOVIS.2005.12

Level Planarity: Transitivity vs. Even Crossings

Guido Brückner[1], Ignaz Rutter[2], and Peter Stumpf[2(✉)]

[1] Faculty of Informatics, Karlsruhe Institute of Technology, Karlsruhe, Germany
brueckner@kit.edu
[2] Department of Computer Science and Mathematics, University of Passau,
Passau, Germany
{rutter,stumpf}@fim.uni-passau.de

Abstract. Recently, Fulek et al. [1–3] have presented Hanani-Tutte results for (radial) level planarity, i.e., a graph is (radial) level planar if it admits a (radial) level drawing where any two (independent) edges cross an even number of times. We show that the 2-SAT formulation of level planarity testing due to Randerath et al. [4] is equivalent to the strong Hanani-Tutte theorem for level planarity [3]. Further, we show that this relationship carries over to radial level planarity, which yields a novel polynomial-time algorithm for testing radial level planarity.

1 Introduction

Planarity of graphs is a fundamental concept for graph theory as a whole, and for graph drawing in particular. Naturally, variants of planarity tailored specifically to directed graphs have been explored. A planar drawing is *upward planar* if all edges are drawn as monotone curves in the upward direction. A special case are *level planar drawings* of level graphs, where the input graph $G = (V, E)$ comes with a level assignment $\ell: V \to \{1, 2, \ldots, k\}$ for some $k \in \mathbb{N}$ that satisfies $\ell(u) < \ell(v)$ for all $(u, v) \in E$. One then asks whether there is an upward planar drawing such that each vertex v is mapped to a point on the horizontal line $y = \ell(v)$ representing the level of v. There are also radial variants of these concepts, where edges are drawn as curves that are monotone in the outward direction in the sense that a curve and any circle centered at the origin intersect in at most one point. Radial level planarity is derived from level planarity by representing levels as concentric circles around the origin.

Despite the similarity, the variants with and without levels differ significantly in their complexity. Whereas testing upward planarity and radial planarity are NP-complete [5], level planarity and radial level planarity can be tested in polynomial time. In fact, linear-time algorithms are known for both problems [6,7]. However, both algorithms are quite complicated, and subsequent research has led to slower but simpler algorithms for these problems [4,8]. Recently also constrained variants of the level planarity problem have been considered [9,10].

One of the simpler algorithms is the one by Randerath et al. [4]. It only considers proper level graphs, where each edge connects vertices on adjacent

T. Biedl and A. Kerren (Eds.): GD 2018, LNCS 11282, pp. 39–52, 2018.
https://doi.org/10.1007/978-3-030-04414-5_3

levels. This is not a restriction, because each level graph can be subdivided to make it proper, potentially at the cost of increasing its size by a factor of k. It is not hard to see that in this case a drawing is fully specified by the vertex ordering on each level. To represent this ordering, define a set of variables $\mathcal{V} = \{uw \mid u, w \in V, u \neq w, \ell(u) = \ell(w)\}$. Randerath et al. observe that there is a trivial way of specifying the existence of a level-planar drawing by the following consistency (1), transitivity (2) and planarity constraints (3):

$$\forall uw \quad \in \mathcal{V} \qquad\qquad\qquad\qquad\qquad : \quad uw \quad\quad \Leftrightarrow \neg \quad wu \quad (1)$$

$$\forall uw, wy \quad \in \mathcal{V} \qquad\qquad\qquad\qquad\quad : \quad uw \wedge wy \quad \Rightarrow \quad uy \quad (2)$$

$$\forall uw, vx \quad \in \mathcal{V} \text{ with } (u,v), (w,x) \in E \text{ independent} \quad : \quad uw \quad\quad \Leftrightarrow \quad vx \quad (3)$$

The surprising result due to Randerath et al. [4] is that the satisfiability of this system of constraints (and thus the existence of a level planar drawing) is equivalent to the satisfiability of a *reduced constraint system* obtained by omitting the transitivity constraints (2). That is, transitivity is irrelevant for the satisfiability. Note that a satisfying assignment of the reduced system is not necessarily transitive, rather Randerath et al. prove that a solution can be made transitive without invalidating the other constraints. Since the remaining conditions 1 and 3 can be easily expressed in terms of 2-SAT, which can be solved efficiently, this yields a polynomial-time algorithm for level planarity.

A very recent trend in planarity research are Hanani-Tutte style results. The (strong) Hanani-Tutte theorem [11,12] states that a graph is planar if and only if it can be drawn so that any two independent edges (i.e., not sharing an endpoint) cross an even number of times. One may wonder for which other drawing styles such a statement is true. Pach and Tóth [13,14] showed that the weak Hanani-Tutte theorem (which requires even crossings for all pairs of edges) holds for a special case of level planarity and asked whether the result holds in general. This was shown in the affirmative by Fulek et al. [3], who also established the strong version for level planarity. Most recently, both the weak and the strong Hanani-Tutte theorem have been established for radial level planarity [1,2].

Contribution. We show that the result of Randerath et al. [4] from 2001 is equivalent to the strong Hanani-Tutte theorem for level planarity.

The key difference is that Randerath et al. consider proper level graphs, whereas Fulek et al. [3] work with graphs with only one vertex per level. For a graph G we define two graphs $G^\star$, G^+ that are equivalent to G with respect to level planarity. We show how to transform a Hanani-Tutte drawing of a graph $G^\star$ into a satisfying assignment for the constraint system of G^+ and vice versa. Since this transformation does not make use of the Hanani-Tutte theorem nor of the result by Randerath et al., this establishes the equivalence of the two results.

Moreover, we show that the transformation can be adapted also to the case of radial level planarity. This results in a novel polynomial-time algorithm for testing radial level planarity by testing satisfiability of a system of constraints that, much like the work of Randerath et al., is obtained from omitting all transitivity constraints from a constraint system that trivially models radial level planarity. Currently, we deduce the correctness of the new algorithm from the strong Hanani-Tutte theorem for radial level planarity [2]. However, also this transformation works both ways, and a new correctness proof of our algorithm in the style of the work of Randerath et al. [4] may pave the way for a simpler proof of the Hanani-Tutte theorem for radial level planarity. We leave this as future work. Omitted proofs, indicated by $(\star)$, can be found in the full version [15].

2   Preliminaries

A *level graph* is a directed graph $G = (V, E)$ together with a *level assignment* $\ell :$ $V \to \{1, 2, \ldots, k\}$ for some $k \in \mathbb{N}$ that satisfies $\ell(u) < \ell(v)$ for all $(u, v) \in E$. If $\ell(u)+1 = \ell(v)$ for all $(u, v) \in E$, the level graph G is *proper*. Two independent edges $(u, v), (w, x)$ are *critical* if $\ell(u) \le \ell(x)$ and $\ell(v) \ge \ell(w)$. Note that any pair of independent edges that can cross in a level drawing of G is a pair of critical edges. Throughout this paper, we consider drawings that may be non-planar, but we assume at all times that no two distinct vertices are drawn at the exact same point, no edge passes through a vertex, and no three (or more) edges cross in a single point. If any two independent edges cross an even number of times in a drawing Γ of G, it is called a *Hanani-Tutte drawing* of G.

For any k-level graph G we now define a *star form* G^* so that every level of G^* consists of exactly one vertex. The construction is similar to the one used by Fulek et al. [3]. Let n_i denote the number of vertices on level i for $1 \le i \le k$. Further, let $v_1, v_2, \ldots, v_{n_i}$ denote the vertices on level i. Subdivide every level i into $2n_i$ sublevels $1^i, 2^i, \ldots, (2n_i)^i$. For $1 \le j \le n_i$, replace vertex v_j by two vertices v_j', v_j'' with $\ell(v_j') = j^i$ and $\ell(v_j'') = n_i + j^i$ and connect them by an edge (v_j', v_j''), referred to as the *stretch edge* $e(v_j)$. Connect all incoming edges of v_j to v_j' instead and connect all outgoing edges of v_j to v_j'' instead. Let $e = (u, v)$ be an edge of G. Then let e^* denote the edge of G^* that connects the endpoint of $e(u)$ with the starting point of $e(v)$. See Fig. 1. Define G^+ as the graph obtained by subdividing the edges of G^* so that the graph becomes proper; again, see Fig. 1. Let $(u, v), (w, x)$ be critical edges in G^*. Define their *limits* in G^+ as $(u', v'), (w', x')$ where u', v' are endpoints or subdivision vertices of (u, v), w', x' are endpoints or subdivision vertices of (w, x) and it is $\ell(u') = \ell(w') = \max(\ell(u), \ell(w))$ and $\ell(v') = \ell(x') = \min(\ell(v), \ell(x))$.

Lemma 1 $(\star)$. *Let G be a level graph. Then*

G *is (radial) level pl.* $\Leftrightarrow G^*$ *is (radial) level pl.* $\Leftrightarrow G^+$ *is (radial) level pl.*

3   Level Planarity

Recall from the introduction that Randerath et al. formulated level planarity of a proper level graph G as a Boolean satisfiability problem $\mathcal{S}'(G)$ on the variables $\mathcal{V} = \{uw \mid u \ne w, \ell(u) = \ell(w)\}$ and the clauses given by Eqs. (1)–(3).

It is readily observed that G is level planar if and only if $\mathcal{S}'(G)$ is satisfiable. Now let $\mathcal{S}(G)$ denote the SAT instance obtained by removing the transitivity

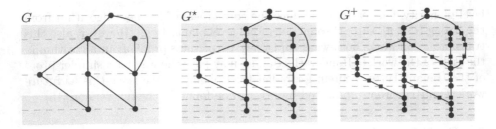

Fig. 1. A level graph G (a) modified to a graph $G^\star$ so as to have only one vertex per level (b) and its proper subdivision G^+ (c).

clauses (2) from $\mathcal{S}'(G)$. Note that it is $(uw \Rightarrow \neg wu) \equiv (\neg uw \vee \neg wu)$ and $(uw \Rightarrow vx) \equiv (\neg uw \vee vx)$, i.e., $\mathcal{S}(G)$ is an instance of 2-SAT, which can be solved efficiently. The key claim of Randerath et al. is that $\mathcal{S}'(G)$ is satisfiable if and only if $\mathcal{S}(G)$ is satisfiable, i.e., dropping the transitivity clauses does not change the satisfiability of $\mathcal{S}'(G)$. In this section, we show that $\mathcal{S}(G)$ is satisfiable if and only if $G^\star$ has a Hanani-Tutte level drawing (Theorem 1). Of course, we do not use the equivalence of both statements to level planarity of G. Instead, we construct a satisfying truth assignment of $\mathcal{S}(G)$ directly from a given Hanani-Tutte level drawing (Lemma 3), and vice versa (Lemma 4). This directly implies the equivalence of the results of Randerath et al. and Fulek et al. (Theorem 1).

The common ground for our constructions is the constraint system $\mathcal{S}'(G^+)$, where a Hanani-Tutte drawing implies a variable assignment that does not necessarily satisfy the planarity constraints (3), though in a controlled way, whereas a satisfying assignment of $\mathcal{S}(G)$ induces an assignment for $\mathcal{S}'(G^+)$ that satisfies the planarity constraints but not the transitivity constraints (2). Thus, in a sense, our transformation trades planarity for transitivity and vice versa.

A (not necessarily planar) drawing Γ of G *induces* a truth assignment φ of $\mathcal{V}$ by defining for all $uw \in \mathcal{V}$ that $\varphi(uw)$ is true if and only if u lies to the left of w in Γ. Note that this truth assignment must satisfy the consistency clauses, but does not necessarily satisfy the planarity constraints. The following lemma describes a relationship between certain truth assignments of $\mathcal{S}(G)$ and crossings in Γ that we use to prove Lemmas 3 and 4.

Lemma 2. *Let $(u, v), (w, x)$ be two critical edges of $G^\star$ and let $(u', v'), (w', x')$ be their limits in G^+. Further, let $\Gamma^\star$ be a drawing of $G^\star$, let Γ^+ be the drawing of G^+ induced by $\Gamma^\star$ and let φ^+ be the truth assignment of $\mathcal{S}(G^+)$ induced by Γ^+. Then (u, v) and (w, x) intersect an even number of times in $\Gamma^\star$ if and only if $\varphi^+(u'w') = \varphi^+(v'x')$.*

Proof. We may assume without loss of generality that any two edges cross at most once between consecutive levels by introducing sublevels if necessary. Let X be a crossing between (u, v) and (w, x) in $G^\star$; see Fig. 2(a). Further, let u_1, w_1 and u_2, w_2 be the subdivision vertices of (u, v) and (w, x) on the levels directly below and above X in $G^\star$, respectively. It is $\varphi^+(u_1w_1) = \neg\varphi^+(u_2w_2)$. In the reverse direction, $\varphi^+(u_1w_1) = \neg\varphi^+(u_2w_2)$ implies that (u, v) and (w, x) cross between the levels $\ell(u_1)$ and $\ell(u_2)$. Due to the definition of limits, any crossing between (u, v) and (w, x) in $G^\star$ must occur between the levels $\ell(u') = \ell(w')$ and $\ell(v') = \ell(x')$. Therefore, it is $\varphi^+(u'w') = \varphi^+(v'x')$ if and only if (u, v) and (w, x) cross an even number of times.   □

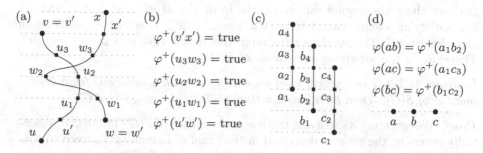

Fig. 2. A Hanani-Tutte drawing (a) induces a truth assignment φ^+ that satisfies $\mathcal{S}(G^+)$ (b), the value where φ^+ differs from ψ^+ is highlighted in red. Using the subdivided stretch edges of G^+ (c), translate φ^+ to a satisfying assignment φ of $\mathcal{S}(G)$ (d). (Color figure online)

Lemma 3. *Let G be a proper level graph and let $\Gamma^\star$ be a Hanani-Tutte drawing of $G^\star$. Then $\mathcal{S}(G)$ is satisfiable.*

Proof. Let Γ^+ be the drawing of G^+ induced by $\Gamma^\star$ and let ψ^+ denote the truth assignment induced by Γ^+. Note that ψ^+ does not necessarily satisfy the crossing clauses. Define φ^+ so that it satisfies all clauses of $\mathcal{S}(G^+)$ as follows.

Let u'', w'' be two vertices of G^+ with $\ell(u'') = \ell(w'')$. If one of them is a vertex in $G^\star$, then set $\varphi^+(u'', w'') = \psi^+(u'', w'')$. Otherwise u'', w'' are subdivision vertices of two edges $(u, v), (w, x) \in E(G^\star)$. If they are independent, then they are critical. In that case their limits $(u', v'), (w', x')$ are already assigned consistently by Lemma 2. Then set $\varphi^+(u''w'') = \psi^+(u'w')$. If $(u, v), (w, x)$ are adjacent, then we have $u = w$ or $v = x$. In the first case, we set $\varphi^+(u''w'') = \psi^+(v'x')$. In the second case, we set $\varphi^+(u''w'') = \psi^+(u'w')$.

Thereby, we have for any critical pair of edges $(u'', v''), (w'', x'') \in E(G^+)$ that $\varphi^+(u''w'') = \varphi^+(v''x'')$ and clearly $\varphi^+(u''w'') = \neg\varphi^+(w''u'')$. Hence, assignment φ^+ satisfies $\mathcal{S}(G^+)$. See Fig. 2 for a drawing Γ^+ (a) and the satisfying assignment of $\mathcal{S}(G^+)$ derived from it (b).

Proceed to construct a satisfying truth assignment φ of $\mathcal{S}(G)$ as follows. Let u, w be two vertices of G with $\ell(u) = \ell(w)$. Then the stretch edges $e(u), e(w)$ in $G^\star$ are critical by construction. Let $(u', u''), (w', w'')$ be their limits in G^+. Set $\varphi(uw) = \varphi^+(u'w')$. Because φ^+ is a satisfying assignment, all crossing clauses of $\mathcal{S}(G^+)$ are satisfied, which implies $\varphi^+(u'w') = \varphi^+(u''w'')$. The same is true for all subdivision vertices of $e(u)$ and $e(w)$ in G^+. Because φ^+ also satisfies the consistency clauses of $\mathcal{S}(G^+)$, this means that φ satisfies the consistency clauses of $\mathcal{S}(G)$. See Fig. 2 for how $\mathcal{S}(G^+)$ is translated from G^+ (c) to G (d). Note that the resulting assignment is not necessarily transitive, e.g., it could be $\varphi(uv) = \varphi(vw) = \neg\varphi(uw)$.

Consider two edges (u, v), (w, x) in G with $\ell(u) = \ell(w)$. Because G is proper, we do not have to consider other pairs of edges. Let (u', u''), (w', w'') be the limits of $e(u), e(w)$ in G^+. Further, let (v', v''), (x', x'') be the limits of $e(v), e(x)$ in G^+. Because there are disjoint directed paths from u' and w' to v' and x' and φ^+ is a satisfying assignment, it is $\varphi^+(u'w') = \varphi^+(v'x')$. Due to the construction of φ described in the previous paragraph, this means that it is $\varphi(uw) = \varphi(vx)$. Therefore, φ is a satisfying assignment of $\mathcal{S}(G)$.   □

Lemma 4. *Let G be a proper level graph together with a satisfying truth assignment φ of $\mathcal{S}(G)$. Then there exists a Hanani-Tutte drawing $\Gamma^\star$ of $G^\star$.*

Proof. We construct a satisfying truth assignment φ^+ of $\mathcal{S}(G^+)$ from φ by essentially reversing the process described in the proof of Lemma 3. Proceed to construct a drawing Γ^+ of G^+ from φ^+ as follows. Recall that by construction, every level of G^+ consists of exactly one non-subdivision vertex. Let u denote the non-subdivision vertex of level i. Draw a subdivision vertex w on level i to the right of u if $\varphi^+(uw)$ is true and to the left of u otherwise. The relative order of subdivision vertices on either side of u can be chosen arbitrarily. Let $\Gamma^\star$ be the drawing of $G^\star$ induced by Γ^+. To see that $\Gamma^\star$ is a Hanani-Tutte drawing, consider two critical edges (u, v), (w, x) of $G^\star$. Let (u', v'), (w', x') denote their limits in G^+. One vertex of u' and v' (w' and x') is a subdivision vertex and the other one is not. Lemma 2 gives $\varphi^+(u'w') = \varphi^+(v'x')$ and then by construction u', w' and v', x' are placed consistently on their respective levels. Moreover, Lemma 2 yields that (u, v) and (w, x) cross an even number of times in $G^\star$. Figure 3 illustrates the construction.   □

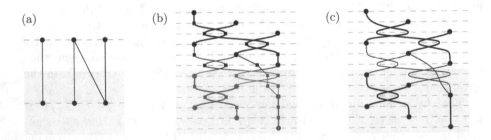

Fig. 3. A proper level graph G together with a satisfying variable assignment φ (a) induces a drawing of G^+ (b), which induces a Hanani-Tutte drawing of $G^\star$ (c).

Theorem 1. *Let G be a proper level graph. Then*
$\mathcal{S}(G)$ *is satisfiable* $\Leftrightarrow$ $G^\star$ *has a Hanani-Tutte level drawing* $\Leftrightarrow$ G *is level planar.*

4   Radial Level Planarity

In this section we present an analogous construction for radial level planarity. In contrast to level planarity, we now have to consider cyclic orders on the levels, and even those may still leave some freedom for drawing the edges between adjacent levels. In the following we first construct a constraint system of radial level planarity for a proper level graph G, which is inspired by the one of Randerath et al. Afterwards, we slightly modify the construction of $G^\star$. Finally, in analogy to the level planar case, we show that a satisfying assignment of our constraint system defines a satisfying assignment of the constraint system of G^+, and that this in turn corresponds to a Hanani-Tutte radial level drawing of $G^\star$.

A Constraint System for Radial Level Planarity. We start with a special case that bears a strong similarity with the level-planar case. Namely, assume that G is a proper level graph that contains a directed path $P = \alpha_1, \ldots, \alpha_k$ that has exactly one vertex α_i on each level i. We now express the cyclic ordering on each level as linear orders whose first vertex is α_i. To this end, we introduce for each level the variables $\mathcal{V}_i = \{\alpha_i uv \mid u, v \in V_i \setminus \{\alpha_i\}\}$, where $\alpha_i uv \equiv$ true means α_i, u, v are arranged clockwise on the circle representing level i. We further impose the following necessary and sufficient *linear ordering constraints* $\mathcal{L}_G(\alpha_i)$.

$$\forall \text{ distinct} \qquad u, v \in V \setminus \{\alpha_i\}: \quad \alpha_i uv \qquad\qquad \Leftrightarrow \neg\ \alpha_i vu \quad (4)$$

$$\forall \text{ pairwise distinct} \quad u, v, w \in V \setminus \{\alpha_i\}: \quad \alpha_i uv \wedge \alpha_i vw \quad \Rightarrow \neg\ \alpha_i uw \quad (5)$$

It remains to constrain the cyclic orderings of vertices on adjacent levels so that the edges between them can be drawn without crossings. For two adjacent levels i and $i + 1$, let $\varepsilon_i = (\alpha_i, \alpha_{i+1})$ be the *reference edge*. Let E_i be the set of edges (u, v) of G with $\ell(u) = i$ that are not adjacent to an endpoint of ε_i. Further $E_i^+ = \{(\alpha_i, v) \in E \setminus \{\varepsilon_i\}\}$ and $E_i^- = \{(u, \alpha_{i+1}) \in E \setminus \{\varepsilon_i\}\}$ denote the edges between levels i and $i + 1$ adjacent to the reference edge ε_i.

In the context of the constraint formulation, we only consider drawings of the edges between levels i and $i + 1$ where any pair of edges crosses at most once and, moreover, ε_i is not crossed. Note that this can always be achieved, independently of the orderings chosen for levels i and $i + 1$. Then, the cyclic orderings of the vertices on the levels i and $i + 1$ determine the drawings of all edges in E_i. In particular, two edges $(u, v), (u', v') \in E_i$ do not intersect if and only if $\alpha_i uu' \Leftrightarrow \alpha_{i+1} vv'$; see Fig. 4(a). Therefore, we introduce constraint (6). For each edge $e \in E_i^+ \cup E_i^-$ it remains to decide whether it is embedded locally to the left or to the right of ε_i. We write $l(e)$ in the former case. Two edges $e \in E_i^-$, $f \in E_i^+$ do not cross if and only if $l(e) \Leftrightarrow \neg l(f)$; see Fig. 4(b). This gives us constraint (7). It remains to forbid crossings between edges in E_i and edges in $E_i^+ \cup E_i^-$. An edge $e = (\alpha_i, v'') \in E_i^+$ and an edge $(u', v') \in E_i$ do not cross if and only if $l(e) \Leftrightarrow \alpha_{i+1} v'v''$; see Fig. 4(c). Crossings with edges $(v, \alpha_{i+1}) \in E_i^-$ can be treated analogously. This yields constraints (8) and (9). We denote the planarity constraints (6)–(9) by $\mathcal{P}_G(\varepsilon_i)$, where $\varepsilon_i = (\alpha_i, \alpha_{i+1})$.

$$\forall \text{ independent } (u,v), (u',v') \in E_i \qquad : \quad \alpha_i uu' \qquad \Leftrightarrow \alpha_{i+1} vv' \quad (6)$$

$$\forall e \in E_i^+, f \in E_i^- \qquad\qquad\qquad : \quad l(e) \qquad\qquad \Leftrightarrow \neg l(f) \quad (7)$$

$$\forall \text{ independent } (\alpha_i, v'') \in E_i^+, (u,v) \in E_i \quad : \quad l(\alpha_i, v'') \qquad \Leftrightarrow \alpha_{i+1} vv'' \quad (8)$$

$$\forall \text{ independent } (u'', \alpha_{i+1}) \in E_i^-, (u,v) \in E_i \quad : \quad l(u'', \alpha_{i+1}) \quad \Leftrightarrow \alpha_i uu'' \quad (9)$$

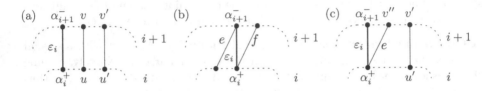

Fig. 4. Illustration of the planarity constraints for radial planarity for the case of two edges in E_i (a), constraint (6); the case of an edge in $e \in E_i^-$ and an edge $f \in E_i^+$ (b), constraint (7); and the case of an edge in E_i and an edge $e \in E_i^+$ (c), constraint (8).

It is not difficult to see that the transformation between Hanani-Tutte drawings and solutions of the constraint system without the transitivity constraints (5) can be performed as in the previous section. The only difference is that one has to deal with edges that share an endpoint with a reference ε_i.

In general, however, such a path P from level 1 to level k does not necessarily exist. Instead, we use an arbitrary reference edge between any two consecutive levels. More formally, we call a pair of sets $A^+ = \{\alpha_1^+, \ldots, \alpha_k^+\}$, $A^- = \{\alpha_1^-, \ldots, \alpha_k^-\}$ *reference sets* for G if we have $\alpha_1^- = \alpha_1^+$ and $\alpha_k^+ = \alpha_k^-$ and for $1 \leq i \leq k$ the *reference vertices* α_i^+, α_i^- lie on level i and for $1 \leq i < k$ graph G contains the *reference edge* $\varepsilon_i = (\alpha_i^+, \alpha_{i+1}^-)$ unless there is no edge between level i and level $i+1$ at all. In that case, we can extend every radial drawing of G by the edge $(\alpha_i^+, \alpha_{i+1}^-)$ without creating new crossings. We may therefore assume that this case does not occur and we do so from now on.

To express radial level planarity, we express the cyclic orderings on each level twice, once with respect to the reference vertex α_i^+ and once with respect to the reference vertex α_i^-. To express planarity between adjacent levels, we use the planarity constraints with respect to the reference edge ε_i. It only remains to specify that, if $\alpha_i^+ \neq \alpha_i^-$, the linear ordering with respect to these reference vertices must be linearizations of the same cyclic ordering. This is expressed by the following *cyclic ordering constraints* $\mathcal{C}_G(\alpha_i^+, \alpha_i^-)$.

$$\forall \text{ distinct } u, \quad v \in V_i \setminus \{\alpha_i^-, \alpha_i^+\} : \quad (\alpha_i^- uv \Leftrightarrow \alpha_i^+ uv) \Leftrightarrow (\alpha_i^- u\alpha_i^+ \Leftrightarrow \alpha_i^- v\alpha_i^+) \quad (10)$$

$$\forall \qquad\qquad v \in V_i \setminus \{\alpha_i^-, \alpha_i^+\} : \qquad\qquad \alpha_i^- v\alpha_i^+ \Leftrightarrow \alpha_i^+ \alpha_i^- v \quad (11)$$

The constraint set $\mathcal{S}'(G, A^+, A^-)$ consists of the linearization constraints $\mathcal{L}_G(\alpha_i^+)$ and $\mathcal{L}_G(\alpha_i^-)$ and the cyclic ordering constraints $\mathcal{C}_G(\alpha_i^+, \alpha_i^-)$

for $i = 1, 2, \ldots, k$ if $\alpha_i^+ \neq \alpha_i^-$, plus the planarity constraints $\mathcal{P}_G(\varepsilon_i)$ for $i = 1, 2, \ldots, k - 1$. This completes the definition of our constraint system.

Theorem 2 ($\star$). *Let G be a proper level graph with reference sets A^+, A^-. Then the constraint system $\mathcal{S}'(G, A^+, A^-)$ is satisfiable if and only if G is radial level planar. Moreover, the radial level planar drawings of G correspond bijectively to the satisfying assignments of $\mathcal{S}'(G, A^+, A^-)$.*

Similar to Sect. 3, we now define a reduced constraints system $\mathcal{S}(G, A^+, A^-)$ obtained from $\mathcal{S}'(G, A^+, A^-)$ by dropping constraint (5). Observe that this reduced system can be represented as a system of linear equations over $\mathbb{F}_2$, which can be solved efficiently. Our main result is that $\mathcal{S}(G, A^+, A^-)$ is satisfiable if and only if G is radial level planar.

Modified Star Form. We also slightly modify the splitting and perturbation operation in the construction of the star form $G^\star$ of G for each level i. This is necessary since we need a special treatment of the reference vertices α_i^+ and α_i^- on each level i. Consider the level i containing the n_i vertices $v_1, \ldots, v_{n_i}$. If $\alpha_i^+ \neq \alpha_i^-$, then we choose the numbering of the vertices such that $v_1 = \alpha_i^-$ and $v_{n_i} = \alpha_i^+$. We replace i by $2n_i - 1$ levels $1^i, 2^i, \ldots, (2n_i - 1)^i$, which is one level less than previously. Similar to before, we replace each vertex v_j by two vertices $\mathrm{bot}(v_j)$ and $\mathrm{top}(v_j)$ with $\ell(\mathrm{bot}(v_j)) = j^i$ and $\ell(\mathrm{top}(v_j)) = (n_i - 1 + j)^i$ and the corresponding stretch edge $(\mathrm{bot}(v_j), \mathrm{top}(v_j))$; see Fig. 5(b). This ensures that the construction works as before, except that the middle level $m_i = j^{n_i}$ contains two vertices, namely $\alpha_i^{+\prime\prime}$ and $\alpha_i^{-\prime}$.

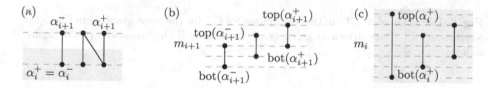

Fig. 5. Illustration of the modified construction of the stretch edges for $G^\star$ for the graph G in (a). The stretch edges for level $i + 1$ where $\alpha_{i+1}^+ \neq \alpha_{i+1}^-$ (b) and for level i where $\alpha_i^+ = \alpha_i^-$ (c).

If, on the other hand, $\alpha_i^+ = \alpha_i^-$, then we choose $v_1 = \alpha_i^+$. But now we replace level i by $2n_i + 1$ levels $1^i, \ldots, (2n_i + 1)^i$. Replace v_1 by vertices $\mathrm{bot}(v_1), \mathrm{top}(v_1)$ with $\ell(\mathrm{bot}(v_1)) = 1^i$ and $\ell(\mathrm{top}(v_1)) = (2n_i + 1)^j$. Replace all other v_j with vertices $\mathrm{bot}(v_j), \mathrm{top}(v_j)$ with $\ell(\mathrm{bot}(v_j)) = j^i$ and $\ell(\mathrm{top}(v_j)) = (n_i + 1 + j)^i$. For all j, we add the stretch edge $(\mathrm{bot}(v_j), \mathrm{top}(v_j))$ as before; see Fig. 5(c). This construction ensures that the stretch edge of $\alpha_i^+ = \alpha_i^-$ starts in the first new level 1^i and ends in the last new level $(2n_i + 1)^i$, and the middle level $m_i = n_i + 1^j$ contains no vertex.

As before, we replace each original edge (u, v) of the input graph G by the edge $(\mathrm{top}(u), \mathrm{bot}(v))$ connecting the upper endpoint of the stretch edge of u

to the lower endpoint of the stretch edge of v. Observe that the construction preserves the properties that for each level i the middle level m_i of the levels that replace i intersects all stretch edges of vertices on level i. Therefore, Lemma 1 also holds for this modified version of G^* and its proper subdivision G^+. For each vertex v of G we use $e(v) = (\mathrm{bot}(v), \mathrm{top}(v))$ to denote its stretch edge.

We define the function L that maps each level j of G^* or G^+ to the level i of G it replaces. For an edge e of G^* and a level i that intersects e, we denote by e_i the subdivision vertex of e at level i in G^+. For two levels i and j that both intersect an edge e of G^*, we denote by e_i^j the path from e_i to e_j in G^+.

Constraint System and Assignment for G^+. We now choose reference sets B^+, B^- for G^+ that are based on the reference sets A^+, A^- for G. Consider a level j of G^* and let $i = L(j)$ be the corresponding level of G. For each level j, define two vertices β_j^+, β_j^-. If $\alpha_i^- = \alpha_i^+$, set $\beta_j^- = \beta_j^+ = e(\alpha_i^-)_j$; see Fig. 6(b). Otherwise, the choice is based on whether j is the middle level $m = m_i$ of the levels $L^{-1}(i)$ that replace level i of G, or whether j lies above or below m. Choose $\beta_m^- = \mathrm{top}(\alpha_i^+)$ and $\beta_m^+ = \mathrm{bot}(\alpha_{i+1}^-)$. For $j < m$, choose $\beta_j^- = \beta_j^+ = e(\alpha_i^-)_j$ and for $j > m$, choose $\beta_j^- = \beta_j^+ = e(\alpha_i^+)_j$; see Fig. 6(c).

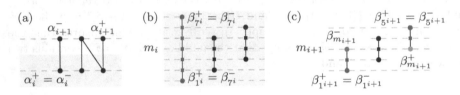

Fig. 6. Definition of β^+, β^- in the assignment for G^+ for the same graph as in Fig. 5(a). Vertices β^+ (β^-) are drawn in green (red), or in blue if they coincide. (Color figure online)

We set B^+ to be the set containing all β_j^+ and likewise for B^-. Our next step is to construct from a satisfying assignment φ of $\mathcal{S}(G, A^+, A^-)$ a corresponding satisfying assignment φ^+ of $\mathcal{S}(G^+, B^+, B^-)$. The construction follows the approach from Lemma 4 and makes use of the fact that G^+ is essentially a stretched and perturbed version of G. Since the construction is straightforward but somewhat technical, we defer it to the full version [15].

Lemma 5 ($\star$). *If $\mathcal{S}(G, A^+, A^-)$ is satisfiable, then $\mathcal{S}(G^+, B^+, B^-)$ is satisfiable.*

Constructing a Hanani-Tutte Drawing. We construct a radial drawing Γ^+ of G^+, from which we obtain the drawing Γ^* of G^* by smoothing the subdivision vertices. Afterwards we show that Γ^* is a Hanani-Tutte drawing.

We construct Γ^+ as follows. Consider a level j of G^+ and let $i = L(j)$ be the original level of G. First assume $j = m_i$ is the middle levels of the levels replacing level i of G. If $\beta_j^- = \beta_j^+$, then we place all vertices of $V_j(G^+)$ in arbitrary order.

Otherwise, we place β_j^- and β_j^+ arbitrarily on the circle representing the level m_i. We then place each vertex $v \in V_j(G^+) \setminus \{\beta_j^-, \beta_j^+\}$ such that β_j^-, v, β_j^+ are ordered clockwise if and only if $\varphi(\beta_j^- v \beta_j^+)$ is true (i.e., we place v on the correct side of β_j^- and β_j^+ and arrange the vertices on both sides of β_j^- and β_j^+ arbitrarily).

Next assume $j \neq m_i$. Then there is exactly one vertex $\xi \in V_j(G^+) \cap V(G^\star)$. If $\xi \in B^-$, then we place all vertices of $V_j(G^+)$ in arbitrary order on the circle representing the level j. Otherwise, we place β_j^- and ξ arbitrarily. We then place any vertex $v \in V_j(G^+) \setminus \{\beta_j^-, \xi\}$ such that β_j^-, ξ, v are ordered clockwise if and only if $\varphi^+(\beta_j^- \xi v)$ is true. Again, we arrange the vertices on either side of β_j^- and ξ arbitrarily. We have now fixed the positions of all vertices and it remains to draw the edges.

Consider two consecutive levels j and $j + 1$ of G^+. We draw the edges in $E_j(G^+)$ such that they do not cross the reference edges in $E(G^+) \cap (B^+ \times B^-)$. We draw an edge $e = (\beta_j^+, x') \in E_j^+(G^+)$ such that it is locally left of (β_j^+, β_j^-) if and only if $\varphi^+(l(e)) = \text{true}$. By reversing the subdivisions of the edges in G^+ we obtain $G^\star$ and along with that we obtain a drawing $\Gamma^\star$ of $G^\star$ from Γ^+.

Let a, b, c be curves or corresponding edges. Then we write $\mathrm{cr}(a, b)$ for the number of crossings between a, b and set $\mathrm{cr}(a, b, c) = \mathrm{cr}(a, b) + \mathrm{cr}(a, c) + \mathrm{cr}(b, c)$. The following lemma is the radial equivalent to Lemma 2 and constitutes our main tool for showing that edges in our drawing cross evenly.

Lemma 6 ($\star$). *Let C_1 and C_2 be distinct concenctric circles and let a, b, c be radially monotone curves from C_1 to C_2 with pairwise distinct start- and endpoints that only intersect at discrete points. Then the start- and endpoints of a, b, c have the same order on C_1 and C_2 if and only if $\mathrm{cr}(a, b, c) \equiv 0 \mod 2$.*

Lemma 7. *The drawing $\Gamma^\star$ is a Hanani-Tutte drawing of $G^\star$.*

Proof. We show that each pair of independent edges of $G^\star$ crosses evenly in $\Gamma^\star$. Of course it suffices to consider critical pairs of edges, since our drawing is radial by construction, and therefore non-critical independent edge pairs cannot cross.

Every edge $(\alpha_i^+, \alpha_{i+1}^-)$ is subdivided into edges of the form $(\beta_j^+, \beta_{j+1}^-)$ and therefore it is not crossed.

Let e, f be two independent edges in $E(G^\star) \setminus (A^+ \times A^-)$ that are critical. Let a and b be the innermost and outermost level shared by e and f.

We seek to use Lemma 6 to analyze the parity of the crossings between e and f. To this end, we construct a curve γ along the edges of the form $(\beta_j^+, \beta_{j+1}^-)$ as follows. For every level j we add a curve c_j between β_j^- and β_j^+ on the circle representing the level j (a point for $\beta_j^- = \beta_j^+$; chosen arbitrarily otherwise). The curve γ is the union of these curves c_j and the curves for the edges of the form $(\beta_j^+, \beta_{j+1}^-)$. Note that γ spans from the innermost level 1 to the outermost level $(2n_k + 1)^k$ with endpoints $\mathrm{bot}(\alpha_1^+)$ and $\mathrm{top}(\alpha_k^-)$.

For any edge $g \in G^\star$, we denote its curve in $\Gamma^\star$ by $c(g)$. For any radial monotone curve c we denote its subcurve between level i and level j by c_i^j (using only one point on circle i and circle j each). We consider the three curves

$g' = \gamma_a^b, e' = c(e)_a^b, f' = c(f)_a^b$. We now distinguish cases based on whether one of the edges e, f starts at the bottom end or ends at the top end of the reference edges on level a or b.

Case 1: We have $e_a, f_a \neq \beta_a^+$ and $e_b, f_b \neq \beta_b^-$. Note that $\mathrm{cr}(e, f, \gamma) = \mathrm{cr}(e, f) + \mathrm{cr}(e, \gamma) + \mathrm{cr}(f, \gamma)$, and therefore $\mathrm{cr}(e, f) \equiv \mathrm{cr}(e, f, \gamma) + \mathrm{cr}(e, \gamma) + \mathrm{cr}(f, \gamma) \mod 2$.

By Lemma 6 we have that the orders of e_a, f_a, β_a^+ and e_b, f_b, β_b^- differ if and only if $\mathrm{cr}(e, f, \gamma) \equiv 1 \mod 2$. That is $\mathrm{cr}(e, f, \gamma) \equiv 0 \mod 2$ if and only if $\varphi^+(\beta_a^+, e_a, f_a) = \varphi^+(\beta_b^-, e_b, f_b)$. We show that $\varphi^+(\beta_a^+, e_a, f_a) = \varphi^+(\beta_b^-, e_b, f_b)$ if and only if $\mathrm{cr}(e, \gamma) + \mathrm{cr}(f, \gamma) \equiv 0 \mod 2$. In either case, $\mathrm{cr}(e, f)$ is even.

Let $a \leq j \leq b - 1$. By construction we have for $\beta_j^- \neq \beta_j^+$ and any other vertex v on level j, that β_j^-, v, β_j^+ are placed clockwise if and only if $\varphi^+(\beta_j^-, v, \beta_j^+)$ is true. Further, since φ^+ satisfies $\mathcal{C}(\beta_j^+, \beta_j^-)$, we have for any other vertex u on level j that β_j^-, u, v and β_j^+, u, v have the same order if and only if β_j^-, v, β_j^+ and β_j^-, u, β_j^+ have the same order, i.e., if and only if u and v lie on the same side of β_j^- and β_j^+. This however, is equivalent to $\mathrm{cr}(e, c_j) + \mathrm{cr}(f, c_j) \equiv 0 \mod 2$.

Since φ^+ satisfies $\mathcal{P}(\delta_j)$ where $\delta_j = (\beta_j^+, \beta_{j+1}^-)$, we have that $\varphi^+(\beta_j^+, e_j, f_j) = \varphi^+(\beta_{j+1}^-, e_{j+1}, f_{j+1})$. We obtain, that $\varphi^+(\beta_j^+, e_j, f_j) = \varphi^+(\beta_{j+1}^-, e_{j+1}, f_{j+1})$ unless $\varphi^+(\beta_{j+1}^-, e_{j+1} f_{j+1}) \neq \varphi^+(\beta_{j+1}^+ e_{j+1} f_{j+1})$ (which requires $\beta_{j+1}^- \neq \beta_{j+1}^+$). This is equivalent to $\mathrm{cr}(e, c_{j+1}) + \mathrm{cr}(f, c_{j+1}) \equiv 1 \mod 2$. Hence, we have $\varphi^+(\beta_a^+ e_a f_a) = \varphi^+(\beta_b^+ e_b f_b)$ if and only if $\sum_{j=a}^{b-1} \mathrm{cr}(c_j, e) + \mathrm{cr}(c_j, f) \equiv 0 \mod 2$ (Note that $\beta_b^- = \beta_b^+$.). Since edges of the form $(\beta_j^+, \beta_{j+1}^-)$ are not crossed, this is equivalent to $\mathrm{cr}(\gamma, e) + \mathrm{cr}(\gamma, f) \equiv 0 \mod 2$. Which we aimed to show. By the above argument we therefore find that $\mathrm{cr}(e, f)$ is even.

Case 2. We do not have $e_a, f_a \neq \beta_a^+$ and $e_b, f_b \neq \beta_b^-$. For example, assume $e_a = \beta_a^+$; the other cases work analogously. We then have $\beta_a^+ = \mathrm{top}(\alpha_i^+)$. This means e originates from an edge in G. Since such edges do not cross middle levels, g' is a subcurve of an original edge ε_i. Especially, we have only three vertices per level between a and b that correspond to γ, e, f.

Let $H \subseteq G^+$ be the subgraph induced by the vertices of $(\varepsilon_i)_a^b, e_a^b, f_a^b$. Then φ^+ satisfies all the constraints of $\mathcal{S}(H, V((\varepsilon_i)_a^b), V((\varepsilon_i)_a^b))$. However, each level of H contains only three vertices, and therefore the transitivity constraints are trivially satisfied, i.e., φ^+ satisfies all the constraints of $\mathcal{S}'(H, V((\varepsilon_i)_a^b), V((\varepsilon_i)_a^b))$. Thus, by Theorem 2, a drawing Γ_H of H according to φ^+ is planar. I.e., we have $\mathrm{cr}_{\Gamma_H}((\varepsilon_i)_a^b, e_a^b, f_a^b) = 0$. Let C_a, C_b be ε-close circles to levels a and b, respectively, that lie between levels a and b. With Lemma 6 we obtain that ε_i, e, f intersect C_a and C_b in the same order.

Note that Γ^+ is drawn according to φ^+ in level a and in level b. We obtain that the curves for ε_i, e, f intersect C_a in the same order in Γ^+ and in Γ_H. The same holds for C_b. Hence, the curves intersect C_a and C_b in the same order in Γ^+. With Lemma 6 we have $\mathrm{cr}_{\Gamma^+}((\varepsilon_i)_a^b, e_a^b, f_a^b) \equiv 0 \mod 2$. Since γ is a subcurve of ε_i and thus not crossed in Γ^+, this yields $\mathrm{cr}_{\Gamma^+}(e_a^b, f_a^b) \equiv 0 \mod 2$. Thus any two independent edges have an even number of crossings. $\square$

As in the level planar case the converse also holds.

Lemma 8 ($\star$). *Let $G^\star$ be a level graph with reference sets A^+, A^- for G^+. If $G^\star$ admits a Hanani-Tutte drawing, then there exists a satisfying assignment φ of $S(G^+, A^+, A^-)$.*

Theorem 3. *Let G be a proper level graph with reference sets A^+, A^-. Then $S(G, A^+, A^-)$ is satisfiable $\Leftrightarrow G^\star$ has a Hanani-Tutte radial level drawing $\Leftrightarrow G$ is radial level planar.*

5 Conclusion

We have established an equivalence of two results on level planarity that have so far been considered as independent. The novel connection has further led us to a new testing algorithm for radial level planarity. Can similar results be achieved for level planarity on a rolling cylinder or on a torus [16]?

References

1. Fulek, R., Pelsmajer, M., Schaefer, M.: Hanani-Tutte for radial planarity. In: Di Giacomo, E., Lubiw, A. (eds.) GD 2015. LNCS, vol. 9411, pp. 99–110. Springer, Cham (2015). https://doi.org/10.1007/978-3-319-27261-0_9

2. Fulek, R., Pelsmajer, M., Schaefer, M.: Hanani-Tutte for radial planarity II. In: Hu, Y., Nöllenburg, M. (eds.) GD 2016. LNCS, vol. 9801, pp. 468–481. Springer, Cham (2016). https://doi.org/10.1007/978-3-319-50106-2_36

3. Fulek, R., Pelsmajer, M.J., Schaefer, M., Štefankovič, D.: Hanani-Tutte, monotone drawings, and level-planarity. In: Pach, J. (ed.) Thirty Essays on Geometric Graph Theory, pp. 263–287. Springer, New York (2013). https://doi.org/10.1007/978-1-4614-0110-0_14

4. Randerath, B., et al.: A satisfiability formulation of problems on level graphs. Electron. Notes Discrete Math. **9**, 269–277 (2001). IICS 2001 Workshop on Theory and Applications of Satisfiability Testing (SAT 2001)

5. Garg, A., Tamassia, R.: On the computational complexity of upward and rectilinear planarity testing. SIAM J. Comput. **31**(2), 601–625 (2002)

6. Bachmaier, C., Brandenburg, F.J., Forster, M.: Radial level planarity testing and embedding in linear time. J. Graph Algorithms Appl. **9**(1), 53–97 (2005)

7. Jünger, M., Leipert, S.: Level planar embedding in linear time. In: Kratochvíyl, J. (ed.) GD 1999. LNCS, vol. 1731, pp. 72–81. Springer, Heidelberg (1999). https://doi.org/10.1007/3-540-46648-7_7

8. Harrigan, M., Healy, P.: Practical level planarity testing and layout with embedding constraints. In: Hong, S.-H., Nishizeki, T., Quan, W. (eds.) GD 2007. LNCS, vol. 4875, pp. 62–68. Springer, Heidelberg (2008). https://doi.org/10.1007/978-3-540-77537-9_9

9. Brückner, G., Rutter, I.: Partial and constrained level planarity. In: Klein, P.N. (ed.) Proceedings of the 28th Annual ACM-SIAM Symposium on Discrete Algorithms (SODA 2017), pp. 2000–2011. SIAM (2017)

10. Klemz, B., Rote, G.: Ordered level planarity, geodesic planarity and bi-monotonicity. In: Frati, F., Ma, K.-L. (eds.) GD 2017. LNCS, vol. 10692, pp. 440–453. Springer, Cham (2018). https://doi.org/10.1007/978-3-319-73915-1_34

11. Chojnacki, C.: Über wesentlich unplättbare Kurven im dreidimensionalen Raume. Fundam. Math. **23**(1), 135–142 (1934)
12. Tutte, W.T.: Toward a theory of crossing numbers. J. Combin. Theory **8**(1), 45–53 (1970)
13. Pach, J., Tóth, G.: Monotone drawings of planar graphs. J. Graph Theory **46**(1), 39–47 (2004)
14. Pach, J., Tóth, G.: monotone drawings of planar graphs. ArXiv:1101.0967 e-prints (2011)
15. Brückner, G., Rutter, I., Stumpf, P.: Level planarity: transitivity vs. even crossings. CoRR abs/1808.09931 (2018)
16. Angelini, P., Da Lozzo, G., Di Battista, G., Frati, F., Patrignani, M., Rutter, I.: Beyond level planarity. In: Hu, Y., Nöllenburg, M. (eds.) GD 2016. LNCS, vol. 9801, pp. 482–495. Springer, Cham (2016). https://doi.org/10.1007/978-3-319-50106-2_37

Short Plane Supports for Spatial Hypergraphs

Thom Castermans[1], Mereke van Garderen[2], Wouter Meulemans[1(✉)],
Martin Nöllenburg[3] [iD], and Xiaoru Yuan[4]

[1] TU Eindhoven, Eindhoven, The Netherlands
{t.h.a.castermans,w.meulemans}@tue.nl
[2] Universität Konstanz, Konstanz, Germany
mereke.van.garderen@uni-konstanz.de
[3] TU Wien, Vienna, Austria
noellenburg@ac.tuwien.ac.at
[4] Peking University, Beijing, China
xiaoru.yuan@pku.edu.cn

Abstract. A graph $G = (V, E)$ is a *support* of a hypergraph $H = (V, S)$ if every hyperedge induces a connected subgraph in G. Supports are used for certain types of hypergraph visualizations. In this paper we consider visualizing *spatial* hypergraphs, where each vertex has a fixed location in the plane. This is the case, e.g., when modeling set systems of geospatial locations as hypergraphs. By applying established aesthetic quality criteria we are interested in finding supports that yield plane straight-line drawings with minimum total edge length on the input point set V. We first show, from a theoretical point of view, that the problem is NP-hard already under rather mild conditions as well as a negative approximability results. Therefore, the main focus of the paper lies on practical heuristic algorithms as well as an exact, ILP-based approach for computing short plane supports. We report results from computational experiments that investigate the effect of requiring planarity and acyclicity on the resulting support length. Further, we evaluate the performance and trade-offs between solution quality and speed of several heuristics relative to each other and compared to optimal solutions.

1 Introduction

A *hypergraph* $H = (V, S)$ is a generalization of a graph, in which each hyperedge in S is a nonempty subset of the vertex set V, that is, $S \subseteq \mathcal{P}(V) \setminus \{\emptyset\}$. Furthermore, we assume here that every element $v \in V$ is in at least one hyperedge $s \in S$. Hypergraphs arise in many domains to model set systems representing clusters, groups or other aggregations. To allow for effective exploration and analysis of such data, visualization is often used. Indeed, drawing hypergraphs relates to set visualization, an active subfield of information visualization (see the recent survey of Alsallakh et al. [3]). Various methods have been developed to visualize set systems for elements fixed in (geo)spatial positions, such as Bubble

© Springer Nature Switzerland AG 2018
T. Biedl and A. Kerren (Eds.): GD 2018, LNCS 11282, pp. 53–66, 2018.
https://doi.org/10.1007/978-3-030-04414-5_4

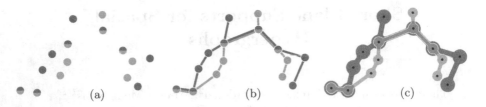

Fig. 1. (a) A set system with colors indicating set membership. (b) The shortest plane support of the corresponding hypergraph. (c) A Kelp-style rendering of the set system. (Color figure online)

Sets [9], LineSets [2], Kelp Diagrams [10] and Kelp Fusion [18]. These methods make different trade-offs between, e.g., Gestalt theory and Tufte's principle of ink minimization [20] to visually convey the set structures; user studies have been performed to analyze the effectiveness of such trade-offs [18].

An important concept to model the drawing of hypergraphs is that of a hypergraph support [14]: a *support* of a hypergraph $H = (V, S)$ is a graph $G = (V, E)$ such that every hyperedge $s \in S$ induces a connected subgraph in G. In other words, for every hyperedge s, the restriction of G to only edges that connect vertices in s, denoted $G[s]$, is connected and spans all vertices in s. Hypergraph supports correspond to a prominent visualization style for geospatial sets, namely that of connecting all elements of a set using colored links, such as seen in Kelp-style diagrams [10,18] (see also Fig. 1) or LineSets [2].

Thus, finding an embedded support that satisfies certain criteria readily translates into a good rendering of the spatial set system. A "good" support should avoid edge crossings, a standard quality criterion in the graph-drawing literature [19]. Moreover, as per Tufte's principle of ink minimization [20], it should have small total edge length. Of course, one may argue that edges of the support that are used by multiple hyperedges do not significantly reduce the "ink" and thus multiplicity should be considered. However, we observe that such edges show co-occurrences of elements and thus have a potential added value in the drawing—user studies that establish the validity of this reasoning are beyond the scope of this paper. The shortest support need not be a tree, but to further build on this idea of co-occurrences, one may want to restrict the support to be acyclic—a support tree.

In many applications, the vertices have some associated (geo)spatial location, thereby prescribing their positions in the drawing of the support. We focus on this case where vertices have fixed positions in the plane and study supports that are embedded using straight-line edges. Figure 2 shows an example on real-world data of restaurants, similar to those used in [18].

Contributions. The contributions of this paper are two-fold: on the one hand we fill some gaps in theoretical knowledge about computing plane supports and support trees; on the other hand, we perform computational experiments to

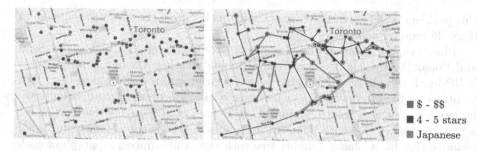

Fig. 2. A set system of restaurants in downtown Toronto: input memberships and locations (left) and a Kelp-style rendering of the shortest plane support (right).

gain more insight into the trade-offs on the complexity of the visual artifact for (implicit) support-based set visualization methods. Our focus is on the latter.

In Sect. 2 we explore computational aspects of the problem and introduce our algorithms. We observe that plane support trees always exist if at least one vertex is contained in all hyperedges, but show that length minimization is NP-hard. Moreover, the natural approach to extend a minimum spanning tree does not even yield a constant-factor approximation. Finally, we present two heuristics, one based on local search, the other on iteratively computing minimum spanning trees, as well as an exact integer linear program (ILP).

In Sect. 3 we describe the results of two computational experiments. The first experiment compares the performance of the two heuristic algorithms in terms of quality and speed. Whereas the local search achieves better quality, the approximation algorithm is faster. The second experiment compares how well these algorithms perform compared to the optimum, computed via the ILP, and investigates the cost in terms of edge length incurred by requiring planarity or acyclicity. The effect of planarity and acyclicity seems to be predictably influenced by the number of hyperedges and the number of incident hyperedges per vertex, but not by the number of vertices. Moreover, the experiment shows that local search often achieves an optimal result.

Related Work. Regarding supports for elements with fixed locations, some results are already known. The results of Bereg et al. [5] imply that existence of a plane support tree for two disjoint hyperedges can be tested in polynomial time; this implies the same result for a plane support. This problem has also been studied in a setting with additional Steiner points [4,11]. Van Goethem et al. [12] enforce a stricter planarity than that of planar supports and investigate the resulting properties for elements on a regular grid, where only neighboring elements can be connected. However, solution length is of no concern in their results.

Without the planarity requirement, existence and length minimization of a (nonplane) support tree for fixed elements can be solved in polynomial time [15,16]. Hurtado et al. [13] show that length minimization of a support for two hyperedges is solvable in polynomial time. However, for three or more hyperedges this problem is NP-hard [1]. We show that this is in fact hard for two hyperedges if we do require planarity.

Planar supports without fixed elements have also received attention. Johnson and Pollak [14] originally showed that deciding whether a planar support exists is NP-hard; various restrictions have since been proven to be NP-hard (e.g., [7]). Contrasting these reductions, our hardness result (Theorem 1) requires only two hyperedges, but uses length minimization. Buchin et al. [7] show that testing for a planar support tree with bounded maximum degree is solvable in polynomial time; testing for a planar support tree such that the induced subgraph of each hyperedge is Hamiltonian can also be done in polynomial time [6].

Various set-visualization methods [2,10,18] implicitly also compute supports, considering various criteria such as length, detour, shape, crossings, and bends.

2   Computing Short Plane Supports

We first describe our theoretical results. Omitted proofs are in the full version [8].

Existence. The observation below gives a sufficient condition for the existence of a plane support tree. Bereg et al. [5] provide a necessary condition for $|S| = 2$, though the problem remains open for $|S| > 2$.

Observation 1. *Consider a hypergraph $H = (V, S)$ with no three vertices in V on a line, such that $V_A = \bigcap_{s \in S} s \neq \emptyset$. Then H has a plane support tree.*

Proof. We use the Euclidean minimum spanning tree on V_A and connect each vertex in $V \setminus V_A$ to the closest one in V_A. This readily yields a support tree; it is plane as no crossings are created when connecting to the closest point in V_A and no overlaps are created in the absence of collinear points.   □

Without a vertex in V_A, one can immediately construct instances that enforce a crossing in any support, e.g., an X-configuration of two disjoint hyperedges.

Approximation. In a support tree the subgraph induced by V_A must be a connected subtree to satisfy the support property for all hyperedges. Next we consider using the above idea to start with an Euclidean minimum spanning tree (EMST) of V_A and extend it to a support tree. Though this leads to an approximation algorithm for two hyperedges [13] if we allow intersections, we show below that the planarity requirement can cause the resulting support length to exceed any constant factor of the length of the shortest plane support tree.

Lemma 1. *There is a family of n-vertex hypergraphs $H = (V, \{r, b\})$ with $V_A = r \cap b \neq \emptyset$ such that any plane support of H that includes an EMST of V_A is a factor $\Theta(|V|)$ longer than the shortest plane support tree.*

Proof (sketch). The family is drawn in Fig. 3. The convex chains force the support with length $\Theta(n) \cdot \ell$ when the EMST on V_A is used. Using a different tree on V_A can give a total of length $\Theta(1) \cdot \ell$.   □

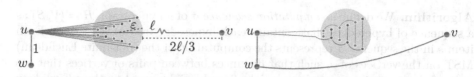

Fig. 3. An n-point instance with approximation ratio $\Theta(n)$ if using an EMST on V_A. All edges are straight-line segments; curvature emphasizes the effect of the convex chain.

Removing vertex w from construction in Fig. 3, we can similarly show that a plane support tree, which now necessarily includes the edge uv, is a factor $\Theta(n)$ longer than a shortest nonplane support tree.

Corollary 1. *There is a family of n-vertex hypergraphs $H = (V, \{r, b\})$ with $V_A = r \cap b \neq \emptyset$ such that any plane support tree of H is a factor $\Theta(n)$ longer than the shortest nonplane support tree.*

Computational Complexity. Unfortunately, finding the shortest plane support and several restricted variants are NP-hard, as captured in the theorem below. It uses a fairly straightforward reduction from planar monotone 3-SAT [17].

Theorem 1. *Let $H = (V, \{r, b\})$ be a hypergraph with vertices V having fixed locations in $\mathbb{R}^2$ and with $r \subseteq b$ or $r \cap b = \emptyset$. It is NP-hard to decide whether H admits a plane support tree with length at most L for some $L > 0$.*

2.1 Iterative Minimum Spanning Trees

Here we focus on computing short supports without requiring planarity. As described by Hurtado et al. [13], EMSTs can be used to find an approximation of the shortest support. In particular, let $H = (V, S)$ be a hypergraph with n vertices and k hyperedges; by computing an EMST for each hyperedge and taking their union, we get a support that is a k-approximation[1] of the shortest support. This algorithm runs in $O(kn \log n)$ time.

Suppose that we compute the EMSTs $T_1, \ldots, T_k$ in that order, for the k hyperedges in S. The final support is the union of these trees: its length is not increased by using an edge in T_i that is already present in some T_j ($j < i$). Hence, we can consider any pair of vertices that is adjacent in $T_1 \cup \ldots \cup T_{i-1}$ to have distance zero, when computing T_i. This heuristically reduces the length of the resulting support (though the approximation ratio remains the same). However, the order in which hyperedges are considered now matters for the result. To alleviate this issue, we iteratively recompute the minimum spanning trees.

---

[1] One can actually do slightly better, by computing spanning trees on the intersection of two hyperedges, yielding roughly a $(0.8k)$-approximation [13].

Algorithm. We define a *computation sequence* σ of a hypergraph $H = (V, S)$ as a sequence of hyperedges that contains each hyperedge in S at least once. Each item s in the sequence σ represents the computation of the (not-quite Euclidean) MST on the vertices of s, such that distances between pairs of vertices that are part of the current support have weight 0 and weight equal to their Euclidean distance otherwise. We use T_s to denote the current MST for hyperedge $s \in S$; the support G is always the union over all T_s. As we compute a spanning tree for each hyperedge, G is a support for H when the algorithm terminates.

Efficiency. Implementing G with adjacency lists, we use $O(nk)$ storage as each of the k trees has $O(n)$ edges. To compute T_s, we use Lemma 2 below to conclude that there are $O(nk)$ candidate edges, ensuring that Prim's MST algorithm runs in $O(nk + n \log n)$ time. To see that we can determine the weight without overhead, consider all vertices to be indexed with numbers from 1 to n. When adding a vertex u to the current tree in Prim's algorithm, we first process the neighbors of u in G (having a weight 0) and mark that these have been processed in an array using the above mentioned vertex index. Only then do we process all other vertices (having weight equal to the Euclidean distance) that are not marked and are not in the current tree. The total algorithm thus takes $O(|\sigma|(nk + n \log n))$ time and $\Theta(nk)$ space.

Lemma 2. *Let P be a point set and $F \subseteq P \times P$. Consider the MST T on P, based on edge weights 0 for edges in F and the Euclidean distance otherwise. Then T is a subset of F and the Euclidean MST on P.*

Properties ($k = 2$). The main question that arises is how long a computation sequence σ must be such that the result *stabilizes*, that is, any sequence that extends σ gives a support that has the same total length. We use G_σ to denote the support resulting from computation sequence σ. Below, we sketch an argument that for $k = 2$, we need to only recompute one hyperedge: sequence $\sigma = \langle r, b, r \rangle$ or $\sigma = \langle b, r, b \rangle$ is sufficient to obtain a stable result. We can compute both sequences and use the result with smallest total edge length.

Lemma 3. *Let $H = (V, \{r, b\})$ be a hypergraph. All computation sequences σ' with $|\sigma'| \geq 4$ have a shorter computation sequence σ with $|\sigma| = 3$ with $G_\sigma = G_{\sigma'}$.*

Proof (sketch). We show that the third computation does not add a new edge with both vertices in $r \cap b$. Hence, the second and fourth computation receive the same input and thus yield the same result. $\qquad\square$

2.2   Local Search

The algorithm described in Sect. 2.1 appears to perform well in practice, as shown in Sect. 3. However, one may wonder whether other commonly employed heuristic approaches outperform it in the experiments. We therefore implement a local-search algorithm, specifically, a hill-climbing heuristic.

Algorithm. This approach assumes that in the given hypergraph $H = (V, S)$, at least one vertex $v \in V$ occurs in all hyperedges $s \in S$ such that Observation 1 applies; let $V_A = \bigcap_{s \in S} s \neq \emptyset$. We need to initialize our hill climbing approach with a valid (plane), easy to find albeit possibly suboptimal solution. Following Observation 1, we obtain this by first calculating an EMST of all vertices in V_A, and subsequently connecting all vertices $v \notin V_A$ to the nearest $v' \in V_A$.

Afterwards, we iteratively execute rounds until no further improvement is gained. Each round consists of checking for each edge in the support if it can be removed, and if the hyperedges using it can be reconnected by (one or more) other edges that have a shorter total length than the removed edge without causing intersections. This check is nontrivial and done in a brute-force manner, improved by caching and pruning. At the end of each round, the edge replacement that reduces the total edge length most is actually executed. More rounds are evaluated until no single edge replacement reduces the total edge length.

As the initial state is a plane support tree, we can also readily enforce acyclicity, or relax the constraints to allow intersections.

2.3   Integer Linear Program

Theorem 1 implies that several variants of computing the shortest plane support are NP-hard. Here we briefly sketch how to obtain an *integer linear programs* (ILP) for a hypergraph $H = (V, S)$, allowing us to leverage effective ILP solvers.

We introduce variables $e_{u,v} \in \{0, 1\}$, indicating whether edge uv is selected for the support. This allows us to represent a graph with fixed vertices. Because the vertex locations are fixed, we can precompute edge lengths $d_{u,v}$ as well as which pairs of edges intersect. This gives the following basic program

$$\text{minimize } \sum_{u,v \in V} d_{u,v} \cdot e_{u,v}$$
$$\text{subject to } \quad e_{u,v} + e_{w,x} \leq 1 \quad \text{for all } u, v, w, x \in V \text{ if edges } uv \text{ and } wx \text{ intersect.}$$

What remains is to ensure that the graph is also a support: we need additional constraints that imply that each hyperedge in S induces a connected subgraph. To this end, we construct a flow tree for each hyperedge s. We pick an arbitrary sink for the hyperedge, $\sigma_s \in s$, that may receive flow, and let the remaining vertices in s generate one unit of flow. To formalize this, we introduce variables $f_{s,u,v} \in \{0, 1, \ldots, |s| - 1\}$ for each $s \in S$ and $u, v \in s$ with $u \neq v$. We now need the following constraints: (a) the incoming flow at σ_s is exactly $|s| - 1$; (b) the outgoing flow at σ_s is zero; (c) except for σ_s, each vertex in s sends out one unit of flow more than it receives; (d) flow can be sent only over selected edges.

(a) $\sum_{u \in s \setminus \{\sigma_s\}} f_{s,u,\sigma_s} = |s| - 1 \quad$ for all $s \in S$

(b) $\qquad\qquad f_{s,\sigma_s,v} = 0 \qquad\qquad$ for all $s \in S, v \in s \setminus \{\sigma_s\}$

(c) $\sum_{v \in s \setminus \{u\}} (f_{s,u,v} - f_{s,v,u}) = 1$ for all $s \in S, u \in s \setminus \{\sigma_s\}$

(d) $\qquad f_{s,u,v} \leq e_{u,v} \cdot (|s| - 1) \qquad$ for all $s \in S, u, v \in s$ with $u \neq v$

Variants. The ILP results in the shortest plane support for H. It can easily be modified to give a shortest (plane or unconstrained) support tree as well as to penalize or admit a limited number of intersections. The latter requires additional variables to indicate whether both edges of a crossing pair are used.

3   Experiments

As discussed above, there are various ways of defining and computing good supports. In this section we discuss several computational experiments that were performed to gain insight into the trade-offs between the different methods and properties. In particular, we use two different setups. First, we exclude optimal but slow algorithms to extensively compare the heuristic algorithms. Second, we include optimal algorithms to answer questions about the effect of requiring planarity or support trees, and to investigate how well heuristic algorithms approximate the optimal solution, albeit on smaller data sets.

Algorithms. We shall study four algorithms under various conditions in these experiments. In particular, we use MSTAPPROXIMATION to refer to the simple approximation algorithm of computing a minimum spanning tree for each hyperedge and then taking their union [13]. We refer to our heuristic improvement as MSTITERATION (Sect. 2.1). Finally, we use LOCALSEARCH to indicate our local search algorithm (Sect. 2.2) and OPT to denote an exact algorithm for computing optimal solutions. The latter two allow four different conditions, by requiring a plane support, a support tree, both (i.e., a plane support tree) or neither (unrestricted). We append P, T, PT and U to denote these conditions.

Data Generation. We generate a random hypergraph $H = (V, S)$ via the procedure described in the full version [8]. Our method ensures that at least one vertex is an element of all hyperedges (necessary for LOCALSEARCH, see Sect. 2.2), and that each hyperedge has at least two vertices. The procedure generates a hypergraph with n vertices, s hyperedges and a degree distribution d according to one of the following scheme:

EVEN   All degrees occur equally frequently.
MID   Degrees are drawn from a normal distribution with a peak on $k/2$.
LOW   Degrees are drawn from a normal distribution with a peak on 1.
HIGH   Degrees are drawn from a normal distribution with a peak on k.

3.1   Experiment 1: Comparison of Heuristics

Here we focus on answering the following three questions: (1) how much does the spanning tree iteration help to reduce the length of the support, compared to computing the minimum spanning trees in isolation; (2) which heuristic algorithm performs best in terms of support length; (3) which heuristic algorithm performs best in terms of computation time?

Setup. For each combination of $n = 20, 40, 60, 80, 100$, $k = 2, 3, 4, 5, 6, 7$ and $d =$ EVEN, MID, LOW, HIGH, we generate 1000 random hypergraphs with n vertices and k hyperedges according to degree distribution scheme d. For each hypergraph, we perform six algorithms: MSTAPPROXIMATION and MSTITERATION as well as LOCALSEARCH U/T/P/PT. This experiment was run on one machine, sequentially in a single thread to also allow for comparison of runtime performance. The machine was an HP ZBook with an Intel Core i7-6700HQ CPU, 24 GB RAM and running Windows 8.1.

Results. We first consider question (1) and compare MSTAPPROXIMATION and MSTITERATION. Since MSTITERATION can only improve upon MSTAPPROXIMATION, we express this as a ratio between 0 and 1. In Fig. 4 we show the results for $n = 20, 60, 100$ (Fig. 10 in the full version [8] provides the chart for all cases). Interestingly, the median gain remains roughly equal as we increase the number of vertices, though the variance becomes lower. Increasing the number of hyperedges gradually increases the relative gain of MSTITERATION. We also observe a dependency on the degree distribution. In particular, MID and EVEN systematically benefit more from iteration than LOW and HIGH. We explain this by observing that in the extreme cases MSTAPPROXIMATION is optimal: if all vertices have degree 1, then the optimal support is simply the union of all (disjoint) minimum spanning trees; if all vertices have degree k, then the optimal support is also simply the minimum spanning tree on the vertices. Difficulties arise when having many vertices that are part of multiple but not all hyperedges. This corresponds to the MID and EVEN schemes.

Let us now turn towards question (2), and consider the resulting support length of the LOCALSEARCH algorithm as well. We omit MSTAPPROXIMATION from these comparisons, since MSTITERATION always performs at least as well. In Fig. 5 we show the results for $n = 40$ and 100 (Fig. 11 in the full version [8] provides the chart for all cases). As one may expect, the length increases gradually with more hyperedges, as the support must use more edges to ensure that each hyperedge induces a connected subgraph. Moreover, we see that LOCALSEARCH U consistently outperforms MSTITERATION. To be exact, this is the case in

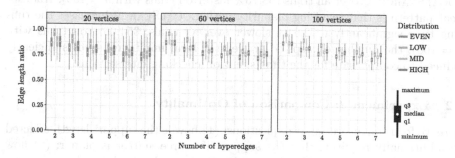

Fig. 4. Ratio of the support length computed by MSTITERATION as a fraction of MSTAPPROXIMATION. Lower values indicate a higher gain of the iteration method.

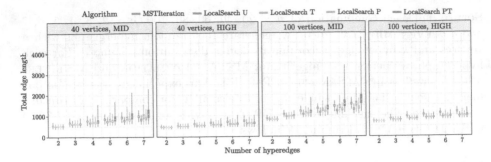

Fig. 5. Support length computed by the algorithms for varying values of n, k and d.

98.5% of all trials; the average ratio of LOCALSEARCH U to MSTITERATION (including those trials in which MSTITERATION performs better) is 0.877, that is, the support length is over 12% shorter on average. The effect of degree distribution also stands out. In LOW and MID, requiring planarity or a support tree has a large effect on the support length, whereas this is not the case in EVEN and HIGH. To explain this, observe that the minimum spanning tree on vertices that are in many or all hyperedges is planar and likely a part of the computed solution; in the EVEN and HIGH cases, there are comparatively many such vertices which can then serve as places to connect the other vertices in the support. In the LOW and MID cases, there are only few such vertices and thus the shortest connections that can be used to connect these to such a "backbone" structure are likely to intersect other connections. Though the number of vertices has little effect on MSTITERATION and LOCALSEARCH U, this does exacerbate the above problem: more vertices leads to a larger increase in support length when we enforce planarity or a support tree.

Finally, we briefly consider question (3) and compare the computation times of the various algorithms (see Fig. 6, or Fig. 12 in the full version [8]). We see that the number of hyperedges impacts the computation only slightly, whereas the number of vertices has a much stronger effect. MSTITERATION clearly outperforms the LOCALSEARCH variants, running on average 95.11% faster than LOCALSEARCH U over all trials (98.73% faster on trials with $n = 100$). Another clear pattern is that requiring planarity with LOCALSEARCH increases the running time significantly (272.64% slower over all trials, 354.06% on trials with $n = 100$); the number of steps to arrive at a local minimum is not sufficiently reduced to compensate for the time spent on checking intersections.

3.2   Experiment 2: Comparison of Optimality

Here we focus on answering two questions: (1) how is the support length affected by additionally requiring that the support is a tree and/or is planar; (2) how well do the heuristic algorithms approximate the optimal solution?

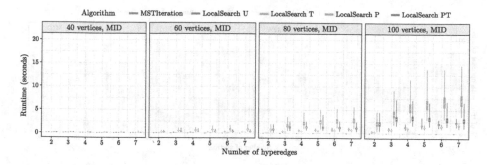

Fig. 6. Computation time of the various algorithms for varying values of n and k.

Setup. For each combination of $n = 10, 15, 20$, $k = 2, 3$ and $d = $ LOW, MID, we generate 1000 random hypergraphs with n vertices, k hyperedges according to degree distribution scheme d. For each hypergraph, we run the LOCALSEARCH U/T/P/PT and compute an optimal solution OPT U/T/P/PT[2]. To obtain a large enough number of trials, these experiments were run on different machines simultaneously and in concurrent threads. As such, we refrain from analyzing algorithm speed in this experiment.

Failed Trials. In about 3.4% of the CPLEX runs for $n = 20$, the computation would run out of memory and therefore not finish successfully. We ran additional trials to compensate, eventually obtaining 1000 successful trials. This likely biases the results for $n = 20$ towards including only the "easier" situations. The full version [8] provides more details including statistics on which cases failed and indicators of the "difficulty" of these cases.

Results. Let us first compare the optimal solutions according to the four different restrictions. In Fig. 7 we show the results. For two hyperedges, we see

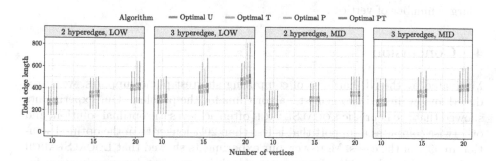

Fig. 7. Support length achieved by OPT in the four conditions U/T/P/PT.

---

[2] For $n = 10, 15$, this is a simple branch and bound algorithm; for $n = 20$ we use the ILP solution, solved with IBM ILOG CPLEX 12.6.3.

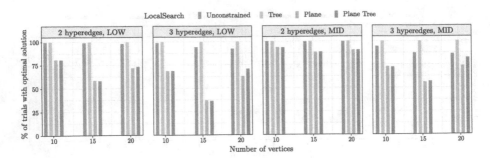

Fig. 8. Percentage of runs of LOCALSEARCH that achieve the optimal solution. Note that LOCALSEARCH T always achieves optimal results.

that there is little to no effect of requiring support trees, but a small worst-case effect for requiring plane supports for the LOW case—the median increases only slightly. For three hyperedges, we see that the effects become slightly larger. Most noticeable is that enforcing support trees has now a slight effect, even for only a few vertices. In terms of plane supports, we see a similar pattern as before, that is, that of an increase particularly in the LOW case, but also some in the MID case. Note that the effects for $n = 20$ are potentially underestimated.

Let us now turn towards how well LOCALSEARCH performs with respect to the optimal solution. Our results indicate that in a majority of the cases, our heuristic actually achieves optimal results (see Fig. 8). For $n = 10, 15$ we see a clear decrease of this percentage for plane supports and trees; we attribute the apparent increase at $n = 20$ to the failed trials. To further see how well LOCALSEARCH performs if it fails to achieve optimal results, we look at the ratio between the support length it achieves and the optimal support length. In all cases, we observe a ratio of less than 1.61. The 90-, 95-, and 99-percentile of this ratio was worst for LOCALSEARCH PT, being 1.05, 1.09, and 1.19, respectively. Again, we have to keep in mind that the data for $n = 20$ likely exclude some more difficult cases and thus the trend in the increasing ratio might extend further for a larger number of vertices.

4   Conclusion

Motivated by the NP-hardness of computing shortest plane supports, we introduced and evaluated two heuristic algorithms for the problem. Our experiments showed that the heuristic LOCALSEARCH often achieves the optimal solution, and otherwise computes a support that is less than 20% longer than the optimal solution in 99% of the cases. Moreover, our experiments showed that LOCALSEARCH performs better than MSTITERATION, which in turn is a k-approximation for k hyperedges. We can also guarantee that LOCALSEARCH (without restrictions) is a k-approximation by initializing it using either MSTAPPROXIMATION or MSTITERATION, though it is not clear whether this change will generally improve the result of LOCALSEARCH. There is a trade-off between speed and support length, where MSTITERATION is better for the former and LOCALSEARCH for the latter. We also observed that the increase in support length caused by additional requirements, depends both on the number of sets and the number of set memberships per element, but this behavior seems predictable and not to depend on the number of elements.

Future Work. From the theoretical side, several questions remain open. For example, can we efficiently decide whether a plane support tree exists? We currently know how to answer this only for two hyperedges (using Observation 1 and [5]). Furthermore, how many iterations do we need for MSTITERATION with more than two hyperedges, to guarantee that the computation stabilizes?

Our experiments indicate that our local search algorithm does not always perform optimally, especially when requiring plane supports. It is, however, based on simple hill climbing. Can we employ better search techniques such as simulated annealing to efficiently find better solutions?

Finally, we chose to generate random hypergraphs for our experiments, as to not depend on particular properties of (geospatial) configurations that may be inherent to some real-world data sets. While this reduces the explanatory power with respect to real-world data sets, it provides us with more insight into the structural problem, unbiased by unknown or hidden structures of real-world data. We leave it to future work to further dive into real-world data sets, to see if similar trends and patterns emerge or more difficult structures arise and to evaluate the impact of the different heuristics on readability.

Acknowledgments. This work started at Dagstuhl seminar 17332 "Scalable Set Visualizations". The authors would like to thank Nathalie Henry Riche for providing the data for Fig. 2. TC was supported by the Netherlands Organisation for Scientific Research (NWO, 314.99.117). MvG received funding from the European Union's Seventh Framework Programme (FP7/2007-2013) under ERC grant agreement n° 319209 (project NEXUS 1492) and the German Research Foundation (DFG) within project B02 of SFB/Transregio 161. WM was partially supported by the Netherlands eScience Centre (NLeSC, 027.015.G02).

References

1. Akitaya, H.A., Löffler, M., Tóth, C.D.: Multi-colored spanning graphs. In: Hu, Y., Nöllenburg, M. (eds.) GD 2016. LNCS, vol. 9801, pp. 81–93. Springer, Cham (2016). https://doi.org/10.1007/978-3-319-50106-2_7
2. Alper, B., Henry Riche, N., Ramos, G., Czerwinski, M.: Design study of LineSets, a novel set visualization technique. IEEE Trans. Vis. Comput. Graph. 17(12), 2259–2267 (2011). https://doi.org/10.1109/TVCG.2011.186
3. Alsallakh, B., Micallef, L., Aigner, W., Hauser, H., Miksch, S., Rodgers, P.: The state of the art of set visualization. Comput. Graph. Forum 35(1), 234–260 (2016). https://doi.org/10.1111/cgf.12722

4. Bereg, S., Fleszar, K., Kindermann, P., Pupyrev, S., Spoerhase, J., Wolff, A.: Colored non-crossing euclidean steiner forest. In: Elbassioni, K., Makino, K. (eds.) ISAAC 2015. LNCS, vol. 9472, pp. 429–441. Springer, Heidelberg (2015). https://doi.org/10.1007/978-3-662-48971-0_37
5. Bereg, S., Jiang, M., Yang, B., Zhu, B.: On the red/blue spanning tree problem. Theor. Comput. Sci. **412**(23), 2459–2467 (2011). https://doi.org/10.1016/j.tcs.2010.10.038
6. Brandes, U., Cornelsen, S., Pampel, B., Sallaberry, A.: Path-based supports for hypergraphs. J. Discrete Algorithms **14**, 248–261 (2012). https://doi.org/10.1016/j.jda.2011.12.009
7. Buchin, K., van Kreveld, M., Meijer, H., Speckmann, B., Verbeek, K.: On planar supports for hypergraphs. J. Graph Algorithms Appl. **15**(4), 533–549 (2011). https://doi.org/10.7155/jgaa.00237
8. Castermans, T., van Garderen, M., Meulemans, W., Nöllenburg, M., Yuan, X.: Short plane supports for spatial hypergraphs. Computing Research Repository, arXiv:1808.09729 (2018)
9. Collins, C., Penn, G., Carpendale, S.: Bubble sets: revealing set relations with isocontours over existing visualizations. IEEE Trans. Vis. Comput. Graph. **15**(6), 1009–1016 (2009). https://doi.org/10.1109/TVCG.2009.122
10. Dinkla, K., van Kreveld, M., Speckmann, B., Westenberg, M.: Kelp diagrams: point set membership visualization. Comput. Graph. Forum **31**(3pt1), 875–884 (2012). https://doi.org/10.1111/j.1467-8659.2012.03080.x
11. Efrat, A., Hu, Y., Kobourov, S.G., Pupyrev, S.: MapSets: visualizing embedded and clustered graphs. J. Graph Algorithms Appl. **19**(2), 571–593 (2015). https://doi.org/10.7155/jgaa.00364
12. van Goethem, A., Kostitsyna, I., van Kreveld, M., Meulemans, W., Sondag, M., Wulms, J.: The painter's problem: covering a grid with colored connected polygons. In: Frati, F., Ma, K.-L. (eds.) GD 2017. LNCS, vol. 10692, pp. 492–505. Springer, Cham (2018). https://doi.org/10.1007/978-3-319-73915-1_38
13. Hurtado, F., et al.: Colored spanning graphs for set visualization. Comput. Geom.: Theory Appl. **68**, 262–276 (2018). https://doi.org/10.1016/j.comgeo.2017.06.006
14. Johnson, D.S., Pollak, H.O.: Hypergraph planarity and the complexity of drawing Venn diagrams. J. Graph Theory **11**(3), 309–325 (1987). https://doi.org/10.1002/jgt.3190110306
15. Klemz, B., Mchedlidze, T., Nöllenburg, M.: Minimum tree supports for hypergraphs and low-concurrency euler diagrams. In: Ravi, R., Gørtz, I.L. (eds.) SWAT 2014. LNCS, vol. 8503, pp. 265–276. Springer, Cham (2014). https://doi.org/10.1007/978-3-319-08404-6_23
16. Korach, E., Stern, M.: The clustering matroid and the optimal clustering tree. Math. Program. **98**(1–3), 385–414 (2003). https://doi.org/10.1007/s10107-003-0410-x
17. Lichtenstein, D.: Planar formulae and their uses. SIAM J. Comput. **11**(2), 329–343 (1982). https://doi.org/10.1137/0211025
18. Meulemans, W., Henry Riche, N., Speckmann, B., Alper, B., Dwyer, T.: KelpFusion: a hybrid set visualization technique. IEEE Trans. Vis. Comput. Graph. **19**(11), 1846–1858 (2013). https://doi.org/10.1109/TVCG.2013.76
19. Purchase, H.: Metrics for graph drawing aesthetics. J. Vis. Lang. Comput. **13**(5), 501–516 (2002). https://doi.org/10.1006/jvlc.2002.0232
20. Tufte, E.: The Visual Display of Quantitative Information. Graphics Press, Cheshire (2001)

Turning Cliques into Paths to Achieve Planarity

Patrizio Angelini[1], Peter Eades[2], Seok-Hee Hong[2], Karsten Klein[3],
Stephen Kobourov[4], Giuseppe Liotta[5], Alfredo Navarra[5],
and Alessandra Tappini[5]($\boxtimes$)

[1] University of Tübingen, Tübingen, Germany
angelini@informatik.uni-tuebingen.de
[2] The University of Sydney, Sydney, Australia
{peter.eades,seokhee.hong}@usyd.edu.au
[3] University of Konstanz, Konstanz, Germany
karsten.klein@uni-konstanz.de
[4] University of Arizona, Tucson, USA
kobourov@cs.arizona.edu
[5] University of Perugia, Perugia, Italy
{giuseppe.liotta,alfredo.navarra}@unipg.it,
alessandra.tappini@studenti.unipg.it

Abstract. Motivated by hybrid graph representations, we introduce
and study the following beyond-planarity problem, which we call h-
CLIQUE2PATH PLANARITY: Given a graph G, whose vertices are partitioned into subsets of size at most h, each inducing a clique, remove
edges from each clique so that the subgraph induced by each subset is
a path, in such a way that the resulting subgraph of G is planar. We
study this problem when G is a simple topological graph, and establish
its complexity in relation to k-planarity. We prove that h-CLIQUE2PATH
PLANARITY is NP-complete even when $h = 4$ and G is a simple 3-plane
graph, while it can be solved in linear time, for any h, when G is 1-plane.

1 Hybrid Representations

A common problem in the visual analysis of real-world networks is that dense
subnetworks create occlusions and hairball-like structures in node-link diagrams
generated by standard layout algorithms, e.g., force-directed methods. On the
other hand, different representations, such as adjacency matrices, are well suited

This work began at the Bertinoro Workshop on Graph Drawing 2018. Research was
partially supported by DFG grant Ka812/17-1 and MIUR-DAAD Joint Mobility Program n.57397196 (PA), by Young Scholar Fund/AFF - Univ. Konstanz (KK), by NSF
grants CCF-1740858 - CCF-1712119 (SK), by project "Algoritmi e sistemi di analisi
visuale di reti complesse e di grandi dimensioni" - Ric. di Base 2018, Dip. Ingegneria -
Univ. Perugia (GL, AT), by project GEO-SAFE n.H2020-691161 (AN).

© Springer Nature Switzerland AG 2018
T. Biedl and A. Kerren (Eds.): GD 2018, LNCS 11282, pp. 67–74, 2018.
https://doi.org/10.1007/978-3-030-04414-5_5

for dense graphs but make neighbor identification and path-tracing more difficult [7,12]. *Hybrid graph representations* combine different representation metaphors in order to exploit their strengths and overcome their drawbacks.

The first example of hybrid representation was the *NodeTrix* model [8], which combines node-link diagrams with adjacency-matrix representations of the denser subgraphs [4,5,8,14]. Another example of hybrid representations are *intersection-link representations* [1]. In this model vertices are geometric objects and edges are either intersections between objects (*intersection edges*), or crossing-free Jordan arcs attaching at their boundary (*link edges*). Different types of objects determine different intersection-link representations.

In [1], *clique-planar* drawings are defined as intersection-link representations in which the objects are isothetic rectangles, and the partition into intersection- and link-edges is given in the input, so that the graph induced by the intersection-edges is composed of a set of vertex-disjoint cliques. The corresponding recognition problem is called CLIQUE-PLANARITY, and it has been proved NP-complete in general and polynomial-time solvable in restricted cases.

We study CLIQUE-PLANARITY when all cliques have bounded size. As proved in [1], the CLIQUE-PLANARITY problem can be reformulated in the terminology of *beyond-planarity* [6,10], as follows. Given a graph $G = (V, E)$ and a partition of its vertex set V into subsets $V_1, \ldots, V_m$ such that the subgraph of G induced by each subset V_i is a clique, the goal is to compute a planar subgraph $G' = (V, E')$ of G by replacing the clique induced by V_i, for each $i = 1, \ldots, m$, with a path spanning the vertices of V_i. We call h-CLIQUE2PATH PLANARITY (for short, h-C2PP) the version of this problem in which each clique has size at most h.

We remark that the version of h-C2PP in which the input graph G is a *geometric graph*, i.e., it is drawn in the plane with straight-line edges, has been recently studied by Kindermann et al. [9] in a different context. The input of their problem is a set of colored points in the plane, and the goal is to decide whether there exist straight-line spanning trees, one for each same-colored point subset, that do not cross each other. Since edges are straight-line, their drawings are determined by the positions of the points, and hence each same-colored point subset can in fact be seen as a straight-line drawing of a clique, from which edges have to be removed so that each clique becomes a tree and the drawing becomes planar. They proved NP-completeness for the case in which the spanning tree must be a path, even when there are at most 4 vertices with the same color. This implies that 4-C2PP for geometric graphs in NP-complete. On the other hand, they provided a linear-time algorithm when there exist at most 3 vertices with the same color, which then extends to 3-C2PP for geometric graphs.

In this paper, we study the version of h-C2PP in which the input graph G is a *simple topological graph*, that is, it is embedded in the plane so that each edge is a Jordan arc connecting its end-vertices; by simple we mean that a Jordan arc does not pass through any vertex, and does not intersect any arc more than once (either with a proper crossing or sharing a common end-vertex); finally, no three arcs pass through the same point. Our main goal is to study the complexity of this problem in relation to the well-studied class of k-*planar graphs*, i.e., those that admit drawings in which each edge has at most k crossings [1,3,6,13].

We observe that the NP-completeness of 4-C2PP for geometric graphs already implies the NP-completeness of 4-C2PP for simple topological graphs; also, though not explicitly mentioned in [9], it is possible to show that the instances produced by that reduction are 4-plane (see [2]). We strengthen this result by proving in Sect. 2 that 4-C2PP is NP-complete even for simple topological 3-plane graphs. On the positive side, we prove in Sect. 3 that the h-C2PP problem for simple topological 1-plane graphs can be solved in linear time for any value of h. We finally remark that the 2-SAT formulation used in [9] to solve 3-C2PP for geometric graphs can be easily extended to solve 3-C2PP for any simple topological graph.

For space reasons, some proofs have been omitted or sketched, and can be found in [2]; the corresponding statements are marked with [*].

2   *NP*-Completeness for Simple Topological 3-Plane Graphs

In this section we prove that the k-C2PP problem remains NP-complete for $k = 4$ even when the input is a simple topological 3-plane graph.

Since the planarity of a simple topological graph can be checked in linear time, the h-C2PP problem for simple topological k-plane graphs belongs to NP for all values of h and k. In the following, we prove the NP-hardness by means of a reduction from the PLANAR POSITIVE 1-IN-3-SAT problem. In this version of the SATISFIABILITY problem, which is known to be NP-complete [11], each variable appears only with its positive literal, each clause has at most three variables, the graph obtained by connecting each variable with all the clauses it belongs to is planar, and the goal is to find a truth assignment in such a way that, for each clause, exactly one of its three variables is set to True. For each 3-clique we use in the reduction, there is a *base edge*, which is crossing-free in the constructed topological graph, while the other two edges always have crossings. We call *left* (*right*) the edge that follows (precedes) the base edge in the clockwise order of the edges along the 3-clique. Also, if an edge e of a clique does not belong to the path replacing the clique, we say that e is *removed*, and that all the crossings involving e in G are *resolved*. For each variable x, let n_x be the number of clauses containing x. We construct a simple topological graph gadget G_x for x, called *variable gadget*; see the left dotted box in Fig. 1(a). This gadget contains $2n_x$ 3-cliques $t_1^x, \ldots, t_{2n_x}^x$, forming a ring, so that the left (right) edge of t_i^x only crosses the left (right) edge of t_{i-1}^x and of t_{i+1}^x, for each $i = 1, \ldots, 2n_x$. Also, gadget G_x contains n_x additional 3-cliques, called $\tau_1^x, \ldots, \tau_{n_x}^x$, so that the right edge of τ_j^x crosses the left edge of t_{2j-1}^x and the right edge of t_{2j}^x, while the left edge of τ_j^x crosses the left edge of t_{2j}^x and the right edge of t_{2j-1}^x. Then, for each clause c, we construct a topological graph gadget G_c, called *clause gadget*, which is composed of a planar drawing of a 4-clique, together with three 3-cliques whose left and right edges cross the edges of the 4-clique as in the right dotted box in Fig. 1(a). In particular, observe that the right (left) edge of each 3-clique crosses exactly one (two) edges of the 4-clique. Every 3-clique in G_c corresponds to one of the three variables of c. Let x be one of such variables; assuming that c is the j-th clause that contains x according to the order of the clauses in the given formula, we connect the 3-clique corresponding to x in the clause gadget G_c to the 3-clique τ_j^x of the variable gadget G_x of x by a chain of 3-cliques of odd length, as in Fig. 1(a).

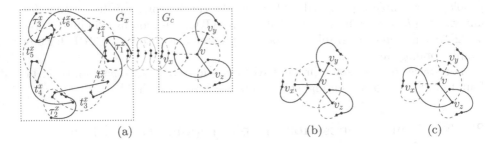

(a)   (b)   (c)

Fig. 1. (a) The variable gadget G_x for a variable x is represented in the left dotted box. The clause gadget for a clause c is represented in the right dotted box. The chain connecting G_x to G_c is represented with lighter colors. The removed edges are dashed red. (b) All variables are **False**. (c) At least two variables are **True**. (Color figure online)

By construction, the resulting simple topological graph G contains cliques of size at most 4, namely one per clause, and hence is a valid instance of 4-C2PP. Also, by collapsing each variable and clause gadget into a vertex, and each chain connecting them into an edge, the resulting graph G' preserves the planarity of the PLANAR POSITIVE 1-IN-3-SAT instance. This implies that the only crossings for each edge of G are with other edges in the gadget it belongs to and, possibly, with the edges of the 3-cliques of a chain. Hence, G is 3-plane. Namely, each base edge is crossing-free; each internal edge of a 4-clique has one crossing; each external edge of a 4-clique has two crossings, and the same is true for the left and right edges of each 3-clique in a chain; finally, the left and right edges of each 3-clique in either a variable or a clause gadget has three crossings.

In the following we prove the equivalence between the original instance of PLANAR POSITIVE 1-IN-3-SAT and the constructed instance G of 4-C2PP. For this, we first give a lemma stating that variable gadgets correctly represent the behavior of a variable; indeed they can assume one out of two possible states in any solution for 4-C2PP. The proof of the next lemma can be found in [2].

Lemma 1 [*]. *Let G_x be the variable gadget for a variable x in G. Then, in any solution for 4-C2PP, either the left edge of each 3-clique τ_j^x, with $j = 1, \ldots, n_x$, is removed, or the right edge of each 3-clique τ_j^x is removed.*

Given Lemma 1, we can associate the truth value of a variable x with the fact that either the left or the right edge of each 3-clique τ_j^x in the variable gadget G_x of G is removed. We use this association to prove the following theorem.

Theorem 1 [*]. *The* 4-C2PP *problem is NP-complete, even for* 3-*plane graphs.*

Proof (sketch). Given an instance of PLANAR POSITIVE 1-IN-3-SAT, we construct an instance G of 4-C2PP in linear time as described above. We prove one direction of the equivalence between the two problems. The other direction follows a similar reasoning. Suppose that there exists a solution for 4-C2PP, i.e., a set of edges of G whose removal resolves all crossings. By Lemma 1, for each variable x either the left or the right edge of each 3-clique τ_j^x in gadget G_x is removed. We assign True (False) to x if the right (left) edge is removed.

We first claim that for each clause c that contains variable x, the right (left) edge of the 3-clique $t_c(x)$ of the clause gadget G_c corresponding to x is removed if and only if the right (left) edge of each 3-clique τ_j^x is removed. Consider the chain that connects $t_c(x)$ with a 3-clique τ_j^x of G_x. For any two consecutive 3-cliques along the chain the left edge of one 3-clique and the right edge of the other 3-clique must be removed. Since the chain has odd length, the truth value of G_x is transferred to the 3-clique $t_c(x)$ of G_c and thus the claim follows.

Consider now a clause c with variables x, y, and z. Let $t_c(x)$, $t_c(y)$, and $t_c(z)$ be the 3-cliques of the clause gadget G_c of c corresponding to x, y, and z, respectively. Let v be the central vertex of the 4-clique of G_c, and let v_x, v_y, v_z be the vertices of this 4-clique lying inside $t_c(x)$, $t_c(y)$, and $t_c(z)$ (see Fig. 1). Assume that v_x, v_y, and v_z appear in this clockwise order around v. We now show that, for exactly one of $t_c(x)$, $t_c(y)$, and $t_c(z)$ the right edge is removed, which implies that exactly one of x, y, and z is True and hence the instance of PLANAR POSITIVE 1-IN-3-SAT is positive. Assume that for each of $t_c(x)$, $t_c(y)$, and $t_c(z)$ the left edge is removed (i.e., all the three variables are set to False), as in Fig. 1(b). The crossings between the right edges of the three 3-cliques and the three edges of triangle (v_x, v_y, v_z) are not resolved. All edges of this triangle should be removed, which is not possible since the remaining edges of the 4-clique do not form a path. Assume now that for at least two of the 3-cliques, say $t_c(x)$ and $t_c(y)$, the right edge is removed (i.e., x and y are set to True), as in Fig. 1(c). Since each edge of triangle (v_x, v_y, v) is crossed by the left edge of one of $t_c(x)$ and $t_c(y)$, by construction, these crossings are not resolved. Hence, all edges of (v_x, v_y, v) should be removed, which is not possible since the remaining edges of the 4-clique do not form a path of length 4. Finally, assume that for exactly one of the 3-cliques, say $t_c(x)$, the right edge is removed (i.e., x is the only one set to True), as in Fig. 1(a). By removing edges (v, v_x), (v_x, v_y), and (v_y, v_z), all crossings are resolved; the remaining edges of the 4-clique form a path of length 4, as desired. □

3 h-CLIQUE2PATH PLANARITY and 1-Planarity

In this section we show that, when the given simple topological graph is 1-plane, problem h-C2PP can be solved in linear time in the size of the input, for any h. We consider all possible simple topological 1-plane cliques and show that the problem can be solved using only local tests, each requiring constant time. Note that $h \leq 6$, since K_6 is the largest 1-planar complete graph [10].

Simple topological 1-plane graphs containing cliques with at most four vertices that cross each other can be constructed, but it is easy to enumerate all these graphs (up to symmetry); see Fig. 2. Note that such graphs involve at most two cliques and that if K_4 has a crossing, combining it with any other clique would violate 1-planarity; see Fig. 2(a) and (b). The next lemma accounts for cliques with five or six vertices.

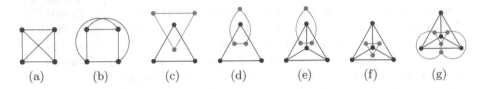

(a)   (b)   (c)   (d)   (e)   (f)   (g)

Fig. 2. All 1-plane graphs involving one or more cliques of type K_3 and K_4.

Lemma 2. *There exists no 1-plane simple topological graph that contains two cliques, one of which with at least five vertices, whose edges cross each other.*

Proof. Consider a simple 1-plane graph G that contains two disjoint cliques K and H, with five and three vertices, respectively. Let K' be the simple plane topological graph obtained from K by replacing each crossing with a dummy vertex. By 1-planarity, every face of K' is a triangle and contains at most one dummy vertex. Suppose, for a contradiction, that there exists a crossing between an edge of K and an edge of H in G. Then there would exist at least a vertex v of H inside a face f of K' and at least one outside f. Since H is a triangle, there must have been two edges that connect vertices inside f to vertices outside f. If f contains one dummy vertex, then two of its edges are not crossed by edges of H, as otherwise G would not be 1-plane. Hence, both the edges that connect vertices inside f to vertices outside f cross the other edge of f, a contradiction. If f contains no dummy vertices, then each edge of f admits one crossing. Let u be the vertex of f that is incident to the two edges crossed by edges of H. Since u has degree 4 in K, it is not possible to draw the third edge of H so that it crosses only one edge of K, which completes the proof.   □

Combining the previous discussion with Lemma 2, we conclude that, for each subgraph of the input graph G that consists either of a combination of at most two cliques of size at most 4, as in Fig. 2, or of a single clique not crossing any other clique, the crossings involving this subgraph (possibly with other edges not belonging to cliques) can only be resolved by removing its edges, which can be checked in constant time. In the next theorem, n denotes the number of vertices.

Theorem 2. *h-C2PP is $O(n)$-time solvable for simple topological 1-plane graphs.*

4   Open Problems

We studied the h-CLIQUE2PATH PLANARITY problem for simple topological k-plane graphs; we proved that this problem is NP-complete for $h = 4$ and $k = 3$, while it is solvable in linear time for every value of h, when $k = 1$. The natural open question is: what is the complexity for simple topological 2-plane graphs?

Kindermann et al. [9] recently proved that problem 4-C2PP is NP-complete for geometric 4-plane graphs. It would be interesting to study this geometric version of the problem for 2-plane and 3-plane graphs.

Finally, note that the version of the h-C2PP problem when the input is an abstract graph (which is equivalent to CLIQUE PLANARITY [1]) is NP-complete when $h \in O(n)$. What if h is bounded by a constant or a sublinear function? We remark that, for $h = 3$, this version of the problem is equivalent to CLUSTERED PLANARITY, when restricted to instances in which the graph induced by each cluster consists of three isolated vertices.

References

1. Angelini, P., Da Lozzo, G., Di Battista, G., Frati, F., Patrignani, M., Rutter, I.: Intersection-link representations of graphs. J. Graph Algorithms Appl. **21**(4), 731–755 (2017). https://doi.org/10.7155/jgaa.00437
2. Angelini, P., Eades, P., Hong, S.-H., Klein, K., Kobourov, S., Liotta, G., Navarra, A., Tappini, A.: Turning cliques into paths to achieve planarity. CoRR 1808.08925v2 (2018). http://arxiv.org/abs/1808.08925v2
3. Bekos, M.A., Kaufmann, M., Raftopoulou, C.N.: On optimal 2- and 3-planargraphs. In: 33rd International Symposium on Computational Geometry, SoCG2017, 4-7 July 2017, Brisbane, Australia, pp. 16:1–16:16 (2017). https://doi.org/10.4230/LIPIcs.SoCG.2017.16
4. Da Lozzo, G., Di Battista, G., Frati, F., Patrignani, M.: Computing NodeTrix representations of clustered graphs. J. Graph Algorithms Appl. **22**(2), 139–176 (2018). https://doi.org/10.7155/jgaa.00461
5. Di Giacomo, E., Liotta, G., Patrignani, M., Tappini, A.: NodeTrix planarity testing with small clusters. In: Frati, F., Ma, K.-L. (eds.) GD 2017. LNCS, vol. 10692, pp. 479–491. Springer, Cham (2018). https://doi.org/10.1007/978-3-319-73915-1_37
6. Didimo, W., Liotta, G., Montecchiani, F.: A survey on graph drawing beyond planarity. CoRR abs/1804.07257 (2018). http://arxiv.org/abs/1804.07257
7. Ghoniem, M., Fekete, J., Castagliola, P.: On the readability of graphs using node-link and matrix-based representations: a controlled experiment and statistical analysis. Inf. Vis. **4**(2), 114–135 (2005). https://doi.org/10.1057/palgrave.ivs.9500092
8. Henry, N., Fekete, J., McGuffin, M.J.: NodeTrix: a hybrid visualization of social networks. IEEE Trans. Vis. Comput. Graph. **13**(6), 1302–1309 (2007). https://doi.org/10.1109/TVCG.2007.70582
9. Kindermann, P., Klemz, B., Rutter, I., Schnider, P., Schulz, A.: The partition spanning forest problem. In: Mulzer, W. (ed.) Proceedings of the 34th European Workshop on Computational Geometry (EuroCG 2018), Berlin (2018, to appear)
10. Kobourov, S.G., Liotta, G., Montecchiani, F.: An annotated bibliography on 1-planarity. Comput. Sci. Rev. **25**, 49–67 (2017). https://doi.org/10.1016/j.cosrev.2017.06.002

11. Mulzer, W., Rote, G.: Minimum-weight triangulation is NP-hard. J. ACM **55**(2), 11:1–11:29 (2008). https://doi.org/10.1145/1346330.1346336

12. Okoe, M., Jianu, R., Kobourov, S.: Revisited experimental comparison of node-link and matrix representations. In: Frati, F., Ma, K.-L. (eds.) GD 2017. LNCS, vol. 10692, pp. 287–302. Springer, Cham (2018). https://doi.org/10.1007/978-3-319-73915-1_23

13. Pach, J., Tóth, G.: Graphs drawn with few crossings per edge. Combinatorica **17**(3), 427–439 (1997). https://doi.org/10.1007/BF01215922

14. Yang, X., Shi, L., Daianu, M., Tong, H., Liu, Q., Thompson, P.: Blockwise human brain network visual comparison using NodeTrix representation. IEEE Trans. Vis. Comput. Graph. **23**(1), 181–190 (2017). https://doi.org/10.1109/TVCG.2016.2598472

Upward Drawings

Universal Slope Sets for Upward Planar Drawings

Michael A. Bekos[1], Emilio Di Giacomo[2(✉)], Walter Didimo[2], Giuseppe Liotta[2], and Fabrizio Montecchiani[2]

[1] Institut für Informatik, Universität Tübingen, Tübingen, Germany
bekos@informatik.uni-tuebingen.de
[2] Dipartimento di Ingegneria, Università degli Studi di Perugia, Perugia, Italy
{emilio.digiacomo,walter.didimo,giuseppe.liotta,
fabrizio.montecchiani}@unipg.it

Abstract. We prove that every set S of Δ slopes containing the horizontal slope is *universal* for 1-bend upward planar drawings of bitonic st-graphs with maximum vertex degree Δ, i.e., every such digraph admits a 1-bend upward planar drawing whose edge segments use only slopes in S. This result is worst-case optimal in terms of the number of slopes, and, for a suitable choice of S, it gives rise to drawings with worst-case optimal angular resolution. In addition, we prove that every such set S can be used to construct 2-bend upward planar drawings of n-vertex planar st-graphs with at most $4n - 9$ bends in total. Our main tool is a constructive technique that runs in linear time.

1 Introduction

Let G be a graph with maximum vertex degree Δ. The *k-bend planar slope number* of G is the minimum number of slopes for the edge segments needed to construct a k-bend planar drawing of G, i.e., a planar drawing where each edge is a polyline with at most $k \geq 0$ bends. Since no more than two edge segments incident to the same vertex can use the same slope, $\lceil \Delta/2 \rceil$ is a trivial lower bound for the k-bend planar slope number of G, irrespectively of k. Besides its theoretical interest, this problem forms a natural extension of two well-established graph drawing models: The *orthogonal* [6,16,18,29] and the *octilinear* drawing models [3,4,7,26], which both have several applications, such as in VLSI and floor-planning [25,30], and in metro-maps and map-schematization [21,27,28]. Orthogonal drawings use only 2 slopes for the edge segments (0 and $\frac{\pi}{2}$), while octilinear drawings use no more than 4 slopes (0, $\frac{\pi}{4}$, $\frac{\pi}{2}$, and $\frac{3\pi}{4}$); consequently, they are limited to graphs with $\Delta \leq 4$ and $\Delta \leq 8$, respectively.

These two drawing models have been generalized to graphs with arbitrary maximum vertex degree Δ by Keszegh et al. [23], who proved that every planar graph admits a 2-bend planar drawing with $\lceil \Delta/2 \rceil$ equispaced slopes. As a witness of the tight connection between the two problems, the result by Keszegh et al. was built upon an older result for orthogonal drawings of degree-4 planar

© Springer Nature Switzerland AG 2018
T. Biedl and A. Kerren (Eds.): GD 2018, LNCS 11282, pp. 77–91, 2018.
https://doi.org/10.1007/978-3-030-04414-5_6

graphs by Biedl and Kant [6]. In the same paper, Keszegh et al. also studied the 1-bend planar slope number and showed an upper bound of 2Δ and a lower bound of $\frac{3}{4}(\Delta-1)$ for this parameter. The upper bound has been recently improved, initially by Knauer and Walczak [24] to $\frac{3}{2}(\Delta-1)$ and subsequently by Angelini et al. [1] to $\Delta-1$. Angelini et al. actually proved a stronger result: Given *any* set S of $\Delta-1$ slopes, every planar graph with maximum vertex degree Δ admits a 1-bend planar drawing whose edge segments use only slopes in S. Any such slope set is hence called *universal* for 1-bend planar drawings. This result simultaneously establishes the best-known upper bound on the 1-bend planar slope number of planar graphs and the best-known lower bound on the angular resolution of 1-bend planar drawings, i.e., on the minimum angle between any two edge segments incident to the same vertex. Indeed, if the slopes in S are equispaced, the resulting drawings have angular resolution at least $\frac{\pi}{\Delta-1}$.

In this paper we study slope sets that are universal for k-bend *upward* planar drawings of directed graphs (or digraphs for short). Recall that in an upward drawing of a digraph G, every edge (u,v) is drawn as a y-monotone non-decreasing curve from u to v. Also, G admits an upward planar drawing if and only if it is a subgraph of a planar st-graph [13,22]. As such drawings are common for representing planar digraphs, they have been extensively studied in the literature (see, e.g., [5,9,15,18,20]). A preliminary result for this setting is due to Di Giacomo et al. [14], who proved that every series-parallel digraph with maximum vertex degree Δ admits a 1-bend upward planar drawing that uses at most Δ slopes, and this bound on the number of slopes is worst-case optimal. Notably, their construction gives rise to drawings with optimal angular resolution $\frac{\pi}{\Delta}$ (but it uses a predefined set of slopes). Upward drawings with one bend per edge and few slopes have also been studied for posets by Czyzowicz et al. [11].

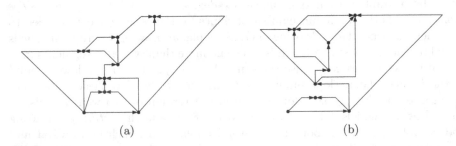

(a)   (b)

Fig. 1. (a) A 1-bend upward planar drawing of a bitonic st-graph, and (b) a 2-bend upward planar drawing of a planar st-graph, both defined on a slope set $S = \{-\frac{\pi}{4}, 0, \frac{\pi}{4}, \frac{\pi}{2}, \pi\}$.

Contribution. We extend the study of universal sets of slopes to upward planar drawings, and present the first constructive technique that works for all planar st-graphs. This technique exploits a linear ordering of the vertices of a planar digraph introduced by Gronemann [19], called *bitonic st-ordering* (see also Sect. 2). We show that any set S of Δ slopes containing the horizontal slope is universal for 1-bend upward planar drawings of degree-Δ planar digraphs having a bitonic *st*-ordering (Sect. 3). We remark that the size of S is worst-case optimal [14] and, if the slopes of S are chosen to be equispaced, the angular resolution of the resulting drawing is at least $\frac{\pi}{\Delta}$ (also optimal); see Fig. 1a for an illustration. We then extend our construction to all planar *st*-graphs by using two bends on a restricted number of edges (Sect. 4). More precisely, we show that, given a set S of Δ slopes containing the horizontal slope, every n-vertex upward planar digraph with maximum vertex degree Δ has a 2-bend upward planar drawing that uses only slopes in S and with at most $4n - 9$ bends in total; see Fig. 1b for an illustration.

For space reasons some proofs are omitted and can be found in [2].

2 Preliminaries

We assume familiarity with common notation and definitions about graphs, drawings, and planarity (see, e.g., [12]). An *upward planar drawing* of a directed

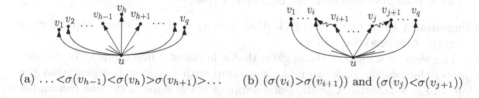

(a) $\ldots < \sigma(v_{h-1}) < \sigma(v_h) > \sigma(v_{h+1}) > \ldots$ (b) $(\sigma(v_i) > \sigma(v_{i+1}))$ and $(\sigma(v_j) < \sigma(v_{j+1}))$

Fig. 2. (a) A bitonic sequence. (b) A forbidden configuration.

simple graph (or *digraph* for short) G is a planar drawing such that each edge of G is drawn as a curve monotonically non-decreasing in the y-direction. An upward drawing is *strict* if its edge curves are monotonically increasing. A digraph is *upward planar* if it admits an upward planar drawing. Note that if a digraph admits an upward drawing then it also admits a strict upward drawing. A digraph is upward planar if and only if it is a subgraph of a *planar st-graph* [13]. Let $G = (V, E)$ be an n-vertex planar *st*-graph, i.e., G is a plane acyclic digraph with a single source s and a single sink t, such that s and t belong to the boundary of the outer face and the edge $(s, t) \in E$ [13]. (Other works do not explicitly require the edge (s, t) to be part of G, see, e.g., [19].) An *st-ordering* of G is a numbering $\sigma : V \to \{1, 2, \ldots, n\}$ such that for each edge $(u, v) \in E$, it holds $\sigma(u) < \sigma(v)$ (which implies $\sigma(s) = 1$ and $\sigma(t) = n$). Every planar *st*-graph has an *st*-ordering, which can be computed in $O(n)$ time (see, e.g., [10]). If u and v are two adjacent vertices of G such that $\sigma(u) < \sigma(v)$, we say that v is a *successor* of u, and u is a *predecessor* of v. Denote by $S(u) = \{v_1, v_2, \ldots, v_q\}$ the sequence of successors of v ordered according to the clockwise circular order of the edges incident to u in the planar embedding of G. The sequence $S(u)$ is *bitonic* if there exists an integer $1 \leq h \leq q$ such that $\sigma(v_1) < \cdots < \sigma(v_{h-1}) < \sigma(v_h) > \sigma(v_{h+1}) > \cdots > \sigma(v_q)$; see Fig. 2a for an illustration. Notice that when $h = 1$ or $h = q$, $S(u)$ is actually a monotonic decreasing or increasing sequence. A *bitonic st-ordering* of G is an st-ordering such that, for every vertex $u \in V$, $S(u)$ is bitonic [19]. A planar st-graph G is a *bitonic st-graph* if it admits a bitonic st-ordering. Deciding whether G is bitonic can be done in linear time both in the fixed [19] and in the variable [8] embedding settings. If G is not bitonic, every st-ordering σ of G contains a forbidden configuration defined as follows. A sequence of successors $S(u)$ of a vertex u forms a *forbidden configuration* if there exist two indices i and j, with $i < j$, such that $\sigma(v_i) > \sigma(v_{i+1})$ and $\sigma(v_j) < \sigma(v_{j+1})$, i.e. there is a path from v_{i+1} to v_i and a path from v_j to v_{j+1}; see Fig. 2b.

Let $G = (V, E)$ be an n-vertex maximal plane graph with vertices u, v, and w on the boundary of the outer face. A *canonical ordering* [17] of G is a linear ordering $\chi = \{v_1 = u, v_2 = v, \ldots, v_n = w\}$ of V, such that for every $3 \leq i \leq n$:

C1: The subgraph G_i induced by $\{v_1, v_2, \ldots, v_i\}$ is 2-connected and internally triangulated, while the boundary of its outer face C_i is a cycle containing (v_1, v_2);

C2: If $i + 1 \leq n$, v_{i+1} belongs to C_{i+1} and its neighbors in G_i form a subpath of the path obtained by removing (v_1, v_2) from C_i.

Computing χ takes $O(n)$ time [17]. Also, χ is *upward* if for every edge (u, v) of a digraph G u precedes v in χ.

The *slope* of a line ℓ is the angle α that a horizontal line needs to be rotated counter-clockwise in order to make it overlap with ℓ. If $\alpha = 0$ we say that the slope of ℓ is *horizontal*. The *slope* of a segment is the slope of the line containing it. Let $S = \{\alpha_1, \ldots, \alpha_h\}$ be a set of h slopes such that $\alpha_i < \alpha_{i+1}$. The slope set S is *equispaced* if $\alpha_{i+1} - \alpha_i = \frac{\pi}{h}$, for $i = 1, \ldots, h - 1$. Consider a *k-bend planar drawing* Γ of a graph G, i.e., a planar drawing in which every edge is mapped to a polyline containing at most $k + 1$ segments. For a vertex v in Γ each slope $\alpha \in S$ defines two different rays that emanate from v and have slope α. If α is horizontal these rays are called *left horizontal* ray and *right horizontal* ray. Otherwise, one of them is the *top* and the other one is the *bottom* ray of v. We say that a ray r_v of a vertex v is *free* if there is no edge attached to v through r_v in Γ. We also say that r_v is *outer* if it is free and the first face encountered when moving from v along r_v is the outer face of Γ. The *slope number* of a k-bend drawing Γ is the number of distinct slopes used for the edge segments of Γ. The *k-bend upward planar slope number* of an upward planar digraph G is the minimum slope number over all k-bend upward planar drawings of G.

3   1-Bend Upward Planar Drawings

Let $G = (V, E)$ be an n-vertex planar st-graph with a bitonic st-ordering $\sigma = \{v_1, v_2, \ldots, v_n\}$; see, e.g., Fig. 3a. We begin by describing an augmentation technique to "transform" σ into an upward canonical ordering of a suitable supergraph $\widehat{G}$ of G. We start from a result by Gronemann [19], whose properties are summarized in the following lemma; see, e.g., Fig. 3b.

Lemma 1 ([19]). *Let $G = (V, E)$ be an n-vertex planar st-graph that admits a bitonic st-ordering $\sigma = \{v_1, v_2, \ldots, v_n\}$. There exists a planar st-graph $G' = (V', E')$ with an st-ordering $\chi = \{v_L, v_R, v_1, v_2, \ldots, v_n\}$ such that: (i) $V' = V \cup \{v_L, v_R\}$; (ii) $E \subset E'$ and $(v_L, v_R) \in E'$; (iii) v_L and v_R are on the boundary of the outer face of G'; (iv) Every vertex of G with less than two predecessors in σ has exactly two predecessors in χ. Also, G' and χ are computed in $O(n)$ time.*

We call G' a *canonical augmentation* of G. Observe that G' always contains the edges (v_L, v_1) and (v_R, v_1) because of Lemma (1). We also insert the edge (v_L, v_n), which is required according to our definition of st-graph; this addition is always possible because v_L and v_n are both on the boundary of the outer face. The next lemma shows that any planar st-graph obtained by triangulating G' admits an upward canonical ordering; see, e.g., Fig. 3c.

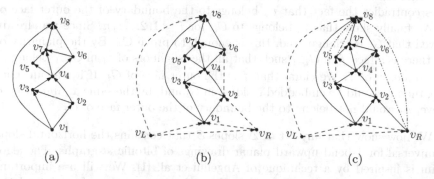

Fig. 3. (a) A bitonic st-graph G with $\sigma = \{v_1, v_2, \ldots, v_8\}$. (b) A canonical augmentation G' of G with $\chi = \{v_L, v_R, v_1, v_2, \ldots, v_8\}$. (c) A planar st-graph $\widehat{G}$ obtained by triangulating G'. χ is an upward canonical ordering of $\widehat{G}$.

Lemma 2. *Let G' be a canonical augmentation of an n-vertex bitonic st-graph G. Every planar st-graph $\widehat{G}$ obtained by triangulating G' has the following properties: (a) it has no parallel edges; (b) $\chi = \{v_L, v_R, v_1, v_2, \ldots, v_n\}$ is an upward canonical ordering.*

Proof. Concerning Property (a), suppose for a contradiction that $\widehat{G}$ has two parallel edges e_1 and e_2 connecting u with v. Let $\mathcal{C}$ be the 2-cycle formed by e_1 and e_2 and let $V_{\mathcal{C}}$ be the set of vertices distinct from u and v that are inside $\mathcal{C}$ in the embedding of $\widehat{G}$. $V_{\mathcal{C}}$ is not empty, as otherwise $\mathcal{C}$ would be a non-triangular face of $\widehat{G}$. Let w be the vertex with the lowest number in χ among those in $V_{\mathcal{C}}$. Since $\widehat{G}$ is planar (in particular e_1 and e_2 are not crossed) and has a single source, it contains a directed path from u to every vertex in $V_{\mathcal{C}}$. Hence, it has an edge from u to w. Also, by assumption, there is no vertex z in $V_{\mathcal{C}}$

such that $\chi(z) < \chi(w)$, which implies that u is the only predecessor of w in χ, a contradiction to Lemma 1(iv). Concerning Property (b), if χ is a canonical ordering of $\widehat{G}$, then χ is actually an upward canonical ordering because it is also an st-ordering. To see that χ is a canonical ordering, observe first that v_L, v_R and v_n are on the boundary of the outer face of $\widehat{G}$ by construction. Denote by $\widehat{G}_i$ the subgraph of $\widehat{G}$ induced by $\{v_L, v_R, v_1, \ldots, v_i\}$ and let $\widehat{C}_i$ be the boundary of its outer face. We first prove by induction on i (for $i = 1, 2, \ldots, n$) that $\widehat{G}_i$ is 2-connected. In the base case $i = 1$, $\widehat{G}_1$ is a 3-cycle and therefore it is 2-connected. In the case $i > 1$, $\widehat{G}_{i-1}$ is 2-connected by induction and v_i has at least two predecessors in $\widehat{G}_{i-1}$ by Lemma 1(iv), thus $\widehat{G}_i$ is 2-connected. We now prove that each $\widehat{G}_i$, for $i = 1, 2, \ldots, n$, is internally triangulated, which concludes the proof of condition **C1** of canonical ordering. Suppose, for a contradiction, that there exists an inner face f that is not a triangle. Since $\widehat{G}$ is triangulated, there exists a vertex v_j, with $j > i$, that is embedded inside f in $\widehat{G}_j$. Since χ is an st-ordering, there is no directed path from v_j to any vertex of f. On the other hand, either $v_j = v_n$ or there is a directed path from v_j to v_n. Both cases contradict the fact that v_n belongs to the boundary of the outer face of $\widehat{G}$. We finally show that v_i belongs to C_i, for $i = 1, 2, \ldots, n$. Since we already proved that $\widehat{G}_i$ is triangulated, this is enough to prove **C2**. By the planarity of $\widehat{G}_i$, there is a face f in $\widehat{G}_{i-1}$ such that all the neighbors of v_i in $\widehat{G}_{i-1}$ belong to the boundary of f. We claim that f is the outer face of $\widehat{G}_i$. If it was an inner face, then v_i would be embedded inside f in $\widehat{G}_i$ and, by the same argument used above, v_n would not belong to the boundary of the outer face of $\widehat{G}$. □

We now show that any set of Δ slopes S that contains the horizontal slope is universal for 1-bend upward planar drawings of bitonic st-graphs. The algorithm is inspired by a technique of Angelini et al. [1]. We will use important additional tools with respect to [1], such as the construction of a triangulated canonical augmentation, extra slopes to draw the edges inserted by the augmentation procedure, and different geometric invariants. Let G be an n-vertex bitonic st-graph with maximum vertex degree Δ; see Fig. 3a. The algorithm first computes a triangulated canonical augmentation $\widehat{G}$ of G; see Figs. 3b and c. We call *dummy edges* all edges that are in $\widehat{G}$ but not in G and *real edges* the edges in $\widehat{G}$ that are also in G. By Lemma 2, $\widehat{G}$ admits an upward canonical ordering $\chi = \{v_L, v_R, v_1, v_2, \ldots, v_n\}$, where χ is an st-ordering such that each vertex distinct from v_L and v_R has at least two predecessors. Let $S = \{\rho_1, \ldots, \rho_\Delta\}$ be any set of Δ slopes, which we call *real slopes*. Let ρ^* be the smallest angle between two slopes in S and let Δ^* be the maximum number of dummy edges incident to a vertex of $\widehat{G}$. For each slope ρ_i ($1 \leq i \leq \Delta$), we add Δ^* *dummy slopes* $\{\delta_1^i, \ldots, \delta_{\Delta^*}^i\}$ such that $\delta_j^i = \rho_i + j \cdot \frac{\rho^*}{\Delta^*+1}$, for $j = 1, 2, \ldots, \Delta^*$. Hence, there are Δ^* dummy slopes between any two consecutive real slopes. We will use the real slopes for the real edges and the dummy slopes for the dummy ones.

Let $\widehat{G}_i$ be the subgraph of $\widehat{G}$ induced by $\{v_L, v_R, v_1, v_2, \ldots, v_i\}$. The algorithm constructs the drawing by adding the vertices according to χ. More precisely, it computes a drawing Γ_i of the digraph $\widehat{G}_i^-$ obtained from $\widehat{G}_i$ by removing the dummy edges (v_L, v_R) and (v_1, v_R), which exist by construction, and (v_R, v_2) if it exists. Let $\widehat{C}_i$ be the boundary of the outer face of $\widehat{G}_i$, and let $\widehat{P}_i$ be the path obtained by removing (v_L, v_R) from $\widehat{C}_i$. For a vertex v of $\widehat{P}_i$, we denote by $d_r(v, i)$ (resp. $d_d(v, i)$) the number of real (resp. dummy) edges incident to v that are not in $\widehat{G}_i$ and by $\widehat{\rho}_j(v, i)$ (resp. $\overset{\frown}{\rho}_j(v, i)$) the j-th outer real top ray in $\widehat{\Gamma}_i$ encountered in clockwise (resp. counterclockwise) order around v starting from the left (resp. right) horizontal ray. For dummy top rays, we define analogously $\overset{\frown}{\delta}_j(v, i)$ and $\widehat{\delta}_j(v, i)$. $\widehat{\Gamma}_i$ satisfies the following invariants:

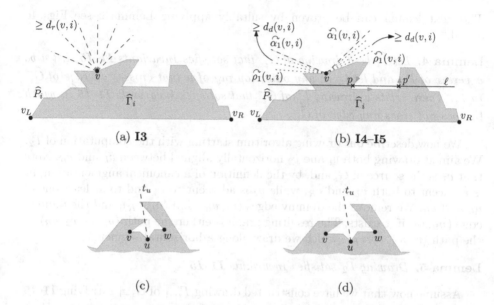

Fig. 4. (a)–(b) Illustration for invariants I3–I5; real rays are dashed, dummy rays are dotted. (c)–(d) Illustration for Lemma 4.

I1 $\widehat{\Gamma}_i$ is a 1-bend upward planar drawing whose real edges use only slopes in $\mathcal{S}$.
I2 Every edge of $\widehat{P}_i$ contains a horizontal segment.
I3 Every vertex v of $\widehat{P}_i$ has at least $d_r(v, i)$ outer real top rays; see Fig. 4a.
I4 Every vertex v of $\widehat{P}_i$ has at least $d_d(v, i)$ outer dummy top rays between $\overset{\frown}{\delta}_1(v, i)$ and $\widehat{\rho}_1(v, i)$ (resp. $\widehat{\delta}_1(v, i)$ and $\overset{\frown}{\rho}_1(v, i)$), including $\overset{\frown}{\delta}_1(v, i)$ (resp. $\widehat{\delta}_1(v, i)$); see Fig. 4b.
I5 Let ℓ be any horizontal line and let p and p' be any two intersection points between ℓ and the polyline representing $\widehat{P}_i$ in $\widehat{\Gamma}_i$; walking along ℓ from left to right, p and p' are encountered in the same order as when walking along $\widehat{P}_i$ from v_L to v_R; see Fig. 4b.

The last vertex v_n is added to $\widehat{\Gamma}_{n-1}$ in a slightly different way and the resulting drawing will satisfy **I1**. The next two lemmas state important properties of any 1-bend upward planar drawing satisfying **I1–I5**. Similar lemmas are proven in [1, Lemmas 2 and 3], but for drawings that satisfy different invariants.

Lemma 3. *Let $\widehat{\Gamma}_i$ be a drawing of $\widehat{G}_i^-$ that satisfies Invariants **I1–I5**. Let (u, v) be any edge of $\widehat{P}_i$ such that u is encountered before v along $\widehat{P}_i$ when going from v_L to v_R, and let λ be a positive number. There exists a drawing $\widehat{\Gamma}_i'$ of $\widehat{G}_i^-$ that satisfies Invariants **I1–I5** and such that: (i) the horizontal distance between u and v is increased by λ; (ii) the horizontal distance between any two other consecutive vertices along $\widehat{P}_i$ is the same as in $\widehat{\Gamma}_i$.*

The next lemma can be proven by suitably applying Lemma 3; see Figs. 4c and d.

Lemma 4. *Let $\widehat{\Gamma}_i$ be a drawing of $\widehat{G}_i^-$ that satisfies Invariants **I1–I5**. Let u be a vertex of $\widehat{P}_i$, and let t_u be any outer top ray of u that crosses an edge of $\widehat{G}_i^-$ in $\widehat{\Gamma}_i$. There exists a drawing $\widehat{\Gamma}_i'$ of $\widehat{G}_i^-$ that satisfies Invariants **I1–I5** in which t_u does not cross any edge of $\widehat{G}_i^-$.*

We now describe our drawing algorithm starting with the computation of $\widehat{\Gamma}_2$. We aim at drawing both v_1 and v_2 horizontally aligned between v_L and v_R. Note that v_1 is the source of G, and, by the definition of a canonical augmentation, v_1 is adjacent to both v_L and v_R, while v_2 is adjacent to v_1 and to at least one of v_L and v_R. We remove the dummy edges (v_1, v_R) and (v_L, v_R), and the dummy edge (v_R, v_2) if it exists. The resulting graph is either the path $\langle v_L, v_1, v_2, v_R \rangle$ or the path $\langle v_L, v_2, v_1, v_R \rangle$, which we draw along a horizontal segment.

Lemma 5. *Drawing $\widehat{\Gamma}_2$ satisfies Invariants **I1–I5**.*

Assume now that we have constructed drawing $\widehat{\Gamma}_{i-1}$ of $\widehat{G}_{i-1}$ satisfying **I1–I5** ($3 \leq i < n$). Let $\{u_1, \ldots, u_q\}$ be the neighbors of the next vertex v_i along $\widehat{P}_{i-1}$. Let t_1 be either $\widehat{\rho}_1(u_1, i-1)$, if (u_1, v_i) is real, or $\widehat{\delta}_1(u_1, i-1)$, if (u_1, v_i) is dummy. Symmetrically, let t_q be either $\widehat{\rho}_1(u_q, i - 1)$, if (u_q, v_i) is real, or $\widehat{\delta}_1(u_q, i-1)$, if (u_q, v_i) is dummy. Let t_j (for $1 < j < q$) be any outer real (resp. dummy) top ray emanating from u_j if (u_j, v_i) is real (resp. dummy). By **I3** all such top rays exist and by Lemma 4 we can assume that none of them crosses $\widehat{\Gamma}_{i-1}$. Let ℓ be a horizontal line above the topmost point of $\widehat{\Gamma}_{i-1}$. Let p_j be the intersection point of t_j and ℓ. We can assume that, for $j = 1, 2, \ldots, q - 1$, p_j is to the left of p_{j+1}. If this is not the case, we can increase the distance between u_j and u_{j+1} so to guarantee that p_j and p_{j+1} appear in the desired order along ℓ; this can be done by applying Lemma 3 with respect to each edge (u_j, u_{j+1}) for a suitable choice of λ; see Figs. 5a and b for an illustration. We will place v_i above ℓ using $q-2$ bottom rays $b_2, b_3, \ldots, b_{q-1}$ of v_i for the segments of the edges (u_j, v_i) $(j = 2, 3, \ldots, q-1)$ incident to v_i such that: (i) b_j ($1 < j < q$) is real (resp. dummy) if (u_j, v_i) is real (resp. dummy); (ii) b_j precedes b_{j+1} in the counterclockwise order around v_i starting from b_2. This choice is possible for the real rays because v_i has $\Delta - 1$ real bottom rays and it has at least one incident real edge not in $\widehat{G}_i$ (otherwise

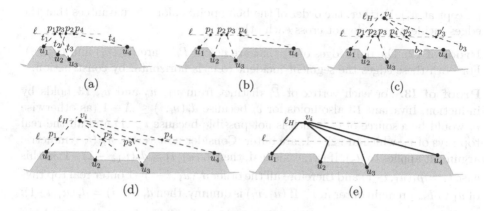

Fig. 5. Addition of vertex v_i.

it would be a sink of G, which is not possible because $i < n$). Concerning the dummy rays, we have at most Δ^* dummy edges incident to v_i and Δ^* dummy bottom rays between any two consecutive real rays. Consider the ray t_1 and choose a point p to the right of t_1 and above ℓ such that placing v_i on p guarantees that $\min_{i=1...q-2}\{x(p'_{i+1}) - x(p'_i)\} > x(p_q) - x(p_1)$, where $p'_1 = p_1$ and $p'_2, p'_3, \ldots, p'_{q-1}$ are the intersection points of the rays $b_2, b_3, \ldots, b_{q-1}$ with the line ℓ (see Fig. 5c). Observe that for a sufficiently large y-coordinate, point p can always be found. We now apply Lemma 3 to each of the edges (u_1, u_2), (u_2, u_3), $\ldots, (u_{q-2}, u_{q-1})$, in this order, choosing $\lambda \geq 0$ so that each p_j is translated to p'_j (for $j = 2, 3, \ldots, q-1$). We finally apply again the same procedure to (u_{q-1}, u_q) so that the intersection point between t_q and the horizontal line ℓ_H passing through v_i is to the right of v_i (see Fig. 5d). After this translation procedure, we can draw the edge (u_1, v_i) (resp. (u_q, v_i)) with a bend at the intersection point between t_1 (resp. t_q) and ℓ_H and therefore using the slope of t_1 (resp. t_q) and the horizontal slope (see Fig. 5e). The edges (u_j, v_i) $(j = 2, 3, \ldots, q-1)$ are drawn with a bend point at $p_j = p'_j$ and therefore using the slopes of t_j and b_j.

Lemma 6. *Drawing $\widehat{\Gamma}_i$, for $i = 3, 4, \ldots, n-1$, satisfies Invariants **I1–I5**.*

Proof. The proof is by induction on $i \geq 3$. $\widehat{\Gamma}_{i-1}$ satisfies Invariants **I1–I5** by Lemma 5 when $i = 3$, and by induction when $i > 3$.

Proof of I1. By construction, each (u_j, v_i) $(j = 1, 2, \ldots, q)$ is drawn as a chain of at most two segments that use real and dummy slopes. In particular, if (u_j, v_i) is real, then it uses real slopes, i.e., slopes in $\mathcal{S}$. By the choice of ℓ, the bend point of (u_j, v_i) has y-coordinate strictly greater than that of u_j and smaller than or equal to that of v_i. Since each (u_j, v_i) is oriented from u_j to v_i (as χ is an upward canonical ordering), the drawing is upward. Concerning planarity, we first observe that $\widehat{\Gamma}_{i-1}$ is planar and it remains planar each time we apply Lemma 3. Also, by Lemma 4 each (u_j, v_i) $(j = 1, 2, \ldots, q)$ does not intersect $\widehat{\Gamma}_{i-1}$

(except at u_j). Further, the order of the bend points along ℓ guarantees that the edges incident to v_i do not cross each other.

Proof of I2. The only edges of $\widehat{P}_i$ that are not in $\widehat{P}_{i-1}$ are (u_1, v_i) and (u_q, v_i). For both these edges the segment incident to v_i is horizontal by construction.

Proof of I3. For each vertex of $\widehat{P}_i$ distinct from u_1, u_q and v_i, **I3** holds by induction. Invariant **I3** also holds for v_i because $d_r(v_i, i) \leq \Delta - 1$ (as otherwise v_i would be a source of G, which is not possible because $i > 1$) and all the real top rays of v_i, which are $\Delta - 1$, are outer. Consider now vertex u_1 (a symmetric argument applies to u_q). If (u_1, v_i) is real, then $d_r(u_1, i) = d_r(u_1, i-1) - 1$; in this case $t_1 = \widehat{\rho}_1(u_1, i-1)$ and therefore all the other $d_r(u_1, i-1) - 1$ outer real top rays of u_1 in $\widehat{\Gamma}_{i-1}$ remain outer in $\widehat{\Gamma}_i$. If (u_1, v_i) is dummy, then $d_r(u_1, i) = d_r(u_1, i-1)$; in this case $t_1 = \widehat{\delta}_1(u_1, i-1)$ and therefore all the $d_r(u_1, i-1)$ outer real top rays of u_1 in $\widehat{\Gamma}_{i-1}$ remain outer in $\widehat{\Gamma}_i$.

Proof of I4. For each vertex of $\widehat{P}_i$ distinct from u_1, u_q and v_i, **I4** holds by induction. **I4** also holds for v_i because $d_d(v_i, i) \leq \Delta^*$ and there are Δ^* dummy top rays between $\widehat{\delta}_1(v_i, i)$ and $\widehat{\rho}_1(v_i, i)$ including $\widehat{\delta}_1(v_i, i)$ (all the top rays of v_i are outer). Analogously, there are Δ^* outer dummy top rays between $\widehat{\delta}_1(v_i, i)$ and $\widehat{\rho}_1(v_i, i)$ including $\widehat{\delta}_1(v_i, i)$. Consider now u_1 (a symmetric argument applies to u_q). If (u_1, v_i) is real, then $d_d(u_1, i) = d_d(u_1, i-1)$; in this case $t_1 = \widehat{\rho}_1(u_1, i-1)$ and there are Δ^* outer dummy top rays between $\widehat{\delta}_1(u_1, i)$ and $\widehat{\rho}_1(u_1, i)$ including $\widehat{\delta}_1(u_1, i)$ (namely, all those between $t_1 = \widehat{\rho}_1(u_1, i-1)$ and $\widehat{\rho}_2(u_1, i-1)$). If (u_1, v_i) is dummy, then $d_d(u_1, i) = d_d(u_1, i-1) - 1$; in this case $t_1 = \widehat{\delta}_1(u_1, i-1)$ and therefore all the other $d_d(u_1, i-1) - 1$ outer dummy top rays of u_1, which by induction were between $\widehat{\delta}_1(u_1, i-1)$ and $\widehat{\rho}_1(u_1, i-1)$, remain outer in $\widehat{\Gamma}_i$.

Proof of I5. Notice that the various applications of Lemma 3 to $\widehat{\Gamma}_{i-1}$ preserve **I5**. Let p and p' be any two intersection points between a horizontal line ℓ and the polyline representing $\widehat{P}_i$ in $\widehat{\Gamma}_i$, with p to the left of p' along ℓ. If p and p' belong to $\widehat{P}_{i-1}$, **I5** holds by induction. If both p and p' belong to the path $\langle u_1, v_i, u_q \rangle$, **I5** holds by construction. If p belongs to $\widehat{P}_{i-1}$ and p' belongs to $\langle u_1, v_i, u_q \rangle$, then p belongs to the subpath of $\widehat{P}_{i-1}$ that goes from v_L to u_1 because the subpath from u_q to v_R is completely to the right of t_q, hence **I5** holds also in this case. If p belongs to $\langle u_1, v_i, u_q \rangle$ and p' belongs to $\widehat{P}_{i-1}$, the proof is symmetric. □

Lemma 7. *G has a 1-bend upward planar drawing Γ using only slopes in $\mathcal{S}$.*

Proof. By Lemma 6, drawing $\widehat{\Gamma}_{n-1}$ satisfies Invariant **I1**–**I5**. We explain how to add the last vertex v_n to obtain a drawing that satisfies Invariant **I1**. Let $\{u_1, \ldots, u_q\}$ be the predecessors of v_n on $\widehat{P}_{n-1}$. Notice that, in this case $u_1 = v_L$ and $u_q = v_R$. Vertex v_n is added to the drawing similarly to all the other vertices added in the previous steps of the algorithm. The only difference is that the number of real incoming edges incident to v_n in $\widehat{\Gamma}_{n-1}$ can be up to Δ. If this is the case, since the real bottom rays are $\Delta - 1$, they are not enough to draw all the real edges incident to v_n. Let j be the smallest index such that (u_j, v_n) is a real edge. We ignore all the dummy edges (u_h, v_n), for $h = 1, 2, \ldots, j-1$, and apply the construction used in the previous steps considering only $\{u_j, u_{j+1}, \ldots, u_q\}$ as predecessors of v_n (notice that such predecessors are at least two because v_n has at least two incident real edges). By ignoring these dummy edges, the segment of the real edge (u_j, v_n) incident to v_n will be drawn using the left horizontal slope. Denote by $\widehat{\Gamma}_n$ the resulting drawing. As in the proof of Lemma 6, we can prove that **I1** holds for $\widehat{\Gamma}_n$ and therefore $\widehat{\Gamma}_n$ is a 1-bend upward planar drawing whose real edges use only slopes in $\mathcal{S}$. The drawing Γ of G is obtained from $\widehat{\Gamma}_n$ by removing all its dummy edges and the two dummy vertices v_L and v_R. □

Lemma 8. *Drawing Γ can be computed in $O(n)$ time.*

Lemmas 7 and 8 are summarized by Theorem 1. Corollary 1 is a consequence of Theorem 1 and of a result in [14].

Theorem 1. *Let $\mathcal{S}$ be any set of $\Delta \geq 2$ slopes including the horizontal slope and let G be an n-vertex bitonic planar st-graph with maximum vertex degree Δ. Graph G has a 1-bend upward planar drawing Γ using only slopes in $\mathcal{S}$, which can be computed in $O(n)$ time.*

Corollary 1. *Every bitonic st-graph with maximum vertex degree $\Delta \geq 2$ has 1-bend upward planar slope number at most Δ, which is worst-case optimal.*

If $\mathcal{S}$ is equispaced, Theorem 1 implies a lower bound of $\frac{\pi}{\Delta}$ on the angular resolution of the computed drawing, which is worst-case optimal [14]. Also, Theorem 1 can be extended to planar st-graphs with $\Delta \leq 3$, as any such digraph can be made bitonic by only rerouting the edge (s, t).

Theorem 2. *Every planar st-graph with maximum vertex degree 3 has 1-bend upward planar slope number at most 3.*

We conclude with the observation that an upward drawing constructed by the algorithm of Theorem 1 can be transformed into a strict upward drawing that uses $\Delta + 1$ slopes rather than Δ. It suffices to replace every horizontal segment oriented from its leftmost (rightmost) endpoint to its rightmost (leftmost) one with a segment having slope ε $(-\varepsilon)$, for a sufficiently small value of $\varepsilon > 0$.

4   2-bend Upward Planar Drawings

We now extend the result of Theorem 1 to non-bitonic planar st-graphs. By adapting a technique of Keszegh et al. [23], one can construct 2-bend upward planar drawings of planar st-graphs using at most Δ slopes. We improve upon this result in two ways: (i) The technique in [23] may lead to drawings with $5n - 11$

bends in total, while we prove that $4n-9$ bends suffice; (ii) It uses a fixed set of Δ slopes (and it is not immediately clear whether it can work with any set of slopes), while we show that any set of Δ slopes with the horizontal one is universal.

Let G be an n-vertex non-bitonic planar st-graph. All forbidden configurations of G can be removed in linear time by subdividing at most $n-3$ edges of G [19]. Let G_b be the resulting bitonic st-graph, called a *bitonic subdivision* of G. Let $\langle u, d, v \rangle$ be a directed path of G_b obtained by subdividing the edge (u, v) of G with the dummy vertex d. We call (u, d) the *lower stub*, and (d, v) the *upper stub* of (u, v). We can prove the existence of an augmentation technique similar to that of Lemma 1, but with an additional property on the upper stubs.

Lemma 9. *Let $G = (V, E)$ be an n-vertex planar st-graph that is not bitonic. Let $G_b = (V_b, E_b)$ be an N-vertex bitonic subdivision of G, with a bitonic st-ordering $\sigma = \{v_1, v_2, \ldots, v_N\}$. There exists a planar st-graph $G' = (V', E')$ with an st-ordering $\chi = \{v_L, v_R, v_1, v_2, \ldots, v_N\}$ such that: (i) $V' = V_b \cup \{v_L, v_R\}$; (ii) $E_b \subset E'$ and $(v_L, v_R) \in E'$; (iii) v_L and v_R are on the boundary of the outer face of G'; (iv) Every vertex of G_b with less than two predecessors in σ has exactly two predecessors in χ. (v) There is no vertex in G' such that its leftmost or its rightmost incoming edge is an upper stub. Also, G' and χ are computed in $O(n)$ time.*

Theorem 3. *Let S be any set of $\Delta \geq 2$ slopes including the horizontal slope and let G be an n-vertex planar st-graph with maximum vertex degree Δ. Graph G has a 2-bend upward planar drawing Γ using only slopes in S, which has at most $4n - 9$ bends in total and which can be computed in $O(n)$ time.*

Proof. We compute a triangulated canonical augmentation $\widehat{G}$ of G by (1) applying Lemma 9 and (2) triangulating the resulting digraph. By Lemma 2, $\widehat{G}$ has an upward canonical ordering χ. The algorithm of Theorem 1 to $\widehat{G}$ would lead to a 3-bend drawing of G (by interpreting every subdivision vertex as a bend). We explain how to modify it to construct a drawing $\widehat{\Gamma}$ of $\widehat{G}$ with at most 2 bends per edge and $4n - 9$ bends in total. Let v_i the next vertex to be added according to χ and let $\{u_1, u_2, \ldots, u_q\}$ its neighbors in $\widehat{P}_{i-1}$. Suppose that u_j is a dummy vertex and that (u_j, v_i) is an upper stub. To save one bend along the edge subdivided by u_j, we draw (u_j, v_i) without bends. By Lemma 9(v), we have that $1 < j < q$. The ray t_j used to draw the segment of (u_j, v_i) incident to u_j can be any outer real top ray; we choose the ray with same slope as the real bottom ray b_j used to draw the segment of (u_j, v_i) incident to v_i. This is possible because all real top rays of u_j are outer (since (u_j, v_i) is the only real outgoing edge of u_j). Hence, edge (u_j, v_i) has no bends. The drawing Γ of G is obtained from $\widehat{\Gamma}$ by removing dummy edges and replacing dummy vertices (except v_L and v_R, which are removed) with bends. Since the upper stubs of subdivided edges has 0 bends, each edge of Γ has at most 2 bends. Let m_1 and m_2 be the number of edges drawn with 1 and 2 bends, respectively; we have $m_2 \leq n - 3$ and $m_1 = m - m_2 \leq 3n - 6 - (n - 3) = 2n - 3$. Thus the total number of bends is at most $2n - 3 + 2(n - 3) = 4n - 9$. Finally, $\widehat{G}$ can be computed in $O(n)$ time (Lemma 9) and the modified drawing algorithm still runs in linear time. $\square$

A planar st-graph with a source/sink of degree Δ requires at least $\Delta - 1$ slopes in any upward planar drawing; thus the gap with Theorem 3 is one unit. Similarly to Theorem 1, Theorem 3 implies a lower bound of $\frac{\pi}{\Delta}$ on the angular resolution of Γ; an upper bound of $\frac{\pi}{\Delta-1}$ can be proven with the same digraph used for the lower bound on the slope number. Finally, Theorem 4 extends the result of Theorem 3 to every upward planar graph using an additional slope.

Theorem 4. *Let S be any set of $\Delta + 1$ slopes including the horizontal slope and let G be an n-vertex upward planar graph with maximum vertex degree $\Delta \geq 2$. Graph G has a 2-bend upward planar drawing using only slopes in S.*

5 Open Problems

(i) Can we draw every planar st-graph with at most one bend per edge (or less than $4n - 9$ in total) and Δ slopes? (ii) What is the 2-bend upward planar slope number of planar st-graphs? Is Δ a tight bound? (iii) What is the straight-line upward planar slope number of upward planar digraphs?

Acknowledgments. Research partially supported by project: "Algoritmi e sistemi di analisi visuale di reti complesse e di grandi dimensioni - Ricerca di Base 2018, Dipartimento di Ingegneria, Università degli Studi di Perugia".

References

1. Angelini, P., Bekos, M.A., Liotta, G., Montecchiani, F.: A universal slope set for 1-bend planar drawings. In: Aronov, B., Katz, M.J. (eds.) SoCG. LIPIcs, vol. 77, pp. 9:1–9:16. Schloss Dagstuhl (2017). https://doi.org/10.4230/LIPIcs.SoCG.2017.9, https://arxiv.org/abs/1703.04283

2. Bekos, M.A., Di Giacomo, E., Didimo, W., Liotta, G., Montecchiani, F.: Universal slope sets for upward planar drawings. ArXiv e-prints abs/1803.09949v2 (2018). https://arxiv.org/abs/1803.09949v2

3. Bekos, M.A., Gronemann, M., Kaufmann, M., Krug, R.: Planar octilinear drawings with one bend per edge. J. Graph Algorithms Appl. **19**(2), 657–680 (2015). https://doi.org/10.7155/jgaa.00369

4. Bekos, M.A., Kaufmann, M., Krug, R.: On the total number of bends for planar octilinear drawings. In: Kranakis, E., Navarro, G., Chávez, E. (eds.) LATIN 2016. LNCS, vol. 9644, pp. 152–163. Springer, Heidelberg (2016). https://doi.org/10.1007/978-3-662-49529-2_12

5. Bertolazzi, P., Di Battista, G., Mannino, C., Tamassia, R.: Optimal upward planarity testing of single-source digraphs. SIAM J. Comput. **27**(1), 132–169 (1998). https://doi.org/10.1137/S0097539794279626

6. Biedl, T.C., Kant, G.: A better heuristic for orthogonal graph drawings. Comput. Geom. **9**(3), 159–180 (1998). https://doi.org/10.1016/S0925-7721(97),00026-6

7. Bodlaender, H.L., Tel, G.: A note on rectilinearity and angular resolution. J. Graph Algorithms Appl. **8**, 89–94 (2004). https://doi.org/10.7155/jgaa.00083

8. Chaplick, S., et al.: Planar L-drawings of directed graphs. In: Frati, F., Ma, K.-L. (eds.) GD 2017. LNCS, vol. 10692, pp. 465–478. Springer, Cham (2018). https://doi.org/10.1007/978-3-319-73915-1_36

9. Chimani, M., Zeranski, R.: Upward planarity testing in practice: SAT formulations and comparative study. ACM J. Exp. Algorithmics **20**, 1.2:1.1–1.2:1.27 (2015). https://doi.org/10.1145/2699875

10. Cormen, T.H., Leiserson, C.E., Rivest, R.L., Stein, C.: Introduction to Algorithms, 3rd edn. MIT Press, Cambridge (2009)

11. Czyzowicz, J., Pelc, A., Rival, I., Urrutia, J.: Crooked diagrams with few slopes. Order **7**(2), 133–143 (1990). https://doi.org/10.1007/BF00383762

12. Di Battista, G., Eades, P., Tamassia, R., Tollis, I.G.: Graph Drawing: Algorithms for the Visualization of Graphs. Prentice-Hall, New Jersey (1999)

13. Di Battista, G., Tamassia, R.: Algorithms for plane representations of acyclic digraphs. Theor. Comput. Sci. **61**, 175–198 (1988). https://doi.org/10.1016/0304-3975(88),90123-5

14. Di Giacomo, E., Liotta, G., Montecchiani, F.: 1-bend upward planar drawings of SP-digraphs. In: Hu, Y., Nöllenburg, M. (eds.) GD 2016. LNCS, vol. 9801, pp. 123–130. Springer, Cham (2016). https://doi.org/10.1007/978-3-319-50106-2_10

15. Didimo, W.: Upward graph drawing. In: Kao, M.Y. (ed.) Encyclopedia of Algorithms. Springer, Heidelberg (2015). https://doi.org/10.1007/978-3-642-27848-8_653-1

16. Duncan, C., Goodrich, M.T.: Planar orthogonal and polyline drawing algorithms. In: Tamassia, R. (ed.) Handbook on Graph Drawing and Visualization. Chapman and Hall/CRC (2013)

17. de Fraysseix, H., Pach, J., Pollack, R.: How to draw a planar graph on a grid. Combinatorica **10**(1), 41–51 (1990). https://doi.org/10.1007/BF02122694

18. Garg, A., Tamassia, R.: On the computational complexity of upward and rectilinear planarity testing. SIAM J. Comput. **31**(2), 601–625 (2001). https://doi.org/10.1137/S0097539794277123

19. Gronemann, M.: Bitonic *st*-orderings for upward planar graphs. In: Hu, Y., Nöllenburg, M. (eds.) GD 2016. LNCS, vol. 9801, pp. 222–235. Springer, Cham (2016). https://doi.org/10.1007/978-3-319-50106-2_18

20. Healy, P., Nikolov, N.S.: Hierarchical drawing algorithms. In: Tamassia, R. (ed.) Handbook on Graph Drawing and Visualization. Chapman and Hall/CRC (2013)

21. Hong, S., Merrick, D., do Nascimento, H.A.D.: Automatic visualisation of metro maps. J. Vis. Lang. Comput. **17**(3), 203–224 (2006). https://doi.org/10.1016/j.jvlc.2005.09.001

22. Kelly, D.: Fundamentals of planar ordered sets. Discrete Math. **63**(2–3), 197–216 (1987). https://doi.org/10.1016/0012-365X(87),90008-2

23. Keszegh, B., Pach, J., Pálvölgyi, D.: Drawing planar graphs of bounded degree with few slopes. SIAM J. Discrete Math. **27**(2), 1171–1183 (2013). https://doi.org/10.1137/100815001

24. Knauer, K., Walczak, B.: Graph drawings with one bend and few slopes. In: Kranakis, E., Navarro, G., Chávez, E. (eds.) LATIN 2016. LNCS, vol. 9644, pp. 549–561. Springer, Heidelberg (2016). https://doi.org/10.1007/978-3-662-49529-2_41

25. Leiserson, C.E.: Area-efficient graph layouts (for VLSI). In: FOCS, pp. 270–281. IEEE (1980). https://doi.org/10.1109/SFCS.1980.13

26. Nöllenburg, M.: Automated drawings of metro maps. Technical report 2005-25, Fakultät für Informatik, Universität Karlsruhe (2005)

27. Nöllenburg, M., Wolff, A.: Drawing and labeling high-quality metro maps by mixed-integer programming. IEEE Trans. Vis. Comput. Graph. **17**(5), 626–641 (2011). https://doi.org/10.1109/TVCG.2010.81

28. Stott, J.M., Rodgers, P., Martinez-Ovando, J.C., Walker, S.G.: Automatic metro map layout using multicriteria optimization. IEEE Trans. Vis. Comput. Graph. **17**(1), 101–114 (2011). https://doi.org/10.1109/TVCG.2010.24

29. Tamassia, R.: On embedding a graph in the grid with the minimum number of bends. SIAM J. Comput. **16**(3), 421–444 (1987). https://doi.org/10.1137/0216030

30. Valiant, L.G.: Universality considerations in VLSI circuits. IEEE Trans. Comput. **30**(2), 135–140 (1981). https://doi.org/10.1109/TC.1981.6312176

Upward Planar Morphs

Giordano Da Lozzo[✉], Giuseppe Di Battista, Fabrizio Frati,
Maurizio Patrignani, and Vincenzo Roselli

Roma Tre University, Rome, Italy
{dalozzo,gdb,frati,patrigna,roselli}@dia.uniroma3.it

Abstract. We prove that, given two topologically-equivalent upward
planar straight-line drawings of an n-vertex directed graph G, there
always exists a morph between them such that all the intermediate drawings of the morph are upward planar and straight-line. Such a morph
consists of $O(1)$ morphing steps if G is a reduced planar st-graph, $O(n)$
morphing steps if G is a planar st-graph, $O(n)$ morphing steps if G is a
reduced upward planar graph, and $O(n^2)$ morphing steps if G is a general
upward planar graph. Further, we show that $\Omega(n)$ morphing steps might
be necessary for an upward planar morph between two topologically-
equivalent upward planar straight-line drawings of an n-vertex path.

1 Introduction

One of the definitions of the word *morph* that can be found in English dictionaries
is "to gradually change into a different image". The Graph Drawing community
defines the morph of graph drawings similarly. Namely, given two drawings Γ_0
and Γ_1 of a graph G, a *morph* between Γ_0 and Γ_1 is a continuously changing
family of drawings of G indexed by time $t \in [0, 1]$, such that the drawing at time
$t = 0$ is Γ_0 and the drawing at time $t = 1$ is Γ_1. Further, the way the Graph
Drawing community adopted the word morph is consistent with its Ancient
Greek root μωρφή, which means "shape" in a broad sense. Namely, if both Γ_0
and Γ_1 have a certain geometric property, it is desirable that all the drawings of
the morph also have the same property. In particular, we talk about a *planar*, a
straight-line, an *orthogonal*, or a *convex morph* if all the intermediate drawings
of the morph are *planar* (edges do not cross), *straight-line* (edges are straight-
line segments), *orthogonal* (edges are polygonal lines composed of horizontal and
vertical segments), or *convex* (the drawings are planar and straight-line, and the
faces are delimited by convex polygons), respectively.

The state of the art on planar morphs covers more than 100 years, starting
from the 1914/1917 works of Tietze [25] and Smith [23]. The seminal papers of
Cairns [13] and Thomassen [24] proved the existence of a planar straight-line
morph between any two topologically-equivalent planar straight-line drawings
of a graph. In the last 10 years, the attention of the research community focused

This research was partially supported by MIUR Project "MODE", by H2020-MSCA-
RISE project "CONNECT", and by MIUR-DAAD JMP N° 34120.

T. Biedl and A. Kerren (Eds.): GD 2018, LNCS 11282, pp. 92–105, 2018.
https://doi.org/10.1007/978-3-030-04414-5_7

on algorithms for constructing planar morphs with few *morphing steps* (see, e.g., [1–7,11,12,21,26]). Each morphing step, sometimes simply called *step*, is a *linear* morph, in which the vertices move along straight-line (possibly distinct) trajectories at uniform speed. A *unidirectional* morph is a linear morph in which the vertex trajectories are all parallel. It is known [2,4] that a planar straight-line morph with a linear number of unidirectional morphing steps exists between any two topologically-equivalent planar straight-line drawings of the same graph, and that this bound is the best possible.

Upward planarity is usually regarded as the natural extension of planarity to directed graphs; see, e.g., [9,10,15,16,18]. A drawing of a directed graph is *upward planar* if it is planar and the edges are represented by curves monotonically increasing in the vertical direction. Despite the importance of upward planarity, up to now, no algorithm has been devised to morph upward planar drawings of directed graphs. This paper deals with the following question: Given two topologically-equivalent upward planar drawings Γ_0 and Γ_1 of an upward planar directed graph G, does an *upward planar straight-line morph* between Γ_0 and Γ_1 always exist? In this paper we give a positive answer to this question.

Problems related to upward planar graphs are usually more difficult than the corresponding problems for undirected graphs. For example, planarity can be tested in linear time [20] while testing upward planarity is NP-complete [18]; all planar graphs admit planar straight-line grid drawings with polynomial area [22] while there are upward planar graphs that require exponential area in any upward planar straight-line grid drawing [17]. Quite surprisingly, we show that, from the morphing point of view, the difference between planarity and upward planarity is less sharp; indeed, in some cases, upward planar straight-line drawings can be morphed even more efficiently than planar straight-line drawings.

More in detail, our results are as follows. Let Γ_0 and Γ_1 be topologically-equivalent upward planar drawings of an n-vertex upward plane graph G. We show algorithms to construct upward planar straight-line morphs between Γ_0 and Γ_1 with the following number of unidirectional morphing steps:

i. $O(1)$ steps if G is a reduced plane st-graph (see Sect. 4);
ii. $O(n)$ steps if G is a plane st-graph (see Sect. 4);
iii. $O(n)$ steps if G is a reduced upward plane graph (see Sect. 5);
iv. $O(n \cdot f(n))$ steps if G is a general upward plane graph, assuming that an $O(f(n))$-step algorithm exists to construct an upward planar morph between any two upward planar drawings of any n-vertex plane st-graph (see Sect. 5). This, together with Result ii., yields an $O(n^2)$-step upward planar morph for general upward plane graphs.

Further, we show (Sect. 3) that there exist two topologically-equivalent upward planar drawings of an n-vertex upward plane path such that any upward planar morph between them consists of $\Omega(n)$ morphing steps.

In order to prove Result i. we devise a technique that allows us to construct a morph in which each morphing step modifies either only the x-coordinates or only the y-coordinates of the vertices. Result ii. builds on the techniques in

[2] and leverages on the arrangement of low-degree vertices in upward planar drawings in order to morph maximal plane st-graphs. We then exploit such morphs for general plane st-graphs. In order to prove Results **iii.** and **iv.** we use an inductive technique for reducing the geometric differences between Γ_0 and Γ_1.

Because of space limitations, some proofs are omitted or sketched. They can be found in the full version of the paper.

2   Preliminaries

We assume familiarity with graph drawing [15] and related concepts.

In this paper we only consider straight-line drawings. Thus, where it leads to no confusion, we will omit the term "straight-line". Let Γ be a drawing of a graph G and let H be a subgraph of G. We denote by $\Gamma[H]$ the restriction of Γ to the vertices and edges of H. Two planar drawings of a connected graph are *topologically equivalent* if they have the same circular order of the edges around each vertex and the same cycle bounding the outer face. A *planar embedding* is an equivalence class of planar drawings. A *plane graph* is a planar graph equipped with a planar embedding. In a planar straight-line drawing an internal face (the outer face) is *strictly convex* if its angles are all smaller (greater) than π. A planar straight-line drawing is *strictly convex* if each face is strictly convex.

A *y-assignment* $y_G : V(G) \to \mathbb{R}$ is an assignment of reals to the vertices of a graph G. A drawing Γ of G *satisfies* y_G if the y-coordinate in Γ of each vertex $v \in V(G)$ is $y_G(v)$. An *x-assignment* x_G for G is defined analogously.

In a directed graph G we denote by uv an edge directed from a vertex u to a vertex v; then v is a *successor* of u, and u is a *predecessor* of v. A *directed path* consists of the edges $u_i u_{i+1}$, for $i = 1, \ldots, n - 1$. The *underlying graph* of G is the undirected graph obtained from G by omitting the directions from its edges. A *transitive edge* in a directed graph G is an edge uv such that G contains a directed path from u to v different from the edge uv. A *reduced* graph is a directed graph that does not contain any transitive edges.

A drawing of a directed graph is *upward planar* if it is planar and each edge uv is drawn as a curve monotonically increasing in the y-direction from u to v. A directed graph is *upward planar* if it admits an upward planar drawing. Consider an upward planar drawing Γ of an upward planar graph G. Let u, v, and w be three vertices consecutive and in this clockwise order along the boundary of a face f of G. We denote by $\angle(u, v, w)$ the angle formed by the (undirected) edges (u, v) and (v, w) in the interior of f. Also, we say that v is a *sink-switch* (*source-switch*) of f if uv and wv (vu and vw) are edges of G. Furthermore, we say that v is a *switch* of G if it is either a sink-switch or a source-switch of some face of Γ. Two switches u and v of a face f are *clockwise* (*counter-clockwise*) *consecutive* if traversing f clockwise (counter-clockwise) no switch is encountered in between u and v. The drawing Γ determines a *large-angle assignment*, that is, a labeling, for each face f and each three clockwise consecutive switches u, v, and w for f of the corresponding angle $\angle(u, v, w)$ as `large`, if it is larger than π in Γ, or `small`, it is smaller than π in Γ [9].

Two upward planar drawings of an upward planar graph G have the same
upward planar embedding if they have the same planar embedding and the same
large-angle assignment. We denote by $\ell(G)$ the number of switches labeled `large`
in G. A combinatorial characterization of upward planar embeddings in terms
of large angles is given in [9]. An *upward plane graph* is an upward planar graph
equipped with an upward planar embedding.

Let Γ_0 and Γ_1 be upward planar drawings of an upward plane graph G. An
upward planar morph is a continuous transformation from Γ_0 to Γ_1 indexed by
time $t \in [0, 1]$ in which the drawing at each time $t \in [0, 1]$ is upward planar.

A *plane st-graph* is an upward plane graph with a single source s and a single
sink t, and with an upward planar embedding in which s and t are incident to
the outer face. A plane st-graph always admits an upward planar straight-line
drawing [16]. A cycle in an upward plane graph is an *st-cycle* if it consists of two
directed paths. A face f of an upward plane graph is an *st-face* if it is delimited
by an st-cycle; the directed paths delimiting an st-face f are called *left* and *right
boundary*, where the edge of the left boundary incident to the source-switch s_f
of f immediately precedes the edge of the right boundary incident to s_f in the
clockwise order of the edges incident to s_f. The following is well-known.

Lemma 1. *An upward plane graph is a plane st-graph iff all its faces are st-faces.*

An internal vertex v of a maximal plane st-graph G is *simple* if the neighbors
of v induce a cycle in the underlying graph of G.

Lemma 2 (Alamdari et al. [2]). *Any maximal plane st-graph contains a simple vertex of degree at most 5.*

3   Slow Morphs and Fast Morphs

We start this section by proving the following lower bound.

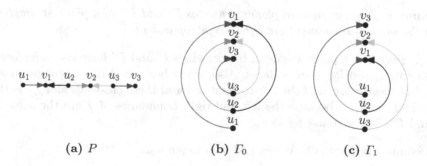

(a) P (b) Γ_0 (c) Γ_1

Fig. 1. Illustration for Theorem 1. (a) P; (b) Γ_0; and (c) Γ_1. For the sake of readability
Γ_0 and Γ_1 have curved edges. However, the x-coordinates of the vertices can be slightly
perturbed in order to make Γ_0 and Γ_1 straight-line.

Theorem 1. *There are two upward planar drawings of an n-vertex upward plane path such that any upward planar morph between them consists of $\Omega(n)$ steps.*

Proof sketch. Assume, for the sake of simplicity, that n is even, and let $n = 2k$. Consider the n-vertex upward plane path P defined as follows (refer to Fig. 1a). The path P contains vertices u_i and v_i, for $i = 1, \ldots, k$, and directed edges $u_i v_i$, for $i = 1, \ldots, k$, and $u_{i+1} v_i$, for $i = 1, \ldots, k-1$. We fix the upward planar embedding of P as in Fig. 1b and c.

Let Γ_0 and Γ_1 be two upward planar straight-line drawings of P in which the bottom-to-top order of the vertices is $u_1, \ldots, u_k, v_k, \ldots, v_1$ (see Fig. 1b) and $u_k, \ldots, u_1, v_1, \ldots, v_k$ (see Fig. 1c), respectively. Let $\langle \Gamma_0 = \Lambda_1, \Lambda_2, \ldots, \Lambda_{h+1} = \Gamma_1 \rangle$ be any upward planar morph from Γ_0 to Γ_1 that consists of h morphing steps. We have the following.

Claim 1.1. *For each $j = 1, \ldots, h+1$, the vertices $u_j, u_{j+1}, \ldots, u_{k-1}, u_k$ appear in this bottom-to-top order in Λ_j.*

By Claim 1.1 and since u_k, u_{k-1} appear in this bottom-to-top order in $\Gamma_1 = \Lambda_{h+1}$, we have that $h + 1 > k - 1$, hence $h \in \Omega(n)$. □

We now establish a tool that will allow us to design efficient algorithms for morphing upward planar drawings. Consider two planar straight-line drawings Γ' and Γ'' of a plane graph G with the same y-assignment. Since the drawings are straight-line and have the same y-assignment, a horizontal line ℓ intersects a vertex or an edge of G in Γ' if and only if it intersects the same vertex or edge in Γ''. We say that Γ' and Γ'' are *left-to-right equivalent* if, for any horizontal line ℓ, for any vertex or edge α of G, and for any vertex or edge β of G such that ℓ intersects both α and β (in Γ' and in Γ''), we have that the intersection of α with ℓ is to the left of the intersection of β with ℓ in Γ' if and only if the intersection of α with ℓ is to the left of the intersection of β with ℓ in Γ''. The definition of *bottom-to-top equivalent* drawings is analogous. We have the following.

Lemma 3. *Any two upward planar drawings Γ' and Γ'' of a plane st-graph G with the same y-assignment are left-to-right equivalent.*

Proof. Since G is a plane st-graph, the drawings Γ' and Γ'' have the same faces. By Lemma1 such faces are st-faces. Also, every horizontal line ℓ crosses an st-face f at most twice, and the left-to-right order of these crossings along ℓ is the same in Γ' and Γ'' because the left and right boundaries of f are the same in Γ' and Γ''. The statement follows. □

Lemma 4 is due to [2]. We extend it in Lemma 5.

Lemma 4 ([2], **Corollary 7.2**). *Consider a unidirectional morph acting on points p, q, and r. If p is on one side of the oriented line through $\overline{qr}$ at the beginning and at the end of the morph, then p is on the same side of the oriented line through $\overline{qr}$ throughout the morph.*

Lemma 5. *Let Γ' and Γ'' be two left-to-right or bottom-to-top equivalent planar drawings of a plane graph. Then the linear morph $\mathcal{M}$ from Γ' to Γ'' is unidirectional and planar.*

Proof. Since Γ' and Γ'' have the same y-assignment (x-assignment), given that they are left-to-right (bottom-to-top) equivalent, it follows that all the vertices move along horizontal (vertical) trajectories. Thus, $\mathcal{M}$ is unidirectional. Also, since Γ' and Γ'' are left-to-right (bottom-to-top) equivalent, each horizontal (vertical) line crosses the same sequence of vertices and edges in both Γ' and Γ''. Thus, by Lemma 4, $\mathcal{M}$ is planar.   □

Lemma 5 allows us to devise a simple morphing technique between any two upward planar drawings Γ_0 and Γ_1 of the same upward plane graph G, when a pair of upward planar drawings of G with special properties can be computed. We say that the pair (Γ_0, Γ_1) is an *hvh-pair* if there exist upward planar drawings Γ_0' and Γ_1' of G such that: (i) Γ_0 and Γ_0' are left-to-right equivalent, (ii) Γ_0' and Γ_1' are bottom-to-top equivalent, and (iii) Γ_1' and Γ_1 are left-to-right equivalent. Our morphing tool is expressed by the following lemma.

Lemma 6 (Fast morph). *Let (Γ_0, Γ_1) be an hvh-pair of upward planar drawings of an upward plane graph G. There is a 3-step upward planar morph from Γ_0 to Γ_1.*

Proof sketch. We define the morph $\mathcal{M}$ as $\langle \Gamma_0, \Gamma_0', \Gamma_1', \Gamma_1 \rangle$. The drawings Γ_0' and Γ_1' exist by hypothesis. Lemma 5 guarantees that $\mathcal{M}$ is unidirectional and planar. We use Lemma 4 to prove that $\mathcal{M}$ is upward.   □

The next lemma will allow us to restrict our attention to biconnected graphs.

Lemma 7. *Let Γ_0 and Γ_1 be two upward planar drawings of an n-vertex upward plane graph G whose underlying graph is connected. There exist upward planar drawings Γ_0' and Γ_1' of an $O(n)$-vertex upward plane graph G' that is a supergraph of G, whose underlying graph is biconnected, and such that $\Gamma_0'[G] = \Gamma_0$ and $\Gamma_1'[G] = \Gamma_1$. Further, if G is reduced or an st-graph, then so is G'.*

Proof sketch. We iteratively apply the following procedure. Consider a cutvertex v of G and two edges that belong to distinct blocks of G and that are consecutive in the circular order of the edges incident to v. Let u and w be the end-vertices of such edges different from v. We add to G a vertex v' and two edges connecting v' with u and w; these edges are oriented as the ones connecting v with u and w, respectively. By placing v' and its incident edges inside the face of G incident to v, u, and w, we obtain an upward plane supergraph of G with one block less than G. Upward planar drawings of this graph extending Γ_0 and Γ_1 can be easily obtained. The repetition of this procedure proves the lemma.   □

4   Plane *st*-Graphs

In this section, we show algorithms for constructing upward planar morphs between upward planar drawings of plane *st*-graphs.

4.1   Reduced Plane st-Graphs

We first consider plane st-graphs without transitive edges. We have the following.

Lemma 8. *Any two upward planar drawings Γ_0 and Γ_1 of a reduced plane st-graph G form an hvh-pair.*

Proof sketch. By Lemma 7 we can assume that G is biconnected. We construct two upward planar drawings Γ_0' and Γ_1' that, together with Γ_0 and Γ_1, satisfy Conditions (i)–(iii) of the definition of hvh-pair. We construct Γ_0' and Γ_1' as follows. First, we draw the left boundary of the outer face of G so that each vertex has the same y-coordinate in Γ_i' as in Γ_i, for $i = 0, 1$. In both Γ_0' and Γ_1' the x-coordinates of all the vertices of this path are 0. Then, we add to Γ_0' and Γ_1' the right boundaries of the st-faces of G one by one, following a topological sorting of the oriented dual graph of G. In Γ_0' (in Γ_1') we assign to the internal vertices of each right boundary the same y-coordinates they have in Γ_0 (Γ_1); since G is reduced, the set of these vertices is non-empty. All the internal vertices of each right boundary get the same x-coordinate, which is used in both Γ_0' and Γ_1'; this x-coordinate is sufficiently large so that no crossing is introduced.   □

Combining Lemma 6 with Lemma 8 we obtain the following result.

Theorem 2. *Let Γ_0 and Γ_1 be any two upward planar drawings of a reduced plane st-graph. There is a 3-step upward planar morph from Γ_0 to Γ_1.*

4.2   General Plane st-Graphs

We now turn our attention to general plane st-graphs. We restate here, in terms of plane st-graphs, a result by Hong and Nagamochi [19] that was originally formulated in terms of hierarchical plane (undirected) graphs.

Theorem 3 ([19], **Theorem 8**). *Consider an internally 3-connected plane st-graph G and let y_G be a y-assignment of the vertices of G such that each vertex v is assigned a value $y_G(v)$ that is greater than those assigned to its predecessors. There exists a strictly-convex upward planar drawing of G satisfying y_G.*

We use Theorem 3 to prove the following theorem, which allows us to restrict our attention to maximal plane st-graphs.

Theorem 4. *Let Γ_0 and Γ_1 be two upward planar drawings of an n-vertex plane st-graph G. Suppose that an algorithm $\mathcal{A}$ exists that constructs an $f(r)$-step upward planar morph between any two upward planar drawings of an r-vertex maximal plane st-graph. Then there exists an $O(f(n))$-step upward planar morph from Γ_0 to Γ_1.*

Proof sketch. By Lemma 7 we can assume that G is biconnected. We augment G to a maximal plane st-graph G^* by inserting a vertex v_f into each face f of G and by inserting a directed edge from the source-switch s_f of f to v_f

and directed edges from v_f to every other vertex incident to f. We define a y-assignment $y^0_{G^*}$ for G^* by setting $y^0_{G^*}(v) = y^0_G(v)$ for each vertex $v \in V(G)$ and by setting, for each vertex $v_f \in V(G^*) \setminus V(G)$, a value for $y^0_{G^*}(v_f)$ that is larger than $y^0_{G^*}(s_f)$ and smaller than $y^0_{G^*}(v)$, for every other vertex v incident to f. We similarly define a y-assignment $y^1_{G^*}$ using the y-coordinates of Γ_1. We use Theorem 3 to construct upward planar drawings Γ^*_0 and Γ^*_1 of G^* satisfying $y^0_{G^*}$ and $y^1_{G^*}$, respectively. By Lemma 3 we have that $\Gamma^*_0[G]$ and Γ_0 ($\Gamma^*_1[G]$ and Γ_1) are left-to-right equivalent. Therefore, by Lemma 5, the linear morph $\mathcal{M}'_0$ from Γ_0 to $\Gamma^*_0[G]$ ($\mathcal{M}'_1$ from $\Gamma^*_1[G]$ to Γ_1) is unidirectional and planar. Such a morph is also upward since both Γ_0 and $\Gamma^*_0[G]$ (Γ_1 and $\Gamma^*_1[G]$) are upward planar and left-to-right equivalent. Then, we apply algorithm $\mathcal{A}$ to construct a morph from Γ^*_0 to Γ^*_1 and restrict such a morph to a morph $\mathcal{M}''$ from $\Gamma^*_0[G]$ to $\Gamma^*_1[G]$. The morph from Γ_0 to Γ_1 is the concatenation of $\mathcal{M}'_0$, $\mathcal{M}''$, and $\mathcal{M}'_1$. $\qquad\square$

The *kernel* of a polygon P is the set of points p inside or on P such that, for any point q on P, the open segment $\overline{pq}$ lies inside P.

Lemma 9 (Convexify). *Let Γ be an upward planar drawing of an internally 3-connected plane st-graph G, let f be an st-face of G, and let P be the polygon representing f in Γ. There exists an upward planar drawing Γ' of G such that the polygon representing the boundary of f is strictly-convex and $\mathcal{M} = \langle \Gamma, \Gamma' \rangle$ is a unidirectional upward planar morph. Further, if v is a vertex incident to f that is in the kernel of P in Γ, then v is in the kernel of the polygon representing the boundary of f throughout $\mathcal{M}$.*

Proof. Denote by y_G the y-assignment for the vertices of G induced by Γ. By Theorem 3, there exists a strictly-convex upward planar drawing Γ' of G satisfying y_G. Thus, by Lemma 3 and since G is a plane st-graph, Γ and Γ' are left-to-right-equivalent drawings. By Lemma 5, the linear morph $\mathcal{M}$ from Γ to Γ' is unidirectional and planar. Since Γ and Γ' are upward, $\mathcal{M}$ is upward as well.

Consider now a vertex v incident to f that is in the kernel of P in Γ. Since the polygon representing the boundary of f in Γ' is strictly-convex, v is also in the kernel of such a polygon. Augment G to a graph G_* by introducing (suitably oriented) edges connecting v to the vertices incident to f that are not already adjacent to v. Since v is in the kernel of the polygon representing the boundary of f both in Γ and in Γ', this results in two left-to-right equivalent upward planar drawings Γ_* and Γ'_* of G_*. By the same arguments used for $\mathcal{M}$, we have that the linear morph $\mathcal{M}_* = \langle \Gamma_*, \Gamma'_* \rangle$ is planar. Hence, v is in the kernel of the polygon representing the boundary of f throughout $\mathcal{M}$. $\qquad\square$

Given two upward planar straight-line drawings Γ_0 and Γ_1 of a maximal plane st-graph G, our strategy for constructing an upward planar morph from Γ_0 to Γ_1 is as follows: (1) we find a simple vertex v of G of degree at most 5; (2) we remove v and its incident edges from G, Γ_0, and Γ_1, obtaining upward planar drawings Γ'_0 and Γ'_1 of an upward plane graph G'; (3) we triangulate G', Γ'_0, and Γ'_1 by inserting edges incident to a former neighbor u of v, obtaining upward planar drawings Γ_0'' and Γ_1'' of a maximal plane st-graph G''; (4) we apply induction in order to construct an upward planar morph $\mathcal{M}''$ from Γ_0'' to Γ_1''; and (5) we remove the edges incident to u that are not in G and insert v and its incident edges in $\mathcal{M}''$, thus obtaining an upward planar morph $\mathcal{M}$ from Γ_0 to Γ_1. In order for this strategy to work, we need u to satisfy certain properties, which are expressed in the upcoming definition of *distinguished neighbor*; further, we need to perform one initial (and one final) unidirectional upward planar morph so to convexify the polygon representing what will be called a *characteristic cycle*.

Let v be a simple vertex with degree at most 5 in a maximal plane st-graph G. Let $G(v)$ be the subgraph of G induced by v and its neighbors. A predecessor u of v in G is a *distinguished predecessor* if, for each predecessor w of v, there is a directed path in $G(v)$ from w to v through u. A successor u of v in G is a *distinguished successor* if, for each successor w of v, there is a directed path in $G(v)$ from v to w through u. A neighbor of v is a *distinguished neighbor* if it is a distinguished predecessor or successor of v. Examples of distinguished neighbors are in Fig. 2.

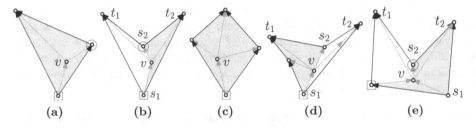

Fig. 2. Distinguished predecessors (enclosed by red squares), distinguished successors (enclosed by red circles), and characteristic cycles (filled yellow). (Color figure online)

Lemma 10. *The vertex v has at most one distinguished predecessor, at most one distinguished successor, and at least one distinguished neighbor.*

We define the *characteristic cycle* $C(v)$ as follows. Let $c_G(v)$ be the subgraph of G induced by the neighbors of v. Since v is simple, the underlying graph of $c_G(v)$ is a cycle. If $c_G(v)$ is an st-cycle, then $C(v) := c_G(v)$; this is always the case if v has degree 3. Otherwise, $c_G(v)$ has two sources s_1 and s_2 and two sinks t_1 and t_2. Suppose that G contains the edges s_1v and vs_2, the cases in which it contains the edges s_2v and vs_1, or t_1v and vt_2, or t_2v and vt_1 are analogous. Suppose also, w.l.o.g., that s_1, t_1, s_2, and t_2 appear in this clockwise order along $c_G(v)$. If v has degree 4, then we define $C(v)$ as the st-cycle composed of the edges s_1v, vs_2, s_2t_2, and s_1t_2. Otherwise, v has degree 5. Consider the directed path $P_1 = (v_1, v, v_2)$, where v_1 (v_2) is the distinguished predecessor (successor) of v or, if such a node does not exist, the source s_1 (s_2). Then P_1 splits $c_G(v)$ into two paths P_2 and P_3 with length 2 and 3, respectively. Cycle $C(v)$ is defined as the st-cycle composed of P_1 and P_3. We have the following structural lemma.

Lemma 11. *The characteristic cycle $C(v)$ is an st-cycle which contains all the distinguished neighbors of v. Further, all the vertices of $c_G(v)$ not belonging to $C(v)$ are adjacent to all the distinguished neighbors of v.*

Characteristic cycles are used in order to prove the following.

Lemma 12. *There is a unidirectional upward planar morph $\langle \Gamma, \Gamma' \rangle$, where in Γ' the distinguished neighbors of v are in the kernel of the polygon representing $c_G(v)$.*

Proof sketch. If $C(v)$ is convex in Γ, then by Lemma 11 the distinguished neighbors of v already are in the kernel of the polygon representing $c_G(v)$. Otherwise, we remove the interior of $C(v)$ and use Lemma 9 to make $C(v)$ convex. Then, we suitably reinsert the interior of $C(v)$ to obtain the desired morph.   □

The following concludes our discussion on maximal plane st-graph.

Theorem 5. *Let Γ_0 and Γ_1 be two upward planar drawings of an n-vertex maximal plane st-graph. There is an $O(n)$-step upward planar morph from Γ_0 to Γ_1.*

Proof sketch. If $n = 3$, then the desired morph is constructed as in Lemma 6. If $n > 3$, then by Lemma 2 a simple vertex v exists in G with degree at most 5. By Lemma 10, v has a distinguished neighbor u. By Lemma 12, unidirectional upward planar morphs $\langle \Gamma_0, \Lambda_0 \rangle$ and $\langle \Lambda_1, \Gamma_0 \rangle$ exist, where Λ_0 and Λ_1 are upward planar drawings of G in which u lies in the kernels of the polygons representing $C(v)$. Remove v from G, Λ_0, and Λ_1, and insert (suitably oriented) edges between u and the former neighbors of v that are not already adjacent to u, thus obtaining upward planar drawings Γ_0'' and Γ_1'' of an $(n-1)$-vertex maximal plane st-graph G''. Recursively compute an upward planar morph $\mathcal{M}''$ from Γ_0'' to Γ_1''. Finally, remove the edges incident to u that are not in G and insert v and its incident edges in $\mathcal{M}''$, obtaining an upward planar morph $\mathcal{M}$ from Λ_0 to Λ_1. This, together with $\langle \Gamma_0, \Lambda_0 \rangle$ and $\langle \Lambda_1, \Gamma_0 \rangle$, provides the desired morph from Γ_0 to Γ_1.   □

We finally get the following.

Corollary 1. *Let Γ_0 and Γ_1 be two upward planar drawings of an n-vertex plane st-graph. There exists an $O(n)$-step upward planar morph from Γ_0 to Γ_1.*

Proof. The statement follows by Lemma 7, Theorem 4, and Theorem 5.   □

5   Upward Plane Graphs

Let G be an upward plane graph, let f be a face of G, and let u, v, and w be three clockwise consecutive switches of f. Also, let v_1 (v_2) be the neighbor of v clockwise preceding (succeeding) v along f, and let u_1 (u_2) be the neighbor of u clockwise preceding (succeeding) u along f. We say that $[u, v, w]$ is a *pocket* for f if $\angle(v_1, v, v_2) = \mathtt{small}$ and $\angle(u_1, u, u_2) = \mathtt{large}$. The following is well-known.

Lemma 13 ([9]). *Let G be an upward plane graph and let f be a face of G that is not an st-face. Then, there exists a pocket $[u, v, w]$ for f.*

Next, we give a lemma that shows how to "simplify" a face of an upward plane graph that is not an *st*-graph, by removing one of its pockets.

Lemma 14. *Let G be an n-vertex (reduced) upward plane graph, let $[u, v, w]$ be a pocket for a face f of G, and let Γ be an upward planar drawing of G. Suppose that an algorithm $\mathcal{A}$ ($\mathcal{A}_R$) exists that constructs an $f(r)$-step ($f_R(r)$-step) upward planar morph between any two upward planar drawings of an r-vertex (reduced) plane st-graph. Then, there exists an $O(f(n))$-step ($O(f_R(n))$-step) upward planar morph from Γ to an upward planar drawing Γ^* of G in which w and u have direct visibility and such that u lies below w, if a directed path exists in f from v to u, and it lies above w, if a directed path exists in f from u to v.*

Proof sketch. Suppose that a directed path p_{vu} exists in f from v to u (see Fig. 3a); the case in which a directed path exists in f from u to v can be treated symmetrically. We first show that there exists an upward planar drawing Γ' of G such that (i) it is possible to add to Γ' an upward planar drawing of two directed paths p' and p'' from u to w that form an *st*-face (see Fig. 3b), and (ii) there exists an $O(f(n))$-step ($O(f_R(n))$-step) upward planar morph $\mathcal{M}'$ from Γ to Γ'. We then show that there exists an upward planar drawing Γ^* of G such that (iii) vertices w and u have direct visibility and u lies below w (see Fig. 3c), and (vi) there exists an $O(f(n))$-step ($O(f_R(n))$-step) upward planar morph $\mathcal{M}^*$ from Γ' to Γ^*. Composing $\mathcal{M}'$ and $\mathcal{M}^*$ yields an upward planar morph from Γ to Γ^*. ☐

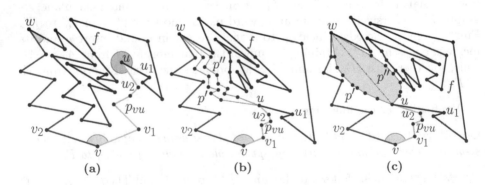

Fig. 3. Illustrations for the proof of Lemma 14.

Theorem 6. *Let Γ_0 and Γ_1 be two upward planar drawings of an n-vertex (reduced) upward plane graph G. Suppose that an algorithm $\mathcal{A}$ ($\mathcal{A}_R$) exists that constructs an $f(r)$-step ($f_R(r)$-step) upward planar morph between any two upward planar drawings of an r-vertex (reduced) plane st-graph. There exists an $O(n \cdot f(n))$-step (an $O(n \cdot f_R(n))$-step) upward planar morph from Γ_0 to Γ_1.*

Proof sketch. By Lemma 7, we can assume that G is biconnected. In order to prove the statement, we show that there exists a $((2\ell(G) + 1) \cdot f(n))$-step (a $((2\ell(G)+1) \cdot f_R(n))$-step) upward planar morph from Γ_0 to Γ_1, if G is a (reduced) upward plane graph. Since $\ell(G) \in O(n)$, the statement follows.

The proof is by induction on $\ell(G)$. In the base case $\ell(G) = 0$ and thus G is a (reduced) plane st-graph. Hence, by applying algorithm $\mathcal{A}$ ($\mathcal{A}_R$) to Γ_0 and Γ_1, we obtain an $f(n)$-step (an $f_R(n)$-step) upward planar morph from Γ_0 to Γ_1.

In the inductive case $\ell(G) > 0$. Then there exists a face f of G that is not an st-face. Thus, by Lemma 13, there exists a pocket $[u, v, w]$ for f. By Lemma 14, we can construct upward planar drawings Γ_0' and Γ_1' of G in which u and w have direct visibility and u lies below w (assuming that a directed path exists in f from v to u, the other case being symmetric), and such that there exists an $f(n)$-step (an $f_R(n)$-step) upward planar morph $\mathcal{M}_{start}$ from Γ_0 to Γ_0' and an $f(n)$-step (an $f_R(n)$-step) upward planar morph $\mathcal{M}_{finish}$ from Γ_1' to Γ_1.

Let G^* be the plane graph obtained from G by splitting f with a directed edge uw. Graph G^* is an upward plane graph whose upward planar embedding is constructed by assigning to each switch in G^* the same label small or large it has in G. Also, $\ell(G^*) = \ell(G) - 1$, since u is not a switch in G^*. Further, G^* is reduced if G is reduced, since there exists no directed path from u to w in G (due to the fact that $[u, v, w]$ is a pocket of f). Let Γ_0^* and Γ_1^* be the planar straight-line drawings of G^* obtained by drawing the directed edge uw as a straight-line segment connecting u and w in Γ_0' and in Γ_1', respectively. It is easy to see that Γ_0^* and Γ_1^* are upward planar drawings of G^*. Therefore, by the inductive hypothesis and since $V(G^*) = V(G)$, we can construct a $((2\ell(G^*)+1) \cdot f(n))$-step (a $((2\ell(G^*)+1) \cdot f_R(n))$-step) upward planar morph from Γ_0^* to Γ_1^*. Observe that, since $G \subset G^*$, restricting each drawing in $\mathcal{M}^*$ to G yields a $((2\ell(G)-1) \cdot f(n))$-step upward planar morph $\mathcal{M}^-$ of G from Γ_0' to Γ_1'. Therefore, by concatenating morphs $\mathcal{M}_{start}$, $\mathcal{M}^-$, and $\mathcal{M}_{finish}$, we obtain a $((2\ell(G) + 1) \cdot f(n))$-step (a $((2\ell(G) + 1) \cdot f_R(n))$-step) upward planar morph of G from Γ_0 to Γ_1. $\qquad\square$

Theorems 2, 4, and 6, imply the following main result.

Theorem 7. *Let Γ_0 and Γ_1 be two upward planar drawings of the same n-vertex (reduced) upward plane graph. There exists an $O(n^2)$-step (an $O(n)$-step) upward planar morph from Γ_0 to Γ_1.*

6   Conclusions and Open Problems

In this paper, we addressed for the first time the problem of morphing upward planar straight-line drawings. We proved that an upward planar morph between any two upward planar drawings of the same upward plane graph always exists. It easy to see that all our algorithms can be implemented in polynomial time.

Several problems remain open. In our opinion the most interesting question is whether an $O(1)$-step upward planar morph between any two upward planar drawings of the same plane st-graph exists. In case of a positive answer, by Theorem 6, an optimal $O(n)$-step upward planar morph would exist between any two upward planar drawings of the same n-vertex upward plane graph. In case of a negative answer, it would be interesting to find broad classes of upward plane graphs that admit upward planar morphs with a sub-linear number of steps. We proved that reduced plane st-graphs have this property and we ask whether the same is true for series-parallel digraphs [8,14].

References

1. Aichholzer, O., et al.: Convexifying polygons without losing visibilities. In: 23rd Canadian Conference on Computational Geometry, CCCG 2011 (2011)
2. Alamdari, S., et al.: How to morph planar graph drawings. SIAM J. Comput. **46**(2), 824–852 (2017)
3. Alamdari, S., et al.: Morphing planar graph drawings with a polynomial number of steps. In: Khanna, S. (ed.) 24th Annual ACM-SIAM Symposium on Discrete Algorithms, SODA 2013, pp. 1656–1667 (2013)
4. Angelini, P., Da Lozzo, G., Di Battista, G., Frati, F., Patrignani, M., Roselli, V.: Morphing planar graph drawings optimally. In: Esparza, J., Fraigniaud, P., Husfeldt, T., Koutsoupias, E. (eds.) ICALP 2014. LNCS, vol. 8572, pp. 126–137. Springer, Heidelberg (2014). https://doi.org/10.1007/978-3-662-43948-7_11
5. Angelini, P., Da Lozzo, G., Frati, F., Lubiw, A., Patrignani, M., Roselli, V.: Optimal morphs of convex drawings. In: Symposium on Computational Geometry, vol. 34 of LIPIcs, pp. 126–140. Schloss Dagstuhl - Leibniz-Zentrum fuer Informatik (2015)
6. Angelini, P., Frati, F., Patrignani, M., Roselli, V.: Morphing planar graph drawings efficiently. In: Wismath, S., Wolff, A. (eds.) GD 2013. LNCS, vol. 8242, pp. 49–60. Springer, Cham (2013). https://doi.org/10.1007/978-3-319-03841-4_5
7. Barrera-Cruz, F., Haxell, P., Lubiw, A.: Morphing planar graph drawings with unidirectional moves. In: Mexican Conference on Discrete Mathematics and Computational Geometry, pp. 57–65 (2013). http://arxiv.org/abs/1411.6185
8. Bertolazzi, P., Cohen, R.F., Di Battista, G., Tamassia, R., Tollis, I.G.: How to draw a series-parallel digraph. Int. J. Comput. Geom. Appl. **4**(4), 385–402 (1994)
9. Bertolazzi, P., Di Battista, G., Liotta, G., Mannino, C.: Upward drawings of triconnected digraphs. Algorithmica **12**(6), 476–497 (1994)
10. Bertolazzi, P., Di Battista, G., Mannino, C., Tamassia, R.: Optimal upward planarity testing of single-source digraphs. SIAM J. Comput. **27**(1), 132–169 (1998)
11. Biedl, T., Lubiw, A., Petrick, M., Spriggs, M.: Morphing orthogonal planar graph drawings. ACM Trans. Algorithms (TALG) **9**(4), 29 (2013)
12. Biedl, T., Lubiw, A., Spriggs, M.J.: Morphing planar graphs while preserving edge directions. In: Healy, P., Nikolov, N.S. (eds.) GD 2005. LNCS, vol. 3843, pp. 13–24. Springer, Heidelberg (2006). https://doi.org/10.1007/11618058_2
13. Cairns, S.S.: Deformations of plane rectilinear complexes. Am. Math. Monthly **51**(5), 247–252 (1944)
14. Cohen, R.F., Di Battista, G., Tamassia, R., Tollis, I.G.: Dynamic graph drawings: Trees, series-parallel digraphs, and planar st-digraphs. SIAM J. Comput. **24**(5), 970–1001 (1995)
15. Di Battista, G., Eades, P., Tamassia, R., Tollis, I.G.: Graph Drawing: Algorithms for the Visualization of Graphs. Prentice-Hall, Upper Saddle River (1999)
16. Di Battista, G., Tamassia, R.: Algorithms for plane representations of acyclic digraphs. Theor. Comput. Sci. **61**, 175–198 (1988)

17. Di Battista, G., Tamassia, R., Tollis, I.G.: Area requirement and symmetry display of planar upward drawings. Discrete Comput. Geom. **7**(4), 381–401 (1992)
18. Garg, A., Tamassia, R.: On the computational complexity of upward and rectilinear planarity testing. SIAM J. Comput. **31**(2), 601–625 (2002)
19. Hong, S., Nagamochi, H.: Convex drawings of hierarchical planar graphs and clustered planar graphs. J. Discrete Alg. **8**(3), 282–295 (2010)
20. Hopcroft, J., Tarjan, R.: Efficient planarity testing. J. ACM **21**(4), 549–568 (1974)
21. Roselli, V.: Morphing and visiting drawings of graphs. Ph.D. thesis, Università degli Studi di Roma "Roma Tre", Dottorato di Ricerca in Ingegneria, Sezione Informatica ed Automazione, XXVI Ciclo (2014)
22. Schnyder, W.: Embedding planar graphs on the grid. In: Proceedings of the First Annual ACM-SIAM Symposium on Discrete Algorithms, SODA 1990, pp. 138–148. Society for Industrial and Applied Mathematics, Philadelphia (1990)
23. Smith, H.L.: On continuous representations of a square upon itself. Ann. Math. **19**(2), 137–141 (1917)
24. Thomassen, C.: Deformations of plane graphs. J. Comb. Theory Ser. B **34**(3), 244–257 (1983)
25. Tietze, H.: Über stetige abbildungen einer quadratfläche auf sich selbst. Rendiconti del Circolo Matematico di Palermo **38**(1), 247–304 (1914)
26. van Goethem, A., Verbeek, K.: Optimal morphs of planar orthogonal drawings. In: 34th International Symposium on Computational Geometry, SoCG 2018, Budapest, Hungary, 11–14 June 2018, pp. 42:1–42:14 (2018)

Visualizing the Template of a Chaotic Attractor

Maya Olszewski[1] ⓘ, Jeff Meder[1] ⓘ, Emmanuel Kieffer[2] ⓘ, Raphaël Bleuse[1] ⓘ,
Martin Rosalie[2] ⓘ, Grégoire Danoy[1(✉)] ⓘ, and Pascal Bouvry[1,2] ⓘ

[1] FSTC/CSC-ILIAS, University of Luxembourg, 6, Avenue de la Fonte,
4364 Esch-sur-Alzette, Luxembourg
{maya.olszewski.001,jeff.meder.001}@student.uni.lu,
{raphael.bleuse,gregoire.danoy,pascal.bouvry}@uni.lu
[2] SnT, University of Luxembourg, 6, Avenue de la Fonte,
4364 Esch-sur-Alzette, Luxembourg
{emmanuel.kieffer,martin.rosalie}@uni.lu

Abstract. Chaotic attractors are solutions of deterministic processes, of
which the topology can be described by templates. We consider templates
of chaotic attractors bounded by a genus–1 torus described by a linking
matrix. This article introduces a novel and unique tool to validate a linking matrix, to optimize the compactness of the corresponding template
and to draw this template. The article provides a detailed description of
the different validation steps and the extraction of an order of crossings
from the linking matrix leading to a template of minimal height. Finally,
the drawing process of the template corresponding to the matrix is saved
in a Scalable Vector Graphics (SVG) file.

Keywords: Chaotic attractor · Template
Linking matrix · Optimization · Visualization

1 Introduction

Resulting of theoretical studies on chaos attractors, applications including
chaotic dynamics can be found in a multitude of domains. Their range goes
from computer science [23], through classical sciences with physical networks
[14], biology and genetics [27] and chemistry with chaotic dynamics in chemical
reactions [8], all the way to electronics and chaos in electronic devices [13] and
even environmental studies on population evolution [5].

Birman and Williams [6] introduce templates as knot-holder to describe the
topological structure of chaotic attractors. The notion of linking matrices to
describe chaotic attractors with integers has been first introduced by Mindlin
et al. in 1990 [18]. The matrix contains the number of torsions and permutations
occurring along the flow of an attractor. The template is a ribbon graph combined with a layering graph. In 1998, Gilmore wrote an extensive survey on the
research on chaotic dynamical systems over the past decade [11], in which one

© Springer Nature Switzerland AG 2018
T. Biedl and A. Kerren (Eds.): GD 2018, LNCS 11282, pp. 106–119, 2018.
https://doi.org/10.1007/978-3-030-04414-5_8

can see various drawings of templates. In his paper, he provides the summary of the topological analysis from dynamical system to template.

The subject of chaotic dynamics studies are promising and on-going. But it clearly misses matrices validation and drawing tools. The research community would benefit from an efficient application that verifies the validity of matrices and draws their corresponding template. The novel tool presented in this paper is publicly available online at https://gitlab.uni.lu/pcog/cate, and aims to fill this gap.

This paper is structured as follows. In Sect. 2 we give an introduction to the problem. Section 3 provides a state-of-the-art analysis in the field of chaotic attractors, focusing on their validation and visualization. In Sect. 4, we first outline our approach to determine the validity of a linking matrix. Secondly, we describe the procedure to get the minimal height of a template and its visualization. In Sect. 5, we present the experimental work and the results in order to validate our proposed approach. Finally, we conclude and outline some directions for future work in Sect. 6.

2   Problem Description

A chaotic attractor is a solution of a dynamic deterministic process that is very sensitive to its initial conditions. The solution will converge to the same global shape (the attractor), independently of the starting position in the basin of attraction. Malasoma [16] proposed a simple differential equations system

$$\begin{cases} \dot{x} = y \\ \dot{y} = z \\ \dot{z} = -\alpha z + xy^2 - x, \end{cases} \tag{1}$$

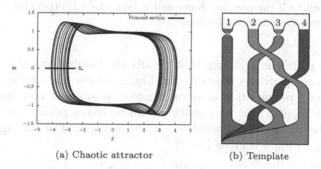

(a) Chaotic attractor   (b) Template

Fig. 1. A representation of a template of a chaotic attractor solution to the Malasoma system (1) for $\alpha = 2.027$. (a) Chaotic attractor with the Poincaré section (see [25] for the definition of this section named S_a). (b) Template of the chaotic attractor from the Poincaré section.

with chaotic dynamics as solutions when $\alpha \in [2.027; 2.08]$. A detailed analysis of the topological properties of the attractors that can be produced by this system has been proposed in [24,25]. For instance, Fig. 1 summarizes some steps of the topological characterization (Poincaré section and template) of a chaotic attractor when $\alpha = 2.027$. In this article, we are considering only attractors bounded by genus–1 torus such as Rössler attractors [26] or Malasoma attractors [16] (Fig. 1a); it does not work for more complex attractors such as Lorenz attractors [15] bounded by a genus–3 torus.

A *template* is a compact branched two-manifold with boundary and smooth expansive semiflow built locally from two types of charts: joining and splitting [10]. It is a figure that represents the topological structure of a chaotic attractor. Since the 1990s there have been two different ways to represent templates with linking matrices that are still used today, as one can see in the recent paper of Gilmore and Rosalie [12], where algorithms are given to switch from one representation to the other. Hereinafter, the representation first given by Melvin and Tufillaro [17] is considered. This representation only requires a linking matrix, and gives a standard representation at the end, where at the bottom of the template the strips are ordered from the back-most on the left to the front-most on the right. This is the representation used for the template shown in Fig. 1. We also use the orientation convention defined by Tufillaro *et al.* [17,18] (Fig. 2).

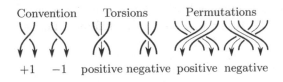

Convention   Torsions   Permutations

+1   −1   positive negative   positive negative

Fig. 2. Convention of representing oriented crossings. The permutation between two branches is positive if the crossing generated is equal to +1, otherwise it is negative. We use the same convention for torsions.

A linking matrix is a matrix that details the number and the direction of crossings in a template. As illustrated in Fig. 2, a torsion is a twist of a branch with itself and a permutation is an exchange of position of two branches. Furthermore, the torsions and permutations can be either positive or negative as defined by the orientation convention shown in Fig. 2. The linking matrix M corresponding to Fig. 1 is given by (2).

$$M = \begin{bmatrix} 1 & 0 & 0 & 0 \\ 0 & 0 & -1 & -1 \\ 0 & -1 & -1 & -1 \\ 0 & -1 & -1 & 0 \end{bmatrix} \tag{2}$$

The diagonal elements in the linking matrix correspond to the torsions. As an example, consider matrix M. The element $M_{1,1} = 1$ represents the number of torsions of branch one of the template from Fig. 1. This branch performs exactly one single positive torsion as indicated by the matrix M. The non-diagonal elements correspond to the number of permutations between the different branches. As an example, $M_{2,4} = -1$ means that branches two and four perform a negative permutation which is depicted by the crossing of the orange and red branch in Fig. 1. It is sufficient to consider the part of the matrix above the diagonal, as it is symmetric.

The linking matrix M is unique but the corresponding template can be drawn in various ways. Some representations can be longer than others. This is why our goal is to find the most concise template. This means that we aim to maximize the number of permutations per level of the template. There might be however several templates with minimum size. In this work we only consider the first template of minimum size generated by the algorithm.

An important remark is that not every matrix corresponds to a valid template of a chaotic attractor. As a chaotic attractor is a solution of a deterministic process and the linking matrix represents it, such a matrix needs to fulfill certain criteria. We will describe the tool we created to verify the validity of a linking matrix, to solve the underlying scheduling problem to find the order of the permutations and to determine the most concise representation of a template. Finally, the tool also renders the solution found.

3   Related Work

The visualization of a template has been addressed in Chap. 5 Sect. 5 of [28] and, according to our best knowledge, the validation of a linking matrix has never been addressed. Usually, this has been done manually by each author. The only comparable project we found is a Mathematica code written by Tufillaro *et al.* [28], which draws templates. Extensive details are available in the Chap. 5 of [28]. It has been used recently in papers written by Barrio *et al.* [2–4]. This implementation, however, only works on older versions of Mathematica. Furthermore, one has to specify as input an explicit order of crossings, which means that it does not find them automatically from a linking matrix, unlike the algorithm presented in this paper. This Mathematica code does not provide a validity verification either, it is purely a tool for drawing "clean" templates.

To the best of our knowledge, such a tool has never been proposed and could be beneficial for the scientific community, as it is not always easy to see whether a matrix is valid or not. Indeed there have been publications with invalid matrices that our tool would have marked as such [18]. Some other papers have presented quite unattractive drawings of templates (eg. Fig. 4 of [1]) and we feel that our tool would provide researchers with an easy and rapid way to solve this problem. Moreover, it can also be used by the community as a tool for building a linking matrix from the linking number numerically obtained during the topological characterization method for attractors bounded by a genus–1 torus (see [11, 21] for details).

4   Linking Matrix and Template of a Chaotic Attractor

In this section, we are going to discuss the approach we developed in order to check the validity of a given linking matrix, to find a corresponding template of minimal height as well as to visualize it. Firstly in Sect. 4.1, we will describe the different validation steps which we are applying on a matrix and justify their necessity. Secondly, Sect. 4.2 explains the tree construction we use in order to minimize the height of the resulting template and the methods we apply for the visualization of the template.

Algorithm 1. Drawing of the template of a linking matrix.

1: verify correct matrix input form
2: verify continuitiy constraints of matrix
3: verify determinism constraints of matrix
4: **if** passed all verification steps **then**:
5:     construct tree
6:     find shortest path in tree
7:     draw template

4.1   Validation of a Linking Matrix

A linking matrix is a topological representation of a chaotic attractor, hence it needs to satisfy certain constraints linked to the attractor. Essentially, a template consists of strips that are stretched, twisted, folded and glued at the bottom over and over again after a clockwise rotation. We remind that we are only considering templates of attractors bounded by a genus–1 torus.

In order to visualize this, one can imagine having a sheet of paper split into several strips. The behavior of those strips is given by the elements of the matrix. If one can deform the paper in such a way that the paper respects the constraints given by the matrix without having to tear it apart, then the matrix corresponds to a valid template. If tears are unavoidable, no valid template exists. If there is a tearing mechanism in the attractor, we are out of the scope because this means that the attractor is at least bounded by a genus–2 torus.

Validation Steps. The steps below evaluate whether or not a linking matrix is valid, i.e., if it corresponds to a chaotic attractor.

First of all, we need to verify that a matrix is of the right form. A valid linking matrix, by definition, has a certain construction. It is square, symmetric and has integers as values [17].

The next three validation steps are constraints on the continuity of the template. Going back to the sheet of paper example, these constraints guarantee that no tears occur. The first of these constraints is linked to the diagonal elements of the matrix. These elements have to respect the condition which dictates that they have to differ by exactly one from their diagonal neighbors. Violating this constraint would result in a discontinuous template. Similar to the diagonal constraint, a linking matrix needs to satisfy the condition which states that an arbitrary value in the matrix cannot differ from the values of all of its neighbors by more than one. Finally, the last continuity constraint is based on the order of the elements on the bottom of the template. From a linking matrix, one needs to be able to obtain a valid order for the template. The order is an array which defines the position of the branches at the bottom of the template after performing the crossings. We obtain this order from the matrix by applying a simple algorithm described in [17]. A valid ordering array contains all branch indexes exactly once. An index being present twice would mean that two branches would end up at the same end position, which is impossible without a tear and therefore would result in an invalid template.

The last two verification steps are linked to the determinism of a chaotic attractor. As stated earlier, chaotic attractors are solutions of dynamic deterministic systems, meaning that from any starting point there is a unique image and no choice is possible. As the template is a topological representation of a chaotic attractor, it also needs to respect its intrinsic properties like determinism. The first of those two verifications consists in checking whether the linking matrix has 2×2 sub-matrices located on its diagonal that are not valid. Up to addition of a global torsion (see [24] for details) there are two 2×2 matrices that are not valid, namely B and C:

$$\left\{ B = \begin{bmatrix} -1 & 0 \\ 0 & 0 \end{bmatrix}, \, C = \begin{bmatrix} 0 & 0 \\ 0 & -1 \end{bmatrix}, \, C + 1 = \begin{bmatrix} 1 & 1 \\ 1 & 0 \end{bmatrix}, \dots \right\}. \tag{3}$$

The set (3) corresponds to matrices that are associated to discontinuous templates. If the matrix has such a sub-matrix on its diagonal, this means that it presents a choice opportunity at some point and violates the determinism condition. Therefore, it is not valid.

Finally, in the second step, which we call planarity check, we verify the order of the end positions of the template. The idea is to take the final positions of the branches at the end of the template, and connect them with arcs in a certain way. Start with 1, and connect it to 2 over the list. Then connect 2 to 3 below the list, 3 to 4 over, and so on. If the arcs cannot be drawn without intersecting, then the matrix is invalid. This is illustrated by Fig. 3, where the left part of the figure corresponds to this verification of the matrix (2), and has no intersections. The right side on the other hand corresponding to matrix N (4) does not pass the test.

$$N = \begin{bmatrix} 0 & 0 & 0 & 0 \\ 0 & 1 & 0 & -1 \\ 0 & 0 & 0 & -1 \\ 0 & -1 & -1 & -1 \end{bmatrix} \tag{4}$$

If this planarity condition was not verified and there was an intersection, the system would have a choice when arriving at this intersection, which would violate the determinism assumption. Therefore, a matrix that does not satisfy this condition cannot correspond to a valid template.

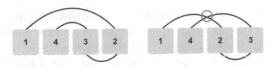

Fig. 3. Planarity check of matrices (2) (left) and (4) (right).

Order of the Validation Steps. The order of the different validation steps is defined in the way described above, we start checking the most general constraints, and then check the most specific ones (Algorithm 1). For example, if a matrix is not square matrix, there is no need to verify specific constraints like the diagonal constraint, as the matrix is not even a linking matrix by definition. The same idea applies to the other criteria.

In literature, there have been publications with invalid matrices that our procedure would have labeled as such. One example would be the first 4 × 4 linking matrix in [18], which gives the matrix with the following diagonal elements: 6, 5, 5 and 4. This matrix would not have passed the validation step which dictates that all elements on the diagonal of a matrix have to differ by one from their diagonal neighbors.

$$K = \begin{bmatrix} 3 & 2 & 2 & 3 \\ 2 & 2 & 2 & 3 \\ 2 & 2 & 3 & 4 \\ 3 & 3 & 4 & 4 \end{bmatrix} \tag{5}$$

For the matrix K (5) the ordering validation step fails because the ordering at the end is given by the array $[2, 2, 3, 3]$, meaning that both strips one and four are on position two and strips two and three are on position three. As this is a problem for continuity, this matrix would not pass the order test. This illustrates that a tool to validate a matrix would facilitate the analysis of linking matrices, as it is not always easy to see whether a matrix is valid or not. A complete example of the validation process can be found in the appendices of the extended version [20].

4.2 Visualization of a Template

Tree Construction. After having verified the validity of a linking matrix, the next step is to generate a visualization of a template with minimal height from a given linking matrix. In order to determine the minimal height of a template, one has to optimize the scheduling of all the crossings between the different branches. For this purpose, we developed an approach where we take as input a valid linking matrix and make use of its permutations to generate a tree graph using a breadth first approach, meaning that we build it level by level.

To do this, we follow Algorithm 2. We derive the initial order from the matrix which represents the root of the tree as a first step. Furthermore, we also retrieve the list of performable permutations between the branches. Beginning at the

---

Algorithm 2. Tree construction

```
1: if validMatrix(matrix) then
2:     init = Node(permutationList, order, father = None) ;
3:     finalOrder = getFinalOrder(matrix)
4:     queue = [init] ;
5:     while queue ≠ ∅ do
6:         node = queue[0] ;
7:         queue = queue[1 :] ;
8:         toExecute = permutationList ∩ allNeighborCombinations(node.order) ;
9:         if toExecute = ∅ and node.order = finalOrder then
10:            setLeaf(node) ;
11:            break ;
12:        for p in toExecute do
13:            newNode =
14:            Node(updatedPermutationList(p), updatedOrder(p), father = node) ;
15:            queue.append(newNode) ;
```

---

root, we simulate the permutations and generate additional nodes which are annotated with an updated order and then added to the tree. For each node created, the list of permutations yet to be performed will differ. Eventually, a node representing a leaf with an empty permutation list and a valid final order will be generated. At this point, the computation of the tree is stopped. By traversing the tree from the root to that leaf, we get the sequence of permutations to execute in order to obtain a template of minimal height. To illustrate this procedure, consider the following 4×4 matrix A (6).

$$A = \begin{bmatrix} -1 & -1 & -1 & -1 \\ -1 & 0 & 0 & 0 \\ -1 & 0 & 1 & 1 \\ -1 & 0 & 1 & 2 \end{bmatrix} \tag{6}$$

From this matrix, we get an initial order where the branches are numbered beginning from 1 to 4. To retrieve the set of permutations to perform, we have to consider the non-diagonal elements of the matrix. For example, the branch with the label 1, has to perform a negative permutation with the branches 2, 3 and 4. There is also a positive permutation between branch 3 and 4. So, we obtain the following list of permutations which needs to be executed $[(1, 2), (1, 3), (1, 4), (3, 4)]$.

To find the permutations which can be performed at this stage, we need to consider our initial order from which we can derive which branches are direct neighbors. For instance, we obtain the following list of neighbor pairs $[(1, 2), (2, 3), (3, 4)]$. By taking the intersection of the neighbor list and the set of permutations to perform, we obtain a set of permutation which are possible to process during the initial stage. By doing so, we can permute branch 1 and 2 or 3 and 4. However, we could also perform both permutations in parallel as performing one of them does not prohibit the other one. As illustrated on top of

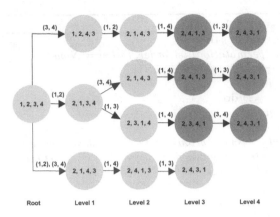

Fig. 4. Final and complete tree for matrix A from (6) including the root and the child nodes generated per level. Each node represents the updated order after each permutation described by the incoming edge. (Color figure online)

Fig. 4, we see the root labeled with the initial order of the branches. After the first set of permutations have been performed, different child nodes are created at level 1. The corresponding order of each child node is obtained by switching the positions of the permuted branches in the initial order of the root.

From the new order of each child node, we try to find a new permutation to perform by defining the neighbor pairs. We then recompute the possible permutations for this iteration. Each iteration will add one or more children to tree and this process is repeated until all permutations have been performed or no new permutation can be computed. However, a node which can no longer perform a permutation while there are still some permutations in the set left to be executed, is not considered valid.

Figure 4 also shows the final tree after all permutations have been performed. The green arrows leading to the green colored leaf denote the shortest path where the labels show the order of execution of the permutations to get to the final order of the template. This will result in a template of shortest possible height. There are also three other possible solutions but they will not reduce the height of the template to a minimum as they perform one additional permutation. However, we stop the computation of building the tree after encountering the first valid leaf, so the red nodes will never be computed. The breadth-first construction of the tree guarantees that the first found solution is the shortest one.

Drawing of the Template. Finally, after verification of the linking matrix and after having found the shortest path in the tree corresponding to the most concise order of crossings, we can now draw the template. To draw the templates as scalable vector graphics, we used python's `swgwrite` module [19].

In order to draw both torsions and permutations, we use a cubic Bézier curve as shape. To illustrate how we use it, consider two points (x_1, y_1) and (x_2, y_2)

and suppose we want to draw this Bézier curve between them, in the same shape as those used in the permutations and torsions. The starting point is given by (x_1, y_1) and we will give the rest of the points relative to this starting point. The relative end point is then given by $(x_2 - x_1, y_2 - y_1)$ and the two relative control points by $(x_1, (y_2 - y_1)/2)$ and $(x_2 - x_1, (y_2 - y_1)/2)$. So the control points are always halfway in height between the two points and straight above respectively below them.

To draw a torsion we first draw one Bézier curve, then add a small white circle in the middle of this curve to *erase* this part. Finally we draw the other Bézier curve. This procedure is illustrated in Fig. 5(a–c). Permutations are drawn in a similar way. The sign of the permutation defines which of the two branches is drawn first, then when the other one is drawn it covers it up as it comes on top of the other one (Fig. 5(d–e)).

(a)   (b)   (c)   (d)   (e)

Fig. 5. An illustration of a positive torsion (a–c) and a positive permutation (d–e) drawing process.

We start by considering the torsions of the matrix and draw all of them. Then we move on to the permutations. They are given by the sequence of edges forming the shortest path of the tree generated by the input matrix. We then draw the rest of the template by levels. At each level, every strip can do one of three actions: do a straight transition, permute left or permute right. The shortest path tells us which two strips should permute. Given this information, it is easy to calculate the coordinates at the next level of each strip and apply the correct transition (Fig. 6).

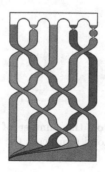

Fig. 6. Template of one linking matrix with five branches and eight permutations.

5   Performance Evaluation

An elementary matrix is a unique linking matrix describing a chaotic mechanism without additional torsions or symmetry properties [22]. Given an input size, Rosalie describes in this article a method to generate all possible elementary linking matrices of such size. We used this method to obtain the 14, 38 and 116 possible elementary matrices with resp. five, six and seven branches (resp. 5×5, 6×6 and 7×7 linking matrices). Figure 7 depicts for each matrix size the distribution of the elementary matrices with respect to the number of permutations to process.

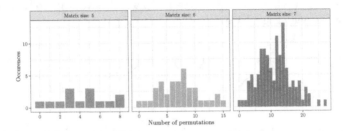

Fig. 7. Distribution of the number of elementary matrices with respect to the number of permutations to process. There are 14 (resp. 38 and 116) matrices of size 5 (resp. 6 and 7).

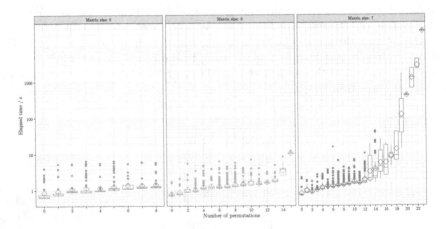

Fig. 8. Elapsed time depending on the number of permutations for the matrices depending on their size. The diamond represents the average value.

The experiments were conducted on a server with an Intel Xeon X7560 processor with a clock speed of 2.27 GHz, and 1024 GB of RAM. Even though this

server is not the fastest available, it is the only one fulfilling the memory requirements (instances required between 25 MB and 400 GB of memory). For the sake
of comparability, all instances have been run on the same machine. For the
complete description of the cluster environment, please refer to https://hpc.uni.
lu/systems/chaos/. We computed the templates of all the elementary matrices
described above. We ran the experiments with version v0.0.1 of the code. For
each input matrix, we measured 30 times the time elapsed to get the template.
The 7 × 7 matrix with 27 permutations ran out of memory and crashed: we
removed it from the graphs. Figures 8 and 9 depict the elapsed computation
time with respect to the number of permutations to process. As expected, we
observe a drastic rise that characterizes a combinatorial explosion in the number
of permutations.

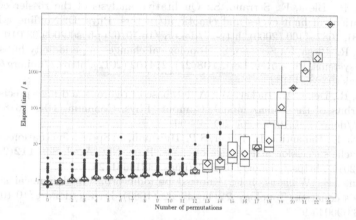

Fig. 9. Elapsed time depending on the number of permutations for the 167 matrices.
The diamond represents the average value.

6   Conclusion

In this paper, we presented a tool which verifies whether a linking matrix corresponds to a topologically valid template. Moreover, our approach computes and
draws a template of minimal height corresponding to this linking matrix. This
is especially interesting for linking matrices with a higher number of crossings.
We believe that this tool could benefit the research community as it eases the
process of verifying the validity of a linking matrix, and quickly draws one of its
matching templates.

A possible extension of our work could be to represent the generated templates as a 3D model in an automated way. One representation of a 3D template
was given by Cross and Gilmore, where they include the torsions as a part of
the global modification [9]. Another visualization was given by Boulant *et al.*

(Fig. 6 of [7]). Such a 3D visualization would allow to be even closer visually to the nature of a chaotic attractor, and thus could provide more intuitive insights.

Acknowledgments. The experiments presented in this paper were carried out using the HPC facilities of the University of Luxembourg [29] (see https://hpc.uni.lu). This work is partially funded by the joint research programme UL/SnT-ILNAS on Digital Trust for Smart ICT.

References

1. Anastassiou, S., Bountis, T., Petalas, Y.G.: On the topology of the Lü attractor and related systems. J. Phys. A: Math. Theor. **41**(48), 485101 (2008). https://doi.org/10.1088/1751-8113/41/48/485101
2. Barrio, R., Blesa, F., Serrano, S.: Qualitative analysis of the rössler equations: bifurcations of limit cycles and chaotic attractors. Phys. D: Nonlinear Phenom. **238**(13), 1087–1100 (2009). https://doi.org/10.1016/j.physd.2009.03.010
3. Barrio, R., Blesa, F., Serrano, S.: Topological changes in periodicity hubs of dissipative systems. Phys. Rev. Lett. **108**(21), 214102 (2012). https://doi.org/10.1103/physrevlett.108.214102
4. Barrio, R., Dena, A., Tucker, W.: A database of rigorous and high-precision periodic orbits of the Lorenz model. Comput. Phys. Commun. **194**, 76–83 (2015). https://doi.org/10.1016/j.cpc.2015.04.007
5. Benincà, E., Ballantine, B., Ellner, S.P., Huisman, J.: Species fluctuations sustained by a cyclic succession at the edge of chaos. Proc. Nat. Acad. Sci. **112**(20), 6389–6394 (2015). https://doi.org/10.1073/pnas.1421968112
6. Birman, J.S., Williams, R.F.: Knotted periodic orbits in dynamical systems–I: Lorenz's equation. Topology **22**(1), 47–82 (1983). https://doi.org/10.1016/0040-9383(83)90045-9
7. Boulant, G., Lefranc, M., Bielawski, S., Derozier, D.: A nonhorseshoe template in a chaotic laser model. Int. J. Bifurcat. Chaos **08**(05), 965–975 (1998). https://doi.org/10.1142/s0218127498000772
8. Budroni, M.A., Calabrese, I., Miele, Y., Rustici, M., Marchettini, N., Rossi, F.: Control of chemical chaos through medium viscosity in a batch ferroin-catalysed Belousov-Zhabotinsky reaction. Phys. Chem. Chem. Phys. **19**(48), 32235–32241 (2017). https://doi.org/10.1039/c7cp06601e
9. Cross, D.J., Gilmore, R.: Dressed return maps distinguish chaotic mechanisms. Phys. Rev. E **87**(1), 012919 (2013). https://doi.org/10.1103/physreve.87.012919
10. Ghrist, R.W., Holmes, P.J., Sullivan, M.C.: Knots and Links in Three-Dimensional Flows. Springer, Berlin (1997). https://doi.org/10.1007/bfb0093387
11. Gilmore, R.: Topological analysis of chaotic dynamical systems. Rev. Mod. Phys. **70**(4), 1455–1529 (1998). https://doi.org/10.1103/revmodphys.70.1455
12. Gilmore, R., Rosalie, M.: Algorithms for concatenating templates. Chaos: Interdisc J. Nonlinear Sci. **26**(3), 033102 (2016). https://doi.org/10.1063/1.4942799
13. Kumar, S., Strachan, J.P., Williams, R.S.: Chaotic dynamics in nanoscale NbO_2 Mott memristors for analogue computing. Nature **548**(7667), 318–321 (2017). https://doi.org/10.1038/nature23307
14. Larger, L., Penkovsky, B., Maistrenko, Y.: Laser chimeras as a paradigm for multistable patterns in complex systems. Nat. Commun. **6**(1), 7752 (2015). https://doi.org/10.1038/ncomms8752

15. Lorenz, E.N.: Deterministic nonperiodic flow. J. Atmos. Sci. **20**(2), 130–141 (1963). https://doi.org/10.1175/1520-0469(1963)020⟨0130:dnf⟩2.0.co;2
16. Malasoma, J.M.: What is the simplest dissipative chaotic jerk equation which is parity invariant? Phys. Lett. A **264**(5), 383–389 (2000). https://doi.org/10.1016/s0375-9601(99)00819-1
17. Melvin, P., Tufillaro, N.B.: Templates and framed braids. Phys. Rev. A **44**, R3419–R3422 (1991). https://doi.org/10.1103/PhysRevA.44.R3419
18. Mindlin, G.B., Hou, X.J., Solari, H.G., Gilmore, R., Tufillaro, N.B.: Classification of strange attractors by integers. Phys. Rev. Lett. **64**(20), 2350–2353 (1990). https://doi.org/10.1103/physrevlett.64.2350
19. Moitzi, M.: svgwrite (Python Library) (2018). https://pypi.org/project/svgwrite/. Accessed 26 May 2018
20. Olszewski, M., et al.: Visualizing the template of a chaotic attractor. arXiv preprint arXiv:1807.11853 (2018)
21. Rosalie, M.: Templates and subtemplates of Rössler attractors from a bifurcation diagram. J. Phys. A: Math. Theor. **49**(31), 315101 (2016). https://doi.org/10.1088/1751-8113/49/31/315101
22. Rosalie, M.: Chaotic mechanism description by an elementary mixer for the template of an attractor. arXiv preprint arXiv:1703.02768 (2017)
23. Rosalie, M., Danoy, G., Chaumette, S., Bouvry, P.: Chaos-enhanced mobility models for multilevel swarms of UAVs. Swarm Evol. Comput. **41**, 36–48 (2018). https://doi.org/10.1016/j.swevo.2018.01.002
24. Rosalie, M., Letellier, C.: Systematic template extraction from chaotic attractors: I. genus-one attractors with an inversion symmetry. J. Phys. A: Math. Theor. **46**(37), 375101 (2013). https://doi.org/10.1088/1751-8113/46/37/375101
25. Rosalie, M., Letellier, C.: Systematic template extraction from chaotic attractors: II. genus-one attractors with multiple unimodal folding mechanisms. J. Phys. A: Math. Theor. **48**(23), 235101 (2015). https://doi.org/10.1088/1751-8113/48/23/235101
26. Rössler, O.: An equation for continuous chaos. Phys. Lett. A **57**(5), 397–398 (1976). https://doi.org/10.1016/0375-9601(76)90101-8
27. Suzuki, Y., Lu, M., Ben-Jacob, E., Onuchic, J.N.: Periodic, quasi-periodic and chaotic dynamics in simple gene elements with time delays. Sci. Rep. **6**(1), 21037 (2016). https://doi.org/10.1038/srep21037
28. Tufillaro, N.B., Abbott, T., Reilly, J.: An Experimental Approach to Nonlinear Dynamics and Chaos. Addison-Wesley, Redwood City (1992)
29. Varrette, S., Bouvry, P., Cartiaux, H., Georgatos, F.: Management of an academic HPC cluster: the UL experience. In: 2014 International Conference on High Performance Computing & Simulation (HPCS). IEEE (2014). https://doi.org/10.1109/hpcsim.2014.6903792

RAC Drawings

On RAC Drawings of Graphs with One Bend per Edge

Patrizio Angelini[✉], Michael A. Bekos, Henry Förster, and Michael Kaufmann

Wilhelm-Schickhard-Institut für Informatik,
Universität Tübingen, Tübingen, Germany
{angelini,bekos,foersth,mk}@informatik.uni-tuebingen.de

Abstract. A k-bend *right-angle-crossing drawing* (or k-bend *RAC drawing*, for short) of a graph is a polyline drawing where each edge has at most k bends and the angles formed at the crossing points of the edges are 90°. Accordingly, a graph that admits a k-bend RAC drawing is referred to as k-bend *right-angle-crossing graph* (or k-bend *RAC*, for short). In this paper, we continue the study of the maximum edge-density of 1-bend RAC graphs. We show that an n-vertex 1-bend RAC graph cannot have more than $5.5n - O(1)$ edges. We also demonstrate that there exist infinitely many n-vertex 1-bend RAC graphs with exactly $5n - O(1)$ edges. Our results improve both the previously known best upper bound of $6.5n - O(1)$ edges and the corresponding lower bound of $4.5n - O(\sqrt{n})$ edges by Arikushi et al. (Comput. Geom. 45(4), 169–177 (2012)).

1 Introduction

A recent research direction in Graph Drawing, which is currently receiving a great deal of attention [26,29,31], focuses on combinatorial and algorithmic aspects for families of graphs that can be drawn on the plane while avoiding specific kinds of edge crossings; see, e.g., [22] for a survey. This direction is informally recognized under the term "beyond planarity". An early work on beyond planarity (and probably the one that initiated this direction in Graph Drawing) is due to Didimo, Eades, and Liotta [21], who introduced and first studied the family of graphs that admit polyline drawings, with few bends per edge, in which the angles formed at the edge crossings are 90°. Their primary motivation stemmed from experiments indicating that the humans' abilities to read and understand drawings of graphs are not affected too much, when the edges cross at large angles [27,28] and the number of bends per edge is limited [34,35]. Their work naturally gave rise to a systematic study of several different variants of these graphs; see, e.g., [7–9,12,18–20,23].

Formally, a k-bend *right-angle-crossing drawing* (or k-bend *RAC drawing*, for short) of a graph is a polyline drawing where each edge has at most k bends and the angles formed at the crossing points of the edges are 90°. Accordingly,

© Springer Nature Switzerland AG 2018
T. Biedl and A. Kerren (Eds.): GD 2018, LNCS 11282, pp. 123–136, 2018.
https://doi.org/10.1007/978-3-030-04414-5_9

a graph that admits a k-bend RAC drawing is referred to as *k-bend right-angle-crossing graph* (or *k-bend RAC*, for short); a 0-bend RAC graph (drawing) is also called a *straight-line RAC graph (drawing)*.

There exist several results for straight-line RAC graphs. Didimo et al. [21] showed that a straight-line RAC graph with n vertices has at most $4n-10$ edges, which is a tight bound, i.e., there exist infinitely many straight-line RAC graphs with n vertices and exactly $4n-10$ edges. These graphs are actually referred to as *optimal* or *maximally-dense* straight-line RAC and are in fact 1-planar [23], i.e., they admit drawings in which each edge is crossed at most once. In general, however, deciding whether a graph is straight-line RAC is NP-hard [8], and remains NP-hard even if the drawing must be upward [7] or 1-planar [12]. Bachmaier et al. [10] and Brandenburg et al. [15] presented interesting relationships between the class of straight-line RAC graphs and subclasses of 1-planar graphs. Variants, in which the vertices are restricted on two parallel lines or on a circle, have been studied by Di Giacomo et al. [18], and by Hong and Nagamochi [25].

An immediate observation emerging from this short literature overview is that the focus has been primarily on the straight-line case; the results for RAC drawings with bends are significantly fewer. Didimo et al. [21] observed that 1- and 2-bend RAC graphs have a sub-quadratic number of edges, while any graph with n vertices admits a 3-bend RAC drawing in $O(n^4)$ area; the required area was improved to $O(n^3)$ by Di Giacomo et al. [19]. Quadratic area for 1-bend RAC drawings can be achieved for subclasses of 1-plane graphs [16]; for general 1-plane graphs the known algorithm may yield 1-bend RAC drawings with super-polynomial area [12]. The best-known upper bounds on the number of edges of 1- and 2-bend RAC graphs are due to Arikushi et al. [9], who showed that these graphs can have at most $6.5n-13$ and $74.2n$ edges, respectively. Arikushi et al. [9] also presented 1- and 2-bend RAC graphs with n vertices, and $4.5n - O(\sqrt{n})$ and $7.83n - O(\sqrt{n})$ edges, respectively. Angelini et al. [7] have shown that all graphs with maximum vertex degree 3 are 1-bend RAC, while those with maximum vertex degree 6 are 2-bend RAC. It is worth noting that the complexity of deciding whether a graph is 1- or 2-bend RAC is still open.

Our Contribution: In this work, we present improved lower and upper bounds on the maximum edge-density of 1-bend RAC graphs. Note that this type of problems is commonly referred to as Turán type, and has been widely studied also in the framework of beyond planarity; see, e.g., [1–5,13,17,24,30,32,33,36]. More precisely, in Sect. 3, we show that an n-vertex 1-bend RAC graph cannot have more than $5.5n - O(1)$ edges, while in Sect. 4 we demonstrate that there exist infinitely many 1-bend RAC graphs with n vertices and exactly $5n - O(1)$ edges. These two results together further narrow the gap between the best-known lower and upper bounds on the maximum edge-density of 1-bend RAC graphs (from $2n$ to $n/2$). Our approach for proving the upper bound in Sect. 3 builds upon the charging technique by Arikushi et al. [9], which we overview in Sect. 2. We discuss open problems in Sect. 5.

2   Overview of the Charging Technique

In this section, we introduce the necessary notation and we describe the most important aspects of the charging technique by Arikushi et al. [9] for bounding the maximum number of edges of a 1-bend RAC graph. Consider an n-vertex 1-bend RAC graph $G = (V, E)$, together with a corresponding 1-bend RAC drawing Γ with the minimum number of crossings. The edges of G are partitioned into two sets E_0 and E_1, based on whether they are crossing-free in Γ (set E_0) or they have at least a crossing (set E_1). Let G_0 and G_1 be the subgraphs of G induced by E_0 and E_1, respectively.

Since G_0 is plane, $|E_0| \leq 3n - 6$ holds. To estimate $|E_1|$, Arikushi et al. consider the graph G'_1 that is obtained from the drawing of G_1, by replacing each crossing point with a dummy vertex; we call G'_1 the *planarization* of the drawing of G_1. Let V'_1, E'_1, and F'_1 be the set of vertices, edges, and faces of G'_1, respectively. Let $\deg(v)$ be the degree of a vertex v of G'_1 and $s(f)$ be the size of a face f of G'_1, that is, the number of edges incident to f. In the charging scheme, every vertex v of G'_1 is initially assigned a charge $ch(v)$ equal to $\deg(v) - 4$, while every face f of G'_1 is initially assigned a charge $ch(f)$ equal to $s(f) - 4$. By Euler's formula, the sum of charges over all vertices and faces of G'_1 is:

$$\sum_{v \in V'_1} (\deg(v) - 4) + \sum_{f \in F'_1} (s(f) - 4) = 2|E'_1| - 4|V'_1| + 2|E'_1| - 4|F'_1| = -8$$

In two subsequent discharging phases, they redistribute the charges in G'_1 so that (i) the total charge remains the same, and (ii) all faces have non-negative charges. In the first discharging phase, for every edge e with one bend, half a unit of charge is passed from each of its two endvertices to the face that is incident to the convex bend of e. Arikushi et al. show that each face of size less than 4 has at least one convex bend, so it receives at least one unit of charge. Hence, after this phase, the only faces that have negative charges are the so-called *lenses*, which have size 2 and only one convex bend (each lens has charge -1). On the other hand, the charge of every vertex $v \in V'_1$ is at least $ch'(v) = \frac{1}{2}\deg(v) - 4$.

In the second discharging phase, Arikushi et al. exploit the crossing minimality of Γ to guarantee the existence of an injective mapping from the lenses to the convex bends incident to faces of G'_1 with size at least 4. Since each such bend yields one additional unit of charge to its incident face, and since this face has already a non-negative charge due to its size, it is possible to move this unit from the face to the mapped lens without introducing faces with negative charge. Hence, after the second phase, the charge $ch''(f)$ of each face $f \in F'_1$ is non-negative (and at least as large as its initial charge, i.e., $ch''(f) \geq ch(f)$). Since $ch''(v) = ch(v)$, $|E_1| \leq 4n - 8$ can be proved as follows:

$$|E_1| - 4n = \sum_{v \in V'_1} \left(\frac{1}{2}\deg(v) - 4 \right) \leq \sum_{v \in V'_1} ch''(v) \leq \sum_{v \in V'_1} ch''(v) + \sum_{f \in F'_1} ch''(f) = -8$$

$$(1)$$

So far, graph G has $|E_0| + |E_1| \leq 7n - 14$ edges. Arikushi et al. improve this bound in a conclusive analysis based on the observation that a triangular face of G_0 cannot contain edges of E_1. Hence, if G_0 contains exactly $3n - 6$ edges, then it is a triangulation, and thus $E_1 = \emptyset$. More in general, they considered how many edges E_1 may contain when G_0 is a graph obtained from a triangulation by removing k edges. Let V_0, E_0, and F_0 be the sets of vertices, edges, and faces of G_0, respectively, and let $d(f)$ be the *degree* of a face $f \in F_0$, i.e., the number of its distinct vertices. Then, by Eq. 1 we have:

$$|E_1| \leq \sum_{f \in F_0; d(f) > 3} (4d(f) - 8) \tag{2}$$

Arikushi et al. proved that the right-hand side of Eq. 2 is at most $8k$. In fact, the removal of any crossing-free edge e leads to one of the following cases.

C.1 if e was a bridge of a face, this yields a face with the same degree, which leaves the right-hand side of Eq. 2 unchanged;

C.2 if e was adjacent to two triangles, this yields a new face f of degree $d(f) = 4$, which can contain at most $4d(f) - 8 = 8$ edges of E_1, which increases the right-hand side of Eq. 2 by 8;

C.3 if e was adjacent to a triangle and to a face of degree $d(f)$ (containing at most $4d(f) - 8$ edges of E_1), this yields a new face of degree at most $d(f) + 1$, which can contain at most $4(d(f) + 1) - 8 = 4d(f) - 4$ edges of E_1, which increases the right-hand side of Eq. 2 by at most 4; finally,

C.4 if e was adjacent to two faces f_1 and f_2 such that $d(f_1), d(f_2) > 3$ (containing at most $4(d(f_1) + d(f_2)) - 16$ edges of E_1), this yields a new face of degree at most $d(f_1) + d(f_2) - 2$, which contains at most $4(d(f_1) + d(f_2) - 2) - 8 = 4(d(f_1) + d(f_2)) - 16$ edges of E_1, leaving the right-hand side of Eq. 2 as is.

Hence, the removal of k uncrossed edges increases the right-hand side of Eq. 2 by at most $8k$. With this observation, Arikushi et al. derived two different upper bounds on the number of edges of G, namely:

$$|E| \leq (3n - 6 - k) + 4n - 8 = 7n - 14 - k \tag{3}$$
$$|E| \leq (3n - 6 - k) + 8k \tag{4}$$

The minimum of the two bounds is maximized when $k = n/2 - 1$, which yields $|E| \leq 6.5n - 13$. Arikushi et al. noticed that the bound of $8k$ is an overestimation, and that possible refinements would lead to improvements of the overall bound.

3   An Improved Upper Bound

In this section, we describe how to improve the analysis of the charging scheme described in Sect. 2 to obtain a better upper bound. W.l.o.g., we assume that G is connected and that $n \geq 5$. Let f be a face of G_0. As in the previous section, we denote by $d(f)$ the degree of f, that is, the number of distinct vertices of f.

Since f is not necessarily simple or connected, the boundary of f is a disjoint set of (not necessarily simple) cycles, which are called *facial walks*; see Fig. 1a. We denote by $\ell(f)$ the length of face f, that is, the number of edges (counted with multiplicities) in all facial walks of f.

Since a vertex v may occur more than once in a facial walk of f, we denote by $m_f(v)$ the number of its occurrences in f minus one (that is, the number of extra occurrences beyond its first). The sum of such extra occurrences over all the vertices of face f is denoted by $m(f)$, that is, $m(f) = \sum_{v \in f} m_f(v)$. Further, we denote by $b(f)$ the number of biconnected components of all facial walks of f. Finally, we assume that an isolated vertex of f (if any) is not a biconnected component of f, and we denote by $i(f)$ the number of isolated vertices of f. It is not difficult to see that $\ell(f) = d(f) + m(f) - i(f)$.

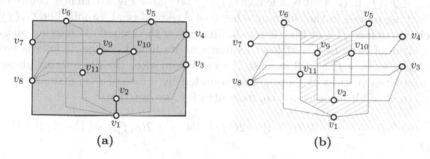

Fig. 1. (a) Illustration of a non-simple, non-connected face f of G_0 (colored in black). The edges of G_1 are colored gray. Face f consists of two facial walks ($w_1 = \langle v_1, v_2, v_1, v_3, v_4, v_5, v_6, v_7, v_8 \rangle$ and $w_2 = \langle v_9, v_{10} \rangle$) and an isolated vertex (v_{11}). Observe that $d(f) = 11$ (as f contains 11 distinct vertices), $\ell(f) = 11$ (as the sum of the lengths of w_1 and w_2 is 11), $m_f(v_1) = 1$ (as v_1 appears twice in w_1), $i(f) = 1$ (as v_{11} is an isolated vertex of f), and $b(f) = 3$ (as w_1 consists of two biconnected components, while w_2 is biconnected). Face f is good, since each of its edges is good. Note that removing edge (v_4, v_7) would make edges (v_5, v_6) and (v_9, v_{10}) not good. (b) The faces of $F_1'(f)$ that are surrounding the three biconnected components of f are tiled in gray.

Let G' be the planarization of the drawing Γ of G. As opposed to G_0, whose faces are not necessarily connected, the faces of G' are in fact connected, since G is connected. Let f be a face of G_0 and let e be any edge incident to f. We say that edge e is *good* for f if and only if there is no other edge e' incident to f such that e and e' are both incident to a face g of G' that lies inside f. Accordingly, face f is called *good* if and only if either all its edges are good for f or if f is a triangle; see Fig. 1a. Note that, if each face of G_0 is good, then every face of the planarization G' is either a triangle of crossing-free edges or contains at most one crossing-free edge, and vice versa. In the next two lemmas, we assume that the faces of G_0 are good; we show later how to guarantee this property. For this, we may need to introduce parallel edges (but no self-loops) in G_0, which however are non-homotopic (each region they define contains at least a vertex). Further, we may need to introduce planar edges with more than one bend; this does not affect the discharging scheme of Arikushi et al. which only considers G_1.

Lemma 1. *Let Γ be a drawing of G such that all faces of G_0 are good. Then, each face f of G_0 contains at most $2d(f) - 2m(f) + 2i(f) + 4b(f) - 8$ edges of G_1.*

Proof. Consider the subgraph $G(f)$ of G which is induced by the interior of f and let $\Gamma(f)$ be the drawing of $G(f)$ derived from Γ. We denote by $G_1(f) = (V_1(f), E_1(f))$ the subgraph of $G(f)$ induced by the set of crossing edges in $\Gamma(f)$, and by $G_1'(f)$ the planarization of $G_1(f)$.

Let $B(f)$ be the set of biconnected components of f and $F_1'(f)$ the set of faces of the drawing of $G_1'(f)$ that is derived from $\Gamma(f)$. Since every edge of f is good, every biconnected component $c \in B(f)$ with length $\ell(c)$ will be surrounded by a face $f_c' \in F_1'(f)$ in G_1' that is of length $\ell(f_c') \geq 2\ell(c)$; see Fig. 1b. Hence, before the discharging phases in the charging scheme of Arikushi et al. (applied on $G_1'(f)$), the charge of face f_c' is at least $2\ell(c) - 4$. Since after the second discharging phase, the charge of each face is at least as much as its initial charge, it follows that the charge of face f_c' is still at least $2\ell(c) - 4$ even after the discharging phases. Since isolated vertices of f are not surrounded by a face of $F_1'(f)$, summing up the charges of all biconnected components of f, we get that

$$\sum_{c \in B(f)} ch''(f_c') \geq \sum_{c \in B(f)} (2\ell(c) - 4) = 2\ell(f) - 4b(f) = 2(d(f) + m(f) - i(f)) - 4b(f)$$

Since, after the second discharging phase, each face has a non-negative charge and the sum of the charges of faces surrounding biconnected components of f is a lower bound for the sum of the charges of all faces in $F_1'(f)$, we get that

$$\sum_{f' \in F_1'(f)} ch''(f') - \sum_{c \in B(f)} ch''(f_c') \geq 0$$

Hence, by refining Eq. 1 we obtain that the number of crossing edges in $G(f)$ can be upper-bounded as follows

$$|E_1(f)| - 4d(f) = \sum_{v \in f} \left(\frac{1}{2} \deg(v) - 4 \right) \leq \sum_{v \in f} ch''(v)$$

$$\leq \sum_{v \in f} ch''(v) + \sum_{f' \in F_1'(f)} ch''(f') - 2(d(f) - m(f) + i(f)) + 4b(f)$$

$$= -8 - 2(d(f) + m(f) - i(f)) + 4b(f)$$

This concludes our proof.   □

In the following lemma, we improve Arikushi et al.'s upper bound on the number of edges of G_1 that G may contain, when the plane subgraph G_0 is obtained from a plane triangulation T by removing k edges, under the assumption that T may contain non-homotopic parallel edges (but no self-loops), and that each face

$f \in F_0$ of G_0 is good. Let $t(f)$ be the minimum number of edges that must be removed from T to obtain f. Similar to Arikushi et al., we preliminarily observe that a face f of G_0 with $t(f) = 0$ cannot contain edges of G_1 in G. If $t(f) = 1$, the only two possible configurations for face f are illustrated in Figs. 2a and b. In both cases, face f can contain at most two crossing edges. If $t(f) = 2$, the only three possible configurations for face f are illustrated in Figs. 2c–2e. Then, face f can contain at most five crossing edges. Let F_0^1 and F_0^2 be the set of faces of G_0 that can be obtained from triangulation T by removing 1 and 2 edges, respectively, that is, $F_0^1 = \{f \in F_0; t(f) = 1\}$ and $F_0^2 = \{f \in F_0; t(f) = 2\}$. By Lemma 1 and the previous observations, we have

$$|E_1| \leq 2|F_0^1| + 5|F_0^2| + \sum_{f \in F_0; t(f) > 2} (2d(f) - 2m(f) + 2i(f) + 4b(f) - 8) \quad (5)$$

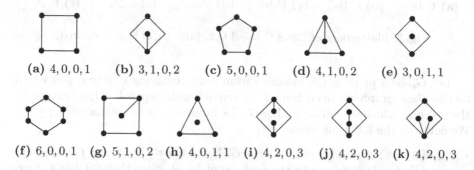

(a) $4, 0, 0, 1$ (b) $3, 1, 0, 2$ (c) $5, 0, 0, 1$ (d) $4, 1, 0, 2$ (e) $3, 0, 1, 1$

(f) $6, 0, 0, 1$ (g) $5, 1, 0, 2$ (h) $4, 0, 1, 1$ (i) $4, 2, 0, 3$ (j) $4, 2, 0, 3$ (k) $4, 2, 0, 3$

Fig. 2. All bounded faces that can be obtained from T by removing (a)–(b) 1 edge, (c)–(e) 2 edges, (f)–(k) 3 edges. The caption of each subfigure indicates the values of $(d(f), m(f), i(f), b(f))$.

In the following lemma, we prove that a slight overestimation of the right-hand side of Eq. 5 is upper-bounded by $\frac{8}{3}k$, which clearly implies that $|E_1| \leq \frac{8}{3}k$.

Lemma 2. *If G_0 is obtained from triangulation T by removing k edges, then:*

$$\frac{8}{3}|F_0^1| + \frac{16}{3}|F_0^2| + \sum_{f \in F_0; t(f) > 2} (2d(f) - 2m(f) + 2i(f) + 4b(f) - 8) \leq \frac{8}{3}k \quad (6)$$

Proof. Our proof is by induction on k and is similar to the corresponding one of Arikushi et al. (Lemma 5 in [9]). In contrast to their proof, we assume that G_0 is obtained from triangulation T by removing edges in a certain order. In particular, we want to avoid the case in which the removal of an edge e results in merging two faces f_1 and f_2 such that $t(f_1), t(f_2) \geq 1$ (refer to Case C.4 in Sect. 2). We guarantee this property as follows. Consider the subgraph D of the dual of T induced by the edges that are dual to those that we have to remove to obtain G_0. We remove the edges in the order in which their dual edges appear in a BFS traversal of each connected component of D. In this way, every inter-level edge in the BFS traversal corresponds to removing an edge that is incident to a triangular face (not visited yet), while each intra-level edge corresponds to removing a bridge from a face that has been created by previously removed edges. In both cases, we avoid merging two faces f_1 and f_2 such that $t(f_1), t(f_2) \geq 1$.

Denote by $\tau(G_0)$ the left-hand side of Eq. 6. In the base of the induction, $k = 0$ holds. In this case, graph G_0 coincides with triangulation T and thus $\tau(G_0) = 0$. In the induction hypothesis, we assume that the lemma holds for $k \geq 0$, and we prove that it also holds for $k' = k + 1$.

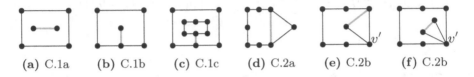

(a) C.1a   (b) C.1b   (c) C.1c   (d) C.2a   (e) C.2b   (f) C.2b

Fig. 3. Illustrations of Cases C.1 and C.2. Edge (u, v) is gray-colored.

Let G'_0 be a plane graph obtained from T by removing k' edges, and let G_0 be the plane graph obtained from T by removing the same k' edges, except for the last one, which we call (u, v). For G_0, by induction, it holds that $\tau(G_0) \leq \frac{8}{3}k$. We consider the following cases:

C.1 Edge (u, v) is a bridge of a face f in G_0 such that $t(f) \geq 3$. Let f' be the face of G'_0 that is obtained by the removal of (u, v). Note that $t(f') \geq 4$. Since (u, v) is a biconnected component of f, it holds that $b(f') = b(f) - 1$. Since (u, v) is a bridge, it also holds that $d(f') = d(f)$. To establish the values of $m(f')$ and $i(f')$, we observe that u, or v, or both may become isolated vertices of G'_0 after the removal of (u, v). We study these cases separately.

  (a) Both u and v become isolated vertices in G'_0; see Fig. 3a. Then $m(f') = m(f)$ and $i(f') = i(f) + 2$. Since $2d(f') - 2m(f') + 2i(f') + 4b(f') - 8 = 2d(f) - 2m(f) + 2(i(f) + 2) + 4(b(f) - 1) - 8 = 2d(f) - 2m(f) + 2i(f) + 4b(f) - 8$, it follows that $\tau(G'_0) = \tau(G_0) \leq \frac{8}{3}k < \frac{8}{3}k'$.

  (b) Exactly one of u and v, say v, becomes an isolated vertex in G'_0; see Fig. 3b. Then $m(f') = m(f) - 1$ and $i(f') = i(f) + 1$. Since $2d(f') - 2m(f') + 2i(f') + 4b(f') - 8 = 2d(f) - 2(m(f) - 1) + 2(i(f) + 1) + 4(b(f) - 1) - 8 = 2d(f) - 2m(f) + 2i(f) + 4b(f) - 8$, it follows that $\tau(G'_0) = \tau(G_0) \leq \frac{8}{3}k < \frac{8}{3}k'$.

  (c) Neither u nor v becomes an isolated vertex in G'_0; see Fig. 3c. Then $m(f') = m(f) - 2$ and $i(f') = i(f)$. Since $2d(f') - 2m(f') + 2i(f') + 4b(f') - 8 = 2d(f) - 2(m(f) - 2) + 2i(f) + 4(b(f) - 1) - 8 = 2d(f) - 2m(f) + 2i(f) + 4b(f) - 8$, it follows that $\tau(G'_0) = \tau(G_0) \leq \frac{8}{3}k < \frac{8}{3}k'$.

C.2 The removal of (u, v) merges a triangular face Δ (that is, $t(\Delta) = 0$) with an adjacent face f of G_0 with $t(f) \geq 3$ into a face f' of G'_0. Note that $t(f') \geq 4$. We consider two cases:

(a) Faces Δ and f share only edge (u, v); see Fig. 3d. Then $d(f') = d(f)+1$, $m(f') = m(f)$, $b(f') = b(f)$, $i(f') = i(f)$. Since $2d(f') - 2m(f') + 2i(f') + 4b(f') - 8 = 2(d(f)+1) - 2m(f) + 2i(f) + 4b(f) - 8 = 2d(f) - 2m(f) + 2i(f) + 4b(f) - 8 + 2$, it follows that $\tau(G_0') = \tau(G_0) + 2 \leq \frac{8}{3}k + 2 < \frac{8}{3}k'$.

(b) Faces Δ and f share at least two edges; see Fig. 3e and f. By removing (u, v), the number of occurrences of the third vertex v' of Δ increases by one and the number of biconnected components increases by one. Then $d(f') = d(f)$, $m(f') = m(f) + 1$, $b(f') = b(f) + 1$, $i(f') = i(f)$. Since $2d(f') - 2m(f') + 2i(f') + 4b(f') - 8 = 2d(f) - 2(m(f) + 1) + 2i(f) + 4(b(f)+1) - 8 = 2d(f) - 2m(f) + 2i(f) + 4b(f) - 8 + 2$, it follows that $\tau(G_0') = \tau(G_0) + 2 \leq \frac{8}{3}k + 2 < \frac{8}{3}k'$.

C.3 The removal of (u, v) yields a face f' of G_0' with $t(f') \in \{1, 2, 3\}$. Note that in the previous cases $t(f') \geq 4$. So, if we rule out this case, then the proof follows. We consider two cases, which correspond to Cases C.1 and C.2 for smaller faces, respectively.

(a) Face f' is obtained by removing a bridge from a face f. Hence, $t(f) = t(f') - 1$ and f' is disconnected. Observe that if $t(f') = 1$, then face f' is not disconnected as can be seen from Fig. 2a and b. Therefore, $t(f') \geq 2$ holds in this subcase.

(b) Face f' is obtained by merging a face f with a triangular face Δ. Hence, $t(f) = t(f') - 1$ holds. Since Δ is triangular, we observe that it does not contribute to $\tau(G_0)$.

In both cases, the face f that is eliminated in order to create face f' is such that $t(f) = t(f') - 1$. We observe that $\tau(G_0')$ is equal to $\tau(G_0)$, plus the contribution of f' to $\tau(G_0')$, minus the contribution of f to $\tau(G_0)$. More precisely: If $t(f') = 1$, then $\tau(G_0') = \tau(G_0) + \frac{8}{3} - 0 \leq \frac{8}{3}k + \frac{8}{3} = \frac{8}{3}k'$; see Fig. 2a–b. If $t(f') = 2$, then $\tau(G_0') = \tau(G_0) + \frac{16}{3} - \frac{8}{3} \leq \frac{8}{3}k + \frac{8}{3} = \frac{8}{3}k'$; see Fig. 2c–e. Otherwise, $t(f') = 3$; see Fig. 2f–k. This implies that $\tau(G_0') \leq \tau(G_0) + (2d(f') - 2m(f') + 2i(f') + 4b(f') - 8) - \frac{16}{3}$. It is easy to verify that $2d(f') - 2m(f') + 2i(f') + 4b(f') - 8 \leq 8$ holds for each of the cases shown in Fig. 2f–k. Hence, $\tau(G_0') \leq \tau(G_0) + \frac{8}{3} \leq \frac{8}{3}k + \frac{8}{3} = \frac{8}{3}k'$.

This concludes the proof.   □

By following a counting similar to Arikushi et al. we obtain a bound on the maximum number of edges of a 1-bend RAC graph with n vertices, when all the faces of G_0 are good. Since planar graphs have at most $3n - 6$ edges even in the presence of non-homotopic parallel edges, the bound is obtained when $7n - 14 - k = 3n - 6 - k + \frac{8}{3}k$, that is, $k = \frac{3}{2}(n - 2)$. This directly implies that in this case $|E| \leq 5.5n - 11$.

In the following, we prove that it is not a loss of generality to assume that all faces of G_0 are good, as otherwise we can augment our graph by adding only crossing-free edges to G (not necessarily drawn with one bend but rather as curves), in such a way that every face of G_0 becomes good. Recall that we denote by G' the planarization of drawing Γ of G.

Assume that there exists a face of G_0 that is not good. Hence, there exist at least two edges belonging to G_0 which are incident to the same face f' in G'. If f' consists exclusively of edges of G_0, then we triangulate f'. Otherwise, we traverse the facial walk of f' starting from any dummy vertex of f' and we connect by a crossing-free edge the first occurring vertex that is incident to an edge of G_0 with the last occurring vertex that is also incident to an edge of G_0. This implies that one of the two faces into which f' is split contains only one crossing-free edge, namely the newly added edge. Note that, in both cases, it is always possible to add the described edges, since we do not require them to be drawn with one bend. Since in both cases, we split a face into smaller faces, this process eventually terminates. At the end, each face is either a triangle of crossing-free edges or contains at most one crossing-free edge. Hence, it is indeed not a loss of generality to assume that all faces of G_0 are good.

We remark that the aforementioned procedure may result in parallel edges or self-loops, which are however non-homotopic by construction. In particular, a self-loop may appear, when the first and the last occurring vertices in the facial walk are identified and form a cut-vertex of G. Note that while Lemma 2 allows non-homotopic parallel edges, it does not allow self-loops. Hence, for self-loops we need to use a different approach. Consider self-loop s. As already mentioned, s is incident to a cut-vertex of G and encloses a part of Γ, which we assume not to contain any other self-loop. Let H_1 and H_2 be the subgraphs of G that are induced by the vertices of G that are in the interior and the exterior of s, respectively. Denote by n_1 and n_2 the number of vertices of H_1 and H_2, respectively, and by m_1 and m_2 their corresponding number of edges. Observe that $n = n_1 + n_2 - 1$. Note that edge s is accounted neither in H_1 nor in H_2. By induction, we may assume that $m_1 \leq 5.5n_1 - 11$ and $m_2 \leq 5.5n_2 - 11$. Hence, graph G (including s) contains at most $5.5(n_1 + n_2) - 22 + 1 = 5.5n - 15.5 \leq 5.5n - 11$ edges. This implies that the upper bound holds even in the presence of self-loops.

We are now ready to state the main theorem of this section.

Theorem 1. *Every n-vertex 1-bend RAC graph has at most $5.5n - 11$ edges.*

4   An Improved Lower Bound

In this section, we present an improved lower bound for the number of edges of 1-bend RAC graphs. Our construction is partially inspired by the corresponding lower bound constructions of 2-planar graphs [14] and fan-planar graphs [30] with maximum density.

Theorem 2. *There exists infinitely many n-vertex 1-bend RAC graphs with exactly $5n - 10$ edges.*

Proof. A central ingredient in our lower bound construction is the dodecahedral graph; see Fig. 4a. This graph admits a straight-line planar drawing in which the outer face is a regular pentagon, and the inner faces can be partitioned into three sets, based on their shape. Namely, the innermost face (shaded in gray in Fig. 4a) is again a regular pentagon, vertically mirrored with respect to the outer one; also, all the faces adjacent to the innermost face have the same shape, which we will describe more precisely later, and the same holds for all the faces adjacent to the outer face. In particular, the drawing of each face is symmetric with respect to the line that is perpendicular to one of its sides (whose length is denoted by a in Fig. 4b) and passes through its opposite vertex (denoted by A in Fig. 4b). Adopting the notation scheme of Fig. 4b, in the following we provide values for the angles and side length ratios to fully describe the shapes of the faces adjacent to the innermost face and to the outer face; for an illustration, refer to Fig. 4a.

(i) The five faces adjacent to the innermost face are realized such that the side of length a is incident to the inner face. Angles α and β are 88° and 100°, respectively. In addition, side-length b is 1.5 times the side-length a.

(ii) The five faces adjacent to the outer face are realized such that the side of length a is incident to the outer face. Angles α and γ are 160° and 54°, respectively. In addition, side-length b is 8.5 times side-length c.

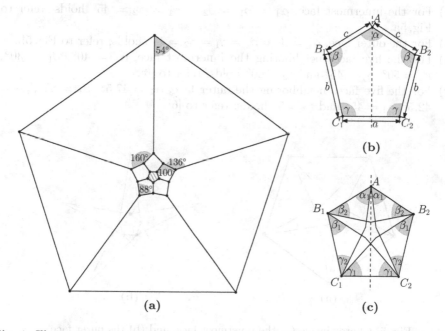

Fig. 4. Illustrations for the lower bound construction: (a) the dodecahedral graph, (b) angles and edge lengths, and (c) crossing configuration.

Consider two copies D_1 and D_2 of this drawing of the dodecahedral graph. Since both the innermost face of D_1 and the outer face of D_2 are drawn as regular pentagons, after scaling the drawing D_2 uniformly and mirroring it vertically, we can construct a drawing of a larger graph by identifying the innermost face of D_1 with the outer face of D_2. This process can be clearly repeated arbitrarily many times. The result is a graph family such that every member of this family admits a straight-line planar drawing, in which each face has one of the shapes described above.

For our lower bound construction, we add five chords in the interior of each face of every member of the above family. Hence, the five vertices that are incident to each face induce a complete graph K_5. In the following, we describe how to draw such chords in the interior of each of the aforementioned faces, based on their shape, so that the resulting drawing is 1-bend RAC. For an illustration of the configuration of the crossing edges in each of these faces refer to Fig. 4c; we will formally define angles $\alpha_1, \beta_1, \beta_2, \gamma_1, \gamma_2$ shortly. Observe that all edges and the formed angles are symmetric with respect to the line through vertex A that is perpendicular to $C_1 C_2$. Also, for every three vertices u, w, and v that are consecutive along the boundary of the face, the chord (u, v) will cross both chords incident to w, making a bend between these two crossings. In the following, we provide values for the angles $\alpha_1, \beta_1, \beta_2, \gamma_1, \gamma_2$ to fully describe the configurations of the crossing edges.

(i) For the innermost face, $\alpha_1 = \beta_1 = \beta_2 = \gamma_1 = \gamma_2 = 45°$ holds; refer to Fig. 5a.
(ii) For the outer face, $\alpha_1 = \beta_1 = \beta_2 = \gamma_1 = \gamma_2 = 45°$ holds; refer to Fig. 5b.
(iii) For the five faces neighboring the innermost face, $\alpha_1 = 40°$, $\beta_1 = 30°$, $\beta_2 = 50°$, $\gamma_1 = 45°$ and $\gamma_2 = 60°$ holds; refer to [6].
(iv) For the five faces neighboring the outer face, $\alpha_1 = 47.5°$, $\beta_1 = 85°$, $\beta_2 = 42.5°$, $\gamma_1 = 45°$ and $\gamma_2 = 5°$ holds; refer to [6].

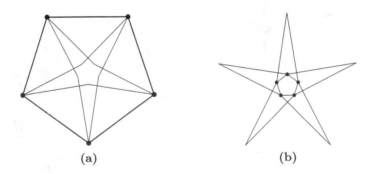

(a)   (b)

Fig. 5. Chords inside (a) the innermost face, and (b) the outer face.

It follows that each graph in the family admits a 1-bend RAC drawing. Let G_n be such a graph with n vertices. Next, we discuss the exact number of edges of graph G_n. Since the crossing-free edges of G_n form a planar graph, whose faces are all of length 5, it follows by Euler's formula that this graph has $\frac{5}{3}(n-2)$ edges and $\frac{2}{3}(n-2)$ faces. Since each of these faces contains five chords, the number of edges of G_n is $\frac{5}{3}(n-2) + 5 \cdot \frac{2}{3}(n-2) = 5n - 10$, and the statement follows. $\square$

5   Conclusions

In this paper, we improved the previously best lower and upper bounds on the number of edges of 1-bend RAC graphs. The gap between our lower and upper bound is approximately $n/2$. A future challenge will be to further narrow this gap. We conjecture that an n-vertex 1-bend RAC graph cannot have more than $5n - 10$ edges (as it is the case for several other classes of beyond planar graphs; see e.g. [11,30,33]). Significantly more difficult seems to be the problem of improving the current best lower and upper bounds on the number of edges of 2-bend RAC graphs, where the gap is significantly wider (approx., $67n$). Closely connected are also complexity related questions; in particular, the characterization and recognition of 1- and 2-bend RAC graphs are still open.

Acknowlegdment. This project was supported by DFG grant KA812/18-1.

References

1. Ackerman, E.: On the maximum number of edges in topological graphs with no four pairwise crossing edges. Discrete Comput. Geom. **41**(3), 365–375 (2009)
2. Ackerman, E.: On topological graphs with at most four crossings per edge. CoRR abs/1509.01932 (2015)
3. Ackerman, E., Keszegh, B., Vizer, M.: On the size of planarly connected crossing graphs. J. Graph Algorithms Appl. **22**(1), 11–22 (2018)
4. Ackerman, E., Tardos, G.: On the maximum number of edges in quasi-planar graphs. J. Comb. Theory, Series A **114**(3), 563–571 (2007)
5. Agarwal, P.K., Aronov, B., Pach, J., Pollack, R., Sharir, M.: Quasi-planar graphs have a linear number of edges. Combinatorica **17**(1), 1–9 (1997)
6. Angelini, P., Bekos, M., Förster, H., Kaufmann, M.: On RACdrawings of graphs with one bend per edge. CoRR 1808.10470 (2018). http://arxiv.org/abs/1808.10470
7. Angelini, P., et al.: On the perspectives opened by right angle crossing drawings. J. Graph Algorithms Appl. **15**(1), 53–78 (2011)
8. Argyriou, E.N., Bekos, M.A., Symvonis, A.: The straight-line RAC drawing problem is NP-hard. J. Graph Algorithms Appl. **16**(2), 569–597 (2012)
9. Arikushi, K., Fulek, R., Keszegh, B., Moric, F., Tóth, C.D.: Graphs that admit right angle crossing drawings. Comput. Geom. **45**(4), 169–177 (2012)
10. Bachmaier, C., Brandenburg, F.J., Hanauer, K., Neuwirth, D., Reislhuber, J.: NIC-planar graphs. Discrete Appl. Math. **232**, 23–40 (2017)
11. Bae, S.W., et al.: Gap-planar graphs. In: Frati, F., Ma, K.-L. (eds.) GD 2017. LNCS, vol. 10692, pp. 531–545. Springer, Cham (2018). https://doi.org/10.1007/978-3-319-73915-1_41
12. Bekos, M.A., Didimo, W., Liotta, G., Mehrabi, S., Montecchiani, F.: On RAC drawings of 1-planar graphs. Theor. Comput. Sci. **689**, 48–57 (2017)
13. Bekos, M.A., Kaufmann, M., Raftopoulou, C.N.: On the density of non-simple 3-planar graphs. In: Hu, Y., Nöllenburg, M. (eds.) GD 2016. LNCS, vol. 9801, pp. 344–356. Springer, Cham (2016). https://doi.org/10.1007/978-3-319-50106-2_27
14. Bekos, M.A., Kaufmann, M., Raftopoulou, C.N.: On optimal 2- and 3-planar graphs. In: Aronov, B., Katz, M.J. (eds.) Symposium on Computational Geometry. LIPIcs, vol. 77, pp. 16:1–16:16, Schloss Dagstuhl - Leibniz-Zentrum fuer Informatik (2017)

15. Brandenburg, F.J., Didimo, W., Evans, W.S., Kindermann, P., Liotta, G., Montec-chiani, F.: Recognizing and drawing IC-planar graphs. Theor. Comput. Sci. **636**, 1–16 (2016)
16. Chaplick, S., Lipp, F., Wolff, A., Zink, J.: 1-bend RAC drawings of NIC-planar graphs in quadratic area. In: Mulzer, W. (ed.) Proceedings of the 34th European Workshop on Computational Geometry (EuroCG 2018). Berlin (2018, to appear)
17. Cheong, O., Har-Peled, S., Kim, H., Kim, H.: On the number of edges of fan-crossing free graphs. Algorithmica **73**(4), 673–695 (2015)
18. Di Giacomo, E., Didimo, W., Eades, P., Liotta, G.: 2-layer right angle crossing drawings. Algorithmica **68**(4), 954–997 (2014)
19. Di Giacomo, E., Didimo, W., Liotta, G., Meijer, H.: Area, curve complexity, and crossing resolution of non-planar graph drawings. Theory Comput. Syst. **49**(3), 565–575 (2011)
20. Didimo, W., Eades, P., Liotta, G.: A characterization of complete bipartite RAC graphs. Inf. Process. Lett. **110**(16), 687–691 (2010)
21. Didimo, W., Eades, P., Liotta, G.: Drawing graphs with right angle crossings. Theor. Comput. Sci. **412**(39), 5156–5166 (2011)
22. Didimo, W., Liotta, G., Montecchiani, F.: A survey on graph drawing beyond planarity. CoRR abs/1804.07257 (2018)
23. Eades, P., Liotta, G.: Right angle crossing graphs and 1-planarity. Discrete Appl. Math. **161**(7–8), 961–969 (2013)
24. Fox, J., Pach, J., Suk, A.: The number of edges in k-quasi-planar graphs. SIAM J. Discrete Math. **27**(1), 550–561 (2013)
25. Hong, S.-H., Nagamochi, H.: Testing full outer-2-planarity in linear time. In: Mayr, E.W. (ed.) WG 2015. LNCS, vol. 9224, pp. 406–421. Springer, Heidelberg (2016). https://doi.org/10.1007/978-3-662-53174-7_29
26. Hong, S., Tokuyama, T.: Algorithmics for beyond planar graphs. NII Shonan Meeting Seminar 089, 27 November–1 December 2016
27. Huang, W.: Using eye tracking to investigate graph layout effects. In: Hong, S., Ma, K. (eds.) APVIS, pp. 97–100. IEEE Computer Society (2007)
28. Huang, W., Eades, P., Hong, S.: Larger crossing angles make graphs easier to read. J. Vis. Lang. Comput. **25**(4), 452–465 (2014)
29. Kaufmann, M., Kobourov, S., Pach, J., Hong, S.: Beyond planargraphs: Algorithmics and combinatorics. Dagstuhl Seminar 16452, 6-11 November 2016
30. Kaufmann, M., Ueckerdt, T.: The density of fan-planar graphs. CoRR abs/1403.6184 (2014)
31. Liotta, G.: Graph drawing beyond planarity: Some results and open problems. SoCG Week, Invited talk (4 July 2017)
32. Pach, J., Radoičić, R., Tardos, G., Tóth, G.: Improving the crossing lemma by finding more crossings in sparse graphs. Discrete Comput. Geom. **36**(4), 527–552 (2006)
33. Pach, J., Tóth, G.: Graphs drawn with few crossings per edge. Combinatorica **17**(3), 427–439 (1997)
34. Purchase, H.C.: Effective information visualisation: a study of graph drawing aesthetics and algorithms. Interact. Comput. **13**(2), 147–162 (2000)
35. Purchase, H.C., Carrington, D.A., Allder, J.: Empirical evaluation of aesthetics-based graph layout. Empir. Softw. Eng. **7**(3), 233–255 (2002)
36. Ringel, G.: Ein Sechsfarbenproblem auf der Kugel. Abh. Math. Sem. Univ. Hamb. **29**, 107–117 (1965)

Compact Drawings of 1-Planar Graphs
with Right-Angle Crossings
and Few Bends

Steven Chaplick[ID], Fabian Lipp[ID], Alexander Wolff[ID], and Johannes Zink[✉]

Lehrstuhl für Informatik I, Universität Würzburg, Würzburg, Germany
{steven.chaplick,fabian.lipp,alexander.wolff,
johannes.zink}@uni-wuerzburg.de
http://www1.informatik.uni-wuerzburg.de/en/staff

Abstract. We study the following classes of beyond-planar graphs: 1-planar, IC-planar, and NIC-planar graphs. These are the graphs that admit a 1-planar, IC-planar, and NIC-planar drawing, respectively. A drawing of a graph is *1-planar* if every edge is crossed at most once. A 1-planar drawing is *IC-planar* if no two pairs of crossing edges share a vertex. A 1-planar drawing is *NIC-planar* if no two pairs of crossing edges share two vertices.

We study the relations of these beyond-planar graph classes to *right-angle crossing* (*RAC*) graphs that admit compact drawings on the grid with few bends. We present four drawing algorithms that preserve the given embeddings. First, we show that every n-vertex NIC-planar graph admits a NIC-planar RAC drawing with at most one bend per edge on a grid of size $\mathcal{O}(n) \times \mathcal{O}(n)$. Then, we show that every n-vertex 1-planar graph admits a 1-planar RAC drawing with at most two bends per edge on a grid of size $\mathcal{O}(n^3) \times \mathcal{O}(n^3)$. Finally, we make two known algorithms embedding-preserving; for drawing 1-planar RAC graphs with at most one bend per edge and for drawing IC-planar RAC graphs straight-line.

1 Introduction

In graph theory and graph drawing, *beyond-planar* graph classes have experienced increasing interest in recent years. A prominent example is the class of *1-planar graphs*, that is, graphs that admit a drawing where each edge is crossed at most once. The 1-planar graphs were introduced by Ringel [18] in 1965; Kobourov et al. [15] surveyed them recently. Another example that has received considerable attention are RAC_k *graphs*, that is, graphs that admit a poly-line drawing where all crossings are at right angles and each edge has at most k bends. The RAC_k graphs were introduced by Didimo et al. [7]. Using right-angle crossings and few bends is motivated by several cognitive studies suggesting a positive correlation between large crossing angles or small curve complexity and the readability of a graph drawing [13,14,17].

The full version of this paper is available on arXiv [4] and the appendices are given therein.

© Springer Nature Switzerland AG 2018
T. Biedl and A. Kerren (Eds.): GD 2018, LNCS 11282, pp. 137–151, 2018.
https://doi.org/10.1007/978-3-030-04414-5_10

We investigate the relationships between (certain subclasses of) 1-planar graphs and RAC_k graphs that admit drawings on a polynomial-size grid. The prior work and our contributions are summarized in Fig. 2. A broader overview of beyond-planar graph classes is given in a recent survey by Didimo et al. [8].

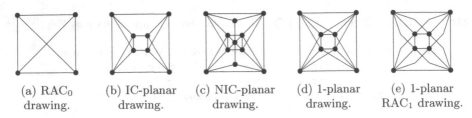

(a) RAC_0 drawing.   (b) IC-planar drawing.   (c) NIC-planar drawing.   (d) 1-planar drawing.   (e) 1-planar RAC_1 drawing.

Fig. 1. Examples of different types of drawings. Figure 1d and e show drawings of the same graph. Figure 1e is taken from the Annotated Bibliography on 1-Planarity [15].

Basic Terminology. A mapping Γ is called a *drawing* of the graph $G = (V, E)$ if each vertex $v \in V$ is mapped to a point in $\mathbb{R}^2$ and each edge uv is mapped to a simple open Jordan curve in $\mathbb{R}^2$ such that the endpoints of this curve are $\Gamma(u)$ and $\Gamma(v)$. For convenience, we will refer to the points and simple open Jordan curves of a drawing as vertices and edges. The topologically connected regions of $\mathbb{R}^2 \setminus \Gamma$ are the *faces* of Γ. The unbounded face of Γ is its *outer* face; the other faces are *inner* faces. Each face defines a circular list of bounding edges (resp. edge sides), which we call its *boundary list*. Two drawings of a graph G are *equivalent* when they have the same set of boundary lists for their inner faces and outer faces. Each equivalence class of drawings of G is an *embedding*. A *k-bend (poly-line) drawing* is a drawing in which every edge is drawn as a connected sequence of at most $k + 1$ line segments. The (up to) k inner vertices of an edge connecting these line segments are called *bend points* or *bends*. A 0-bend drawing is more commonly referred to as a *straight-line* drawing. A *drawing on the grid of size $w \times h$* is a drawing where every vertex, bend point, and crossing point has integer coordinates in the range $[0, w] \times [0, h]$. In any drawing we require that vertices, bends, and crossings are pairwise distinct points. A drawing is *1-planar* if every edge is crossed at most once. A 1-planar drawing is *independent-crossing planar* (*IC-planar*) if no two pairs of crossing edges share a vertex. A 1-planar drawing is *near-independent-crossing planar* (*NIC-planar*) if any two pairs of crossing edges share at most one vertex. A drawing is *right-angle-crossing* (*RAC*) if (*i*) it is a poly-line drawing, (*ii*) no more than two edges cross in the same point, and (*iii*) in every crossing point the edges intersect at right angles. We further specialize the notion of RAC drawings. A drawing is RAC_k if it is RAC and k-bend; it is RAC^{poly} if it is RAC and on a grid whose size is polynomial in its number of vertices. Examples for IC-planar, NIC-planar, 1-planar, and RAC drawings are given in Fig. 1. The *planar, 1-planar, NIC-planar, IC-planar*, and RAC_k graphs are the graphs that admit a crossing-free, 1-planar, NIC-planar,

IC-planar, and RAC_k drawing, respectively. More specifically, RAC_k^{poly} is the set of graphs that admit a RAC_k^{poly} drawing. A *plane, 1-plane, NIC-plane,* and *IC-plane* graph is a graph given with a specific planar, 1-planar, NIC-planar, and IC-planar embedding, respectively. In a 1-planar embedding the edge crossings are known and they are stored as if they were vertices. We will denote an embedded graph by $(G, \mathcal{E})$ where G is the graph and $\mathcal{E}$ is the embedding of this graph. For a point p in the plane, let $x(p)$ and $y(p)$ denote its x- and y-coordinate, respectively. Given two points p and q, we denote the straight-line segment connecting them by $\overline{pq}$ and its length, the Euclidean distance of p and q, by $\|\overline{pq}\|$.

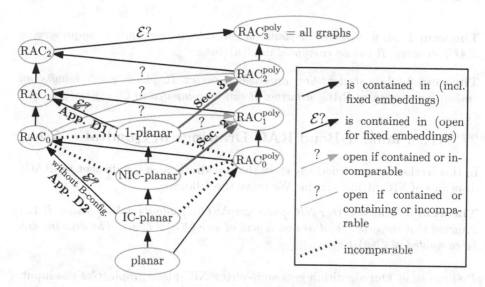

Fig. 2. Relating some classes of (beyond-)planar graphs and RAC graphs. Our main results are the containment relationships indicated by the thick blue arrows (Color figure online).

Previous Work. In the diagram in Fig. 2, we give an overview of the relationships between classes of 1-planar graphs and RAC_k graphs. Clearly, the planar graphs are a subset of the IC-planar graphs, which are a subset of the NIC-planar graphs, which are a subset of the 1-planar graphs. It is well known that every plane graph can be drawn with straight-line edges on a grid of quadratic size [10,19]. Every IC-planar graph admits an IC-planar RAC_0 drawing but not necessarily in polynomial area [3]. Moreover, there are graphs in RAC_0^{poly} that are not 1-planar [9] and, therefore, also not IC-planar. The class of RAC_0 graphs is incomparable with the classes of NIC-planar graphs [1] and 1-planar graphs [9]. Bekos et al. [2] showed that every 1-planar graph admits a 1-planar RAC_1 drawing, but their recursive drawings may need exponential area. Every graph admits a RAC_3 drawing in polynomial area, but this does not hold if a given embedding of the graph must be preserved [7].

Our Contributions. We contribute four new results; two main results and two adaptations of prior results. First, we constructively show that every NIC-plane graph admits a RAC_1 drawing in quadratic area; see Sect. 2. This improves upon a side result by Liotta and Montecchiani [16], who showed that every IC-plane graph admits a RAC_2 drawing on a grid of quadratic size. Second, we constructively show that every 1-plane graph admits a RAC_2 drawing in polynomial area; see Sect. 3. Beside these two main results, we show how to preserve a given embedding when computing RAC drawings. Precisely, we show Theorem 1 in Appendix D.1 by adapting an algorithm of Bekos et al. [2] and we show Theorem 2 in Appendix D.2 by adapting an algorithm of Brandenburg et al. [3].

Theorem 1. *Any n-vertex 1-plane graph admits an embedding-preserving RAC_1 drawing. It can be computed in $\mathcal{O}(n)$ time.*

Theorem 2. *Any straight-line drawable n-vertex IC-plane graph admits an embedding-preserving RAC_0 drawing. It can be computed in $\mathcal{O}(n^3)$ time.*

2   NIC-Planar 1-Bend RAC Drawings in Quadratic Area

In this section we constructively show that quadratic area is sufficient for RAC_1 drawings of NIC-planar graphs. We prove the following.

Theorem 3. *Any n-vertex NIC-plane graph $(G, \mathcal{E})$ admits a NIC-planar RAC_1 drawing that respects $\mathcal{E}$ and lies on a grid of size $\mathcal{O}(n) \times \mathcal{O}(n)$. The drawing can be computed in $\mathcal{O}(n)$ time.*

Preprocessing. Our algorithm gets an n-vertex NIC-plane graph $(G, \mathcal{E})$ as input. We first aim to make $(G, \mathcal{E})$ biconnected and planar so that we can draw it using the algorithm by Harel and Sardas [11]. Around each crossing in $\mathcal{E}$, we insert up to four dummy edges to obtain *empty kites*. A *kite* is a K_4 that is embedded such that (i) every vertex lies on the boundary of the outer face, and (ii) there is exactly one crossing, which does not lie on the boundary of the outer face. A kite K as a subgraph of a graph H is said to be *empty* if there is no edge of $H \backslash K$ that is on an inner face of K or crosses edges of K. Inserting a dummy edge could create a pair of parallel edges. If this happens, we subdivide the original edge participating in this pair by a dummy vertex (see the transition from Fig. 3a – b). Note that we never create parallel dummy edges since G is NIC-planar. After this, we remove both crossing edges from each empty kite and obtain *empty quadrangles* (see Fig. 3c). We store each such empty quadrangle in a list Q. At the end of the preprocessing, we make the resulting plane graph biconnected via, e.g., the algorithm of Hopcroft and Tarjan [12]. Since each empty quadrangle is contained in a biconnected component, no edges are inserted into it. Let $(G', \mathcal{E}')$ be the resulting plane biconnected graph.

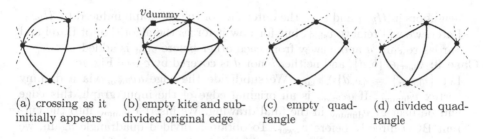

(a) crossing as it initially appears   (b) empty kite and sub-divided original edge   (c) empty quad-rangle   (d) divided quad-rangle

Fig. 3. Modifying the crossings and computing the BCO.

Drawing Step. Now, we draw a graph that we obtain from $(G', \mathcal{E}')$ by first producing a *biconnected canonical ordering (BCO)*[1]. We use the algorithm by Harel and Sardas [11], which is a generalization of the algorithm of Chrobak and Payne [5], which in turn is based on the shift algorithm of de Fraysseix et al. [10]. The algorithm of Harel and Sardas consists of two phases. Given a plane biconnected graph H, in the first phase a BCO Π of the vertices in H is computed. In the second phase, H is drawn according to Π on a grid of size $(2|V(H)| - 4) \times (|V(H)| - 2)$. Unlike the classical shift algorithm, the algorithm of Harel and Sardas computes the (biconnected) canonical ordering *bottom-up*, which we will exploit here. Let $\Pi_k = (v_1, \ldots, v_k)$ be a partial BCO of H after step k, and let H_k be the plane subgraph of H induced by Π_k. We say that a vertex u is *covered* by v_k if u is on the boundary of the outer face of H_{k-1}, but not on that of H_k.

We perform the following additional operations when we compute the BCO $\hat{\Pi}$. Whenever we reach an empty quadrangle $q = (a, b, c, d)$ of the list Q for the first time, i.e., when the first vertex of q—say a—is added to the BCO, we insert an edge inside q from a to the vertex opposite a in q, that is, to c. We call the resulting structure a *divided quadrangle* (see Fig. 3d). In two special cases, we perform further modifications of the graph. They will help us to guarantee a correct reinsertion of the crossing edges in the next step of the algorithm. Namely, when we encounter the last vertex $v_{\text{last}} \in \{b, c, d\}$ of q, we distinguish three cases.

Case 1: $v_{\text{last}} = c$ (see Fig. 4a). Here, no operations are performed.

Case 2: $v_{\text{last}} \in \{b, d\}$, and the other of $\{b, d\}$ is covered by c (see Fig. 4b).

We insert a dummy vertex v_{shift}, which we call *shift vertex*, into the current BCO directly before v_{last} and make it adjacent to a and c. Observe that, if v_{shift} is the k-th vertex in $\hat{\Pi}$, this still yields a valid BCO since v_{shift} has two

---

[1] BCOs are a generalization of canonical orderings that assume only biconnectivity (instead of triconnectivity). In a BCO of a plane graph H, the subgraph H_k of H induced by $v_1, \ldots, v_k$ is connected, the edge $v_1 v_2$ lies on the boundary of the outer face and all vertices in $H - H_k$ lie within the outer face of H_k. For $k > 2$, the vertex v_k has one or more neighbors in H_{k-1}. If v_k has exactly one neighbor u in H_{k-1}, then it has a legal support on the outer face of H_{k-1}, i.e., in the circular order of adjacent vertices around u, it follows or precedes a vertex in H_{k-1}.

neighbors in $\hat{\Pi}_{k-1}$ and is on the outer face of the subgraph induced by $\hat{\Pi}_{k-1}$. Later, we will remove v_{shift}, but for now it forces the algorithm of Harel and Sardas to shift a and c away from each other before v_{last} is added.

Case 3: $v_{last} \in \{b, d\}$, and neither b nor d is covered by c (see Fig. 4c).

Let $\{v_{lower}\} = \{b, d\} \setminus v_{last}$. We subdivide the edge av_{lower} via a dummy vertex v_{dummy}. If av_{lower} is an original edge of the input graph, this edge will be bent at v_{dummy} in the final drawing. We insert v_{dummy} into the current BCO directly before v_{lower}. To obtain a divided quadrangle again, we insert the dummy edge av_{lower}, which we will remove before we reinsert the crossing edges. This will give us some extra space inside the triangle $(a, v_{dummy}, v_{lower})$ for a bend point. Inserting v_{dummy} as k-th vertex into $\hat{\Pi}$ keeps $\hat{\Pi}$ valid since v_{dummy} uses the support edge incident to a that would have been covered by v_{lower} otherwise. Then, v_{lower} has at least two neighbors in $\hat{\Pi}_k$, namely a and v_{dummy}.

We draw the resulting plane biconnected $\hat{n}$-vertex graph $(\hat{G}, \hat{\mathcal{E}})$ according to its BCO $\hat{\Pi}$ via the algorithm by Harel and Sardas and obtain a crossing-free drawing $\hat{\Gamma}$. We do not modify the actual drawing phase.

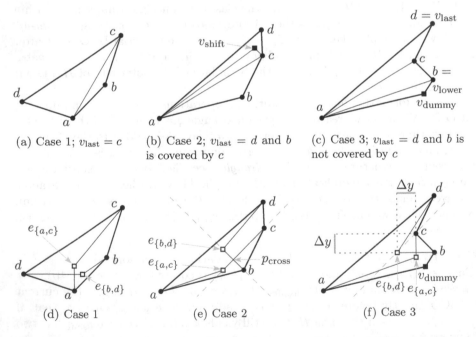

(a) Case 1; $v_{last} = c$

(b) Case 2; $v_{last} = d$ and b is covered by c

(c) Case 3; $v_{last} = d$ and b is not covered by c

(d) Case 1

(e) Case 2

(f) Case 3

Fig. 4. Divided quadrangles produced in the three cases of the drawing step (a)–(c) and the crossing edges after the reinsertion step (d)–(f) in our algorithm. For orientation, lines with slope 1 or −1 are dashed violet. (Color figure online)

Postprocessing (Reinserting the Crossing Edges). We refine the underlying grid of $\hat{\Gamma}$ by a factor of 2 in both dimensions. Let $q = (a, b, c, d)$ be a quadrangle in Q, where a is the first and v_{last} the last vertex in $\hat{\Pi}$ among the vertices in q. From q, we first remove the chord edge ac and obtain an empty quadrangle. Then, we distinguish three cases for reinserting the crossing edges that we removed in the preprocessing. These are the same cases as in the description of the modified computation of the BCO before. In this case distinction we omit some lengthy but straight-forward calculations; see Zink's master's thesis [21] for the details.

Case 1: $v_{last} = c$ (see Fig. 4a).

Since c is adjacent to a, b, and d in $\hat{G}$, it has the largest y-coordinate among the vertices in q. Assume that $y(d)$ is smaller or equal to $y(b)$ since the other case is symmetric. An example of a quadrangle in this case before and after the reinsertion of the crossing edges is given in Fig. 4a and d, respectively. We will have a crossing point at $(x(a), y(d))$. To this end, we insert the edge ac with a bend at $e_{ac} = (x(a), y(d)+1)$ and we insert the edge bd with a bend at $e_{bd} = (x(a)+1, y(d))$. Clearly the crossing is at a right angle. Observe that q is convex since c is the last drawn vertex of q and c is adjacent to b, a, and d in this circular order in the embedding and observe that both bend points lie inside q. Therefore, it follows that both crossing edges lie completely inside q.

Case 2: $v_{last} \in \{b, d\}$, and the other of $\{b, d\}$ is covered by c (see Fig. 4b).

Assume that $y(d) > y(b)$; the other case is symmetric. An example of a quadrangle in this case before and after the reinsertion of the crossing edges is given in Fig. 4b and e, respectively. We remove v_{shift} in addition to removing the edge ac. We define the crossing point $p_{cross} = (x_{cross}, y_{cross})$ as the intersection point of the lines with slope 1 and -1 through c and b, respectively. The coordinates of this crossing point are $x_{cross} = (x(c) - y(c) + x(b) + y(b))/2$ and $y_{cross} = (-x(c) + y(c) + x(b) + y(b))/2$. Since we refined the grid by a factor of 2 in each dimension, the above coordinates are both integers. We place the two bend points onto the same lines at the closest grid points that are next to p_{cross}, i.e., we draw the edge ac with a bend point at $e_{ac} = (x_{cross} - 1, y_{cross} - 1)$ and we insert the edge bd with a bend point at $e_{bd} = (x_{cross} - 1, y_{cross} + 1)$. We do not intersect or touch the edge ad because we shifted a far enough away from c by the extra shift due to v_{shift}. Moreover, the points e_{ac} and p_{cross} on the line with slope 1 through c are inside the empty quadrangle q since b is covered by c (then b is below the line with slope 1 through c) and $y(b)$ is at most equal to $y(e_{ac})$.

Case 3: $v_{last} \in \{b, d\}$, and neither b nor d is covered by c (see Fig. 4c).

Assume that $y(d) > y(b)$; again, the other case is symmetric. An example of a quadrangle in this case before and after the reinsertion of the crossing edges is given in Figs. 4c and f, respectively. Note that the edge ab is a dummy edge, which we inserted during the computation of $\hat{\Pi}$, and next to this edge, there is the path $av_{dummy}b$. This path is the former edge ab. We will reinsert the edges ac and bd such that they cross in $(x(c), y(b))$. We will bend the edge bd on the line with slope 1 through c at $y = y(b)$ because from this point we always "see" d inside q. So, we define $x_{bend} := x(c) - \Delta y$ with $\Delta y := y(c) - y(b)$. First, we remove the dummy edge ab. Second, we insert the edge ac with a bend point at $e_{ac} = (x(c), y(b) - 1)$. Third, we insert the edge bd with a bend point at $e_{bd} = (x_{\text{bend}}, y(b))$. Note that e_{ac} might be below the straight-line segment $\overline{ab}$ since a could have been shifted far away from c. However, e_{ac} cannot be on or below the path $av_{\text{dummy}}b$ because $y(v_{\text{dummy}}) < y(e_{ac})$ and the slope of the line segment $\overline{v_{\text{dummy}}b}$ is either greater than 1 or negative. Therefore, the crossing edges ac and bd lie completely inside the pentagonal face $(a, v_{\text{dummy}}, b, c, d)$.

Result. After we have reinserted the crossing edges into each quadrangle of Q, we remove all dummy edges and transform the remaining dummy vertices to bend points. The resulting drawing Γ is a RAC_1 drawing that preserves the embedding of the NIC-plane input graph $(G, \mathcal{E})$. In Appendix A (p. 15), we bound the size of the grid that our drawings need, as follows.

Lemma 4. *Every vertex, bend point, and crossing point of the drawing returned by our algorithm lies on a grid of size at most $(16n - 32) \times (8n - 16)$.*

The shift algorithm of Harel and Sardas runs in linear time [11]. Also, our additional operations can be performed in linear time [21]. This proves Theorem 3. We give a full example of a NIC-plane RAC_1 drawing generated by a Java implementation of our algorithm in Figs. 9 and 10 in Appendix B.

3   1-Planar 2-Bend RAC Drawings in Polynomial Area

In this section we constructively prove the following.

Theorem 5. *Any n-vertex 1-plane graph $(G, \mathcal{E})$ admits a 1-planar RAC_2 drawing that respects $\mathcal{E}$ and lies on a grid of size $\mathcal{O}(n^3) \times \mathcal{O}(n^3)$. The drawing can be computed in $\mathcal{O}(n)$ time.*

The idea of our algorithm is to draw a slightly modified, planarized version of the 1-plane input graph with a variant of the shift algorithm (by Harel and Sardas [11]) and then "manually" redraw the crossing edges so that they cross at right angles and have at most two bends each. The difficulty is to find grid points for the bend points and the crossings so that the redrawn edges do not touch or cross the surrounding edges drawn by the shift algorithm. To this end, we refine our grid and place the middle part of each crossing edge onto a horizontal or vertical grid line so that the edge crossings are at right angles.

Preprocessing. Our algorithm gets an n-vertex 1-plane graph $(G, \mathcal{E})$ as input. First, we planarize G by replacing each crossing point by a vertex (see Fig. 5a). We will refer to them as *crossing vertices*. Second, we enclose each crossing vertex by a *subdivided kite*, which is an empty kite where the four boundary edges are subdivided by a vertex (see Fig. 5b). We use subdivided kites instead of empty kites to maintain the embedding and to avoid adding parallel edges. Third, we make the graph biconnected using, e.g., the algorithm of Hopcroft and Tarjan [12]. Note that we do not insert edges into inner faces of subdivided kites because all vertices and edges of a subdivided kite are in the same biconnected

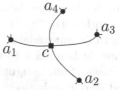
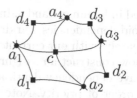
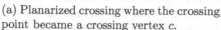

(a) Planarized crossing where the crossing   (b) Enclosing the crossing vertex c by a
point became a crossing vertex c.   subdivided kite.

Fig. 5. A crossing point is replaced by a crossing vertex c and we insert four 2-paths of two dummy edges and a dummy vertex to induce a subdivided kite at each crossing. The vertices d_1, d_2, d_3, and d_4 are the dummy vertices of these 2-paths.

component. After these three steps, we have a biconnected plane graph $(G', \mathcal{E}')$. We draw $(G', \mathcal{E}')$ using the algorithm of Harel and Sardas [11]. This algorithm returns a crossing-free straight-line drawing Γ' of $(G', \mathcal{E}')$, whose vertices lie on a grid of size $(2n' - 4) \times (n' - 2)$, where n' is the number of vertices of G'.

Assignment of Edges to Axis-Parallel Half-Lines. For each crossing vertex c there are four incident edges in G'. They correspond to two edges of G. Consider the circular order around c in $(G', \mathcal{E}')$. The first and the third edge incident to c correspond to one edge in $(G, \mathcal{E})$; symmetrically, the second and fourth incident edge correspond to one edge. To obtain a RAC drawing from this, we redraw each of the four edges around c. Consider an edge ac from a vertex a of the subdivided kite to the crossing vertex c. This edge is then redrawn with a bend point b that lies on an axis-parallel line through c. For an example how a crossing in Γ' is replaced by a RAC crossing, see the transition from Fig. 8a to f. In order to obtain a right-angle crossing, we bijectively assign the four incident edges to the four axis-parallel half-lines originating in c. We call such a mapping an *assignment*. We do not take an arbitrary assignment, but take care to avoid extra crossings with edges that are redrawn or previously drawn. We call an assignment A *valid* if there is a way to redraw each edge e with one bend so that the bend point of e lies on the half-line $A(e)$ and the resulting drawing is plane.

To ensure that our valid assignment can be realized on a small grid, we introduce further criteria. We say that an edge e_1 *depends* on another edge e_2 with respect to an assignment A if e_2 lies in the angular sector between e_1 and the half-line $A(e_1)$. In Fig. 6a, for example, the edge e_3 depends on e_4 and e_2 depends on e_1, but e_1 and e_4 do not depend on any edge. We call edges (such as e_1 and e_4) that do not depend on other edges *independent*. We define the *dependency depth* of an assignment to be the largest integer k with $0 \leq k \leq 3$ such that there is a chain of $k+1$ edges $e_1, e_2, \ldots, e_{k+1}$ incident to c such that e_1 depends on e_2 and $\ldots$ and e_k depends on e_{k+1}, but there is no such chain of $k + 2$ edges. For example, in Fig. 6a, b, and c, the assignment has a dependency depth of 1, whereas in Fig. 6d, the assignment has a dependency depth of 0. Showing that there is a valid assignment of dependency depth at most 1 will imply the existence of an appropriate set of grid points for the bend points as formalized in Lemmas 7 and 8. In fact, as we will see in the discussion below, if we could avoid dependencies, our drawing would fit on a grid of size $\mathcal{O}(n^2) \times \mathcal{O}(n^2)$. Unfortunately, with our current approach this seems to be unavoidable.

We now construct an assignment that we will show in Lemma 6 to be valid and to have dependency depth at most 1. The four cases of our assignment are given in order of priority. Note that, in Cases 1 and 2, our assignment always contains dependencies; see Fig. 6a and b. Note further that it is enough to specify the assignment of one edge; the remaining assignment is determined since the circular orders of the edges and the assigned half-lines must be the same.

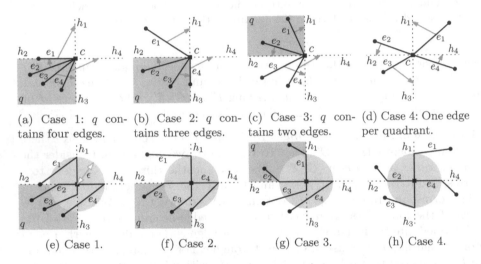

(a) Case 1: q contains four edges. (b) Case 2: q contains three edges. (c) Case 3: q contains two edges. (d) Case 4: One edge per quadrant.

(e) Case 1.　(f) Case 2.　(g) Case 3.　(h) Case 4.

Fig. 6. The four cases of our assignment procedure: (a)–(d) indicate the assignment with orange arrows and show that the dependency depth is always at most 1, (e)–(f) show that the assignment is valid; the radius of the light blue disk is ϵ.

Case 1: There is a quadrant q that contains all four incident edges; see Fig. 6a. Take the two "inner" edges in q and assign them to the two half-lines that bound q, while keeping the circular order.

Case 2: There is a quadrant q that contains three incident edges; see Fig. 6b. Consider the edge outside q, say e_1, and assign it to the closest half-line h_i that does not bound q.

Case 3: There is a quadrant q that contains two incident edges; see Fig. 6c. Assign the incident edges in q to their closest half-lines.

Case 4: Each quadrant contains exactly one incident edge; see Fig. 6d. Assign each edge to its closest half-line in counter-clockwise direction.

See also Appendix C, where we prove the following lemma on p. 16.

Lemma 6. *Our assignment procedure returns a valid assignment with dependency depth at most 1.*

Note that Lemma 6 already gives us a RAC_2 drawing of the input graph, but in order to get a (good) bound on the grid size of the drawing, we have to place the bend points on a grid that is as coarse as possible, but still fine enough to provide us with grid points where we need them: on the half-lines emanating from the crossing vertices. This is what the remainder of this section is about.

(a) available polygon   (b) triangle for valid edge placement given points p and q

Fig. 7. Example of an available polygon in which we determine the points p and q and with them the triangle for valid edge placement and the line segment $\overline{qc}$.

Placement of Bend Points on the Grid. In Γ', we have a drawing of a subdivided kite for every crossing in the 1-plane input graph. It is an octagon with a central crossing vertex c of degree four in its interior. For an example, see Fig. 8a. We will redraw the straight-line edges between c and its four adjacent vertices as 1-bend edges according to the assignment A computed in the previous step. The segment of such a 1-bend edge ac that ends at c will lie on the axis-parallel half-line $A(ac)$. If we pair and concatenate the 1-bend edges that enter c from opposite sides, we obtain two 2-bend edges and a right-angle crossing in c; see Fig. 8f. It remains to show how the bend points for the edges are placed on the grid. We proceed as follows.

First, we determine for each edge ac incident to a crossing vertex c the available region into which we can redraw ac with a bend b on $A(ac)$. The region between $\overline{ac}$ and the half-line $A(ac)$ inside the subdivided kite defines an *available polygon*. Examples of such an available polygon are given in Figs. 7a and 8b. Note that the available polygons might overlap (as they do once in Fig. 8b). Observe that there is only a triangle inside each available polygon in which the new line segment $\overline{ab}$ can be placed. Such a *triangle for valid edge placement* is determined by a, c and a corner point p of the available polygon. The point p is the corner point (excluding a and c) for which the angle between $\overline{ac}$ and $\overline{ap}$ inside the available polygon is the smallest. These triangles for valid edge placement are depicted in Figs. 7b and 8c. Again, they might overlap. Observe that in such a triangle, the angle at a cannot become arbitrarily small because every determining point lies on a grid point. Let q be the intersection point of the line through $\overline{ap}$ and the half-line $A(ac)$. One can see q as the projection of p onto $A(ac)$ seen from a. Note that we have a degenerated case if $a \in A(ac)$. Then, the available polygon has no area and equals the line segment $\overline{ac}$. In this case let $a = p = q$. Moreover, note that p can be equal to q because the intersection of

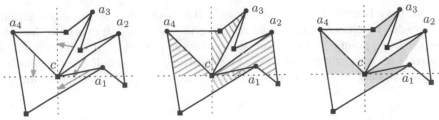

(a) A subdivided kite. The assignment of edges to half-lines is indicated by arrows.

(b) Available polygons for each pair of edge and assigned half-line.

(c) Triangles for valid edge placement.

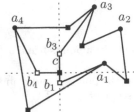

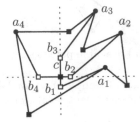

(d) After the insertion of the bend points of the three independent edges.

(e) Available polygon and triangle for valid edge placement for the edge a_2c which depends on a_1c.

(f) Result after the insertion of the bend point b_2.

Fig. 8. Transformation from a planarized crossing to a RAC_2 crossing.

$A(ac)$ and an edge of the subdivided kite is also a corner point of the available polygon. This is the only case where p may not be a grid point.

We will place the bend point b onto the line segment $\overline{qc}$, but observe that the triangles for valid edge placement of two edges e_1 and e_2 might overlap if e_1 depends on e_2 in A. To solve this, we first draw the independent edges, then recompute the available polygons and the triangles for valid edge placement for the other edges, and finally draw those edges. Remember that our assignment procedure returns only assignments with dependency depth at most 1. Let Γ' be drawn on a grid of size $\tilde{n} \times \tilde{n}$. We refine the grid by a factor of $\tilde{n}$ in each dimension. The next step in our algorithm relies on the following lemma (which we prove in Appendix C, p. 19).

An important tool in our analysis will be the so-called *Farey sequence* [20] of order $\tilde{n} - 1$, which is the sequence of all reduced fractions from 0 to 1 with numerator and denominator being positive integers bounded by $\tilde{n} - 1$.

Lemma 7. *For any independent edge ac, the interior of the line segment $\overline{qc}$ contains at least one grid point of the refined $\tilde{n}^2 \times \tilde{n}^2$ grid.*

Using Lemma 7, we pick for each independent edge any grid point of $\overline{qc}$, place a bend point b on it, and replace the segment $\overline{ac}$ by the two segments $\overline{ab}$ and $\overline{bc}$. In Fig. 8c, the edges a_1c, a_3c, and a_4c are independent, but a_2c depends on a_1c.

We again refine the grid by a factor of $\tilde{n}$ in each dimension. The grid size is now $\tilde{n}^3 \times \tilde{n}^3$. For the remaining edges incident to a crossing vertex c, we compute new available polygons and triangles for valid edge placement since we need to take the 1-bend edges into account that were inserted in the previous step. Now the following lemma (proved in Appendix C, p. 22) yields grid points for the bend points of the remaining edges.

Lemma 8. *After having redrawn the independent edges, the interior of the line segment $\overline{qc}$ of each edge depending on an independent edge contains at least one grid point of the refined $\tilde{n}^3 \times \tilde{n}^3$ grid.*

For each remaining edge incident to a crossing vertex c we pick any grid point of its line segment $\overline{qc}$ and place a bend point b on it. Again, we replace $\overline{ac}$ by the two line segments $\overline{ab}$ and $\overline{bc}$.

Result. Finally, we remove the dummy edges and dummy vertices that bound the subdivided kites and interpret the crossing vertices as crossing points. We return the resulting RAC_2 drawing Γ. It is drawn on a grid of size $(8n'^3 - 48n'^2 + 96n' - 64) \times (4n'^3 - 24n'^2 + 48n' - 32)$, where n' is the number n of vertices of G plus 5 times the number of crossings $\mathrm{cr}(\mathcal{E})$ in $\mathcal{E}$. Note that $\mathrm{cr}(\mathcal{E}) \leq n - 2$ for 1-plane graphs [6]. If we ignore the bend points, the drawing is on a grid of size $(2n' - 4) \times (n' - 2)$, i.e., its size is quadratic. Again, the algorithm by Harel and Sardas [11] and our modification run in linear time. Therefore, we conclude the correctness of Theorem 5.

4   Conclusion and Open Questions

We have shown that any n-vertex NIC-plane graph admits a $\mathrm{RAC}_1^{\mathrm{poly}}$ drawing in $\mathcal{O}(n^2)$ area and that any n-vertex 1-plane graph admits a $\mathrm{RAC}_2^{\mathrm{poly}}$ drawing in $\mathcal{O}(n^6)$ area. We have also shown how to adjust two existing algorithms for drawing certain 1-planar graphs such that their embedding is preserved. More precisely, we have proved that any 1-plane graph admits a RAC_1 drawing. This answers an open question explicitly asked by the authors of the original algorithm [2]. We have also proved that any straight-line drawable IC-plane graph admits a RAC_0 drawing, where the original algorithm did not necessarily preserve the embedding [3]. The diagram in Fig. 2 leaves some open questions. Does any 1-planar graph admit a $\mathrm{RAC}_1^{\mathrm{poly}}$ drawing? Can we draw any graph in RAC_0 with only right-angle crossings in polynomial area when we allow one or two bends per edge? What is the relationship between RAC_1 and $\mathrm{RAC}_2^{\mathrm{poly}}$? Can we compute $\mathrm{RAC}_2^{\mathrm{poly}}$ drawings of 1-plane graphs in $o(n^6)$ area?

References

1. Bachmaier, C., Brandenburg, F.J., Hanauer, K., Neuwirth, D., Reislhuber, J.: NIC-planar graphs. Discrete Appl. Math. **232**, 23–40 (2017). https://doi.org/10.1016/j.dam.2017.08.015
2. Bekos, M.A., Didimo, W., Liotta, G., Mehrabi, S., Montecchiani, F.: On RAC drawings of 1-planar graphs. Theor. Comput. Sci. **689**, 48–57 (2017). https://doi.org/10.1016/j.tcs.2017.05.039
3. Brandenburg, F.J., Didimo, W., Evans, W.S., Kindermann, P., Liotta, G., Montecchiani, F.: Recognizing and drawing IC-planar graphs. Theor. Comput. Sci. **636**, 1–16 (2016). https://doi.org/10.1016/j.tcs.2016.04.026
4. Chaplick, S., Lipp, F., Wolff, A., Zink, J.: Compact drawings of 1-planar graphs with right-angle crossings and few bends. Arxiv report (2018). http://arxiv.org/abs/1806.10044v4
5. Chrobak, M., Payne, T.H.: A linear-time algorithm for drawing a planar graph on a grid. Inf. Process. Lett. **54**(4), 241–246 (1995). https://doi.org/10.1016/0020-0190(95)00020-D
6. Czap, J., Hudák, D.: On drawings and decompositions of 1-planar graphs. Electr. J. Comb. **20**(2), 54 (2013). http://www.combinatorics.org/ojs/index.php/eljc/article/view/v20i2p54
7. Didimo, W., Eades, P., Liotta, G.: Drawing graphs with right angle crossings. Theor. Comput. Sci. **412**(39), 5156–5166 (2011). https://doi.org/10.1016/j.tcs.2011.05.025
8. Didimo, W., Liotta, G., Montecchiani, F.: A survey on graph drawing beyond planarity. Arxiv report (2018). http://arxiv.org/abs/1804.07257
9. Eades, P., Liotta, G.: Right angle crossing graphs and 1-planarity. Discrete Appl. Math. **161**(7–8), 961–969 (2013). https://doi.org/10.1016/j.dam.2012.11.019
10. de Fraysseix, H., Pach, J., Pollack, R.: How to draw a planar graph on a grid. Combinatorica **10**(1), 41–51 (1990). https://doi.org/10.1007/BF02122694
11. Harel, D., Sardas, M.: An algorithm for straight-line drawing of planar graphs. Algorithmica **20**(2), 119–135 (1998). https://doi.org/10.1007/PL00009189
12. Hopcroft, J.E., Tarjan, R.E.: Algorithm 447: efficient algorithms for graph manipulation. Commun. ACM **16**(6), 372–378 (1973). https://doi.org/10.1145/362248.362272
13. Huang, W., Eades, P., Hong, S.: Larger crossing angles make graphs easier to read. J. Vis. Lang. Comput. **25**(4), 452–465 (2014). https://doi.org/10.1016/j.jvlc.2014.03.001
14. Huang, W., Hong, S., Eades, P.: Effects of crossing angles. In: Proceedings IEEE VGTC Pacific Visualization Symposium (PacificVis 2008), pp. 41–46 (2008). https://doi.org/10.1109/PACIFICVIS.2008.4475457
15. Kobourov, S.G., Liotta, G., Montecchiani, F.: An annotated bibliography on 1-planarity. Comput. Sci. Rev. **25**, 49–67 (2017). https://doi.org/10.1016/j.cosrev.2017.06.002
16. Liotta, G., Montecchiani, F.: L-visibility drawings of IC-planar graphs. Inf. Process. Lett. **116**(3), 217–222 (2016). https://doi.org/10.1016/j.ipl.2015.11.011
17. Purchase, H.: Which aesthetic has the greatest effect on human understanding? In: DiBattista, G. (ed.) GD 1997. LNCS, vol. 1353, pp. 248–261. Springer, Heidelberg (1997). https://doi.org/10.1007/3-540-63938-1_67
18. Ringel, G.: Ein Sechsfarbenproblem auf der Kugel. Abh. Math. Seminar Univ. Hamburg **29**(1–2), 107–117 (1965)

19. Schnyder, W.: Embedding planar graphs on the grid. In: Johnson, D.S. (ed.) Proceedings 1st ACM-SIAM Symposium Discrete Algorithms (SODA 1990), pp. 138–148 (1990). http://dl.acm.org/citation.cfm?id=320176.320191
20. Wikipedia contributors: Farey sequence—Wikipedia, the free encyclopedia (2018). https://en.wikipedia.org/w/index.php?title=Farey_sequence&oldid=844932264. Accessed 8 June 2018
21. Zink, J.: 1-planar RAC drawings with bends. Master's thesis, Institut für Informatik, Universität Würzburg (2017). http://www1.pub.informatik.uni-wuerzburg.de/pub/theses/2017-zink-master.pdf

Drawing Subcubic 1-Planar Graphs with Few Bends, Few Slopes, and Large Angles

Philipp Kindermann[1]([⊠])([iD]), Fabrizio Montecchiani[2], Lena Schlipf[3], and André Schulz[3]

[1] University of Waterloo, Waterloo, Canada
philipp.kindermann@uwaterloo.ca
[2] Università degli Studi di Perugia, Perugia, Italy
fabrizio.montecchiani@unipg.it
[3] FernUniversität in Hagen, Hagen, Germany
{lena.schlipf,andre.schulz}@fernuni-hagen.de

Abstract. We show that the 1-planar slope number of 3-connected cubic 1-planar graphs is at most 4 when edges are drawn as polygonal curves with at most 1 bend each. This bound is obtained by drawings whose vertex and crossing resolution is at least $\pi/4$. On the other hand, if the embedding is fixed, then there is a 3-connected cubic 1-planar graph that needs 3 slopes when drawn with at most 1 bend per edge. We also show that 2 slopes always suffice for 1-planar drawings of subcubic 1-planar graphs with at most 2 bends per edge. This bound is obtained with vertex resolution $\pi/2$ and the drawing is RAC (crossing resolution $\pi/2$). Finally, we prove lower bounds for the slope number of straight-line 1-planar drawings in terms of number of vertices and maximum degree.

1 Introduction

A graph is 1-*planar* if it can be drawn in the plane such that each edge is crossed at most once. The notion of 1-planarity naturally extends planarity and received considerable attention since its first introduction by Ringel in 1965 [33], as witnessed by recent surveys [14,27]. Despite the efforts made in the study of 1-planar graphs, only few results are known concerning their geometric representations (see, e.g., [1,4,7,11]). In this paper, we study the existence of 1-planar drawings that simultaneously satisfy the following properties: edges are polylines using few bends and few distinct slopes for their segments, edge crossings occur at large angles, and pairs of edges incident to the same vertex form large angles. For example, Fig. 1d shows a 1-bend drawing of a 1-planar graph (i.e., a drawing in which each edge is a polyline with at most one bend) using 4 distinct slopes, such that edge crossings form angles at least $\pi/4$, and the angles formed by edges incident to the same vertex are at least $\pi/4$. In what follows, we briefly recall known results concerning the problems of computing polyline drawings with few bends and few slopes or with few bends and large angles.

Related Work. The *k-bend (planar) slope number* of a (planar) graph G with maximum vertex degree Δ is the minimum number of distinct edge slopes needed

© Springer Nature Switzerland AG 2018
T. Biedl and A. Kerren (Eds.): GD 2018, LNCS 11282, pp. 152–166, 2018.
https://doi.org/10.1007/978-3-030-04414-5_11

to compute a (planar) drawing of G such that each edge is a polyline with at most k bends. When $k = 0$, this parameter is simply known as the *(planar) slope number* of G. Clearly, if G has maximum vertex degree Δ, at least $\lceil \Delta/2 \rceil$ slopes are needed for any k. While there exist non-planar graphs with $\Delta \geq 5$ whose slope number is unbounded with respect to Δ [3,32], Keszegh et al. [24] proved that the planar slope number is bounded by $2^{O(\Delta)}$. Several authors improved this bound for subfamilies of planar graphs (see, e.g., [21,26,28]).

Concerning k-bend drawings, Angelini et al. [2] proved that the 1-bend planar slope number is at most $\Delta - 1$, while Keszegh et al. [24] proved that the 2-bend planar slope number is $\lceil \Delta/2 \rceil$ (which is tight). Special attention has been paid in the literature to the slope number of *(sub)cubic* graphs, i.e., graphs having vertex degree (at most) 3. Mukkamala and Pálvölgyi showed that the four slopes $\{0, \frac{\pi}{4}, \frac{\pi}{2}, \frac{3\pi}{4}\}$ suffice for every cubic graph [31]. For planar graphs, Kant and independently Dujmović et al. proved that cubic 3-connected planar graphs have planar slope number 3 disregarding the slopes of three edges on the outer face [15,22], while Di Giacomo et al. [13] proved that the planar slope number of subcubic planar graphs is 4. We also remark that the slope number problem is related to orthogonal drawings, which are planar and with slopes $\{0, \frac{\pi}{2}\}$ [16], and with octilinear drawings, which are planar and with slopes $\{0, \frac{\pi}{4}, \frac{\pi}{2}, \frac{3\pi}{4}\}$ [5]. All planar graphs with $\Delta \leq 4$ (except the octahedron) admit 2-bend orthogonal drawings [6,29], and planar graphs admit octilinear drawings without bends if $\Delta \leq 3$ [13,22], with 1 bend if $\Delta \leq 5$ [5], and with 2 bends if $\Delta \leq 8$ [24].

Of particular interest for us is the k-bend *1-planar slope number* of 1-planar graphs, i.e., the minimum number of distinct edge slopes needed to compute a 1-planar drawing of a 1-planar graph such that each edge is a polyline with at most $k > 0$ bends. Di Giacomo et al. [12] proved an $O(\Delta)$ upper bound for the 1-planar slope number ($k = 0$) of outer 1-planar graphs, i.e., graphs that can be drawn 1-planar with all vertices on the external boundary.

Finally, the *vertex resolution* and the *crossing resolution* of a drawing are defined as the minimum angle between two consecutive segments incident to the same vertex or crossing, respectively (see, e.g., [17,20,30]). A drawing is *RAC (right-angle crossing)* if its crossing resolution is $\pi/2$. Eades and Liotta proved that 1-planar graphs may not have straight-line RAC drawings [18], while Chaplick et al. [8] and Bekos et al. [4] proved that every 1-planar graph has a 1-bend RAC drawing that preserves the embedding.

Our Contribution. We prove upper and lower bounds on the k-bend 1-planar slope number of 1-planar graphs, when $k \in \{0, 1, 2\}$. Our results are based on techniques that lead to drawings with large vertex and crossing resolution.

In Sect. 3, we prove that every 3-connected cubic 1-planar graph admits a 1-bend 1-planar drawing that uses at most 4 distinct slopes and has both vertex and crossing resolution $\pi/4$. In Sect. 4, we show that every subcubic 1-planar graph admits a 2-bend 1-planar drawing that uses at most 2 distinct slopes and has both vertex and crossing resolution $\pi/2$. These bounds on the number of slopes and on the vertex/crossing resolution are clearly worst-case optimal. In Sect. 5.1, we give a 3-connected cubic 1-plane graph for which any embedding-preserving 1-bend drawing uses at least 3 distinct slopes. The lower bound holds even if we are allowed to change the outer face. In Sect. 5.2, we present 2-connected subcubic 1-plane graphs with n vertices such that any embedding-preserving straight-line drawing uses $\Omega(n)$ distinct slopes, and 3-connected 1-plane graphs with maximum degree $\Delta \geq 3$ such that any embedding-preserving straight-line drawing uses at least $9(\Delta - 1)$ distinct slopes, which implies that at least 18 slopes are needed if $\Delta = 3$.

Preliminaries can be found in Sect. 2, while open problems are in Sect. 6.

2 Preliminaries

We only consider *simple* graphs with neither self-loops nor multiple edges. A *drawing* Γ of a graph G maps each vertex of G to a point of the plane and each edge to a simple open Jordan curve between its endpoints. We always refer to *simple* drawings where two edges can share at most one point, which is either a common endpoint or a proper intersection. A drawing divides the plane into topologically connected regions, called *faces*; the infinite region is called the *outer face*. For a planar (i.e., crossing-free) drawing, the boundary of a face consists of vertices and edges, while for a non-planar drawing the boundary of a face may also contain crossings and parts of edges. An *embedding* of a graph G is an equivalence class of drawings of G that define the same set of faces and the same outer face. A *(1-)plane graph* is a graph with a fixed (1-)planar embedding. Given a 1-plane graph G, the *planarization* G^* of G is the plane graph obtained by replacing each crossing of G with a *dummy vertex*. To avoid confusion, the vertices of G^* that are not dummy are called *real*. Moreover, we call *fragments* the edges of G^* that are incident to a dummy vertex. The next lemma will be used in the following and can be of independent interest, as it extends a similar result by Fabrici and Madaras [19]. The proof is given in the full version [25].

Lemma 1. *Let $G = (V, E)$ be a 1-plane graph and let G^* be its planarization. We can re-embed G such that each edge is still crossed at most once and (i) no cutvertex of G^* is a dummy vertex, and (ii) if G is 3-connected, then G^* is 3-connected.*

A drawing Γ is *straight-line* if all its edges are mapped to segments, or it is *k-bend* if each edge is mapped to a chain of segments with at most $k > 0$ bends. The *slope* of an edge segment of Γ is the slope of the line containing this segment. For convenience, we measure the slopes by their angle with respect to the x-axis. Let $\mathcal{S} = \{\alpha_1, \ldots, \alpha_t\}$ be a set of t distinct slopes. The *slope number* of a k-bend drawing Γ is the number of distinct slopes used for the edge segments of Γ. An edge segment of Γ uses the *north (N) port* (*south (S) port*) of a vertex v if it has slope $\pi/2$ and v is its bottommost (topmost) endpoint. We can define analogously the *west (W)* and *east (E)* ports with respect to the slope 0, the *north-west (NW)* and *south-east (SE)* ports with respect to slope $3\pi/4$, and the *south-west (SW)* and *north-east (NE)* ports with respect to slope $\pi/4$. Any such port is *free* for v if there is no edge that attaches to v by using it.

We will use a decomposition technique called *canonical ordering* [23]. Let $G = (V, E)$ be a 3-connected plane graph. Let $\delta = \{\mathcal{V}_1, \ldots, \mathcal{V}_K\}$ be an ordered partition of V, that is, $\mathcal{V}_1 \cup \cdots \cup \mathcal{V}_K = V$ and $\mathcal{V}_i \cap \mathcal{V}_j = \emptyset$ for $i \neq j$. Let G_i be the subgraph of G induced by $\mathcal{V}_1 \cup \cdots \cup \mathcal{V}_i$ and denote by C_i the outer face of G_i. The partition δ is a canonical ordering of G if: (i) $\mathcal{V}_1 = \{v_1, v_2\}$, where v_1 and v_2 lie on the outer face of G and $(v_1, v_2) \in E$. (ii) $\mathcal{V}_K = \{v_n\}$, where v_n lies on the outer face of G, $(v_1, v_n) \in E$. (iii) Each C_i ($i > 1$) is a cycle containing (v_1, v_2). (iv) Each G_i is 2-connected and internally 3-connected, that is, removing any two interior vertices of G_i does not disconnect it. (v) For each $i \in \{2, \ldots, K-1\}$, one of the following conditions holds: (a) $\mathcal{V}_i$ is a *singleton* v^i that lies on C_i and has at least one neighbor in $G \setminus G_i$; (b) $\mathcal{V}_i$ is a *chain* $\{v_1^i, \ldots, v_l^i\}$, both v_1^i and v_l^i have exactly one neighbor each in C_{i-1}, and $v_2^i, \ldots, v_{l-1}^i$ have no neighbor in C_{i-1}. Since G is 3-connected, each v_j^i has at least one neighbor in $G \setminus G_i$.

Let v be a vertex in $\mathcal{V}_i$, then its neighbors in G_{i-1} (if G_{i-1} exists) are called the *predecessors* of v, while its neighbors in $G \setminus G_i$ (if G_{i+1} exists) are called the *successors* of v. In particular, every singleton has at least two predecessors and at least one successor, while every vertex in a chain has either zero or one predecessor and at least one successor. Kant [23] proved that a canonical ordering of G always exists and can be computed in $O(n)$ time; the technique in [23] is such that one can arbitrarily choose two adjacent vertices u and w on the outer face so that $u = v_1$ and $w = v_2$ in the computed canonical ordering.

An n-vertex *planar st-graph* $G = (V, E)$ is a plane acyclic directed graph with a single source s and a single sink t, both on the outer face [10]. An *st-ordering* of G is a numbering $\sigma : V \to \{1, 2, \ldots, n\}$ such that for each edge $(u, v) \in E$, it holds $\sigma(u) < \sigma(v)$ (thus $\sigma(s) = 1$ and $\sigma(t) = n$). For an *st*-graph, an *st*-ordering can be computed in $O(n)$ time (see, e.g., [9]) and every biconnected undirected graph can be oriented to become a planar *st*-graph (also in linear time).

3   1-Bend Drawings of 3-Connected Cubic 1-Planar Graphs

Let G be a 3-connected 1-plane cubic graph, and let G^* be its planarization. We can assume that G^* is 3-connected (else we can re-embed G by Lemma 1). We choose as outer face of G a face containing an edge (v_1, v_2) whose vertices are both real (see Fig. 1a). Such a face exists: If G has n vertices, then G^* has fewer than $3n/4$ dummy vertices because G is subcubic. Hence we find a face in G^* with more real than dummy vertices and hence with two consecutive real vertices. Let $\delta = \{\mathcal{V}_1, \ldots, \mathcal{V}_K\}$ be a canonical ordering of G^*, let G_i be the graph obtained by adding the first i sets of δ and let C_i be the outer face of G_i.

Note that a real vertex v of G_i can have at most one successor w in some set $\mathcal{V}_j$ with $j > i$. We call w an *L-successor* (resp., *R-successor*) of v if v is the leftmost (resp., rightmost) neighbor of $\mathcal{V}_j$ on C_i. Similarly, a dummy vertex x of G_i can have at most two successors in some sets $\mathcal{V}_j$ and $\mathcal{V}_l$ with $l \geq j > i$. In both cases, a vertex v of G_i having a successor in some set $\mathcal{V}_j$ with $j > i$ is called *attachable*. We call v *L-attachable* (resp., *R-attachable*) if v is attachable

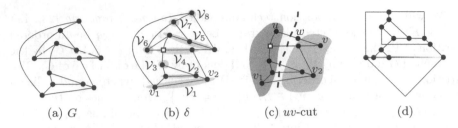

(a) G (b) δ (c) uv-cut   (d)

Fig. 1. (a) A 3-connected 1-plane cubic graph G; (b) a canonical ordering δ of the planarization G^* of G—the real (dummy) vertices are black points (white squares); (c) the edges crossed by the dashed line are a uv-cut of G_5 with respect to (u, w)—the two components have a yellow and a blue background, respectively; (d) a 1-bend 1-planar drawing with 4 slopes of G (Color figure online)

and has no L-successor (resp., R-successor) in G_i. We will draw an upward edge at u with slope $\pi/4$ (resp., $3\pi/4$) only if it is L-attachable (resp., R-attachable).

Let u and v be two vertices of C_i, for $i > 1$. Denote by $P_i(u, v)$ the path of C_i having u and v as endpoints and that does not contain (v_1, v_2). Vertices u and v are *consecutive* if they are both attachable and if $P_i(u, v)$ does not contain any other attachable vertex. Given two consecutive vertices u and v of C_i and an edge e of C_i, a *uv-cut of G_i with respect to e* is a set of edges of G_i that contains both e and (v_1, v_2) and whose removal disconnects G_i into two components, one containing u and one containing v (see Fig. 1c). We say that u and v are *L-consecutive* (resp., *R-consecutive*) if they are consecutive, u lies to the left (resp., *right*) of v on C_i, and u is L-attachable (resp., R-attachable).

We construct an embedding-preserving drawing Γ_i of G_i, for $i = 2, \ldots, K$, by adding one by one the sets of δ. A drawing Γ_i of G_i is *valid*, if:

P1 It uses only slopes in the set $\{0, \frac{\pi}{4}, \frac{\pi}{2}, \frac{3\pi}{4}\}$;

P2 It is a 1-bend drawing such that the union of any two edge fragments that correspond to the same edge in G is drawn with (at most) one bend in total.

A valid drawing Γ_K of G_K will coincide with the desired drawing of G, after replacing dummy vertices with crossing points.

Construction of Γ_2. We begin by showing how to draw G_2. We distinguish two cases, based on whether $\mathcal{V}_2$ is a singleton or a chain, as illustrated in Fig. 2.

Construction of Γ_i, for $2 < i < K$. We now show how to compute a valid drawing of G_i, for $i = 3, \ldots, K - 1$, by incrementally adding the sets of δ.

We aim at constructing a valid drawing Γ_i that is also *stretchable*, i.e., that satisfies the following two more properties; see Fig. 3. These two properties will be useful to prove Lemma 2, which defines a standard way of stretching a drawing by lengthening horizontal segments.

P3 The edge (v_1, v_2) is drawn with two segments s_1 and s_2 that meet at a point p. Segment s_1 uses the SE port of v_1 and s_2 uses the SW port of v_2. Also, p is the lowest point of Γ_i, and no other point of Γ_i is contained by the two lines that contain s_1 and s_2.

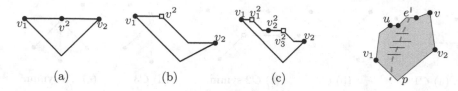

Fig. 2. Construction of Γ_2: (a) $\mathcal{V}_2$ is a real singleton;   **Fig. 3.** Γ_i is stretchable.
(b) $\mathcal{V}_2$ is a dummy singleton; (c) $\mathcal{V}_2$ is a chain.

P4 For every pair of consecutive vertices u and v of C_i with u left of v on C_i, it holds that (a) If u is L-attachable (resp., v is R-attachable), then the path $P_i(u, v)$ is such that for each vertical segment s on this path there is a horizontal segment in the subpath before s if s is traversed upwards when going from u to v (resp., from v to u); (b) if both u and v are real, then $P_i(u, v)$ contains at least one horizontal segment; and (c) for every edge e of $P_i(u, v)$ such that e contains a horizontal segment, there exists a uv-cut of G_i with respect to e whose edges all contain a horizontal segment in Γ_i except for (v_1, v_2), and such that there exists a y-monotone curve that passes through all and only such horizontal segments and (v_1, v_2).

Lemma 2. *Suppose that Γ_i is valid and stretchable, and let u and v be two consecutive vertices of C_i. If u is L-attachable (resp., v is R-attachable), then it is possible to modify Γ_i such that any half-line with slope $\pi/4$ (resp., $3\pi/4$) that originates at u (resp., at v) and that intersects the outer face of Γ_i does not intersect any edge segment with slope $\pi/2$ of $P_i(u, v)$. Also, the modified drawing is still valid and stretchable.*

Proof Sketch. Crossings between such half-lines and vertical segments of $P_i(u, v)$ can be solved by finding suitable uv-cuts and moving everything on the right/left side of the cut to the right/left. The full proof is given in the full version [25]. □

Let P be a set of ports of a vertex v; the *symmetric* set of ports P' of v is the set of ports obtained by mirroring P at a vertical line through v. We say that Γ_i is *attachable* if the following two properties also apply.

P5 At any attachable real vertex v of Γ_i, its N, NW, and NE ports are free.
P6 Let v be an attachable dummy vertex of Γ_i. If v has two successors, there are four possible cases for its two used ports, illustrated with two *solid* edges in Fig. 4a–d. If v has only one successor not in Γ_i, there are eight possible cases for its three used ports, illustrated with two solid edges plus one *dashed* or one *dotted* edge in Fig. 4a–e.

Observe that Γ_2, besides being valid, is also stretchable and attachable by construction (see also Fig. 2). Assume that G_{i-1} admits a valid, stretchable, and attachable drawing Γ_{i-1}, for some $2 \le i < K - 1$; we show how to add the next set $\mathcal{V}_i$ of δ so to obtain a drawing Γ_i of G_i that is valid, stretchable and attachable. We distinguish between the following cases.

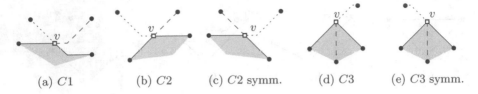

| (a) $C1$ | (b) $C2$ | (c) $C2$ symm. | (d) $C3$ | (e) $C3$ symm. |

Fig. 4. Illustration for **P6**. If v has two successors not in Γ_i, then the edges connecting v to its two neighbors in Γ_i are solid. If v has one successor in Γ_i, then the edge between v and this successor is dashed or dotted.

Case 1. $\mathcal{V}_i$ is a singleton, i.e., $\mathcal{V}_i = \{v^i\}$. Note that if v^i is real, it has two neighbors on C_{i-1}, while if it is dummy, it can have either two or three neighbors on C_{i-1}. Let u_l and u_r be the first and the last neighbor of v^i, respectively, when walking along C_{i-1} in clockwise direction from v_1. We will call u_l (resp., u_r) the *leftmost predecessor* (resp., *rightmost predecessor*) of v^i.

Case 1.1. Vertex v^i is real. Then, u_l and u_r are its only two neighbors in C_{i-1}. Each of u_l and u_r can be real or dummy. If u_l (resp., u_r) is real, we draw (u_l, v^i) (resp., (u_r, v^i)) with a single segment using the NE port of u_l and the SW port of v^i (resp., the NW port of u_r and the SE port of v^i). If u_l is dummy and has two successors not in Γ_{i-1}, we distinguish between the cases of Fig. 4 as shown in Fig. 5. The symmetric configuration of $C3$ is only used for connecting to u_r.

If u_l is dummy and has one successor not in Γ_{i-1}, we distinguish between the various cases of Fig. 4 as indicated in Fig. 6. Observe that $C1$ requires a local reassignment of one port of u_l. The edge (u_r, v^i) is drawn by following a similar case analysis. Vertex v^i is then placed at the intersection of the lines passing through the assigned ports, which always intersect by construction. In particular, the S port is only used when u_l has one successor, but the same situation cannot occur when drawing (u_r, v^i). Otherwise, there is a path of C_{i-1} from u_l via its successor x on C_{i-1} to u_r via its successor y on C_{i-1}. Note that $x = y$ is possible but $x \neq u_r$. Since the first edge on this path goes from a predecessor to a successor and the last edge goes from a successor to a predecessor, there has to be a vertex z without a successor on the path; but then u_l and u_r are not consecutive. To avoid crossings between Γ_{i-1} and the new edges (u_l, v^i) and (u_r, v^i), we apply Lemma 2 to suitably stretch the drawing. In particular, possible crossings can occur only with vertical edge segments of $P_{i-1}(u_l, u_r)$,

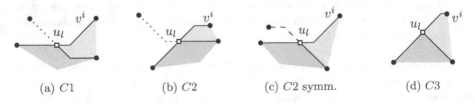

| (a) $C1$ | (b) $C2$ | (c) $C2$ symm. | (d) $C3$ |

Fig. 5. A real singleton when u_l is dummy with two successors not in Γ_{i-1}

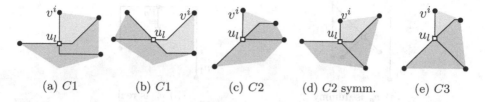

(a) $C1$ (b) $C1$ (c) $C2$ (d) $C2$ symm. (e) $C3$

Fig. 6. Some cases for the addition of a real singleton when u_l is dummy with one successor not in Γ_{i-1}

because when walking along $P_{i-1}(u_l, u_r)$ from u_l to u_r we only encounter a (possibly empty) set of segments with slopes in the range $\{3\pi/4, \pi/2, 0\}$, followed by a (possibly empty) set of segments with slopes in the range $\{\pi/2, \pi/4, 0\}$.

Case 1.2. Vertex v^i is dummy. By 1-planarity, the two or three neighbors of v^i on C_{i-1} are all real. If v^i has two neighbors, we draw (u_l, v^i) and (u_r, v^i) as shown in Fig. 7a, while if v^i has three neighbors, we draw (u_l, v^i) and (u_r, v^i) as shown in Fig. 7b. Analogous to the previous case, vertex v^i is placed at the intersection of the lines passing through the assigned ports, which always intersect by construction, and avoiding crossings between Γ_{i-1} and the new edges (u_l, v^i) and (u_r, v^i) by applying Lemma 2. In particular, if v^i has three neighbors on C_{i-1}, say u_l, w, and u_r, by P4 there is a horizontal segment between u_l and w, as well as between w and u_r. Thus, Lemma 2 can be applied not only to resolve crossings, but also to find a suitable point where the two lines with slopes $\pi/4$ and $3\pi/4$ meet along the line with slope $\pi/2$ that passes through w.

Case 2. $\mathcal{V}_i$ is a chain, i.e., $\mathcal{V}_i = \{v_1^i, v_2^i, \ldots, v_l^i\}$. We find a point as if we had to place a vertex v whose leftmost predecessor is the leftmost predecessor of v_1^i and whose rightmost predecessor is the rightmost predecessor of v_l^i. We then draw the chain slightly below this point by using the same technique used to draw $\mathcal{V}_2$. Again, Lemma 2 can be applied to resolve possible crossings.

We formally prove the correctness of our algorithm in the full version [25].

Lemma 3. *Drawing Γ_{K-1} is valid, stretchable, and attachable.*

Construction of Γ_K. We now show how to add $\mathcal{V}_K = \{v_n\}$ to Γ_{K-1} so as to obtain a valid drawing of G_K, and hence the desired drawing of G after replacing dummy vertices with crossing points. Recall that (v_1, v_n) is an edge of G by the definition of canonical ordering. We distinguish whether v_n is real or dummy;

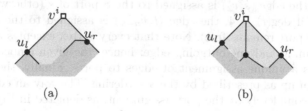

(a) (b)

Fig. 7. Illustration for the addition of a dummy singleton

(a) v_n is dummy   (b) v_n is real

Fig. 8. Illustration for the addition of $\mathcal{V}_k$

the two cases are shown in Fig. 8. Note that if v_n is dummy, its four neighbors are all real and hence their N, NW, and NE ports are free by P5. If v_n is real, it has three neighbors in Γ_{K-1}, v_1 is real by construction, and the S port can be used to attach with a dummy vertex. Finally, since Γ_{K-1} is attachable, we can use Lemma 2 to avoid crossings and to find a suitable point to place v_n. A complete drawing is shown in Fig. 1d.

The theorem follows immediately by the choice of the slopes.

Theorem 1. *Every 3-connected cubic 1-planar graph admits a 1-bend 1-planar drawing with at most 4 distinct slopes and angular and crossing resolution $\pi/4$.*

4   2-Bend Drawings

Liu et al. [29] presented an algorithm to compute orthogonal drawings for planar graphs of maximum degree 4 with at most 2 bends per edge (except the octahedron, which requires 3 bends on one edge). We make use of their algorithm for biconnected graphs. The algorithm chooses two vertices s and t and computes an st-ordering of the input graph. Let $V = \{v_1, \ldots, v_n\}$ with $\sigma(v_i) = i$, $1 \le i \le n$. Liu et al. now compute an embedding of G such that v_2 lies on the outer face if $\deg(s) = 4$ and v_{n-1} lies on the outer face if $\deg(t) = 4$; such an embedding exists for every graph with maximum degree 4 except the octahedron.

The edges around each vertex $v_i, 1 \le i \le n$, are assigned to the four ports as follows. If v_i has only one outgoing edge, it uses the N port; if v_i has two outgoing edges, they use the N and E port; if v_i has three outgoing edges, they use the N, E, and W port; and if v_i has four outgoing edges, they use all four ports. Symmetrically, the incoming edges of v_i use the S, W, E, and N port, in this order. The edge (s, t) (if it exists) is assigned to the W port of both s and t. If $\deg(s) = 4$, the edge (s, v_2) is assigned to the S port of s (otherwise the port remains free); if $\deg(t) = 4$, the edge (t, v_{n-1}) is assigned to the N port of t (otherwise the port remains free). Note that every vertex except s and t has at least one incoming and one outgoing edge; hence, the given embedding of the graph provides a unique assignment of edges to ports. Finally, they place the vertices bottom-up as prescribed by the st-ordering. The way an edge is drawn is determined completely by the port assignment, as depicted in Fig. 9.

Let $G = (V, E)$ be a subcubic 1-plane graph. We first re-embed G according to Lemma 1. Let G^* be the planarization of G after the re-embedding. Then, all

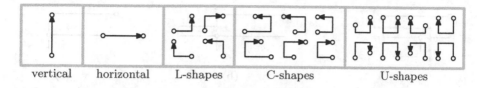

Fig. 9. The shapes to draw edges

cutvertices of G^* are real vertices, and since they have maximum degree 3, there is always a bridge connecting two 2-connected components. Let $G_1, \ldots, G_k$ be the 2-connected components of G, and let G_i^* be the planarization of $G_i, 1 \leq i \leq k$. We define the *bridge decomposition tree* $\mathcal{T}$ of G as the graph having a node for each component G_i of G, and an edge (G_i, G_j), for every pair G_i, G_j connected by a bridge in G. We root $\mathcal{T}$ in G_1. For each component $G_i, 2 \leq i \leq k$, let u_i be the vertex of G_i connected to the parent of G_i in $\mathcal{T}$ by a bridge and let u_1 be an arbitrary vertex of G_1. We will create a drawing Γ_i for each component G_i with at most 2 slopes and 2 bends such that u_i lies on the outer face.

To this end, we first create a drawing Γ_i^* of G_i^* with the algorithm of Liu et al. [29] and then modify the drawing. Throughout the modifications, we will make sure that the following invariants hold for the drawing Γ_i^*.

(I1) Γ_i^* is a planar orthogonal drawing of G_i^* and edges are drawn as in Fig. 9;
(I2) u_i lies on the outer face of Γ_i^* and its N port is free;
(I3) every edge is y-monotone from its source to its target;
(I4) every edge with 2 bends is a C-shape, there are no edges with more bends;
(I5) if a C-shape ends in a dummy vertex, it uses only E ports; and
(I6) if a C-shape starts in a dummy vertex, it uses only W ports.

Lemma 4. *Every G_i^* admits a drawing Γ_i^* that satisfies invariants (I1)–(I6).*

Proof Sketch. We choose $t = u_i$ and some real vertex s and use the algorithm by Liu et al. to draw G_i. Since s and t are real, there are no U-shapes. Since no real vertex can have an outgoing edge at its W port or incoming edge at its E port, the invariants follow. The full proof is given in the full version [25]. □

We now iteratively remove the C-shapes from the drawing while maintaining the invariants. We make use of a technique similar to the stretching in Sect. 3. We lay an orthogonal y-monotone curve S through our drawing that intersects no vertices. Then we stretch the drawing by moving S and all features that lie right of S to the right, and stretching all points on S to horizontal segments. After this stretch, in the area between the old and the new position of S, there are only horizontal segments of edges that are intersected by S. The same operation can be defined symmetrically for an x-monotone curve that is moved upwards.

Lemma 5. *Every G_i admits an orthogonal 2-bend drawing such that u_i lies on the outer face and its N port is free.*

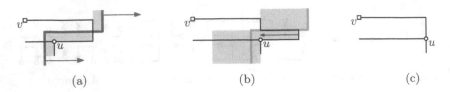

Fig. 10. Proof of Lemma 5, Case 1

Proof Sketch. We start with a drawing Γ_i^* of G_i^* that satisfies invariants (I1)–(I6), which exists by Lemma 4. By (I2), u_i lies on the outer face and its N port is free. If no dummy vertex in Γ_i^* is incident to a C-shape, by (I4) all edges incident to dummy vertices are drawn with at most 1 bend, so the resulting drawing Γ_i of G_i is an orthogonal 2-bend drawing. Otherwise, there is a C-shape between a real vertex u and a dummy vertex v. We show how to eliminate this C-shape without introducing new ones while maintaining all invariants.

We prove the case that (u, v) is directed from u to v, so by (I5) it uses only E ports; the other case is symmetric. We do a case analysis based on which ports at u are free. We show one case here and the rest in the full version [25].
Case 1. The N port at u is free; see Fig. 10. Create a curve S as follows: Start at some point p slightly to the top left of u and extend it downward to infinity. Extend it from p to the right until it passes the vertical segment of (u, v) and extend it upwards to infinity. Place the curve close enough to u and (u, v) such that no vertex or bend point lies between S and the edges of u that lie right next to it. Then, stretch the drawing by moving S to the right such that u is placed below the top-right bend point of (u, v). Since S intersected a vertical segment of (u, v), this changes the edge to be drawn with 4 bends. However, now the region between u and the second bend point of (u, v) is empty and the N port of u is free, so we can make an L-shape out of (u, v) that uses the N port at u. This does not change the drawing style of any edge other than (u, v), so all the invariants are maintained and the number of C-shapes is reduced by one.   □

Finally, we combine the drawings Γ_i to a drawing Γ of G. Recall that every cutvertex is real and two biconnected components are connected by a bridge. Let G_j be a child of G_i in the bridge decomposition tree. We have drawn G_j with u_j on the outer face and a free N port. Let v_i be the neighbor of u_j in G_i. We choose one of its free ports, rotate and scale Γ_j such that it fits into the face of that port, and connect u_j and v_i with a vertical or horizontal segment. Doing this for every biconnected component gives an orthogonal 2-bend drawing of G.

Theorem 2. *Every subcubic 1-plane graph admits a 2-bend 1-planar drawing with at most 2 distinct slopes and both angular and crossing resolution $\pi/2$.*

5   Lower Bounds for 1-Plane Graphs

5.1   1-Bend Drawings of Subcubic Graphs

Theorem 3. *There exists a subcubic 3-connected 1-plane graph such that any embedding-preserving 1-bend drawing uses at least 3 distinct slopes. The lower bound holds even if we are allowed to change the outer face.*

Proof. Let G be the K_4 with a planar embedding. The outer face is a 3-cycle, which has to be drawn as a polygon Π with at least four (nonreflex) corners. Since we allow only one bend per edge, one of the corners of Π has to be a vertex of G. The vertex in the interior has to connect to this corner, however, all of its free ports lie on the outside. Thus, no drawing of G is possible.   □

5.2   Straight-Line Drawings

The full proofs for this section are given in the full version [25].

Theorem 4. *There exist 2-regular 2-connected 1-plane graphs with n vertices such that any embedding-preserving straight-line drawing uses $\Omega(n)$ distinct slopes.*

Proof Sketch. Let G_k be the graph given by the cycle $a_1 \ldots, a_{k+1}, b_{k+1}, \ldots, b_1, a_1$ and the embedding shown in Fig. 11a. Walking along the path $a_1, \ldots, a_{k+1}$, we find that the slope has to increase at every step.   □

Lemma 6. *There exist 3-regular 3-connected 1-plane graphs such that any embedding-preserving straight-line drawing uses at least 18 distinct slopes.*

Proof Sketch. Consider the graph depicted in Fig. 11b. We find that the slopes of the edges $(a_i, b_i), (a_i, c_i), (c_i, d_i), (c_i, e_i), (e_i, d_i), (e_i, a_{i+1})$ have to be increasing in this order for every $i = 1, 2, 3$.   □

Theorem 5. *There exist 3-connected 1-plane graphs such that any embedding-preserving straight-line drawing uses at least $9(\Delta - 1)$ distinct slopes.*

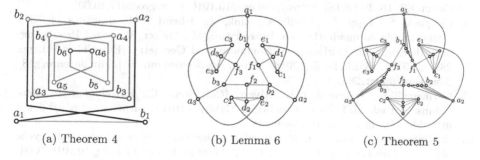

(a) Theorem 4   (b) Lemma 6   (c) Theorem 5

Fig. 11. The constructions for the results of Sect. 5

Proof Sketch. Consider the graph depicted in Fig. 11c. The degree of a_i, c_i, and e_i is Δ. We repeat the proof of Lemma 6, but observe that the slopes of the $9(\Delta-3)$ added edges lie between the slopes of $(a_i, b_i), (a_i, c_i), (c_i, e_i)$, and (e_i, a_{i+1}).   □

6 Open Problems

The research in this paper gives rise to interesting questions, among them: (1) Is it possible to extend Theorem 1 to all subcubic 1-planar graphs? (2) Can we drop the embedding-preserving condition from Theorem 3? (3) Is the 1-planar slope number of 1-planar graphs bounded by a function of the maximum degree?

References

1. Alam, M.J., Brandenburg, F.J., Kobourov, S.G.: Straight-line grid drawings of 3-connected 1-planar graphs. In: Wismath, S., Wolff, A. (eds.) GD 2013. LNCS, vol. 8242, pp. 83–94. Springer, Cham (2013). https://doi.org/10.1007/978-3-319-03841-4_8

2. Angelini, P., Bekos, M.A., Liotta, G., Montecchiani, F.: A universal slope set for 1-bend planar drawings. In: Aronov, B., Katz, M.J. (eds.) Proceedings of 33rd International Symposium on Computational Geometry (SoCG 2017). LIPIcs, vol. 77, pp. 9:1–9:16. Schloss Dagstuhl (2017). https://doi.org/10.4230/LIPIcs.SoCG.2017.9

3. Barát, J., Matousek, J., Wood, D.R.: Bounded-degree graphs have arbitrarily large geometric thickness. Electr. J. Comb. **13**(1), 1–14 (2006). http://www.combinatorics.org/Volume_13/Abstracts/v13i1r3.html

4. Bekos, M.A., Didimo, W., Liotta, G., Mehrabi, S., Montecchiani, F.: On RAC drawings of 1-planar graphs. Theor. Comput. Sci. **689**, 48–57 (2017). https://doi.org/10.1016/j.tcs.2017.05.039

5. Bekos, M.A., Gronemann, M., Kaufmann, M., Krug, R.: Planar octilinear drawings with one bend per edge. J. Graph Algorithms Appl. **19**(2), 657–680 (2015). https://doi.org/10.7155/jgaa.00369

6. Biedl, T., Kant, G.: A better heuristic for orthogonal graph drawings. Comput. Geom. Theory Appl. **9**(3), 159–180 (1998). https://doi.org/10.1016/S0925-7721(97)00026-6

7. Brandenburg, F.J.: T-shape visibility representations of 1-planar graphs. Comput. Geom. **69**, 16–30 (2018). https://doi.org/10.1016/j.comgeo.2017.10.007

8. Chaplick, S., Lipp, F., Wolff, A., Zink, J.: 1-bend RAC drawings of NIC-planar graphs in quadratic area. In: Korman, M., Mulzer, W. (eds.) Proceedings of 34th European Workshop on Computational Geometry (EuroCG 2018). pp. 28:1–28:6. FU Berlin, Berlin (2018). https://conference.imp.fu-berlin.de/eurocg18/download/paper_28.pdf

9. Cormen, T.H., Leiserson, C.E., Rivest, R.L., Stein, C.: Introduction to Algorithms, 3rd ed. MIT Press, Cambridge (2009). https://mitpress.mit.edu/books/introduction-algorithms-third-edition

10. Di Battista, G., Tamassia, R.: Algorithms for plane representations of acyclic digraphs. Theor. Comput. Sci. **61**, 175–198 (1988). https://doi.org/10.1016/0304-3975(88)90123-5

11. Di Giacomo, E., et al.: Ortho-polygon visibility representations of embedded graphs. Algorithmica **80**(8), 2345–2383 (2018). https://doi.org/10.1007/s00453-017-0324-2

12. Di Giacomo, E., Liotta, G., Montecchiani, F.: Drawing outer 1-planar graphs with few slopes. J. Graph Algorithms Appl. **19**(2), 707–741 (2015). https://doi.org/10.7155/jgaa.00376

13. Di Giacomo, E., Liotta, G., Montecchiani, F.: Drawing subcubic planar graphs with four slopes and optimal angular resolution. Theor. Comput. Sci. **714**, 51–73 (2018). https://doi.org/10.1016/j.tcs.2017.12.004

14. Didimo, W., Liotta, G., Montecchiani, F.: A survey on graph drawing beyond planarity. arxiv report arXiv:1804.07257 (2018)

15. Dujmović, V., Eppstein, D., Suderman, M., Wood, D.R.: Drawings of planar graphs with few slopes and segments. Comput. Geom. **38**(3), 194–212 (2007). https://doi.org/10.1016/j.comgeo.2006.09.002

16. Duncan, C., Goodrich, M.T.: Planar orthogonal and polyline drawing algorithms. In: Tamassia, R. (ed.) Handbook on Graph Drawing and Visualization. Chapman and Hall/CRC, Boca Raton (2013). http://cs.brown.edu/people/rtamassi/gdhandbook/chapters/orthogonal.pdf

17. Duncan, C.A., Kobourov, S.G.: Polar coordinate drawing of planar graphs with good angular resolution. J. Graph Algorithms Appl. **7**(4), 311–333 (2003). https://doi.org/10.7155/jgaa.00073

18. Eades, P., Liotta, G.: Right angle crossing graphs and 1-planarity. Discrete Appl. Math. **161**(7–8), 961–969 (2013). https://doi.org/10.1016/j.dam.2012.11.019

19. Fabrici, I., Madaras, T.: The structure of 1-planar graphs. Discrete Math. **307**(7–8), 854–865 (2007). https://doi.org/10.1016/j.disc.2005.11.056

20. Formann, M., et al.: Drawing graphs in the plane with high resolution. SIAM J. Comput. **22**(5), 1035–1052 (1993). https://doi.org/10.1137/0222063

21. Jelinek, V., Jelinková, E., Kratochvil, J., Lidický, B., Tesar, M., Vyskocil, T.: The planar slope number of planar partial 3-trees of bounded degree. Graphs Comb. **29**(4), 981–1005 (2013). https://doi.org/10.1007/s00373-012-1157-z

22. Kant, G.: Hexagonal grid drawings. In: Mayr, E.W. (ed.) WG 1992. LNCS, vol. 657, pp. 263–276. Springer, Heidelberg (1993). https://doi.org/10.1007/3-540-56402-0_53

23. Kant, G.: Drawing planar graphs using the canonical ordering. Algorithmica **16**(1), 4–32 (1996). https://doi.org/10.1007/BF02086606

24. Keszegh, B., Pach, J., Pálvölgyi, D.: Drawing planar graphs of bounded degree with few slopes. SIAM J. Discrete Math. **27**(2), 1171–1183 (2013). https://doi.org/10.1137/100815001

25. Kindermann, P., Montecchiani, F., Schlipf, L., Schulz, A.: Drawing subcubic 1-planar graphs with few bends, few slopes, and large angles. arxiv report arXiv:1808.08496 (2018)

26. Knauer, K.B., Micek, P., Walczak, B.: Outerplanar graph drawings with few slopes. Comput. Geom. **47**(5), 614–624 (2014). https://doi.org/10.1016/j.comgeo.2014.01.003

27. Kobourov, S.G., Liotta, G., Montecchiani, F.: An annotated bibliography on 1-planarity. Comput. Sci. Reviews **25**, 49–67 (2017). https://doi.org/10.1016/j.cosrev.2017.06.002

28. Lenhart, W., Liotta, G., Mondal, D., Nishat, R.I.: Planar and plane slope number of partial 2-trees. In: Wismath, S., Wolff, A. (eds.) GD 2013. LNCS, vol. 8242, pp. 412–423. Springer, Cham (2013). https://doi.org/10.1007/978-3-319-03841-4_36

29. Liu, Y., Morgana, A., Simeone, B.: A linear algorithm for 2-bend embeddings of planar graphs in the two-dimensional grid. Discrete Appl. Math. **81**(1–3), 69–91 (1998). https://doi.org/10.1007/978-3-319-03841-4_36
30. Malitz, S.M., Papakostas, A.: On the angular resolution of planar graphs. SIAM J. Discrete Math. **7**(2), 172–183 (1994). https://doi.org/10.1137/S0895480193242931
31. Mukkamala, P., Pálvölgyi, D.: Drawing cubic graphs with the four basic slopes. In: van Kreveld, M., Speckmann, B. (eds.) GD 2011. LNCS, vol. 7034, pp. 254–265. Springer, Heidelberg (2012). https://doi.org/10.1007/978-3-642-25878-7_25
32. Pach, J., Pálvölgyi, D.: Bounded-degree graphs can have arbitrarily large slope numbers. Electr. J. Comb. **13**(1), 1–4 (2006). http://www.combinatorics.org/Volume_13/Abstracts/v13i1n1.html
33. Ringel, G.: Ein Sechsfarbenproblem auf der Kugel. Abh. Math. Semin. Univ. Hambg. **29**(1–2), 107–117 (1965). https://doi.org/10.1007/BF02996313

Best Paper Track 2

Aesthetic Discrimination of Graph Layouts

Moritz Klammler[1], Tamara Mchedlidze[1](✉), and Alexey Pak[2]

[1] Karlsruhe Institute of Technology, 76131 Karlsruhe, Germany
moritz@klammler.eu, mched@iti.uka.de
[2] Fraunhofer Institute of Optronics, System Technologies and Image Exploitation,
Fraunhoferstraße 1, 76131 Karlsruhe, Germany
alexey.pak@iosb.fraunhofer.de

Abstract. This paper addresses the following basic question: given two layouts of the same graph, which one is more aesthetically pleasing? We propose a neural network-based discriminator model trained on a labeled dataset that decides which of two layouts has a higher aesthetic quality. The feature vectors used as inputs to the model are based on known graph drawing quality metrics, classical statistics, information-theoretical quantities, and two-point statistics inspired by methods of condensed matter physics. The large corpus of layout pairs used for training and testing is constructed using force-directed drawing algorithms and the layouts that naturally stem from the process of graph generation. It is further extended using data augmentation techniques. Our model demonstrates a mean prediction accuracy of 96.48%, outperforming discriminators based on stress and on the linear combination of popular quality metrics by a small but statistically significant margin.

The full version of the paper including the appendix with additional illustrations is available at https://arxiv.org/abs/1809.01017.

Keywords: Graph drawing · Graph drawing aesthetics
Machine learning · Neural networks · Graph drawing syndromes

1 Introduction

What makes a drawing of a graph aesthetically pleasing? This admittedly vague question is central to the field of Graph Drawing which has over its history suggested numerous answers. Borrowing ideas from Mathematics, Physics, Arts, etc., many researchers have tried to formalize the elusive concept of aesthetics.

In particular, dozens of formulas collectively known as *drawing aesthetics* (or, more precisely, *quality metrics* [6]) have been proposed that attempt to capture in a single number how beautiful, readable and clear a drawing of an abstract graph is. Of those, simple metrics such as the number of edge crossings, minimum crossing angle, vertex distribution or angular resolution parameters, are obviously incapable *per se* of providing the ultimate aesthetic statement.

© Springer Nature Switzerland AG 2018
T. Biedl and A. Kerren (Eds.): GD 2018, LNCS 11282, pp. 169–184, 2018.
https://doi.org/10.1007/978-3-030-04414-5_12

Advanced metrics may represent, for example, the energy of a corresponding system of physical bodies [5,9]. This approach underlies many popular graph drawing algorithms [39] and often leads to pleasing results in practice. However, it is known that low values of energy or stress do not always correspond to the highest degree of symmetry [43] which is an important aesthetic criterion [30].

Another direction of research aims to narrow the scope of the original question to specific application domains, focusing on the purpose of a drawing or possible user actions it may facilitate (*tasks*). The target parameters – readability and the clarity of representation – may be assessed via user performance studies. However, even in this case such aesthetic notions as symmetry still remain important [30]. In general, aesthetically pleasing designs are known to positively affect the apparent and the actual usability [25,41] of interfaces and induce positive mental states of users, enhancing their problem-solving abilities [8].

In this work, we offer an alternative perspective on the aesthetics of graph drawings. First, we address a slightly modified question: "Of two given drawings of the same graph, which one is more aesthetically pleasing?". With that, we implicitly admit that "the ultimate" quality metric may not exist and one can hope for at most a (partial) ordering. Instead of a metric, we therefore search for a binary *discriminator function* of graph drawings. As limited as it is, it could be useful for practical applications such as picking the best answer out of outputs of several drawing algorithms or resolving local minima in layout optimization.

Second, like Huang et al. [13], we believe that by combining multiple metrics computed for each drawing, one has a better chance of capturing complex aesthetic properties. We thus also consider a "meta-algorithm" that aggregates several "input" metrics into a single value. However, unlike the recipe by Huang et al., we do not specify the form of this combination *a priori* but let an artificial neural network "learn" it based on a sample of labeled training data. In the recent years, machine learning techniques have proven useful in such aesthetics-related tasks as assessing the appeal of 3D shapes [4] or cropping photos [24]. Our network architecture is based on a so-called *Siamese neural network* [3] – a generic model specifically designed for binary functions of same-kind inputs.

Finally, we acknowledge that any simple or complex input metric may become crucial to the answer in some cases that are hard to predict *a priori*. We therefore implement as many input metrics as we can and relegate their ranking to the model. In addition to those known from the literature, we implement a few novel metrics inspired by statistical tools used in Condensed Matter Physics and Crystallography, which we expect to be helpful in capturing the symmetry, balance, and salient structures in large graphs. These metrics are based on so-called *syndromes* – variable-size multi-sets of numbers computed for a graph or its drawing (e.g. vertex coordinates or pairwise distances). In order to reduce these heterogeneous multi-sets to a fixed-size *feature vector* (input to the discriminator model), we perform a *feature extraction* process which may involve steps such as creating histograms or performing regressions.

In our experiments, our discriminator model outperforms the known (metric-based) algorithms and achieves an average accuracy of 96.48% when identifying the "better" graph drawing out of a pair. The project source code including the data generation procedure is available online [20].

The remainder of this paper is structured as follows. In Sect. 2 we briefly overview the state-of-the-art in quantifying graph layout aesthetics. Section 4 discusses the used syndromes of aesthetic quality, Sect. 5 feature extraction, and Sect. 6 the discriminator model. The dataset used in our experiments is described in Sect. 7. The results and the comparisons with the known metrics are presented in Sect. 8. Section 9 finalizes the paper and provides an outlook for future work.

2   Related Work

According to empirical studies, graph drawings that maximize one or several quality metrics are more aethetically pleasing and easier to read [12,13,28,31,42]. For instance, in their seminal work, Purchase et al. have established [30] that higher numbers of edge crossings and bends as well as lower levels of symmetry negatively influence user performance in graph reading tasks.

Many graph drawing algorithms attempt to optimize multiple quality metrics. As one way to combine them, Huang et al. [13] have used a weighted sum of "simple" metrics, effects of their interactions (see Purchase [29] or Huang and Huang [16]), and error terms to account for possible measurement errors.

In another work, Huang et al. [15] have empirically demonstrated that their "aggregate" metric is sensitive to quality changes and is correlated with the human performance in graph comprehension tasks. They have also noticed that the dependence of aesthetic quality on input quality metrics can be non-linear (e.g. a quadratic relationship better describes the interplay between crossing angles and drawing quality [14]). Our work extends this idea as we allow for arbitrary non-linear dependencies implemented by an artificial neural network.

In evolutionary graph drawing approaches, several techniques have been suggested to "train" a *fitness function*[1] from the user's responses as a composition of several known quality metrics. Masui [23] modeled the fitness function as a linear combination in which the weights are obtained via genetic programming from the pairs of "good" and "bad" layouts provided by users. The so-called co-evolution was used by Barbosa and Barreto [1] to evolve the weights of the fitness function in parallel with a drawing population in order to match the ranking made by users. Spönemann and others [37] suggested two alternative techniques. In the first one, the user directly chooses the weights with a slider. In the second, they select good layouts from the current population and the weights are adjusted according to the selection. Rosete-Suarez [32] determined the relative importance of individual quality metrics based on user inputs. Several machine learning-based approaches to graph drawing are described by dos Santos Vieira et al. [33]. Recently, Kwon et al. [22] presented a novel work on topological similarity of graphs. Their goal was to avoid expensive computations of graph layouts and their quality measures. The resulting system was able to sketch a graph in different layouts and estimate corresponding quality measures.

---

[1] Objective function in genetic algorithms that summarizes optimization goals.

3   Definitions

In this paper we consider general simple graphs $G = (V, E)$ where $V = V(G)$ and $E = E(G)$ are the vertex and edge sets of G with $|V| = n$ and $|E| = m$. A *drawing* or *layout* of a graph is its graphical representation where vertices are drawn as points or small circles, and the edges as straight line segments. Vertex positions in a drawing are denoted by $\boldsymbol{p}^k = (p_1^k, p_2^k)^{\mathrm{T}}$ for $k = 1, \ldots, n$ and their set $P = \{\boldsymbol{p}^k\}_{k=1}^n$. Furthermore, we use $\mathrm{dist}_G(u, v)$ to denote the *graph-theoretical distance* – the length of the shortest path between vertices u and v in G – and $\mathrm{dist}_\Gamma(u, v)$ for the Euclidean distance between u and v in the drawing $\Gamma(G)$.

4   Quality Syndromes of Graph Layouts

A *quality syndrome* of a layout Γ is a multi-set of numbers sharing an interpretation that are known or suspected to correlate with the aesthetic quality (e.g. all pairwise angles between incident edges in Γ). In the following we describe several syndromes (implemented in our code) inspired by popular quality metrics and common statistical tools. The list is by no means exhaustive, nor do we claim syndromes below as necessary or independent. Our model accepts any combination of syndromes; better choices remain to be systematically investigated.

PRINVEC1 **and** PRINVEC2. The two principal axes of the set P. If we define a covariance matrix $C = \{c_{ij}\}$, $c_{ij} = \frac{1}{n} \sum_{k=1}^n (p_i^k - \overline{p_i})(p_j^k - \overline{p_j}))$, $i, j \in \{1, 2\}$, where $\overline{p_i} = \frac{1}{n} \sum_{k=1}^n p_i^k$ are the mean values over each dimension, then PRINVEC1 and PRINVEC2 will be its eigenvectors.

PRINCOMP1 **and** PRINCOMP2. Projections of vertex positions onto $\boldsymbol{v}_1 =$ PRINVEC1 and $\boldsymbol{v}_2 =$ PRINVEC2, that is, $\{\langle (\boldsymbol{p}^j - \overline{\boldsymbol{p}}), \boldsymbol{v}_i \rangle\}_{j=1}^n$ for $i \in \{1, 2\}$ where $\langle \cdot, \cdot \rangle$ denotes the scalar product.

ANGULAR. Let $A(v)$ denote the sequence of edges incident to a vertex v, appearing in a clockwise order around it in Γ. Let $\alpha(e_i, e_j)$ denote the clockwise angle between edges e_i and e_j incident to the same vertex. This syndrome is then defined as $\bigcup_{v \in V(G)} \{\alpha(e_i, e_j) : e_i, e_j \text{ are consecutive in } A(v)\}$.

EDGE_LENGTH. $\bigcup_{(u,v) \in E(G)} \{\mathrm{dist}_\Gamma(u, v)\}$ is the set of edge lengths in Γ.

RDF_GLOBAL. $\bigcup_{u \neq v \in V(G)} \{\mathrm{dist}_\Gamma(u, v)\}$ contains distances between all vertices in the drawing. The concept of a *radial distribution function (RDF)* [7] (the distribution of RDF_GLOBAL) is borrowed from Statistical Physics and Crystallography and characterizes the regularity of molecular structures. In large graph layouts it captures regular, periodic and symmetric patterns in the vertex positions.

RDF_LOCAL(d). $\bigcup_{u \neq v \in V(G)} \{\mathrm{dist}_\Gamma(u, v) : \mathrm{dist}(u, v) \leq d\}$ is the set of distances between vertices such that the graph-theoretical distance between them is bounded by $d \in \mathbb{N}$. In our implementation, we compute RDF_LOCAL(2^i) for $i \in \{0, \ldots, \lceil \log_2(D) \rceil\}$ where D is the diameter of G. RDF_LOCAL(d) in a sense interpolates between EDGE_LENGTH $(d = 1)$ and RDF_GLOBAL $(d \to \infty)$.

TENSION. $\bigcup_{u \neq v \in V(G)} \{\text{dist}_\Gamma(u,v)/\text{dist}_G(u,v)\}$ are the ratios of Euclidean and graph-theoretical distances computed for all vertex pairs. TENSION is motivated by and is related to the well-known stress function [17].

Note that before computing the quality syndromes, we *normalize* all layouts so that the center of gravity of V is at the origin and the mean edge length is fixed in order to remove the effects of scaling and translation (but not rotation).

5   Feature Vectors

The sizes of quality syndromes are in general graph- and layout-dependent. A neural network, however, requires a fixed-size input. A collection of syndromes is condensed to this *feature vector* via *feature extraction*. Our approach to this step relies on several auxiliary definitions. Let $S = \{x_i\}_{i=1}^p$ be a syndrome with p entries. By S^μ we denote the arithmetic mean and by S^ρ the root mean square of S. We also define a *histogram sequence* $S^\beta = \frac{1}{p}(S_1, \ldots, S_\beta)$ – normalized counts in a histogram built over S with β bins. The *entropy* [36] of S^β is defined as

$$\mathscr{E}(S^\beta) = - \sum_{i=1}^p \log_2(S_i) S_i. \tag{1}$$

We expect the entropy, as a measure of disorder, to be related to the aesthetic quality of a layout and convey important information to the discriminator.

   The entropy $\mathscr{E}(S^\beta)$ is sensitive to the number of bins β (cf. Fig. 1). In order to avoid influencing the results via arbitrary choices of β, we compute it for $\beta = 8, 16, \ldots, 512$. After that, we perform a linear regression of $\mathscr{E}(S^\beta)$ as a function of $\log_2(\beta)$. Specifically, we find S^η and S^σ such that $\sum_\beta (S^\sigma \log_2 \beta + S^\eta - \mathscr{E}(S^\beta))^2$ is minimized. The parameters (intercept S^η and slope S^σ) of this regression no longer depend on the histogram size and are used as feature vector components. Figure 1 illustrates that the dependence of $\mathscr{E}(S^\beta)$ on $\log_2(\beta)$ is indeed often close to linear and the regression provides a decent approximation.

   A discrete histogram over S can be generalized to a continuous *sliding average*

$$S^F(x) = \frac{\sum_{i=1}^p F(x, x_i)}{\int_{-\infty}^{+\infty} dy \ \sum_{i=1}^p F(y, x_i)}. \tag{2}$$

A natural choice for the kernel $F(x,y)$ is the Gaussian $F_\sigma(x,y) = \exp\left(-\frac{(x-y)^2}{2\sigma^2}\right)$. By analogy to Eq. 1, we may now define the *differential entropy* [36] as

$$\mathscr{D}(S^{F_\sigma}) = - \int_{-\infty}^{+\infty} dx \ \log_2(S^{F_\sigma}(x)) \ S^{F_\sigma}(x). \tag{3}$$

This entropy via kernel function still depends on parameter σ (the filter width). Computing $\mathscr{D}(S^{F_\sigma})$ for multiple σ values as we do for $\mathscr{E}(S^\beta)$ is too expensive. Instead, we have found that using Scott's Normal Reference Rule [35] as a heuristic to fix σ yields satisfactory results, and allows us to define $S^\epsilon = \mathscr{D}(S^{F_\sigma})$.

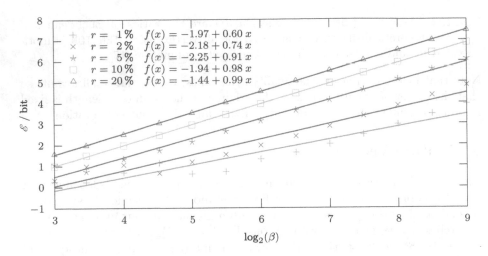

Fig. 1. Entropy $\mathcal{E} = \mathcal{E}(S^\beta)$ computed for histogram sequences S^β defined for different numbers of histogram bins β. Different markers (colors) correspond to several layouts of a regular grid-like graph, progressively distorted according to the parameter r. The dependence of $\mathcal{E}$ on $\log_2(\beta)$ is well approximated by a linear function. Both intercept and slope show a strong correlation with the levels of distortion r. (Color figure online)

Using these definitions, for the most complex syndrome RDF_LOCAL(d) we introduce RDF_LOCAL – a 30-tuple containing the arithmetic mean, root mean square and the differential entropy of RDF_LOCAL(2^i) for $i \in (0, \ldots, 9)$. With that[2], $\text{RDF\_LOCAL} = \big(\text{RDF\_LOCAL}(2^i)^\mu, \text{RDF\_LOCAL}(2^i)^\rho, \text{RDF\_LOCAL}(2^i)^\epsilon\big)_{i=0}^9$.

Finally, we assemble the 57-dimensional[3] feature vector for a layout Γ as

$$F_{\text{layout}}(\Gamma) = \text{PRINVEC1} \cup \text{PRINVEC2} \cup \text{RDF\_LOCAL} \cup \bigcup_S (S^\mu, S^\rho, S^\eta, S^\sigma)$$

where S ranges over PRINCOMP1, PRINCOMP2, ANGULAR, EDGE_LENGTH, RDF_GLOBAL and TENSION.

In addition, the discriminator model receives the trivial properties of the underlying graph as the second 2-dimensional vector $F_{\text{graph}}(G) = (\log(n), \log(m))$.

6   Discriminator Model

Feature extractors such as those introduced in the previous section reduce an arbitrary graph G and its arbitrary layout Γ to fixed-size vectors $F_{\text{graph}}(G)$ and

---

[2] Values $i < 10$ are sufficient as no graph in our dataset has a diameter exceeding 2^9.

[3] The size is one less than expected from the explanation above because we do not include the arithmetic mean for EDGE_LENGTH as it is constant (due to the layout normalization mentioned earlier) and therefore non-informative.

$F_{\text{layout}}(\Gamma)$. Given a graph G and a pair of its alternative layouts Γ_a and Γ_b, the discriminator function DM receives the feature vectors $v_a = F_{\text{layout}}(\Gamma_a)$, $v_b = F_{\text{layout}}(\Gamma_b)$ and $v_G = F_{\text{graph}}(G)$ and outputs a scalar value

$$t = \text{DM}(v_G, v_a, v_b) \in [-1, 1]. \tag{4}$$

The interpretation is as follows: if $t < 0$, then the model believes that Γ_a is "prettier" than Γ_b; if $t > 0$, then it prefers Γ_b. Its magnitude $|t|$ encodes the confidence level of the decision (the higher $|t|$, the more solid the answer).

For the implementation of the function DM we have chosen a practically convenient and flexible model structure known as *Siamese neural networks*, originally proposed by Bromley and others [3] that is defined as

$$\text{DM}(v_G, v_a, v_b) = \text{GM}(\sigma_a - \sigma_b, v_G) \tag{5}$$

where $\sigma_a = \text{SM}(v_a)$ and $\sigma_b = \text{SM}(v_b)$. The *shared model* SM and the *global model* GM are implemented as multi-layer neural networks with a simple structure shown in Fig. 2. The network was implemented using the *Keras* [18] framework with the *TensorFlow* [40] library as back-end.

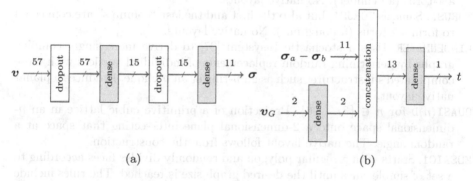

Fig. 2. Structure of the neural networks $\text{SM}(v)$ (a) and $\text{GM}(\sigma_a - \sigma_b, v_G)$ (b). Shaded blocks denote standard network layers, and the numbers on the arrows denote the dimensionality of the respective representations.

The SM network (Fig. 2(a)) consists of two "dense" (fully-connected) layers, each preceded by a "dropout" layer (discarding 50% and 25% of the signals, respectively). Dropout is a stochastic regularization technique intended to avoid overfitting that was first proposed by Srivastava and others [38].

In the GM network (Fig. 2(b)), the graph-related feature vector v_G is passed through an auxiliary dense layer, and concatenated with the difference signal $(\sigma_a - \sigma_b)$ obtained from the output vectors of SM for the two layouts. The final dense layer produces the scalar output value. The first and the auxiliary layers use linear activation functions, the hidden layer uses ReLU [11] and the final layer hyperbolic tangent activation. Following the standard practice, the inputs to the network are normalized by subtracting the mean and dividing by the standard deviation of the feature computed over the complete dataset.

In total, the DM model has 1 066 free parameters, trained via stochastic gradient descent-based optimization of the mean squared error (MSE) loss function.

7   Training and Testing Data

For training, all machine learning methods require datasets representing the variability of possible inputs. Our DM model needs a dataset containing graphs, their layouts, and known aesthetic orderings of layout pairs. We have assembled such a dataset using two types of sources. First, we used the collections of the well-known graph archives ROME, NORTH and RANDDAG which are published on graphdrawing.org as well as the NIST's "Matrix Market" [2].

Second, we have generated random graphs using the algorithms listed below. As a by-product, some of them produce layouts that stem naturally from the generation logic. We refer to these as *native* layouts (see [19] for details).

GRID. Regular $n \times m$ grids. Native layouts: regular rectangular grids.

TORUS1. Same as GRID, but the first and the last "rows" are connected to form a 1-torus (a cylinder). No native layouts.

TORUS2. Same as TORUS1, but also the first and the last "columns" are connected to form a 2-torus (a doughnut). No native layouts.

LINDENMAYER. Uses a stochastic L-system [27] to derive increasingly complex graphs by performing random replacements of individual vertices with more complicated substructures such as an n-ring or an n-clique. Produces a planar native layout.

QUASI⟨n⟩D for $n \in \{3, \ldots, 6\}$. Projection of a primitive cubic lattice in an n-dimensional space onto a 2-dimensional plane intersecting that space at a random angle. The native layout follows from the construction.

MOSAIC1. Starts with a regular polygon and randomly divides faces according to a set of simple rules until the desired graph size is reached. The rules include adding a vertex connected to all vertices of the face; subdividing each edge and adding a vertex that connects to each subdivision vertex; subdividing each edge and connecting them to a cycle. The native layout follows from the construction.

MOSAIC2. Applies a randomly chosen rule of MOSAIC1 to every face, with the goal of obtaining more symmetric graphs.

BOTTLE. Constructs a graph as a three-dimensional mesh over a random solid of revolution. The native layout is an axonometric projection.

For each graph, we have computed force-directed layouts using the FM^3 [10] and stress-minimization [17] algorithms. We assume these and native layouts to be generally aesthetically pleasing and call them all *proper* layouts of a graph.

Furthermore, we have generated a priori un-pleasing (*garbage*) layouts as follows. Given a graph $G = (V, E)$, we generate a random graph $G' = (V', E')$ with $|V'| = |V|$ and $|E'| = |E|$ and compute a force-directed layout for G'.

The coordinates found for the vertices V' are then assigned to V. We call these "phantom" layouts due to the use of a "phantom" graph G'. We find that phantom layouts look less artificial than purely random layouts when vertex positions are sampled from a uniform or a normal distribution. This might be due to the fact that G and G' have the same density and share some beneficial aspects of the force-directed method (such as mutual repelling of nodes).

For training and testing of the discriminator model we need a corpus of labeled pairs – triplets (Γ_a, Γ_b, t) where Γ_a and Γ_a are two different layouts for the same graph and $t \in [-1, 1]$ is a value indicating the relative aesthetic quality of Γ_a and Γ_b. A negative (positive) value for t expresses that the quality of Γ_a is superior (inferior) compared to Γ_b and the magnitude of t expresses the confidence of this prediction. We only use pairs with sufficiently large $|t|$.

As manually-labelled data were unavailable, we have fixed the values of t as follows. First, we paired a proper and a garbage layout of a graph. The assumption is that the former is always more pleasing (i.e. $t = \pm 1$). Second, in order to obtain more nuanced layout pairs and to increase the amount of data, we have employed the well-known technique of *data augmentation* as follows.

Layout Worsening: Given a proper layout Γ, we apply a transformation designed to gradually reduce its aesthetic quality that is modulated by some parameter $r \in [0, 1]$, resulting in a transformed layout Γ'_r. By varying the degree r of the distortion, we may generate a sequence of layouts ordered by their anticipated aesthetic value: a layout with less distortion is expected to be more pleasing than a layout with more distortion when starting from a presumably decent layout. We have implemented the following worsening techniques. PERTURB: add Gaussian noise to each node's coordinates. FLIP_NODES: swap coordinates of randomly selected node pairs. FLIP_EDGES: same as FLIP_NODES but restricted to connected node pairs. MOVLSQ: apply an affine deformation based on moving least squares suggested (although for a different purpose) by Schaefer et al. [34]. In essence, all vertices are shifted according to some smoothly varying coordinate mapping.

Layout Interpolation: As the second data augmentation technique, we linearly interpolated the positions of corresponding vertices between the proper and garbage layouts of the same graph. The resulting label t is then proportional to the difference in the interpolation parameter.

In total, using all the methods described above, we have been able to collect a database of about 36 000 labeled layout pairs.

8   Evaluation

The performance of the discriminator model was evaluated using *cross-validation* with 10-fold random subsampling [21]. In each round, 20% of graphs (with all their layouts) were chosen randomly and were set aside for testing, and the model was trained using the remaining layout pairs. Of N labeled pairs used for testing, in each round we computed the number $N_{correct}$ of pairs for which the model properly predicted the aesthetic preference, and derived the accuracy (success rate) $A = N_{correct}/N$. The standard deviation of A over the 10 runs was taken as the uncertainty of the results. With the average number of test samples of $N = 7415$, the eventual success rate was $A = (96.48 \pm 0.85)\%$.

8.1   Comparison with Other Metrics

In order to assess the relative standing of the suggested method, we have implemented two known aesthetic metrics (*stress* and the *combined metric* by Huang et al. [15]) and evaluated them over the same dataset. The metric values were trivially converted to the respective discriminator function outputs.

Stress $\mathcal{T}$ of a layout Γ of a simple connected graph $G = (V, E)$ was defined by Kamada and Kawai [17] as

$$\mathcal{T}(\Gamma) = \sum_{i=1}^{n-1} \sum_{j=i+1}^{n} k_{ij} \left(\text{dist}_\Gamma(v_i, v_j) - L \, \text{dist}_G(v_i, v_j)\right)^2 , \tag{6}$$

where L denotes the desirable edge length and $k_{ij} = K/\text{dist}_G(v_i, v_j)^2$ is the strength of a "spring" attached to v_i and v_j. The constant K is irrelevant in the context of discriminator functions and can be set to any value.

As observed by Welch and Kobourov [43], the numeric value of stress depends on the layout scale via the constant L in the Eq. 6 which complicates comparisons. Their suggested solution was for each layout to find L that minimizes $\mathcal{T}$ (e.g. using binary search). In our implementation, we applied a similar technique based on fitting and minimizing a quadratic function to the stress computed at three scales. We refer to this quantity as STRESS.

The combined metric proposed by Huang et al. [15] (referred to as COMB) is a weighted average of four simpler quality metrics: the number of edge crossings (CC), the minimum crossing angle between any two edges in the drawing (CR), the minimum angle between two adjacent edges (AR), and the standard deviation computed over all edge lengths (EL).

The average is computed over the so-called z-scores of the above metrics. Each z-score is found by subtracting the mean and dividing by the standard deviation of the metric for all layouts of a given graph to be compared with each other. More formally, let G be a graph and $\Gamma_1, \ldots, \Gamma_k$ be its k layouts to be compared pairwise. Let $M(\Gamma_i)$ be the value of metric M for Γ_i and μ_M and σ_M be the mean and the standard deviation of $M(\Gamma_i)$ for $i \in \{1, \ldots, k\}$. Then

$$z_M^{(i)} = \frac{M(\Gamma_i) - \mu_M}{\sigma_M} \tag{7}$$

is the z-score for metric M and layout Γ_i. The combined metric then is

$$\text{COMB}(\Gamma_j) = \sum_M w_M \, z_M^{(j)}. \tag{8}$$

The weights w_M were found via Nelder-Mead maximization [26] of the prediction accuracy over the training dataset[4].

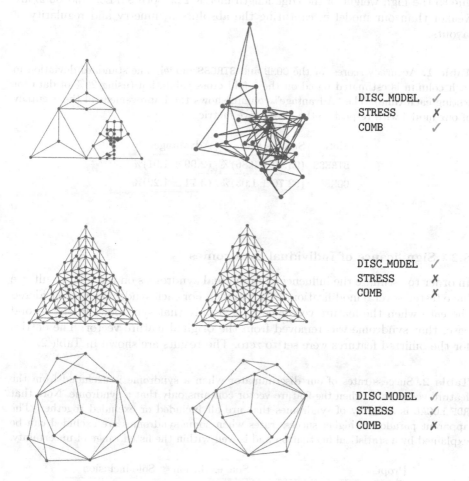

Fig. 3. Examples where our discriminator model (DISC_MODEL) succeeds ($\checkmark$) and the competing metrics fail ($\times$) to predict the answer correctly. In each row, the layout on the left is expected to be superior compared to the one on the right.

The accuracy of the stress-based and the combined model-based discriminators is shown in Table 1. In most cases, our model outperforms these algorithms by a comfortable margin. Figure 3 provides examples of mis-predictions. By inspecting such cases, we notice that STRESS often fails to guess the aesthetics of (almost) planar layouts that contain both very short and very long edges (such behavior may also be inferred from the definition of STRESS). We observe

---

[4] The obtained weights are: $w_{EL} = +0.4803 \pm 0.0855$, $w_{CC} = +0.4679 \pm 0.1069$, $w_{CR} = -0.0431 \pm 0.0315$, $w_{AR} = -0.0087 \pm 0.0072$.

that there are planar graphs, such as nested triangulations, for which this property is unavoidable in planar drawings. The mis-predictions of COMB seem to be due to the high weight of the edge length metric EL. Both STRESS and COMB are weaker than our model in capturing the absolute symmetry and regularity of layouts.

Table 1. Accuracy scores for the COMB and STRESS model. The standard deviation in each column is estimated based on the 5-fold cross-validation (using 20% of data for testing each time). The "Advantage" column shows the improvement in the accuracy of our model with respect to the alternative metric.

Metric	Success rate	Advantage
STRESS	$(93.49 \pm 0.86)\%$	$(2.99 \pm 1.01)\%$
COMB	$(92.76 \pm 1.03)\%$	$(3.71 \pm 1.22)\%$

8.2 Significance of Individual Syndromes

In order to estimate the influence of individual syndromes on the final result, we have tested several modifications of our model. For each syndrome, we considered the case when the feature vector contained only that syndrome. In the second case, that syndrome was removed from the original feature vector. The entries for the omitted features were set to zero. The results are shown in Table 2.

Table 2. Success rates of our discriminator when a syndrome is excluded from the feature vector, and when the feature vector contains only that a syndrome. Note that RDF_LOCAL is a family of syndromes that are all included or excluded together. The apparent paradox of higher success rates when some syndromes are excluded can be explained by a statistical fluctuation and is well within the listed range of uncertainty.

Property	Sole exclusion	Sole inclusion
PRINCOMP1	$(96.37 \pm 0.84)\%$	$(55.51 \pm 6.50)\%$
PRINCOMP2	$(96.20 \pm 0.76)\%$	$(61.08 \pm 5.24)\%$
EDGE_LENGTH	$(96.33 \pm 0.59)\%$	$(71.65 \pm 3.38)\%$
ANGULAR	$(96.40 \pm 0.34)\%$	$(77.79 \pm 6.06)\%$
RDF_GLOBAL	$(95.92 \pm 0.94)\%$	$(86.37 \pm 3.43)\%$
TENSION	$(96.83 \pm 0.31)\%$	$(89.78 \pm 0.95)\%$
RDF_LOCAL	$(90.04 \pm 2.04)\%$	$(94.78 \pm 1.60)\%$
Baseline using all properties	$(96.48 \pm 0.85)\%$	

As can be observed, the dominant contribution to the accuracy of the model is due to the RDF-based properties RDF_LOCAL and RDF_GLOBAL. The exclusion of other syndromes does not significantly change the results (they agree within the estimated uncertainty). However, the sole inclusion of these syndromes still performs better than random choice. This suggests that there is a considerable overlap between the aesthetic aspects captured by various syndromes. Further analysis is needed to identify the nature and the magnitude of these correlations.

9   Conclusion

In this paper we propose a machine learning-based discriminator model that selects the more aesthetically pleasing drawing from a pair of graph layouts. Our model picks the "better" layout in more than 96% cases and outperforms known stress-based and linear combination-based models. To the best of our knowledge, this is the first application of machine learning methods to this question. Previously, such techniques have proven successful in a range of complex issues involving aesthetics, prior knowledge, and unstated rules in object recognition, industrial design, and digital arts. As our model uses a simple network architecture, investigating the performance of more complex networks is warranted.

Previous efforts were focused on determining the aesthetic quality of a layout as a weighted average of individual quality metrics. We extend these ideas and findings in the sense that we do not assume any particular form of dependency between the overall aesthetic quality and the individual quality metrics.

Going beyond simple quality metrics, we define quality syndromes that capture arrays of information about graphs and layouts. In particular, we borrow the notion of RDF from Statistical Physics and Crystallography; RDF-based features demonstrate the strongest potential in extracting the aesthetic quality of a layout. We expect RDFs (describing the microscopic structure of materials) to be the most relevant for large graphs. It is tempting to investigate whether further tools from physics can be useful in capturing drawing aesthetics.

From multiple syndromes, we construct fixed-size feature vectors using common statistical tools. Our feature vector does not contain any information on crossings or crossing angles, nevertheless its performance is superior with respect to the weighted averages-based model which accounts for both. It would be interesting to investigate whether including these and other features further improves the performance of the neural network-based model.

In order to train and evaluate the model, we have assembled a relatively large corpus of labeled pairs of layouts, using available and generated graphs and exploiting the assumption that layouts produced by force-directed algorithms and native graph layouts are aesthetically pleasing and that disturbing them reduces the aesthetic quality. We admit that this study should ideally be repeated with human-labeled data. However, this requires that a dataset be collected with a size similar to ours, which is a challenging task. Creating such a dataset may become a critically important accomplishment in the graph drawing field.

References

1. Barbosa, H.J.C., Barreto, A.M.S.: An interactive genetic algorithm with coevolution of weights for multiobjective problems. In: Spector, L., Goodman, E.D., Wu, A., Langdon, W.B., Voigt, H.M. (eds.) An Interactive Genetic Algorithm With Co-evolution of Weights for Multiobjective Problems, GECCO 2001, pp. 203–210. Morgan Kaufmann Publishers Inc. (2001)
2. Boisvert, R.F., Pozo, R., Remington, K., Barrett, R.F., Dongarra, J.J.: Matrix market: a web resource for test matrix collections. In: Boisvert, R.F. (ed.) Quality of Numerical Software: Assessment and enhancement, pp. 125–137. Springer, Heidelberg (1997). https://doi.org/10.1007/978-1-5041-2940-4_9
3. Bromley, J., Guyon, I., LeCun, Y., Säckinger, E., Shah, R.: Signature verification using a "Siamese" time delay neural network. In: Advances in Neural Information Processing Systems, pp. 737–744 (1994). https://doi.org/10.1142/S0218001493000339
4. Dev, K., Villar, N., Lau, M.: Polygons, points, or voxels?: Stimuli selection for crowdsourcing aesthetics preferences of 3d shape pairs. In: Gooch, B., Gingold, Y.I., Winnemoeller, H., Bartram, L., Spencer, S.N. (eds.) Proceedings of the symposium on Computational Aesthetics, CAE 2017, Los Angeles, California, USA, pp. 2:1–2:7. ACM (2017). https://doi.org/10.1145/3092912.3092918
5. Eades, P.: A heuristic for graph drawing. Congr. Numer. **24**, 149–160 (1984)
6. Eades, P., Hong, S., Nguyen, A., Klein, K.: Shape-based quality metrics for large graph visualization. J. Graph Algorithms Appl. **21**(1), 29–53 (2017). https://doi.org/10.7155/jgaa.00405
7. Findenegg, G.H., Hellweg, T.: Statistische Thermodynamik, 2nd edn. Springer, Darmstadt (2015). https://doi.org/10.1007/978-3-642-37872-0
8. Fredrickson, B.L.: What good are positive emotions. Rev. Gen. Psychol. **2**, 300–319 (1998)
9. Fruchterman, T.M.J., Reingold, E.M.: Graph drawing by force-directed placement. Softw. Pract. Exper. **21**(11), 1129–1164 (1991). https://doi.org/10.1002/spe.4380211102
10. Hachul, S., Jünger, M.: Drawing large graphs with a potential-field-based multilevel algorithm. In: Pach, J. (ed.) GD 2004. LNCS, vol. 3383, pp. 285–295. Springer, Heidelberg (2005). https://doi.org/10.1007/978-3-540-31843-9_29
11. Hahnloser, R.H.R., Sarpeshkar, R., Mahowald, M.A., Douglas, R.J., Seung, H.S.: Digital selection and analogue amplification coexist in a cortex-inspired silicon circuit. Nature **405**, 947–951 (2000). https://doi.org/10.1038/35016072
12. Huang, W., Eades, P.: How people read graphs. In: Hong, S. (ed.) Asia-Pacific Symposium on Information Visualisation, APVIS 2005, pp. 51–58, Australia, Sydney (2005)
13. Huang, W., Eades, P., Hong, S.H., Lin, C.C.: Improving multiple aesthetics produces better graph drawings. J. Vis. Lang. Comput. **24**(4), 262–272 (2013). https://doi.org/10.1016/j.jvlc.2011.12.002
14. Huang, W., Hong, S., Eades, P.: Effects of crossing angles. In: IEEE VGTC Pacific Visualization Symposium 2008, PacificVis 2008, Kyoto, Japan, pp. 41–46 (2008). https://doi.org/10.1109/PACIFICVIS.2008.4475457
15. Huang, W., Huang, M.L., Lin, C.C.: Evaluating overall quality of graph visualizations based on aesthetics aggregation. Inf. Sci. **330**, 444–454 (2016). https://doi.org/10.1016/j.ins.2015.05.028, SI: Visual Info Communication

16. Huang, W., Huang, M.: Exploring the relative importance of crossing number and crossing angle. In: Dai, G., et al. (eds.) Proceedings of the 3rd International Symposium on Visual Information Communication, VINCI 2010, pp. 1–8, ACM (2010). https://doi.org/10.1145/1865841.1865854
17. Kamada, T., Kawai, S.: An algorithm for drawing general undirected graphs. Inf. Process. Lett. **31**(1), 7–15 (1989). https://doi.org/10.1016/0020-0190(89)90102-6
18. Keras. https://keras.io/
19. Klammler, M.: Aesthetic value of graph layouts: investigation of statistical syndromes for automatic quantification. Master's thesis, Karlsruhe Institute of Technology (2018). http://klammler.eu/msc/
20. Klammler, M., et al.: Source code for aesthetic discrimination of graph layouts. https://github.com/5gon12eder/msc-graphstudy
21. Kohavi, R.: A study of cross-validation and bootstrap for accuracy estimation and model selection. In: Proceedings of the 14thInternational Joint Conference on Artificial Intelligence, IJCAI 1995, Montréal Québec, Canada, vol. 2, pp. 1137–1145. Morgan Kaufmann (1995)
22. Kwon, O.H., Crnovrsanin, T., Ma, K.L.: What would a graph look like in this layout? A machine learning approach to large graph visualization. IEEE Trans. Vis. Comput. Graph. **24**(1), 478–488 (2018). https://doi.org/10.1109/TVCG.2017.2743858
23. Masui, T.: Evolutionary learning of graph layout constraints from examples. In: Szekely, P.A. (ed.) Proceedings of the 7th Annual ACM Symposium on User Interface Software and Technology, pp. 103–108. UIST 1994. ACM (1994). https://doi.org/10.1145/192426.192468
24. Nishiyama, M., Okabe, T., Sato, Y., Sato, I.: Sensation-based photo cropping. In: Gao, W., et al. (eds.) Proceedings of the 17th ACM International Conference on Multimedia. pp. 669–672. MM 2009, ACM (2009). https://doi.org/10.1145/1631272.1631384
25. Norman, D.A.: Emotion & design: attractive things work better. Interactions **9**(4), 36–42 (2002). https://doi.org/10.1145/543434.543435
26. Press, W., Teukolsky, S., Vetterling, W., Flannery, B.: Numerical Recipes: The Art of Scientific Computing, 3 edn. Cambridge University Press (2007)
27. Prusinkiewicz, P., Lindenmayer, A.: The Algorithmic Beauty of Plants. Springer, New York (1990). https://doi.org/10.1007/978-1-4613-8476-2
28. Purchase, H.: Which aesthetic has the greatest effect on human understanding? In: DiBattista, G. (ed.) GD 1997. LNCS, vol. 1353, pp. 248–261. Springer, Heidelberg (1997). https://doi.org/10.1007/3-540-63938-1_67
29. Purchase, H.C.: Performance of layout algorithms: comprehension, not computation. J. Vis. Lang. Comput. **9**(6), 647–657 (1998). https://doi.org/10.1006/jvlc.1998.0093
30. Purchase, H.C., Cohen, R.F., James, M.: Validating graph drawing aesthetics. In: Brandenburg, F.J. (ed.) GD 1995. LNCS, vol. 1027, pp. 435–446. Springer, Heidelberg (1996). https://doi.org/10.1007/BFb0021827
31. Purchase, H.C., Hamer, J., Nöllenburg, M., Kobourov, S.G.: On the usability of Lombardi graph drawings. In: Didimo, W., Patrignani, M. (eds.) GD 2012. LNCS, vol. 7704, pp. 451–462. Springer, Heidelberg (2013). https://doi.org/10.1007/978-3-642-36763-2_40
32. Rosete-Suarez, A., Sebag, M., Ochoa-Rodriguez, A.: A study of evolutionary graph drawing: laboratoire de Recherche en Informatique (LRI), Universite Paris-Sud XI, p. 1228. Technical report (1999)

33. dos Santos Vieira, R., do Nascimento, H.A.D., da Silva, W.B.: The application of machine learning to problems in graph drawing – a literature review. In: Proceedings of the 7th International Conference on Information, Process, and Knowledge Management,eKNOW 2015, Lisbon, Portugal, pp. 112–118 (2015)

34. Schaefer, S., McPhail, T., Warren, J.: Image deformation using moving least squares. ACM Trans. Graph. **25**(3), 533–540 (2006). https://doi.org/10.1145/1141911.1141920

35. Scott, D.W.: On optimal and data-based histograms. Biometrika **66**(3), 605–610 (1979). https://doi.org/10.1093/biomet/66.3.605

36. Shannon, C.E.: A mathematical theory of communication. Bell Syst. Tech. J. **27**(4), 623–656 (1948). https://doi.org/10.1002/j.1538-7305.1948.tb00917.x

37. Spönemann, M., Duderstadt, B., von Hanxleden, R.: Evolutionary meta layout of graphs. In: Dwyer, T., Purchase, H., Delaney, A. (eds.) Diagrams 2014. LNCS (LNAI), vol. 8578, pp. 16–30. Springer, Heidelberg (2014). https://doi.org/10.1007/978-3-662-44043-8_3

38. Srivastava, N., Hinton, G., Krizhevsky, A., Sutskever, I., Salakhutdinov, R.: Dropout: a simple way to prevent neural networks from overfitting. J. Mach. Learn. Res. **15**(1), 1929–1958 (2014)

39. Tamassia, R.: Handbook of Graph Drawing and Visualization. Discrete Mathematics and Its Applications. CRC Press (2013)

40. Tensor flow. https://tensorflow.org/

41. Tractinsky, N., Katz, A.S., Ikar, D.: What is beautiful is usable. Interact. Comput. **13**(2), 127–145 (2000). https://doi.org/10.1016/S0953-5438(00)00031-X

42. Ware, C., Purchase, H.C., Colpoys, L., McGill, M.: Cognitive measurements of graph aesthetics. Inf. Vis. **1**(2), 103–110 (2002). https://doi.org/10.1057/palgrave.ivs.9500013

43. Welch, E., Kobourov, S.: Measuring symmetry in drawings of graphs. Comput. Graph. Forum **36**(3), 341–351 (2017). https://doi.org/10.1111/cgf.13192

Orders

A Flow Formulation for Horizontal Coordinate Assignment with Prescribed Width

Michael Jünger[1], Petra Mutzel[2], and Christiane Spisla[2]($\boxtimes$)

[1] University of Cologne, Cologne, Germany
mjuenger@informatik.uni-koeln.de
[2] TU Dortmund University, Dortmund, Germany
{petra.mutzel,christiane.spisla}@cs.tu-dortmund.de

Abstract. We consider the coordinate assignment phase of the well known Sugiyama framework for drawing directed graphs in a hierarchical style. The extensive literature in this area has given comparatively little attention to a prescribed width of the drawing. We present a minimum cost flow formulation that supports prescribed width and optionally other criteria like lower and upper bounds on the distance of neighboring nodes in a layer or enforced vertical edge segments. In our experiments we demonstrate that our approach can compete with state-of-the-art algorithms.

Keywords: Hierarchical drawings · Coordinate assignment
Minimum cost flow · Prescribed drawing width

1 Introduction

The Sugiyama framework [12] is a popular approach for drawing directed graphs. It layouts the graph in a hierarchical manner and works in five phases: Cycle removal, layer assignment, crossing minimization, coordinate assignment and edge routing. If the graph is not already acyclic, some edges are reversed to prepare the graph for the next phase. Then each node is assigned to a layer so that all edges point from top to bottom. After that the orderings of the nodes within each layer are determined. In the coordinate assignment phase that we consider here, the exact positions of the nodes are fixed. Finally the edges are layouted, e.g., as straight lines. A good overview over the different phases of the framework can be found in [9].

After the nodes are assigned to layers and the orderings of the nodes within their layers are fixed, the task of the coordinate assignment phase is to compute x-coordinates for all nodes. There are several, sometimes contradicting, objectives in this phase, e.g., short edges, minimum distance between neighboring nodes, straight edges, balanced positions of the nodes between their neighbors in adjacent layers, and few bend points of edges that cross multiple layers. The

© Springer Nature Switzerland AG 2018
T. Biedl and A. Kerren (Eds.): GD 2018, LNCS 11282, pp. 187–199, 2018.
https://doi.org/10.1007/978-3-030-04414-5_13

criterion "short edges" can be handled by exact algorithms as well as fast heuristics that give pleasant results, possibly also considering other aesthetic criteria.

When it comes to the width of the drawing one usually tries to restrict the maximum number of nodes in one layer, see e.g. [5]. Long edges, i.e. edges that span more than two layers, are often split into paths with one dummy node on each intermediate layer. Healy and Nikolov [8] present a branch-and-cut approach to compute a layering that takes the influence of the number of dummy nodes on the width into account. Jabrayilov et al. [10] do the same in a mixed integer program that treats the first two phases of the Sugiyama framework simultaneously. But still, the maximum number of nodes in one layer does not necessarily define the actual width of the final drawing, as illustrated in Fig. 1. The main objective of most methods for the coordinate assignment phase is "short edges", which often leads to small drawings, but the width of the final layout is not directly addressed.

Fig. 1. In the left picture the horizontal edge length is $k - 3$ and the width is 1, in the right picture the horizontal edge length is 0 and the width is $k - 2$, where k is the number of layers.

There may be further requirements for the final drawing, such as an aspect ratio in order to make optimal use of the drawing area, or a maximum distance between two nodes on the same layer if they are semantically related. A common request is that inner segments of long edges are drawn as vertical straight lines in order to improve readability.

Related Work. Sugiyama et al. [12] present a quadratic programming formulation that has a combination of two asthetic criteria as objective function, short edges (closeness to adjacent nodes) and a balanced layout (positioning nodes close to the barycenter of their upper and lower neighbors). Gansner et al. [7] give a simpler formulation in which they replace quadratic terms of the form $(x_v - x_u)^2$ by $|x_v - x_u|$ and leave out the balance terms. The coordinate assignment problem can be interpreted as an instance of the layer assignment problem, and they suggest to apply the network simplex algorithm to an auxiliary graph to obtain a drawing with minimum horizontal edge length. Given an initial layout, some heuristics sweep through the layers and try to shift the nodes to better positions depending on the fixed x-coordinates of their neighbors in adjacent layers, see e.g. [6,11,12]. Two fast heuristics that compute coordinates from scratch are presented by Buchheim et al. [3] and by Brandes and Köpf [2]. Both algorithms draw inner segments of long edges straight and aim for a balanced layout with short edges.

Our Contribution. We formulate the coordinate assignment problem as a minimum cost flow problem that can be solved efficiently. Within this formulation we can fix the maximum width of the final drawing as well as a maximum and minimum horizontal distance between nodes in the same layer and we can enforce straightness to some edges. We compute x-coordinates such that the total horizontal edge length is minimized subject to these further constraints.

2   Notation and Preliminaries

Let $G = (V, E)$ be a directed graph with $|V| = n$ nodes and $|E| = m$ edges. For a directed edge $e = (u, v)$ we denote the start node of e with $start(e) = u$ and the target node of e with $target(e) = v$. A *path* P from u to v of length k is a set of edges $\{e_i = (v_i, v_{i+1}) \mid i = 1, \dots, k$ where $u = v_1$ and $v = v_{k+1}\}$. We also write $u \xrightarrow{*} v$. If $v_{k+1} = v_1$ it is called a *cycle*. A graph is called a *directed acyclic graph (DAG)* if it has no cycles. A *layering* $\mathcal{L}$ of a graph assigns every $v \in V$ a *layer* L_i, such that $i < j$ holds for every edge $e = (u, v)$ with $\mathcal{L}(u) = L_i$ and $\mathcal{L}(v) = L_j$. The layering is called *proper* if $\mathcal{L}(v) = \mathcal{L}(u) + 1$ for every edge (u, v), i.e., the layers of every pair of adjacent nodes are consecutive. An edge that violates the latter property is called a *long edge*. Every graph with a layering can be transformed into a graph with a proper layering by subdividing every long edge into a chain of edges. We denote with $|\mathcal{L}|$ the number of layers and with $|L_i|$ the number of nodes in layer L_i.

An *ordering* ord defines a partial ordering on the nodes of G. For every layer L_i it assigns each node in L_i a number $1 \leq j \leq |L_i|$ and we write $u < v$ if $ord(u) < ord(v)$. We denote with v_j^i the j-th node in layer L_i.

Given a graph G with a layering $\mathcal{L}$ and an ordering ord the *horizontal coordinate assignment problem (HCAP)* asks for x-coordinates for every node, so that $x(u) < x(v)$ if $u < v$. We will restrict ourselves to integer coordinates. The *horizontal length* of an edge $e = (u, v)$ is defined as $length(e) = |x(v) - x(u)|$ and the *total horizontal edge length* is $length(E) = \sum_{e \in E} length(e)$. The *width* of the assignment is $\max_{v \in V} x(v) - \min_{v \in V} x(v)$. Unless otherwise stated, we mean the horizontal length whenever we talk about the length of an edge.

$HCAP_{minEL}$ is the variant of HCAP in which we also want to minimize the total horizontal edge length.

We assume familiarity with minimum cost flows. Ahuja et al. [1] give a good overview. Let $N = (V_N, E_N)$ be a directed graph with a super source s and a super sink t, so for all other nodes the amount of incoming flow equals the amount of outgoing flow. We have lower and upper bounds on the edges and a cost function $cost : E_N \to \mathbb{R}$. Let f be a feasible flow. For a subset of nodes $V' \subseteq V_N \setminus \{s, t\}$ we denote with $f(V') = \sum_{v \in V'} \sum_{e=(v,w)} f(e) = \sum_{v \in V'} \sum_{e=(u,v)} f(e)$ the flow through V'. For s we define $f(s)$ to be the total amount of flow leaving s. For a subset of edges $E' \subset E_N$ we denote with $f(E') = \sum_{e \in E'} f(e)$ the flow over E' and with $cost(E') = \sum_{e \in E'} cost(e)$ the cost of E' and with $cost_f = \sum_{e \in E_N} f(e) \cdot cost(e)$ the total cost of f.

3   Network Flow Formulation

In this section we describe the construction of a network for the horizontal coordinate assignment problem. Given a minimum cost flow in this network we show how to obtain x-coordinates for all nodes such that the total horizontal edge length is minimized. By a simple modification we can compute x-coordinates that give us minimum total horizontal edge length with respect to a given maximum width of the drawing. The basic idea is that flow represents horizontal distance and we send flow from top to bottom through the layers.

3.1   Network Construction

Let $G = (V, E)$ be a DAG with a proper layering $\mathcal{L}$ and an ordering and let $N = (V_N, E_N)$ be the minimum cost flow network. For now let us assume that neighboring nodes on a layer should have an equal minimum distance of one and that we have no further requirements concerning the edges.

For every layer L_i with $i \in \{1, \ldots, |\mathcal{L}|\}$ we add nodes $w_0^i, w_1^i, \ldots, w_{|L_i|}^i$ and $z_0^i, z_1^i, \ldots, z_{|L_i|}^i$ to N. Imagine the node w_j^i placed above the layer L_i and between v_j^i and v_{j+1}^i (w_0^i is placed at the left end and $w_{|L_i|}^i$ at the right end of the layer). The nodes z_j^i are placed in the same way below layer L_i. Although we do not have a drawing of G at this moment we can still use terms like "above" and "below" because the layering gives us a vertical ordering of the nodes of G and we can talk about "left" and "right" because of the given ordering of the nodes in each layer. Since we are placing the nodes w_j^i and z_j^i "between" the nodes v_j^i and v_{j+1}^i we want to extend the "$<$" relation to give a partial ordering on $V \cup V_N$ in the following way: $w_0^i < v_1^i < w_1^i < v_2^i < \cdots < v_{|L_i|}^i < w_{|L_i|}^i$ and $z_0^i < v_1^i < z_1^i < v_2^i < \cdots < v_{|L_i|}^i < z_{|L_i|}^i$. We connect w_j^i to z_j^i with an edge a_j^i that has a lower bound of one and an upper bound of ∞ and a cost of zero. The flow over these edges will define the distance between v_j^i and v_{j+1}^i. We denote the set of these edges with A. Figure 2(a) shows an example.

For every layer L_i with $i \in \{1, \ldots, |\mathcal{L}|\}$ and every $j \in \{0, \ldots, |L_i| - 1\}$ we add edges $\overrightarrow{bw}_j^i = (w_j^i, w_{j+1}^i)$, $\overleftarrow{bw}_j^i = (w_{j+1}^i, w_j^i)$, $\overrightarrow{bz}_j^i = (z_j^i, z_{j+1}^i)$ and $\overleftarrow{bz}_j^i = (z_{j+1}^i, z_j^i)$ to N. The lower bound of these edges is zero and the upper bound is ∞. The cost of these network edges equals the number of graph edges they "cross over". That means, the cost of $\overleftarrow{bw}_j^i$ and $\overrightarrow{bw}_j^i$ equals the number of incoming graph edges of node v_j^i and the cost of $\overleftarrow{bz}_j^i$ and $\overrightarrow{bz}_j^i$ equals the number of outgoing graph edges of v_j^i, see Fig. 2(a). Positive flow over one of these edges will cause the crossed-over graph edges to have positive horizontal length. We call the set of these edges B.

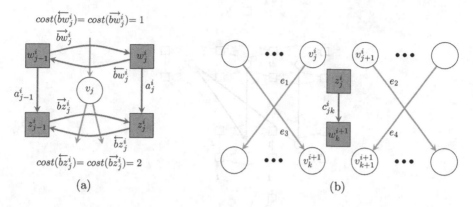

Fig. 2. Illustration of edges of the sets (a) A, B and (b) C. Nodes of G are white circles, nodes of N are green rectangles. Edges of G are gray, edges of N are green. (Color figure online)

Now we connect the nodes of neighboring layers. We could add edges between every z_j^i and every w_k^{i+1}, but we want to keep the number of edges between layers as small as possible. We add edges only in special situations and will show later that this suffices for correctness. For every layer L_i with $i \in \{1, \ldots, |\mathcal{L}| - 1\}$ we add edges $c_{00}^i = (z_0^i, w_0^{i+1})$ and $c_{|L_i||L_{i+1}|}^i = (z_{|L_i|}^i, w_{|L_{i+1}|}^{i+1})$ to the network with a lower bound of zero, an upper bound of ∞ and a cost of zero. Additionally we add edges $c_{jk}^i = (z_j^i, w_k^{i+1})$ if there exist $e_1, e_2, e_3, e_4 \in E$ with $start(e_1) = v_j^i$, $start(e_2) = v_{j'}^i$, where $v_{j'}^i$ is the next node to the right of v_j^i with an outgoing edge and $target(e_3) = v_k^{i+1}$, $target(e_4) = v_{k'}^{i+1}$, where $v_{k'}^{i+1}$ is the next node to the right of v_k^{i+1} with an incoming edge and the following conditions holds: $start(e_3) \leq start(e_1) < start(e_2) \leq start(e_4)$ and $target(e_1) \leq target(e_3) < target(e_4) \leq target(e_2)$. We call this situation a *hug* between z_j^i and w_k^{i+1}. These edges get a lower bound of zero, an upper bound of ∞, and the cost equals the number of graph edges they cross over: $cost(c_{jk}^i) = |\{e = (v_p^i, v_q^i) \in E \mid p \leq j \wedge q \geq k'$ or $p \geq j' \wedge q \leq k\}|$. Like the edges of B, flow on edges of this kind will cause horizontal length and we denote the set of all c_{jk}^i by C. Figure 2(b) illustrates a hug situation.

Finally we add a super source s and a super sink t to the network. We connect s with every $w_j^1, j \in \{1, \ldots, |L_1|\}$ and t with every $z_k^{|\mathcal{L}|}, k \in \{1, \ldots, |L_{|\mathcal{L}|}|\}$. These edges get a lower bound of zero, an upper bound of ∞ and a cost of zero. Figure 3 shows a complete example network. If it is clear from the context which layer or which node is meant, we omit the node subscripts and superscripts.

3.2   Obtaining Coordinates and Correctness

Let f be a feasible flow in the network described above. We observe that $f(a_j^i) = f(w_j^i) = f(z_j^i)$ since a_j^i is the only outgoing edge of w_j^i and the only incoming

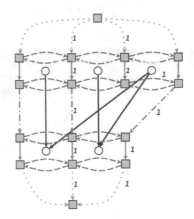

Fig. 3. An example network with underlying graph. Edges of the set A are red (solid), edges of the set B are blue (dashed) and edges of the set C are purple (dashdotted). The numbers along the edges denote the flow, unlabeled edges carry no flow. (Color figure online)

edge of z_j^i. We define the x-coordinate of a node v_j^i as

$$x(v_j^i) := \sum_{l=0}^{j-1} f(a_l^i) = \sum_{l=0}^{j-1} f(w_l^i) = \sum_{l=0}^{j-1} f(z_l^i). \tag{1}$$

Together with $y(v_j^i) = i$ we get an induced drawing with a feasible coordinate assignment, because for every v_j, v_k within the same layer $x(v_j) < x(v_k)$ if and only if $v_j < v_k$ (since the amount of flow over edges a $\in A$ is always positive).

Now we want to explain the correspondence between the cost of a flow f and the total horizontal edge length of the resulting drawing. The intuition is, that if flow is sent from the right of $start(e)$ to the left of $target(e)$ for some edge e, then $target(e)$ is "pushed" to the right because of the additional flow on the left. This results in a horizontal expansion of e. We define for an edge $e = (u, v) \in E$

$$\overrightarrow{E}(e) := \{bw \in B \mid start(bw) < v \;\wedge\; target(bw) > v\}$$
$$\cup \; \{bz \in B \mid start(bz) < u \;\wedge\; target(bz) > u\}$$
$$\cup \; \{c \in C \mid start(c) < u \;\wedge\; target(c) > v\}$$

as the set of network edges that start to the left of e and end to the right of e, thus cross over e from left to right. Analogously the set of network edges that cross over a graph edge from right to left is

$$\overleftarrow{E}(e) := \{bw \in B \mid start(bw) > v \;\wedge\; target(bw) < v\}$$
$$\cup \; \{bz \in B \mid start(bz) > u \;\wedge\; target(bz) < u\}$$
$$\cup \; \{c \in C \mid start(c) > u \;\wedge\; target(c) < v\}.$$

We make the following observations:

Property 1. $cost(g) = |\{e \in E \mid g \in \overrightarrow{E}(e)\}| + |\{e \in E \mid g \in \overleftarrow{E}(e)\}| \; \forall g \in B \cup C.$

Property 2. *Because of the flow conservation rule we have*
$\sum_{j=0}^{|L_i|} f(w_j^i) = \sum_{j=0}^{|L_i|} f(z_j^i)$ *for all* $i \in \{1, \ldots, |\mathcal{L}|\}$ *and*
$\sum_{j=0}^{|L_i|} f(w_j^i) = \sum_{j=0}^{|L_k|} f(w_j^k) = f(s)$ *for all* $i, k \in \{1, \ldots, |\mathcal{L}|\}.$

Property 3. *The width of the induced drawing is*
$\max_{1 \leq i \leq |\mathcal{L}|} \left(\sum_{j=1}^{|L_i|-1} f(w_j^i) \right) \leq f(s).$

Property 4. *Let* $e = (v_j^i, v_k^{i+1})$ *be an edge. Then*
$\sum_{w_l^{i+1} < v_k^{i+1}} f(w_l^{i+1}) = \sum_{z_l^i < v_j^i} f(z_l^i) + f(\overrightarrow{E}(e)) - f(\overleftarrow{E}(e)).$

The last property is illustrated in Fig. 4. The total flow that reaches all w_j^{i+1} that are to left of $target(e)$ comes from the z_j^i that are to the left of $start(e)$ and from nodes that are to the right of $target(e)$ or $start(e)$. Flow from the latter nodes has to pass over e from right to left. Flow from a node z_j^i that is to the left of $start(e)$ and does not enter one of the w_j^{i+1} left of $target(e)$ has to pass over e from left to right.

Lemma 1. *For a feasible flow f and the induced drawing* $cost_f \geq length(E)$ *holds.*

Proof. Let $e = (v_j^i, v_k^{i+1})$ be an edge of G. The length of e is $length(e) = |x(v_j^i) - x(v_k^{i+1})|$ and together with (1) we have

$$
length(e) = \left| \sum_{l=0}^{j-1} f(z_l^i) - \sum_{l=0}^{k-1} f(w_l^{i+1}) \right|
$$

$$
= \left| \sum_{z_l^i < start(e)} f(z_l^i) - \sum_{w_l^{i+1} < target(e)} f(w_l^{i+1}) \right|
$$

$$
= \left| f(\overrightarrow{E}(e)) - f(\overleftarrow{E}(e)) \right| \quad \text{(by Property 4).}
$$

Therefore we have for the total edge length

$$
length(E) = \sum_{e \in E} \left| f(\overrightarrow{E}(e)) - f(\overleftarrow{E}(e)) \right|
$$

$$
\leq \sum_{e \in E} \left(\left| f(\overrightarrow{E}(e)) \right| + \left| f(\overleftarrow{E}(e)) \right| \right)
$$

$$
= \sum_{e \in E} \left(f(\overrightarrow{E}(e)) + f(\overleftarrow{E}(e)) \right)
$$

$$
= \sum_{g \in E_N} f(g) \cdot |\{e \in E \mid g \in \overrightarrow{E}(e)\} \cup \{e \in E \mid g \in \overleftarrow{E}(e)\}|
$$

$$
= \sum_{g \in E_N} f(g) \cdot cost(g) \quad \text{(by Property 1)}
$$

$$
= cost_f.
$$

$\square$

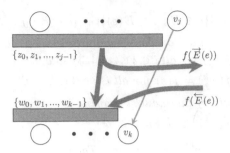

Fig. 4. Illustration of Property 4. The rectangles represent all network nodes to the left of v_j and v_k, respectively. The thick arrows represent the flow of several edges.

Lemma 2. *Let Γ be a drawing of G. There exists a flow f that induces Γ and whose cost is equal to the total edge length of Γ.*

Proof. If necessary, we set $x(v) := x(v) - \min_{v \in V} x(v)$ so that the smallest x-coordinate is zero. That gives us an equivalent drawing. We construct the flow f as follows: Let ω be the width of Γ. We send ω units of flow from s to t, so that the k-th unit takes the path $P_k = s \xrightarrow{*} w_{j_1}^1 \xrightarrow{*} w_{j_2}^2 \xrightarrow{*} \cdots \xrightarrow{*} w_{j_{|\mathcal{L}|}}^{|\mathcal{L}|} \xrightarrow{*} t$, where $w_{j_i}^i$ is chosen so that $x(v_{j_i+1}^i) \geq k$ and $x(v_{j_i}^i) < k$ ($w_{j_i}^i = w_0^i$, if $x(v_1^i) \geq k$ and $w_{j_i}^i = w_{|L_i|}^i$, if $x(v_{|L_i|}^i) < k$). That means we send the k-th unit through the k-th "column" of Γ. This is always possible, because of the subpaths $w_{j_i}^i \to z_{j_i}^i \xrightarrow{*} z_0^i \to w_0^{i+1} \xrightarrow{*} w_{j_{i+1}}^{i+1}$. So for every v there are $x(v)$ units of flow that pass by to the left of v, thus giving us correct coordinates for all nodes.

We define $E_k^i := \{e = (v_j^i, v_l^{i+1}) \in E \mid x(v_j^i) < k \text{ and } x(v_l^{i+1}) \geq k\} \cup \{e \in E \mid x(v_j^i) \geq k \text{ and } x(v_l^{i+1}) < k\}$, i.e. all edges that cross over the k-th column between L_i and L_{i+1}. We show that there exists a path P_k that produces the same cost as the number of graph edges that cross over the k-th column in total, that is $cost(P_k) = \sum_{i=1}^{|\mathcal{L}|-1} |E_k^i|$. Then we have $\sum_{k=1}^{\omega} cost(P_k) = \sum_{k=1}^{\omega} \sum_{i=1}^{|\mathcal{L}|-1} |E_k^i| = length(E)$ and we have proven the lemma.

It suffices to focus on the subpath P_k^i from $z = z_{j_i}^i$ to $w = w_{j_{i+1}}^{i+1}$ between two consecutive layers. Notice that network edges (s, w^1), (w^i, z^i) and $(z^{|\mathcal{L}|}, t)$ do not contribute to the cost of the flow. For better readability we denote the nodes of L_i with u_j and the nodes of L_{i+1} with v_j and we omit the superscripts. If not stated otherwise we use z_j for z_j^i and w_j for w_j^{i+1}. We construct $P' = P_k^i$ so that $cost(P') = |E'| = |E_k^i|$.

Case 1: There exists no edge e with $start(e) < z$ and $target(e) < w$.
That means every edge e with $start(e) < z$ has $target(e) > w$, and if $target(e) < w$ then $start(e) > z$. Then we set $P' = z \to z_{j_i-1} \xrightarrow{*} z_0 \to w_0 \to w_1 \xrightarrow{*} w$. For every $u_j < z$ with p outgoing edges P' uses exactly one $\overleftarrow{bz}$ with cost p. All these edges are in E'. For every $v_j < w$ with q incoming edges we use exactly one $\overrightarrow{bw}$

with cost q. Again these edges are in E'. So $cost(P') = |E'|$, since there are no other edges in E'.

Case 2: There exists no edge e with $start(e) > z$ and $target(e) > w$.

Arguing like in Case 1, we set $P' = z \xrightarrow{*} z_{|L_i|} \rightarrow w_{|L_{i+1}|} \xrightarrow{*} w$. As before the cost of P' equals $|E'|$.

Case 3: There exists an edge e_l with $start(e_l) < z$ and $target(e_l) < w$ and another edge e_r with $start(e_r) > z$ and $target(e_r) > w$.

Let e_l be the edge with the biggest $x(start(e))$ of all edges e with $start(e) < z$ and $target(e) < w$, and let e_r be the edge with the smallest $x(start(e))$ of all edges e with $start(e) > z$ and $target(e) > w$.

Case 3.1: There is at least one node u' with outgoing edges and $start(e_l) < u' < z$.

Let $u_g = start(e_l)$ and $v_{g'} = target(e_l)$. We know $v_{g'} < w$. Let $v_{h'}$ be the first node to the right of $v_{g'}$ with an edge $e_{r'} = (u_h, v_{h'})$ and $u_h > u_g$. Such a node does exist, since we have e_r. Notice that $v_{h'}$ might be to the right of w.

Then we have a hug: Set $e_1 = e_l$, set e_2 to one outgoing edge of u_{g+1} (or the next node to the right of u_g, which has an outgoing edge), $e_4 = e_{r'}$ and set e_3 to one incoming edge of $v_{h'-1}$ (or the next one to the left of $v_{h'}$), see Fig. 5. Notice that e_1 may coincide with e_3 and e_2 with e_4.

We have $start(e_3) \leq start(e_1)$, because we chose $e_1 = e_l$ with the biggest $x(start(e))$ and $v_{h'}$ is the first node to the right of $v_{g'}$ with an adjacent node to the right of u_g. So every node between $v_{g'} = target(e_1)$ and $v_{h'}$, including $v_{h'-1} = target(e_3)$, can only have adjacent nodes to the left of $u_g = start(e_1)$. It is clear that $start(e_1) < start(e_2)$ and $start(e_2) \leq start(e_4)$, since $start(e_4) = u_h > u_g$. By choice of e_1, e_3 and e_4 $target(e_1) \leq target(e_3) < target(e_4)$ holds. We know that $target(e_2) > w$ because there is at least one node between $start(e_1)$ and z whose outgoing edges have to end to the right of w because of the choice of e_1. If $target(e_4) > target(e_2)$ then e_2 would have been chosen for $e_{r'}$ and therefore for e_4. So $target(e_4) \leq target(e_2)$ also holds. So there exists $c_{g(h'-1)} \in E_N$ and we set $P' = z \xrightarrow{*} z_g \rightarrow w_{h'-1} \xrightarrow{*} w$.

Now for the cost. A subset of E' are the edges e with $z_g < start(e) < z$ and $target(e) > w$, which are covered by the $\overleftarrow{bz}$ of P'.

We have two options. First, if $w_{h-1} > w$ then all edges e with $start(e) < z_g$ and $target(e) > w_{h'-1}$ are covered by $c_{g(h'-1)}$ and the remaining edges e with $start(e) < z_g$ and $w < target(e) < w_{h'-1}$ are covered by the $\overleftarrow{bw}$ of P'. Edges e with $start(e) > z > u_g$ and $target(e) < w < w_{h'}$ are also covered by $c_{g(h'-1)}$. There cannot be any edge e with $z < start(e)$ and $w < target(e) < w_{h'-1}$ or $z_g < start(e) < z$ and $target(e) < w$, which would be crossed over by two different edges of P', due to the choice of edges e_1 to e_4.

Second, if $w_{h-1} < w$ then $c_{g(h'-1)}$ covers all edges e with $start(e) < z_g$ and $target(e) > w > w_{h'-1}$ and all edges e with $start(e) > z > z_g$ and $target(e) < w_{h'-1} < w$. Edges e with $start(e) > z$ and $w_{h'-1} < target(e) < w$ are covered by the $\overrightarrow{bw}$. Again there are no edges that are crossed over twice by P' due to the choice of e_1 to e_4. And there are no edges in E' that are not covered by some edge of P'.

Case 3.2: $start(e_l)$ is the next node to the left of z with outgoing edges, but there is at least one node u' with outgoing edges and $start(e_r) > u' > z$. This case is analogous to Case 3.1.

Case 3.3: $start(e_l)$ is the next node to the left of z with outgoing edges and $start(e_r)$ is the next node to the right of z with outgoing edges.
Let $e_l = (u_g, v_{g'})$ and $v_{h'}$ be the first node right of $v_{g'}$ with an adjacent node $u_h > u_g$. Again we have a hug. Set $e_1 = e_l$, $e_2 = e_r$, e_3 to an incoming edge of $v_{h'-1}$ (or a lower node, if necessary) and $e_4 = (u_h, v_{h'})$.

With the same arguments as in Case 3.1 we convince ourselves that e_1, e_2, e_3 and e_4 are indeed a hug and we have $c_{j_i(h'-1)}$. We set $P' = z \rightarrow w_{h'-1} \overset{*}{\rightarrow} w$. As before $cost(P') = |E'|$. □

Theorem 1. *A minimum cost flow in the network described above solves* $HCAP_{minEL}$.

Proof. Lemma 1 and Lemma 2. □

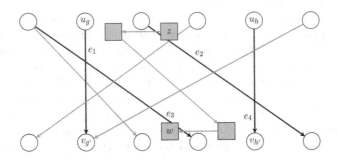

Fig. 5. Case 3.1. Only relevant network nodes and edges are depicted. Edges that participate in the hug are black.

For controlling the maximum width of the drawing we make use of Property 3, which states that the width of the drawing is at most the flow leaving s. We can add an additional node s' and an edge (s, s') to N and replace all edges of the form (s, w_j^1) with (s', w_j^1). Now we can limit the maximum width of the drawing by setting the upper bound of (s, s') to an appropriate value.

Further constraints can be modelled by manipulating the network. By adjusting the lower and upper bounds of edges $a \in A$ we can realize minimum and maximum distances between two neighboring nodes on the same layer. By removing every $g \in \overleftarrow{E}(e) \cup \overrightarrow{E}(e)$ from the network, we can enforce the edge e to be drawn vertically.

4   Experimental Results

In our experiment we want to demonstrate that we are able to restrict the width of the drawing without paying too much in terms of total (horizontal) edge length and time.

We implemented the algorithm from Sect. 3, which we will call MCF within the Open Graph Drawing Framework [4] (OGDF) and used the OGDF network simplex software to solve the minimum cost flow problem. We also implemented the approach of Gansner et al. [7] (Gansner) that also uses the network simplex algorithm. Additionally we use three other OGDF methods: an ILP that also takes balancing the nodes between their neighbors into account (LP), the algorithm of Buchheim, Jünger and Leipert [3] (BJL) and the algorithm of Brandes and Köpf [2] (BK). All algorithms draw inner segments of long edges as vertical lines, since this is generally desirable for good readability. MCF is configured to compute a layout with minimum edge length with respect to minimum possible width and Gansner computes coordinates that minimize the total edge length regardless of width. We used a subset of the AT&T graphs from www.graphdrawing.org/data.html consisting of 1277 graphs with 10 to 100 nodes as our test set.

The test was run on an Intel Xeon E5-2640v3 2.6 GHz CPU with 128 GB RAM.

Figures 6, 7, and 8 show the results. The whiskers in Figs. 6 and 7 cover 95% of the data and outliers are omitted for better readability. Figure 8 shows absolute values for MCF and Fig. 9 displays three example drawings.

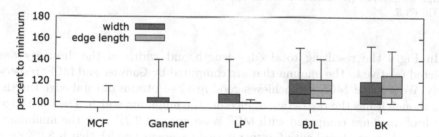

Fig. 6. Width and total edge length produced by MCF, Gansner, LP, BJL and BK relative to minimum width, resp. edge length.

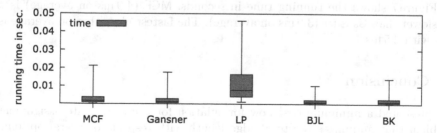

Fig. 7. Running time for MCF, Gansner, LP, BJL and BK.

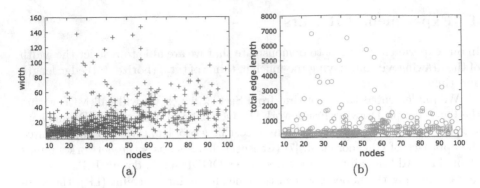

Fig. 8. Absolute values of (a) width and (b) total edge length for MCF.

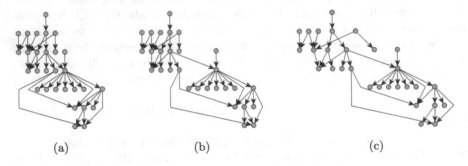

Fig. 9. Example drawings of a graph with 29 nodes and 33 edges. (a) MCF: width: 9, edge length: 58. (b) Gansner: width: 13, edge length: 54. (c) BK: width: 16.5, edge length: 63.5.

In Fig. 6 the resulting total edge length and width of the drawings are depicted relative to the minima that are computed by Gansner and MCF, respectively. We see that MCF still achieves good results in terms of total edge length, even though it has the restriction of meeting the minimum width. The total edge length of drawings computed with MCF is on average 2.2% over the minimum, while drawings produced with Gansner have on average a width that is 8.9% over the minimum. In an extreme example with minimum width 1, Gansner results in width 15.

Figure 7 shows the running time in seconds. MCF (4.9 ms on average) is a bit slower than Gansner (3.9 ms on average). The fastest algorithm on average is BJL with 2.5 ms.

5   Conclusion

We presented a minimum cost flow formulation for the coordinate assignment problem that minimizes the total edge length with respect to several optional criteria like the maximum width or lower and upper bounds on the distance of neighboring nodes in a layer. In our experiments we showed that our approach can compete with state-of-the-art algorithms.

References

1. Ahuja, R.K., Magnanti, T.L., Orlin, J.B.: Network Flows - Theory, Algorithms and Applications. Prentice Hall, Upper Saddle River (1993)
2. Brandes, U., Köpf, B.: Fast and simple horizontal coordinate assignment. In: Mutzel, P., Jünger, M., Leipert, S. (eds.) GD 2001. LNCS, vol. 2265, pp. 31–44. Springer, Heidelberg (2002). https://doi.org/10.1007/3-540-45848-4_3
3. Buchheim, C., Jünger, M., Leipert, S.: A fast layout algorithm for k-level graphs. In: Marks, J. (ed.) GD 2000. LNCS, vol. 1984, pp. 229–240. Springer, Heidelberg (2001). https://doi.org/10.1007/3-540-44541-2_22
4. Chimani, M., Gutwenger, C., Jünger, M., Klau, G.W., Klein, K., Mutzel, P.: The open graph drawing framework (OGDF). In: Tamassia, R. (ed.) Handbook on Graph Drawing and Visualization, pp. 543–569. Chapman and Hall/CRC, Boca Raton (2013)
5. Coffman, E.G., Graham, R.L.: Optimal scheduling for two-processor systems. Acta Informatica $1(3)$, 200–213 (1972). https://doi.org/10.1007/BF00288685
6. Eades, P., Lin, X., Tamassia, R.: An algorithm for drawing a hierarchical graph. Int. J. Comput. Geom. Appl. $6(2)$, 145–156 (1996). https://doi.org/10.1142/S0218195996000101
7. Gansner, E.R., Koutsofios, E., North, S.C., Vo, K.P.: A technique for drawing directed graphs. Softw. Eng. $19(3)$, 214–230 (1993). https://doi.org/10.1109/32.221135
8. Healy, P., Nikolov, N.S.: A branch-and-cut approach to the directed acyclic graph layering problem. In: Goodrich, M.T., Kobourov, S.G. (eds.) GD 2002. LNCS, vol. 2528, pp. 98–109. Springer, Heidelberg (2002). https://doi.org/10.1007/3-540-36151-0_10
9. Healy, P., Nikolov, N.S.: Hierarchical drawing algorithms. In: Tamassia, R. (ed.) Handbook on Graph Drawing and Visualization, pp. 409–453. Chapman and Hall/CRC, Boca Raton (2013)
10. Jabrayilov, A., Mallach, S., Mutzel, P., Rüegg, U., von Hanxleden, R.: Compact layered drawings of general directed graphs. In: Hu, Y., Nöllenburg, M. (eds.) GD 2016. LNCS, vol. 9801, pp. 209–221. Springer, Cham (2016). https://doi.org/10.1007/978-3-319-50106-2_17
11. Sander, G.: A fast heuristic for hierarchical Manhattan layout. In: Brandenburg, F.J. (ed.) GD 1995. LNCS, vol. 1027, pp. 447–458. Springer, Heidelberg (1996). https://doi.org/10.1007/BFb0021828
12. Sugiyama, K., Tagawa, S., Toda, M.: Methods for visual understanding of hierarchical system structures. IEEE Trans. Syst. Man Cybern. $11(2)$, 109–125 (1981). https://doi.org/10.1109/TSMC.1981.4308636

The Queue-Number of Posets of Bounded Width or Height

Kolja Knauer[1]([✉]), Piotr Micek[2], and Torsten Ueckerdt[3]

[1] Aix Marseille Univ, Université de Toulon, CNRS, LIS, Marseille, France
kolja.knauer@lis-lab.fr
[2] Faculty of Mathematics and Computer Science, Theoretical Computer Science
Department, Jagiellonian University, Kraków, Poland
piotr.micek@tcs.uj.edu.pl
[3] Karlsruhe Institute of Technology (KIT), Institute of Theoretical Informatics,
Karlsruhe, Germany
torsten.ueckerdt@kit.edu

Abstract. Heath and Pemmaraju [9] conjectured that the queue-number of a poset is bounded by its width and if the poset is planar then also by its height. We show that there are planar posets whose queue-number is larger than their height, refuting the second conjecture. On the other hand, we show that any poset of width 2 has queue-number at most 2, thus confirming the first conjecture in the first non-trivial case. Moreover, we improve the previously best known bounds and show that planar posets of width w have queue-number at most $3w - 2$ while any planar poset with 0 and 1 has queue-number at most its width.

1 Introduction

A *queue layout* of a graph consists of a total ordering on its vertices and an assignment of its edges to *queues*, such that no two edges in a single queue are nested. The minimum number of queues needed in a queue layout of a graph G is called its *queue-number* and denoted by $\mathrm{qn}(G)$.

To be more precise, let G be a graph and let L be a linear order on the vertices of G. We say that the edges $uv, u'v' \in E(G)$ are *nested* with respect to L if $u < u' < v' < v$ or $u' < u < v < v'$ in L. Given a linear order L of the vertices of G, the edges $u_1v_1, \ldots, u_kv_k$ of G form a *rainbow* of size k if $u_1 < \cdots < u_k < v_k < \cdots < v_1$ in L. Given G and L, the edges of G can be partitioned into k queues if and only if there is no rainbow of size $k + 1$ in L, see [10].

The queue-number was introduced by Heath and Rosenberg in 1992 [10] as an analogy to book embeddings. Queue layouts were implicitly used before and have

K. Knauer was supported by ANR projects GATO: ANR-16-CE40-0009-01 and DISTANCIA: ANR-17-CE40-0015.
P. Micek was partially supported by the National Science Center of Poland under grant no. 2015/18/E/ST6/00299.

T. Biedl and A. Kerren (Eds.): GD 2018, LNCS 11282, pp. 200–212, 2018.
https://doi.org/10.1007/978-3-030-04414-5_14

applications in fault-tolerant processing, sorting with parallel queues, matrix computations, scheduling parallel processes, and communication management in distributed algorithm (see [8,10,13]).

Perhaps the most intriguing question concerning queue-numbers is whether planar graphs have bounded queue-number.

Conjecture 1 (Heath and Rosenberg [10]).
The queue-number of planar graphs is bounded by a constant.

In this paper we study queue-numbers of posets. The parameter was introduced in 1997 by Heath and Pemmaraju [9] and the main idea is that given a poset one should lay it out respecting its relation. Two elements a, b of a poset are called *comparable* if $a < b$ or $b < a$, and *incomparable*, denoted by $a \parallel b$, otherwise. Posets are visualized by their *diagrams*: Elements are placed as points in the plane and whenever $a < b$ in the poset, and there is no element c with $a < c < b$, there is a curve from a to b going upwards (that is y-monotone). We denote this case as $a \prec b$. The diagram represents those relations which are essential in the sense that they are not implied by transitivity, also known as *cover relations*. The undirected graph implicitly defined by such a diagram is the *cover graph* of the poset. Given a poset P, a *linear extension* L of P is a linear order on the elements of P such that $x <_L y$, whenever $x <_P y$. (Throughout the paper we use a subscript on the symbol $<$, if we want to emphasize which order it represents.) Finally, the *queue-number of a poset* P, denoted by $\operatorname{qn}(P)$, is the smallest k such that there is a linear extension L of P for which the resulting linear layout of G_P contains no $(k+1)$-rainbow. Clearly we have $\operatorname{qn}(G_P) \leq \operatorname{qn}(P)$, i.e., the queue-number of a poset is at least the queue-number of its cover graph. It is shown in [9] that even for *planar posets*, that is posets admitting crossing-free diagrams, there is no function f such that $\operatorname{qn}(P) \leq f(\operatorname{qn}(G_P))$ (Fig. 1).

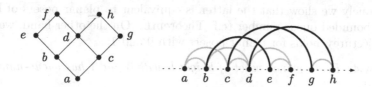

Fig. 1. A poset and a layout with two queues (gray and black). Note that the order of the elements on the spine is a linear extension of the poset.

Heath and Pemmaraju [9] investigated the maximum queue-number of several classes of posets, in particular with respect to bounded width (the maximum number of pairwise incomparable elements) and height (the maximum number of pairwise comparable elements). A set with every two elements being comparable is a *chain*. A set with every two distinct elements being incomparable is an *antichain*. They proved that if $\operatorname{width}(P) \leq w$, then $\operatorname{qn}(P) \leq w^2$. The lower bound is attained by *weak orders*, i.e., chains of antichains and is conjectured to be the upper bound as well:

Conjecture 2 (Heath and Pemmaraju [9]).
Every poset of width w has queue-number at most w.

Furthermore, they made a step towards this conjecture for planar posets: if a planar poset P has width$(P) \leq w$, then qn$(P) \leq 4w - 1$. For the lower bound side they provided planar posets of width w and queue-number $\lceil \sqrt{w} \rceil$.

We improve the bounds for planar posets and get the following:

Theorem 1. *Every planar poset of width w has queue-number at most $3w - 2$. Moreover, there are planar posets of width w and queue-number w.*

As an ingredient of the proof we show that posets without certain subdivided crowns satisfy Conjecture 2 (c.f. Theorem 5). This implies the conjecture for interval orders and planar posets with (unique minimum) 0 and (unique maximum) 1 (c.f. Corollary 2). Moreover, we confirm Conjecture 2 for the first non-trivial case $w = 2$:

Theorem 2. *Every poset of width 2 has queue-number at most 2.*

An easy corollary of this is that all posets of width w have queue-number at most $w^2 - w + 1$ (c.f. Corollary 1).

Another conjecture of Heath and Pemmaraju concerns planar posets of bounded height:

Conjecture 3 (Heath and Pemmaraju [9]).
Every planar poset of height h has queue-number at most h.

We show that Conjecture 3 is false for the first non-trivial case $h = 2$:

Theorem 3. *There is a planar poset of height 2 with queue-number at least 4.*

Furthermore, we establish a link between a relaxed version of Conjectures 3 and 1, namely we show that the latter is equivalent to planar posets of height 2 having bounded queue-number (c.f. Theorem 6). On the other hand, we show that Conjecture 3 holds for planar posets with 0 and 1:

Theorem 4. *Every planar poset of height h with 0 and 1 has queue-number at most $h - 1$.*

Organization of the paper. In Sect. 2 we consider general (not necessarily planar) posets and give upper bounds on their queue-number in terms of their width, such as Theorem 2. In Sect. 3 we consider planar posets and bound the queue-number in terms of the width, both from above and below, i.e., we prove Theorem 1. In Sect. 4 we give a counterexample to Conjecture 3 by constructing a planar poset with height 2 and queue-number at least 4. Here we also argue that proving *any* upper bound on the queue-number of such posets is equivalent to proving Conjecture 1. Finally, we show that Conjecture 3 holds for planar posets with 0 and 1 and that for every h there is a planar poset of height h and queue-number $h - 1$ (c.f. Proposition 3).

2 General Posets of Bounded Width

By Dilworth's Theorem [3], the width of a poset P coincides with the smallest integer w such that P can be decomposed into w chains of P. Let us derive Proposition 1 of Heath and Pemmaraju [9] from such a chain partition.

Proposition 1. *For every poset P, if* $\mathrm{width}(P) \leq w$ *then* $\mathrm{qn}(P) \leq w^2$.

Proof. Let P be a poset of width w and $C_1, \ldots, C_w$ be a chain partition of P. Let L be any linear extension of P and $a <_L b <_L c <_L d$ with $a \prec d$ and $b \prec c$. Note that we must have either $a \parallel b$ or $c \parallel d$. If follows that if $a \in C_i$, $b \in C_j$, $c \in C_k$, and $d \in C_\ell$, then $(i, \ell) \neq (j, k)$. As there are only w^2 ordered pairs (x, y) with $x, y \in [w]$, we can conclude that every nesting set of covers has cardinality at most w^2. □

Note that in the above proof L is *any* linear extension and that without choosing the linear extension L carefully, upper bound w^2 is best-possible. Namely, if $P = \{a_1, \ldots, a_k, b_1, \ldots, b_k\}$ with comparabilities $a_i < b_j$ for all $1 \leq i, j \leq k$, then P has width k and the linear extension $a_1 < \ldots < a_k < b_k < \ldots < b_1$ creates a rainbow of size k^2.

We continue by showing that every poset of width 2 has queue-number at most 2, that is, we prove Theorem 2.

Proof (Theorem 2). Let P be a poset of width 2 and minimum element 0 and C_1, C_2 be a chain partition of P. Note that the assumption of the minimum causes no loss of generality, since a 0 can be added without increasing the width nor decreasing the queue-number. Any linear extension L of P partitions the ground set X naturally into inclusion-maximal sets of elements, called *blocks*, from the same chain in $\{C_1, C_2\}$ that appear consecutively along L, see Fig. 2. We denote the blocks by $B_1, \ldots, B_k$ according to their appearance along L. We say that L is *lazy* if for each $i = 2, \ldots, k$, each element $x \in B_i$ has a relation to some element $y \in B_{i-1}$. A linear extension L can be obtained by picking any minimal element $m \in P$, put it into L, and recurse on $P \setminus \{m\}$. Lazy linear extensions (with respect to C_1, C_2) can be constructed by the same process where additionally the next element is chosen from the same chain as the element before, if possible. Note that the existence of a 0 is needed in order to ensure the property of laziness with respect to B_2.

Now we shall prove that in a lazy linear extension no three covers are pairwise nesting. So assume that $a \prec b$ is any cover and that $a \in B_i$ and $b \in B_j$. As L is lazy, b is comparable to some element in B_{j-1} (if $j \geq 2$) and all elements in $B_1, \ldots, B_{j-2}$ (if $j \geq 3$). With $a \prec b$ being a cover, it follows from L being lazy that $i \in \{j-2, j-1, j\}$. If $i = j$, then no cover is nested under $a \prec b$. If $i = j-1$, then no cover $c \prec d$ is nested above $a \prec b$: either $c \in B_i$ and $d \in B_j$ and hence $c \prec d$ is not a cover, or both endpoints would be inside the same chain, i.e., c, d are the last and first element of B_{j-2} and B_j or B_i and B_{i+2}, respectively. This implies $c <_L a <_L d <_L b$ or $a <_L c <_L b <_L r$, respectively, and $c \prec d$ cannot nest above $a \prec b$. If $i = j-2$, then no cover is nested above $a \prec b$. Thus, either

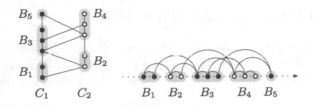

Fig. 2. A poset of width 2 with a 0 and a chain partition C_1, C_2 and the blocks $B_1, \ldots, B_5$ induced by a lazy linear extension with respect to C_1, C_2.

no cover is nested below $a \prec b$, or no cover is nested above $a \prec b$, or both. In particular, there is no three nesting covers and $qn(P) \leq 2$.   □

Corollary 1. *Every poset of width w has queue-number at most $w^2 - 2\lfloor w/2 \rfloor$.*

Proof. We take any chain partition of size w and pair up chains to obtain a set S of $\lfloor w/2 \rfloor$ disjoint pairs. Each pair from S induces a poset of width at most 2, which by Theorem 2 admits a linear order with at most two nesting covers. Let L be a linear extension of P respecting all these partial linear extensions.

Now, following the proof of Proposition 1 any cover can be labeled by a pair (i, j) corresponding to the chains containing its endpoint. Thus, in a set of nesting covers any pair appears at most once, but for each i, j such that $(i, j) \in S$ only two of the four possible pairs can appear simultaneously in a nesting. This yields the upper bound.   □

For an integer $k \geq 2$ we define a *subdivided k-crown* as the poset P_k as follows. The elements of P_k are $\{a_1, \ldots, a_k, b_1, \ldots, b_k, c_1, \ldots, c_k\}$ and the cover relations are given by $a_i \prec b_i$ and $b_i \prec c_i$ for $i = 2, \ldots, k$, $a_i \prec c_{i-1}$ for $i = 1, \ldots, k-1$, and $a_1 \prec c_k$; see the left of Fig. 3. We refer to the covers of the form $a_i \prec c_j$ as the *diagonal covers* and we say that a poset P has an *embedded P_k* if P contains $3k$ elements that induce a copy of P_k in P with all diagonal covers of that copy being covers of P.

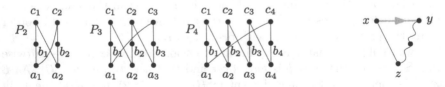

Fig. 3. Left: The posets P_2, P_3, and P_4. Right: The existence of an element z with cover relation $z \prec x$ and non-cover relation $z < y$ gives rise to a gray edge from x to y.

Theorem 5. *If P is a poset that for no $k \geq 2$ has an embedded P_k, then the queue-number of P is at most the width of P.*

Proof. Let P be any poset. For this proof we consider the cover graph G_P of P as a directed graph with each edge xy directed from x to y if $x \prec y$ in P. We call these edges the *cover edges*. Now we augment G_P to a directed graph G by introducing for some incomparable pairs $x \parallel y$ a directed edge. Specifically, we add a directed edge from x to y if there exists a z with $z < x, y$ in P where $z \prec x$ is a cover relation and $z < y$ is not a cover relation; see the right of Fig. 3. We call these edges the *gray edges* of G.

Now we claim that if G has a directed cycle, then P has an embedded subdivided crown. Clearly, every directed cycle in G has at least one gray edge. We consider the directed cycles with the fewest gray edges and among those let $C = [c_1, \dots, c_\ell]$ be one with the fewest cover edges. First assume that C has a cover edge (hence $\ell \geq 3$), say $c_1 c_2$ is a gray edge followed by a cover edge $c_2 c_3$. Consider the element z with cover relation $z \prec c_1$ and non-cover relation $z < c_2$ in P. By $z < c_2 \prec c_3$ we have a non-cover relation $z < c_3$ in P. Now if $c_1 \parallel c_3$ in P, then G contains the gray edge $c_1 c_3$ (see Fig. 4(a)) and $[c_1, c_3, \dots, c_\ell]$ is a directed cycle with the same number of gray edges as C but fewer cover edges, a contradiction. On the other hand, if $c_1 < c_3$ in P (note that $c_3 < c_1$ is impossible as $z \prec c_1$ is a cover), then there is a directed path Q of cover edges from c_1 to c_3 (see Fig. 4(b)) and $C + Q - \{c_1 c_2, c_2 c_3\}$ contains a directed cycle with fewer gray edges than C, again a contradiction.

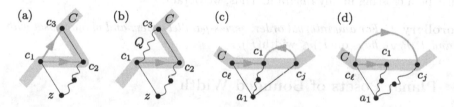

Fig. 4. Illustrations for the proof of Theorem 5.

Hence $C = [c_1, \dots, c_\ell]$ is a directed cycle consisting solely of gray edges. Note that by the first paragraph $\{c_1, \dots, c_\ell\}$ is an antichain in P. For $i = 2, \dots, \ell$ let a_i be the element of P with cover relation $a_i \prec c_{i-1}$ and non-cover relation $a_i < c_i$, as well as a_1 with cover relation $a_1 \prec c_\ell$ and non-cover relation $a_1 < c_1$. As $\{c_1, \dots, c_\ell\}$ is an antichain and $a_i < c_i$ holds for $i = 1, \dots, \ell$, we have $\{c_1, \dots, c_\ell\} \cap \{a_1, \dots, a_\ell\} = \emptyset$. Let us assume that $a_1 < c_j$ in P for some $j \neq 1, \ell$. If $a_1 \prec c_j$ is a cover relation, then there is a gray edge $c_j c_1$ in G (see Fig. 4(c)) and the cycle $[c_1, \dots, c_j]$ is shorter than C, a contradiction. If $a_1 < c_j$ is a non-cover relation, then there is a gray edge $c_\ell c_j$ in G (see Fig. 4(d)) and the cycle $[c_j, \dots, c_\ell]$ is shorter than C, again a contradiction.

Hence, the only relations between $a_1, \dots, a_\ell$ and $c_1, \dots, c_\ell$ are cover relations $a_1 \prec c_\ell$ and $a_i \prec c_{i-1}$ for $i = 2, \dots, \ell$ and the non-cover relations $a_i < c_i$ for $i = 1, \dots, \ell$. Hence $a_1, \dots, a_\ell$ are pairwise distinct. Moreover, $\{a_1, \dots, a_\ell\}$ is an antichain in P since the only possible relations among these elements are of the form $a_1 < a_\ell$ or $a_i < a_{i-1}$, which would contradict that $a_1 \prec c_\ell$ and $a_i \prec c_{i-1}$ are cover relations. Finally, we pick for every $i = 1, \ldots, \ell$ an element b_i with $a_i < b_i < c_i$, which exists as $a_i < c_i$ is a non-cover relation. Together with the above relations between $a_1, \ldots, a_\ell$ and $c_1, \ldots, c_\ell$ we conclude that $b_1, \ldots, b_\ell$ are pairwise distinct and these 3ℓ elements induce a copy of P_ℓ in P with all diagonal covers in that copy being covers of P.

Thus, if P has no embedded P_k, then the graph G we constructed has no directed cycles, and we can pick L to be any topological ordering of G. As $G_P \subseteq G$, L is a linear extension of P. For any two nesting covers $x_2 <_L x_1 <_L y_1 <_L y_2$ we have $x_1 \parallel x_2$ or $y_1 \parallel y_2$ or both, since $x_2 \prec y_2$ is a cover. However, if $x_2 < x_1$ in P, then there would be a gray edge from y_2 to y_1 in G, contradicting $y_1 <_L y_2$ and L being a topological ordering of G. We conclude that $x_1 \parallel x_2$ and the left endpoints of any rainbow form an antichain, proving $\mathrm{qn}(P) \leq \mathrm{width}(P)$.   □

Let us remark that several classes of posets have no embedded subdivided crowns, e.g., graded posets, interval orders (since these are $2 + 2$-free, see [6]), or (quasi-)series-parallel orders (since these are N-free, see [7]). Here, $2 + 2$ and N are the four-element posets defined by $a < b, c < d$ and $a < b, c < d, c < b$, respectively. Also note that while subdivided crowns are planar posets, no planar poset with 0 and 1 has an embedded k-crown. Indeed, already looking at the subposet induced by the k-crown and the 0 and the 1, it is easy to see that there must be a crossing in any diagram. Thus, we obtain:

Corollary 2. *For any interval order, series-parallel order, and planar poset with 0 and 1, P we have $\mathrm{qn}(P) \leq \mathrm{width}(P)$.*

3   Planar Posets of Bounded Width

Heath and Pemmaraju [9] show that the largest queue-number among planar posets of width w lies between $\lceil \sqrt{w} \rceil$ and $4w - 1$. Here we improve the lower bound to w and the upper bound to $3w - 2$.

Proposition 2. *For each w there exists a planar poset Q_w with 0 and 1 of width w and queue-number w.*

Proof. We shall define Q_w recursively, starting with Q_1 being any chain. For $w \geq 2$, Q_w consists of a *lower copy* P and a disjoint *upper copy* P' of Q_{w-1}, three additional elements a, b, c, and the following cover relations in between:

- $a \prec x$, where x is the 0 of P
- $y \prec x'$, where y is the 1 of P and x' is the 0 of P'
- $y' \prec c$, where y' is the 1 of P'
- $a \prec b \prec c$

It is easily seen that all cover relations of P and P' remain cover relations in Q_w, and that Q_w is planar, has width w, a is the 0 of Q_w, and c is the 1 of Q_w. See Fig. 5 for an illustration.

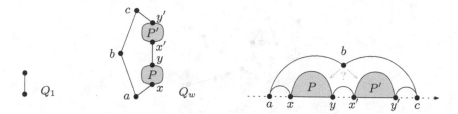

Fig. 5. Recursively constructing planar posets Q_w of width w and queue-number w. Left: Q_1 is a two-element chain. Middle: Q_w is defined from two copies P, P' of Q_{w-1}. Right: The general situation for a linear extension of Q_w.

To prove that $\mathrm{qn}(Q_w) = w$ we argue by induction on w, with the case $w = 1$ being immediate. Let L be any linear extension of Q_w. Then a is the first element in L and c is the last. Since $y \prec x'$, all elements in P come before all elements of P'. Now if in L the element b comes after all elements of P, then P is nested under cover $a \prec b$, and if b comes before all elements of P', then P' is nested under cover $b \prec c$. We obtain w nesting covers by induction on P in the former case, and by induction on P' in the latter case. This concludes the proof. $\quad\square$

Next we prove Theorem 1, namely that the maximum queue-number of planar posets of width w lies between w and $3w - 2$.

Proof (Theorem 1). By Proposition 2 some planar posets of width w have queue-number w. So it remains to consider an arbitrary planar poset P of width w and show that P has queue-number at most $3w - 2$. To this end, we shall add some relations to P, obtaining another planar poset Q of width w that has a 0 and 1, with the property that $\mathrm{qn}(P) \le \mathrm{qn}(Q) + 2w - 2$. Note that this will conclude the proof, as by Corollary 2 we have $\mathrm{qn}(Q) \le w$.

Given a planar poset P of width w, there are at most w minima and at most w maxima. Hence there are at most $2w - 2$ extrema that are not on the outer face. For each such extremum x – say x is a minimum – consider the unique face f with an obtuse angle at x. We introduce a new relation $y < x$, where y is a smallest element at face f, see Fig. 6. Note that this way we introduce at most $2w - 2$ new relations, and that these can be drawn y-monotone and crossing-free by carefully choosing the other element in each new relation. Furthermore, every inner face has a unique source and unique sink.

Now consider a cover relation $a \prec_P b$ that is not a cover relation in the new poset Q. For the corresponding edge e from a to b in Q there is one face f with unique source a and unique sink b. Now either way the other edge in f incident to a or to b must be one of the $2w - 2$ newly inserted edges, see again Fig. 6. This way we assign $a \prec b$ to one of $2w - 2$ queues, one for each newly inserted edge. Every such queue contains either at most one edge or two incident edges, i.e., a nesting is impossible, no matter what linear ordering is chosen later.

We create at most $2w - 2$ queues to deal with the cover relations of P that are not cover relations of Q and spend another w queues for Q dealing with the remaining cover relations of P. Thus, $\mathrm{qn}(P) \le \mathrm{qn}(Q) + 2w - 2 \le 3w - 2$. $\quad\square$

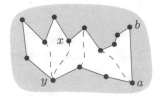

Fig. 6. Inserting new relations (dashed) into a face of a plane diagram. Note that relation $a < b$ is a cover relation in P but not in Q.

4   Planar Posets of Bounded Height

Recall Conjecture 3, which states that every planar poset of height h has queue-number at most h. In the following, we give a counterexample to this conjecture:

Proof (Theorem 3). Consider the graph G that is constructed as follows: Start with $K_{2,10}$ with bipartition classes $\{a_1, a_2\}$ and $\{b_1, \ldots, b_{10}\}$. For every $i = 1, \ldots, 9$ add four new vertices $c_{i,1}, \ldots, c_{i,4}$, each connected to b_i and b_{i+1}. The resulting graph G has 46 vertices, is planar and bipartite with bipartition classes $X = \{b_1, \ldots, b_{10}\}$ and $Y = \{a_1, a_2\} \cup \{c_{i,j} \mid 1 \le i \le 9, 1 \le j \le 4\}$. See Fig. 7.

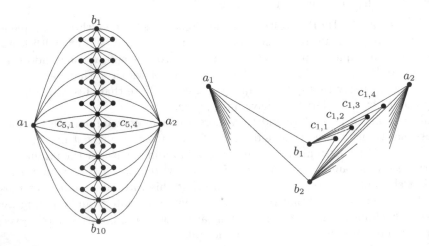

Fig. 7. A planar poset P of height 2 and queue-number at least 4. Left: The cover graph G_P of P. Right: A part of a planar diagram of P.

Let P be the poset arising from G by introducing the relation $x < y$ for every edge xy in G with $x \in X$ and $y \in Y$. Clearly, P has height 2 and hence the cover relations of P are exactly the edges of G. Moreover, by a result of Moore [12] (see also [2]) P is planar because G is planar, also see the right of Fig. 7.

We shall argue that $\mathrm{qn}(P) \ge 4$. To this end, let L be any linear extension of P. Without loss of generality we have $a_1 <_L a_2$. Note that since in P one bipartition class of G is entirely below the other, any 4-cycle in G gives a 2-rainbow. Let b_{i_1}, b_{i_2} be the first two elements of X in L, b_{j_1}, b_{j_2} be the last two such elements. As $|X| = 10$ there exists $1 \leq i \leq 9$ such that $\{i, i+1\} \cap \{i_1, i_2, j_1, j_2\} = \emptyset$, i.e., we have $b_{i_1}, b_{i_2} <_L b_i, b_{i+1} <_L b_{j_1}, b_{j_2} <_L a_1 <_L a_2$, where we use that a_1 and a_2 are above all elements of X in P.

Now consider the elements $C = \{c_{i,1}, \ldots, c_{i,4}\}$ that are above b_i and b_{i+1} in P. As $|C| \geq 4$, there are two elements c_1, c_2 of C that are both below a_1, a_2 in L, or both between a_1 and a_2 in L, or both above a_1, a_2 in L. Consider the 2-rainbow R in the 4-cycle $[c_1, b_i, c_2, b_{i+1}]$. In the first case R is nested below the 4-cycle $[a_1, b_{i_1}, a_2, b_{i_2}]$, in the second case the cover $b_{j_1} \prec a_1$ is nested below R and R is nested below the cover $b_{i_1} \prec a_2$, and in the third case 4-cycle $[a_1, b_{j_1}, a_2, b_{j_2}]$ is nested below R. As each case results in a 4-rainbow, we have $\mathrm{qn}(P) \geq 4$. □

Even though Conjecture 3 has to be refuted in its strongest meaning, it might hold that planar posets of height h have queue-number $O(h)$, or at least bounded by some function $f(h)$ in terms of h, or at least that planar posets of height 2 have bounded queue-number. As it turns out, all these statements are equivalent, and in turn equivalent to Conjecture 1.

Theorem 6. *The following statements are equivalent:*

(i) Planar graphs have queue-number $O(1)$ (Conjecture 1).
(ii) Planar posets of height h have queue-number $O(h)$.
(iii) Planar posets of height h have queue-number at most $f(h)$ for a function f.
(iv) Planar posets of height 2 have queue-number $O(1)$.
(v) Planar bipartite graphs have queue-number $O(1)$.

Proof.   (i)⇒(ii) Pemmaraju proves in his thesis [14] (see also [4]) that if G is a graph, π is a vertex ordering of G with no $(k+1)$-rainbow, $V_1, \ldots, V_m$ are color classes of any proper m-coloring of G, and π' is the vertex ordering with $V_1 <_{\pi'} \cdots <_{\pi'} V_m$, where within each V_i the ordering of π is inherited, then π' has no $(2(m-1)k+1)$-rainbow. So if P is any poset of height h, its cover graph G_P has $\mathrm{qn}(G_P) \leq c$ by (i) for some global constant $c > 0$. Splitting P into h antichains $A_1, \ldots, A_h$ by iteratively removing all minimal elements induces a proper h-coloring of G_P with color classes $A_1, \ldots, A_h$. As every vertex ordering π' of G with $A_1 <_{\pi'} \cdots <_{\pi'} A_h$ is a linear extension of P, it follows by Pemmaraju's result that $\mathrm{qn}(P) \leq 2(h-1)\,\mathrm{qn}(G_P) \leq 2ch$, i.e., $\mathrm{qn}(P) \in O(h)$.

(ii) ⇒(iii)⇒(iv) These implications are immediate.

(iv)⇒(v) Moore proves in his thesis [12] (see also [2]) that if G is a planar and bipartite graph with bipartition classes A and B, and P_G is the poset on element set $A \cup B = V(G)$ where $x < y$ if and only if $x \in A, y \in B, xy \in E(G)$, then P_G is a planar poset of height 2. As G is the cover graph of P_G, we have $\mathrm{qn}(G) \leq \mathrm{qn}(P_G) \leq c$ for some constant $c > 0$ by (iv), i.e., $\mathrm{qn}(G) \in O(1)$.

(v)⇒(i) This is a result of Dujmović and Wood [5].   □

Finally, we show that Conjecture 3 holds for planar posets with 0 and 1.

Proof (Theorem 4). Let P be a planar poset with 0 and 1. Then P has dimension at most two [1], i.e., it can be written as the intersection of two linear extensions of P. A particular consequence of this is, that there is a well-defined dual poset P^* in which two distinct elements x, y are comparable in P if and only if they are incomparable in P^*. Poset P^* reflects a "left of"-relation for each incomparable pair $x \parallel y$ in P in the following sense: Any maximal chain C in P corresponds to a 0-1-path Q in G_P, which splits the elements of $P \setminus C$ into those left of Q and those right of Q. Now $x <_{P^*} y$ if and only if x is left of the path for every maximal chain containing y (equivalently y is right of the path for every maximal chain containing x). Due to planarity, if $a \prec b$ is a cover in P and C is a maximal chain containing neither a nor b, then a and b are on the same side of the path Q corresponding to C. In particular, if for $x, y \in C$ we have $a <_{P^*} x$ and $b \parallel y$, then b and y are comparable in P^*, but if $y <_{P^*} b$ we would get a crossing of C and $a \prec b$. Also see the left of Fig. 8. We summarize:

$(\star)$ If $a \prec b$, $a <_{P^*} x$ for some $x \in C$ and $b \parallel y$ for some $y \in C$, then $b <_{P^*} y$.

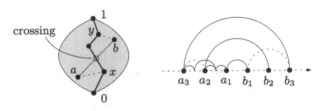

Fig. 8. Left: Illustration of $(\star)$: If $a <_{P^*} x$, $b \parallel y$, $x < y$, and $a \prec b$ is a cover, then $b <_{P^*} y$ due to planarity. Right: If $a_3 <_L a_2 <_L a_1 <_L b_1 <_L b_2 <_L b_3$ is a 3-rainbow with $a_2, a_3 < a_1$, then $a_3 < a_2$.

Now let L be the *leftmost* linear extension of P, i.e., the unique linear extension L with the property that for any $x \parallel y$ in P we have $x <_L y$ if and only if $x < y$ in P^*. Assume that $a_2 <_L a_1 <_L b_1 <_L b_2$ is a pair of nesting covers $a_1 \prec b_1$ below $a_2 \prec b_2$. Then $a_1 \parallel a_2$ (hence $a_2 <_{P^*} a_1$) or $b_1 \parallel b_2$ (hence $b_1 <_{P^*} b_2$) or both. Observe that the latter case is impossible, as for any maximal chain C containing $a_1 \prec b_1$ we would have $a_2 <_{P^*} a_1$ with $a_1 \in C$ and $b_1 <_{P^*} b_2$ with $b_1 \in C$, contradicting $(\star)$. So the nesting of $a_1 \prec b_1$ below $a_2 \prec b_2$ is either of type A with $a_2 < a_1$, or of type B with $b_1 < b_2$. See Fig. 9.

Now consider the case that cover $a_2 \prec b_2$ is nested below another cover $a_3 \prec b_3$, see the right of Fig. 8. Then also $a_1 \prec b_1$ is nested below $a_3 \prec b_3$ and we claim that if both, the nesting of $a_1 \prec b_1$ below $a_2 \prec b_2$ as well as the nesting of $a_1 \prec b_1$ below $a_3 \prec b_3$, are of type A (respectively type B), then also the nesting of $a_2 \prec b_2$ below $a_3 \prec b_3$ is of type A (respectively type B). Indeed, assuming type B, we would get $a_3 <_{P^*} a_2$ and $b_1 <_{P^*} b_3$, which together with any maximal chain C containing $a_2 < a_1 < b_1$ contradicts $(\star)$.

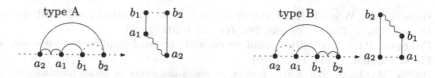

Fig. 9. A nesting of $a_1 \prec b_1$ below $a_2 \prec b_2$ of type A (left) and type B (right).

Finally, let $a_k <_L \cdots <_L a_1 <_L b_1 <_L \cdots <_L b_k$ be any k-rainbow and let $I = \{i \in [k] \mid a_i < a_1\}$, i.e., for each $i \in I$ the nesting of $a_1 \prec b_1$ below $a_i \prec b_i$ is of type A. Then we have just shown that the nesting of $a_j \prec b_j$ below $a_i \prec b_i$ is of type A whenever $i, j \in I$ and of type B whenever $i, j \notin I$. Hence, the set $\{a_i \mid i \in I\} \cup \{a_1, b_1\} \cup \{b_i \mid i \notin I\}$ is a chain in P of size $k+1$, and thus $k \leq h-1$. It follows that P has queue-number at most $h-1$, as desired. $\square$

The proof of the following can be found in the arXiv version of the present paper, [11].

Proposition 3. *For each h there exists a planar poset Q_h of height h and queue-number $h-1$.*

5   Conclusions

We studied the queue-number of (planar) posets of bounded height and width. Two main problems remain open: bounding the queue-number by the width and bounding it by a function of the height in the planar case, where the latter is equivalent to the central conjecture in the area of queue-numbers of graphs. For the first problem the biggest class known to satisfy it are posets without the embedded the subdivided k-crowns for $k \geq 2$ as defined in Sect. 2. Note, that proving it for $k \geq 3$ would imply that Conjecture 2 holds for all 2-dimensional posets, which seems to be a natural next step.

Let us close the paper by recalling another interesting conjecture from [9], which we would like to see progress in:

Conjecture 4. (Heath and Pemmaraju [9]).
Every planar poset on n elements has queue-number at most $\lceil \sqrt{n} \rceil$.

References

1. Baker, K.A., Fishburn, P.C., Roberts, F.S.: Partial orders of dimension 2. Networks **2**(1), 11–28 (1972)
2. Di Battista, G., Liu, W.-P., Rival, I.: Bipartite graphs, upward drawings, and planarity. Inf. Process. Lett. **36**(6), 317–322 (1990)
3. Dilworth, R.P.: A decomposition theorem for partially ordered sets. Ann. Math. **2**(51), 161–166 (1950)
4. Dujmović, V., Pór, A., Wood, D.R.: Track layouts of graphs. Discret. Math. Theor. Comput. Sci. **6**(2), 497–522 (2004)

5. Dujmović, V., Wood, D.R.: Stacks queues and tracks: layouts of graph subdivisions. Discret. Math. Theor. Comput. Sci. **7**(1), 155–202 (2005)
6. Fishburn, P.C.: Intransitive indifference with unequal indifference intervals. J. Math. Psychol. **7**, 144–149 (1970)
7. Habib, M., Jegou, R.: N-free posets as generalizations of series-parallel posets. Discret. Appl. Math. **12**, 279–291 (1985)
8. Heath, L.S., Leighton, F.T., Rosenberg, A.L.: Comparing queues and stacks as machines for laying out graphs. SIAM J. Discret. Math. **5**(3), 398–412 (1992)
9. Heath, L.S., Pemmaraju, S.V.: Stack and queue layouts of posets. SIAM J. Discret. Math. **10**(4), 599–625 (1997)
10. Heath, L.S., Rosenberg, A.L.: Laying out graphs using queues. SIAM J. Comput. **21**(5), 927–958 (1992)
11. Knauer, K., Micek, P., Ueckerdt, T.: The queue-number of posets of bounded width or height. arXiv:1806.04489v4 (2018). https://arxiv.org/abs/1806.04489v4
12. Moore, J.I.: Graphs and partially ordered sets. Ph.D. thesis, University of South Carolina (1975)
13. Nešetřil, J., de Mendez, P.O., Wood, D.R.: Characterisations and examples of graph classes with bounded expansion. Eur. J. Comb. **33**(3), 350–373 (2012). https://doi.org/10.1016/j.ejc.2011.09.008
14. Pemmaraju, S.V.: Exploring the powers of stacks and queues via graph layouts. Ph.D. thesis, Virginia Polytechnic Institute & State University, Blacksburg, Virginia (1992)

Queue Layouts of Planar 3-Trees

Jawaherul Md. Alam[1], Michael A. Bekos[2], Martin Gronemann[3(✉)],
Michael Kaufmann[2], and Sergey Pupyrev[1]

[1] Department of Computer Science, University of Arizona, Tucson, USA
jawaherul@gmail.com, spupyrev@gmail.com
[2] Institut für Informatik, Universität Tübingen, Tübingen, Germany
{bekos,mk}@informatik.uni-tuebingen.de
[3] Institut für Informatik, Universität zu Köln, Köln, Germany
gronemann@informatik.uni-koeln.de

Abstract. A *queue layout* of a graph G consists of a *linear order* of the
vertices of G and a partition of the edges of G into *queues*, so that no two
independent edges of the same queue are nested. The *queue number* of G
is the minimum number of queues required by any queue layout of G. In
this paper, we continue the study of the queue number of planar 3-trees.
As opposed to general planar graphs, whose queue number is not known
to be bounded by a constant, the queue number of planar 3-trees has
been shown to be at most seven. In this work, we improve the upper
bound to five. We also show that there exist planar 3-trees, whose queue
number is at least four; this is the first example of a planar graph with
queue number greater than three.

1 Introduction

In a *queue* layout [12], the vertices of a graph are restricted to a line and the
edges are drawn at different half-planes delimited by this line, called *queues*.
The task is to find a linear order of the vertices along the underlying line and
a corresponding assignment of the edges of the graph to the queues, so that no
two independent edges of the same queues are nested; see Fig. 1. Recall that
two edges are called *independent* if they do not share an endvertex. The *queue
number* of a graph is the smallest number of queues that are required by any
queue layout of the graph. Note that queue layouts form the "dual" concept of
stack layouts [14], which do not allow two edges of the same stack to cross.

Apart from the intriguing theoretical interest, queue layouts find applications
in several domains [2,11,15,20]. As a result, they have been studied extensively
over the years [3,5,9,10,12,16–21]. An important open problem in this area is
whether the queue number of *planar* graphs is bounded by a constant. A positive
answer to this problem would have several important implications, e.g., (i) that
every n-vertex planar graph admits a $\mathcal{O}(1) \times \mathcal{O}(1) \times \mathcal{O}(n)$ straight-line grid
drawing [22], (ii) that every Hamiltonian bipartite planar graph admits a 2-
layer drawing and an edge-coloring of bounded size, such that edges of the same

This work is supported by the DFG grant Ka812/17-1 and DAAD project 57419183.

T. Biedl and A. Kerren (Eds.): GD 2018, LNCS 11282, pp. 213–226, 2018.
https://doi.org/10.1007/978-3-030-04414-5_15

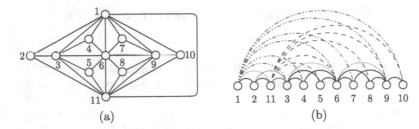

Fig. 1. (a) The Goldner-Harary planar 3-tree, and (b) a 5-queue layout of it produced by our algorithm, in which edges of different queues are colored differently. (Color figure online)

color do not cross [8], and (iii) that the queue number of k-planar graphs is also bounded by a constant [9]. The best-known upper bound is due to Dujmović [4], who showed that the queue number of an n-vertex planar graph is at most $\mathcal{O}(\log n)$ (improving upon an earlier bound by Di Battista et al. [3]).

It is worth noting that many subclasses of planar graphs have bounded queue number. Every tree has queue number one [12], outerplanar graphs have queue number at most two [11], and series-parallel graphs have queue number at most three [18]. Surprisingly, planar 3-trees have queue number at most seven [21], although they were conjectured to have super-constant queue number by Pemmaraju [16]. As a matter of fact, every graph that admits a 1-queue layout is planar with at most $2n - 3$ edges; however, testing this property is $\mathcal{NP}$-complete [11]; for a survey refer to [9].

Our Contribution. In Sect. 2, we improve the upper bound on the queue number of planar 3-trees from seven [21] to five; recall that a planar 3-tree is a triangulated plane graph G with $n \geq 3$ vertices, such that G is either a 3-cycle, if $n = 3$, or has a vertex whose deletion gives a planar 3-tree with $n - 1$ vertices, if $n > 3$. In Sect. 3, we show that there exist planar 3-trees, whose queue number is at least four, thus strengthening a corresponding result of Wiechert [21] for general (that is, not necessarily planar) 3-trees. We stress that our lower bound is also the best known for planar graphs. Table 1 puts our results in the context of existing bounds. We conclude in Sect. 4 with open problems.

Preliminaries. For a pair of distinct vertices u and v, we write $u \prec v$, if u precedes v in a linear order. We also write $[v_1, v_2, \ldots, v_k]$ to denote that v_i precedes v_{i+1} for all $1 \leq i < k$. Assume that F is a set of $k \geq 2$ independent edges (s_i, t_i) with $s_i \prec t_i$, for all $1 \leq i \leq k$. If the linear order is $[s_1, \ldots, s_k, t_k, \ldots, t_1]$, then we say that F is a k-*rainbow*, while if the linear order is $[s_1, \ldots, s_k, t_1, \ldots, t_k]$, we say that F is a k-*twist*. The edges of F form a k-*necklace*, if $[s_1, t_1, \ldots, s_k, t_k]$; see Fig. 2a. A preliminary result for queue layouts is the following.

Lemma 1 (Heath and Rosenberg [12]). *A linear order of the vertices of a graph admits a k-queue layout if and only if there exists no $(k + 1)$-rainbow.*

Table 1. Queue numbers of various subclasses of planar graphs

Graph class	Upper bound		Lower bound	
	Old	New	Old	New
Tree	1 [12]		1 [12]	
Outerplanar	2 [11]		2 [12]	
Series-parallel	3 [18]		3 [21]	
Planar 3-tree	7 [21]	**5** [Theorem 1]	3 [21]	**4** [Theorem 2]
Planar	$\mathcal{O}(\log n)$ [4]		3 [21]	**4** [Theorem 2]

Central in our approach is also the following construction by Dujmović et al. [7] for internally-triangulated outerplane graphs; for an illustration see Figs. 2b–c.

Lemma 2 (Dujmović, Pór, Wood [7]). *Every internally-triangulated outerplane graph, G, admits a straight-line outerplanar drawing, $\Gamma(G)$, such that the y-coordinates of vertices of G are integers, and the absolute value of the difference of the y-coordinates of the endvertices of each edge of G is either one or two. Furthermore, the drawing can be used to construct a 2-queue layout of G.*

Let $\langle u, v, w \rangle$ be a face of a drawing $\Gamma(G)$ produced by the construction of Lemma 2, where G is an internally triangulated outerplane graph. Up to renaming of the vertices of this face, we may assume that $|y(u) - y(v)| = |y(u) - y(w)| = 1$, $|y(v) - y(w)| = 2$ and $y(v) > y(w)$. We refer to vertex u as to the *anchor* of the face $\langle u, v, w \rangle$ of $\Gamma(G)$; v and w are referred to as *top* and *bottom*, respectively. It is easy to verify that drawing $\Gamma(G)$ can be converted to a 2-queue layout of G as follows: (i) for any two distinct vertices u and v of G, $u \prec v$, if and only if the y-coordinate of u is strictly greater than the one of v, or the y-coordinate of u is equal to the one of v, and u is to the left of v in $\Gamma(G)$, (ii) edge (u, v) is assigned to the first (second) queue if and only if the absolute value of the difference of the y-coordinates of u and v is one (two, respectively) in $\Gamma(G)$.

Finally, let $\langle u, v, w \rangle$ and $\langle u', v', w' \rangle$ be two faces of $\Gamma(G)$, such that u and u' are their anchors, v and v' are their top vertices, and w and w' are their bottom vertices. If u and u' are distinct and $u \prec u'$ in the 2-queue layout, then $v \prec v'$ (if $v \neq v'$) and $w \prec w'$ (if $w \neq w'$). The property clearly holds, if u and u' do not have the same y-coordinate. Otherwise, the property holds, since $\Gamma(G)$ is planar.

2   The Upper Bound

In this section, we prove that the queue number of every planar 3-tree is at most five. Our approach is inspired by the algorithm of Wiechert [21] to compute 7-queue layouts for general (not necessarily planar) 3-trees. To reduce the number of required queues in the produced layouts, we make use of structural properties of the input graph. In particular, we put the main ideas of the algorithm of

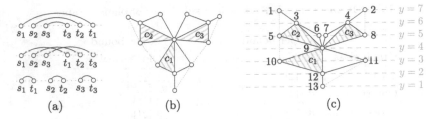

Fig. 2. (a) 3-rainbow, 3-twist and 3-necklace (from top to bottom); (b) an internally-triangulated outerplane graph G_0; the dotted-gray edges are added to make it biconnected; its gray-shaded faces contain components c_1, c_2 and c_3 of G_1; (c) the drawing $\Gamma(G_0)$ by Lemma 2; the vertex-labels indicate the linear order of its 2-queue layout; the anchor vertices of faces $\langle 9, 10, 12 \rangle$, $\langle 3, 5, 9 \rangle$ and $\langle 4, 8, 9 \rangle$ are 10, 5, 8, respectively.

Wiechert [21] into a *peeling-into-levels* approach (see, e.g., [23]), according to which the vertices and the edges of the input graph are partitioned as follows: (i) vertices incident to the outerface are at level zero, (ii) vertices incident to the outerface of the graph induced by deleting all vertices of levels $0, \ldots, i - 1$ are at level i, (iii) edges between same-level vertices are called *level edges*, and (iv) edges between vertices of different levels are called *binding edges*.

To keep the description simple, we first show how to compute a 5-queue layout of a planar 3-tree G, assuming that G has only two levels. Then, we extend our approach to more than two levels. We conclude by discussing the differences between the approach of Wiechert [21] and ours; we also describe which properties of planar 3-trees we exploited to reduce the required number of queues.

The Two-Level Case. We start with the (intuitively easier) case in which the given planar 3-tree G consists of two levels, L_0 and L_1. Since we use this case as a tool to cope with the general case of more than two levels, we consider a slightly more general scenario. In particular, we make the following assumptions (see Fig. 2b): (A.1) the graph G_0 induced by the vertices of level L_0 is outerplane and internally-triangulated, and (A.2) each connected component of the graph G_1 induced by the vertices of level L_1 is outerplane and resides within a (triangular) face of G_0. Without loss of generality we may also assume that G_0 is biconnected, as otherwise we can augment it to being biconnected by adding (level-L_0) edges without affecting its outerplanarity. Note that in a planar 3-tree, graph G_0 is simply a triangle (and not an outerplane graph, as we have assumed), and as a result G_1 is a single outerplane component. Our algorithm maintains the following invariants:

I.1 the linear order is such that all vertices of L_0 precede all vertices of L_1;
I.2 the level edges use two queues, $\mathcal{Q}_0$ and $\mathcal{Q}_1$;
I.3 the binding edges use three queues, $\mathcal{Q}_2$, $\mathcal{Q}_3$, and $\mathcal{Q}_4$.

In the following lemma, we show how to determine a (partial) linear order of the vertices of levels L_0 and L_1 that satisfies the first two invariants of our algorithm.

Lemma 3. *There is an order of vertices of level L_0 and a partial order of vertices of level L_1 such that I.1 and I.2 are satisfied.*

Proof. To compute an order that satisfies I.1, we construct two orders, one for the vertices of level L_0 (that satisfies I.2) and one for the vertices of level L_1 (that also satisfies I.2), and then we concatenate them so that the vertices of L_0 precede the vertices of L_1.

To compute an order of the vertices of L_0 satisfying I.2, we apply Lemma 2, as by our initial assumption A.1, graph G_0 is internally-triangulated and outer-plane. Thus, I.2 is satisfied for the vertices of level L_0. To compute an order of the vertices of L_1 satisfying I.2, we apply Lemma 2 individually for every connected component of G_1, which can be done by our initial assumption A.2. Then the resulting orders are concatenated (as defined by next Lemma 4). Since for every two connected components of G_1, all vertices of the first one either precede or follow all vertices of the second one, we can use the same two queues (denoted by $\mathcal{Q}_0$ and $\mathcal{Q}_1$ in I.2) for all the vertices of L_1. Therefore, I.2 is satisfied.   □

Next, we complete the order of the vertices of G, in a way that the binding edges between L_0 and L_1 require at most three additional queues so as to satisfy I.3.

Lemma 4. *Given the linear order of the vertices of level L_0 and the partial order of the vertices of level L_1 produced by Lemma 3, there is a total order of the vertices of L_0 and L_1 that extends their partial orders and an assignment of the binding edges between L_0 and L_1 into three queues such that I.3 is satisfied.*

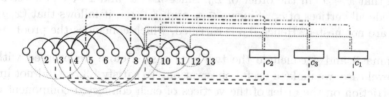

Fig. 3. The 5-queue layout for the graph in Fig. 2; since $5 \prec 8$ and $8 \prec 10$ in the order of the vertices of level L_0 as seen in Fig. 2, c_2 precedes c_3, and c_3 precedes c_1. (Color figure online)

Proof. Consider a connected component c of G_1. By our initial assumption A.2, component c resides within a triangular face $\langle u, v, w \rangle$ of G_0. Let u, v and w be the anchor, top and bottom vertices of the face, respectively. We assign the binding edges incident to u to queue $\mathcal{Q}_2$, the ones incident to v to queue $\mathcal{Q}_3$ and the ones incident to w to queue $\mathcal{Q}_4$; see the blue, red, and green edges in Fig. 3.

Next we describe how to compute the relative order of the connected components of G_1. Let c and c' be two such components. By our initial assumption A.2, c and c' reside within two triangular faces $\langle u, v, w \rangle$ and $\langle u', v', w' \rangle$ of G_0. Assume that u and u' are the anchors of the two faces, v, v' are top and w, w' are bottom vertices. If $u \neq u'$, then c precedes c' if and only if $u \prec u'$ in the order of L_0.

If $u = u'$, we have $v \neq v'$ or $w \neq w'$. If $v \neq v'$, then c precedes c' if and only if $v \prec v'$ in the order of L_0. Otherwise (that is, $u = u'$ and $v = v'$), c precedes c' if and only if $w \prec w'$ in the order of L_0. We claim that for the resulting order of L_1, I.3 is satisfied, that is, no two edges of each of $\mathcal{Q}_2$, $\mathcal{Q}_3$ and $\mathcal{Q}_4$ are nested.

We start our proof with $\mathcal{Q}_2$. Consider two independent edges $(x, y) \in \mathcal{Q}_2$ and $(x', y') \in \mathcal{Q}_2$, where $x, x' \in L_0$ and $y, y' \in L_1$ (see the blue edges in Fig. 3 incident to 5 and 8). By construction of $\mathcal{Q}_2$, x and x' are anchors of two different faces f_x and $f_{x'}$ of G_0 (see the faces of Fig. 2c that contain c_2 and c_3). Without loss of generality we assume that $x \prec x'$ in the order of L_0. Then, the two components c_y and $c_{y'}$ of G_1, that reside within f_x and $f_{x'}$ and contain y and y', are such that all vertices of c_y precede all vertices of $c_{y'}$ (in Fig. 3, $x = 5$ precedes $y = 8$; thus, $c_y = c_2$ precedes $c_{y'} = c_3$). Since $y \in c_y$ and $y' \in c_{y'}$, edges (x, y) and (x', y') do not nest.

We continue our proof with $\mathcal{Q}_3$ (the proof for $\mathcal{Q}_4$ is similar). Let (x, y) and (x', y') be two independent edges of $\mathcal{Q}_3$, where $x, x' \in L_0$ and $y, y' \in L_1$ (see the red edges in Fig. 3 incident to 3 and 4). By construction of $\mathcal{Q}_3$, x and x' are the top vertices of two different faces f_x and $f_{x'}$ of G_0 (see the faces of Fig. 2c that contain c_2 and c_3). Let c_y and $c_{y'}$ be the components of G_1 that reside within f_x and $f_{x'}$ and contain y and y'. Finally, let u and u' be the anchors of f_x and $f_{x'}$, respectively. Suppose first that $u \neq u'$ and assume that $u \prec u'$ in the order of L_0. Since $u \prec u'$, it follows that $x \prec x'$ and that all vertices of c_y precede all vertices of $c_{y'}$ (in Fig. 3, $u = 5$ precedes $u' = 8$, which implies that $x = 3$ precedes $x' = 4$; thus, $c_y = c_2$ precedes $c'_y = c_3$). Since $y \in c_y$ and $y' \in c_{y'}$, it follows that (x, y) and (x', y') are not nested. Suppose now that $u = u'$ and assume that $x \prec x'$ in the order of L_0. Since $u = u'$ and $x \prec x'$, all vertices of c_y precede all vertices of $c_{y'}$. Since $y \in c_y$ and $y' \in c_{y'}$, it follows that (x, y) and (x', y') are not nested. Hence, I.3 is satisfied, which concludes the proof.   □

Lemmas 3 and 4 conclude the two-level case. Before we proceed with the multi-level case, we make a useful observation. To satisfy I.3, we did not impose any restriction on the order of the vertices of each connected component of G_1 (any order that satisfies I.2 for level L_1 would be suitable for us, that is, not necessarily the one constructed by Lemma 2). What we fixed, was the relative order of these components. We are now ready to proceed to the multi-level case.

The Multi-level Case. We now consider the general case, in which our planar 3-tree G consists of more than two levels, say $L_0, L_1, \ldots, L_\lambda$ with $\lambda \geq 2$. Let G_i be the subgraph of G induced by the vertices of level L_i; $i = 0, 1, \ldots, \lambda$. The connected components of each graph G_i are internally-triangulated outerplane graphs that are not necessarily biconnected: Clearly, this holds for G_0, which is a triangle. Assuming that for some $i = 1, \ldots, \lambda$, graph G_{i-1} has the claimed property, we observe that each connected component of G_i resides within a facial triangle of G_{i-1}. Since each non-empty facial triangle of G_{i-1} in G induces a planar 3-tree [13], the claim follows by observing that the removal of the outer face of a planar 3-tree yields a plane graph, whose outer vertices induce an internally-triangulated outerplane graph.

For the recursive step of our algorithm, assume that for some $i = 0, \ldots, \lambda - 1$ we have a 5-queue layout for each of the connected components of the graph H_{i+1} induced by the vertices of $L_{i+1}, \ldots, L_\lambda$, that satisfies the following invariants.

M.1 the linear order is such that all vertices of L_j precede all vertices of L_{j+1} for every $j = i + 1, \ldots, \lambda - 1$;

M.2 the level edges of $L_{i+1}, \ldots, L_\lambda$ use two queues, $\mathcal{Q}_0$ and $\mathcal{Q}_1$;

M.3 for every $j = i + 1, \ldots, \lambda - 1$, the binding edges between L_j and L_{j+1} use three queues, $\mathcal{Q}_2$, $\mathcal{Q}_3$, and $\mathcal{Q}_4$.

Based on these layouts, we show how to construct a 5-queue layout (satisfying M.1–M.3) for each of the connected components of the graph H_i induced by the vertices of $L_i, \ldots, L_\lambda$. Let C_i be such a component. By definition, C_i is delimited by a connected component c_i of G_i which is internally-triangulated and outerplane. If none of the faces of c_i contains a connected component of H_{i+1}, then we compute a 2-queue layout of it using Lemma 2. Consider now the more general case, in which some of the faces of c_i contain connected components of H_{i+1}. By M.1–M.3, we have computed 5-queue layouts for all the connected components, say $d_1, \ldots, d_k$, of H_{i+1} that reside within the faces of c_i.

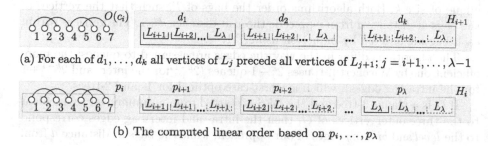

(a) For each of $d_1, \ldots, d_k$ all vertices of L_j precede all vertices of L_{j+1}; $j = i+1, \ldots, \lambda-1$

(b) The computed linear order based on $p_i, \ldots, p_\lambda$

Fig. 4. Illustrations for the proof of Theorem 1.

We proceed by applying the two-level algorithm to the subgraph of C_i induced by the vertices of c_i and the vertices incident to the outer faces of $d_1, \ldots, d_k$. By the last observation we made in the two-level case, this will result in: (a) a linear order $\mathcal{O}(c_i)$ of the vertices of c_i, (b) a relative order of the components $d_1, \ldots, d_k$, (c) an assignment of the (level-L_i) edges of c_i into $\mathcal{Q}_0$ and $\mathcal{Q}_1$, and (d) an assignment of the binding edges between c_i and each of $d_1, \ldots, d_k$ into $\mathcal{Q}_2$, $\mathcal{Q}_3$ and $\mathcal{Q}_4$. Up to renaming, we assume that $d_1, \ldots, d_k$ is the computed order of these components; see Fig. 4a.

By (c) and (d), all edges of C_i are assigned to $\mathcal{Q}_0, \ldots, \mathcal{Q}_4$, since the edges of $d_1, \ldots, d_k$ have been recursively assigned to these queues. Next, we partition the order of vertices of C_i into $\lambda - i + 1$ disjoint intervals, say $p_i, \ldots, p_\lambda$, such that p_μ precedes p_ν if and only if $\mu \prec \nu$. All the (level-L_i) vertices of c_i are contained in p_i in the order $\mathcal{O}(c_i)$ by (a). For $j = i+1, \ldots, \lambda$, p_j contains the vertices of L_j of each of the components $d_1, \ldots, d_k$, such that the vertices of L_j of d_μ precede the vertices of L_j of d_ν if and only if $\mu \prec \nu$; see Fig. 4b. The proof that M.1–M.3 are satisfied can be found in the full version [1]. We summarize in the following.

Theorem 1. *Every planar 3-tree has queue number at most 5.*

We note here that queue layouts are closely related to track layouts; for definitions refer to [7]. The following result follows immediately from a known result by Dujmović, Morin, Wood [6]; see the full version [1] for details.

Corollary 1. *The track number of a planar 3-tree is at most 4000.*

Differences with Wiechert's Algorithm. Wiechert's algorithm [21] builds upon a previous algorithm by Dujmović et al. [6]. Both yield queue layouts for general k-trees, using the breadth-first search (BFS) starting from an arbitrary vertex r of G. For each $d > 0$ and each connected component C induced by the vertices at distance d from r, create a node (called *bag*) "containing" all vertices of C; two bags are adjacent if there is an edge of G between them. For a k-tree, the result is a tree of bags T, called *tree-partition*, so that (P.1) every node of T induces a connected $(k-1)$-tree, and (P.2) for each non-root node $x \in T$, if $y \in T$ is the parent of x, then the vertices in y having a neighbor in x form a clique of size k. Both algorithms order the bags of T, such that the vertices of the bags at distance d from r precede those at distance $d+1$. The vertices within each bag are ordered by induction using P.1.

The algorithms differ in the way the edges are assigned to queues; the more efficient one by Wiechert [21] uses $2^k - 1$ queues (2^{k-1} for the inter- and $2^{k-1}+1$ for the intra-bag edges), which is worst-case optimal for 1- and 2-trees.

If G is a planar 3-tree and the BFS is started from a dummy vertex incident to the three outervertices of G, then the intra- and inter-bag edges correspond to the *level* and *binding* edges of our approach, while the bags at distance d from r in T correspond to different connected components of level d.

To reduce the number of queues, we observed that in G (i) every node of T induces a connected outerplanar graph, while (ii) each clique of size three by P.2 is a triangular face of G. By the first observation, we reduced the number of queues for intra-bag edges; by the second, we combined orders from different bags more efficiently.

3   The Lower Bound

In the following, we prove that the queue number of planar 3-trees is at least four. To this end, we will define recursively a subgraph of a planar 3-tree G and we will show that it contains at least one 4-rainbow in any ordering. Starting with a set of T independent edges (s_i, t_i) with $1 \le i \le T$ and T to be determined later, we connect their endpoints to two unique vertices, say A and B, which we assume to be neighboring. We refer to these edges as (s, t)-*edges*.

As a next step, we *stellate* each triangle $\langle A, s_i, t_i \rangle$ with a vertex x_i, that is, we introduce vertex x_i and connect it to A, s_i, and t_i. Symmetrically, we also

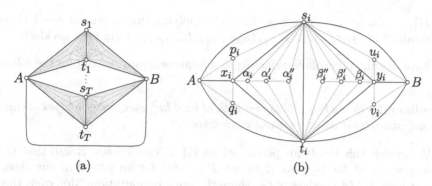

Fig. 5. Construction of graph G_T: Each gray subgraph in (a) corresponds to a copy of the graph of (b).

stellate each triangle $\langle B, s_i, t_i \rangle$ with a vertex y_i. Afterwards, we add one more level, that is, we stellate each of the triangles $\langle A, s_i, t_i \rangle$, $\langle B, s_i, t_i \rangle$, $\langle A, x_i, s_i \rangle$, $\langle A, x_i, t_i \rangle$, $\langle B, y_i, s_i \rangle$ and $\langle B, y_i, t_i \rangle$ with vertices α_i, β_i, p_i, q_i, u_i and v_i, respectively; see Fig. 5b. We further stellate $\langle s_i, t_i, \alpha_i \rangle$ with α_i' and then $\langle s_i, t_i, \alpha_i' \rangle$ with α_i''. Symmetrically, we stellate $\langle s_i, t_i, \beta_i \rangle$ with β_i' and $\langle s_i, t_i, \beta_i' \rangle$ with β_i''.

Let G_T be the graph constructed so far. We refer to vertices A and B as the *poles* of G_T and we assume that G_T admits a 3-queue layout $\mathcal{Q}$. By symmetry, we may assume that $A \prec B$ and that $s_i \prec t_i$ for each edge (s_i, t_i). Consider a single edge (s_i, t_i) and the relative order of its endvertices to A and B. Then, there exist six possible permutations: (P.1) $s_i \prec A \prec B \prec t_i$, (P.2) $A \prec s_i \prec B \prec t_i$, (P.3) $s_i \prec A \prec t_i \prec B$, (P.4) $A \prec B \prec s_i \prec t_i$, (P.5) $s_i, \prec t_i \prec A \prec B$, and (P.6) $A \prec s_i \prec t_i \prec B$.

By the pigeonhole principle and by setting $T = 6l$, we may claim that at least one of the permutations P.1–P.6 applies to at least l edges. We will show that if too many (s, t)-edges share one of the permutations P.1–P.5, then there exists a 4-rainbow, contradicting the fact that $\mathcal{Q}$ is a 3-queue layout for G_T. This implies that if T is large enough, then for at least one (s, t)-edge of G_T permutation P.6 applies. Based on this fact, we describe later how to augment the graph that we have constructed so far using a recursive construction such that we can also rule out permutation P.6. Thereby, proving the claimed lower bound of four. We start with an auxiliary lemma.

Lemma 5. *In every queue that contains r^2 independent edges, there exists either an r-twist or an r-necklace.*

Proof. Assume that no r-twist exists, as otherwise the lemma holds. We will prove the existence of an r-necklace. Let $(s_1, t_1), \dots, (s_{r^2}, t_{r^2})$ be the r^2 independent edges. Assume w.l.o.g. that $s_i \prec s_{i+1}$ for each $i = 1, \dots, r^2 - 1$. Consider the edge (s_1, t_1). Since s_1 is the first vertex in the order and no two edges nest, each vertex t_i, with $i > 1$, is to the right of t_1. Since no r-twist exists, vertex s_r is to the right of t_1. Thus, (s_1, t_1) and (s_r, t_r) do not cross. The removal of

$(s_1, t_1), \ldots, (s_{r-1}, t_{r-1})$ makes s_r first. By applying this argument $r - 1$ times, we obtain that $(s_1, t_1), (s_r, t_r), \ldots (s_{(r-1)^2+1}, t_{(r-1)^2+1})$ form an r-necklace. □

Applying the pigeonhole principle to a k-queue layout, we obtain the following.

Corollary 2. *Every k-queue layout with at least kr^2 independent edges contains at least one r-twist or at least one r-necklace.*

We exploit this result for permutations P.1–P.6 as follows. Recall that Q is a 3-queue layout for G_T. So, if we set $T = 18r^2$ for an $r > 0$ of our choice, then at least $3r^2$ (s,t)-edges of G_T share the same permutation. Moreover, these edges are by construction independent. Therefore, by Corollary 2 at least r of them form a necklace or a twist (while also sharing the same permutation). In the following, we show that if r (s,t)-edges, say w.l.o.g. $(s_1, t_1), \ldots, (s_r, t_r)$, form a necklace or a twist (for an appropriate choice of r) and simultaneously share one of the permutations P.1–P.5, then a 4-rainbow is inevitably induced, which contradicts the fact that Q is a 3-queue layout. We consider each case separately.

Case P.1: Let $r = 8$. It suffices to consider the case, in which $(s_1, t_1), \ldots, (s_8, t_8)$ form a twist, since in general for $r > 1$ the necklace case is impossible. Hence, the order is $[s_1 \ldots s_8 A B t_1 \ldots t_8]$. We show that x_4 always yields a 4-rainbow; Fig. 6 shows the three subcases arising when x_4 is such that $x_4 \prec B$ holds. Clearly, each yields a 4-rainbow. Since we did not use the edge (x_4, A), by symmetry, a 4-rainbow is also obtained when $B \prec x_4$.

Case P.2: As in the previous case, we set $r = 8$ and we only consider the case, in which $(s_1, t_1), \ldots, (s_8, t_8)$ form a twist, since the necklace case is again impossible. Hence, the order is $[A s_1 \ldots s_8 B t_1 \ldots t_8]$. One may verify that placing x_4 and x_5 to the left of t_8 always results in a 4-rainbow (see the full version [1] for details). For the case in which x_4 and x_5 are preceded by t_8, we distinguish between if $x_4 \prec x_5$ holds or not. Both result in a 4-rainbow.

Case P.3: This case can be ruled out like Case P.2 due to symmetry.

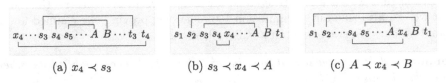

(a) $x_4 \prec s_3$ (b) $s_3 \prec x_4 \prec A$ (c) $A \prec x_4 \prec B$

Fig. 6. Illustration for the Case P.1 when $x_4 \prec B$ holds.

Case P.4: Let $r = 10$. We distinguish two subcases based on whether the edges $(s_1, t_1), \ldots, (s_{10}, t_{10})$ form a twist or a necklace (in contrast to the previous case, here both cases are possible).

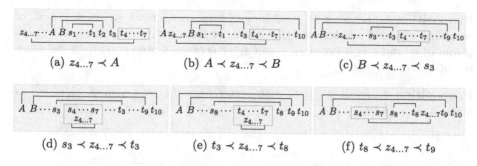

Fig. 7. Illustration for the Case P.4 when $z_{4...7} \prec t_9$ holds.

We start with the twist case. Hence, the order is $[ABs_1 \ldots s_{10}t_1 \ldots t_{10}]$. Let $Z_{4...7} = \{x_4, \ldots, x_7\} \cup \{y_4, \ldots y_7\}$ and let $z_{4...7}$ be any element of $Z_{4...7}$. Similar to the previous case, we sweep from left to right and rule out easy subcases. However, we have to ensure that we do not use any edge from $z_{4...7}$ to A or B in order to keep the roles of x_i and y_i interchangeable. Figure 7 shows that we may assume that $t_9 \prec z_{4...7}$, that is, all $x_4, \ldots, x_7$ and $y_4, \ldots, y_7$ are preceded by t_9.

Next, we show that we can always construct a 3-rainbow spanning (s_8, t_8), which then yields the desired 4-rainbow. Let us take a closer look at the ordering of the 8 vertices in $Z_{4...7}$. To prevent the creation of a 3-rainbow that spans (s_8, t_8), we claim that the ordering has to comply with two requirements: (R.1) the indices of the first 7 elements of $Z_{4...7}$ are non-decreasing, and (R.2) for the last 7 elements of $Z_{4...7}$, it must hold that all x precede all y. Assume to the contrary, that R.1 does not hold. Hence, there exists a pair of vertices, say w.l.o.g $x_j \prec x_i$, with $i < j$ and x_i is not the last element of $Z_{4...7}$. Then, $[s_i \ldots s_j \ldots x_j \ldots x_i]$ forms a 2-rainbow and together with the last element of $Z_{4...7}$ that is adjacent to either A or B, we obtain a 3-rainbow spanning (s_8, t_8); a contradiction. Assume now that R.2 does not hold. Then, there exists a pair $y_i \prec x_j$ with y_i not being the first element. Let the first element be x_l. Then, $[A \ldots B \ldots s_l \ldots x_l \ldots y_i \ldots x_j]$ is a 3-rainbow spanning (s_8, t_8); a contradiction.

Now, we show that R.1 and R.2 cannot simultaneously hold, which implies the existence of a 4-rainbow. Consider the last element of $Z_{4...7}$. Assume that R.1 and R.2 both hold. By R.2, we may deduce that the last three elements of $Z_{4...7}$ belong to $\{y_4, \ldots y_7\}$. Let them be y_i, y_j, y_ℓ as they appear from left to right. Then, by R.1 we have that $i < j$. Consider now x_j. By R.1, $y_i \prec x_j$ must hold. This contradicts the fact that y_i, y_j, y_ℓ are the last three elements of $Z_{4...7}$.

We continue with the necklace case. Here, the order is $[ABs_1t_1 \ldots s_{10}t_{10}]$. We make several observations about the ordering in the form of propositions; their formal proofs can be found in the full version [1].

Proposition 1. *Let w be a neighbor of s_i and t_i for $3 \leq i \leq 8$. Then, either $s_{i-1} \prec w \prec t_{i+1}$ holds, or $s_{10} \prec w$.*

Proposition 2. *Let w and z be two vertices that form a K_4 with s_i and t_i, for $3 \leq i \leq 8$. Then, at least one of the following holds: $s_{10} \prec w$ or $s_{10} \prec z$.*

Proposition 3. *Let w, z be neighbors of both s_i, t_i, for $3 \leq i \leq 8$. Then, at most one of w and z is between s_{i-1} and s_i or between t_i and t_{i-1}. Furthermore, if one of w and z is between s_{i-1} and s_i or between t_i and t_{i-1}, then the other is not between s_i and t_i.*

Proposition 4. *For $4 \leq i \leq 8$, each vertex from the set $\{x_i, y_i, p_i, q_i, u_i, v_i\}$ is between s_{i-1} and t_{i+1}.*

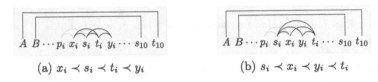

(a) $x_i \prec s_i \prec t_i \prec y_i$ (b) $s_i \prec x_i \prec y_i \prec t_i$

Fig. 8. Contradiction for placing $x_i, y_i, p_i, q_i, u_i, v_i$ in range (s_{i-1}, t_{i+1}), $4 \leq i \leq 8$.

By Proposition 4, for $4 \leq i \leq 8$, each vertex from $\{s_i, t_i, x_i, y_i, p_i, q_i, u_i, v_i\}$ is in (s_{i-1}, t_{i+1}). Then, the edges between these vertices cannot form a 2-rainbow, as otherwise this 2-rainbow along with the two edges (A, t_{10}) and (B, s_{10}) would form a 4-rainbow. Assume w.l.o.g. that $x_i \prec y_i$. Then, by Proposition 3, one of the following two conditions hold: (i) $x_i \prec s_i \prec t_i \prec y_i$, (ii) $s_i \prec x_i \prec y_i \prec t_i$; see Fig. 8. In both cases, p_i must precede both x_i and s_i, as otherwise either $(p_i, s_i), (x_i, t_i)$, or $(p_i, x_i), (s_i, t_i)$ would form a 2-rainbow; see Fig. 8. But then there is no valid position for q_i without creating a 2-rainbow in either case, resulting together with (A, t_{10}) and (B, s_{10}) in a 4-rainbow.

Case P.5: This case can be ruled out like Case P.4 due to symmetry.

From the above case analysis it follows that if r is at least 10 (which implies that T is at least 1,800), then for at least one (s, t)-edge of G_T permutation P.6 applies, that is, there exists $1 \leq i_0 \leq T$ such that $A \prec s_{i_0} \prec t_{i_0} \prec B$. Notice that the edges (A, B) and (s_{i_0}, t_{i_0}) form a 2-rainbow.

We proceed by augmenting graph G_T as follows. For each edge (s_i, t_i) of G_T, we introduce a new copy of G_T, which has s_i and t_i as poles. Let G'_T be the augmented graph and let $(s'_1, t'_1), \ldots, (s'_T, t'_T)$ be the (s, t)-edges of the copy of graph G_T in G'_T corresponding to the edge (s_{i_0}, t_{i_0}) of the original graph G_T. Then, by our arguments above there exists $1 \leq i'_0 \leq T$ such that $s_{i_0} \prec s'_{i'_0} \prec t'_{i'_0} \prec t_{i_0}$. Hence, the edges (A, B), (s_{i_0}, t_{i_0}) and $(s'_{i'_0}, t'_{i'_0})$ form a 3-rainbow, since $A \prec s_{i_0} \prec t_{i_0} \prec B$ holds. If we apply the same augmentation procedure to graph G'_T, then we guarantee that the resulting graph G''_T, which is clearly a subgraph of a planar 3-tree, has inevitably a 4-rainbow. Hence, either G_T does not admit a 3-queue layout, as we initially assumed, or G''_T does not admit a 3-queue layout. In both cases, Theorem 2 follows.

Theorem 2. *There exist planar 3-trees that have queue number at least 4.*

4   Conclusions

In this work, we presented improved bounds on the queue number of planar 3-trees. Three main open problems arise from our work. The first one concerns the exact upper bound on the queue number of planar 3-trees. Does there exist a planar 3-tree, whose queue number is five (as our upper bound) or the queue number of every planar 3-tree is four (as our lower bound example)? The second problem is whether the technique that we developed for planar 3-trees can be extended so to improve the upper bound for the queue number of general (that is, non-planar) k-trees, which is currently exponential in k [21]. Finally, the third problem is the central question in the area. Is the queue number of general planar graphs (that is, that are not necessarily planar 3-trees) bounded by a constant?

References

1. Alam, J.M., Bekos, M.A., Gronemann, M., Kaufmann, M., Pupyrev, S.: Queue layouts of planar 3-trees. CoRR abs/1808.10841 (2018)
2. Bhatt, S.N., Chung, F.R.K., Leighton, F.T., Rosenberg, A.L.: Scheduling tree-dags using FIFO queues: a control-memory trade-off. J. Parallel Distrib. Comput. **33**(1), 55–68 (1996)
3. Di Battista, G., Frati, F., Pach, J.: On the queue number of planar graphs. SIAM J. Comput. **42**(6), 2243–2285 (2013)
4. Dujmović, V.: Graph layouts via layered separators. J. Comb. Theory Ser. B **110**, 79–89 (2015)
5. Dujmović, V., Frati, F.: Stack and queue layouts via layered separators. J. Graph Algorithms Appl. **22**(1), 89–99 (2018)
6. Dujmović, V., Morin, P., Wood, D.R.: Layout of graphs with bounded tree-width. SIAM J. Comput. **34**(3), 553–579 (2005)
7. Dujmović, V., Pór, A., Wood, D.R.: Track layouts of graphs. Discret. Math. Theor. Comput. Sci. **6**(2), 497–522 (2004)
8. Dujmović, V., Wood, D.R.: Tree-partitions of k-trees with applications in graph layout. In: Bodlaender, H.L. (ed.) WG 2003. LNCS, vol. 2880, pp. 205–217. Springer, Heidelberg (2003). https://doi.org/10.1007/978-3-540-39890-5_18
9. Dujmović, V., Wood, D.R.: Stacks, queues and tracks: layouts of graph subdivisions. Discret. Math. Theor. Comput. Sci. **7**(1), 155–202 (2005)
10. Hasunuma, T.: Laying out iterated line digraphs using queues. In: Liotta, G. (ed.) GD 2003. LNCS, vol. 2912, pp. 202–213. Springer, Heidelberg (2004). https://doi.org/10.1007/978-3-540-24595-7_19
11. Heath, L.S., Leighton, F.T., Rosenberg, A.L.: Comparing queues and stacks as mechanisms for laying out graphs. SIAM J. Discrete Math. **5**(3), 398–412 (1992)
12. Heath, L.S., Rosenberg, A.L.: Laying out graphs using queues. SIAM J. Comput. **21**(5), 927–958 (1992)
13. Mondal, D., Nishat, R.I., Rahman, M.S., Alam, M.J.: Minimum-area drawings of plane 3-trees. J. Graph Algorithms Appl. **15**(2), 177–204 (2011)
14. Ollmann, T.: On the book thicknesses of various graphs. In: Hoffman, F., Levow, R., Thomas, R. (eds.) Southeastern Conference on Combinatorics, Graph Theory and Computing. Congressus Numerantium, vol. VIII, p. 459 (1973)

15. Pach, J., Thiele, T., Tóth, G.: Three-dimensional grid drawings of graphs. In: DiBattista, G. (ed.) GD 1997. LNCS, vol. 1353, pp. 47–51. Springer, Heidelberg (1997). https://doi.org/10.1007/3-540-63938-1_49
16. Pemmaraju, S.V.: Exploring the powers of stacks and queues via graph layouts. Ph.D. thesis, Virginia Tech (1992)
17. Pupyrev, S.: Mixed linear layouts of planar graphs. In: Frati, F., Ma, K.-L. (eds.) GD 2017. LNCS, vol. 10692, pp. 197–209. Springer, Cham (2018). https://doi.org/10.1007/978-3-319-73915-1_17
18. Rengarajan, S., Veni Madhavan, C.E.: Stack and queue number of 2-trees. In: Du, D.-Z., Li, M. (eds.) COCOON 1995. LNCS, vol. 959, pp. 203–212. Springer, Heidelberg (1995). https://doi.org/10.1007/BFb0030834
19. Shahrokhi, F., Shi, W.: On crossing sets, disjoint sets, and pagenumber. J. Algorithms **34**(1), 40–53 (2000)
20. Tarjan, R.E.: Sorting using networks of queues and stacks. J. ACM **19**(2), 341–346 (1972)
21. Wiechert, V.: On the queue-number of graphs with bounded tree-width. Electr. J. Comb. **24**(1) (2017). P1.65
22. Wood, D.R.: Queue layouts, tree-width, and three-dimensional graph drawing. In: Agrawal, M., Seth, A. (eds.) FSTTCS 2002. LNCS, vol. 2556, pp. 348–359. Springer, Heidelberg (2002). https://doi.org/10.1007/3-540-36206-1_31
23. Yannakakis, M.: Embedding planar graphs in four pages. J. Comput. Syst. Sci. **38**(1), 36–67 (1989)

Crossings

Crossing Minimization in Perturbed Drawings

Radoslav Fulek[1]([✉]) and Csaba D. Tóth[2,3]

[1] Institute of Science and Technology, Klosterneuburg, Austria
radoslav.fulek@gmail.com
[2] California State University Northridge, Los Angeles, CA, USA
[3] Tufts University, Medford, MA, USA
cdtoth@eecs.tufts.edu

Abstract. Due to data compression or low resolution, nearby vertices and edges of a graph drawing may be bundled to a common node or arc. We model such a "compromised" drawing by a piecewise linear map $\varphi : G \to \mathbb{R}^2$. We wish to perturb φ by an arbitrarily small $\varepsilon > 0$ into a proper drawing (in which the vertices are distinct points, any two edges intersect in finitely many points, and no three edges have a common interior point) that minimizes the number of crossings. An ε-perturbation, for every $\varepsilon > 0$, is given by a piecewise linear map $\psi_\varepsilon : G \to \mathbb{R}^2$ with $\|\varphi - \psi_\varepsilon\| < \varepsilon$, where $\|.\|$ is the uniform norm (i.e., sup norm).

We present a polynomial-time solution for this optimization problem when G is a cycle and the map φ has no **spurs** (i.e., no two adjacent edges are mapped to overlapping arcs). We also show that the problem becomes NP-complete (i) when G is an arbitrary graph and φ has no spurs, and (ii) when φ may have spurs and G is a cycle or a union of disjoint paths.

Keywords: Map approximation · C-planarity · Crossing number

1 Introduction

A graph $G = (V, E)$ is a 1-dimensional simplicial complex. A continuous piecewise linear map $\varphi : G \to \mathbb{R}^2$ maps the vertices in V into points in the plane, and the edges in E to piecewise linear arcs between the corresponding vertices. However, several vertices may be mapped to the same point, and two edges may be mapped to overlapping arcs. This scenario arises in applications in cartography, clustering, and visualization, due to data compression, graph semantics, or low resolution. Previous research focused on determining whether such a map φ can be "perturbed" into an embedding. Specifically, a continuous piecewise linear map $\varphi : G \to M$ is a **weak embedding** if, for every $\varepsilon > 0$, there is an

Research supported in part by the NSF awards CCF-1422311 and CCF-1423615, and the Austrian Science Fund (FWF): M2281-N35.

T. Biedl and A. Kerren (Eds.): GD 2018, LNCS 11282, pp. 229–241, 2018.
https://doi.org/10.1007/978-3-030-04414-5_16

embedding $\psi_\varepsilon : G \to M$ with $\|\varphi - \psi_\varepsilon\| < \varepsilon$, where $\|.\|$ is the uniform norm (i.e., sup norm). Recently, Fulek and Kynčl [1] gave a polynomial-time algorithm for recognizing weak embeddings, and the running time was subsequently improved to $O(n \log n)$ for simplicial maps by Akitaya et al. [2]. Note, however, that only planar graphs admit embeddings and weak embeddings. In this paper, we extend the concept of ε-perturbations to nonplanar graphs, and seek a perturbation with the minimum number of crossings.

A continuous map $\varphi : G \to M$ of a graph G to a 2-manifold M is a **drawing** if (i) the vertices in V are mapped to distinct points in M, (ii) each edge is mapped to a Jordan arc between two vertices without passing through any other vertex, and (iii) any two edges intersect in finitely many points. A **crossing** between two edges, $e_1, e_2 \in E$, is defined as an intersection point between the relative interiors of the arcs $\varphi(e_1)$ and $\varphi(e_2)$. For a piecewise linear map $\varphi : G \to \mathbb{R}^2$, let $\mathrm{cr}(\varphi)$ be the minimum nonnegative integer k such that for every $\varepsilon > 0$, there exists a drawing $\psi_\varepsilon : G \to \mathbb{R}^2$ with $\|\varphi - \psi_\varepsilon\| < \varepsilon$ and k crossings, see Fig. 1 for an illustration.

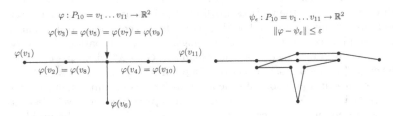

Fig. 1. An example for a map $\varphi : G \to \mathbb{R}^2$, where $G = P_{10}$, i.e., a path of length 10, with $\mathrm{cr}(\varphi) = 1$ (left); and a perturbation ψ_ε witnessing that $\mathrm{cr}(\varphi) \leq 1$ (right).

It is clear that φ is a weak embedding if and only if $\mathrm{cr}(\varphi) = 0$. Note also that if $e_1, e_2 \in E$ and the arcs $\varphi(e_1)$ and $\varphi(e_2)$ cross transversely at some point $p \in \mathbb{R}^2$, then $\psi_\varepsilon(e_1)$ and $\psi_\varepsilon(e_2)$ also cross in the ε-neighborhood of p for any sufficiently small $\varepsilon > 0$. An ε-perturbation may, however, remove tangencies and partial overlaps between edges.

The problem of determining $\mathrm{cr}(\varphi)$ for a given map $\varphi : G \to \mathbb{R}^2$ is NP-complete: In the special case that $\varphi(G)$ is a single point, $\mathrm{cr}(\varphi)$ equals the crossing number of G, and it is NP-complete to find the crossing number of a given graph [3] (even if G is a planar graph plus one edge [4]).

In this paper, we focus on the special case that G is a cycle. A series of recent papers [5–7] show that weak embeddings can be recognized in $O(n \log n)$ time. Chang et al. [6] identified two features of a map $\varphi : G \to \mathbb{R}^2$ that are difficult to handle: A **spur** is a vertex whose incident edges are mapped to the same arc or overlapping arcs, and a **fork** is a vertex mapped to the relative interior of the image of some nonincident edge (a vertex may be both a fork and a spur). We prove the following results.

Theorem 1. *Given a cycle $G = (V, E)$ and a piecewise linear map $\varphi : G \to \mathbb{R}^2$, where G has n vertices and the image $\varphi(G)$ is a plane graph with m vertices, then $cr(\varphi)$ can be computed*

1. *in $O((m + n) \log(m + n))$ time if φ has neither spurs nor forks,*
2. *in $O((mn) \log(mn))$ time if φ has no spurs.*

As noted above, the problem of determining $cr(\varphi)$ is NP-complete when G is an arbitrary graph (even if φ is a constant map). We show that the problem remains NP-complete if G is a cycle and we drop the condition that φ has no spurs.

Theorem 2. *Given $k \in \mathbb{N}$ and a piecewise linear map $\varphi : G \to \mathbb{R}^2$, it is NP-complete to decide whether $cr(\varphi) \leq k$ if $\varphi : G \to \mathbb{R}^2$ may have spurs and*

1. *G is a cycle, or*
2. *G is a union of disjoint paths.*

Related Previous Work. Finding efficient algorithms for the recognition of weak embeddings $\varphi : G \to M$, where G is an arbitrary graph, was posed as an open problem in [5–7]. The first polynomial-time solution for the general version follows from a recent variant [1] of the Hanani-Tutte theorem [8,9], which was conjectured by Skopenkov [10] in 2003 and in a slightly weaker form already by Repovš and Skopenkov [11] in 1998. Weak embeddings of graphs also generalize various graph visualization models such as **strip planarity** [12] and **level planarity** [13]; and can be seen as a special case [14] of the notoriously difficult **cluster-planarity** (for short, **c-planarity**) [15,16], whose tractability remains elusive today.

Organization. We start in Sect. 2 with preliminary observations that show that determining $cr(\varphi)$ is a purely combinatorial problem, which can be formulated without metric inequalities. We describe and analyse a recognition algorithm, proving Theorem 1 in Sect. 3. We prove NP-hardness by a reduction from 3SAT in Sect. 4, and conclude in Sect. 5. Omitted proofs are available in the Appendix.

2 Preliminaries

We rely on techniques introduced in [1,5,6,17], and complement them with additional tools to keep track of edge crossings. A piecewise linear function $\varphi : G \to \mathbb{R}^2$ is a composition $\varphi = \gamma \circ \lambda$, where $\lambda : G \to H$ is a continuous map from G to a graph H (i.e., a 1-dimensional simplicial complex) and $\gamma : H \to \mathbb{R}^2$ is a drawing of H. We may further assume, by subdividing the edges of G if necessary, that the map $\lambda : G \to H$ is **simplicial**, that is, it maps vertices to vertices and edges to edges; and $\gamma : H \to \mathbb{R}^2$ is a straight-line drawing of H, where each edge in $E(H)$ is mapped to a line segment. To distinguish the graphs G and H in our terminology, G has **vertices** $V(G)$ and **edges** $E(G)$, and H has **clusters** $V(H)$ and **pipes** $E(H)$.

A perturbation ψ_ε of φ lies in the ε-neighborhood of $\varphi(G)$. We define suitable neighborhoods for the graph H, and the image $\gamma(H) = \varphi(G)$. For the graph H and its drawing $\gamma : H \to \mathbb{R}^2$, we define the **neighborhood** $\mathcal{N} \subset \mathbb{R}^2$ as the union of regions N_u and N_{uv} for every $u \in V(H)$ and $uv \in E(G)$, respectively, as follows. Let $\varepsilon_0 > 0$ be a sufficiently small constant specified below. For every $u \in V(H)$, let N_u be the closed disk of radius ε_0 centered at $\gamma(u)$. For every edge $uv \in E(H)$, let N_{uv} be the set of points at distance at most ε_0^2 from $\gamma(uv)$ that lie in the interior of neither N_u nor N_v. Let $\varepsilon_0 > 0$ be so small that for every triple $\{u, v, w\} \subset V(H)$, the disk N_u is disjoint from both N_v and N_{vw}, and the regions N_{uv} and N_{uw} are disjoint from each other. (Note, however, that regions N_{uv} and $N_{u'v'}$ may intersect if the line segments $\gamma(uv)$ and $\gamma(u'v')$ cross.)

Such $\varepsilon_0 > 0$ exists due to piecewise linearity of φ and by compactness. (Indeed, consider the intersection $B_{u,v}$ and $B_{u,w}$ of the boundary of N_u with that of N_{uv} and N_{uw}, respectively. Taking ε_0 sufficiently small, we assume that $N_u \cap \gamma(uv)$ and $N_u \cap \gamma(uw)$ are line segments meeting in u at some angle $\alpha \leq \pi$. We require $\varepsilon_0 < \frac{1}{\pi}\alpha$ since we need $\varepsilon_0^2 < \frac{1}{\pi}\varepsilon_0\alpha$ for $B_{u,v}$ and $B_{u,w}$ to be disjoint, and hence N_{uv} and N_{uw}.) By definition, an ε-perturbation of $\varphi = \gamma \circ \lambda$ lies in the neighborhood $\mathcal{N}$ for all $\varepsilon \in (0, \varepsilon_0^2)$.

For the graph H and its drawing $\gamma : H \to \mathbb{R}^2$, we also define the **thickening** $\mathcal{H}$, $H \subset \mathcal{H}$, as a 2-dimensional manifold with boundary as follows. For every $u \in V(H)$, create a topological disk D_u, and for every edge $uv \in E(H)$, create a rectangle R_{uv}. For every D_u and R_{uv}, fix an arbitrary orientation of ∂D_u and ∂R_{uv}, respectively. Partition the boundary of ∂D_u into $\deg(u)$ arcs, and label them by $A_{u,v}$, for all $uv \in E(H)$, in the cyclic order around ∂D_u determined by the rotation of u in the the drawing $\gamma(G)$. The manifold $\mathcal{H}$ is obtained by identifying two opposite sides of every rectangle R_{uv} with $A_{u,v}$ and $A_{v,u}$ via an orientation preserving homeomorphism. Note that there is a natural map $\Gamma : \mathcal{H} \to \mathcal{N}$ such that $\Gamma|_H = \gamma$; Γ is a homeomorphism between D_u and N_u for every $u \in V(H)$; and Γ maps R_{uv} to N_{uv} for every $uv \in E(H)$.

We reformulate a problem instance $\varphi : G \to \mathbb{R}^2$ as two functions $\lambda : G \to H$ and $\gamma : H \to \mathbb{R}^2$, where G and H are abstract graphs, λ is a simplicial map and γ is a straight-line drawing of H. A **perturbation** of the map $\varphi = \gamma \circ \lambda$ is a drawing $\psi = \Gamma \circ \Lambda$, where $\Lambda : G \to \mathcal{H}$ is a drawing of G on $\mathcal{H}$ with the following properties:

(P1) for every vertex $a \in V(G)$, $\Lambda(a) \in D_{\lambda(a)}$,
(P2) for every edge $ab \in E(G)$, $\Lambda(ab) \subset D_{\lambda(a)} \cup R_{\lambda(a)\lambda(b)} \cup D_{\lambda(b)}$ such that it crosses the boundary of the disks $D_{\lambda(a)}$ and $D_{\lambda(b)}$ precisely once, and
(P3) all crossing between arcs $\Lambda(e)$, $e \in E(G)$, lie in the disks D_u, $u \in V(H)$;

and $\Gamma : \mathcal{H} \to \mathbb{R}^2$ maps the disk D_u injectively into $\mathcal{N}_u$ for all $u \in V(H)$, and rectangle R_{uv} into N_{uv} for all $uv \in E(H)$ (however the rectangles R_{uv} and $R_{u'v'}$ may be mapped to crossing neighborhoods N_{uv} and $N_{u'v'}$ for two independent edges $uv, u'v' \in E(H)$).

Combinatorial Representation. Properties (P1)–(P3) allow for a combinatorial representation of the drawing $\Lambda : G \to \mathcal{H}$: For every pipe $uv \in E(H)$, let π_{uv} be a total order of the edges in $\lambda^{-1}[uv] \subseteq E(G)$ in $R_{\lambda(a)\lambda(b)}$; and let $\pi_\Lambda = \{\pi_{uv} : uv \in E(H)\}$ the collection of these total orders. In fact, we can assume that $\Lambda(G)$ consists of straight-line segments in every rectangle R_{uv}, and every disk D_u. The number of crossings in each disk D_u is determined by the cyclic order of the segment endpoints along ∂D_u. Thus the number of crossings in all disk D_u, $u \in V(H)$ is determined by π_Λ.

Two Types of Crossings. The reformulation of the problem allows us to distinguish two types of crossings in a piecewise-linear map $\varphi : G \to \mathbb{R}^2$: edge-crossings in the neighborhoods N_u, $u \in V(H)$, and crossings between edges mapped to two pipes that cross each other.

The number of crossings between the edges of G inside a disk N_u, $u \in V(H)$, is the same as the number of crossings in D_u, since Γ is injective on D_u. We denote the total number of such crossings by

$$\mathrm{cr}_1(\lambda) = \min_\Lambda \left(\sum_{u \in V(H)} \mathrm{CR}_\Lambda(u) \right),$$

where $\mathrm{CR}_\Lambda(u)$ is the number of crossings of the drawing $\Lambda(G)$ in the disk D_u.

Let the **weight** of a pipe $e \in E(H)$ be the number of edges of G mapped to e, that is, $w(e) := |\lambda^{-1}[e]|$. If the arcs $\gamma(e_1)$ and $\gamma(e_2)$ cross in the plane, for some $e_1, e_2 \in E(H)$, then every edge in $\lambda^{-1}[e_1]$ crosses all edges in $\lambda^{-1}[e_2]$. The total number of crossings between the edges of G attributed to the crossings between pipes is

$$\mathrm{cr}_2(\gamma, \lambda) = \sum_{\{e_1, e_2\} \in C} w(e_1)w(e_2),$$

where C is the multiset of pipe pairs $\{e_1, e_2\}$ such that $\gamma(e_1)$ and $\gamma(e_2)$ cross. It is now clear that

$$\mathrm{cr}(\gamma \circ \lambda) = \mathrm{cr}_1(\lambda) + \mathrm{cr}_2(\gamma, \lambda). \tag{1}$$

The operations in Sect. 3 successively modify an instance $\varphi = \gamma \circ \lambda$ until H becomes a cycle. In this case, it is easy to determine $\mathrm{cr}_2(\gamma, \lambda)$, which is a consequence of the following folklore lemma.

Lemma 1 [18, Lemma 1.12]. *If $G = C_n$ and $H = C_k$ and $\lambda : G \to H$ is a simplicial map without spurs, where the cycle G winds around the cycle H precisely n/k times, then $cr_1(\lambda) = \frac{n}{k} - 1$.*

3 Cycles Without Spurs

Let $G = C_n$ be a cycle with n vertices, and H an arbitrary abstract graph, $\lambda : G \to H$ a simplicial map that does not map any two consecutive edges of G to the same edge in H, and $\gamma : H \to \mathbb{R}^2$ a straight-line drawing. In this section, we prove that $\mathrm{cr}(\gamma \circ \lambda)$ is invariant under the so-called ClusterExpansion and

PipeExpansion operations. (Similar operations for weak embeddings have been introduced in [1,5,6,17].) We show that a sequence of $O(n)$ operations produces an instance in which H is a cycle, where we can easily determine both $cr_1(\lambda)$ and $cr_2(\gamma, \lambda)$, hence $cr(\gamma \circ \lambda)$.

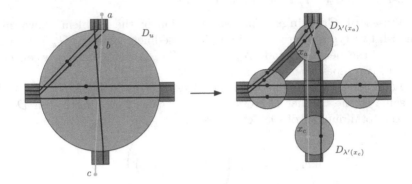

Fig. 2. ClusterExpansion(u).

ClusterExpansion(u). See Fig. 2 for an illustration. (1) Let D_u be a sufficiently small disk centered at $\gamma(u)$ that intersects only the images of pipes incident to u. (2) Subdivide every pipe $uv \in E(H)$ incident to u with a new cluster y_v, let $\gamma(y_v) := \partial D_u \cap \gamma(uv)$. (3) Subdivide every edge $ab \in E(G)$ such that $\lambda(b) = u$ with a new vertex x_a such that $\lambda(x_a) = y_{\lambda(a)}$. (4) For every vertex $b \in \lambda^{-1}[u]$, and any two neighbors x_a and x_c, insert an edge $x_a x_c$ in G, insert a pipe $\lambda(x_a)\lambda(x_c)$ in H if it is not already present, and draw this pipe in the plane as a straight-line segment between $\gamma(\lambda(x_a))$ and $\gamma(\lambda(x_c))$. (5) Delete cluster u from H, and delete all vertices in $\lambda^{-1}[u]$ from G. (6) Return the resulting instance by $\lambda' : G' \to H'$ and $\gamma' : H' \to \mathbb{R}^2$.

Lemma 2. *If G is a cycle, $\lambda : G \to H$ has no spur, and $u \in V(H)$, then ClusterExpansion(u) produces an instance where G' is a cycle, $\lambda' : G' \to H'$ has no spur, and $cr(\gamma \circ \lambda) = cr(\gamma' \circ \lambda')$.*

We remark that $cr(\gamma \circ \lambda)$ is invariant under the ClusterExpansion(u) operation even in the presence of spurs, however the proof is somewhat simpler in the absence spurs, and Lemma 2 also establishes that ClusterExpansion(u) does not create new spurs.

Pipe Expansion. A cluster $u \in V(H)$ is a *base* of an incident pipe uv if every vertex in $\lambda^{-1}[u]$ is incident to an edge in $\lambda^{-1}[uv]$. A pipe $uv \in E(H)$ is *safe* if both u and v are bases of uv. The following operation is defined on safe pipes. See Fig. 2 for an illustration. (We note that our algorithm would be correct even if PipeExpansion(uv) were defined on all pipes, unlike the result in [2], since λ does not contain spurs. We restrict this operation to safe pipe to simplify the runtime analysis.)

PipeExpansion(uv). (1) Let D_{uv} be a sufficiently narrow ellipse with foci at $\gamma(u)$ and $\gamma(v)$ that intersects only the images of pipes incident to u and v. (2) Subdivide every pipe $e \in E(H)$ incident to u or v with a new cluster y_e, let $\gamma(y_e) := \partial D_{uv} \cap \gamma(e)$. (3) Subdivide every edge $ab \in E(G)$ such that $\lambda(a) \notin \{u, v\}$ and $\lambda(b) \in \{u, v\}$ with a new vertex x_a such that $\lambda(x) = y_{\lambda(ab)}$. (4) For every edge $bc \in \lambda^{-1}[uv]$, and the two neighbors x_a and x_d of b and c, respectively, insert an edge $x_a x_d$ in G, insert a pipe $\lambda(x_a)\lambda(x_d)$ in H if it is not already present, and draw this pipe in the plane as a straight-line segment between $\gamma(\lambda(x_a))$ and $\gamma(\lambda(x_d))$. (5) Delete clusters u and v from H, and delete all vertices in $\lambda^{-1}[uv]$ from G. (6) Return the resulting instance by $\lambda' : G' \to H'$ and $\gamma' : H' \to \mathbb{R}^2$ (Fig. 3).

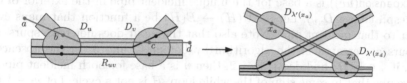

Fig. 3. PipeExpansion(uv) for a safe pipe uv.

Lemma 3. *If G is a cycle, $\lambda : G \to H$ has no spur, and $uv \in E(H)$ is a safe pipe, then PipeExpansion(uv) produces an instance where G' is a cycle, $\lambda' : G' \to H'$ has no spur, and $cr(\gamma \circ \lambda) = cr(\gamma' \circ \lambda')$.*

We remark that Lemma 3 holds even for uv that is not safe, provided that $\lambda : G \to H$ has no spur.

Main Algorithm. Given an instance $\lambda : G \to H$ and $\gamma : H \to \mathbb{R}^2$, we apply the two operations defined above as follows.

Algorithm 1. Input: (G, H, λ, γ)
$U_0 \longleftarrow V(H)$
for *every $u \in U_0$* **do**
  $\lfloor$ ClusterExpansion(u)
while *there is a safe pipe $uv \in E(H)$ such that* $\deg_H(u) \geq 3$ *or* $\deg_H(v) \geq 3$ **do**
  $\lfloor$ PipeExpansion(uv)
$uv \longleftarrow$ an arbitrary edge in $E(H)$.
return $cr_2(\gamma, \lambda) + |\lambda^{-1}[uv]| - 1$.

Lemma 4. *Algorithm 1 terminates.*

Proof. By Lemmas 2 and 3, $\lambda : G \to H$ has no spurs in any step of the algorithm. It is enough to show that the while loop of Algorithm 1 terminates. We define the potential function $\Phi(G, H) = |E(G)| - |E(H)|$, and show that $\Phi(G, H) \geq 0$

and it decreases in every invocation of PipeExpansion(uv). Since G is a cycle and λ has no spur, every edge in $\lambda^{-1}[uv]$ is adjacent to one edge in some other pipe incident to u and one edge in some other pipe incident to v. Each of these edges contributes to one edge in $E(G')$ inside the ellipse D_{uv}. Since uv is safe, G' has no other new edges. Consequently, $|E(G')| = |E(G)|$. Since $\deg_H(u) \geq 3$ or $\deg_H(v) \geq 3$, PipeExpansion(uv) replaces the clusters u and v with at least 3 clusters, each of which is incident to at least one pipe in the ellipse D_{uv}. Consequently, $|E(H')| > |E(H)|$, and so $\Phi(G, H) > \Phi(G', H')$, as claimed. □

Lemma 5. *At the end of the while loop of Algorithm 1, H is a cycle.*

Proof. It is enough to show that if H is not a cycle in the while loop of Algorithm 1, then there is a safe pipe $uv \in E(H)$ such that $\deg_H(u) \geq 3$ or $\deg_H(v) \geq 3$. Observe that every cluster created by ClusterExpansion(u) (resp., PipeExpansion(uv)) is a base for the unique incident pipe in the exterior of disk D_u (resp., ellipse D_{uv}). Let $s : V(H) \to E(H)$ be a function that maps every cluster to that incident pipe. Note also that the input does not have spurs, and no spurs are created in the algorithm by Lemmas 2 and 3. In the absence of spurs, if $u \in V(H)$ and $\deg_H(u) = 2$, then u is a base for both incident pipes.

Assume that in some step of the while loop, H is not a cycle. Let $v_1 \in V(H)$ be an arbitrary cluster such that $\deg_H(v_1) \geq 3$. Construct a maximal simple path $(v_1, v_2, \ldots, v_\ell)$ incrementally such that $s(v_i) = v_i v_{i+1}$ for $i = 1, 2, \ldots \ell$. If the path encounters a cluster v_i where $s(v_i) = s(v_{i-1})$, then the pipe $v_{i-1}v_i$ is safe. Similarly, if $\deg_H(v_{i+1}) = 2$, then $v_i v_{i+1}$ is safe. Otherwise, the path ends with a repeated cluster: $s(v_\ell) = v_\ell v_i$, for some $1 \leq i < \ell - 1$, and so we obtain a cycle $(v_i, v_{i+1}, \ldots, v_\ell)$ of at least 3 vertices. Let v_j, $i \leq j \leq \ell$, be the cluster created in the most recent ClusterExpansion(u) or PipeExpansion(uv) operation. Then $s(v_j)$ is a pipe in the exterior of a disk D_u or an ellipse D_{uv}. Hence, the pipe $v_{j-1}v_j$ is in the interior of D_u or D_{uv}, moreover v_j and v_{j-1} were created by the same operation. However, this implies $s(v_{j-1}) \neq v_{j-1}v_j$, contradicting the assumption that $(v_i, v_{i+1}, \ldots, v_\ell)$ is a cycle. We conclude that the path finds a safe pipe before any cluster repeats. □

Lemma 6. *Algorithm 1 returns* $\mathrm{cr}(\gamma \circ \lambda)$.

Proof. By (1), $\mathrm{cr}(\gamma \circ \lambda) = \mathrm{cr}_1(\lambda) + \mathrm{cr}_2(\gamma, \lambda)$. Here $\mathrm{cr}_2(\gamma, \lambda)$ can be computed by a line sweep of the drawing $\gamma(H)$. By Lemmas 1 and 5, at the end of the algorithm, $\mathrm{cr}_1(\lambda) = |\lambda^{-1}[uv]| - 1$ for an arbitrary edge $uv \in E(H)$. By Lemmas 2 and 3, $\mathrm{cr}(\gamma \circ \lambda)$ is invariant in the operations, so the algorithm reports $\mathrm{cr}(\gamma \circ \lambda)$ for the input instance. □

Running Time. The efficient implementation of our algorithm relies on the following data structures. For every cluster $u \in V(H)$ we maintain the set of vertices of $V(G)$ in $\lambda^{-1}[u]$. For every pipe $uv \in E(H)$, we maintain $\lambda^{-1}[uv] \subset E(G)$, the weight $w(uv) = |\lambda^{-1}[uv]|$, and the sum of weights of all pipes that cross uv, that we denote by $W(uv)$. Then we have $\mathrm{cr}_2(\gamma, \lambda) = \frac{1}{2}\sum_{uv \in E(H)} w(uv)W(uv)$. We also maintain the current value of $\mathrm{cr}_2(\gamma, \lambda)$. We further maintain indicator variables that support checking the conditions of the while loop in Algorithm 1: (i) whether the cluster is a base for the pipe, (ii) whether a cluster has degree 2, and (iii) whether a pipe is safe.

Lemma 7. *With the above data structures, Algorithm 1 runs in $O((M + R)\log M)$ time, where $M = |E(H)| + |E(G)|$ and $R = cr(\gamma \circ \lambda) < M^2$.*

4   NP-Completeness in the Presence of Spurs

In this section, we prove Theorem 2. In a problem instance, we are given a simplicial map $\lambda : G \to H$, a straight-line drawing $\gamma : H \to \mathbb{R}^2$, and a nonnegative integer K, and ask whether $cr(\gamma \circ \lambda) \leq K$.

Lemma 8. *The above problem is in NP.*

Proof. A feasible drawing $\Gamma \circ \Lambda : G \to \mathbb{R}^2$ with $cr(\Gamma \circ \Lambda) \leq K$ can be witnessed by a combinatorial representation of Λ. Specifically, we can determine $cr_2(\gamma, \lambda)$ by computing the weight of each pipe $uv \in E(H)$ in $O(|E(G)| + |E(H)|)$ time, and finding all edge-crossings in the drawing $\gamma(H)$ in $O(|E(H)| \log |E(H)|)$ time. Given a combinatorial representation of a drawing $\Lambda : G \to \mathcal{H}$, we can determine the number of crossings at all nodes $u \in V(H)$ in $O(\sum_{u \in V(H)} |\lambda^{-1}[u]|) = O(|E(G)|)$ time.   □

We prove NP-hardness by a reduction from 3SAT. Let Φ be a boolean formula in 3CNF with a set $\mathcal{X} = \{x_1, \ldots, x_n\}$ of variables and a set $\mathcal{C} = \{c_1, \ldots, c_m\}$ of clauses. We construct graphs G and H, a simplicial map $\lambda : G \to H$, a straight-line drawing $\gamma : H \to \mathbb{R}^2$, and an integer $K \in \mathbb{N}$ such that $cr(\gamma \circ \lambda) \leq K$ if and only if Φ is satisfiable.

First Construction: Disjoint Union of Paths. Refer to Fig. 4.

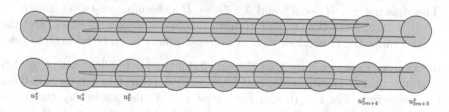

Fig. 4. Two embeddings of G_x. Top: P_1^x is above P_3^x. Bottom: P_1^x is below P_3^x.

Construction of H and $\gamma : H \to \mathbb{R}^2$. For every variable $x \in \mathcal{X}$, create a path $H_x = (u_3^x, u_4^x, \ldots, u_{5m+5}^x)$.

For $i = 1, \ldots, m$, the i-th clause $c_i \in \mathcal{C}$ is associated to at most three (negated or non-negated) variables, say, $x, y, z \in \mathcal{X}$. Identify the clusters $u_{5i+\ell}^x = u_{5i+\ell}^y = u_{5i+\ell}^z$ for $\ell = 0, 1, 2, 3$ and we denote the resulting clusters also by $u_{5i+\ell}$ and associate them with clause c_i. Add two new clusters v_i an w_i, and two new pipes $v_i u_{5i+1}^x$ and $w_i u_{5i+2}^x$. This completes the description of H.

For every $i = 1, \ldots, m$, we map clusters $u_{5i}, \ldots, u_{5i+3}$ to integer points $5i, \ldots, 5i + 3$ on the x-axis. The two additional clusters, v_i and w_i, are mapped to points $\gamma(v_i) = (5i + 1, 1)$ and $\gamma(w_i) = (5i + 2, -1)$, above and below the x-axis. The remaining clusters and pipes of H_x, $x \in \mathcal{X}$, are mapped to integer points in the horizonal line $y = j + 1$. Specifically, $\gamma(u_i^{x_j}) = (i, j + 1)$, for $3 \le i \le 5m + 5$, except for clusters $u_i^{x_j}$ that have been merged and incorporated in clause gadgets.

Observation 1. *For every $x \in \mathcal{X}$, $\gamma(H_x)$ is an x-monotone polygonal path in the plane. This ensures, in particular, that if $c_i \in \mathcal{C}$ contains variables x, y, and z, then the pipes of H_x, H_y, and H_z that enter u_{5i} and exit u_{5i+3} appear in reverse ccw order in the rotation of u_{5i} and u_{5i+3}, respectively.*

Construction of G and $\lambda : G \to H$. For each clause $c_i \in \mathcal{C}$, create a path G_i of 4 vertices mapped to $(v_i, u_{5i+1}, u_{5i+2}, w_i)$. For each variable $x \in \mathcal{X}$, create a path G_x as follows. First create a path of $15m + 5$ vertices as a concatenation of three paths: P_1^x, P_2^x, and P_3^x, which are mapped to $(u_3^x, \ldots, u_{5m+4}^x)$, $(u_{5m+4}^x, \ldots, u_4^x)$, and $(u_4^x, \ldots, u_{5m+5}^x)$, respectively. We shall modify P_1^x and P_3^x within each cluster. Regardless of these local modifications, in every embedding of G_x, the path P_2^x lies between P_1^x and P_3^x. The truth value of variable x is encoded by the above-below relationship between P_1^x and P_3^x (Fig. 4(a–b)).

Each pair $(x, c_i) \in \mathcal{X} \times \mathcal{C}$, where a literal x or $\overline{x}$ appears in c_i, corresponds to the subpath $(u_{5i}, \ldots, u_{5i+3})$ of H_x. Suppose that a subpath $A \subset P_1^x$ and $B \subset P_3^x$ are mapped to this subpath. To simplify notation, we assume that A and B are *directed* from u_{5i} to u_{5i+3}.

Refer to Fig. 5. If c_0 contains the non-negated x, then replace A on P_x^1 with a subpath mapped to $A' = (u_{5i}, u_{5i+1}, u_{5i+2}, u_{5i+3}, u_{5i+2}, u_{5i+1}, u_{5i+2}, u_{5i+3})$ and B with a subpath mapped to $B' = (u_{5i}, u_{5i+1}, u_{5i+2}, u_{5i+1}, u_{5i}, u_{5i+1}, u_{5i+2}, u_{5i+3})$. If c_0 contains the negated $\overline{x}$ then replace A with B', and B with A'. This completes the definition of G.

The drawing $\gamma : H \to \mathbb{R}^2$ and $\lambda : G \to H$ determine $cr_2(\gamma, \lambda)$. Let $K = cr_2(\gamma, \lambda) + 13m$. Note that G and H have $O(mn)$ vertices and edges, and the drawing γ maps the clusters in $V(H)$ to integer points in an $O(m) \times O(n)$ grid.

Equivalence. First, we show that the satisfiability of Φ implies that $cr(\gamma, \lambda) \le K$. Assume that Φ is satisfiable, and let $\tau : \mathcal{X} \to \{\text{true}, \text{false}\}$ be a satisfying truth assignment. Fix $\varepsilon \in (0, \varepsilon_0)$. For every $x \in \mathcal{X}$, denote by $\mathcal{N}_x$ the union of disks N_u and N_{uv} for all clusters $v \in V(H_x)$ and pipes $uv \in E(H_x)$; and similarly let $\mathcal{N}_i$ be the union of such regions for the path $(u_{5i}, \ldots, u_{5i+3})$ in H. For every $x \in \mathcal{X}$, incrementally, embed the path G_x in $\mathcal{N}_x$ as follows: each edge is an x-monotone Jordan arc; if $\tau(x) = \text{true}$, then P_1^x lies above P_3^x; otherwise P_3^x lies above P_1^x. If a clause c_i contains variables $x, y, z \in \mathcal{X}$, we also ensure that the embeddings of G_z, G_y, and G_y are pairwise disjoint within $\mathcal{N}_i$. This is possible by Observation 1. Finally, for $i = 1, \ldots, m$, embed the path G_i as follows. Assume that c_i contains the variables $x, y, z \in \mathcal{X}$, where x corresponds to a true literal in c_i. Then $\Gamma(G_i)$ starts from $\gamma(v_i)$ along the vertical line $x = 5i + 1$ until it crosses the arc $\Gamma(P_2^x)$, then follows $\Gamma(P_2^x)$ to the vertical line $x = 5i + 2$,

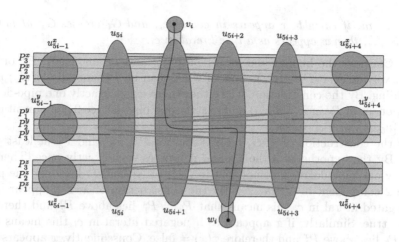

Fig. 5. A clause gadget for $c_i = (x \lor y \lor z)$, where $\tau(x) = \tau(z) =$ false and $\tau(y) =$ true. The neighborhood of the four middle "vertically prolonged" clusters and pipes between them forms $\mathcal{N}_i$.

and continues to $\gamma(w_i)$ along that line. Note that $\Gamma(P_2^x)$ crosses only 3 edges in $\Gamma(G_x)$, and 5 edges in $\Gamma(G_y)$ and $\Gamma(G_z)$. So there are 13 crossings in $\mathcal{N}_i$ for $i = 1, \ldots, m$; and the total number of crossings is $\mathrm{cr}_2(\gamma, \lambda) + 13m$, as required.

Second, we show that $\mathrm{cr}(\gamma, \lambda) \leq K$ implies that Φ is satisfiable by constructing a satisfying assignment. Consider functions $\Lambda : G \to \mathcal{H}$ and $\Gamma : \mathcal{H} \to \mathbb{R}^2$ such that $\Gamma \circ \Lambda : G \to \mathbb{R}^2$ is a drawing in which $\mathrm{cr}(\Gamma \circ \Lambda) \leq K$. Note that $\mathrm{cr}_2(\gamma, \lambda)$ crossings are unavoidable due to edge-crossings in the drawing $\gamma(H)$. Hence, by the definition of K, there are at most $13m$ crossings in the neighborhoods of clusters. We show that (1) there must be precisely 13 crossings in each neighborhood $\mathcal{N}_i$, (2) $\Gamma \circ \Lambda(G_x)$ is an embedding for every $x \in \mathcal{X}$, and (3) the embeddings of G_x, for all $x \in \mathcal{X}$, jointly encode a satisfying truth assignment for Φ. (1) and (2) is established by the following lemma.

Lemma 9. *Let $i \in \{1, \ldots, m\}$ and let $x, y, z \in \mathcal{X}$ be the three variables in c_i. In $\Gamma \circ \Lambda$, there are at least 13 crossings in neighborhood $\mathcal{N}_i$, and equality is possible only if none of the drawings $\Gamma \circ \Lambda(G_x)$, $x \in \mathcal{X}$, has self-crossings in $\mathcal{N}_i$, and at least one of G_x, G_y and G_z is crossed exactly 3 times by G_i.*

By Lemma 9, $\mathrm{cr}_1(\lambda) \leq 13m$ implies that $\Gamma \circ \Lambda$ defines an embedding of G_x, for all $x \in \mathcal{X}$, in each region $\mathcal{N}_i$, $i = 1, \ldots, m$. Consequently, $\Gamma \circ \Lambda$ defines an embedding of G_x in $\mathbb{R}^2$ for all $x \in \mathcal{X}$. In every embedding $\Gamma \circ \Lambda(G_x)$, for $x \in \mathcal{X}$, either P_1^x lies above P_2^x, or vice versa. We can now define a truth assignment $\tau : \mathcal{X} \to \{\text{true}, \text{false}\}$ such that for every $x \in \mathcal{X}$, $\tau(x) =$ true if and only if P_1^x lies above P_2^x in $\Gamma \circ \Lambda(G_x)$.

Lemma 10. *Assume that $\Gamma \circ \Lambda(G_x)$ is an embedding for every $x \in \mathcal{X}$, which determines the truth assignment $\tau : \mathcal{X} \to \{\text{true}, \text{false}\}$ described above. For every*

$i = 1, \ldots, m$, *if variable x appears in clause c_i, and G_i crosses G_x at most 3 times in $\mathcal{N}_i$, then x appears as a true literal in c_i.*

Proof. Consider the highest and lowest path P_h and P_l among P_1^x, P_2^x or P_3^x, respectively, in $\mathcal{N}_i \cap \Gamma \circ \Lambda(G_x)$, none of which can be P_2^x since $\Gamma \circ \Lambda(G_x)$ is an embedding. By the construction of λ, either there exists exactly one pipe-degree 2 component of P_h in $\lambda^{-1}[u_{5i+1}]$ and exactly one pipe-degree 2 component of P_l in $\lambda^{-1}[u_{5i+2}]$, or vice versa.

By the construction of λ, G_i crosses each of P_1^x, P_2^x, and P_3^x at least once in $\mathcal{N}_i$. By the hypothesis of the lemma, it crosses each exactly once. Then P_h has only one pipe-degree 2 component in $\lambda^{-1}[u_{5i+1}]$, and P_l has only one pipe-degree 2 component in $\lambda^{-1}[u_{5i+2}]$. By the construction of λ, if x appears as a non-negated literal in c_i this means that $P_h = P_1^x$ lies above P_2^x and therefore $\tau(x) = $ true. Similarly, if x appears as a negated literal in c_i this means that $P_3^x = P_h$ lies above P_2^x and therefore $\tau(x) = $ false. Consequently, x appears as a true literal in c_i and that concludes the proof. □

Since $\mathrm{cr}_1(\lambda) \leq 13m$, for every $i = 1, \ldots, m$, there are exactly 13 crossings in $\mathcal{N}_i$ by Lemma 9. Moreover, by Lemma 9 the drawing $\Gamma \circ \Lambda(G_x)$ is an embedding for every $x \in \mathcal{X}$, and in every c_i for one its variables x the drawing of G_x is crossed by G_i exactly 3 times. By Lemma 10, the assignment τ makes at least one literal in each clause c_i of Φ true. We conclude that Φ is satisfiable, as required. This completes the proof of NP-hardness.

Second Construction: Cycle. In our first construction, G is a disjoint union of paths, and for every path endpoint $a \in V(G)$, a is the only vertex mapped to the cluster $\lambda(a) \in V(H)$. This property allows us to expand the construction as follows. We augment G into a cycle $\overline{G}$ by adding a perfect matching M_G connecting the path endpoints, and we augment H with the corresponding matching between the clusters $M_H = \{\lambda(a)\lambda(b) : ab \in M_G\}$, and for every new pipe $uv \in M_H$ draw a polygonal arc $\gamma(uv)$ between $\gamma(u)$ and $\gamma(v)$ that does not pass through the image of any other cluster (but may cross images of other pipes). The augmentation does not change $\mathrm{cr}_1(\lambda)$, and we can easily compute the increase in $\mathrm{cr}_2(\gamma, \lambda)$ due to new crossings. Consequently, finding $\mathrm{cr}(\gamma \circ \lambda)$ remains NP-hard.

5 Conclusions

Motivated by recent efficient algorithms that can decide whether a piecewise linear map $\varphi : G \to \mathbb{R}^2$ can be perturbed into an embedding, we investigate the problem of computing the minimum number of crossings in a perturbation. We have described an efficient algorithm when G is a cycle and φ has no spurs (Theorem 1); and the problem becomes NP-hard if G is an arbitrary graph, or if G is a cycle but φ may have spurs (Theorem 2). However, perhaps one can minimize the number of crossings efficiently under milder assumptions. We formulate one promising scenario as follows: Is there a polynomial-time algorithm that finds $\mathrm{cr}(\gamma \circ \lambda)$ when $\lambda^{-1}[u]$ is a planar graph (resp., an edgeless graph) for every cluster $u \in V(H)$ and λ has no spurs?

References

1. Fulek, R., Kynčl, J.: Hanani-Tutte for approximating maps of graphs. In: Proceedings of the 34th Symposium on Computational Geometry (SoCG). LIPIcs, vol. 99, pp. 39:1–39:15. Dagstuhl, Germany (2018)
2. Akitaya, H.A., Fulek, R., Tóth, C.D.: Recognizing weak embeddings of graphs. In: Proceedings of the 29th ACM-SIAM Symposium on Discrete Algorithms (SODA), pp. 274–292. SIAM (2018)
3. Garey, M.R., Johnson, D.S.: Crossing number is NP-complete. SIAM J. Algebr. Discret. Methods **4**(3), 312–316 (1982)
4. Cabello, S., Mohar, B.: Adding one edge to planar graphs makes crossing number and 1-planarity hard. SIAM J. Comput. **42**(5), 1803–1829 (2013)
5. Akitaya, H.A., Aloupis, G., Erickson, J., Tóth, C.D.: Recognizing weakly simple polygons. Discret. Comput. Geom. **58**(4), 785–821 (2017)
6. Chang, H.C., Erickson, J., Xu, C.: Detecting weakly simple polygons. In: Proceedings of the 26th ACM-SIAM Symposium on Discrete Algorithms (SODA), pp. 1655–1670 (2015)
7. Cortese, P.F., Di Battista, G., Patrignani, M., Pizzonia, M.: On embedding a cycle in a plane graph. Discret. Math. **309**(7), 1856–1869 (2009)
8. Hanani, H.: Über wesentlich unplättbare Kurven im drei-dimensionalen Raume. Fundam. Math. **23**, 135–142 (1934)
9. Tutte, W.T.: Toward a theory of crossing numbers. J. Comb. Theory **8**, 45–53 (1970)
10. Skopenkov, M.: On approximability by embeddings of cycles in the plane. Topol. Appl. **134**(1), 1–22 (2003)
11. Repovš, D., Skopenkov, A.B.: A deleted product criterion for approximability of maps by embeddings. Topol. Appl. **87**(1), 1–19 (1998)
12. Angelini, P., Da Lozzo, G., Di Battista, G., Frati, F.: Strip planarity testing for embedded planar graphs. Algorithmica **77**(4), 1022–1059 (2017)
13. Jünger, M., Leipert, S., Mutzel, P.: Level planarity testing in linear time. In: Whitesides, S.H. (ed.) GD 1998. LNCS, vol. 1547, pp. 224–237. Springer, Heidelberg (1998). https://doi.org/10.1007/3-540-37623-2_17
14. Angelini, P., Lozzo, G.D.: Clustered planarity with pipes. In: Hong, S.H. (ed.) Proceedings of the 27th International Symposium on Algorithms and Computation (ISAAC). LIPIcs, vol. 64, pp. 13:1–13:13. Schloss Dagstuhl (2016)
15. Feng, Q.-W., Cohen, R.F., Eades, P.: How to draw a planar clustered graph. In: Du, D.-Z., Li, M. (eds.) COCOON 1995. LNCS, vol. 959, pp. 21–30. Springer, Heidelberg (1995). https://doi.org/10.1007/BFb0030816
16. Feng, Q.-W., Cohen, R.F., Eades, P.: Planarity for clustered graphs. In: Spirakis, P. (ed.) ESA 1995. LNCS, vol. 979, pp. 213–226. Springer, Heidelberg (1995). https://doi.org/10.1007/3-540-60313-1_145
17. Cortese, P.F., Di Battista, G., Patrignani, M., Pizzonia, M.: Clustering cycles into cycles of clusters. J. Graph Algorithms Appl. **9**(3), 391–413 (2005)
18. Hass, J., Scott, P.: Intersections of curves on surfaces. Isr. J. Math. **51**(1), 90–120 (1985)

The Number of Crossings in Multigraphs
with No Empty Lens

Michael Kaufmann[1], János Pach[2,3], Géza Tóth[3], and Torsten Ueckerdt[4(✉)]

[1] Wilhelm-Schickard-Institut für Informatik, Universität Tübingen,
Tübingen, Germany
mk@informatik.uni-tuebingen.de
[2] EPFL, Lausanne, Switzerland
pach@cims.nyu.edu
[3] Rényi Institute, Budapest, Hungary
toth.geza@renyi.mta.hu
[4] Institute of Theoretical Informatics, Karlsruhe Institute of Technology (KIT),
Karlsruhe, Germany
torsten.ueckerdt@kit.edu

Abstract. Let G be a multigraph with n vertices and $e > 4n$ edges, drawn in the plane such that any two parallel edges form a simple closed curve with at least one vertex in its interior and at least one vertex in its exterior. Pach and Tóth [5] extended the Crossing Lemma of Ajtai *et al.* [1] and Leighton [3] by showing that if no two adjacent edges cross and every pair of nonadjacent edges cross at most once, then the number of edge crossings in G is at least $\alpha e^3/n^2$, for a suitable constant $\alpha > 0$. The situation turns out to be quite different if nonparallel edges are allowed to cross any number of times. It is proved that in this case the number of crossings in G is at least $\alpha e^{2.5}/n^{1.5}$. The order of magnitude of this bound cannot be improved.

1 Introduction

In this paper, multigraphs may have parallel edges but no loops. A topological graph (or multigraph) is a graph (multigraph) G drawn in the plane with the property that every vertex is represented by a point and every edge uv is represented by a curve (continuous arc) connecting the two points corresponding to the vertices u and v. We assume, for simplicity, that the points and curves are in "general position", that is, (a) no vertex is an interior point of any edge; (b) any pair of edges intersect in at most finitely many points; (c) if two edges share an interior point, then they properly cross at this point; and (d) no 3 edges cross at the same point. Throughout this paper, every multigraph G is a topological multigraph, that is, G is considered with a fixed drawing that is given from the context. In notation and terminology, we then do not distinguish between the vertices (edges) and the points (curves) representing them. The number of crossing points in the considered drawing of G is called its *crossing number*, denoted

© Springer Nature Switzerland AG 2018
T. Biedl and A. Kerren (Eds.): GD 2018, LNCS 11282, pp. 242–254, 2018.
https://doi.org/10.1007/978-3-030-04414-5_17

by $\mathrm{cr}(G)$. (i.e., $\mathrm{cr}(G)$ is defined for topological multigraphs rather than abstract multigraphs.)

The classic "crossing lemma" of Ajtai, Chvátal, Newborn, Szemerédi [1] and Leighton [3] gives an asymptotically best-possible lower bound on the crossing number in any n-vertex e-edge topological graph without loops or parallel edges, provided $e > 4n$.

Theorem A (Crossing Lemma, Ajtai *et al.* [1] and Leighton [3]). *There is an absolute constant $\alpha > 0$, such that for any n-vertex e-edge topological graph G we have*

$$\mathrm{cr}(G) \geq \alpha \frac{e^3}{n^2}, \qquad provided \ e > 4n.$$

In general, the Crossing Lemma does not hold for topological multigraphs with parallel edges, as for every n and e there are n-vertex e-edge topological multigraphs G with $\mathrm{cr}(G) = 0$. Székely proved the following variant for multigraphs by restricting the edge multiplicity, that is the maximum number of pairwise parallel edges, in G to be at most m.

Theorem B (Székely [6]). *There is an absolute constant $\alpha > 0$ such that for any $m \geq 1$ and any n-vertex e-edge multigraph G with edge multiplicity at most m we have*

$$\mathrm{cr}(G) \geq \alpha \frac{e^3}{mn^2}, \qquad provided \ e > 4mn.$$

Most recently, Pach and Tóth [5] extended the Crossing Lemma to so-called branching multigraphs. We say that a topological multigraph is

- *separated* if any pair of parallel edges form a simple closed curve with at least one vertex in its interior and at least one vertex in its exterior,
- *single-crossing* if any pair of edges cross at most once (that is, edges sharing k endpoints, $k \in \{0, 1, 2\}$, may have at most $k + 1$ points in common), and
- *locally starlike* if no two adjacent edges cross (that is, edges sharing k endpoints, $k \in \{1, 2\}$, may not cross).

A topological multigraph is *branching* if it is separated, single-crossing and locally starlike. Note that the edge multiplicity of a branching multigraph may be as high as $n - 2$.

Theorem C (Pach and Tóth [5]). *There is an absolute constant $\alpha > 0$ such that for any n-vertex e-edge branching multigraph G we have*

$$\mathrm{cr}(G) \geq \alpha \frac{e^3}{n^2}, \qquad provided \ e > 4n.$$

In this paper we generalize Theorem C by showing that the Crossing Lemma holds for all topological multigraphs that are separated and locally starlike, but not necessarily single-crossing. We shall sometimes refer to the separated condition as the multigraph having "no empty lens," where we remark that here a lens is bounded by two entire edges, rather than general edge segments as sometimes defined in the literature. We also prove a Crossing Lemma variant for separated (and not necessarily locally starlike) multigraphs, where however the term $\alpha\frac{e^3}{n^2}$ must be replaced by $\alpha\frac{e^{2.5}}{n^{1.5}}$. Both results are best-possible up to the value of constant α.

Theorem 1. *There is an absolute constant $\alpha > 0$ such that for any n-vertex e-edge topological multigraph G with $e > 4n$ we have*

(i) $\mathrm{cr}(G) \geq \alpha\frac{e^3}{n^2}$, if G is separated and locally starlike.
(ii) $\mathrm{cr}(G) \geq \alpha\frac{e^{2.5}}{n^{1.5}}$, if G is separated.

Moreover, both bounds are best-possible up to the constant α.

We prove Theorem 1 in Sect. 3. Our arguments hold in a more general setting, which we present in Sect. 2. In Sect. 4 we use this general setting to deduce other known Crossing Lemma variants, including Theorem B. We conclude the paper with some open questions in Sect. 5.

2   A Generalized Crossing Lemma

In this section we consider general drawing styles and propose a generalized Crossing Lemma, which will subsume all Crossing Lemma variants mentioned here. A *drawing style* D is a predicate over the collection of all topological drawings, i.e., for each topological drawing of a multigraph G we specify whether G is in drawing style D or not. We say that G is a multigraph in drawing style D when G is a topological multigraph whose drawing is in drawing style D.

In order to prove our generalized Crossing Lemma, we follow the line of arguments of Pach and Tóth [5] for branching multigraphs. Their main tool is a bisection theorem for branching drawings, which easily generalizes to all separated drawings. We generalize their definition as follows.

Definition 1 (D-bisection width). *For a drawing style D the D-bisection width $\mathrm{b}_D(G)$ of a multigraph G in drawing style D is the smallest number of edges whose removal splits G into two multigraphs, G_1 and G_2, in drawing style D with no edge connecting them such that $|V(G_1)|, |V(G_2)| \geq n/5$.*

We say that a drawing style is *monotone* if removing edges retains the drawing style, that is, for every multigraph G in drawing style D and any edge removal, the resulting multigraph with its inherited drawing from G is again in drawing style D. Note that we require a monotone drawing style to be retained only after removing edges, but not necessarily after removing vertices. For example, the branching drawing style is in general not maintained after removing a vertex, since a closed curve formed by a pair of parallel edges might become empty.

Given a topological multigraph G, we call any operation of the following form a *vertex split*: (1) Replace a vertex v of G by two vertices v_1 and v_2 and (2) by locally modifying the edges in a small neighborhood of v, connect each edge in G incident to v to either v_1 or v_2 in such a way that no new crossing is created. We say that a drawing style is *split-compatible* if performing vertex splits retains the drawing style, that is, for every multigraph G in drawing style D and any vertex split, the resulting multigraph with its inherited drawing from G is again in drawing style D.

We are now ready to state our main result.

Theorem 2 (Generalized Crossing Lemma). *Suppose D is a monotone and split-compatible drawing style, and that there are constants $k_1, k_2, k_3 > 0$ and $b > 1$ such that each of the following holds for every n'-vertex e'-edge multigraph G' in drawing style D:*

(P1) If $\operatorname{cr}(G') = 0$, then the edge count satisfies $e' \leq k_1 \cdot n'$.
(P2) The D-bisection width satisfies $b_D(G') \leq k_2 \sqrt{\operatorname{cr}(G') + \Delta(G') \cdot e' + n'}$.
(P3) The edge count satisfies $e' \leq k_3 n'^b$.

Then there exists an absolute constant $\alpha > 0$ such that for any n-vertex e-edge multigraph G in drawing style D we have

$$\operatorname{cr}(G) \geq \alpha \frac{e^{x(b)+2}}{n^{x(b)+1}}, \qquad provided \ e > (k_1 + 1)n,$$

where $x(b) := 1/(b-1)$ and $\alpha = \alpha_b \cdot k_2^{-2} \cdot k_3^{-x(b)}$ for some constant α_b depending only on b.

Lemma 1. *If there exist for arbitrarily large n multigraphs in drawing style D with n vertices and $e = \Theta(n^b)$ edges such that any two edges cross at most a constant number of times, then the bound in Theorem 2 is asymptotically tight.*

Proof. Consider such an n-vertex e-edge multigraph in drawing style D. Clearly, there are at most $O(e^2) = O(n^{2b})$ crossings, while Theorem 2 gives with $x(b) = 1/(b-1)$ that there are at least

$$\Omega\left(\frac{e^{x(b)+2}}{n^{x(b)+1}}\right) = \Omega\left(\frac{e^{x(b)+2}}{n^{b \cdot x(b)}}\right) = \Omega\left(\frac{n^{b \cdot x(b)+2b}}{n^{b \cdot x(b)}}\right) = \Omega\left(n^{2b}\right)$$

crossings. □

2.1 Proof of Theorem 2

Proof Idea. Before proving Theorem 2, let us sketch the rough idea. Suppose, for a contradiction, that G is a multigraph in drawing style D with fewer than $\alpha \frac{e^{x(b)+2}}{n^{x(b)+1}}$ crossings, for a constant α to be defined. First, we conclude from **(P1)** that G must have many edges. Then, by **(P2)**, the D-bisection width of G is small, and thus we can remove few edges from the drawing to obtain two smaller multigraphs, G_1 and G_2, both also in drawing style D, which we call parts. We then repeat splitting each large enough part into two parts each, again using **(P2)**. Note that each part has at most 4/5 of the vertices of the corresponding part in the previous step. We continue until all parts are smaller than a carefully chosen threshold. As we removed relatively few edges during this decomposition algorithm, the final parts still have a lot of edges, while having few vertices each. This will contradict **(P3)** and hence complete the proof.

Now, let us start with the proof of Theorem 2. We define an absolute constant

$$\alpha := \frac{1}{2^{2x(b)+14}} \cdot \frac{1}{k_2^2} \cdot \frac{1}{k_3^{x(b)}} \tag{1}$$

Now let $\tilde{G}$ be a fixed multigraph in drawing style D with $\tilde{n}$ vertices and $\tilde{e} > (k_1 + 1)\tilde{n}$ edges. Let G' be an edge-maximal subgraph of $\tilde{G}$ on vertex set $V(\tilde{G})$ such that the inherited drawing of G' has no crossings. Since D is monotone, G' is in drawing style D. Hence, by **(P1)**, for the number e' of edges in G' we have $e' \leq k_1 \cdot n' = k_1 \cdot \tilde{n}$. Since G' is edge-maximal crossing-free, each edge in $E(\tilde{G}) - E(G')$ has at least one crossing with an edge in $E(G')$. Thus

$$\mathrm{cr}(\tilde{G}) \geq \tilde{e} - e' \geq \tilde{e} - k_1\tilde{n} > \tilde{n}. \tag{2}$$

In case $(k_1 + 1)\tilde{n} < \tilde{e} \leq \beta\tilde{n}$ for $\beta := \alpha^{-1/(x(b)+2)}$, we get

$$\mathrm{cr}(\tilde{G}) \overset{(2)}{>} \tilde{n} \geq \alpha \cdot \frac{\tilde{e}^{x(b)+2}}{\tilde{n}^{x(b)+1}},$$

as desired. To prove Theorem 2 in the remaining case $\tilde{e} > \beta\tilde{n}$ we use proof by contradiction. Therefore assume that the number of crossings in $\tilde{G}$ satisfies

$$\mathrm{cr}(\tilde{G}) < \alpha \cdot \frac{\tilde{e}^{x(b)+2}}{\tilde{n}^{x(b)+1}}.$$

Let d denote the average degree of the vertices of $\tilde{G}$, that is, $d = 2\tilde{e}/\tilde{n}$. For every vertex $v \in V(\tilde{G})$ whose degree, $\deg(v, \tilde{G})$, is larger than d, we perform $\lceil \deg(v, \tilde{G})/d \rceil - 1$ vertex splits so as to split v into $\lceil \deg(v, \tilde{G})/d \rceil$ vertices, each of degree at most d. At the end of the procedure, we obtain a multigraph G with $e = \tilde{e}$ edges, $n < 2\tilde{n}$ vertices, and maximum degree $\Delta(G) \leq d = 2\tilde{e}/\tilde{n} < 4e/n$. Moreover, as D is split-compatible, G is in drawing style D. For the number of crossings in G, we have

$$\mathrm{cr}(G) = \mathrm{cr}(\tilde{G}) < \alpha \cdot \frac{\tilde{e}^{x(b)+2}}{\tilde{n}^{x(b)+1}} < 2^{x(b)+1}\alpha \cdot \frac{e^{x(b)+2}}{n^{x(b)+1}}. \tag{3}$$

Moreover, recall that

$$e > \beta\tilde{n} > \beta\frac{n}{2} \qquad \text{for } \beta = \frac{1}{\alpha^{1/(x(b)+2)}}. \tag{4}$$

We break G into smaller parts, according to the following procedure. At each step the parts form a partition of the entire vertex set $V(G)$.

DECOMPOSITION ALGORITHM

STEP 0.
▷ **Let** $G^0 = G, G_1^0 = G, M_0 = 1, m_0 = 1$.

Suppose that we have already executed STEP i, and that the resulting graph G^i consists of M_i parts, $G_1^i, G_2^i, \ldots, G_{M_i}^i$, each in drawing style D and having at most $(4/5)^i n$ vertices. Assume without loss of generality that each of the first m_i parts of G^i has at least $(4/5)^{i+1}n$ vertices and the remaining $M_i - m_i$ have fewer. Letting $n(G_j^i)$ denote the number of vertices of the part G_j^i, we have

$$(4/5)^{i+1}n(G) \leq n(G_j^i) \leq (4/5)^i n(G), \qquad 1 \leq j \leq m_i. \qquad (5)$$

Hence,

$$m_i \leq (5/4)^{i+1}. \qquad (6)$$

STEP $i + 1$.
▷ **If**

$$(4/5)^i < \frac{1}{(2k_3)^{x(b)}} \cdot \frac{e^{x(b)}}{n^{x(b)+1}}, \qquad (7)$$

then STOP.
▷ **Else**, for $j = 1, 2, \ldots, m_i$, delete $b_D(G_j^i)$ edges from G_j^i, as guaranteed by **(P2)**, such that G_j^i falls into two parts, each of which is in drawing style D and contains at most $(4/5)n(G_j^i)$ vertices. Let G^{i+1} denote the resulting graph on the original set of n vertices.

Clearly, each part of G^{i+1} has at most $(4/5)^{i+1}n$ vertices.

Suppose that the DECOMPOSITION ALGORITHM terminates in STEP $k + 1$. If $k > 0$, then

$$(4/5)^k < \frac{1}{(2k_3)^{x(b)}} \cdot \frac{e^{x(b)}}{n^{x(b)+1}} \leq (4/5)^{k-1}. \qquad (8)$$

First, we give an upper bound on the total number of edges deleted from G. Using Cauchy-Schwarz inequality, we get for any nonnegative numbers $a_1, \ldots, a_m$,

$$\sum_{j=1}^m \sqrt{a_j} \leq \sqrt{m \sum_{j=1}^m a_j}, \qquad (9)$$

and thus obtain that, for any $0 \le i \le k$,

$$\sum_{j=1}^{m_i} \sqrt{\mathrm{cr}(G_j^i)} \overset{(9)}{\le} \sqrt{m_i \sum_{j=1}^{m_i} \mathrm{cr}(G_j^i)} \overset{(6)}{\le} \sqrt{(5/4)^{i+1}} \sqrt{\mathrm{cr}(G)}$$

$$\overset{(3)}{<} \sqrt{(5/4)^{i+1}} \sqrt{2^{x(b)+1}\alpha \cdot \frac{e^{x(b)+2}}{n^{x(b)+1}}}. \tag{10}$$

Letting $e(G_j^i)$ and $\Delta(G_j^i)$ denote the number of edges and maximum degree in part G_j^i, respectively, we obtain similarly

$$\sum_{j=1}^{m_i} \sqrt{\Delta(G_j^i) \cdot e(G_j^i) + n(G_j^i)} \overset{(9)}{\le} \sqrt{m_i \left(\sum_{j=1}^{m_i} \Delta(G_j^i) \cdot e(G_j^i) + n(G_j^i) \right)}$$

$$\overset{(6)}{\le} \sqrt{(5/4)^{i+1}} \sqrt{\Delta(G) \cdot e + n} \le \sqrt{(5/4)^{i+1}} \sqrt{\frac{4e}{n} e + n}$$

$$< \sqrt{(5/4)^{i+1}} \sqrt{\frac{5e^2}{n}} < \sqrt{(5/4)^{i+1}} \frac{3e}{\sqrt{n}}, \tag{11}$$

where we used in the last line the fact that $n < e$.

Using a partial sum of a geometric series we get

$$\sum_{i=0}^{k} (\sqrt{5/4})^{i+1} = \frac{(\sqrt{5/4})^{k+2} - 1}{\sqrt{5/4} - 1} - 1 < \frac{(\sqrt{5/4})^3}{\sqrt{5/4} - 1} \cdot (\sqrt{5/4})^{k-1} < 12 \cdot (\sqrt{5/4})^{k-1} \tag{12}$$

Thus, as each G_j^i is in drawing style D and hence **(P2)** holds for each G_j^i, the total number of edges deleted during the decomposition procedure is

$$\sum_{i=0}^{k} \sum_{j=1}^{m_i} \mathrm{b}_D(G_j^i) \le k_2 \sum_{i=0}^{k} \sum_{j=1}^{m_i} \sqrt{\mathrm{cr}(G_j^i) + \Delta(G_j^i) \cdot e(G_j^i) + n(G_j^i)}$$

$$\le k_2 \left(\sum_{i=0}^{k} \sum_{j=1}^{m_i} \sqrt{\mathrm{cr}(G_j^i)} + \sum_{i=0}^{k} \sum_{j=1}^{m_i} \sqrt{\Delta(G_j^i) \cdot e(G_j^i) + n(G_j^i)} \right)$$

$$\overset{(10),(11)}{\le} k_2 \left(\sum_{i=0}^{k} \sqrt{(5/4)^{i+1}} \right) \left(\sqrt{2^{x(b)+1}\alpha \cdot \frac{e^{x(b)+2}}{n^{x(b)+1}}} + \frac{3e}{\sqrt{n}} \right)$$

$$\overset{(12)}{<} k_2 \cdot 12 \sqrt{(5/4)^{k-1}} \left(\sqrt{2^{x(b)+1}\alpha \cdot \frac{e^{x(b)+2}}{n^{x(b)+1}}} + \frac{3e}{\sqrt{n}} \right)$$

$$\overset{(8)}{<} k_2 \cdot 12 \sqrt{(2k_3)^{x(b)} \cdot \frac{n^{x(b)+1}}{e^{x(b)}}} \left(\sqrt{2^{x(b)+1}\alpha \cdot \frac{e^{x(b)+2}}{n^{x(b)+1}}} + \frac{3e}{\sqrt{n}} \right)$$

$$< k_2 \cdot 36 \cdot \sqrt{k_3^{x(b)}} \left(2^{x(b)} \sqrt{\alpha} e + \sqrt{\frac{2^{x(b)} n^{x(b)}}{e^{x(b)-2}}} \right)$$

$$\overset{(4)}{<} k_2 \cdot 36 \cdot \sqrt{k_3^{x(b)}} \cdot 2^{x(b)} \left(\sqrt{\alpha} + \sqrt{\frac{1}{\beta^{x(b)}}} \right) e$$

$$\overset{(4)}{=} k_2 \cdot 36 \cdot \sqrt{k_3^{x(b)}} \cdot 2^{x(b)} \left(\sqrt{\alpha} + \sqrt{\alpha^{\frac{x(b)}{x(b)+2}}} \right) e < k_2 \cdot \sqrt{k_3^{x(b)}} \cdot 2^{x(b)+6} \sqrt{\alpha} e \overset{(1)}{=} \frac{e}{2}.$$

$$(13)$$

By (13) the DECOMPOSITION ALGORITHM removes less than half of the edges of G if $k > 0$. Hence, the number of edges of the graph G^k obtained in the final step of this procedure satisfies

$$e(G^k) > \frac{e}{2}. \tag{14}$$

(Note that this inequality trivially holds if the algorithm terminates in the very first step, i.e., when $k = 0$.)

Next we shall give an upper bound on $e(G^k)$ that contradicts (14). The number of vertices of each part G_j^k of G^k satisfies

$$n(G_j^k) \leq (4/5)^k n \overset{(8)}{<} \left(\frac{1}{(2k_3)^{x(b)}} \cdot \frac{e^{x(b)}}{n^{x(b)+1}} \right) n = \left(\frac{e}{2 \cdot k_3 \cdot n} \right)^{x(b)}, \quad 1 \leq j \leq M_k.$$

Hence

$$n(G_j^k)^{b-1} < \left(\frac{e}{2 \cdot k_3 \cdot n} \right)^{x(b)(b-1)} = \frac{e}{2 \cdot k_3 \cdot n},$$

since $x(b) = 1/(b-1)$ and hence $x(b)(b-1) = 1$.

As G_j^k is in drawing style D, (P3) holds for G_j^k and we have

$$e(G_j^k) \leq k_3 \cdot n(G_j^k)^b < k_3 \cdot n(G_j^k) \cdot \frac{e}{2 \cdot k_3 \cdot n} = n(G_j^k) \cdot \frac{e}{2n}.$$

Therefore, for the total number of edges of G^k we have

$$e(G^k) = \sum_{j=1}^{M_k} e(G_j^k) < \frac{e}{2n} \sum_{j=1}^{M_k} n(G_j^k) = \frac{e}{2},$$

contradicting (14). This completes the proof of Theorem 2. □

3 Separated Multigraphs

We derive our Crossing Lemma variants for separated multigraphs (Theorem 1) from the generalized Crossing Lemma (Theorem 2) presented in Sect. 2. Let us denote the separated drawing style by D_{sep} and the separated and locally star-like drawing style by $D_{\text{loc-star}}$. In order to apply Theorem 2, we shall find for $D = D_{\text{sep}}, D_{\text{loc-star}}$ (1) the largest number of edges in a crossing-free n-vertex multigraph in drawing style D, (2) an upper bound on the D-bisection width of multigraphs in drawing style D, and (3) an upper bound on the number of edges in any n-vertex multigraph in drawing style D.

As for crossing-free multigraphs D_{sep} and $D_{\text{loc-star}}$ are equivalent to the branching drawing style, we can rely on the following Lemma of Pach and Tóth.

Lemma 2 (Pach and Tóth [5]). *Any n-vertex crossing-free branching multigraph, $n \geq 3$, has at most $3n - 6$ edges.*

Corollary 1. *Any n-vertex crossing-free multigraph in drawing style D_{sep} or $D_{\text{loc}-\text{star}}$, $n \geq 3$, has at most $3n - 6$ edges.*

Also we can derive the bounds on the D-bisection width from the corresponding bound for the branching drawing style due to Pach and Tóth.

Lemma 3 (Pach and Tóth [5]). *For any multigraph G in the branching drawing style D with n vertices of degrees $d_1, d_2, \ldots, d_n$, and with $\operatorname{cr}(G)$ crossings, the D-bisection width of G satisfies*

$$b_D(G) \leq 22 \sqrt{\operatorname{cr}(G) + \sum_{i=1}^{n} d_i^2 + n}.$$

Lemma 4. *For $D = D_{\text{sep}}, D_{\text{loc}-\text{star}}$ any multigraph G in the drawing style D with n vertices, e edges, maximum degree $\Delta(G)$, and with $\operatorname{cr}(G)$ crossings, the D-bisection width of G satisfies*

$$b_D(G) \leq 44 \sqrt{\operatorname{cr}(G) + \Delta(G) \cdot e + n}.$$

Proof. Let G be a multigraph in drawing style D. Suppose there is a simple closed curve γ formed by parts of only two edges e_1 and e_2, which does not have a vertex in its interior. This can happen between two consecutive crossings of e_1 and e_2, or for $D \neq D_{\text{loc}-\text{star}}$ between a common endpoint and a crossing of e_1 and e_2. Further assume that the interior of γ is inclusion-minimal among all such curves, and note that this implies that an edge crosses e_1 along γ if and only if it crosses e_2 along γ. Say e_1 has at most as many crossings along γ as e_2. We then reroute the part of e_2 on γ very closely along the part of e_1 along γ so as to reduce the number of crossings between e_1 and e_2. The rerouting does not introduce new crossing pairs of edges. Hence, the resulting multigraph is again in drawing style D and has at most as many crossings as G. Similarly, we proceed when γ has no vertex in its exterior.

Thus, we can redraw G to obtain a multigraph G' in drawing style D with $\operatorname{cr}(G') \leq \operatorname{cr}(G)$, such that introducing a new vertex at each crossing of G' creates a crossing-free multigraph that is separated, i.e., in drawing style D. Now, using precisely the same proof as the proof of its special case Lemma 3 in [5], we can show that

$$b_D(G') \leq 22 \sqrt{\operatorname{cr}(G') + \sum_{i=1}^{n} d_i^2 + n},$$

where $d_1, \ldots, d_n$ denote the degrees of vertices in G'. Thus with

$$\sum_{i=1}^{n} d_i^2 \leq \Delta(G) \sum_{i=1}^{n} d_i \leq 2\Delta(G) \cdot e$$

the result follows.   □

Finally, let us bound the number of edges in crossing-free multigraphs. Again, we can utilize the result of Pach and Tóth for the branching drawing style.

Lemma 5 (Pach and Tóth [5]). *For any n-vertex e-edge, $n \geq 3$, multigraph of maximum degree $\Delta(G)$ in the branching drawing style we have $\Delta(G) \leq 2n - 4$ and $e \leq n(n-2)$, and both bounds are best-possible.*

Lemma 6. *For any n-vertex e-edge multigraph in drawing style D of maximum degree $\Delta(G)$ we have*

(i) $\Delta(G) \leq (n-1)(n-2)$ and $e \leq \binom{n}{2}(n-2)$ if $D = D_{\text{sep}}$,
(ii) $\Delta(G) \leq 2n - 4$ and $e \leq n(n-2)$ if G if $D = D_{\text{loc-star}}$.

Moreover, each bound is best-possible.

Proof. Let G be a fixed n-vertex, $n \geq 3$, e-edge crossing-free multigraph in drawing style D.

(i) Let $D = D_{\text{sep}}$. Clearly, every set of pairwise parallel edges contains at most $n - 2$ edges, since every lens has to contain a vertex different from the two endpoints of these edges. This gives $\Delta(G) \leq (n-1)(n-2)$ and $e \leq n\Delta(G)/2 = \binom{n}{2}(n-2)$. To see that these bounds are tight, consider n points in the plane with no four points on a circle. Then it is easy to draw between any two points $n-2$ edges as circular arcs such that the resulting multigraph (which has $\binom{n}{2}(n-2)$ edges) is in separating drawing style.

(ii) Let $D = D_{\text{loc-star}}$. Consider any fixed vertex v in G and remove all edges not incident to v. The resulting multigraph is branching and hence by Lemma 5 v has at most $2n-4$ incident edges. Thus $\Delta(G) \leq 2n-4$ and $e \leq n\Delta(G)/2 = n(n-2)$. By Lemma 5, these bounds are tight, even for the more restrictive branching drawing style. $\square$

We are now ready to prove that drawing styles $D_{\text{loc-star}}$ and D_{sep} fulfill the requirements of the generalized Crossing Lemma (Theorem 2), which lets us prove Theorem 1.

Proof (Proof of Theorem 1). Let $D = D_{\text{loc-star}}$ for (i) and $D = D_{\text{sep}}$ for (ii). Clearly, these drawing styles are monotone, i.e., maintained when removing edges, as well as split-compatible. So it remains to determine the constants $k_1, k_2, k_3 > 0$ and $b > 1$ such that **(P1)**, **(P2)**, and **(P3)** hold for D.

(P1) holds with $k_1 = 3$ for $D = D_{\text{loc-star}}, D_{\text{sep}}$ by Corollary 1. **(P2)** holds with $k_2 = 44$ for $D = D_{\text{sep}}$ by Lemma 4, which implies the same for $D = D_{\text{loc-star}}$. **(P3)** holds with $k_3 = 1$ and $b = 3$ for $D = D_{\text{sep}}$ by Lemma 6(i), and with $k_3 = 1$ and $b = 2$ for $D = D_{\text{loc-star}}$ by Lemma 6(ii).

For $b = 2$ we have $x(b) = 1/(b-1) = 1$. Thus Theorem 2 for $D = D_{\text{loc-star}}$ gives an absolute constant $\alpha > 0$ such that for every n-vertex e-edge separated and locally starlike multigraph we have $\text{cr}(G) \geq \alpha e^{x(b)+2}/n^{x(b)+1} = \alpha e^3/n^2$, provided $e > (k_1 + 1)n = 4n$. Moreover, by Lemma 6(ii) there are separated multigraphs with n vertices and $\Theta(n^2)$ edges, any two of which cross at most once. Hence, the term e^3/n^2 is best-possible by Lemma 1.

For $b = 3$ we have $x(b) = 1/(b-1) = 0.5$. Thus Theorem 2 for $D = D_{\text{sep}}$ gives an absolute constant $\alpha > 0$ such that for every n-vertex e-edge separated multigraph we have $\operatorname{cr}(G) \geq \alpha e^{x(b)+2}/n^{x(b)+1} = \alpha e^{2.5}/n^{1.5}$, provided $e > (k_1 + 1)n = 4n$. Moreover, by Lemma 6(i) there are separated multigraphs with n vertices and $\Theta(n^3)$ edges, any two of which cross at most twice. Hence, the term $e^{2.5}/n^{1.5}$ is best-possible by Lemma 1. $\square$

4   Other Crossing Lemma Variants

We use the generalized Crossing Lemma (Theorem 2) to reprove existing variants of the Crossing Lemma due to Székely and Pach, Spencer, Tóth, respectively.

4.1   Low Multiplicity

Here we consider for fixed $m \geq 1$ the drawing style D_m which is characterized by the absence of $m + 1$ pairwise parallel edges. In particular, any n-vertex multigraph G in drawing style D_m has at most $m\binom{n}{2}$ edges, i.e., **(P3)** holds for D_m with $b = 2$ and $k_3 = m$. Moreover, if G is crossing-free on n vertices and e edges, then $e \leq 3mn$, i.e., **(P1)** holds for D_m with $k_1 = 3m$.

Finally, we claim that **(P2)** holds for D_m with k_2 being independent of m. To this end, let G be any n-vertex e-edge multigraph in drawing style D_m. As already noted by Székely [6], we can reroute all but one edge in each bundle in such a way that in the resulting multigraph G' every lens is empty, no two adjacent edges cross, and $\operatorname{cr}(G') \leq \operatorname{cr}(G)$. (Simply route every edge very closely to its parallel copy with the fewest crossings.) Clearly, G' has drawing style D_m.

Now, we place a new vertex in each lens of G', giving a multigraph G'' with $n'' \leq n+e$ vertices and $e'' = e$ edges, which is in the separated drawing style D. By Lemma 4, there is an absolute constant k such that

$$b_D(G'') \leq k\sqrt{\operatorname{cr}(G'') + \Delta(G'') \cdot e'' + n''}.$$

As $b_{D_m}(G) \leq b_D(G'')$, $\operatorname{cr}(G'') = \operatorname{cr}(G') \leq \operatorname{cr}(G)$, $\Delta(G'') = \Delta(G)$, and $\Delta(G)+1 \leq 2\Delta(G)$ we conclude that

$$b_{D_m}(G) \leq 2k\sqrt{\operatorname{cr}(G) + \Delta(G) \cdot e + n}.$$

In other words, **(P2)** holds for drawing style D_m with an absolute constant $k_2 = 2k$ that is independent of m.

Note that for $b = 2$, we have $x(b) = 1$. We conclude with Theorem 2 that there is an absolute constant α' such that for every m and every n-vertex e-edge multigraph G in drawing style D_m we have

$$\operatorname{cr}(G) \geq \alpha' \cdot \frac{1}{k_3^{x(b)}} \cdot \frac{e^{x(b)+2}}{n^{x(b)+1}} = \alpha' \cdot \frac{e^3}{mn^2}, \qquad \text{provided } e > (3m+1)n,$$

which is the statement of Theorem B.

4.2   High Girth

Theorem D (Pach, Spencer, Tóth [4]). *For any $r \geq 1$ there is an absolute constant $\alpha_r > 0$ such that for any n-vertex e-edge graph G of girth larger than $2r$ we have*

$$\mathrm{cr}(G) \geq \alpha_r \cdot \frac{e^{r+2}}{n^{r+1}}, \qquad provided\ e > 4n.$$

Here we consider for fixed $r \geq 1$ the drawing style D_r which is characterized by the absence of cycles of length at most $2r$. In particular, any multigraph G in drawing style D_r has neither loops nor multiple edges. Hence **(P1)** holds for drawing style D_r with $k_1 = 3$. Secondly, drawing style D_r is more restrictive than the branching drawing style and thus also **(P2)** holds for D_r. Moreover, any n-vertex graph in drawing style D_r has $O(n^{1+1/r})$ edges [2], i.e., **(P3)** holds for D_r with $b = 1+1/r$. Finally, D_r is obviously a monotone and split-compatible drawing style.

Thus with $x(b) = 1/(b-1) = r$, Theorem 2 immediately gives

$$\mathrm{cr}(G) \geq \alpha_r \cdot \frac{e^{r+2}}{n^{r+1}}, \qquad provided\ e > 4n$$

for any n-vertex e-edge multigraph in drawing style D_r, which is the statement of Theorem D.

5   Conclusions

Let G be a topological multigraph with n vertices and $e > 4n$ edges. We have shown that $\mathrm{cr}(G) \geq \alpha e^3/n^2$ if G is separated and locally starlike, which generalizes the result for branching multigraphs [5], which are additionally single-crossing. Moreover, if G is only separated, then the lower bound drops to $\mathrm{cr}(G) \geq \alpha e^{2.5}/n^{1.5}$, which is tight up to the constant factor, too. It remains open to determine a best-possible Crossing Lemma for separated and single-crossing multigraphs. This would follow from our generalized Crossing Lemma (Theorem 2), where the missing ingredient is the determination of the smallest b such that every separated and single-crossing multigraph G on n vertices has $O(n^b)$ edges. It is easy to see that the maximum degree $\Delta(G)$ may be as high as $(n-1)(n-2)$, but we suspect that any such G has $O(n^2)$ edges.

Acknowledgements. This project initiated at the Dagstuhl seminar 16452 "Beyond-Planar Graphs: Algorithmics and Combinatorics," November 2016. We would like to thank all participants, especially Stefan Felsner, Vincenzo Roselli, and Pavel Valtr, for fruitful discussions.

References

1. Ajtai, M., Chvátal, V., Newborn, M.M., Szemerédi, E.: Crossing-free subgraphs. North-Holland Math. Stud. **60**(C), 9–12 (1982)
2. Alon, N., Hoory, S., Linial, N.: The Moore bound for irregular graphs. Graphs Comb. **18**(1), 53–57 (2002)
3. Leighton, T.: Complexity Issues in VLSI. Foundations of Computing Series. MIT Press, Cambridge (1983)
4. Pach, J., Spencer, J., Tóth, G.: New bounds on crossing numbers. Discret. Comput. Geom. **24**(4), 623–644 (2000)
5. Pach, J., Tóth, G.: A crossing lemma for multigraphs. In: Speckmann, B., Tóth, C.D. (eds.) 34th International Symposium on Computational Geometry (SoCG 2018). Leibniz International Proceedings in Informatics (LIPIcs), vol. 99, pp. 65:1–65:13. Schloss Dagstuhl-Leibniz-Zentrum fuer Informatik, Dagstuhl (2018)
6. Székely, L.A.: Crossing numbers and hard Erdős problems in discrete geometry. Comb. Prob. Comput. **6**(3), 353–358 (1997)

Crossing Numbers and Stress of Random Graphs

Markus Chimani[1]([⊠])(iD), Hanna Döring[2], and Matthias Reitzner[2]

[1] Theoretical CS, Institute of Computer Science, Uni Osnabrück,
Osnabrück, Germany
markus.chimani@uni-osnabrueck.de
[2] Stochastics, Institute of Mathematics, Uni Osnabrück, Osnabrück, Germany
{hanna.doering,matthias.reitzner}@uni-osnabrueck.de

Abstract. Consider a random geometric graph over a random point process in $\mathbb{R}^d$. Two points are connected by an edge if and only if their distance is bounded by a prescribed distance parameter. We show that projecting the graph onto a two dimensional plane is expected to yield a constant-factor crossing number (and rectilinear crossing number) approximation. We also show that the crossing number is positively correlated to the stress of the graph's projection.

1 Introduction

An undirected abstract graph G_0 consists of vertices and edges connecting vertex pairs. An *injection* of G_0 into $\mathbb{R}^d$ is an injective map from the vertices of G_0 to $\mathbb{R}^d$, and edges onto curves between their corresponding end points but not containing any other vertex point. For $d \geq 3$, we may assume that distinct edges do not share any point (other than a common end point). For $d = 2$, we call the injection a *drawing*, and it may be necessary to have points where curves *cross*. A drawing is *good* if no pair of edges crosses more than once, nor meets tangentially, and no three edges share the same crossing point. Given a drawing D, we define its crossing number $\mathrm{cr}(D)$ as the number points where edges cross. The crossing number $\mathrm{cr}(G_0)$ of the graph itself is the smallest $\mathrm{cr}(D)$ over all its good drawings D. We may restrict our attention to the *rectilinear crossing number* $\overline{\mathrm{cr}}(G_0)$, where edge curves are straight lines; note that $\overline{\mathrm{cr}}(G_0) \geq \mathrm{cr}(G_0)$.

The crossing number and its variants have been studied for several decades, see, e.g., [30], but still many questions are widely open. We know the crossing numbers only for very few graph classes; already for $\mathrm{cr}(K_n)$, i.e., on complete graphs with n vertices, we only have conjectures, and for $\overline{\mathrm{cr}}(K_n)$ not even them. Since deciding $\mathrm{cr}(G_0)$ is NP-complete [15] (and $\overline{\mathrm{cr}}$ even ∃ℝ-complete [4]), several attempts for approximation algorithms have been undertaken. The problem does not allow a PTAS unless P = NP [6]. For general graphs, we currently do not know whether there is an α-approximation for any constant α. However, we can achieve constant ratios for dense graphs [14] and for bounded pathwidth graphs [3]. Other strong algorithms deal with graphs of maximum bounded

© Springer Nature Switzerland AG 2018
T. Biedl and A. Kerren (Eds.): GD 2018, LNCS 11282, pp. 255–268, 2018.
https://doi.org/10.1007/978-3-030-04414-5_18

degree and achieve either slightly sublinear ratios [13], or constant ratios for further restrictions such as embeddability on low-genus surfaces [16–18] or a bounded number of graph elements to remove to obtain planarity [7,9,10,12].

We will make use of the *crossing lemma*, originally due to [2,25][1]: There are constants[2] $d \geq 4, c \geq \frac{1}{64}$ such that any abstract graph G_0 on n vertices and $m \geq dn$ edges has $\mathrm{cr}(G_0) \geq cm^3/n^2$. In particular for (dense) graphs with $m = \Theta(n^2)$, this yields the asymptotically tight maximum of $\Theta(m^2)$ crossings.

Random Geometric Graphs (RGGs). We always consider a *geometric* graph G as input, i.e., an abstract graph G_0 together with a straight-line injection into $\mathbb{R}^d$, for some $d \geq 2$; we identify the vertices with their points. For a 2-dimensional plane L, the postfix operator $|_L$ denotes the projection onto L.

Given a set of points V in $\mathbb{R}^d$, the *unit-ball graph* (*unit-disk graph* if $d = 2$) is the geometric graph using V as vertices that has an edge between two points iff balls of radius 1 centered at these points touch or overlap. Thus, points are adjacent iff their distance is ≤ 2. In general, we may use arbitrary threshold distances $\delta > 0$. We are interested in *random* geometric graphs *(RGGs)*, i.e., when using a Poisson point process to obtain V for the above graph class 2.

Stress. When drawing (in particular large) graphs with straight lines in practice, *stress* is a well-known and successful concept, see, e.g., [5,20,21]: let G be a geometric graph, d_0, d_1 two distance functions on vertex pairs—(at least) the latter of which depends on an injection—and w weights. We have:

$$\mathrm{stress}(G) := \sum_{v_1, v_2 \in V(G), v_1 \neq v_2} w(v_1, v_2) \cdot (d_0(v_1, v_2) - d_1(v_1, v_2))^2. \qquad (1)$$

In a typical scenario, G is injected into $\mathbb{R}^2$, d_0 encodes the graph-theoretic distances (number of edges on the shortest path) or some given similarity matrix, and d_1 is the Euclidean distance in $\mathbb{R}^2$. Intuitively, in a drawing of 0 (or low) stress, the vertices' geometric distances d_1 are (nearly) identical to their "desired" distance according to d_0. A typical weight function $w(v_1, v_2) := d_0(v_1, v_2)^{-2}$ softens the effect of "bad" geometric injections for vertices that are far away from each other anyhow. It has been observed *empirically* that low-stress drawings *tend* to be visually pleasing and to have a low number of crossings, see, e.g., [8,22]. While it may seem worthwhile to approximate the crossing number by minimizing a drawing's stress, there is no sound mathematical basis for this approach.

There are different ways to find (close to) minimal-stress drawings in 2D [5]. One way is multidimensional scaling, cf. [20], where we start with an injection of an abstract graph G_0 into some high-dimensional space $\mathbb{R}^d$ and asking for a projection of it onto $\mathbb{R}^2$ with minimal stress. It should be understood that Euclidean distances in a unit-ball graph in $\mathbb{R}^d$ by construction closely correspond to the graph-theoretic distances. In fact, for such graphs it seems reasonable to

---

[1] Incidentally, the lemma allows an intriguingly elegant proof using stochastics [1].

[2] The currently best constants $d = 7, c = \frac{1}{20}$ are due to [19].

use the distances in $\mathbb{R}^d$ as the given metrics d_0, and seek an injection into $\mathbb{R}^2$—whose resulting distances form d_1—by means of projection.

Contribution. We consider RGGs for large t and investigate the mean, variance, and corresponding law of large numbers both for their rectilinear crossing number and their minimal stress when projecting them onto the plane. We also prove, for the first time, a positive correlation between these two measures.

While our technical proofs make heavy use of stochastic machinery (several details of which have to be deferred to the arXiv version [11]), the consequences are very algorithmic: We give a surprisingly simple algorithm that yields an *expected constant* approximation ratio for random geometric graphs even in the pure abstract setting. In fact, we can state the algorithm already now; the remainder of this paper deals with the proof of its properties and correctness:

Given a random geometric graph G in $\mathbb{R}^d$ (see below for details), we pick a *random* 2-dimensional plane L in $\mathbb{R}^d$ to obtain a straight-line drawing $G|_L$ that yields a crossing number approximation both for $\overline{\mathrm{cr}}(G_0)$ and for $\mathrm{cr}(G_0)$.

Throughout this paper, we prefer to work within the setting of a Poisson point process because of the strong mathematical tools from the Malliavin calculus that are available in this case. It is straightforward to de-Poissonize our results: this yields asymptotically the same results—even with the same constants—for n uniform random points instead of a Poisson point process; we omit the details.

2 Notations and Tools from Stochastic Geometry

Let $W \subset \mathbb{R}^d$ be a convex set of volume $\mathrm{vol}_d(W) = 1$. Choose a Poisson distributed random variable n with parameter t, i.e., $\mathbb{E}n = t$. Next choose n points $V = \{v_1, \ldots, v_n\}$ independently in W according to the uniform distribution. Those points form a Poisson point process V in W of intensity t. A Poisson point process has several nice properties, e.g., for disjoint subsets $A, B \subset W$, the sets $V \cap A$ and $V \cap B$ are independent (thus also their size is independent). Let $V_{\neq}^k$, $k \geq 1$, be the set of all ordered k-tuples over V with pairwise distinct elements. We will consider V as the vertex set of a geometric graph G for the distances parameter $(\delta_t)_{t>0}$ with edges $E = \{\{u, v\} \mid u, v \in V, u \neq v, \|u - v\| \leq \delta_t\}$, i.e., we have an edge between two distinct points if and only if their distance is at most δ_t. Such *random geometric graphs (RGG)* have been extensively investigated, see, e.g., [27,29], but nothing is known about the stress or crossing number of its underlying abstract graph G_0.

A *U-statistic* $U(k, f) := \sum_{\mathbf{v} \in V_{\neq}^k} f(\mathbf{v})$ is the sum over $f(\mathbf{v})$ for all k-tuples $\mathbf{v}$. Here, f is a measurable non-negative real-valued function, and $f(\mathbf{v})$ only depends on $\mathbf{v}$ and is independent of the rest of V. The number of edges in G is a U-statistic as $m = \frac{1}{2} \sum_{v,u \in V, v \neq u} \mathbb{1}(\|v - u\| \leq \delta_t)$. Likewise, the stress of a geometric graph as well as the crossing number of a straight-line drawing is a U-statistic, using 2- and 4-tuples of V, respectively. The well-known multivariate Slivnyak-Mecke formula tells us how to compute the expectation $\mathbb{E}_V$ over all realizations of the Poisson process V; for U-statistics we have, see [31, Cor. 3.2.3]:

$$\mathbb{E}_V \sum_{(v_1,\ldots,v_k)\in V_{\neq}^k} f(v_1,\ldots,v_k) = t^k \int_{W^k} f(v_1,\ldots,v_k)\, dv_1 \cdots dv_k. \qquad (2)$$

We already know $\mathbb{E}_V n = \mathbb{E}_V |V| = t$. Solving the above formula for the expected number of edges, we obtain

$$\mathbb{E}_V m = \mathbb{E}_V |E| = \frac{\kappa_d}{2}\, t^2 \delta_t^d + O(t^2 \delta_t^{d+1}\, \mathrm{surf}(W)), \qquad (3)$$

where $\kappa_d = \mathrm{vol}_d(B_d)$ is the volume of the unit ball B_d in $\mathbb{R}^d$, and $\mathrm{surf}(W)$ the surface area of W. For n and m, central limit theorems and concentration inequalities are well known as $t \to \infty$, see, e.g., [27,29].

The expected degree $\mathbb{E}_V \deg(v)$ of a typical vertex v is approximately of order $\kappa_d\, t\, \delta_t^d$ (this can be made precise using Palm distributions). This naturally leads to three different asymptotic *regimes* as introduced in Penrose's book [27]:

- in the *sparse regime* we have $\lim_{t\to\infty} t\, \delta_t^d = 0$, thus $\mathbb{E}_V \deg(v)$ tends to zero;
- in the *thermodynamic regime* we have $\lim_{t\to\infty} t\, \delta_t^d = c > 0$, thus $\mathbb{E}_V \deg(v)$ is asymptotically constant;
- in the *dense regime* we have $\lim_{t\to\infty} t\, \delta_t^d = \infty$, thus $\mathbb{E}_V \deg(v) \to \infty$.

Observe that in standard graph theoretic terms, the *thermodynamic regime* leads to *sparse graphs*, i.e., via (3) we obtain $\mathbb{E}_V m = \Theta(t) = \Theta(\mathbb{E}_V n)$. Similarly, the dense regime—together with $\delta_t \to c$—leads to *dense graphs*, i.e., $\mathbb{E}_V m = \Theta(t^2) = \Theta((\mathbb{E}_V n)^2)$. Recall that to employ the crossing lemma, we want $m \geq 4n$. Also, the lemma already shows that any good (straight-line) drawing of a dense graph G_0 already gives a constant-factor approximation for $\mathrm{cr}(G_0)$ (and $\overline{\mathrm{cr}}(G_0)$). In the following we thus assume a constant $0 < c \leq t\, \delta_t^d$ and $\delta_t \to 0$, i.e., $m = o(n^2)$.

The Slivnyak-Mecke formula is a classical tool to compute expectations and will thus be used extensively throughout this paper. Yet, suitable tools to compute variances came up only recently. They emerged in connection with the development of the Malliavin calculus for Poisson point processes [23,26]. An important operator for functions $g(V)$ of Poisson point processes is the *difference* (also called *add-one-cost*) operator,

$$D_v g(V) := g(V \cup \{v\}) - g(V),$$

which considers the change in the function value when adding a single further point v. We know that there is a Poincaré inequality for Poisson functionals [23, 32], yielding the upper bound in (4) below. On the other hand, the isometry property of the Wiener-Itô chaos expansion [24] of an (square integrable) L^2-function $g(V)$ leads to the lower bound in (4):

$$t \int_W (\mathbb{E}_V D_v g(V))^2\, dv \;\leq\; \mathrm{Var}_V g(V) \;\leq\; t \int_W \mathbb{E}_V (D_v g(V))^2\, dv. \qquad (4)$$

Often, in particular in the cases we are interested in in this paper, the bounds are sharp in the order of t and often even sharp in the occurring constant. This is due to the fact that the Wiener-Itô chaos expansion, the Poincaré inequality, and the lower bound are particularly well-behaved for Poisson U-statistics [28].

3   Rectilinear Crossing Number of an RGG

Let $\mathcal{L}$ be the set of all two-dimensional linear planes and $L \in \mathcal{L}$ be a random plane chosen according to a (uniform) Haar probability measure on $\mathcal{L}$. The drawing $G_L := G|_L$ is the projection of G onto L. Let $[u, v]$ denote the segment between vertex points $u, v \in V$ if their distance is at most δ_t and $\emptyset$ otherwise. The rectilinear crossing number of G_L is a U-statistic of order 4:

$$\overline{\mathrm{cr}}(G_L) = \frac{1}{8} \sum_{(v_1, v_2, v_3, v_4) \in V_{\neq}^4} \mathbb{1}([v_1, v_2]|_L \cap [v_3, v_4]|_L \neq \emptyset).$$

Keep in mind that even for the best possible projection we only obtain $\min_{L \in \mathcal{L}} \overline{\mathrm{cr}}(G|_L) \geq \overline{\mathrm{cr}}(G_0)$. To analyze $\mathbb{E}_V \min_{L \in \mathcal{L}} \overline{\mathrm{cr}}(G|_L)$ is more complicated than $\mathbb{E}_{L,V} \overline{\mathrm{cr}}(G|_L)$; fortunately, we will not require it.

3.1   The Expectation of the Rectilinear Crossing Numbers

For the expectation with respect to the underlying Poisson point process the Slivnyak-Mecke formula (2) gives

$$\mathbb{E}_V \overline{\mathrm{cr}}(G_L) = \frac{1}{8} t^4 \int_W \underbrace{\int_{W^3} \mathbb{1}([v_1, v_2]|_L \cap [v_3, v_4]|_L \neq \emptyset) \, dv_4 dv_3 dv_2}_{=:I_W(v_1)} \, dv_1.$$

Let c_d be the constant given by the expectation of the event that two independent edges cross. In this paper's arXiv version [11, Appendix A], we prove in Proposition 15 that $c_d \leq 2\pi \kappa_d^2$, that $\frac{I_W(v_1)}{\delta_t^{2d+2}}$ is bounded by c_d times the volume of the maximal $(d - 2)$-dimensional section of W, and that

$$\lim_{\delta_t \to 0} \frac{I_W(v_1)}{\delta_t^{2d+2}} = c_d \mathrm{vol}_{d-2}((v_1 + L^\perp) \cap W), \tag{5}$$

where $L^\perp$ is the $d - 2$ dimensional hyperplane perpendicular to L. Using the dominated convergence theorem of Lebesgue and Fubini's theorem we obtain

$$\lim_{t \to \infty} \frac{\mathbb{E}_V \overline{\mathrm{cr}}(G_L)}{t^4 \delta_t^{2d+2}} = \frac{1}{8} c_d \int_W \mathrm{vol}_{d-2}((v_1 + L^\perp) \cap W) \, dv_1$$

$$= \frac{1}{8} c_d \int_{W|_L} \int_{(v_1^L + L^\perp) \cap W} \mathrm{vol}_{d-2}((v_1^L + L^\perp) \cap W) \, dv_1^{L^\perp} dv_1^L$$

$$= \frac{1}{8} c_d \underbrace{\int_{W|_L} \mathrm{vol}_{d-2}((v_1^L + L^\perp) \cap W)^2 \, dv_1^L}_{=:I^{(2)}(W,L)}.$$

Theorem 1. *Let G_L be the projection of an RGG onto a two-dimensional plane L. Then, as $t \to \infty$ and $\delta_t \to 0$,*

$$\mathbb{E}_V \overline{\text{cr}}(G_L) = \frac{1}{8} c_d \, t^4 \delta_t^{2d+2} \, I^{(2)}(W, L) + o(\delta_t^{2d+2} t^4).$$

For unit-disk graphs, i.e., $d = 2$, the choice of L is unique and the projection superfluous. There the expected crossing number is asymptotically $\frac{c_2}{8} t^4 \delta_t^6$ and thus of order $\Theta(m^3/n^2)$ which is asymptotically optimal as witnessed by the crossing lemma. In general, the expectation is of order

$$t^4 \delta_t^{2d+2} = \Theta\left(\frac{m^3}{n^2} \left(\frac{m}{n^2} \right)^{\frac{2-d}{d}} \right).$$

The extra factor m/n^2 can be understood as the probability that two vertices are connected via an edge, thus measures the "density" of the graph.

3.2 The Variance of the Rectilinear Crossing Numbers

By the variance inequalities (4) for functionals of Poisson point processes we are interested in the moments of the difference operator of the crossing numbers:

$$\mathbb{E}_V D_v \overline{\text{cr}}(G_L) = \frac{1}{8} \mathbb{E}_V \sum_{(v_2,\dots,v_4) \in V_{\neq}^3} \mathbb{1}([v, v_2]|_L \cap [v_3, v_4]|_L \neq \emptyset) = \frac{1}{8} t^3 I_W(v) \quad (6)$$

$$\mathbb{E}_V (D_v \overline{\text{cr}}(G_L))^2 = \mathbb{E}_V \left(\frac{1}{8} \sum_{(v_2,\dots,v_4) \in V_{\neq}^3} \mathbb{1}([v, v_2]|_L \cap [v_3, v_4]|_L \neq \emptyset) \right)^2 \quad (7)$$

Plugging (7) into the Poincaré inequality (4) gives

$$\mathbb{V}\text{ar}_V \overline{\text{cr}}(G_L) \le \frac{1}{64} t \int_W \mathbb{E}_V \left(\sum_{(v_2,\dots,v_4) \in V_{\neq}^3} \mathbb{1}([v, v_2]|_L \cap [v_3, v_4]|_L \neq \emptyset) \right)^2 dv.$$

Using calculations from integral geometry (see this paper's arXiv version [11, Appendix B]), there is a constant $0 < c_d' \le 2\pi \kappa_d c_d$ (given by the expectation of the event that two pairs of independent edges cross) such that

$$\mathbb{V}\text{ar}_V \overline{\text{cr}}(G_L) \le \frac{1}{64} \left(c_d^2 + \frac{c_d'}{t\delta_t^d} \right) t^7 \delta_t^{4d+4} \int_W \text{vol}_{d-2}((v + L^\perp) \cap W)^2 (1 + o(1)) dv$$

$$+ O(\max\{t^6 \delta_t^{4d+2}, t^5 \delta_t^{3d+2}, t^4 \delta_t^{2d+2}\}).$$

We use that $t\delta_t^d \ge c > 0$, assume $d \ge 3$, and use Fubini's theorem again.

$$\lim_{t \to \infty} \frac{\mathbb{V}\text{ar}_V \overline{\text{cr}}(G_L)}{t^7 \delta_t^{4d+4}} \le \frac{1}{64} \left(c_d^2 + c_d' \lim_{t \to \infty} \frac{1}{t\delta_t^d} \right) \underbrace{\int_{W|L} \text{vol}_{d-2}((v + L^\perp) \cap W)^3 \, dv}_{=:I^{(3)}(W,L)}.$$

On the other hand, (6) and the lower bound in (4) gives in our case

$$\mathbb{V}\mathrm{ar}_V \overline{\mathrm{cr}}(G_L) \geq t \int\limits_W (\mathbb{E}_V D_v \overline{\mathrm{cr}}(G_L))^2 \, dv$$

$$\geq \frac{1}{64} t^7 \int\limits_W I_W(v)^2 \, dv = \frac{1}{64} c_d^2 \, t^7 \delta_t^{4d+4} \, I^{(3)}(W, L)(1 + o(1)).$$

Thus our bounds have the correct order and, in the dense regime where $t \delta_t^d \to \infty$, are even sharp. Using $0 < c_d' \leq 2\pi \kappa_d c_d$ we obtain:

Theorem 2. *Let G_L be the projection of an RGG in $\mathbb{R}^d$, $d \geq 3$, onto a two-dimensional plane L. Then, as $t \to \infty$ and $\delta_t \to 0$,*

$$\frac{1}{64} c_d^2 I^{(3)}(W, L) \leq \lim_{t \to \infty} \frac{\mathbb{V}\mathrm{ar}_V \overline{\mathrm{cr}}(G_L)}{t^7 \delta_t^{4d+4}} \leq \frac{1}{64} (c_d^2 + 2\pi \kappa_d c_d \lim_{t \to \infty} \frac{1}{t \delta_t^d}) I^{(3)}(W, L).$$

Theorems 1 and 2 show for the standard deviation

$$\sigma(\overline{\mathrm{cr}}(G_L)) = \sqrt{\mathbb{V}\mathrm{ar}_V \overline{\mathrm{cr}}(G_L)} = \Theta(t^4 \delta_t^{2d+2} \, t^{-\frac{1}{2}}) = \Theta(\mathbb{E}_V \overline{\mathrm{cr}}(G_L) \, (\mathbb{E}_V n)^{-\frac{1}{2}}),$$

which is smaller than the expectation by a factor $(\mathbb{E}_V n)^{-\frac{1}{2}} = t^{-\frac{1}{2}}$. Or, equivalently, the coefficient of variation $\frac{\sigma(\overline{\mathrm{cr}}(G_L))}{\mathbb{E}_V \overline{\mathrm{cr}}(G_L)}$ is of order $t^{-\frac{1}{2}}$. As $t \to \infty$, our bounds on the expectation and variance together with Chebychev's inequality lead to

$$\mathbb{P}\left(\left| \frac{\overline{\mathrm{cr}}(G_L)}{t^4 \delta_t^{2d+2}} - \frac{\mathbb{E}_V \overline{\mathrm{cr}}(G_L)}{t^4 \delta_t^{2d+2}} \right| \geq \varepsilon \right) \leq \frac{\mathbb{V}\mathrm{ar}_V \overline{\mathrm{cr}}(G_L)}{t^8 \delta_t^{4d+4} \varepsilon^2} \to 0.$$

Corollary 3 (Law of Large Numbers). *For given L, the normalized random crossing number converges in probability (with respect to the Poisson point process V) as $t \to \infty$,*

$$\frac{\overline{\mathrm{cr}}(G_L)}{t^4 \delta_t^{2d+2}} \to \frac{1}{8} c_d I^{(2)}(W, L).$$

Until now we fixed a plane L and computed the variance with respect to the random points V. Theorems 1 and 2 allow to compute the expectation and variance with respect to V *and* a randomly chosen plane L. For the expectation we obtain from Theorem 1 and by Fubini's theorem

$$\mathbb{E}_{L,V} \overline{\mathrm{cr}}(G_L) = \frac{1}{8} c_d \, t^4 \delta_t^{2d+2} \int\limits_{\mathcal{L}} I^{(2)}(W, L) \, dL + o(t^4 \delta_t^{2d+2}), \tag{8}$$

as $t \to \infty$ and $\delta_t \to 0$, where dL denotes integration with respect to the Haar measure on $\mathcal{L}$. For simplicity we assume in the following that $\lim_{t \to \infty} (t \delta_t^d)^{-1} = 0$. We use the variance decomposition $\mathbb{V}\mathrm{ar}_{L,V} X = \mathbb{E}_L \mathbb{V}\mathrm{ar}_V X + \mathbb{V}\mathrm{ar}_L \mathbb{E}_V X$. By

$$\mathbb{E}_L \mathbb{V}\mathrm{ar}_V \overline{\mathrm{cr}}(G_L) = \frac{1}{64} c_d^2 t^7 \delta_t^{4d+4} \int_{\mathcal{L}} I^{(3)}(W,L)\, dL \ + o(t^7 \delta_t^{4d+4}), \quad \text{and}$$

$$\mathbb{V}\mathrm{ar}_L \mathbb{E}_V \overline{\mathrm{cr}}(G_L) = \mathbb{E}_L (\mathbb{E}_V \overline{\mathrm{cr}}(G_L))^2 - (\mathbb{E}_{L,V} \overline{\mathrm{cr}}(G_L))^2$$

$$= \frac{1}{64} c_d^2 t^8 \delta_t^{4d+4} \left[\int_{\mathcal{L}} I^{(2)}(W,L)^2\, dL - \left(\int_{\mathcal{L}} I^{(2)}(W,L) dL \right)^2 \right] + o(t^8 \delta_t^{4d+4})$$

we obtain

$$\mathbb{V}\mathrm{ar}_{L,V} \overline{\mathrm{cr}}(G_L) \ = \frac{1}{64} c_d^2 t^8 \delta_t^{4d+4} \left[\int_{\mathcal{L}} I^{(2)}(W,L)^2 dL - \left(\int_{\mathcal{L}} I^{(2)}(W,L) dL \right)^2 \right]$$

$$+ o(t^8 \delta_t^{4d+4}). \tag{9}$$

Hölder's inequality implies that the term in brackets is positive as long as $I^{(2)}(W,L)$ is not a constant function.

3.3   The Rotation Invariant Case

If W is the ball B of unit volume and thus V is *rotation invariant*, then $I^{(2)}(B,L) = I^{(2)}(B)$ is a constant function independent of L, and the leading term in (9) is vanishing. From (8) we see that in this case the expectation is independent of L.

$$\mathbb{E}_V \overline{\mathrm{cr}}(G_L) = \mathbb{E}_L \mathbb{E}_V \overline{\mathrm{cr}}(G_L) = t^4 \delta_t^{2d+2} I^{(2)}(B) + o(t^4 \delta_t^{2d+2})$$

For the variance this implies $\mathbb{V}\mathrm{ar}_L \mathbb{E}_V \overline{\mathrm{cr}}(G_L) = 0$, and hence

$$\mathbb{V}\mathrm{ar}_{L,V} \overline{\mathrm{cr}}(G_L) = \mathbb{E}_L \mathbb{V}\mathrm{ar}_V \overline{\mathrm{cr}}(G_L) = \frac{1}{64} c_d^2 t^7 \delta_t^{4d+4} I^{(3)}(B) + o(t^7 \delta_t^{4d+4}).$$

In this case the variance $\mathbb{V}\mathrm{ar}_{L,V}$ is of the order t^{-1}—and thus surprisingly significantly—smaller than in the general case.

Theorem 4. *Let G_L be the projection of an RGG in the ball $B \subset \mathbb{R}^d$, $d \geq 3$, onto a two-dimensional uniformly chosen random plane L. Then*

$$\mathbb{E}_{L,V} \overline{\mathrm{cr}}(G_L) = \frac{1}{8} c_d\, t^4 \delta_t^{2d+2} I^{(2)}(B) + o(t^4 \delta_t^{2d+2}) \quad \text{and}$$

$$\mathbb{V}\mathrm{ar}_{L,V} \overline{\mathrm{cr}}(G_L) = \frac{1}{64} c_d^2 t^7 \delta_t^{4d+4} I^{(3)}(B) + o(t^7 \delta_t^{4d+4}),$$

as $t \to \infty$, $\delta_t \to 0$ and $t\delta_t^d \to \infty$.

Again, Chebychev's inequality immediately yields a law of large numbers which states that with high probability the crossing number of G_L in a random direction is very close to $\frac{1}{8} c_d\, t^4 \delta_t^{2d+2} I^{(2)}(B)$.

Corollary 5 (Law of Large Numbers). *Let G_L be the projection of an RGG in $B \subset \mathbb{R}^d$, $d \geq 3$, onto a random two-dimensional plane L. Then the normalized random crossing number converges in probability (with respect to the Poisson point process V and to L), as $t \to \infty$,*

$$\frac{\overline{cr}(G_L)}{t^4 \delta_t^{2d+2}} \to \frac{1}{8} c_d I^{(2)}(B).$$

As known by the crossing lemma, the optimal crossing number is of order $\frac{m^3}{n^2}$. In our setting this means that we are looking for the optimal direction of projection which leads to a crossing number of order $t^4 \delta_t^{3d}$, much smaller than the expectation $\mathbb{E}_V \overline{cr}(G_L)$. Chebychev's inequality shows that if $W = B$ it is difficult to find this optimal direction and to reach this order of magnitude; using $\delta_t \to 0$ in the last step we have:

$$\mathbb{P}_{L,V}(\overline{cr}(G_L) \leq ct^4 \delta_t^{3d}) \leq \mathbb{P}_{L,V}(|\overline{cr}(G_L) - \mathbb{E}_{L,V} \overline{cr}(G_L)| \geq \mathbb{E}_{L,V} \overline{cr}(G_L) - ct^4 \delta_t^{3d})$$

$$\leq \frac{\mathrm{Var}_{L,V} \overline{cr}(G_L)}{(\mathbb{E}_{L,V} \overline{cr}(G_L) - ct^4 \delta_t^{3d})^2} = O(t^{-1}).$$

Hence a computational naïve approach of minimizing the crossing numbers by just projecting onto a sample of random planes seems to be expensive. This suggests to combine the search for an optimal choice of the direction of projection with other quantities of the RGG. It is a long standing assumption in graph drawing that there is a connection between the crossing number and the stress of a graph. Therefore the next section is devoted to investigations concerning the stress of RGGs.

4   The Stress of an RGG

According to (1) we define the stress of G_L as

$$\mathrm{stress}(G, G_L) := \frac{1}{2} \sum_{(v_1, v_2) \in V_{\neq}^2} w(v_1, v_2)(d_0(v_1, v_2) - d_L(v_1, v_2))^2,$$

where $w(v_1, v_2)$ a positive weight-function and d_0 resp. d_L are the distances between v_1 and v_2, resp $v_1|_L$ and $v_2|_L$. As $\overline{cr}(G)$, stress is a U-statistic, but now of order two. Using the Slivnyak-Mecke formula, it is immediate that

$$\mathbb{E}_V \mathrm{stress}(G, G_L) = \frac{1}{2} t^2 \underbrace{\int_{W^2} w(v_1, v_2)(d_0(v_1, v_2) - d_L(v_1, v_2))^2 dv_1 dv_2}_{=:S^{(1)}(W, L)}.$$

For the variance, the Poincaré inequality (4) implies

$$\mathrm{Var}_V \mathrm{stress}(G, G_L) \leq t \int\limits_W \mathbb{E}_V (D_v(\mathrm{stress}(G, G_L)))^2 dv$$

$$= \frac{1}{4} t \int\limits_W \mathbb{E}_V \left(\sum_{v_1 \in V} w(v, v_1)(d_0(v, v_1) - d_L(v, v_1))^2 \right)^2 dv$$

$$= \frac{1}{4} t^3 \underbrace{\int\limits_{W^3} \prod_{i=1}^{2} \left(w(v, v_i)(d_0(v, v_i) - d_L(v, v_i))^2 \right) dv_1 dv_2 dv}_{=:S^{(2)}(W, L)}$$

$$+ \frac{1}{4} t^2 \int\limits_{W^2} w(v, v_1)^2 (d_0(v, v_1) - d_L(v, v_1))^4 dv_1 dv.$$

Hence the standard deviation of the stress is smaller than the expectation by a factor $t^{-\frac{1}{2}}$ and thus the stress is concentrated around its mean. Again the computation of the lower bound for the variance in (4) is asymptotically sharp.

$$\mathrm{Var}_V \mathrm{stress}(G, G_L) \geq t \int\limits_W (\mathbb{E}_V D_v(\mathrm{stress}(G, G_L)))^2 dv$$

$$= \frac{1}{4} t \int\limits_W \left(\mathbb{E}_V \sum_{v_1 \in V} w(v, v_1)(d_0(v, v_1) - d_L(v, v_1))^2 \right)^2 dv = \frac{1}{4} t^3 S^{(2)}(W, L).$$

Theorem 6. *Let G_L be the projection of an RGG in $\mathbb{R}^d$, $d \geq 3$, onto a two-dimensional plane L. Then*

$$\mathbb{E}_V \mathrm{stress}(G, G_L) = \frac{1}{2} t^2 S^{(1)}(W, L) \quad and$$

$$\mathrm{Var}_V \mathrm{stress}(G, G_L) = \frac{1}{4} t^3 S^{(2)}(W, L) + O(t^2).$$

The discussions from Sects. 3.2 and 3.3 lead to analogous results for the stress of the RGG. Using Chebychev's inequality we could derive a law of large numbers. Taking expectations with respect to a uniform plane L we obtain:

$$\mathbb{E}_{L,V} \mathrm{stress}(G, G_L) = \frac{1}{2} t^2 \int\limits_{\mathcal{L}} S^{(1)}(W, L) dL,$$

$$\mathrm{Var}_{L,V} \mathrm{stress}(G, G_L) = \frac{1}{4} t^4 \left[\int\limits_{\mathcal{L}} S^{(1)}(W, L)^2 dL - \left(\int\limits_{\mathcal{L}} S^{(1)}(W, L) dL \right)^2 \right] + O(t^3).$$

Again, the term in brackets is only vanishing if $W = B$. In this case

$$\mathrm{Var}_{L,V} \mathrm{stress}(G, G_L) = \mathbb{E}_L \mathrm{Var}_V \mathrm{stress}(G, G_L) = \frac{1}{4} t^3 S^{(2)}(B) + O(t^2).$$

5 Correlation Between Crossing Number and Stress

It seems to be widely conjectured that the crossing number and the stress should be positively correlated. Yet it also seems that a rigorous proof is still missing. It is the aim of this section to provide the first proof of this conjecture, in the case where the graph is a random geometric graph.

Clearly, by the definition of $\overline{cr}$ and stress we have

$$D_v \, \overline{cr}(G_L) \geq 0 \text{ and } D_v \, \text{stress}(G, G_L) \geq 0,$$

for all v and all realizations of V. Such a functional F satisfying $D_v(F) \geq 0$ is called increasing. The Harris-FKG inequality for Poisson point processes [23] links this fact to the correlation of $\overline{cr}(G_L)$ and $\text{stress}(G, G_L)$.

Theorem 7. *Because stress and $\overline{cr}$ are increasing we have*

$$\mathbb{E}_V \overline{cr}(G_L)\text{stress}(G, G_L) \geq \mathbb{E}_V \overline{cr}(G_L) \, \mathbb{E}_V \text{stress}(G, G_L),$$

and thus the correlation is positive.

We immediately obtain that the covariance is positive and is of order at most

$$\text{Cov}_V\left(\overline{cr}(G_L), \text{stress}(G_L)\right) \leq \sqrt{\text{Var}_V \overline{cr}(G_L) \, \text{Var}_V \text{stress}(G, G_L)}$$

$$\leq \frac{1}{16}c_d\left(1 + \frac{2\pi\kappa_d}{c_d} \lim_{t\to\infty} \frac{1}{t\delta_t^d}\right)^{\frac{1}{2}} t^5 \delta_t^{2d+2} I^{(3)}(W, L)^{\frac{1}{2}} S^{(2)}(W, L)^{\frac{1}{2}} + o(t^5 \delta_t^{2d+2}).$$

In [11, Appendix C] we use Mehler's formula to prove a lower bound:

$$\text{Cov}_V\left(\overline{cr}(G_L), \text{stress}(G, G_L)\right) \geq \frac{t^5}{16} \int_{W^2} I_W(v)w(v, v_1)(d_0(v, v_1) - d_L(v, v_1))^2 dv_1 dv.$$

We combine this bound with (5), divide by the standard deviations from Theorems 2 and 6 and obtain the asymptotics for the correlation coefficient:

Theorem 8. *Let G_L be the projection of an RGG in $\mathbb{R}^d$, $d \geq 3$, onto a two-dimensional plane L. Then*

$$\lim_{t\to\infty} \text{Corr}_V\left(\overline{cr}(G_L), \text{stress}(G, G_L)\right)$$

$$\geq \frac{\int_{W^2} \text{vol}_{d-2}((v + L^\perp) \cap W)w(v, v_1)(d_0(v, v_1) - d_L(v, v_1))^2 dv_1 dv}{\left(1 + \frac{2\pi\kappa_d}{c_d} \lim_{t\to\infty} \frac{1}{t\delta_t^d}\right)^{\frac{1}{2}} I^{(3)}(W, L)^{\frac{1}{2}} S^{(2)}(W, L)^{\frac{1}{2}}}.$$

It can be shown that this bound is even tight and asymptotically gives the correct correlation coefficient.

5.1   The Rotation Invariant Case

In principle the bounds for the covariance in the Poisson point process V given above can be used to compute covariance bounds in L and V when L is not fixed but random. For this we could use the covariance decomposition

$$\operatorname{Cov}_{L,V}(X,Y) = \mathbb{E}_L \operatorname{Cov}_V(X,Y) + \operatorname{Cov}_L(\mathbb{E}_V X, \mathbb{E}_V Y).$$

Here we concentrate again on the case when $W = B$ is the ball of unit volume and thus V is rotation invariant. Then $\operatorname{Cov}_L(\mathbb{E}_V \overline{\operatorname{cr}}(G_L), \mathbb{E}_V \operatorname{stress}(G_L)) = 0$, and as an immediate consequence of Theorem 8 we obtain

Corollary 9. *Let G_L be the projection of an RGG in $B \subset \mathbb{R}^d$, $d \geq 3$, onto a two-dimensional random plane L. Then the correlation between the crossing number and the stress of the RGG is positive with*

$$\lim_{t \to \infty} \operatorname{Corr}_{L,V}(\overline{\operatorname{cr}}(G_L), \operatorname{stress}(G, G_L))$$

$$\geq \frac{\int_{B^2} \operatorname{vol}_{d-2}(v + L^\perp) \cap B) w(v, v_1)(d_0(v, v_1) - d_L(v, v_1))^2 dv_1 dv}{(1 + \frac{2\pi\kappa_d}{c_d} \lim_{t \to \infty} \frac{1}{t\delta_t^d})^{\frac{1}{2}} I^{(3)}(B)^{\frac{1}{2}} S^{(2)}(B)^{\frac{1}{2}}}.$$

In particular, the correlation does not vanish as $t \to \infty$. This gives the first proof we are aware of, that there is a strict positive correlation between the crossing number and the stress of a graph. Hence, at least for RGGs, the method to optimize the stress to obtain good crossing numbers can be supported by rigorous mathematics.

6   Consequences and Conclusion

Apart from providing precise asymptotics for the crossing numbers of drawings of random geometric graphs, the main findings are the positive covariance and the non-vanishing correlation between the stress and the crossing number of the drawing of a random geometric graph. Of interest would be whether $\operatorname{Cov}_L(\overline{\operatorname{cr}}(G_L), \operatorname{stress}(G, G_L)) > 0$ for arbitrary graphs G. Yet there are simple examples of graphs G where this is wrong. Yet we could ask in a slightly weaker form whether at least $\mathbb{E}_V \operatorname{Cov}_L(\overline{\operatorname{cr}}(G_L), \operatorname{stress}(G_L)) > 0$, but we have not been able to prove that.

We may coarsely summarize the gist of all the above findings algorithmically in the context of crossing number approximation, ignoring precise numeric terms that can be found above. We yield the first (expected) crossing number approximations for a rich class of randomized graphs:

Corollary 10. *Let G be a random geometric graph in $\mathbb{R}^2$ (unit-disk graph) as defined above. With high probability, the number of crossings in its natural straight-line drawing is at most a constant factor away from $\operatorname{cr}(G_0)$ and $\overline{\operatorname{cr}}(G_0)$.*

Corollary 11. *Let G be a random geometric graph in $\mathbb{R}^d$ (unit-ball graph) as defined above. We obtain a straight-line drawing D by projecting it onto a randomly chosen 2D plane. With high probability, the number of crossings in D is at most a factor α away from $\mathrm{cr}(G_0)$ and $\overline{\mathrm{cr}}(G_0)$. Thereby, α is only dependent on the graph's density.*

Corollary 12. *Let G be a random geometric graph and use its natural distances in $\mathbb{R}^d$ as input for stress minimization. The stress is positively correlated to the crossing number. Loosely speaking, a drawing of G with close to minimal stress is expected to yield a close to minimal number of crossings.*

References

1. Aigner, M., Ziegler, G.M.: Proofs from THE BOOK, 4th edn. Springer, Heidelberg (2009). https://doi.org/10.1007/978-3-642-00856-6
2. Ajtai, M., Chvátal, V., Newborn, M., Szemerédi, E.: Crossing-free subgraphs. In: Theory and Practice of Combinatorics, North-Holland Mathematics Studies, North-Holland, vol. 60, pp. 9–12 (1982)
3. Biedl, T.C., Chimani, M., Derka, M., Mutzel, P.: Crossing number for graphs with bounded pathwidth. In: Proceedings of International Symposium on Algorithms and Computation (ISAAC) 2017, pp. 13:1–13:13. LIPIcs 92 (2017)
4. Bienstock, D.: Some provably hard crossing number problems. Discrete Comput. Geom. **6**(5), 443–459 (1991)
5. Brandes, U., Pich, C.: An experimental study on distance-based graph drawing. In: Tollis, I.G., Patrignani, M. (eds.) GD 2008. LNCS, vol. 5417, pp. 218–229. Springer, Heidelberg (2009). https://doi.org/10.1007/978-3-642-00219-9_21
6. Cabello, S.: Hardness of approximation for crossing number. Discrete Comput. Geom. **49**(2), 348–358 (2013)
7. Cabello, S., Mohar, B.: Crossing number and weighted crossing number of near-planar graphs. Algorithmica **60**(3), 484–504 (2011). https://doi.org/10.1007/s00453-009-9357-5
8. Chimani, M., et al.: People prefer less stress and fewer crossings. In: Duncan, C., Symvonis, A. (eds.) Proceedings of International Symposium on Graph Drawing (GD) 2014. LNCS, vol. 8871, pp. 523–524. Springer, Heidelberg (2014)
9. Chimani, M., Hliněný, P.: Inserting multiple edges into a planar graph. In: Proceedings of International SYmposium on Computational Geometry (SoCG) 2016, pp. 30:1–30:15. LIPIcs 51 (2016)
10. Chimani, M., Hliněný, P., Mutzel, P.: Vertex insertion approximates the crossing number for apex graphs. Eur. J. Comb. **33**, 326–335 (2012)
11. Chimani, M., Döring, H., Reitzner, M.: Crossing numbers and stress of random graphs. eprint arXiv:1808.07558v1 (2018)
12. Chimani, M., Hliněný, P.: A tighter insertion-based approximation of the crossing number. J. Comb. Optim. **33**(4), 1183–1225 (2017)
13. Chuzhoy, J.: An algorithm for the graph crossing number problem. In: Proceedings of ACM Symposium on Theory of Computing (STOC) 2011, pp. 303–312. ACM (2011)
14. Fox, J., Pach, J., Suk, A.: Approximating the Rectilinear crossing number. In: Hu, Y., Nöllenburg, M. (eds.) GD 2016. LNCS, vol. 9801, pp. 413–426. Springer, Cham (2016). https://doi.org/10.1007/978-3-319-50106-2_32

15. Garey, M.R., Johnson, D.S.: Crossing number is NP-complete. SIAM J. Algebr. Discrete Methods **4**(3), 312–316 (1983)
16. Gitler, I., Hliněný, P., Leaño, J., Salazar, G.: The crossing number of a projective graph is quadratic in the face-width. Electron. J. Comb. **29**, 219–233 (2007)
17. Hliněný, P., Chimani, M.: Approximating the crossing number of graphs embeddable in any orientable surface. In: Proceedings of ACM-SIAM Symposium on Discrete Algorithms (SODA) 2010, pp. 918–927 (2010)
18. Hliněný, P., Salazar, G.: Approximating the crossing number of toroidal graphs. In: Tokuyama, T. (ed.) ISAAC 2007. LNCS, vol. 4835, pp. 148–159. Springer, Heidelberg (2007). https://doi.org/10.1007/978-3-540-77120-3_15
19. de Klerk, E., Maharry, J., Pasechnik, D.V., Richter, R.B., Salazar, G.: Improved bounds for the crossing numbers of $K_{m,n}$ and K_n. SIAM J. Discrete Math. **20**(1), 189–202 (2006)
20. Klimenta, M., Brandes, U.: Graph drawing by classical multidimensional scaling: new perspectives. In: Didimo, W., Patrignani, M. (eds.) GD 2012. LNCS, vol. 7704, pp. 55–66. Springer, Heidelberg (2013). https://doi.org/10.1007/978-3-642-36763-2_6
21. Kobourov, S.G.: Force-directed drawing algorithms. In: Handbook of Graph Drawing and Visualization, pp. 383–408. CRC Press (2013)
22. Kobourov, S.G., Pupyrev, S., Saket, B.: Are crossings important for drawing large graphs? In: Duncan, C., Symvonis, A. (eds.) GD 2014. LNCS, vol. 8871, pp. 234–245. Springer, Heidelberg (2014). https://doi.org/10.1007/978-3-662-45803-7_20
23. Last, G., Penrose, M.: Fock space representation, chaos expansion and covariance, inequalities for general Poisson processes. Probab. Theory Relat. Fields **150**, 663–690 (2011)
24. Last, G., Penrose, M.: Lectures on the Poisson Process. Cambridge University Press, Cambridge (2017)
25. Leighton, F.T.: Complexity Issues in VLSI: Optimal Layouts for the Shuffle-exchange Graph and Other Networks. MIT Press, Cambridge (1983)
26. Peccati, G., Reitzner, M. (eds.): Stochastic Analysis for Poisson Point Processes: Malliavin Calculus, Wiener-Itô Chaos Expansions and Stochastic Geometry. Bocconi & Springer Series, vol. 7. Springer, Switzerland (2016). https://doi.org/10.1007/978-3-319-05233-5. Bocconi University Press, Milan
27. Penrose, M.: Random Geometric Graphs (Oxford Studies in Probability). Oxford University Press, Oxford (2004)
28. Reitzner, M., Schulte, M.: Central limit theorems for U-statistics of poisson point processes. Ann. Probab. **41**, 3879–3909 (2013)
29. Reitzner, M., Schulte, M., Thäle, C.: Limit theory for the Gilbert graph. Adv. Appl. Math. **88**, 26–61 (2017)
30. Schaefer, M.: Crossing Numbers of Graphs. CRC-Press, Boca Raton (2017)
31. Schneider, R., Weil, W.: Stochastic and Integral Geometry. Probability and Its Applications. Springer, Heidelberg (2008). https://doi.org/10.1007/978-3-540-78859-1
32. Wu, L.: A new modified logarithmic Sobolev inequality for Poisson point processes and several applications. Probab. Theory Relat. Fields **118**, 427–438 (2000)

Crossing Angles

A Heuristic Approach Towards Drawings of Graphs with High Crossing Resolution

Michael A. Bekos, Henry Förster(✉), Christian Geckeler, Lukas Holländer,
Michael Kaufmann, Amadäus M. Spallek, and Jan Splett

Institut für Informatik, Universität Tübingen, Tübingen, Germany
{bekos,foersth,geckeler,mk}@informatik.uni-tuebingen.de
{jan-lukas.hollaender,amadaeus.spallek,
jan.splett}@student.uni-tuebingen.de

Abstract. The *crossing resolution* of a non-planar drawing of a graph
is the value of the minimum angle formed by any pair of crossing edges.
Recent experiments have shown that the larger the crossing resolution
is, the easier it is to read and interpret a drawing of a graph. However,
maximizing the crossing resolution turns out to be an NP-hard problem
in general and only heuristic algorithms are known that are mainly based
on appropriately adjusting force-directed algorithms.

In this paper, we propose a new heuristic algorithm for the crossing resolution maximization problem and we experimentally compare
it against the known approaches from the literature. Our experimental
evaluation indicates that the new heuristic produces drawings with better crossing resolution, but this comes at the cost of slightly higher aspect
ratio, especially when the input graph is large.

1 Introduction

In Graph Drawing, there exists a rich literature and a wide range of techniques
for drawing planar graphs; see, e.g., [11,28,34]. However, drawing a non-planar
graph, and in particular when it does not have some special structure (e.g., degree
restriction), is a difficult and challenging task, mainly due to the edge crossings
that negatively affect the drawing's quality [39]. As a result, the established
techniques are significantly fewer (e.g., crossing minimization heuristics [22,40],
energy-based layout algorithms [20,24]); for an overview refer to [13,36,41].

In this context, Huang et al. [31,32] a decade ago introduced some important
experimental evidence, that edge crossings may not negatively affect the drawing's quality too much (and hence the human's ability to read and interpret it),
when the angles formed by the crossing edges are large. In other words, while
prior to these experiments it was commonly accepted that mainly the number of
crossings is the most important parameter for judging the quality of a non-planar
graph drawing, it turned out that the types of edge crossings also matter. As a
result, a new and prominent research direction was initiated, recognized under

T. Biedl and A. Kerren (Eds.): GD 2018, LNCS 11282, pp. 271–285, 2018.
https://doi.org/10.1007/978-3-030-04414-5_19

(a)

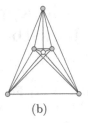
(b)

Fig. 1. (a) A RAC drawing of the complete graph K_5, and (b) a drawing of the complete graph K_6, whose crossing resolution is arbitrarily close to $90°$.

the term "beyond planarity" [30, 35, 37], which focuses on graphs and their properties, when different constraints on the types of edges crossings are imposed; refer to [16] for a recent survey.

The value of the minimum angle formed by any two crossing edges in a drawing is referred to as its *crossing resolution*; the crossing resolution of a graph is defined as the maximum crossing resolution over all its drawings. Clearly, the crossing resolution of a non-planar graph is at most $90°$, while a graph that admits a drawing with crossing resolution $90°$ is called *right-angle-crossing* (*RAC*) graph; see Fig. 1. Notably, RAC graphs are sparse with at most $4n - 10$ edges [15], while deciding whether a graph is RAC is NP-hard [4].

The latter result is an indication that the problem of finding drawings with high crossing resolution might also be difficult, even though, formally, its complexity has not been settled yet for values of the crossing resolution smaller than $90°$. Also, the literature is significantly more limited, when restricting the crossing resolution to be smaller than $90°$, as also evidenced by Sect. 2.

From a practical point of view, we are only aware of two methods that aim at drawings with high crossing resolution; both of them are adjustments of force-directed algorithms [20]. The first one is due to Huang et al. [33], while the second one is due to Argyriou et al. [5]. Common in both algorithms is that they apply appropriate forces on the endvertices of every pair of crossing edges. Each of them uses a different way to compute (the direction and the magnitude of) the forces, but the underlying idea of both is the same: the smaller the crossing angles are, the larger are the magnitudes of the forces applied at their endvertices.

In this work, we approach the crossing resolution maximization problem from a different perspective. We suggest a simple and intuitive randomization method, which, in a sense, mimics the way a human would try to increase the crossing resolution of a drawing. How would one increase the crossing resolution of a given drawing? First, she would try to identify the pair of edges that define the crossing resolution of the drawing (we call them *critical* edges); then, she would try to move an endvertex of this pair (which we choose at random), hoping that by this move the crossing resolution will increase. Of course, we cannot consider all possible positions for the vertex to be moved. Instead, we consider a small set of randomly generated ones. If there exists a position among them, that does not lead to a reduction of the crossing resolution, we move the vertex to this position.

In general, randomization is a technique that has not been deeply examined in Graph Drawing, as it seems difficult to even speculate about the expected quality of the produced drawings; a notable exception is the randomized approach by Goldschmidt and Takvorian [27] for computing large planar subgraphs. Since we also could not provide any theoretical guarantee on the expected quality of the produced drawings, we followed a more practical approach. We implemented our algorithm and the force-directed ones of [5] and [33], and we experimentally compared them on standard benchmark graphs. Our evaluation indicates that our method significantly outperforms the force-directed ones [5,33] in terms of crossing resolution, but this comes at the cost of slightly worse running time for large and dense graphs. Analogous results are obtained, when our algorithm and the ones of [5] and [33] are adjusted to maximize the *angular resolution* (i.e., the minimum value of the angle between any two adjacent edges [23]) or the *total resolution* (i.e., the minimum of the angular and the crossing resolution [5]).

Preliminaries: Unless otherwise specified, in this paper we consider simple undirected graphs. Let $G = (V, E)$ be such a graph. The degree of vertex $u \in V$ of G is denoted by $d(u)$. The degree $d(G)$ of graph G is defined as the maximum degree of its vertices, i.e., $d(G) = \max_{u \in V} d(u)$. Given a drawing $\Gamma(G)$ of G, we denote by $p(u) = (x_u, y_u)$ the position of vertex $u \in V$ of G in $\Gamma(G)$.

Structure of the Paper: The remainder of this paper is structured as follows. Section 2 overviews related works. Our algorithm is presented in detail in Sect. 3 and is experimentally evaluated against the ones of Huang et al. [33] and Argyriou et al. [5] in Sect. 4, where we also discuss our insights from this project. In [9], we provide experimental results on grid restricted drawings, on more test sets and on the graphs from the Graph Drawing Competition in 2017.

2   Related Work

As already mentioned, the study of the crossing resolution maximization problem has mainly focused on its optimal case, i.e., on the study of RAC graphs. An n-vertex RAC graph has at most $4n - 10$ edges [15], while deciding whether a graph is RAC is NP-hard [4]. The maximally-dense RAC graphs are 1-planar [21], i.e., they can be drawn with at most one crossing per edge. Actually, several relationships between the class of RAC graphs and subclasses of 1-planar graphs are known [7,10]. Deciding, however, whether a 1-planar graph is RAC is NP-hard [8]. Note that the problem of finding RAC drawings has also been studied in the presence of bends [2,6,15,26] and by imposing restrictions on the degree [3], the structure [14] and the drawing [25,29] of the graph. The results are fewer, when the right-angle constraint is relaxed. Dujmovic et al. [19] proved that an n-vertex graph with crossing resolution at least α radians, has at most $(3n - 6)\pi/\alpha$ edges. Corresponding density results are also known in the presence of bends [1,26].

An immediate observation emerging from the above overview is that the focus has been primarily on theoretical aspects of the problem. Most of the approaches that could be useful in practice are based on force-directed techniques [13,20]. COWA is a system that supports conceptual web site traffic analysis [17]; its algorithmic core is a force-directed heuristic to compute simultaneous embeddings of two non-planar graphs with high crossing resolution. Didimo et al. [18] describe topology-driven force-directed heuristics to achieve good trade-offs in terms of number of edge crossings, crossing resolution, and geodesic edge tendency; the obtained drawings, however, are not straight-line. For straight-line drawings, Nguyen et al. [38] suggest a quadratic-program to increase the crossing angles of circular drawings. Of more general scope are the already mentioned force-directed algorithms of Argyriou et al. [5] and Huang et al. [33].

3   Description of Our Heuristic Approach

In this section, we describe our heuristic for obtaining drawings with high crossing resolution. The input of our heuristic consists of a graph G and an initial drawing Γ_0 of G with crossing resolution $c(\Gamma_0)$. We assume that no two edges of G overlap in Γ_0, i.e., $c(\Gamma_0) > 0$. A circular drawing or a drawing obtained by applying a force-directed algorithm on G clearly meets this precondition.

Our algorithm is iterative and at each iteration performs some operations that are mainly based on randomization. At the i-th iteration, we assume that we have computed a drawing Γ_{i-1} of crossing resolution $c(\Gamma_{i-1}) \geq c(\Gamma_0)$. In other words, we assume, as an invariant for our algorithm, that the crossing resolution cannot be decreased at some iteration. Then, a vertex of Γ_{i-1} is chosen arbitrarily at random based on the so-called *vertex-pool*, which may contain: (i) either all vertices of Γ_{i-1}, or (ii) a prespecified subset of the vertices of Γ_{i-1}, called *critical*.

Intuitively, the critical vertices are the endpoints of the edges that define the crossing resolution of drawing Γ_{i-1}. To formally define them, we first need to introduce the notion of critical edge-pairs. A pair of edges e and e' is called *critical* in Γ_{i-1}, if e and e' cross in Γ_{i-1} and the minimum angle that is formed at their crossing point is equal to $c(\Gamma_{i-1})$. The set of critical vertices of Γ_{i-1} is then defined by the four endvertices of each critical edge-pair.

The role of critical vertices is central in our algorithm[1]: By appropriately changing the location of a critical vertex or of a vertex in the neighbourhood of the critical vertices, we naturally expect to improve the crossing resolution of the current drawing. We turned this observation into an algorithmic implementation through a probabilistic random selection procedure, so that the vertices at graph-distance i from the ones of the vertex-pool have higher probability for selection than the corresponding ones at distance j in the graph, when $0 \leq i < j$. So, if the vertex-pool contains only critical vertices, then the closer a vertex is to the

---

[1] If the focus is not on the critical vertices for a large graph, then our algorithm will need a large number of iterations to converge to a good solution, because it is simply very unlikely to select to move one of the vertices that define the crossing resolution.

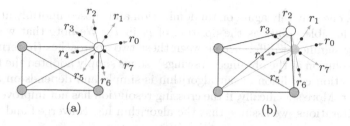

Fig. 2. Illustration of an iteration step of our algorithm: (a) The chosen vertex is the white one; the computed rays $r_0, \ldots, r_7$ have been rotated by $8°$; the black-colored points along these rays are points $\pi_0, \ldots, \pi_7$; among them, π_4 yields the best solution. (b) The resulting drawing after moving the vertex at position π_2.

critical vertices, the more likely it is to be chosen. Otherwise, the vertex-pool contains all vertices and each vertex can be chosen with the same probability.

What we quickly realized from our practical analysis, is that the crossing resolution of the initial drawing improves rapidly during the first iterations of the algorithm. However, by focusing only at the critical vertices, it is highly possible that the algorithm will get trapped to some local maxima after a number of iterations. So, special care is needed to avoid these bottlenecks, especially when the input graph is large. We will discuss ways to avoid them later in this section.

So far, we have described the main idea of our algorithm, which at each iteration chooses uniformly at random a vertex of the current drawing to move (based on the content of the vertex-pool), so to improve the crossing resolution. Next, we describe how to compute its new position in the next drawing. Note that our method resembles probabilistic hill climbing approaches.

Let v_i be the vertex of Γ_{i-1} that has been chosen to be moved at the i-th iteration. To compute the position of v_i in the next drawing Γ_i, we consider a set of ρ rays $r_0, r_1, \ldots, r_{\rho-1}$ that all emanate from $p(v_i)$ in Γ_{i-1}, such that the angle formed by ray r_j, with $j = 0, 1, \ldots, \rho-1$, and the horizontal axis equals to $2j\pi/\rho$, where $\rho > 0$ is an integer parameter of the algorithm. These rays are then rotated by an angle that is chosen uniformly at random in the interval $[0, 2\pi]$; see Fig. 2. The position of vertex v_i in Γ_i will eventually be along one of the rays $r_0, r_1, \ldots, r_{\rho-1}$. More precisely, for each ray r_i we choose a distance value δ_i uniformly at random from the interval $[\delta_{min}, \delta_{max}]$, where δ_{min} and δ_{max} are two positive parameters of the algorithm. For each $j = 0, 1, \ldots, \rho-1$, a new point π_j is obtained by translating $p(u)$ along r_j by a distance δ_j; point π_j is *feasible*, if the crossing resolution of the drawing obtained by placing vertex v_i at π_j and by keeping all other vertices of G in their positions in Γ_{i-1} is at least as large as the crossing resolution of Γ_{i-1}, and there is no vertex of Γ_{i-1} at π_j.

If none of the points π_j, with $j = 0, 1, \ldots, \rho-1$ is feasible, then the position of v_i in Γ_i is $p(v_i)$, i.e., same as in Γ_{i-1}, since $c(\Gamma_i) \geq c(\Gamma_{i-1})$ must hold. If there is one or more feasible points, then one may consider two different approaches to determine the position of v_i in Γ_i. The most natural is to choose the feasible point that maximizes the crossing resolution of the obtained drawing. As an alternative, one may rely again on randomization and chose uniformly at random one of the feasible points as the position of v_i in Γ_i. We note that we did not observe any significant difference between these two approaches (in terms of the crossing resolution of the obtained drawings), so we simply adopted the first one. The termination condition of our algorithm is simple and depends on an input parameter τ. More specifically, if the crossing resolution has not improved during the last τ iterations, we assume that the algorithm has converged and we stop.

Avoiding Local Maxima. To avoid getting trapped to locally optimal solutions, we mainly investigated two approaches, which are both parametrizable by two input parameters ζ and ζ'. The first mimics the human behaviour. What would one do to escape from a locally optimal solution? She would stop trying to move the endvertices of the edges defining the crossing resolution; she would rather start moving "irrelevant" vertices hoping that by doing so a better solution will be easier to be computed afterwards. Our algorithm is mimicking this idea as follows: (i) if during the last ζ iterations the crossing resolution has not been improved, then the vertex-pool becomes *wider* by including all the vertices, and the algorithm is executed with this vertex-pool for ζ' iterations; (ii) afterwards, the vertex-pool switches back to the critical vertices. While this approach turned out to be effective for smaller graphs, for graphs with more than 100 vertices, it was not so efficient; in most iterations with the wider vertex-pool, the embedding could not change in a beneficial way for the algorithm to proceed.

Our second approach is based on parameters ρ, δ_{min} and δ_{max} of the algorithm. Our idea was that if the algorithm gets trapped to a locally optimal solution, then a "drastic" or "sharp" move may help to escape. We turned this idea into an algorithmic implementation as follows: (i) if during the last ζ iterations the crossing resolution has not been improved, we double the values of ρ, δ_{min} and δ_{max}, and the algorithm is executed with these values for ζ' iterations; (ii) afterwards, ρ, δ_{min} and δ_{max} switch back to this initial value. This approach may lead to drawings with larger area, but this is "expected", as it turns out that drawings with high crossing resolution may require large area [2,10].

Complexity Issues. A factor that highly affects the efficiency of our algorithm is the computation of the crossing points of the edges and the corresponding angles at these points. Given a drawing, a naïve approach to compute its crossings requires $O(m^2)$ time, which can be improved by a plane-sweep technique to $O(m \log m + c)$ time, where m and c denote the number of edges and crossings.

Instead of computing all crossing points and the corresponding angles for each candidate position of each iteration, we adopted a different approach for determining the set of feasible candidate positions, which turned out to be quite efficient in practice. Recall that we denoted by v_i the vertex chosen at the i-th iteration step, and by $\pi_0, \ldots, \pi_{\rho-1}$ the candidate positions to move v_i. Let $e_0, \ldots, e_{d_i-1}$ be the edges incident to v_i, where $d_i = deg(v_i)$. Next, for each edge e_k with $k = 0, \ldots, d_i - 1$ we compute the crossings and the corresponding crossing angles of e_k with all other edges in Γ_{i-1}. Let ϕ_i be the minimum crossing angle computed; this is our reference angle. Also, for each candidate position π_j with $j = 0, \ldots, \rho - 1$, and for each edge e_k with $k = 0, \ldots, d_i - 1$, we compute the crossings and the corresponding crossing angles of e_k with all other edges of the drawing, assuming that v_i is at π_j. Let χ_j be the minimum crossing angle computed with this approach, when v_i is at position π_j. Clearly, π_j is feasible only if $\chi_j \geq \phi_i$. Note that the complexity of this approach is $O(deg(v_i)m) = O(nm)$.

3.1   Some Interesting Variants

In general, aesthetically pleasant drawings of graphs are usually the result of compromising between different aesthetic criteria. Towards this direction, we discuss in this section interesting variants of our algorithm, which are motivated by the following observation that we made while working on this project (see Sect. 4): Drawings that are optimised only in terms of the crossing resolution tend to have bad aspect ratio and poor angular resolution.

Aspect Ratio. It was easy to instruct our algorithm to prevent producing drawings with aspect ratio either higher than the one of the starting layout or higher than a given input value. What we simply had to do was to reject candidate positions, which violate this precondition.

Total Resolution. Similarly as above, we could adjust our algorithm to yield drawings with high total resolution by simply taking into account also the angular resolution of the drawing. In particular, if the total resolution of the drawing is defined by its angular resolution, then the way we compute the critical vertices of this drawing has to change; the critical vertices must be the endvertices of the pairs of edges that define the angular resolution. Also, at each iteration of our algorithm we have to ensure that the total resolution does not decrease. We do so by rejecting candidate positions which yield a reduced total resolution.

Angular Resolution. As it is the case with the force-directed algorithms of Huang et al. [33] and Argyriou et al. [5], our algorithm can be also restricted to maximize only the angular resolution (by neglecting its crossing resolution). We already described in the previous paragraph the necessary changes in the definition of the critical vertices and the rule according to which a candidate position is rejected (i.e., when it yields a drawing with a reduced angular resolution).

Grid Drawings. Our algorithm, as it has been described so far, does not necessarily produce grid drawings, i.e., drawings in which the vertices are at integer coordinates. However, it can be easily adjusted to produce such drawings. More precisely, if we round the candidate positions computed at each iteration of our algorithm to their closest grid points and use these grid points as candidates for the next position of the vertex to be moved, then the obtained drawing will be grid (assuming, of course, that the starting drawing is grid). One can even bound the size of the grid, by rejecting candidate grid positions outside the bounds. In [9], we report experimental results on this variant.

4   Experimental Evaluation

In this section, we present the results of our experimental evaluation. For comparison purposes, apart from our algorithm, we also implemented the force-directed algorithms of Argyriou et al. [5] and Huang et al. [33]. The implementations[2] were in Java using yFiles [42]. The experiment was performed on a Linux laptop with four cores at 2.4 GHz and 8 GB RAM. As a test set for our experiment, we used the non-planar Rome graphs [12], which form a collection of around 8.100 benchmark graphs; in [9], we also report on the AT&T graphs.

The experiment was performed as follows. Initially, each Rome graph was laid out using the SmartOrganic layouter of yFiles [42]. Starting from this layout, every graph was drawn with (i) our algorithm, (ii) our algorithm restricted not to violate the aspect ratio of the initial layout, and the force-directed algorithms (iii) by Argyriou et al. and (iv) by Huang et al. Since all algorithms of the experiment can easily be adjusted to maximize only the crossing resolution, or only the angular resolution or both (by maximizing the total resolution), we adjusted each of the algorithms to maximize exclusively the corresponding measures; see Figs. 3, 4 and 5. In our algorithm, this can be achieved by modifying appropriately the content of the vertex-pool (as we saw in Sect. 3.1), while in the algorithms of Argyriou et al. and of Huang et al. by switching on only the forces that maximize the corresponding properties under measure (note that, each of these two algorithms has a different set of forces to maximize the crossing and the angular resolution, such that together they maximize the total resolution). The reported results are on average across different drawings with same number of vertices. Finally, we mention that for our algorithm, we chose $\delta_{max} = \frac{1}{2} \max\{w, h\}$, where w and h are the width and the height of the initial drawing, respectively, $\delta_{min} = \frac{1}{100}\delta_{max}$ and $\rho = 10$.

Crossing Resolution. Our results for the crossing resolution are summarized in Fig. 3. Here, each algorithm was adjusted to maximize exclusively the crossing resolution (i.e., by ignoring the drawing's angular resolution). It is immediate to see that our algorithm outperforms all other ones in terms of the crossing resolution of the produced drawings, when we do not impose any restriction on the aspect ratio of the computed drawings; refer to the solid-black curve, denoted as *Unrestricted*, in Fig. 3a. The variant of our algorithm, which does not violate the aspect ratio of the initial layout, leads to drawings with slightly smaller crossing resolution; refer to the solid-gray curve, denoted as *AR-restricted*, in Fig. 3a. Finally, the two force-directed algorithms seem to produce drawings with worse crossing resolution; refer to the dotted-gray and dotted-black curves of Fig. 3a (by Argyriou et al. and by Huang et al., respectively).

While our unrestricted algorithm produces drawings with better crossing resolution, this comes at a cost of drastically increased aspect ratio (see Fig. 3b), which, however, is still better that the corresponding aspect ratio of the drawings produced by the algorithm of Argyriou et al. For the latter algorithm, it seems that the forces due to the angles formed at the crossings outperform the corresponding spring forces, which try to keep the lengths of the edges short. Going back to our unrestricted algorithm, its behaviour is up to a certain degree expected, mainly due to the fact that there is no control on the lengths of the edges. On the other hand, the restricted variant of our algorithm, which does

---

[2] Our implementation is available on request from the authors.

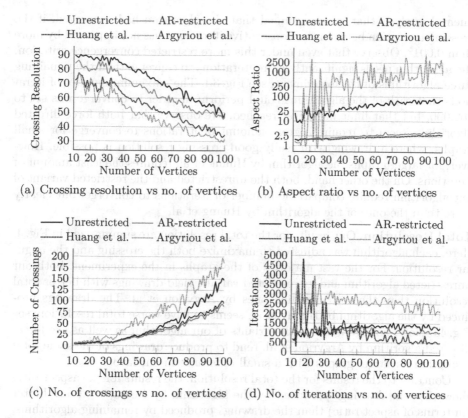

Fig. 3. Experimental results on the crossing resolution for the Rome graphs.

not allow the aspect ratio to increase, has more or less comparable performance (in terms of aspect ratio) with the one of Huang et al.

Regarding the number of crossings, the restricted variant of our algorithm and the force-directed algorithm of Huang et al. yield drawings with comparable number of crossings, which at the same time is significantly smaller than the number of crossings produced by the two other algorithms; see Fig. 3c.

A different behaviour can be observed in the number of iterations, which are required by the algorithms to converge; refer to Fig. 3d. We note here that we used different criteria to determine whether the algorithms of our experiment had converged. For our algorithms and for the force-directed algorithm by Huang et al., we assumed that the algorithm had converged, if the crossing resolution between 500 consecutive iterations was not improved by more than 0.001°. For the algorithm by Argyriou et al., we decided to use a much more restricted convergence criterion, because the produced layouts can change vastly between consecutive iterations. We made this choice mainly to have "comparable" number of iterations among the algorithms of the experiment. In this direction, we adopted the convergence criterion that the authors used in their previous experimental analysis that is, we assumed that the algorithm had converged, if the crossing resolution between two consecutive iterations was not improved by more than 0.001°. Observe that even under this more restricted convergence criterion, the algorithm needs significantly more iterations to converge than the remaining three algorithms of the experiment; see Fig. 3d. The maximum number of iterations that each of the algorithms could perform in order to converge was set to 100.000, but that limit was never reached. We observe that both force-directed algorithms seem to require a great amount of iterations to converge for small graphs, where a drawing with really good crossing resolution is possible. However, for larger graphs the algorithm by Huang et al. requires the least amount of iterations. On the other hand, both the unrestricted and the restricted variant of our algorithm require comparable number of iterations to converge, but clearly more than the ones of the algorithm by Huang et al.

Total Resolution. Our results for the total resolution are summarized in Fig. 4. Here, each algorithm was adjusted to maximize both the crossing and the angular resolution. For the vast majority of the graphs in the experiment, both our unrestricted algorithm and its restricted variant yield drawings with better total resolution than the corresponding ones by Argyriou et al. The drawings produced by the algorithm by Huang et al. seems to have worse total resolution; see Fig. 4a. Note, however, that both variants of our algorithm as well as the force-directed algorithm by Argyriou et al. tend to produce drawings of the same total resolution for larger graphs with a small difference in our favor.

Contrary to the results for the total resolution, the results for the aspect ratio show that the drawings produced by the algorithm by Huang et al. are better (in terms of aspect ratio) than the drawings produced by remaining algorithms; see Fig. 4b. More concretely, the drawings produced by the restricted variant of our algorithm have slightly worse aspect ratios. Then, the ones produced by the force-directed algorithm by Argyriou et al. follow. Again, we observe that our unrestricted algorithm leads to drawings with very high aspect ratio.

The restricted variant of our algorithm and the algorithm by Huang et al. yield drawings with the least number of crossings; see Fig. 4c. Comparable but slightly worse (in terms of the number of crossings) are the drawings produced by the force-directed algorithm by Argyriou et al. Our unrestricted algorithm seems to require the largest number of crossings, which turn out to be notably higher than the corresponding ones of the other three algorithms.

On the negative side, both the unrestricted and the restricted variant of our algorithm require more iterations than the force-directed algorithm by Huang et al.; see Fig. 4d. Recall, however, that the latter algorithm is clearly outperformed by both our variants in term of total resolution. The algorithm by Argyriou et al. clearly requires the highest number of iterations (especially for large graphs). We note that the convergence criterion was the same as for the crossing resolution; however, the measured quality was (not the crossing but) the total resolution.

Angular Resolution. We conclude the analysis of our experimental evaluation with the results for the angular resolution; see Fig. 5. Here, each algorithm was adjusted to maximize only the angular resolution (i.e., by ignoring the drawing's

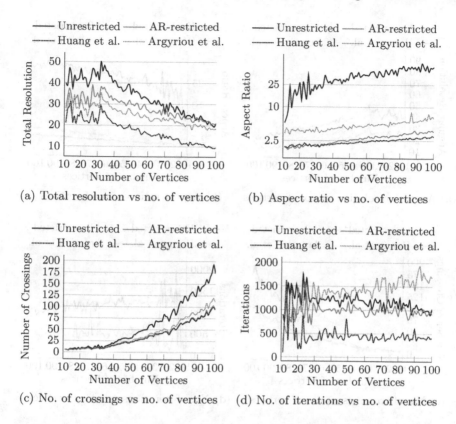

Fig. 4. Experimental results on the total resolution for the Rome graphs.

crossing resolution). A notable observation is that, for small graphs the best results are achieved by the algorithm by Argyriou et al., while for medium-size graphs by our unrestricted algorithm; see Fig. 5a. For large graphs, the two algorithms tend to have the same performance. The restricted variant of our algorithm yields drawings with slightly worse angular resolution. The algorithm by Huang et al. is outperformed by all algorithms of the experiment.

The results for the aspect ratio, the number of crossings and the required number of iterations are very similar with corresponding ones for the total resolution; see Fig. 5b–d. This observation suggests that, for most of the graphs of our experiment, the angular resolution dominates the crossing resolution (and thus is the one defining the total resolution) in the constructed drawings, which explains the similarity in the reported results. The small differences result from the fact that the crossing resolution cannot be entirely neglected.

Discussion. While working on this project, we made some useful observations and obtained some interesting insights. In particular, there is a recent hypothesis (also supported by experiments) that drawings, in which the crossing angles, are large are easy to read and understand. We observed that drawings that are

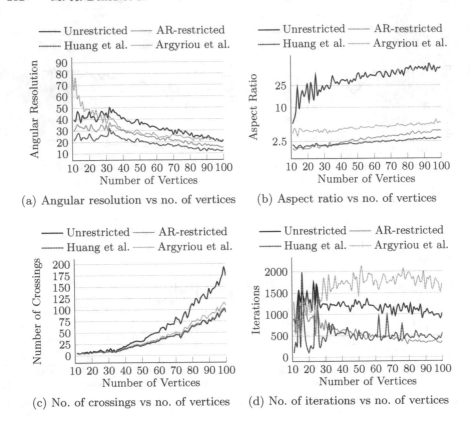

(a) Angular resolution vs no. of vertices   (b) Aspect ratio vs no. of vertices

(c) No. of crossings vs no. of vertices   (d) No. of iterations vs no. of vertices

Fig. 5. Experimental results on the angular resolution for the Rome graphs.

optimized only in terms of the crossing angles might be arbitrarily bad and may have several undesired properties. In particular, in these drawings it was very common to have adjacent edges to run almost in parallel and vertices to be very close to each other. Hence, angular resolution and aspect ratio were often poor. The additional restrictions that we imposed regarding the angular resolution and the aspect ratio helped significantly improving the readability of the drawings, without loosing too much of their quality in terms of the crossing resolution.

We conclude by noting that our motivation to work with this problem was our participation to GD2017 contest, where we performed miserably using a force-directed algorithm; for details see [9]. As our evaluation shows, the performance of such algorithms is good, only when several aesthetic criteria are taken into account; our new approach is definitely more promising than our previous as evidenced by our experiments. The framework that we developed seems to be quite adaptable to optimize or to take into account also other desired aesthetic properties of a drawing.

Acknowledgments. This project was supported by DFG grant KA812/18-1. The authors would also like to acknowledge Simon Wegendt and Jessica Wolz for implementing the first version of our prototype.

References

1. Ackerman, E., Fulek, R., Tóth, C.D.: Graphs that admit polyline drawings with few crossing angles. SIAM J. Discrete Math. **26**(1), 305–320 (2012). https://doi.org/10.1137/100819564

2. Angelini, P., et al.: On the perspectives opened by right angle crossing drawings. J. Graph Algorithms Appl. **15**(1), 53–78 (2011). https://doi.org/10.7155/jgaa.00217

3. Angelini, P., et al.: Large angle crossing drawings of planar graphs in subquadratic area. In: Márquez, A., Ramos, P., Urrutia, J. (eds.) EGC 2011. LNCS, vol. 7579, pp. 200–209. Springer, Heidelberg (2012). https://doi.org/10.1007/978-3-642-34191-5_19

4. Argyriou, E.N., Bekos, M.A., Symvonis, A.: The straight-line RAC drawing problem is NP-hard. J. Graph Algorithms Appl. **16**(2), 569–597 (2012). https://doi.org/10.7155/jgaa.00274

5. Argyriou, E.N., Bekos, M.A., Symvonis, A.: Maximizing the total resolution of graphs. Comput. J. **56**(7), 887–900 (2013). https://doi.org/10.1093/comjnl/bxs088

6. Arikushi, K., Fulek, R., Keszegh, B., Moric, F., Tóth, C.D.: Graphs that admit right angle crossing drawings. Comput. Geom. **45**(4), 169–177 (2012). https://doi.org/10.1016/j.comgeo.2011.11.008

7. Bachmaier, C., Brandenburg, F.J., Hanauer, K., Neuwirth, D., Reislhuber, J.: NIC-planar graphs. Discrete Appl. Math. **232**, 23–40 (2017). https://doi.org/10.1016/j.dam.2017.08.015

8. Bekos, M.A., Didimo, W., Liotta, G., Mehrabi, S., Montecchiani, F.: On RAC drawings of 1-planar graphs. Theor. Comput. Sci. **689**, 48–57 (2017). https://doi.org/10.1016/j.tcs.2017.05.039

9. Bekos, M.A., et al.: A heuristic approach towards drawings of graphs with high crossing resolution. CoRR, abs/xxxx.yyyy (2018). https://arxiv.org/abs/xxxx.yyyy

10. Brandenburg, F.J., Didimo, W., Evans, W.S., Kindermann, P., Liotta, G., Montecchiani, F.: Recognizing and drawing IC-planar graphs. Theor. Comput. Sci. **636**, 1–16 (2016). https://doi.org/10.1016/j.tcs.2016.04.026

11. de Fraysseix, H., Pach, J., Pollack, R.: How to draw a planar graph on a grid. Combinatorica **10**(1), 41–51 (1990). https://doi.org/10.1007/BF02122694

12. Di Battista, G., Didimo, W.: GDToolkit. In: Tamassia, R. (ed), Handbook on Graph Drawing and Visualization, pp. 571–597. Chapman and Hall/CRC (2013)

13. Di Battista, G., Eades, P., Tamassia, R., Tollis, I.G.: Graph Drawing: Algorithms for the Visualization of Graphs. Prentice-Hall, Upper Saddle River (1999)

14. Didimo, W., Eades, P., Liotta, G.: A characterization of complete bipartite RAC graphs. Inf. Process. Lett. **110**(16), 687–691 (2010). https://doi.org/10.1016/j.ipl.2010.05.023

15. Didimo, W., Eades, P., Liotta, G.: Drawing graphs with right angle crossings. Theor. Comput. Sci. **412**(39), 5156–5166 (2011). https://doi.org/10.1016/j.tcs.2011.05.025

16. Didimo, W., Liotta, G., Montecchiani, F.: A survey on graph drawing beyond planarity. CoRR, abs/1804.07257 arXiv:1803.03705 (2018)

17. Didimo, W., Liotta, G., Romeo, S.A.: Graph visualization techniques for conceptual web site traffic analysis. In: IEEE PacificVis, pp. 193–200. IEEE Computer Society (2010). https://doi.org/10.1109/PACIFICVIS.2010.5429593
18. Didimo, W., Liotta, G., Romeo, S.A.: Topology-driven force-directed algorithms. In: Brandes, U., Cornelsen, S. (eds.) GD 2010. LNCS, vol. 6502, pp. 165–176. Springer, Heidelberg (2011). https://doi.org/10.1007/978-3-642-18469-7_15
19. Dujmovic, V., Gudmundsson, J., Morin, P., Wolle, T.: Notes on large angle crossing graphs. Chicago J. Theor. Comput. Sci. (2011). http://cjtcs.cs.uchicago.edu/articles/CATS2010/4/contents.html
20. Eades, P.: A heuristic for graph drawing. Congressus Numerantium **42**, 149–160 (1984)
21. Eades, P., Liotta, G.: Right angle crossing graphs and 1-planarity. Discrete Appl. Math. **161**(7–8), 961–969 (2013). https://doi.org/10.1016/j.dam.2012.11.019
22. Eades, P., Wormald, N.C.: Edge crossings in drawings of bipartite graphs. Algorithmica **11**(4), 379–403 (1994). https://doi.org/10.1007/BF01187020
23. Formann, M., et al.: Drawing graphs in the plane with high resolution. SIAM J. Comput. **22**(5), 1035–1052 (1993). https://doi.org/10.1137/0222063
24. Fruchterman, T.M.J., Reingold, E.M.: Graph drawing by force-directed placement. Softw. Pract. Exper. **21**(11), 1129–1164 (1991). https://doi.org/10.1002/spe.4380211102
25. Giacomo, E.D., Didimo, W., Eades, P., Liotta, G.: 2-layer right angle crossing drawings. Algorithmica **68**(4), 954–997 (2014). https://doi.org/10.1007/s00453-012-9706-7
26. Giacomo, E.D., Didimo, W., Liotta, G., Meijer, H.: Area, curve complexity, and crossing resolution of non-planar graph drawings. Theory Comput. Syst. **49**(3), 565–575 (2011). https://doi.org/10.1007/s00224-010-9275-6
27. Goldschmidt, O., Takvorian, A.: An efficient graph planarization two-phase heuristic. Networks **24**(2), 69–73 (1994). https://doi.org/10.1002/net.3230240203
28. Gutwenger, C., Mutzel, P.: Planar polyline drawings with good angular resolution. In: Whitesides, S.H. (ed.) GD 1998. LNCS, vol. 1547, pp. 167–182. Springer, Heidelberg (1998). https://doi.org/10.1007/3-540-37623-2_13
29. Hong, S.-H., Nagamochi, H.: Testing full outer-2-planarity in linear time. In: Mayr, E.W. (ed.) WG 2015. LNCS, vol. 9224, pp. 406–421. Springer, Heidelberg (2016). https://doi.org/10.1007/978-3-662-53174-7_29
30. Hong, S., Tokuyama, T.: Algorithmics for beyond planar graphs. NII Shonan Meeting Seminar 089, 27 November–1 December 2016
31. Huang, W.: Using eye tracking to investigate graph layout effects. In: Hong, S., Ma, K. (eds.) APVIS, pp. 97–100. IEEE Computer Society (2007). https://doi.org/10.1109/APVIS.2007.329282
32. Huang, W., Eades, P., Hong, S.: Larger crossing angles make graphs easier to read. J. Vis. Lang. Comput. **25**(4), 452–465 (2014). https://doi.org/10.1016/j.jvlc.2014.03
33. Huang, W., Eades, P., Hong, S., Lin, C.: Improving multiple aesthetics produces better graph drawings. J. Vis. Lang. Comput. **24**(4), 262–272 (2013). https://doi.org/10.1016/j.jvlc.2011.12.002
34. Kant, G.: Drawing planar graphs using the canonical ordering. Algorithmica **16**(1), 4–32 (1996). https://doi.org/10.1007/BF02086606
35. Kaufmann, M., Kobourov, S., Pach, J., Hong, S.: Beyond planar graphs: Algorithmics and combinatorics. Dagstuhl Seminar **16452**(November), 6–11 (2016)
36. Kaufmann, M., Wagner, D. (eds.): Drawing Graphs, Methods and Models. LNCS, vol. 2025. Springer, Heidelberg (2001). https://doi.org/10.1007/3-540-44969-8

37. Liotta, G.: Graph drawing beyond planarity: some results and open problems. SoCG Week, Invited talk, 4th July 2017
38. Nguyen, Q., Eades, P., Hong, S.-H., Huang, W.: Large crossing angles in circular layouts. In: Brandes, U., Cornelsen, S. (eds.) GD 2010. LNCS, vol. 6502, pp. 397–399. Springer, Heidelberg (2011). https://doi.org/10.1007/978-3-642-18469-7_40
39. Purchase, H.C.: Effective information visualisation: a study of graph drawing aesthetics and algorithms. Interact. Comput. **13**(2), 147–162 (2000)
40. Sugiyama, K., Tagawa, S., Toda, M.: Methods for visual understanding of hierarchical system structures. IEEE Trans. Syst. Man Cybern. **11**(2), 109–125 (1981)
41. Tamassia, R. (ed.): Handbook on Graph Drawing and Visualization. Chapman and Hall/CRC, Boca Raton (2013)
42. Wiese, R., Eiglsperger, M., Kaufmann, M.: yFiles - visualization and automatic layout of graphs. In: Jünger, M., Mutzel, P., et al. (eds.) Graph Drawing Software, pp. 173–191. Springer, Heidelberg (2004). https://doi.org/10.1007/978-3-642-18638-7_8

A Greedy Heuristic for Crossing-Angle Maximization

Almut Demel, Dominik Dürrschnabel, Tamara Mchedlidze,
Marcel Radermacher[(✉)], and Lasse Wulf

Department of Computer Science, Karlsruhe Institute of Technology,
Karlsruhe, Germany
mched@iti.uka.de, radermacher@kit.edu

Abstract. The crossing angle of a straight-line drawing Γ of a graph $G = (V, E)$ is the smallest angle between two crossing edges in Γ. Deciding whether a graph G has a straight-line drawing with a crossing angle of $90°$ is $\mathcal{NP}$-hard [1]. We propose a simple heuristic to compute a drawing with a large crossing angle. The heuristic greedily selects the best position for a single vertex in a random set of points. The algorithm is accompanied by a speed-up technique to compute the crossing angle of a straight-line drawing. We show the effectiveness of the heuristic in an extensive empirical evaluation. Our heuristic was clearly the winning algorithm (CoffeeVM) in the Graph Drawing Challenge 2017 [6].

1 Introduction

The *crossing angle* cr-$\alpha(\Gamma)$ of a straight-line drawing Γ is defined to be the minimum over all angles created by two crossing edges in Γ. The 24th edition of the annual *Graph Drawing Challenge*, held during the Graph Drawing Symposium, posed the following problem: Given a graph G, compute a straight-line drawing Γ on an integer grid that has a large crossing angle. In this paper we present a greedy heuristic that starts with a carefully chosen initial drawing and repeatedly moves a vertex v to a random point p if this increases the crossing angle of Γ. This heuristic was the winning algorithm of the GD Challenge'17 [6].

Related Works. A drawing of a graph is called RAC if its minimum crossing angle is $90°$. Deciding whether a graph has a straight-line RAC drawing is an $\mathcal{NP}$-hard problem [1]. Giacomo et al. [13] proved that every straight-line drawing of a complete graph with at least 12 vertices has a crossing angle of $\Theta(\pi/n)$. Didimo et al. [7] have shown that every n-vertex graph that admits a straight-line RAC drawing has at most $4n - 10$ edges. This bound is tight, since there is an infinite family of graphs with $4n - 10$ edges that have straight-line RAC drawings. Moreover they proved that every graph has a RAC drawing with three

Work was partially supported by grant WA 654/21-1 of the German Research Foundation (DFG).

T. Biedl and A. Kerren (Eds.): GD 2018, LNCS 11282, pp. 286–299, 2018.
https://doi.org/10.1007/978-3-030-04414-5_20

bends per edge. Arikishu et al. [3] showed that any n-vertex graph that admits a RAC drawing with one bend or two bends per edge has at most $6.5n$ and $74.2n$ edges, respectively. For an overview over further results on RAC drawings we refer to [8]. Dujmović et al. [9] introduced the concept of αAC graphs. A graph is αAC if it admits a drawing with crossing angle of at least α. For $\alpha > \pi/3$, αAC graphs are *quasiplanar* graphs, i.e., graphs that admit a drawing without three mutually crossing edges, and thus have at most $6.5n - 20$ edges. Moreover, every n-vertex αAC graph with $\alpha \in (0, \pi/2)$ has at most $(\pi/\alpha)(3n - 6)$ edges. Besides the theoretical work on this topic, there are a few force-directed approaches that optimize the crossing angle in drawings of arbitrary graphs [2,14], see Sect. 2.1.

Contribution. We introduce a heuristic to increase the crossing angle in a given straight-line drawing Γ (Sect. 3). The heuristic is accompanied by a speed-up technique to compute the pair of crossing edges in Γ that create the smallest crossing angle. In Sect. 4 we give an extensive evaluation of our heuristic. The evaluation is driven by three main research questions: (i) What is a good parametrization of our heuristic? (ii) Does our heuristic improve the crossing angle of a given initial drawing? (iii) What is a good choice for an initial drawing?

2 Preliminaries

Let Γ be a straight-line drawing of a graph $G = (V, E)$. Denote by n and m the number of vertices and edges of G, respectively. Let e and e' be two distinct edges of G. If e and e' have an interior intersection in Γ, the function cr-$\alpha(\Gamma, e, e')$ denotes the smallest angle formed by e and e' in Γ. In case that e and e' do not intersect, we define cr-$\alpha(\Gamma, e, e')$ to be $\pi/2$. The *local crossing angle of a vertex* v is defined as the minimum angle of the edges incident to v, i.e., cr-$\alpha(\Gamma, v) = \min_{e, uv \in E, e \neq uv}$ cr-$\alpha(\Gamma, e, uv)$. The *crossing angle* of a drawing Γ is defined as cr-$\alpha(\Gamma) = \min_{e, e' \in E, e \neq e'}$ cr-$\alpha(\Gamma, e, e')$. Let Δx and Δy be the difference of the x-coordinates and the y-coordinates of the endpoints of e in a drawing Γ. The *slope of* e is the angle between e and the x-axis, i.e. slope$(\Gamma, e) = \arctan(\Delta y/\Delta x)$ if $\Delta x \neq 0$ and slope$(\Gamma, e) = -\pi/2$ otherwise.

2.1 Force-Directed Approaches

In general, force-directed algorithms [10,11] compute for each vertex v of a graph $G = (V, E)$ a force F_v. A new drawing Γ' is obtained from a drawing Γ by

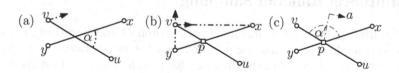

Fig. 1. Sketches of the force (a) $F_{\cos}(v)$, (b) $F_{\text{cage}}(v)$ and (c) $F_{\text{ang}}(v)$.

Algorithm 1. Random Sampling

Input : Initial drawing Γ, number of levels $L \in \mathbb{N}$, number of samples $T \in \mathbb{N}$,
            scaling factor $b \in (0, 1)$, side length $s > 0$
Output : Drawing Γ

1 **while** *stopping criteria* **do**
2 $\quad$ $(e_1, e_2) \leftarrow$ crossing edges with smallest crossing angle in Γ
3 $\quad$ $v \leftarrow$ random vertex in $e_1 \cup e_2$
4 $\quad$ **for** $i \leftarrow 1$ **to** L **do**
5 $\quad\quad$ $R^i \leftarrow$ square centered at $\Gamma[v]$ with side length $s \cdot b^{i-1}$
6 $\quad\quad$ **for** 1 **to** T **do**
7 $\quad\quad\quad$ $q \leftarrow$ uniform random position in R^i
8 $\quad\quad\quad$ **if** cr-$\alpha(\Gamma[v \mapsto q], v) >$ cr-$\alpha(\Gamma, v)$ **then**
9 $\quad\quad\quad\quad$ $\Gamma[v] \leftarrow q$

---

displacing every vertex v according to the force F_v. Classically, the force F_v is a linear combination of repelling and attracting forces, i.e., all pairs of vertices repel each other, and incident vertices attract each other. It is easy to integrate new forces into this generic system, e.g., in order to increase the crossing angle. For this purpose, Huang et al. [14] introduced the *cosine force* $F_{\cos}$. The force-directed approach considered by Argyriou et al. [2] uses two forces, F_{cage} and F_{ang}, to increase the crossing angle. In the following we will describe each force.

Let $\overrightarrow{xy}$ denote the unit length vector from x to y. Let uv, xy be two crossing edges in Γ and let α be the angle as depicted in Fig. 1(a) and let p denote the intersection point of uv and xy, see Fig. 1. The cosine force for v is defined as $F_{\cos}(v) = k_{\cos} \cdot \cos \alpha \cdot \overrightarrow{yx}$, where $k_{\cos}$ is a positive constant.

The force $F_{\text{cage}}(v)$ is a compound of two forces $F_{\text{cage}}(v, x)$ and $F_{\text{cage}}(v, y)$. Let l_{ab} denote the distance between two points a and b. Let $l_{vx}^{\star}$ be the length of the edge vx in a triangle vxp with side length l_{vp} and l_{xp}, and a right angle at the point p. Then, $F_{\text{cage}}(v, x) = k_{\text{cage}} \cdot \log(l_{vx}/l_{vx}^{\star})\overrightarrow{vx}$, where k_{cage} is positive constant. The force $F_{\text{cage}}(v, y)$ is defined symmetrically.

Again the force $F_{\text{ang}}(v)$ is a compound of the forces $F_{\text{ang}}(v, x)$ and $F_{\text{ang}}(v, y)$. Consider the unit vector a that is perpendicular to the bisector of $\overrightarrow{uv}$ and $\overrightarrow{yx}$, refer Fig. 1c. Further, let α' be the angle between the $\overrightarrow{uv}$ and $\overrightarrow{yx}$. Then the force $F_{\text{ang}}(v, x)$ is defined as $k_{\text{ang}} \cdot \text{sign}(\alpha' - \pi/2) \cdot |\pi/2 - \alpha'|/\alpha' \cdot a$ where k_{ang} is a positive constant. The force $F_{\text{ang}}(v, y)$ is defined correspondingly.

3   Multilevel Random Sampling

Our algorithm starts with a drawing Γ of a graph G and iteratively improves the crossing angle of Γ by moving a vertex to a better position, i.e., we locally optimize the crossing angle of the drawing; for pseudocode refer to Algorithm 1. For this purpose we greedily select a vertex v with a minimal crossing angle cr-$\alpha(\Gamma, v)$. More precisely, let e and e' be two edges with a minimal crossing angle in Γ. We set v randomly to be an endpoint of e and e'. We iteratively improve the crossing angle of v by sampling a set S of T points within a square R and by moving v to the position $p \in S$ that induces a maximal local crossing angle. We repeat this process $L \in \mathbb{N}^+$ times and decrease the size of R in each iteration.

More formally, denote by $\Gamma[v \mapsto p]$ the drawing obtained from Γ by moving v to the point $p = (p_x, p_y) \in \mathbb{R}^2$. Let $R^i(p) = [p_x - s \cdot b^i/2, p_y - s \cdot b^i/2] \times [p_x + s/2, p_y + s \cdot b^i/2] \subset \mathbb{R}^2$ be a square centered at the point p with a *scaling factor* $b \in (0,1)$ and *initial side length* $s > 0$. Let p^0 be the position of v in Γ and let $S^0 \subset R^0(p^0)$ be a set of T points in $R^0(p^0)$ chosen uniformly at random. Let p^i be a point in $S^{i-1} \cup \{p^{i-1}\}$ that maximizes cr-$\alpha(\Gamma[v \mapsto p^i], v)$. We obtain a new sample S^i by randomly selecting T points within the square $R^i(p^i)$. Since cr-$\alpha(\Gamma[v \mapsto p^i], v) = \max_{uv \in E, e \in E \setminus \{uv\}}$ cr-$\alpha(\Gamma[v \mapsto p^i], uv, e)$, the function can be evaluated in $O(\deg(v)|E|)$ time.

3.1   Fast Minimum Angle Computation

The running time of the random sampling approach relies on computing in each iteration a pair of edges creating the minimum crossing angle cr-$\alpha(\Gamma)$. More formally, we are looking for a pair of distinct edges $e, f \in E$ that have a minimal crossing angle in a straight-line drawing Γ, i.e., cr-$\alpha(\Gamma, e, f) = $ cr-$\alpha(\Gamma)$. The well known sweep-line algorithm [4] requires $O((n+k) \log(n+k))$ time to report all k intersecting edges in Γ. In general the number of intersecting edges can be $\Omega(m^2)$, but we are only interested in a single pair that forms the minimal crossing angle. Therefore, we propose an algorithm, which uses the slopes of the edges in Γ to rule out pairs of edges, which cannot form the minimum angle.

Assume that we already found two intersecting edges forming a small angle of size $\delta > 0$. We set $t := \lfloor \pi/\delta \rfloor$ and distribute the edges into t buckets $B_0, \ldots, B_{t-1}$ such that bucket B_i contains exactly the edges e with $i\pi/t \leq$ slope$(\Gamma, e) + \pi/2 < (i+1)\pi/t$. Then each bucket covers an interval of size $\pi/\lfloor \pi/\delta \rfloor \geq \delta$. Thus, if there exist edges e, f with cr-$\alpha(\Gamma, e, f) < \delta$, they belong to the same or to the adjacent buckets (modulo t). Overall, we consider all pairs of edges in $B_i \cup B_{i+1 \pmod{t}}$, $i = 1, \ldots t$, and find the pair forming the smallest crossing angle. To find this pair we could apply a sweep-line algorithm to the set $B_i \cup B_{i+1}$. In general this set can contain $\Omega(m)$ edges. Thus, in worst case we would not gain a speed up in comparison to a sweep-line algorithm applied to Γ. On the other hand, in practice we expect the number of edges in a bucket to be small. If we assume this number to be a constant, the overall running time of the exhaustive check is linear in m and does not depend on the number of crossings.

Implementation Details. In the case that the slopes in Γ are uniformly distributed, we expect the number of edges in a bucket to decrease with an decreasing estimate δ. We set the value δ to be the minimal crossing angle of the r longest edges in Γ. In our implementation we set r to be 50 if the graph contains at most 5000 edges, otherwise it is 300.

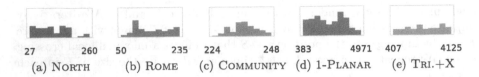

27   260   50   235   224   248   383   4971   407   4125
(a) NORTH   (b) ROME   (c) COMMUNITY (d) 1-PLANAR   (e) TRI.+X

Fig. 2. The distribution of the sum of number of vertices and edges per graph class. The plot is scaled such that a bar of full height would contain 40 graphs.

4   Experimental Evaluation

The RANDOM SAMPLING heuristic has several parameters which allow for many different configurations. In Sect. 4.4, we investigate the influence of the configuration on the crossing angle of the drawing computed by the random sampling approach. We investigate the question of whether the RANDOM SAMPLING approach improves the crossing angle of a given drawing. Our evaluation in Sect. 4.5 answers the question affirmatively. Moreover, we expect that the crossing angle of the drawing computed by the random sampling approach depends on the choice of the initial drawing. We show that this is indeed the case (Sect. 4.6). We close the evaluation with a short running time analysis in Sect. 4.7. Our evaluation is based on a selection of artificial and real world graphs (Sect. 4.1), several choices of the initial drawing, see Sect. 4.2, and a specific way to compare two drawing algorithms (Sect. 4.3).

Setup. All experiments were conducted on a single core of an AMD Opteron Processor 6172 clocked at 2.1 GHz. The server is equipped with 256 GB RAM. All algorithms were compiled with g++-4.8.5 with optimization mode -O3.

4.1   Benchmark Graphs

We evaluate the heuristic on the following graph classes, either purely synthetic or with a structure resembling real-world data. Figure 2 shows the size distribution of these graphs. The color of each class is used consistently throughout the paper.

Real World. The classes *Rome* and *North* (AT&T)[1] are the non-planar subsets of the corresponding well known benchmark sets, respectively. From each graph class we picked 100 graphs uniformly at random. The COMMUNITY graphs are generated with the LFR-GENERATOR [17] implemented in NETWORKIT [19]. These graphs resemble social networks with a community structure.

Artificial. For each artificial graph we picked the number n of vertices uniformly at random between 100 and 1000. The TRIANGULATION+X class contains randomly generated n-vertex triangulations with an additional set of x edges.

---

[1]   http://graphdrawing.org/data.html.

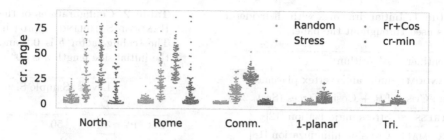

Fig. 3. Crossing angles of the initial drawings.

The number x is picked uniformly at random between $0.1n$ and $0.15n$. The endpoints of the additional edge are picked uniformly at random, as well.

The class 1-PLANAR consists of graphs that admit drawings where every edge has at most one crossing. We used a *geometric* and *topological* procedure to generate these graphs. For the former consider a random point set P of n points. Let $e_1, \ldots e_k$ be a random permutation of all pairs of points in P. Let $G_0 = (P, \emptyset)$. If the drawing $G_{i-1} + e_i$ induced by P is simple and 1-planar, we define G_i to be this graph, otherwise we set $G_i = G_{i-1}$. We construct the topological 1-PLANAR graphs based on a random planar triangulation G generated with OGDF [5]. Let v be a random vertex of G and let v, x, u, y be an arbitrary 4-cycle. We add uv to G if $G + uv$ is 1-planar. The process is repeated x times, for a random number $x \in [0.3n, 0.4n]$. In contrast to the experimental work on crossing minimization in book embeddings [16], we did not observe that our heuristic performs differently on the topological and geometric 1-PLANAR graphs. Hence, we merge the two classes into a single class. Thus, in total the 1-PLANAR graphs contain 200 graphs compared to 100 in the other graph classes.

4.2   Initial Drawings

In our evaluation we consider four initial drawings of each benchmark graph; refer to Table 1. A random point set P of size n induces a RANDOM drawing of an n-vertex graph. The FR+COS drawings are generated by applying our implementation of the force-directed method of Fruchtermann and Reingold [11] to the RANDOM drawings with the additional $F_{\cos}$ force (Sect. 2.1). We applied the stress majorization [5,12] implementation of the OPEN GRAPH DRAWING FRAMEWORK (OGDF) to RANDOM in order to obtain the STRESS drawings. The CR-SMALL drawings are computed with the heuristic introduced by Rader-macher et al. [18] in order to decrease the number of crossings in straight-line drawings. They showed that the heuristic computes drawings with significantly less crossings than drawings computed by stress majorization. Unfortunately, within a feasible amount of time we were not able to compute CR-SMALL draw-ing for the classes 1-PLANAR and TRIANGULATION+X.

A point in Fig. 3 corresponds to the crossing angle of an initial drawing. The plot is categorized by graph class. The RANDOM drawings have the smallest

Table 1. Initial drawings with their identi-
fiers used throughout the paper.

Identifier	Algorithm
RANDOM	uni. rand. vertex placement
FR+COS	FR + Cosine forces (Sect. 2.1)
STRESS	Stress majorization [12]
CR-SMALL	Crossing minimization [18]

Table 2. Configurations of the
RANDOM SAMPLING approach.
The scaling factor b is 0.2 and
the initial side length s is 10^5.

	Levels	Sample Size
	L	T
SLOPPY	3	50
MEDIUM	4	175
PRECISE	5	400

crossing angles. The STRESS drawings have a larger crossing angle than CR-SMALL and overall, FR+COS drawings tend to have the largest crossing angle.

We point out that in contrast to the evaluation of Argyriou et al. [2], our implementation of the force-directed method with F_{cage} and F_{ang} produces drawings with smaller crossing angles than with F_{cos}. Thus, we do not consider these drawings in our evaluation.

4.3   Differences Between Paired Drawings

In order to compare the performance of two algorithms on multiple graphs and to investigate by how much one of the algorithms outperforms the other, we employ the following machinery. We denote by $\Gamma\{G\}$ the set of all drawings of G. Let $\mathcal{G} = \{G_1, G_2, \ldots, G_k\}$ be a family of (non-planar) graphs. We refer to a set $\Lambda = \{\Gamma_1, \ldots, \Gamma_k\}$ as a *family of drawings* of $\mathcal{G}$ where $\Gamma_i \in \Gamma\{G_i\}$. Let Λ^1 and Λ^2 be two families of drawings of $\mathcal{G}$. Let $\mathcal{F}$ be a subset of $\mathcal{G}$. We say Λ^1 *outperforms* Λ^2 *on* $\mathcal{F}$ if and only if for all $G_i \in \mathcal{F}$ the inequality cr-$\alpha(\Gamma_i^1) >$ cr-$\alpha(\Gamma_i^2)$ holds. If Λ^1 outperforms Λ^2 on $\mathcal{F}$ then Λ^1 has an *advantage of* $\Delta > 0$ *on* $\mathcal{F}$ if for all $G_i \in \mathcal{F}$ the inequality cr-$\alpha(\Gamma_i^1) >$ cr-$\alpha(\Gamma_i^2) + \Delta$ holds. For a finite set $\mathcal{G}$, we say $\mathcal{F}$ has *relative size at least* $p \in [0,1]$ if $|\mathcal{F}| \geq p \cdot |\mathcal{G}|$.

In order to compare two families of drawings we plot the advantage as a function of p; refer to Fig. 5. For each value p the plot contains 5 five bars, each corresponding to a graph class. The height of the bars correspond to advantages Δ for a set of relative size p. A caption of a figure in the form of A vs B indicates that if Δ is positive, B has advantage Δ over A. Correspondingly, if Δ is negative, A has an advantage of $-\Delta$ over B. Thus, Fig. 5 shows that for $p = 0.1$, for each graph class there is a subset $\mathcal{F}$ of relative size 0.1, i.e., $\mathcal{F}$ contains at least 10 graphs, such that the set SLOPPY has an advantage of Δ over PRECISE on $\mathcal{F}$. In greater detail, SLOPPY has an advantage of 7.9° over PRECISE on the NORTH graphs, 12.9° on the ROME graphs, 11.5° on the COMMUNITY graphs, 1.2° on the 1-PLANAR graphs and 1.2° on the TRIANGULATION+X graphs. On the other side, PRECISE has an advantage of 12.9° over SLOPPY on at least 10 NORTH graphs, 15.7° on the ROME graphs, 13.8° on the COMMUNITY graphs, 1.1° on the 1-PLANAR graphs and 0.4° on the TRIANGULATION+X graphs. Note that only for $p < 0.5$ there can be two disjoint subsets $\mathcal{F}_1, \mathcal{F}_2$ of a graph class

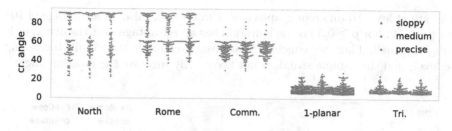

Fig. 4. Performance of different configurations

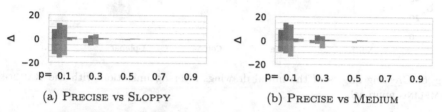

(a) PRECISE vs SLOPPY   (b) PRECISE vs MEDIUM

Fig. 5. Comparison of the SLOPPY configuration to the MEDIUM and PRECISE configuration. The colors indicate the graph as indicated by Fig. 2.

of relative size p such that PRECISE has an advantage over SLOPPY on $\mathcal{F}_1$ and SLOPPY has an advantage over PRECISE on $\mathcal{F}_2$.

4.4 Parametrization of the Random Sampling Approach

The RANDOM SAMPLING approach introduced in Sect. 3 has four different parameters, the number of levels L, the size of the sample T, the initial side length s and the scaling factor b, that allows for many different configurations. With an increasing number T of samples, we expect to obtain a larger crossing angle in each iteration to the cost of an increasing running time. If we allow each configuration the same running time, it is unclear whether it is beneficial to increase the number of iterations or to increase the number of samples (T) and levels (L) per iteration. This motivates the following question: does the crossing angle of a drawing of an n-vertex graph computed by the random sampling approach within a given time limit t_n increase with an increasing number of samples and levels? We choose to set the time limit t_n to n seconds. This allows for at least $1.6 \cdot n$ iterations on our benchmark instances. Since the parametrization space is infeasibly large, we evaluate three exemplary configurations, SLOPPY, MEDIUM and PRECISE; see Table 2.

The plot in Fig. 4 does not indicate that the distributions of the crossing angle differ across different configurations significantly . With the plot in Fig. 5 we can confirm this observation. For each configuration there is only a small subset of each class such that the configuration has an advantage over the other configurations. For example, for the ROME graphs there exist at least 10 graphs such that SLOPPY has an advantage of $10°$ over PRECISE. On the other hand, there are at least 10 different graphs such PRECISE has also an advantage of 10° over SLOPPY. For $p \geq 0.5$ no configuration has an advantage over the other, or it negligibly small. Thus, we conclude that given a common time limit, increasing the levels and the sample size does not necessarily increase the crossing angle.

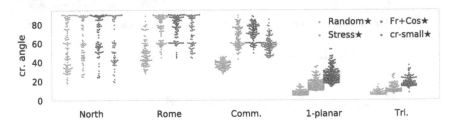

Fig. 6. Crossing angles of the initial drawings after optimization with the RANDOM SAMPLING approach.

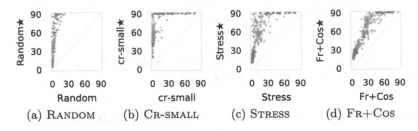

(a) RANDOM   (b) CR-SMALL   (c) STRESS   (d) FR+COS

Fig. 7. Initial crossing angle vs the final crossing angle. The plots show the crossing angles of the classes NORTH, COMMUNITY and 1-PLANAR.

4.5 Improvement of the Crossing Angles

In this section we investigate whether the RANDOM SAMPLING approach is able to improve the crossing angle of a given drawing within $2n$ iterations. Given the same number of iterations, it is most-likely that we obtain a larger crossing angle of a drawing if we increase the number of samples. Thus, we use the PRECISE configuration for the evaluation of the above question. We refer to the drawings after the application of the RANDOM SAMPLING approach as RANDOM*, FR+COS*, STRESS* and CR-SMALL*, respectively.

The plots in Figs. 3 and 6 indicate that the RANDOM SAMPLING approach indeed improves the crossing angle of the initial drawings. Figure 7 shows the relationship between the crossing angle of the initial drawing and the final drawing. For the purpose of clarity, the plot only shows drawings of the classes NORTH, COMMUNITY and 1-PLANAR. The plots shows that the RANDOM SAMPLING approach considerably improves the crossing angle of the initial drawing. In case of the NORTH graphs there are a few graphs that have an improvement of at least

70°. There are at least 10 drawings in RANDOM whose crossing angle is improved by at least 75° . For all real world graph classes and all initial layouts there are 70 graphs in each class, such that the final drawing has an advantage of over 25°.

For TRIANGULATION+X, FR+COS* has an advantage of at least 11° over FR+COS on at least 90 TRIANGULATION+X. For the remaining initial layouts the corresponding advantage is at most 7.6°. Considering the 1-PLANAR graphs the corresponding advantages are 14° and 9.7°. This indicates that within $2n$ iterations a large initial crossing angle helps to further improve the crossing angle of 1-PLANAR and TRIANGULATION+X graphs. Overall we observe that the 1-PLANAR and TRIANGULATION+X classes are rather difficult to optimize. This can either be a limitation of our heuristic or the crossing angle of these graphs are indeed small. Unfortunately, we are not aware of meaningful upper and lower bounds on the crossing angle of straight-line drawing of these graphs. Nevertheless, we can conclude that our heuristic indeed improves the initial crossing angle. To which extend our heuristic is able to increase crossing angle of a drawing depends on the graph class and on the initial drawing itself.

4.6   Effect of the Initial Drawing

The RANDOM SAMPLING approach iteratively improves the crossing angle of a given drawing. Given a different drawing of the same graph the heuristic might be able to compute a drawing with a larger crossing angle. Hence, we investigate whether the choice of the initial drawing influences the crossing angle of a drawing obtained by the RANDOM SAMPLING approach with $2n$ iterations.

For all graph classes, except from NORTH, it is apparent from Fig. 6 that the drawings in the set RANDOM* have noticeably smaller crossing angles compared to the remaining drawings. This meets our expectations, since the initial RANDOM drawings presumably has many crossings [15] and thus is likely to have many small crossing angles; compare the initial crossing angles plotted in Fig. 3.

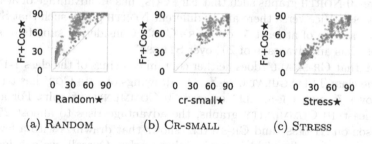

(a) RANDOM   (b) CR-SMALL   (c) STRESS

Fig. 8. Comparison of the initial layout.

The plot in Fig. 6 suggests that the set FR+COS* contains drawings with the largest crossing angles. In order to corroborate this claim, Fig. 8 shows crossing angles obtained by different algorithms. It shows that except for one graph, each drawing in FR+COS* has a larger crossing angle than the corresponding drawing in RANDOM*. Figure 8b and c suggest that FR+COS* overall contains more drawings with a larger crossing angle compared to STRESS* and CR-SMALL*. With the help of Fig. 9 we are able to quantify the number of graphs above the diagonal and the difference of the crossing angles. Figure 9a shows that for the graph classes 1-PLANAR and TRIANGULATION+X, there are each at least 90 of 100 graphs whose drawings in FR+COS* have a crossing angle larger then the corresponding drawing in STRESS*, i.e., FR+COS* has an advantage of 4.5° degrees over STRESS*.

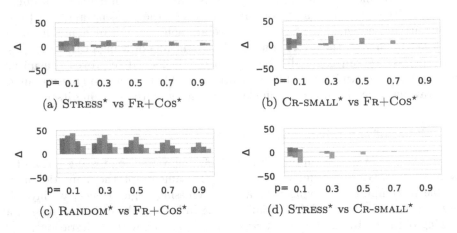

(a) STRESS* vs FR+COS*   (b) CR-SMALL* vs FR+COS*

(c) RANDOM* vs FR+COS*   (d) STRESS* vs CR-SMALL*

Fig. 9. Comparison of the crossing angle of the final drawings.

There are at least 50 1-PLANAR graphs such that the FR+COS* has an advantage of 10° over STRESS*. At least 50 COMMUNITY graphs have drawings in FR+COS* with an advantage of 5° over the corresponding drawings in STRESS*. There are 10 NORTH graphs such that FR+COS* has an advantage of at least 5° over STRESS*. Vice versa there are 10 different NORTH graph such that STRESS* has an advantage of at least 5° over FR+COS*. Considering subsets of size 10, FR+COS* has an advantage of 20° over STRESS*.

Recall that CR-SMALL* does neither contain drawings of the class 1-PLANAR nor of the class TRIANGULATION+X. The drawings of FR+COS* has an advantage of over 7° over CR-SMALL* on over 70 COMMUNITY graphs. For a subset with at least 10 COMMUNITY graphs, the advantage rises to almost 25°. The comparison on STRESS* and CR-SMALL* shows that drawings with a few crossings do not necessarily yield larger crossing angles. Overall, we conclude that the RANDOM SAMPLING approach computes the largest crossing angle when applied to the FR+COS drawings. This is plausible, since the crossing angles of the initial crossing angles are already good. As shown in the previous section, depending on the graph class, there is a large improvement in the crossing angle, if we start with such an initial drawing. In further investigations we were able to show that the advantages of FR+COS* decreases comparably to STRESS* with

$4n$ iterations. However, doubling the iterations does not entirely cover the gap between the crossing angles of the initial drawings.

4.7   Note on the Running Time

In this section we shortly evaluate the running time of our algorithm on all our graphs. For this purpose, we applied two implementations of the RANDOM SAMPLING heuristic to the RANDOM drawings. The SWEEP implementation uses a sweep-line algorithm to compute the pair of crossing edges that create the smallest crossing. BUCKET uses the algorithm described in Sect. 3.1. We employ the speed-up technique only for graphs with at least 1000 edges, we refer to these graphs as *large*. Figure 10 plots the running time per iteration for n-vertex graphs. The median and the second 3-quantile of the running time on the large graphs are highlighted. BUCKET has an average running time of 391 ms per iteration on the large graphs and SWEEP has an average running time of 500 ms. On all graph BUCKET requires on average 328 ms per iteration.

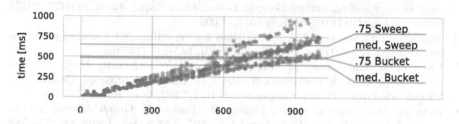

Fig. 10. Average Running time per iteration vs the number of vertices.

5   Conclusion

We designed and evaluated a simple heuristic to increase the crossing angle in a straight-line drawing of a graph. On real world networks our heuristic is able to compute larger crossing angles than on artificial networks. This can either be a limitation of our heuristic or the crossing angle of our artificial graph classes are small. We are not aware of lower and upper bounds of the crossing

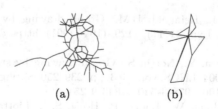

(a)   (b)

Fig. 11. (a) STRESS drawing of a Rome graph. (b) Drawing after optimizing the crossing angle. The ratio between the longest and shortest edges is large.

angle of these graphs. Thus, investigating such bounds of the 1-PLANAR and TRIANGULATION+X graphs is an interesting theoretical question.

Figure 11 shows that our heuristic does not necessarily compute readable drawings. Nevertheless, parts of the RANDOM SAMPLING heuristic are easily exchangeable. For example, the objective function can be replaced by a linear combination of number of crossing and the crossing angle. Thus, future work can be concerned with adapting the RANDOM SAMPLING approach with the aim to compute readable drawings.

References

1. Argyriou, E.N., Bekos, M.A., Symvonis, A.: The straight-line RAC drawing problem is NP-hard. J. Graph Algorithms Appl. **16**(2), 569–597 (2012). https://doi.org/10.7155/jgaa.00274
2. Argyriou, E.N., Bekos, M.A., Symvonis, A.: Maximizing the total resolution of graphs. Comput. J. **56**(7), 887–900 (2013). https://doi.org/10.1093/comjnl/bxs088
3. Arikushi, K., Fulek, R., Keszegh, B., Morić, F., Tóth, C.D.: Graphs that admit right angle crossing drawings. Comput. Geom. Theory Appl. **45**(4), 169–177 (2012). https://doi.org/10.1016/j.comgeo.2011.11.008
4. Bentley, J.L., Ottmann, T.A.: Algorithms for reporting and counting geometric intersections. IEEE Trans. Comput. **C–28**(9), 643–647 (1979)
5. Chimani, M., Gutwenger, C., Jünger, M., Klau, G.W., Klein, K., Mutzel, P.: The Open Graph Drawing Framework (OGDF). In: Handbook of Graph Drawing and Visualization, pp. 543–569. Chapman and Hall/CRC, Boca Raton (2013)
6. Devanny, W., Kindermann, P., Löffler, M., Rutter, I.: Graph drawing contest report. In: Frati, F., Ma, K.-L. (eds.) GD 2017. LNCS, vol. 10692, pp. 575–582. Springer, Cham (2018). https://doi.org/10.1007/978-3-319-73915-1_44
7. Didimo, W., Eades, P., Liotta, G.: Drawing graphs with right angle crossings. Theor. Comput. Sci. **412**(39), 5156–5166 (2011). https://doi.org/10.1016/j.tcs.2011.05.025
8. Didimo, W., Liotta, G.: The crossing-angle resolution in graph drawing. In: Pach, J. (ed.) Thirty Essays on Geometric Graph Theory, pp. 167–184. Springer, New York (2013). https://doi.org/10.1007/978-1-4614-0110-0_10
9. Dujmović, V., Gudmundsson, J., Morin, P., Wolle, T.: Notes on large angle crossing graphs. In: Viglas, T., Potanin, A. (eds.) Proceedings of the 16th Symposium on Computing: The Australasian Theory (CATS 2010), pp. 19–24. Australian Computer Society (2010)
10. Eades, P.: A heuristic for graph drawing. Congressus Numerantium **42**, 149–160 (1984)
11. Fruchterman, T.M.J., Reingold, E.M.: Graph drawing by force-directed placement. Softw.: Pract. Exp. **21**(11), 1129–1164 (1991). https://doi.org/10.1002/spe.4380211102
12. Gansner, E.R., Koren, Y., North, S.: Graph drawing by stress majorization. In: Pach, J. (ed.) GD 2004. LNCS, vol. 3383, pp. 239–250. Springer, Heidelberg (2005). https://doi.org/10.1007/978-3-540-31843-9_25
13. Giacomo, E.D., Didimo, W., Eades, P., Hong, S.H., Liotta, G.: Bounds on the crossing resolution of complete geometric graphs. Discret. Appl. Math. **160**(1), 132–139 (2012). https://doi.org/10.1016/j.dam.2011.09.016

14. Huang, W., Eades, P., Hong, S.H., Lin, C.C.: Improving force-directed graph drawings by making compromises between aesthetics. In: Hundhausen, C.D., Pietriga, E., Díaz, P., Rosson, M.B. (eds.) Proceedings of the IEEE Symposium on Visual Languages and Human-Centric Computing (VL/HCC2010), pp. 176–183. IEEE Computer Society (2010). https://doi.org/10.1109/VLHCC.2010.32
15. Huang, W., Huang, M.: Exploring the relative importance of crossing number and crossing angle. In: Wang, H., Yuan, X., Tao, L., Chen, W. (eds.) Proceedings of the 3rd International Symposium on Visual Information Communication (VINCI 2010), pp. 10:1–10:8. ACM (2010). https://doi.org/10.1145/1865841.1865854
16. Klawitter, J., Mchedlidze, T., Nöllenburg, M.: Experimental evaluation of book drawing algorithms. In: Frati, F., Ma, K.-L. (eds.) GD 2017. LNCS, vol. 10692, pp. 224–238. Springer, Cham (2018). https://doi.org/10.1007/978-3-319-73915-1_19
17. Lancichinetti, A., Fortunato, S., Radicchi, F.: Benchmark graphs for testing community detection algorithms. Phys. Rev. E **78**, 046110 (2008). https://doi.org/10.1103/PhysRevE.78.046110
18. Radermacher, M., Reichard, K., Rutter, I., Wagner, D.: A geometric heuristic for rectilinear crossing minimization. In: Pagh, R., Venkatasubramanian, S. (eds.) Proceedings of the 20th Workshop on Algorithm Engineering and Experiments (ALENEX 2018), pp. 129–138 (2018). https://doi.org/10.1137/1.9781611975055.12
19. Staudt, C.L., Sazonovs, A., Meyerhenke, H.: NetworKit: a tool suite for large-scale complex network analysis. Netw. Sci. **4**(4), 508–530 (2016). https://doi.org/10.1017/nws.2016.20

Contact Representations

Recognition and Drawing of Stick Graphs

Felice De Luca[1]([✉]), Md Iqbal Hossain[1], Stephen Kobourov[1], Anna Lubiw[2], and Debajyoti Mondal[3]

[1] Department of Computer Science, University of Arizona, Tucson, USA
{felicedeluca,hossain}@email.arizona.edu, kobourov@cs.arizona.edu
[2] Cheriton School of Computer Science, University of Waterloo, Waterloo, Canada
alubiw@uwaterloo.ca
[3] Department of Computer Science, University of Saskatchewan, Saskatoon, Canada
dmondal@cs.usask.ca

Abstract. A *Stick graph* is an intersection graph of axis-aligned segments such that the left end-points of the horizontal segments and the bottom end-points of the vertical segments lie on a "ground line", a line with slope -1. It is an open question to decide in polynomial time whether a given bipartite graph G with bipartition $A \cup B$ has a Stick representation where the vertices in A and B correspond to horizontal and vertical segments, respectively. We prove that G has a Stick representation if and only if there are orderings of A and B such that G's bipartite adjacency matrix with rows A and columns B excludes three small 'forbidden' submatrices. This is similar to characterizations for other classes of bipartite intersection graphs.

We present an algorithm to test whether given orderings of A and B permit a Stick representation respecting those orderings, and to find such a representation if it exists. The algorithm runs in time linear in the size of the adjacency matrix. For the case when only the ordering of A is given, we present an $O(|A|^3|B|^3)$-time algorithm. When neither ordering is given, we present some partial results about graphs that are, or are not, Stick representable.

1 Introduction

Let $\mathcal{O}$ be a set of geometric objects in the Euclidean plane. The *intersection graph* of $\mathcal{O}$ is a graph where each vertex corresponds to a distinct object in $\mathcal{O}$, and two vertices are adjacent if and only if the corresponding objects intersect. Recognition of intersection graphs that arise from different types of geometric objects such as segments, rectangles, discs, intervals, etc., is a classic problem in combinatorial geometry. Some of these classes, such as interval graphs [2], can be recognized in polynomial-time, whereas many others are NP-hard [4,25,27]. There are many beautiful results that characterize intersection classes in terms of a vertex ordering without certain forbidden patterns, and recently, Hell *et al.* [20] unified many previous results by giving a general polynomial time recognition algorithm for all cases of small forbidden patterns.

© Springer Nature Switzerland AG 2018
T. Biedl and A. Kerren (Eds.): GD 2018, LNCS 11282, pp. 303–316, 2018.
https://doi.org/10.1007/978-3-030-04414-5_21

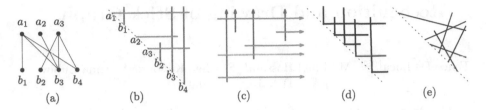

a_1 a_2 a_3

b_1 b_2 b_3 b_4

(a)   (b)   (c)   (d)   (e)

Fig. 1. (a) A bipartite graph $G = (A \cup B, E)$. (b) A Stick representation of G. (c)–(e) Illustration for different types of intersection representations. (c) A *2DOR* representation. (d) A *Hook* representation. (e) A grounded segment representation.

In this paper we study a class of bipartite intersection graphs called *Stick graphs*. A *Stick graph* is an intersection graph of axis-aligned segments with the property that the left end-points of horizontal segments and the bottom end-points of vertical segments all lie on a *ground line*, ℓ, which we take, without loss of generality, to be a line of slope -1. See Fig. 1(a)–(b). It is an open problem to recognize Stick graphs in polynomial time [8].

Stick graphs lie between two well-studied classes of bipartite intersection graphs. First of all, they are a subset of the *grid intersection graphs* (GIG) [19]—intersection graphs of horizontal and vertical segments in the plane—which are NP-complete to recognize [25]. When all the horizontal segments extend rightward to infinity and the vertical segments extend upward to infinity, we obtain the subclass of *2-directional orthogonal ray* (2DOR) graphs (e.g., see Fig. 1(c)), which can be recognized in polynomial time [30]. It is easy to show that every 2DOR graph is a Stick graph—truncate each ray at a ground line placed above and to the right of every intersection point (and then flip the picture upside-down). Thus the class of Stick graphs lies strictly between these two classes.

What the two classes (GIG and 2DOR) have in common is a nice characterization in terms of vertex orderings. A bipartite graph G with vertex bipartition $A \cup B$ can be represented as a *bipartite adjacency matrix*, $M(G)$ with rows and columns corresponding to A and B, respectively, and a 1 in row i, column j, if (i, j) is an edge. Both GIG graphs and 2DOR graphs can be characterized as graphs G for which $M(G)$ has a row and column ordering without certain 'forbidden' submatrices. (Details below.) Many other bipartite intersection graphs can be similarly characterized in terms of forbidden submatrices, see [24].

One of our main results is a similar characterization of Stick graphs. Specifically, we will prove that a bipartite graph G with vertex bipartition $A \cup B$ has a Stick representation with vertices of A corresponding to horizontal segments and vertices of B corresponding to the vertical segments if and only if there is an ordering of A and an ordering of B such that $M(G)$ has no submatrix of the following form, where $*$ stands for either 0 or 1:

$$\begin{bmatrix} * & 1 & * \\ * & 0 & 1 \\ 1 & * & * \end{bmatrix} \quad \begin{bmatrix} 1 & * \\ 0 & 1 \\ 1 & * \end{bmatrix} \quad \begin{bmatrix} * & 1 & * \\ 1 & 0 & 1 \end{bmatrix}$$

Although this characterization does not (yet) give us a polynomial time algorithm to recognize Stick graphs, it allows us to make some progress. Given a bipartite graph G with vertex bipartition $A \cup B$, we want to know if G has a Stick representation with A and B corresponding to horizontal and vertical segments, respectively. It is easy to show that a solution to this problem is completely determined by a total ordering σ of the vertices of G corresponding to the order (from left to right) in which the segments touch the ground line. A natural way to tackle the recognition of Stick graphs is as a hierarchy of problems, each (possibly) more difficult than the next:

(i) **Fixed As and Bs**: In this case an ordering, σ_a, of the vertices in A and an ordering, σ_B, of the vertices in B are given, and the output ordering σ must respect these given orderings. Because of our forbidden submatrix characterization, this problem can be solved in polynomial time.

(ii) **Fixed As**: In this case only the ordering σ_A is given.

(iii) **General Stick graphs**: In this case, neither σ_A nor σ_B is given, i.e., there is no restriction on the ordering of the vertices.

Our Results: We give an algorithm with run-time $O(|A||B|)$ for problem (i). This is faster than naively looking for the forbidden submatrices. (And in fact, we use our algorithm to prove the forbidden submatrix characterization). Furthermore, the algorithm will find a Stick representation when one exists.

We give an algorithm for problem (ii) with run time $O(|A|^3|B|^3)$ that uses the forbidden submatrix characterization and reduces the problem to 2-Satisfiability. For problem (iii), recognizing Stick graphs, we give some conditions that ensure a graph is a Stick graph, and some conditions that ensure a graph is not a Stick graph.

Related Work: We now review the research related to the recognition of intersection graphs, in particular those that are bipartite.

Interval graphs, i.e., intersection graph of horizontal intervals on the real line, can be recognized in linear time [2, 13]. Bipartite interval graphs with a fixed bipartition are known as *interval bigraphs* (IBG) [14, 28], and can be recognized in polynomial time [28]. In contrast to the interval graphs, no linear-time recognition algorithm is known for IBG.

Many bipartite graph classes have been characterized in terms of forbidden submatrices of the graph's bipartite adjacency matrix, and a rich body of research examines when the rows and columns of a matrix can be permuted to avoid forbidden submatrices [24]. For example, a graph G is chordal bipartite if and only if $M(G)$ can be permuted to avoid the matrix γ_1 in Fig. 2 [24], which led to a polynomial-time algorithm [26]. G is a bipartite permutation graph if and only if $M(G)$ can be permuted to avoid γ_1, γ_2, and γ_3 [11].

A graph is a *two-directional orthogonal ray* (2DOR) graph if it admits an intersection representation of upward and rightward rays [30,31]. A graph is a 2DOR graph if and only if its incidence matrix admits a permutation of its rows and columns that avoids γ_1 and γ_2 [30].

$$\gamma_1 = \begin{bmatrix} 1 & 0 \\ 1 & 1 \end{bmatrix} \quad \gamma_2 = \begin{bmatrix} 1 & 0 \\ 0 & 1 \end{bmatrix} \quad \gamma_3 = \begin{bmatrix} 1 & 1 \\ 0 & 1 \end{bmatrix} \quad \gamma_4 = \begin{bmatrix} 1 & 0 & 1 \\ * & 1 & * \end{bmatrix} \quad \gamma_5 = \begin{bmatrix} * & 1 & * \\ 1 & 0 & 1 \\ * & 1 & * \end{bmatrix}$$

Fig. 2. Forbidden submatrices, where $*$ stands for either 0 or 1.

There is a linear-time algorithm to recognize 2DOR graphs [12,30]. If there are 3 or 4 allowed directions for the rays, then the graphs are called 3DOR or 4DOR graphs, respectively. Felsner *et al.* [16] showed that if the direction (right, left, up, or down) for each vertex is given, then the existence of a 4DOR representation respecting the given directions can be decided in polynomial time. If the horizontal elements are segments and the vertical elements are rays, then the corresponding intersection graphs are called *SegRay* graphs [7–9,23]. A graph G is a SegRay graph if and only if $M(G)$ can be permuted to avoid γ_4 [10].

The time-complexity questions for 3DOR, 4DOR and SegRay are all open.

The class of *segment graphs* contains the graphs that can be represented as intersections of segments (with arbitrary slopes and intersection angles). Every planar graph has a segment intersection representation [6]. Restricting to axis-aligned segments gives rise to *grid intersection graphs* (GIG) [25]. A bipartite graph is a GIG graph if and only if its incidence matrix admits a permutation of its rows and columns that avoids γ_5 [19]. If all the segments must have the same length, then the graphs are known as *unit grid intersection graphs* (UGIG) [27]. The recognition problem is NP-complete for both GIG [25] and UGIG [27]. We note that 4DOR is a subset of UGIG but Stick is not [8].

Researchers have examined further restrictions on GIG. For example, the graphs that admit a GIG representation with the additional constraint that all the segments must intersect (or be "stabbed by") a ground line form the *stabbable grid intersection* (StabGIG) graph class [8].

Another class of intersection graphs that restricts the objects on a ground line is defined in terms of *hooks*. A *hook* consists of a center point on the ground line together with an incident vertical segment and horizontal segment above the ground line. *Hook* graphs are intersection graphs of hooks [5,21,32], e.g., see Fig. 1(d). Hook graphs are also known as *max point-tolerance graphs* [5] and *heterozygosity graphs* [18]. The bipartite graphs that admit a Hook representation are called BipHook [8]. The complexities of recognizing the classes StabGIG, BipHook, and Stick are all open [8]. Chaplick *et al.* [8] examined the containment relations of these graph classes.

Grounded segment representations are a generalization of Stick representations, where the segments can have arbitrary slopes, e.g., see Fig. 1(e). Note that the segments are still restricted to lie on the same side of the ground line. Cardinal *et al.* [4] showed that the problem of deciding whether a graph admits a grounded segment representation is $\exists\mathbb{R}$-complete. We refer to [3,4] for other related classes such as outersegment and outerstring graphs, and for the study of their containment relations.

The following table summarizes the time complexities of recognizing different classes of bipartite intersection graphs, where n and m are the sizes of the two vertex sets of the bipartition.

Graph class	Time complexity	Ref
Chordal Bipartite Graphs	$O((n + m)^2)$, or $|E| \log(n + m)$	[26,33]
Bipartite Permutation Graphs	$O(nm)$-time	[34]
2-Directional Ray Graphs (2DOR)	$O(nm)$-time	[12,30]
3- or 4-Directional Ray Graphs (3DOR, 4DOR)	Open	[12,30]
4-DOR with given directions for vertices	$f(n, m)$-time[a]	[16]
3-DOR with a given bipartition $(A \cup B)$, and an ordering for As, i.e., vertical rays	$O((n + m)^2)$-time	[16]
Grid Intersection Graphs (GIG)	NP-complete	[25]
Unit Grid Intersection Graphs (UGIG)	NP-complete	[27]
Grounded Segment Intersection Graphs	$\exists \mathbb{R}$-complete	[4]
StabGIG, SegRay, Hook, BipHook and Stick Graphs	Open	[8,22]

[a]Multiplication time for two $(n + m) \times (n + m)$ matrices

2   Fixed As and Bs

In this section we study Stick representations of graphs with a fixed bipartition of the vertices and fixed vertex orderings for each vertex set. We call this problem Stick$_{AB}$, defined formally as follows.

Problem: STICK REPRESENTATION WITH FIXED AS AND BS (STICK$_{AB}$)
Input: A bipartite graph $G = (A \cup B, E)$, an ordering σ_A of the vertices in A, and an ordering σ_B of the vertices in B.
Question: Does G admit a Stick representation such that the ith horizontal segment on the ground line ℓ corresponds to the ith vertex of σ_A and the jth vertical segment on ℓ corresponds to the jth vertex of σ_B?

We first present an $O(|A||B|)$-time algorithm for Stick$_{AB}$. A Stick representation is totally determined by the order σ of the segments' intersection with the ground line (details in the proof of Lemma 1). Thus the idea of the algorithm is to impose some ordering constraints between the vertices of A and B based on some submatrices of the adjacency matrix of G. We show that the required

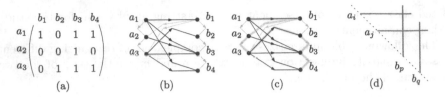

Fig. 3. (a) The incidence matrix M for the graph of Fig. 1(a). (b) The directed graph H. (c) A total order $(a_1, b_1, a_2, a_3, b_2, b_3, b_4)$ of the vertices in H. The corresponding Stick drawing is in Fig. 1(b). (d) Illustration of a forbidden ordering if $m_{j,p} = 0$.

Stick representation exists if and only if there exists a total order σ of $(A \cup B)$ that satisfies the constraints and preserves the given orderings σ_A and σ_B. We now describe the details.

Assume that $\sigma_A = (a_1, \ldots, a_n)$ and $\sigma_B = (b_1, \ldots, b_m)$. Let M be the ordered bipartite adjacency matrix of A and B, i.e., M has rows $a_1, \ldots, a_n$ and columns $b_1, \ldots, b_m$, where the entry $m_{i,p}$, i.e., the entry at the ith row and pth column, is 1 or 0 depending on whether a_i and b_p are adjacent or not, as illustrated in Fig. 3(a).

We start with the constraints $a_{i-1} \prec a_i$, where $2 \le i \le n$, and $b_{p-1} \prec b_p$, where $2 \le p \le m$ to enforce the given orderings σ_A and σ_B. We now add some more constraints, as follows.

C_1: If an entry $m_{i,p}$ is 1, then add the constraint $a_i \prec b_p$, e.g., see the black edges in Fig. 3(b).

C_2: If M contains an ordered submatrix $\begin{matrix} & b_p\ b_q \\ a_i \\ a_j \end{matrix} \begin{bmatrix} 1 & * \\ 0 & 1 \end{bmatrix}$, then add the constraint

$b_p \prec a_j$. For example, see the gray edges in Fig. 3(b).

We now test whether the set of constraints is consistent. Consider a directed graph H with vertex set $(A \cup B)$, where each constraint corresponds to a directed edge (Fig. 3(b)). Then the set of constraints is consistent if and only if H is acyclic, and the following lemma claims that this occurs if and only if the graph admits a Stick representation.

Lemma 1. *G admits a Stick representation respecting σ_A and σ_B if and only if H is acyclic, i.e., the constraints are consistent.*

Proof. We first show that the constraints are necessary. Every constraint between two vertices of the same set is implied by σ_A or σ_B. For Condition C_1, observe that a horizontal segment a_i can intersect a vertical segment b_p only if a_i precedes b_p, i.e., we must have $a_i \prec b_p$. For Condition C_2, we already have $a_i \prec a_j, b_p \prec b_q, a_i \prec b_p, a_j \prec b_q$. If we assume that $a_j \prec b_p$, then we have $a_i \prec a_j \prec b_p \prec b_q$, and to reach the vertical segment b_q, a_j would intersect b_p. Since $m_{j,p} = 0$, this intersection is forbidden. Figure 3(d) illustrates this scenario. Therefore, we must have the constraint $b_p \prec a_j$. Since all the constraints are necessary, if G admits an intersection representation, then the set of constraints is consistent.

We now prove the converse. Suppose the set of constraints is consistent. Take a total order of $A \cup B$ which is consistent with all the constraints, e.g., see Fig. 1(c). This is a "topological order" of H. Initiate the drawing of the corresponding orthogonal segments in this order on the ground line ℓ. This determines the y-coordinate of every $a \in A$ and the x-coordinate of every $b \in B$. For each vertex $a \in A$, let $\max_B(a)$ be the neighbor of a in G with the largest index. We extend the horizontal segment corresponding to a to the right until the x-coordinate of $\max_B(a)$. Similarly, for each vertex $b \in B$, let $\min_A(b)$ be the neighbor of b in G with the minimum index. We extend the vertical segment corresponding to b upward until the y-coordinate of $\min_A(b)$.

We must show that the resulting drawing does not contain any forbidden intersection. Suppose by contradiction that the segments of a_j and b_p intersect, but they are not adjacent in G, i.e., $m_{j,p} = 0$. We now have $a_j \prec b_p$, and the entries $b_q = \max_B(a_j)$ and $a_i = \min_A(b_p)$ give the submatrix described in Condition C_2, thus the constraint $b_p \prec a_j$ applies, a contradiction. □

An algorithm to solve Stick$_{AB}$ follows immediately, and can be implemented in linear time in the size of the adjacency matrix M.

Theorem 1. *There is an $O(|A||B|)$-time algorithm to decide the Stick$_{AB}$ problem, and construct a Stick representation if one exists.*

Proof. The algorithm was given above: We construct the directed graph H from the 0-1 matrix M and test if H is acyclic. This correctly decides Stick$_{AB}$ by Lemma 1. Furthermore, if H is acyclic, then we can construct a Stick representation as specified in the proof of Lemma 1. Pseudocode for the algorithm is given in the full version [15].

The matrix M has size $O(nm)$ where $n = |A|$ and $m = |B|$, and the graph H has $n + m$ vertices and $O(nm)$ edges. We can test acyclicity of a graph and find a topological ordering in linear time. Also, the construction of the Stick representation is clearly doable in linear time.

Thus we only need to give details on constructing H in time $O(nm)$. We can construct the edges of H that correspond to σ_A and σ_B in time $O(n + m)$. The edges arising from constraints C_1 correspond to the 1's in the matrix M, so we can construct them in $O(nm)$ time. The edges arising from constraints C_2 correspond to some of the 0's in the matrix M. Specifically, a 0 in position $m_{j,p}$ gives a C_2 constraint $b_p \prec a_j$ if and only if there is a 1 in row j to the right of the 0 and a 1 in column p above the 0. We can flag the 0's that have a 1 to their right by scanning each row of M from right to left. Similarly, we can flag the 0's that have a 1 above them by scanning each column of M from bottom to top. These scans take time $O(nm)$. Finally, if a 0 in M has both flags, then we add the corresponding edge to H. The total time is $O(nm)$. □

Lemma 1 also yields a forbidden submatrix characterization for Stick$_{AB}$.

Theorem 2. *An instance of Stick$_{AB}$ with graph $G = (A \cup B, E)$ has a solution if and only if G's ordered adjacency matrix M has no ordered submatrix of the following form:*

$$P_1 = \begin{array}{c} \\ a_i \\ a_j \\ a_k \end{array}\begin{array}{c} b_p\ b_q\ b_r \\ \begin{bmatrix} * & 1 & * \\ * & 0 & 1 \\ 1 & * & * \end{bmatrix} \end{array}, \ P_2 = \begin{array}{c} \\ a_i \\ a_j \\ a_k \end{array}\begin{array}{c} b_p\ b_q \\ \begin{bmatrix} 1 & * \\ 0 & 1 \\ 1 & * \end{bmatrix} \end{array}, \ P_3 = \begin{array}{c} \\ a_i \\ a_j \end{array}\begin{array}{c} b_p\ b_q\ b_r \\ \begin{bmatrix} * & 1 & * \\ 1 & 0 & 1 \end{bmatrix} \end{array}.$$

Observe that P_2 and P_3 are special cases of P_1 with $p=q$ and $j=k$, respectively.

Proof. We will use the graph H that we constructed above and used in Lemma 1. By Lemma 1, the theorem statement is equivalent to the statement that M has a submatrix P_1, P_2 or P_3 if and only if H has a directed cycle.

We first show that if the matrix M has one of the ordered submatrices P_1, P_2, P_3 then H has a directed cycle. For P_1, the cycle in H is $a_k \prec b_p$ (by C_1), $b_p \prec b_q$ (by σ_B), $b_q \prec a_j$ (by C_2), $a_j \prec a_k$ (by σ_A). For P_2, the cycle is $b_p \prec a_j$ (by C_2), $a_j \prec a_k$ (by σ_A), $a_k \prec b_p$ (by C_1). For P_3, the cycle is $b_q \prec a_j$ (by C_2), $a_j \prec b_p$ (by C_1), $b_p \prec b_q$ (by σ_B).

To prove the other direction, suppose that H has a directed cycle O. We will show that M has one of the submatrices P_1, P_2, P_3. Let b_q be the rightmost vertex of O in σ_B, and let (b_q, z) be the outgoing edge of b_q in O. Since b_q is the rightmost vertex of O in σ_B, z must be a vertex a_j of A. The constraint $b_q \prec a_j$ can only be added by C_2. Therefore, we must have the configuration $\begin{array}{c} a_i \\ a_j \end{array}\begin{array}{c} b_q\ b_r \\ \begin{bmatrix} 1 & * \\ 0 & 1 \end{bmatrix} \end{array}$ The path can now continue from a_j following zero or more A vertices,

but to complete the cycle, it eventually needs to reach a vertex b_p of B. Since b_q is the rightmost in σ_B, b_p must appear either before b_q or coincide with b_q. First suppose that $b_p \neq b_q$. If the outgoing edge of a_j is (a_j, b_p), then we obtain the configuration P_3. Otherwise, the path visits several vertices of A and then visits b_p, and we thus obtain the configuration P_1.

Suppose now that $b_p = b_q$. In this case the outgoing edge of a_j cannot be (a_j, b_p), because such an edge can only be added by C_1, which would imply $m_{j,p} = m_{j,q} = 1$, violating the configuration above. If the path visits several vertices of A and then visits $b_p(= b_q)$, then there must be a 1 in the qth column below the jth row. We thus obtain the configuration P_2.   □

Bipartite Graphs Representable for All Orderings: The above forbidden submatrix characterization allows us to characterize the bipartite graphs $G = (A \cup B, E)$ that have a Stick representation for *every* possible ordering of A and B. Observe that the forbidden submatrices P_2, P_3, P_1 correspond, respectively, to the bipartite graphs shown in Fig. 4(a)–(c). We can construct $2^2 = 4$ graphs from Fig. 4(a) based on whether each of the dotted edges is present or not. Similarly, we can construct $2^2 = 4$ graphs from Fig. 4(b), and $2^5 = 32$ graphs from Fig. 4(b). Let $\mathcal{H}$ be the set that consists of these 40 graphs. From Theorem 2 we immediately obtain:

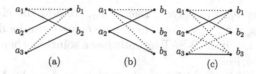

Fig. 4. The forbidden subgraphs for Theorem 3. Dotted edges are optional.

Theorem 3. *A bipartite graph $G = (A \cup B, E)$ admits a Stick representation for every possible ordering of A and B if and only if G does not contain any graph of $\mathcal{H}$.*

3   Fixed As

In this section we study the Stick representation problem when the ordering of only the vertices in A is given. A formal description of the problem, which we call Stick_A, is as follows.

Problem: STICK REPRESENTATION WITH FIXED AS (STICK_A)
Input: A bipartite graph $G = (A \cup B, E)$, and a vertex-ordering σ_A of A.
Question: Does G admit a Stick representation such that the ith horizontal segment on the ground line corresponds to the ith vertex of σ_A?

We give a polynomial-time algorithm for Stick_A. The idea is to use the forbidden submatrix characterization for Stick_{AB} (Theorem 2). We need an ordering of the B vertices that, together with the given ordering σ_A, avoids the forbidden submatrices P_1, P_2, P_3. We will express the conditions for the ordering of the B vertices as a 2-SAT formula, i.e., a CNF (conjunctive normal form) formula where each clause contains at most two literals. 2-SAT can be solved in polynomial time [1].

Theorem 4. *There is an algorithm with run-time $O(|A|^3|B|^3)$ to decide the Stick_A problem, and construct a Stick representation if one exists.*

Proof. For each pair of vertices v, w of G, we create variables $p_{v \prec w}$ and $p_{w \prec v}$ (representing the ordering of segments v and w on the ground line). We will enforce $p_{v \prec w} = \neg p_{w \prec v}$ by adding clauses $(\neg p_{v \prec w} \vee \neg p_{w \prec v}) \wedge (p_{v \prec w} \vee p_{w \prec v})$. (One variable would suffice, but it is notationally easier to have both.) We first set the truth values of all the variables involving two vertices of A based on σ_A. We then add a few other clauses based on P_1, P_2, P_3, as follows.

For every b_p, b_q, b_r giving rise to P_1, we add the clauses $(\neg p_{b_q \prec b_r} \vee p_{b_q \prec b_p})$ and $(\neg p_{b_p \prec b_q} \vee p_{b_r \prec b_q})$. The first clause means that if $b_q \prec b_r$, then to avoid P_1, we must have $b_q \prec b_p$. Similarly, the second clause means if $b_p \prec b_q$, then to avoid P_1, we must have $b_r \prec b_q$. These clauses ensure that if the SAT formula has a solution, then no configuration of the form P_1 can arise.

For every b_p, b_q giving rise to P_2, we set $p_{b_q \prec b_p}$ to true. This would avoid any forbidden configuration of the form P_2 in a solution of the 2-SAT formula.

Finally, for every b_p, b_q, b_r giving rise to P_3, we add the clauses $(\neg p_{b_q \prec b_r} \vee p_{b_q \prec b_p})$ and $(\neg p_{b_p \prec b_q} \vee p_{b_r \prec b_q})$. Note that these clauses can be interpreted in the same way as for P_1, i.e., if the 2-SAT formula has a solution, then no configuration of the form P_3 can arise.

Let F be the resulting 2-SAT formula, which can be solved in linear time in the input size [1], i.e., $O((|A| + |B|)^2)$ time. If F does not have a solution, then there does not exist any ordering of the Bs that avoids the forbidden patterns. Thus G does not admit the required Stick representation. If F has a solution, then there exists an ordering σ_B of Bs that together with σ_A avoids all the forbidden patterns. By Theorem 2, G admits the required Stick representation, and it can be constructed from σ_A and σ_B using Theorem 1.

Thus the time complexity of the algorithm is dominated by the time to construct the 2-SAT formula, which is $O(|A|^3|B|^3)$. Pseudocode for the algorithm is given in the full version [15]. □

Bipartite Graphs Representable for All A Orderings: We also considered the class of bipartite graphs $G = (A \cup B, E)$ such that for every ordering of the vertices of A there exists a Stick representation. We will call this the Stick$_{\forall A}$ class. Although we do not have a characterization of the Stick$_{\forall A}$ class, we describe some positive and negative instances below in Remark 1 and Remark 2, with proofs in the full version [15].

Remark 1. Any bipartite graph $G = (A \cup B, E)$ with at most three vertices in A belongs to the Stick$_{\forall A}$ class.

Remark 2. A graph does not belong to Stick$_{\forall A}$ if its bipartite adjacency matrix contains the submatrix $\begin{matrix} a_1 \\ a_2 \\ a_3 \\ a_4 \end{matrix} \begin{bmatrix} 1 & * \\ 0 & 1 \\ 1 & 0 \\ * & 1 \end{bmatrix}$. (Here the columns are unordered.)

4   Stick Graphs

In this section we examine general Stick representations, i.e., we do not impose any constraints on the ordering of the vertices.

Problem: STICK REPRESENTATION
Input: A bipartite graph $G = (A \cup B, E)$.
Question: Does G admit a Stick representation such that the vertices in A and B correspond to horizontal and vertical segments, respectively?

It is an open question to find a polynomial time algorithm for the above problem of recognizing Stick graphs.

We give some positive instances (Remarks 3–4) and some negative instances (Remark 5). The proofs of all but the first remark are included in the full version [15]. We need a few definitions to state the remarks, as follows.

A matrix has the *simultaneous consecutive ones* property if the rows and columns can be permuted so that the 1's in each row and each column appear consecutively [29]. A *one-sided drawing* of a planar bipartite graph $G = (A \cup B, E)$ is a planar straight-line drawing of G, where all vertices in A lie on the x-axis, and the vertices of B lie strictly above the x-axis [17].

Remark 3. Let $G = (A \cup B, E)$ be a bipartite graph and let M be its adjacency matrix, where the rows and columns correspond to As and Bs, respectively. If M has the simultaneous consecutive ones property, then G admits a Stick representation, which can be computed in $O(|A||B|)$ time.

Proof. One can determine whether M has the simultaneous consecutive ones property in $O(|A||B|)$ time [29], and if so, then one can construct such a matrix M' within the same time complexity.

We now show how to construct the Stick representation from M'. For each row (resp., column), we draw a horizontal (resp., vertical) segment starting from the rightmost (resp., topmost) 1 entry. We extend the horizontal segments to the left and vertical segments downward such that they touch a ground line ℓ.

Let the resulting drawing be D, which may contain many unnecessary crossings. However, for each unnecessary crossing, we can follow the segments involved in the crossings upward and rightward to find two distinct 1 entries. Since the matrix has the simultaneous ones property, the violated entries in each row (column) must lie consecutively at the left end of the row (bottom end of the column). Therefore, one can find a $(+x, -y)$-monotone path P that separates the violated entries from the rest of the matrix.

Let $b_1, b_2, \ldots, b_k$ be the bend points creating $90°$ angles towards ℓ. To compute the required Stick representation, we remove these bends one after another, as follows. Consider the topmost bend point b_i. Imagine a Cartesian coordinate system with origin at b_i. Move the rows above b_i and columns to the right of b_i towards the upward and rightward directions, respectively. It is straightforward to observe that one now can construct a ground line ℓ' through b_i such that the violated entries lie in the region below the path determined by $b_{i+1}, \ldots, b_k$. $\square$

Remark 4. Let $G = (A \cup B, E)$ be an n-vertex bipartite graph that admits a one-sided planar drawing. Then G is a Stick graph, and its Stick representation can be computed in $O(n^2)$ time.

Remark 5. Let H be the graph obtained by deleting a perfect matching from a complete bipartite graph $K_{4,4}$. Any graph $G = (A \cup B, E)$ containing H as an induced subgraph does not admit a Stick representation. Since H is a planar graph, not all planar bipartite graphs are Stick graphs.

5   Open Problems

We conclude the paper with the following open problems.

Open Problem 1. What is the complexity of recognizing Stick graphs? Is the problem NP-complete? By Theorem 2 the problem is equivalent to ordering the rows and columns of a 0-1 matrix to exclude the 3 forbidden submatrices given in the Theorem statement. Note that these forbidden submatrices involve 5 or 6 rows and columns (vertices of the graph) so the results of Hell et al. [20], which apply to patterns of at most 4 vertices in a bipartite graph, do not provide a polynomial time algorithm.

One possible approach using 3-SAT is as follows. Given a bipartite graph $G = (A \cup B, E)$, one can create a 3-SAT formula Φ such that Φ is satisfiable if and only if G admits a Stick representation, as follows. For each pair of vertices v, w of G, create variables $p_{v \prec w}$ and $p_{w \prec v}$ (representing the ordering of v and w on the ground line), and add clauses $(\neg p_{v \prec w} \vee \neg p_{w \prec v}) \wedge (p_{v \prec w} \vee p_{w \prec v})$ to enforce $p_{v \prec w} = \neg p_{w \prec v}$. Now express the conditions C1 and C2 from Sect. 2 as 3-SAT clauses.

Φ_1: (Condition C1.) If $m_{i,p} = 1$, then set $p_{a_i \prec b_p} = 1$.

Φ_2: (Condition C2.) We must express the condition that if the ordered submatrix

$$\begin{array}{c} \\ a_i \\ a_j \end{array} \begin{array}{c} b_p\ b_q \\ \begin{bmatrix} 1 & * \\ 0 & 1 \end{bmatrix} \end{array} \text{ exists, then } p_{b_p \prec a_j} = 1. \text{ Thus, if } m_{i,p} = 1, m_{jq} = 1 \text{ and } m_{j,p} = 0,$$

then we add the clause $(\neg p_{a_i \prec a_j} \vee \neg p_{b_p \prec b_q} \vee \neg p_{a_j \prec b_p})$.

Φ_3: For each triple u, v, w of vertices, add the clause $(\neg p_{u \prec v} \vee \neg p_{v \prec w} \vee p_{u \prec w})$. Intuitively, these are transitivity constraints, which would ensure a total ordering on the ground line.

It is not difficult to show that the 3-SAT Φ is satisfiable if and only if G admits the required intersection representation. However, since Φ contains $O(n^2)$ variables, using known SAT-solvers would not be faster than a naive algorithm that simply guesses the order of the segments along the ground line. Therefore, an interesting direction for future research would be to find a 3-SAT formulation with a linear number of variables.

Open Problem 2. Can we improve the time complexity of the recognition algorithm for graphs with fixed As?

Acknowledgments. The research of A. Lubiw and D. Mondal is supported in part by the Natural Sciences and Engineering Research Council of Canada (NSERC). Also, this work is supported in part by NSF grants CCF-1423411 and CCF-1712119.

References

1. Aspvall, B., Plass, M.F., Tarjan, R.E.: A linear-time algorithm for testing the truth of certain quantified Boolean formulas. Inf. Process. Lett. **8**(3), 121–123 (1979)
2. Booth, K.S., Lueker, G.S.: Testing for the consecutive ones property, interval graphs, and graph planarity using PQ-tree algorithms. J. Comput. Syst. Sci. **13**(3), 335–379 (1976)
3. Cabello, S., Jejcic, M.: Refining the hierarchies of classes of geometric intersection graphs. Electr. J. Comb. **24**(1), 33 (2017)

4. Cardinal, J., Felsner, S., Miltzow, T., Tompkins, C., Vogtenhuber, B.: Intersection graphs of rays and grounded segments. In: Bodlaender, H.L., Woeginger, G.J. (eds.) WG 2017. LNCS, vol. 10520, pp. 153–166. Springer, Cham (2017). https://doi.org/10.1007/978-3-319-68705-6_12

5. Catanzaro, D., et al.: Max point-tolerance graphs. Discrete Appl. Math. **216**, 84–97 (2017)

6. Chalopin, J., Gonçalves, D.: Every planar graph is the intersection graph of segments in the plane. In: Proceedings of the Forty-First Annual ACM Symposium on Theory of Computing, pp. 631–638. ACM (2009)

7. Chan, T.M., Grant, E.: Exact algorithms and APX-hardness results for geometric packing and covering problems. Comput. Geom. **47**(2, Part A), 112–124 (2014)

8. Chaplick, S., Felsner, S., Hoffmann, U., Wiechert, V.: Grid intersection graphs and order dimension. Order **35**(2), 363–391 (2018)

9. Chaplick, S., Cohen, E., Morgenstern, G.: Stabbing polygonal chains with rays is hard to approximate. In: Canadian Conference on Computational Geometry (CCCG) (2013)

10. Chaplick, S., Hell, P., Otachi, Y., Saitoh, T., Uehara, R.: Intersection dimension of bipartite graphs. In: Gopal, T.V., Agrawal, M., Li, A., Cooper, S.B. (eds.) TAMC 2014. LNCS, vol. 8402, pp. 323–340. Springer, Cham (2014). https://doi.org/10.1007/978-3-319-06089-7_23

11. Chen, L., Yesha, Y.: Efficient parallel algorithms for bipartite permutation graphs. Networks **23**(1), 29–39 (1993)

12. Cogis, O.: On the Ferrers dimension of a digraph. Discrete Math. **38**(1), 47–52 (1982)

13. Corneil, D.G., Olariu, S., Stewart, L.: The LBFS structure and recognition of interval graphs. SIAM J. Discrete Math. **23**(4), 1905–1953 (2009)

14. Das, S., Sen, M., Roy, A.B., West, D.B.: Interval digraphs: an analogue of interval graphs. J. Graph Theory **13**(2), 189–202 (1989)

15. De Luca, F., Hossain, M.I., Kobourov, S., Lubiw, A., Mondal, D.: Recognition and Drawing of Stick Graphs. arXiv preprint arXiv:1808.10005, August 2018

16. Felsner, S., Mertzios, G.B., Musta, I.: On the recognition of four-directional orthogonal ray graphs. In: Chatterjee, K., Sgall, J. (eds.) MFCS 2013. LNCS, vol. 8087, pp. 373–384. Springer, Heidelberg (2013). https://doi.org/10.1007/978-3-642-40313-2_34

17. Fößmeier, U., Kaufmann, M.: Nice drawings for planar bipartite graphs. In: Bongiovanni, G., Bovet, D.P., Di Battista, G. (eds.) CIAC 1997. LNCS, vol. 1203, pp. 122–134. Springer, Heidelberg (1997). https://doi.org/10.1007/3-540-62592-5_66

18. Halldórsson, B.V., Aguiar, D., Tarpine, R., Istrail, S.: The clark phase-able sample size problem: long-range phasing and loss of heterozygosity in GWAS. In: Berger, B. (ed.) RECOMB 2010. LNCS, vol. 6044, pp. 158–173. Springer, Heidelberg (2010). https://doi.org/10.1007/978-3-642-12683-3_11

19. Hartman, I.B.A., Newman, I., Ziv, R.: On grid intersection graphs. Discrete Math. **87**(1), 41–52 (1991)

20. Hell, P., Mohar, B., Rafiey, A.: Ordering without forbidden patterns. In: Schulz, A.S., Wagner, D. (eds.) ESA 2014. LNCS, pp. 554–565. Springer, Heidelberg (2014). https://doi.org/10.1007/978-3-662-44777-2_46

21. Hixon, T.: Hook graphs and more: some contributions to geometric graph theory. Master's thesis, Technische Universität Berlin (2013)

22. Hoffmann, U.: Intersection graphs and geometric objects in the plane. Master's thesis, Technische Universität Berlin (2016)

23. Katz, M.J., Mitchell, J.S., Nir, Y.: Orthogonal segment stabbing. Comput. Geom. **30**(2), 197–205 (2005)
24. Klinz, B., Rudolf, R., Woeginger, G.J.: Permuting matrices to avoid forbidden submatrices. Discrete Appl. Math. **1**(60), 223–248 (1995)
25. Kratochvíl, J.: A special planar satisfiability problem and a consequence of its NP-completeness. Discrete Appl. Math. **52**(3), 233–252 (1994)
26. Lubiw, A.: Doubly lexical orderings of matrices. SIAM J. Comput. **16**(5), 854–879 (1987)
27. Mustaţă, I., Pergel, M.: Unit grid intersection graphs: Recognition and properties. arXiv preprint arXiv:1306.1855 (2013)
28. Müller, H.: Recognizing interval digraphs and interval bigraphs in polynomial time. Discrete Appl. Math. **78**(1), 189–205 (1997)
29. Oswald, M., Reinelt, G.: The simultaneous consecutive ones problem. Theor. Comput. Sci. **410**(21–23), 1986–1992 (2009)
30. Shrestha, A.M.S., Tayu, S., Ueno, S.: On orthogonal ray graphs. Discrete Appl. Math. **158**(15), 1650–1659 (2010)
31. Soto, J.A., Telha, C.: Jump number of two-directional orthogonal ray graphs. In: Günlük, O., Woeginger, G.J. (eds.) IPCO 2011. LNCS, vol. 6655, pp. 389–403. Springer, Heidelberg (2011). https://doi.org/10.1007/978-3-642-20807-2_31
32. Soto, M., Caro, C.T.: p-Box: a new graph model. Discrete Math. Theor. Comput. Sci. **17**(1), 169 (2015)
33. Spinrad, J.P.: Doubly lexical ordering of dense 0–1 matrices. Inf. Process. Lett. **45**(5), 229–235 (1993)
34. Spinrad, J.P., Brandstädt, A., Stewart, L.: Bipartite permutation graphs. Discrete Appl. Math. **18**(3), 279–292 (1987)

On Contact Graphs of Paths on a Grid

Zakir Deniz[1], Esther Galby[2(✉)], Andrea Munaro[2], and Bernard Ries[2]

[1] Department of Mathematics, Duzce University, Duzce, Turkey
zakirdeniz@duzce.edu.tr
[2] Department of Informatics, Decision Support & Operations Research,
University of Fribourg, Fribourg, Switzerland
{esther.galby,andrea.munaro,bernard.ries}@unifr.ch

Abstract. In this paper we consider *Contact graphs of Paths on a Grid* (*CPG graphs*), i.e. graphs for which there exists a family of interiorly disjoint paths on a grid in one-to-one correspondence with their vertex set such that two vertices are adjacent if and only if the corresponding paths touch at a grid-point. Our class generalizes the well studied class of VCPG graphs (see [1]). We examine CPG graphs from a structural point of view which leads to constant upper bounds on the clique number and the chromatic number. Moreover, we investigate the recognition and 3-colorability problems for B_0-CPG, a subclass of CPG. We further show that CPG graphs are not necessarily planar and not all planar graphs are CPG.

1 Introduction

Asinowski et al. [3] introduced the class of *vertex intersection graphs of paths on a grid*, referred to as *VPG graphs*. An undirected graph $G = (V, E)$ is called a *VPG graph* if one can associate a path on a grid with each vertex such that two vertices are adjacent if and only if the corresponding paths intersect on at least one grid-point. It is not difficult to see that the class of VPG graphs coincides with the class of string graphs, i.e. intersection graphs of curves in the plane (see [3]).

A natural restriction which was forthwith considered consists in limiting the number of *bends* (i.e. 90° turns at a grid-point) that the paths may have: an undirected graph $G = (V, E)$ is a B_k-*VPG graph*, for some integer $k \geq 0$, if one can associate a path on a grid having at most k *bends* with each vertex such that two vertices are adjacent if and only if the corresponding paths intersect on at least one grid-point. Since their introduction, B_k-VPG have been extensively studied (see [2,3,5,7–9,14,15,18–20]).

A notion closely related to intersection graphs is that of *contact graphs*. Such graphs can be seen as a special type of intersection graphs of geometrical objects in which these objects are not allowed to have common interior points but only to touch each other. Contact graphs of various types of objects have been studied in the literature (see, e.g., [1,10,11,21–23]). In this paper, we consider *Contact graphs of Paths on a Grid* (*CPG graphs* for short) which are defined as follows.

© Springer Nature Switzerland AG 2018
T. Biedl and A. Kerren (Eds.): GD 2018, LNCS 11282, pp. 317–330, 2018.
https://doi.org/10.1007/978-3-030-04414-5_22

(a) Allowed contacts.   (b) Forbidden contact.

Fig. 1. Examples of types of contact between two paths (the endpoints of a path are marked by an arrow).

A graph G is a *CPG graph* if the vertices of G can be represented by a family of interiorly disjoint paths on a grid, two vertices being adjacent in G if and only if the corresponding paths touch, i.e. share a grid-point which is an endpoint of at least one of the two paths (see Fig. 1). Note that this class is hereditary, i.e. closed under vertex deletion. Similarly to VPG, a B_k-CPG graph is a CPG graph admitting a representation in which each path has at most k bends. Clearly, any B_k-CPG graph is also a B_k-VPG graph.

Aerts and Felsner [1] considered a similar family of graphs, namely those admitting a *Vertex Contact representation of Paths on a Grid* (*VCPG* for short). The vertices of such graphs can be represented by a family of interiorly disjoint paths on a grid, but the adjacencies are defined slightly differently: two vertices are adjacent if and only if the endpoint of one of the corresponding paths touches an interior point of the other corresponding path (observe that this is equivalent to adding the constraint forbidding two paths from having a common endpoint, i.e. contacts as in Fig. 1a on the right). This class has been considered by other authors as well (see [6,7,14,19,24]).

It is not difficult to see that graphs admitting a VCPG are planar (see [1]) and it immediately follows from the definition that those graphs are CPG graphs. This containment is in fact strict even when restricted to planar CPG graphs, as there exist, in addition to nonplanar CPG graphs, planar graphs which are CPG but do not admit a VCPG.

To the best of our knowledge, the class of CPG graphs has never been studied in itself and our present intention is to provide some structural properties (see Sect. 3). By considering a specific weight function on the vertices, we provide upper bounds on the number of edges in CPG graphs as well as on the clique number and the chromatic number (see Sect. 3). In particular, we show that B_0-CPG graphs are 4-colorable and that 3-COLORABILITY restricted to B_0-CPG is NP-complete (see Sect. 5). We further prove that recognizing B_0-CPG graphs is NP-complete. Additionally, we show that the classes of CPG graphs and planar graphs are incomparable (see Sect. 4).

2   Preliminaries

Throughout this paper, all considered graphs are undirected, finite and simple. For any graph theoretical notion not defined here, we refer the reader to [13].

Let $G = (V, E)$ be a graph with vertex set V and edge set E. The *degree* of a vertex $v \in V$, denoted by $d(v)$, is the number of neighbors of v in G. A graph G is *k-regular* if the degree of every vertex in G is $k \geq 0$. A *clique* (resp. *stable set*) in G is a set of pairwise adjacent (resp. nonadjacent) vertices. The graph obtained from G by deleting a vertex $v \in V$ is denoted by $G - v$. For a given graph H, G is *H-free* if it contains no induced subgraph isomorphic to H.

As usual, K_n (resp. C_n) denotes the complete graph (resp. chordless cycle) on n vertices and $K_{m,n}$ denotes the complete bipartite graph with bipartition (V_1, V_2) such that $|V_1| = m$ and $|V_2| = n$. Given a graph G, the *line graph of* G, denoted by $L(G)$, is the graph such that each vertex v_e in $L(G)$ corresponds to an edge e in G and two vertices are adjacent in $L(G)$ if and only if their corresponding edges in G have a common endvertex.

A graph G is *planar* if it can be drawn in the plane without crossing edges; such a drawing is then called a *planar embedding* of G. A planar embedding divides the plane into several regions referred to as *faces*. A planar graph is *maximally planar* if adding any edge renders it nonplanar. A maximally planar graph has exactly $2n - 4$ faces, where n is the number of vertices in the graph. A graph H is a *minor* of a graph G, if H can be obtained from G by deleting edges and vertices and by contracting edges. It is well-known that a graph is planar if and only if it does not contain K_5 or $K_{3,3}$ as a minor [13].

A *coloring* of a graph G is a mapping $\mathbf{c}$ associating with every vertex u an integer $\mathbf{c}(u)$, called a *color*, such that $\mathbf{c}(v) \neq \mathbf{c}(u)$ for every edge uv. If at most k distinct colors are used, $\mathbf{c}$ is called a *k-coloring*. The smallest integer k such that G admits a k-coloring is called the *chromatic number* of G, denoted by $\chi(G)$.

Consider a rectangular grid $\mathcal{G}$ where the horizontal lines are referred to as *rows* and the vertical lines as *columns*. The grid-point lying on row x and column y is denoted by (x, y). An *interior point* of a path P on $\mathcal{G}$ is a point belonging to P and different from its endpoints; the *interior* of P is the set of all its interior points. A graph $G = (V, E)$ is *CPG* if there exists a collection $\mathcal{P}$ of interiorly disjoint paths on a grid $\mathcal{G}$ such that $\mathcal{P}$ is in one-to-one correspondence with V and two vertices are adjacent in G if and only if the corresponding paths touch; if every path in $\mathcal{P}$ has at most k bends, G is B_k-CPG. The pair $\mathcal{R} = (\mathcal{G}, \mathcal{P})$ is a *CPG representation* of G, and more specifically a *k-bend CPG representation* if every path in $\mathcal{P}$ has at most k bends. In the following, the path representing some vertex u in a CPG representation $\mathcal{R}$ of a graph G is denoted by $P_u^{\mathcal{R}}$, or simply P_u if it is clear from the context.

Let $G = (V, E)$ be a CPG graph and $\mathcal{R} = (\mathcal{G}, \mathcal{P})$ be a CPG representation of G. A grid-point p is of *type I* if it corresponds to an endpoint of four paths in $\mathcal{P}$ (see Fig. 2a), and of *type II* if it corresponds to an endpoint of two paths in $\mathcal{P}$ and an interior point of a third path in $\mathcal{P}$ (see Fig. 2b).

For any grid-point p, we denote by $\tau(p)$ the number of edges in the subgraph induced by the vertices whose corresponding paths contain or have p as an endpoint. Note that this subgraph is a clique and so $\tau(p) = \binom{j}{2}$ if j paths touch at grid-point p.

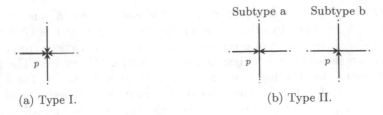

(a) Type I.   (b) Type II.

Fig. 2. Two types of grid-points.

For any path P, we denote by $\mathring{P}$ (resp. $\partial(P)$) the interior (resp. endpoints) of P. For a vertex $u \in V$, we define the *weight of u with respect to $\mathcal{R}$*, denoted by $w_u^{\mathcal{R}}$ or simply w_u if it is clear from the context, as follows. Let q_u^i $(i = 1, 2)$ be the endpoints of the corresponding path P_u in $\mathcal{P}$ and consider, for $i = 1, 2$,

$$w_u^i = |\{P \in \mathcal{P} \mid q_u^i \in \mathring{P}\}| + \frac{1}{2} \cdot |\{P \in \mathcal{P} \mid P \neq P_u \text{ and } q_u^i \in \partial(P)\}|.$$

Then $w_u = w_u^1 + w_u^2$.

Observation 1. *Let $G = (V, E)$ be a CPG graph and $\mathcal{R} = (\mathcal{G}, \mathcal{P})$ be a CPG representation of G. For any vertex $u \in V$ and $i = 1, 2$, $w_u^i \leq \frac{3}{2}$ where equality holds if and only if q_u^i is a grid-point of type I or II.*

Indeed, the contribution of q_u^i to w_u^i is maximal if all four grid-edges containing q_u^i are used by paths of $\mathcal{P}$, which may only happen when q_u^i is a grid-point of type I or II.

Remark. In fact, we have $w_u^i \in \{0, \frac{1}{2}, 1, \frac{3}{2}\}$ for any vertex $u \in V$ and $i = 1, 2$.

Observation 2. *Let $G = (V, E)$ be a CPG graph and $\mathcal{R} = (\mathcal{G}, \mathcal{P})$ be a CPG representation of G. Then*

$$|E| \leq \sum_{u \in V} w_u,$$

where equality holds if and only if all paths of $\mathcal{P}$ pairwise touch at most once.

Indeed, if $uv \in E$, we may assume that either an endpoint of P_u touches the interior of P_v, or P_u and P_v have a common endpoint. In the first case, the edge uv is fully accounted for in the weight of u, and in the second case, the edge uv is accounted for in both w_u and w_v by one half. The characterization of equality then easily follows.

3   Structural Properties of CPG Graphs

In this section, we investigate CPG graphs from a structural point of view and present some useful properties which we will further exploit.

Lemma 1. *A CPG graph is either 6-regular or has a vertex of degree at most 5.*

Proof. If $G = (V, E)$ is a CPG graph and $\mathcal{R}$ is a CPG representation of G, by combining Observations 1 and 2, we obtain

$$\sum_{u \in V} d(u) = 2|E| \leq 2 \sum_{u \in V} w_u \leq 2 \sum_{u \in V} \left(\frac{3}{2} + \frac{3}{2} \right) = 6|V|. \qquad \square$$

Remark. We can show that there exists an infinite family of 6-regular CPG graphs. Due to lack of space, this proof is here omitted but can be found in the full version [12].

For B_1-CPG graphs, we can strengthen Lemma 1 as follows.

Proposition 1. *Every B_1-CPG graph has a vertex of degree at most 5.*

Proof. Let $G = (V, E)$ be a B_1-CPG graph and $\mathcal{R}$ be a 1-bend CPG representation of G. Denote by p the upper-most endpoint of a path among the left-most endpoints in $\mathcal{R}$, and by P_x (with $x \in V$) an arbitrary path having p as an endpoint. Since $\mathcal{R}$ is a 1-bend CPG representation, no path uses the grid-edge on the left of p, for otherwise p would not be a left-most endpoint. Therefore, p contributes to the weight of x with respect to $\mathcal{R}$ by at most 1 and, by Observations 1 and 2, we have

$$\sum_{u \in V} d(u) = 2|E| \leq 2(w_x + \sum_{u \neq x} w_u) \leq 6|V| - 1,$$

which implies the existence of a vertex of degree at most 5. $\qquad \square$

A natural question that arises when considering CPG graphs is whether they may contain large cliques. It immediately follows from Observation 2 that CPG graphs cannot contain K_n, for $n \geq 8$. This can be further improved as shown in the next result.

Theorem 1. *CPG graphs are K_7-free.*

Proof. Since the class of CPG graphs is hereditary, it is sufficient to show that K_7 is not a CPG graph. Suppose, to the contrary, that K_7 is a CPG graph and consider a CPG representation $\mathcal{R} = (\mathcal{G}, \mathcal{P})$ of K_7. Observe first that the weight of every vertex with respect to $\mathcal{R}$ must be exactly $2 \cdot 3/2$, as otherwise by Observation 1, we would have $\sum_{u \in V} w_u < 3|V| = 21 = |E|$ which contradicts Observation 2. This implies in particular that every grid-point corresponding to an endpoint of a path is either of type I or II. Furthermore, any two paths must touch at most once, for otherwise by Observation 2, $|E| < \sum_{u \in V} w_u = 3|V| = |E|$. Hence, if we denote by P_I (resp. P_{II}) the set of grid-points of type I (resp. type II), then since $\tau(p) = 6$ for all $p \in P_I$ and $\tau(p) = 3$ for all $p \in P_{II}$, we have that $6|P_I| + 3|P_{II}| = 21$, which implies $|P_{II}| \neq 0$. Suppose that there exists a path P_u having one endpoint corresponding to a grid-point of type I and the other corresponding to a grid-point of type II. Since the corresponding vertex u has degree 6, P_u must then properly contain an endpoint of another path which, as first observed, necessarily corresponds to a grid-point of type II. But vertex u would then have degree $3 + 2 + 2$ as no two paths touch more than once, a contradiction. Hence, every path has both its endpoints of the same type. But then, $|P_I| = 0$; indeed, if there exists a path having both its endpoints of type I, since no two paths touch more than once, this implies that every path has both its endpoints of type I, i.e. $|P_{II}| = 0$, a contradiction. Now, if we consider each grid-point of type II as a vertex and connect any two such vertices when the corresponding grid-points belong to a same path, then we obtain a planar embedding of a 4-regular graph on 7 vertices. But this contradicts the fact that every 4-regular graph on 7 vertices contains $K_{3,3}$ as a minor (a proof of this result can be found in the full version [12]).   □

However, CPG graphs may contain cliques on 6 vertices as shown in Proposition 2. Due to lack of space, its proof is omitted here and can be found in the full version [12].

Proposition 2. K_6 *is in* B_2-*CPG* $\setminus B_1$-*CPG*.

We conclude this section with a complexity result pointing towards the fact that there may not be a polynomial characterization of B_0-CPG graphs. Let us first introduce rectilinear planar graphs: a graph G is *rectilinear planar* if it admits a rectilinear planar drawing, i.e. a drawing mapping each edge to a horizontal or vertical segment.

Theorem 2. RECOGNITION *is* NP-*complete for* B_0-*CPG graphs*.

Proof. We show that a graph G is rectilinear planar if and only if its line graph $L(G)$ is B_0-CPG. As RECOGNITION for rectilinear planar graphs was shown to be NP-complete in [17], this concludes the proof. Suppose G is a rectilinear planar graph and let $\mathcal{D}$ be the collection of horizontal and vertical segments in a rectilinear planar drawing of G. It is not difficult to see that the contact graph of $\mathcal{D}$ is isomorphic to $L(G)$. Conversely, assume that $L(G)$ is a B_0-CPG graph and consider a 0-bend CPG representation $\mathcal{R} = (\mathcal{G}, \mathcal{P})$ of $L(G)$. Since $L(G)$ is $K_{1,3}$-free [4], every path in $\mathcal{P}$ has at most two contact points. Thus, by eventually shortening paths, we may assume that contacts only happen at endpoints of paths. Therefore, $\mathcal{R}$ induces a rectilinear planar drawing of G, where each vertex corresponds to a contact point in $\mathcal{R}$ and each edge is mapped to its corresponding path in $\mathcal{P}$.   □

4   Planar CPG Graphs

In this section, we focus on planar graphs and their relation with CPG graphs. In particular, we show that not every planar graph is CPG and not all CPG graphs are planar.[1]

---

[1] We can further show that not all CPG graphs are 1-planar as $K_7 - E(K_3)$ is CPG but not 1-planar [25].

   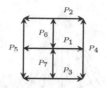

(a) A nonplanar graph G.   (b) A 0-bend CPG representation of G.

Fig. 3. A B_0-CPG graph containing $K_{3,3}$ as a minor (contract the edge e).

Lemma 2. *If G is a CPG graph for which there exists a CPG representation containing no grid-point of type I or II.a, then G is planar. In particular, if G is a triangle-free CPG graph, then G is planar.*

Proof. Let $G = (V, E)$ be a CPG graph for which there exists a CPG representation $\mathcal{R}$ containing no grid-point of type I or II.a. By considering each path of $\mathcal{R}$ as a curve in the plane, it follows that G is a curve contact graph having a representation (namely $\mathcal{R}$) in which any point in the plane belongs to at most three curves. Furthermore, whenever a point in the plane belongs to the interior of a curve $\mathcal{C}$ and corresponds to an endpoint of two other curves, then those two curves lie on the same side of $\mathcal{C}$ (recall that there is no grid-point of type II.a). Hence, it follows from Proposition 2.1 in [21] that G is planar.

If G is a triangle-free CPG graph, then no CPG representation of G contains grid-points of type I or II.a. Hence, G is planar. ⊓

Remark. Since $K_{3,3}$ is a triangle-free nonplanar graph, it follows from Lemma 2 that $K_{3,3}$ is not CPG. Therefore, CPG graphs are $K_{3,3}$-free. Observe however that for any $k \geq 0$, B_k-CPG is not a subclass of planar graphs as there exist B_0-CPG graphs which are not planar (see Fig. 3).

It immediately follows from [7] that all triangle-free planar graphs are B_1-CPG; hence, we have the following corollary.

Corollary 1. *If a graph G is triangle-free, then G is planar if and only if G is B_1-CPG.*

The next result allows us to detect planar graphs that are not CPG.

Lemma 3. *Let $G = (V, E)$ be a planar graph. If G is a CPG graph, then G has at most $4|V| - 2f + 4$ vertices of degree at most 3, where f denotes the number of faces of G. In particular, if G is maximally planar, then G has at most 12 vertices of degree at most 3.*

Proof. Let $G = (V, E)$ be a planar CPG graph and $\mathcal{R} = (\mathcal{G}, \mathcal{P})$ a CPG representation of G. Denote by U the subset of vertices in G of degree at most 3. If a path P_u, with $u \in U$, touches every other path in $\mathcal{P}$ at most once, then, since at least one endpoint of P_u is then not a grid-point of type I or II, the weight of

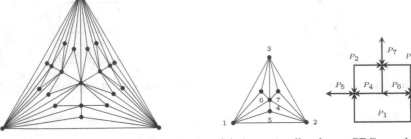

(a) A non CPG maximally planar graph.   (b) A maximally planar CPG graph.

Fig. 4. Two maximally planar graphs.

u with respect to $\mathcal{R}$ is at most $3/2 + 1$. Thus, if we assume that this is the case for all paths whose corresponding vertex is in U, we have by Observation 2

$$|E| \leq \left(\frac{3}{2} + 1\right)|U| + 3(|V| - |U|) = 3|V| - \frac{|U|}{2}.$$

On the other hand, if there exists $u \in U$ such that P_u touches some path more than once, then the above inequality still holds as the corresponding edge is already accounted for. Using the fact that $f = |E| - |V| + 2$ (Euler's formula), we obtain the desired upper bound. Moreover, if G is maximally planar, then $f = 2|V| - 4$ and so $|U| \leq 12$. $\square$

Remark. In Fig. 4a, we give an example of a maximally planar graph which is not CPG due to Lemma 3. It is constructed by iteratively adding a vertex in a triangular face, starting from the triangle, so that it has exactly 13 vertices of degree 3. There exist however maximally planar graphs which are CPG (see Fig. 4b). Note that maximally planar graphs do not admit a VCPG [1].

5   Coloring CPG Graphs

In this section, we provide tight upper bounds on the chromatic number of B_k-CPG graphs for different values of k and investigate the 3-COLORABILITY problem for CPG graphs. The proof of the following result is an easy exercise left to the reader (see the full version [12]).

Theorem 3. *CPG graphs are 6-colorable.*

Remark. Since K_6 is B_2-CPG, this bound is tight for B_k-CPG graphs with $k \geq 2$. We leave as an open problem whether this bound is also tight for B_1-CPG graphs (note that it is at least 5 since K_5 is B_1-CPG).

Theorem 4. B_0-*CPG graphs are 4-colorable. Moreover, K_4 is a 4-chromatic B_0-CPG graph.*

Proof. Let G be a B_0-CPG graph and $\mathcal{R} = (\mathcal{G}, \mathcal{P})$ a 0-bend CPG representation of G. Denote by $\mathcal{L}$ (resp. $\mathcal{C}$) the set of rows (resp. columns) of $\mathcal{G}$ on which lies at least one path of $\mathcal{P}$. Since the representation contains no bend, if A is a row in $\mathcal{L}$ (resp. column in $\mathcal{C}$), then the set of vertices having their corresponding path on A induces a collection of disjoint paths in G. If $B \neq A$ is another row in $\mathcal{L}$ (resp. column in $\mathcal{C}$), then no path in A touches a path in B. Hence, it suffices to use two colors to color the vertices having their corresponding path in a row of $\mathcal{L}$ and two other colors to color the vertices having their corresponding path in a column of $\mathcal{C}$ to obtain a proper coloring of G.   □

It immediately follows from a result in [22] that the 3-COLORABILITY problem is NP-complete in CPG, even if the graph admits a representation in which each grid-point belongs to at most two paths. We conclude this section by a strengthening of this result.

Theorem 5. 3-COLORABILITY *is* NP-*complete in* B_0-*CPG.*

Proof. We exhibit a polynomial reduction from 3-COLORABILITY restricted to planar graphs of maximum degree 4, which was shown to be NP-complete in [16].

Let $G = (V, E)$ be a planar graph of maximum degree 4. It follows from [26] that G admits a grid embedding where each vertex is mapped to a grid-point and each edge is mapped to a grid-path with at most 4 bends, in such a way that all paths are interiorly disjoint (such an embedding can be obtained in linear time). Denote by $\mathcal{D} = (\mathcal{V}, \mathcal{E})$ such an embedding, where $\mathcal{V}$ is the set of grid-points in one-to-one correspondence with V and $\mathcal{E}$ is the set of grid-paths in one-to-one correspondence with E. For any vertex $u \in V$, we denote by (x_u, y_u) the grid-point in $\mathcal{V}$ corresponding to u and by P_u^N (resp. P_u^S) the path of $\mathcal{E}$, if any, having (x_u, y_u) as an endpoint and using the grid-edge above (resp. below) (x_u, y_u). For any edge $e \in E$, we denote by P_e the path in $\mathcal{E}$ corresponding to e. We construct from $\mathcal{D}$ a 0-bend CPG representation $\mathcal{R}$ in such a way that the corresponding graph G' is 3-colorable if and only if G is 3-colorable.

By eventually adding rows and columns to the grid, we may assume that the interior of each path P in $\mathcal{E}$ is surrounded by an empty region, i.e. no path $P' \neq P$ or grid-point of $\mathcal{V}$ lies in the interior of this region. In the following, we denote this region by $\mathcal{R}_P$ (delimited by red dashed lines in every subsequent figure) and assume, without loss of generality, that it is always large enough for the following operations.

We first associate with every vertex $u \in V$ a *vertical path* P_u containing the grid-point (x_u, y_u) as follows. If P_u^N (resp. P_u^S) is not defined, the top (resp. lower) endpoint of P_u is $(x_u, y_u + \varepsilon)$ (resp. $(x_u, y_u - \varepsilon)$) for a small enough ε so that the segment $[(x_u, y_u), (x_u, y_u + \varepsilon)]$ (resp. $[(x_u, y_u), (x_u, y_u - \varepsilon)]$) touches no path of $\mathcal{E}$. If P_u^N has at least one bend, then the top endpoint of P_u lies at the border of $\mathcal{R}_{P_u^N}$ on column x_u (see Fig. 5a). If P_u^N has no bend, then the top endpoint of P_u lies at the middle of P_u^N (see Fig. 5b). Similarly, we define the lower endpoint of P_u according to P_u^S: if P_u^S has at least one bend, then the lower endpoint of P_u lies at the border of $\mathcal{R}_{P_u^S}$ on column x_u, otherwise it lies at the middle of P_u^S.

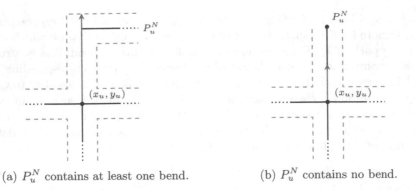

(a) P_u^N contains at least one bend.   (b) P_u^N contains no bend.

Fig. 5. Constructing the path P_u corresponding to vertex u (in blue). (Color figure online)

(a) A path containing one interior vertical segment.   (b) A path containing two interior vertical segments.

Fig. 6. Replacing interior vertical segments by 0-bend paths (in blue). (Color figure online)

For any path P of $\mathcal{E}$ with at least two bends, an *interior vertical segment of P* is a vertical segment of P containing none of its endpoints (note that since every path in $\mathcal{E}$ has at most 4 bends, it may contain at most two interior vertical segments). We next replace every interior segment of P by a slightly longer vertical path touching the border of $\mathcal{R}_P$ (see Fig. 6).

We finally introduce two gadgets H (see Fig. 7) and H', where H' is the subgraph of H induced by $\{b, c, 4, 5, 6, 7, 8, 9, 10\}$, as follows. Denote by $\mathcal{P}'$ the set of vertical paths introduced so far and by V' the set of vertices of the contact graph of $\mathcal{P}'$. Observe that V' contains a copy of V and that two vertices are adjacent in the contact graph of $\mathcal{P}'$ if and only if they are both copies of vertices in V and the path P of $\mathcal{P}$ corresponding to the edge between these two copies is a vertical path with no bend. Now, along each path P_{uv} of $\mathcal{P}$ such that the vertical paths P_u and P_v of $\mathcal{P}'$ do not touch, we add gadgets H and H' as follows. Let $P_1, \ldots, P_k$ be the vertical paths of $\mathcal{P}'$ encountered in order when going along P_{uv} from (x_u, y_u) to (x_v, y_v) and let u_j be the vertex of V' corresponding to P_j, for $1 \leq j \leq k$. Note that P_1 (resp. P_k) is the path corresponding to vertex

$u = u_1$ (resp. $v = u_k$) and that P_j, for $2 \leq j \leq k - 1$, is a path corresponding to an interior vertical segment of P_{uv} (this implies in particular that $k \leq 4$). We add the gadget H' in between u_1 and u_2 by identifying u_1 with b and u_2 with c. Moreover, for any $2 \leq j \leq k - 1$, we add the gadget H in between u_j and u_{j+1} by identifying u_j with b and u_{j+1} with a (see Fig. 8 where $k = 4$ and each box labeled H (resp. H') means that gadget H (resp. H') has been added by identifying the vertex lying to the left of the box to b and the vertex lying on the right of the box to a (resp. c)).

The resulting graph G' remains B_0-CPG. Indeed, we may add 0-bend CPG representations of the gadgets H and H' inside $\mathcal{R}_{P_{uv}}$ and at different heights so that they do not touch any other such gadget, as shown in Fig. 9. In the full version [12], we give a local example of the resulting 0-bend CPG representation $\mathcal{R}$.

We now show that G is 3-colorable if and only if G' is. To this end, we prove the following.

Claim 1

- In any 3-coloring $\mathbf{c}$ of H', we have $\mathbf{c}(b) \neq \mathbf{c}(c)$.
- In any 3-coloring $\mathbf{c}$ of H, we have $\mathbf{c}(a) = \mathbf{c}(b)$ and $\mathbf{c}(b) \neq \mathbf{c}(c)$.

Proof. Let $\mathbf{c} \colon \{a, b, c, 1, 2, 3, 4, 5, 6, 7, 8, 9, 10\} \to \{blue, red, green\}$ be a 3-coloring of H and assume without loss of generality that $\mathbf{c}(b) = blue$. Clearly, at least two vertices among 4, 6 and 8 have the same color. If vertices 4, 6 and 8 all have the same color, say *red*, then either $\mathbf{c}(7) = blue$ and $\mathbf{c}(9) = green$, or $\mathbf{c}(7) = green$ and $\mathbf{c}(9) = blue$. Therefore, $\{\mathbf{c}(5), \mathbf{c}(10)\} = \{blue, green\}$ and since c is adjacent to all three colors, we then obtain a contradiction. Now if vertices 4 and 8 have the same color, say *red*, then vertex 6 has color *green* and both 7 and 9 have color *blue*, a contradiction. Hence, either $\mathbf{c}(4) = \mathbf{c}(6) \neq \mathbf{c}(8)$ or $\mathbf{c}(8) = \mathbf{c}(6) \neq \mathbf{c}(4)$. By symmetry, we may assume that vertices 4 and 6 have the same color, say *red*, and that vertex 8 has color *green*. This implies that vertex 7 has color *green*, vertices 9 and 5 have color *blue* and vertex 10 has color *red*; but then, $\mathbf{c}(c) = green \neq \mathbf{c}(b)$. This proves the first point of the claim. Observe that each coloring of b and c with distinct colors can be extended to a 3-coloring of H' and H.

As for the second point, since vertices 4 and 6 have color *red*, both 1 and 2 must have color *green*, and since vertex 8 has color *green*, vertex 3 must have color *red*. Consequently, $\mathbf{c}(a) = blue = \mathbf{c}(b)$. ◇

We finally conclude the proof of Theorem 5. By Claim 1, if $\mathbf{c}$ is a 3-coloring of G' then, for any path P_{uv} of $\mathcal{P}$, we have $\mathbf{c}(u_1) \neq \mathbf{c}(u_2)$ and $\mathbf{c}(u_2) = \mathbf{c}(u_i)$ for all $3 \leq i \leq k$. Hence, $\mathbf{c}$ induces a 3-coloring of G. Conversely, it is easy to see that any 3-coloring of G can be extended to a 3-coloring of G'. □

Fig. 7. The gadget H (left) and a 0-bend CPG representation of it (right).

Fig. 8. Adding gadgets H and H'.

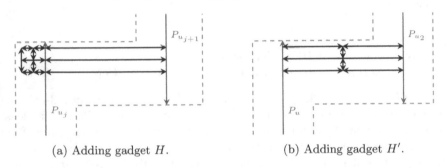

(a) Adding gadget H.   (b) Adding gadget H'.

Fig. 9. Locally adding gadgets to control the color of the vertices.

6   Conclusion

We conclude by stating the following open questions:

1. Are B_1-CPG graphs 5-colorable?
2. Can we characterize those planar graphs which are CPG?
3. Is RECOGNITION NP-complete for B_k-CPG graphs with $k > 0$?

References

1. Aerts, N., Felsner, S.: Vertex contact graphs of paths on a grid. In: Kratsch, D., Todinca, I. (eds.) WG 2014. LNCS, vol. 8747, pp. 56–68. Springer, Cham (2014). https://doi.org/10.1007/978-3-319-12340-0_5
2. Alcón, L., Bonomo, F., Mazzoleni, M.P.: Vertex intersection graphs of paths on a grid: characterization within block graphs. Graphs Comb. **33**(4), 653–664 (2017)
3. Asinowski, A., Cohen, E., Golumbic, M.C., Limouzy, V., Lipshteyn, M., Stern, M.: Vertex intersection graphs of paths on a grid. J. Graph Algorithms Appl. **16**, 129–150 (2012)

4. Beineke, L.W.: Characterizations of derived graphs. J. Comb. Theory **9**(2), 129–135 (1970)
5. Chaplick, S., Cohen, E., Stacho, J.: Recognizing some subclasses of vertex intersection graphs of 0-bend paths in a grid. In: Kolman, P., Kratochvíl, J. (eds.) WG 2011. LNCS, vol. 6986, pp. 319–330. Springer, Heidelberg (2011). https://doi.org/10.1007/978-3-642-25870-1_29
6. Chaplick, S., Kobourov, S.G., Ueckerdt, T.: Equilateral L-contact graphs. In: Brandstädt, A., Jansen, K., Reischuk, R. (eds.) WG 2013. LNCS, vol. 8165, pp. 139–151. Springer, Heidelberg (2013). https://doi.org/10.1007/978-3-642-45043-3_13
7. Chaplick, S., Ueckerdt, T.: Planar graphs as VPG-graphs. In: Didimo, W., Patrignani, M. (eds.) GD 2012. LNCS, vol. 7704, pp. 174–186. Springer, Heidelberg (2013). https://doi.org/10.1007/978-3-642-36763-2_16
8. Cohen, E., Golumbic, M.C., Ries, B.: Characterizations of cographs as intersection graphs of paths on a grid. Discrete Appl. Math. **178**, 46–57 (2014)
9. Cohen, E., Golumbic, M.C., Trotter, W.T., Wang, R.: Posets and VPG graphs. Order **33**(1), 39–49 (2016)
10. de Castro, N., Cobos, F.J., Dana, J.C., Márquez, A., Noy, M.: Triangle-free planar graphs as segments intersection graphs. In: Kratochvíyl, J. (ed.) GD 1999. LNCS, vol. 1731, pp. 341–350. Springer, Heidelberg (1999). https://doi.org/10.1007/3-540-46648-7_35
11. de Fraysseix, H., de Mendez, P.O.: Representations by contact and intersection of segments. Algorithmica **47**(4), 453–463 (2007)
12. Deniz, Z., Munaro, A., Galby, E., Ries, B.: On contact graphs of paths on a grid. arXiv:1803.03468 (2018)
13. Diestel, R.: Graph Theory. GTM, vol. 173. Springer, Heidelberg (2017). https://doi.org/10.1007/978-3-662-53622-3
14. Felsner, S., Knauer, K., Mertzios, G.B., Ueckerdt, T.: Intersection graphs of L-shapes and segments in the plane. Discrete Appl. Math. **2016**, 48–55 (2016)
15. Francis, M., Lahiri, A.: VPG and EPG bend-numbers of Halin graphs. Discrete Appl. Math. **215**, 95–105 (2016)
16. Garey, M.R., Johnson, D.S., Stockmeyer, L.: Some simplified NP-complete graph problems. Theor. Comput. Sci. **1**(3), 237–267 (1976)
17. Garg, A., Tamassia, R.: On the computational complexity of upward and rectilinear planarity testing. SIAM J. Comput. **31**, 601–625 (2001)
18. Golumbic, M.C., Ries, B.: On the intersection graphs of orthogonal line segments in the plane: characterizations of some subclasses of chordal graphs. Graphs Comb. **29**(3), 499–517 (2013)
19. Gonçalves, D., Isenmann, L., Pennarun, C.: Planar graphs as L-intersection or L-contact graphs. In: Proceedings of the 29th Annual ACM-SIAM Symposium on Discrete Algorithms. SODA 2018, pp. 172–184. Society for Industrial and Applied Mathematics (2018)
20. Heldt, D., Knauer, K., Ueckerdt, T.: On the bend-number of planar and outerplanar graphs. Discrete Appl. Math. **179**(C), 109–119 (2014)
21. Hliněný, P.: Classes and recognition of curve contact graphs. J. Comb. Theory Ser. B **74**(1), 87–103 (1998)
22. Hliněný, P.: The maximal clique and colourability of curve contact graphs. Discrete Appl. Math. **81**(1), 59–68 (1998)
23. Hliněný, P.: Contact graphs of line segments are NP-complete. Discrete Math. **235**(1–3), 95–106 (2001)

24. Kobourov, S., Ueckerdt, T., Verbeek, K.: Combinatorial and geometric properties of planar Laman graphs. In: Proceedings of the 24th Annual ACM-SIAM Symposium on Discrete Algorithms. SODA 2013, pp. 1668–1678. Society for Industrial and Applied Mathematics (2013)
25. Korzhik, V.P.: Minimal non-1-planar graphs. Discrete Math. **308**(7), 1319–1327 (2008)
26. Tamassia, R., Tollis, I.G.: Planar grid embedding in linear time. IEEE Trans. Circuits Syst. **36**(9), 1230–1234 (1989)

Specialized Graphs and Trees

On the Area-Universality
of Triangulations

Linda Kleist$^{(\boxtimes)}$

Technische Universität Berlin, Berlin, Germany
kleist@math.tu-berlin.de

Abstract. We study straight-line drawings of planar graphs with prescribed face areas. A plane graph is *area-universal* if for every area assignment on the inner faces, there exists a straight-line drawing realizing the
prescribed areas.

For triangulations with a special vertex order, we present a sufficient
criterion for area-universality that only requires the investigation of one
area assignment. Moreover, if the sufficient criterion applies to one plane
triangulation, then all embeddings of the underlying planar graph are
also area-universal. To date, it is open whether area-universality is a
property of a plane or planar graph.

We use the developed machinery to present area-universal families of
triangulations. Among them we characterize area-universality of accordion graphs showing that area-universal and non-area-universal graphs
may be structural very similar.

Keywords: Area-universality · Triangulation · Planar graph
Face area

1   Introduction

By Fary's theorem [11,20,22], every plane graph has a straight-line drawing. We
are interested in straight-line drawings with the additional property that the
face areas correspond to prescribed values. Particularly, we study *area-universal*
graphs for which all prescribed face areas can be realized by a straight-line
drawing. Usually, in a *planar* drawing, no two edges intersect except in common
vertices. It is worthwhile to be slightly more generous and allow *crossing-free*
drawings, i.e., drawings that can be obtained as the limit of a sequence of planar
straight-line drawings. Note that a crossing-free drawing of a triangulation is not
planar (*degenerate*) if and only if the area of at least one face vanishes. Moreover,
we consider two crossing-free drawings of a plane graph as *equivalent* if the cyclic
order of the incident edges at each vertex and the outer face coincide.

For a plane graph G, we denote the set of faces by F, and the set of inner
faces by F'. An *area assignment* is a function $\mathcal{A}: F' \to \mathbb{R}_{\geq 0}$. We say G is *area-
universal* if for every area assignment $\mathcal{A}$ there exists an equivalent crossing-free

© Springer Nature Switzerland AG 2018
T. Biedl and A. Kerren (Eds.): GD 2018, LNCS 11282, pp. 333–346, 2018.
https://doi.org/10.1007/978-3-030-04414-5_23

drawing where every inner face $f \in F'$ has area $\mathcal{A}(f)$. We call such a drawing $\mathcal{A}$-*realizing* and the area assignment $\mathcal{A}$ *realizable*.

Related Work. Biedl and Ruiz Velázquez [6] showed that planar partial 3-trees, also known as subgraphs of *stacked triangulations* or *Apollonian networks*, are area-universal. In fact, every subgraph of a plane area-universal graph is area-universal. Ringel [19] gave two examples of graphs that have drawings where all face areas are of equal size, namely the octahedron graph and the icosahedron graph. Thomassen [21] proved that plane 3-regular graphs are area-universal. Moreover, Ringel [19] showed that the octahedron graph is not area-universal. Kleist [15] generalized this result by introducing a simple counting argument which shows that no Eulerian triangulation, different from K_3, is area-universal. Moreover, it is shown in [15] that every 1-subdivision of a plane graphs is area-universal; that is, every area assignment of a plane graph has a realizing polyline drawing where each edge has at most one bend. Evans et al. [10,17] present classes of area-universal plane quadrangulations. In particular, they verify the conjecture that plane bipartite graphs are area-universal for quadrangulations with up to 13 vertices. Particular graphs have also been studied: It is known that the square grid [9] and the unique triangulation on seven vertices [4] are area-universal. Moreover, non-area-universal triangulations on up to ten vertices have been investigated in [13].

The computational complexity of the decision problem of area-universality for a given graph was studied by Dobbins et al. [7]. The authors show that this decision problem belongs to UNIVERSAL EXISTENTIAL THEORY OF THE REALS ($\forall \exists \mathbb{R}$), a natural generalization of the class EXISTENTIAL THEORY OF THE REALS ($\exists \mathbb{R}$), and conjecture that this problem is also $\forall \exists \mathbb{R}$-complete. They show hardness of several variants, e.g., the analogue problem of volume universality of simplicial complexes in three dimensions.

In a broader sense, drawings of planar graphs with prescribed face areas can be understood as *cartograms*. Cartograms have been intensely studied for duals of triangulations [1,3,5,14] and in the context of rectangular layouts, dissections of a rectangle into rectangles [8,12,23]. For a detailed survey of the cartogram literature, we refer to [18].

Our Contribution. In this work we present three characterizations of area-universal triangulations. We use these characterizations for proving area-universality of certain triangulations. Specifically, we consider triangulations with a vertex order, where (most) vertices have at least three neighbors with smaller index, called *predecessors*. We call such an order a *p-order*. For triangulations with a p-order, the realizability of an area assignment reduces to finding a real root of a univariate polynomial. If the polynomial is surjective, we can guarantee area-universality. In fact, this is the only known method to prove the area-universality of a triangulation besides the simple argument for plane 3-trees relying on K_4.

We discover several interesting facts. First, to guarantee area-universality it is enough to investigate one area assignment. Second, if the polynomial is surjective for one plane graph, then it is for every embedding of the underlying planar graph. Consequently, the properties of one area assignment can imply the area-universality of all embeddings of a planar graph. This may indicate that area-universality is a property of planar graphs.

We use the method to prove area-universality for several graph families including accordion graphs. To obtain an *accordion graph* from the plane octahedron graph, we introduce new vertices of degree 4 by subdividing an edge of the central triangle. Figure 1 presents four examples of accordion graphs. Surprisingly, the insertion of an even number of vertices yields a non-area-universal graph while the insertion of an odd number of vertices yields an area-universal graph. Accordions with an even number of vertices are Eulerian and thus not area-universal [15]. Consequently, area-universal and non-area-universal graphs may have a very similar structure. (In [17], we use the method to classify small triangulations with p-orders on up to ten vertices.)

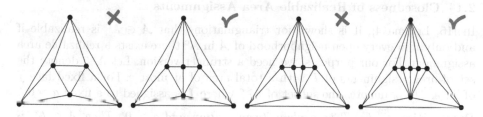

Fig. 1. Examples of accordion graphs. A checkmark indicates area-universality and a cross non-area-universality.

Organization. We start by presenting three characterizations of area-universality of triangulations in Sect. 2. In Sect. 3, we turn our attention to triangulations with p-orders and show how the analysis of one area assignment can be sufficient to prove area-universality of all embeddings of the given triangulation. Then, in Sect. 4, we apply the developed method to prove area-universality for certain graph families; among them we characterize the area-universality of accordion graphs. We end with a discussion and a list of open problems in Sect. 5. For omitted proofs consider the appendices of the full version [16].

2   Characterizations of Area-Universal Triangulations

Throughout this section, let T be a plane triangulation on n vertices. A straight-line drawing of T can be encoded by the $2n$ vertex coordinates, and hence, by a point in the Euclidean space $\mathbb{R}^{2n}$. We call such a vector of coordinates a *vertex placement* and denote the set of all vertex placements encoding crossing-free drawings by $\mathcal{D}(T)$; we also write $\mathcal{D}$ if T is clear from the context.

It is easy to see that an $\mathcal{A}$-realizing drawing of a triangulation can be transformed by an affine linear map into an $\mathcal{A}$-realizing drawing where the outer face corresponds to any given triangle of correct total area $\Sigma\mathcal{A} := \sum_{f \in F'} \mathcal{A}(f)$, where F' denotes the set of inner faces as before.

Lemma 1. [15, Obs. 2] *A plane triangulation T with a realizable area assignment $\mathcal{A}$, has an $\mathcal{A}$-realizing drawing within every given outer face of area $\Sigma\mathcal{A}$.*

Likewise, affine linear maps can be used to scale realizing drawings by any factor. For any positive real number $\alpha \in \mathbb{R}$ and area assignment $\mathcal{A}$, let $\alpha\mathcal{A}$ denote the *scaled area assignment* of $\mathcal{A}$ where $\alpha\mathcal{A}(f) := \alpha \cdot \mathcal{A}(f)$ for all $f \in F'$.

Lemma 2. *Let $\mathcal{A}$ be an area assignment of a plane graph and $\alpha > 0$. The scaled area assignment $\alpha\mathcal{A}$ is realizable if and only if $\mathcal{A}$ is realizable.*

For a plane graph and $c > 0$, let $\mathbb{A}^c$ denote the set of area assignments with a total area of c. Lemma 2 directly implies the following property.

Lemma 3. *Let $c > 0$. A plane graph is area universal if all area assignments in $\mathbb{A}^c$ are realizable.*

2.1   Closedness of Realizable Area Assignments

In [15, Lemma 4], it is shown for triangulations that $\mathcal{A} \in \mathbb{A}^c$ is realizable if and only if in every open neighborhood of $\mathcal{A}$ in $\mathbb{A}^c$ there exists a realizable area assignment. For our purposes, we need a stronger version. Let $\mathbb{A}^{\leq c}$ denote the set of area assignments of T with a total area of at most c. For a fixed face f of T, $\mathbb{A}^{\leq c}|_{f \rightarrow a}$ denotes the subset of $\mathbb{A}^{\leq c}$ where f is assigned to a fixed $a > 0$.

Proposition 1. *Let T be a plane triangulation and $c > 0$. Then $\mathcal{A} \in \mathbb{A}^c$ is realizable if and only if for some face f with $\mathcal{A}(f) > 0$ every open neighborhood of $\mathcal{A}$ in $\mathbb{A}^{\leq 2c}|_{f \rightarrow \mathcal{A}(f)}$ contains a realizable area assignment.*

Intuitively, Proposition 1 enables us not to worry about area assignments with bad but unlikely properties. In particular, area-universality is guaranteed by the realizability of a dense subset of $\mathbb{A}^c$. Moreover, this stronger version allows to certify the realizability of an area assignment by realizable area assignments with slightly different total areas. The proof of Proposition 1 goes along the same lines as in [15, Lemma 4]; it is based on the fact that the set of drawings of T with a fixed face f and a total area of at most $2c$ is compact.

2.2   Characterization by 4-Connected Components

For a plane triangulation T, a *4-connected component* is a maximal 4-connected subgraph of T. Moreover, we call a triangle t of T *separating* if at least one vertex of T lies inside t and at least one vertex lies outside t; in other words, t is not a face of T.

Proposition 2. *A plane triangulation T is area-universal if and only if every 4-connected component of T is area-universal.*

Proof (Sketch). The proof is based on the fact that a plane graph G with a separating triangle t is area-universal if and only if G_E, the induced graph by t and its exterior, and G_I, the induced graph by t and its interior, are area-universal. In particular, Lemma 1 allows us to combine realizing drawings of G_E and G_I to a drawing of G.

Remark. Note that a plane 3-tree has no 4-connected component. (Recall that K_4 is 3-connected and a graph on $n > 4$ vertices is 4-connected if and only if it has no separating triangle.) This is another way to see their area-universality.

2.3 Characterization by Polynomial Equation System

Dobbins et al. [7, Proposition 1] show a close connection of area-universality and equation systems: For every plane graph G with area assignment $\mathcal{A}$ there exists a polynomial equation system $\mathcal{E}$ such that $\mathcal{A}$ is realizable if and only if $\mathcal{E}$ has a real solution. Here we strengthen the statement for triangulations, namely it suffices to guarantee the face areas; these imply all further properties such as planarity and the equivalent embedding. To do so, we introduce some notation.

A plane graph G induces an orientation of the vertices of each face. For a face f given by the vertices $v_1, \ldots, v_k$, we say f is *counter clockwise (ccw)* if the vertices $v_1, \ldots, v_k$ appear in ccw direction on a walk on the boundary of f; otherwise f is *clockwise (cw)*. Moreover, the function $\mathrm{AREA}(f, D)$ measures the area of a face f in a drawing D. For a ccw triangle t with vertices v_1, v_2, v_3, we denote the coordinates of v_i by (x_i, y_i). Its area in D is given by the determinant

$$\mathrm{Det}(v_1, v_2, v_3) := \det \big(c(v_1), c(v_2), c(v_3)\big) = 2 \cdot \mathrm{AREA}(t, D), \qquad (1)$$

where $c(v_i) := (x_i, y_i, 1)$. Since the (complement of the) outer face f_o has area $\Sigma\mathcal{A}$ in an $\mathcal{A}$-realizing drawing, we define $\mathcal{A}(f_o) := \Sigma\mathcal{A}$. For a set of faces $\tilde{F} \subset F$, we define the *area equation system of $\tilde{F}$* as

$$\mathrm{AEQ}(T, \mathcal{A}, \tilde{F}) := \{\mathrm{Det}(v_i, v_j, v_k) = \mathcal{A}(f) \mid f \in \tilde{F}, f =: (v_i, v_j, v_k) \text{ ccw}\}.$$

For convenience, we omit the factor of 2 in each *area equation*. Therefore, without mentioning it any further, we usually certify the realizability of $\mathcal{A}$ by a $1/2\mathcal{A}$-realizing drawing. That is, if we say a triangle has area a, it may have area $1/2a$. Recall that, by Lemma 2, consistent scaling has no further implications.

Proposition 3. *Let T be a triangulation, $\mathcal{A}$ an area assignment, and f a face of T. Then $\mathcal{A}$ is realizable if and only if $\mathrm{AEQ}(T, \mathcal{A}, F \setminus \{f\})$ has a real solution.*

The key idea is that a (scaled) vertex placement of an $\mathcal{A}$-realizing drawing is a real solution of $\mathrm{AEQ}(T, \mathcal{A}, F \setminus \{f\})$ and vice versa. The main task is to guarantee crossing-freeness of the induced drawing; it follows from the following neat fact.

Lemma 4. *Let D be a vertex placement of a triangulation T where the orientation of each inner face in D coincides with the orientation in T. Then D represents a crossing-free straight-line drawing of T.*

A proof of Lemma 4 can be found in [2, in the end of the proof of Lemma 4.2]. An alternative proof relies on the properties of the determinant, in particular, on the fact that for any vertex placement D the area of the triangle formed by its outer vertices evaluates to

$$\mathrm{AREA}(f_o, D) = \sum_{f \in F'} \mathrm{AREA}(f, D). \qquad (2)$$

Equation (2) shows that for every face $f \in F'$, the equation systems
$\mathrm{AEQ}(T, \mathcal{A}, F')$ and $\mathrm{AEQ}(T, \mathcal{A}, F \setminus \{f\})$ are equivalent. This fact is also used for
Proposition 3.

Remark 1. In fact, Lemma 4 and Proposition 3 generalize to *inner triangula-tions*, i.e., 2-connected plane graphs where every inner face is a triangle.

3   Area-Universality of Triangulations with p-orders

We consider planar triangulations with the following property: An order of the
vertices $(v_1, v_2, \ldots, v_n)$, together with a set of *predecessors* $\mathrm{pred}(v_i) \subset N(v_i)$ for
each vertex v_i, is a *p-order* if the following conditions are satisfied:

- $\mathrm{pred}(v_i) \subseteq \{v_1, v_2, \ldots, v_{i-1}\}$, i.e., the predecessors of v_i have an index $< i$,
- $\mathrm{pred}(v_1) = \emptyset$, $\mathrm{pred}(v_2) = \{v_1\}$, $\mathrm{pred}(v_3) = \mathrm{pred}(v_4) = \{v_1, v_2\}$, and
- for all $i > 4$: $|\mathrm{pred}(v_i)| = 3$, i.e., v_i has exactly three predecessors.

Note that $\mathrm{pred}(v_i)$ specifies a subset of preceding neighbors. Moreover, a p-order
is defined for a planar graph independent of a drawing. We usually denote a
p-order by $\mathcal{P}$ and state the order of the vertices; the predecessors are then implic-itly given by $\mathrm{pred}(v_i)$. Figure 2 illustrates a p-order.

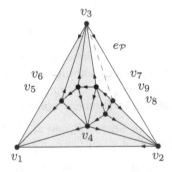

i	$\mathrm{pred}(v_i)$
5	$\{v_1, v_3, v_4\}$
6	$\{v_3, v_4, v_5\}$
7	$\{v_3, v_4, v_6\}$
8	$\{v_2, v_4, v_7\}$
9	$\{v_2, v_7, v_8\}$

Fig. 2. A plane 4-connected triangulation with a p-order $\mathcal{P}$. In an almost realizing
vertex placement constructed with $\mathcal{P}$, all face areas are realized except for the two
faces incident to the unoriented (dashed) edge $e_{\mathcal{P}}$ of $\mathcal{O}_{\mathcal{P}}$ (Lemma 8).

We pursue the following one-degree-of-freedom mechanism to construct real-izing drawings for a plane triangulation T with a p-order $(v_1, v_2, \ldots, v_n)$ and an
area assignment $\mathcal{A}$:

- Place the vertices v_1, v_2, v_3 at positions realizing the area equation of the
  face $v_1 v_2 v_3$. Without loss of generality, we set $v_1 = (0, 0)$ and $v_2 = (1, 0)$.
- Insert v_4 such that the area equation of face $v_1 v_2 v_4$ is realized; this is fulfilled
  if y_4 equals $\mathcal{A}(v_1 v_2 v_4)$ while $x_4 \in \mathbb{R}$ is arbitrary. The value x_4 is our variable.

- Place each remaining vertex v_i with respect to its predecessors $\text{pred}(v_i)$ such that the area equations of the two incident face areas are respected; the coordinates of v_i are *rational* functions of x_4.
- Finally, all area equations are realized except for two special faces f_a and f_b. Moreover, the face area of f_a is a *rational* function $\mathfrak{f}$ of x_4.
- If $\mathfrak{f}$ is *almost surjective*, then there is a vertex placement D respecting all face areas and orientations, i.e., D is a real solution of $\text{AEQ}(T, \mathcal{A}, F)$.
- By Proposition 3, D guarantees the realizability of $\mathcal{A}$.
- If this holds for *enough* area assignments, then T is area-universal.

3.1   Properties of p-orders

A p-order $\mathcal{P}$ of a plane triangulation T induces an orientation $\mathcal{O}_\mathcal{P}$ of the edges: For $w \in \text{pred}(v_i)$, we orient the edge from v_i to w, see also Fig. 2. By Proposition 2, we may restrict our attention to 4-connected triangulations. We note that 4-connectedness is not essential for our method but yields a cleaner picture.

Lemma 5. *Let T be a planar 4-connected triangulation with a p-order $\mathcal{P}$. Then $\mathcal{O}_\mathcal{P}$ is acyclic, $\mathcal{O}_\mathcal{P}$ has a unique unoriented edge $e_\mathcal{P}$, and $e_\mathcal{P}$ is incident to v_n.*

It follows that the p-order encodes all but one edge which is easy to recover. Therefore, the p-order of a planar triangulation T encodes T. In fact, T has a p-order if and only if there exists an edge e such that $T - e$ is 3-degenerate.

Convention. Recall that a drawing induces an orientation of each face. We follow the convention of stating the vertices of inner faces ccw and of the outer face in cw direction. This convention enables us to switch between different plane graphs of the same planar graph without changing the order of the vertices. To account for our convention, we redefine $\mathcal{A}(f_o) := -\Sigma \mathcal{A}$ for the outer face f_o. Then, for different embeddings, only the right sides of the AEQs change.

The next properties can be proved by induction and are shown in Fig. 3.

Lemma 6. *Let T be a plane 4-connected triangulation with a p-order $\mathcal{P}$ specified by $(v_1, v_2, \ldots, v_n)$ and let T_i denote the subgraph of T induced by $\{v_1, v_2, \ldots, v_i\}$. For $i \geq 4$,*

- *T_i has one 4-face and otherwise only triangles,*
- *T_{i+1} can be constructed from T_i by inserting v_{i+1} in the 4-face of T_i, and*
- *the three predecessors of v_i can be named (p_F, p_M, p_L) such that $p_F p_M v_i$ and $p_M p_L v_i$ are (ccw inner and cw outer) faces of T_i.*

Remark 2. For every (non-equivalent) plane graph T' of T, the three predecessors (p_F, p_M, p_L) of v_i in T' and T coincide.

Remark 3. Lemma 6 can be used to show that the number of 4-connected planar triangulations on n vertices with a p-order is $\Omega(2^n/n)$.

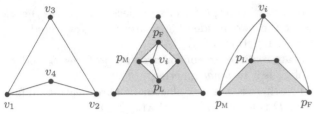

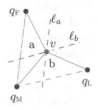

Fig. 3. Illustration of Lemma 6: (a) T_4, (b) v_i is inserted in an inner 4-face, (c) v_i is inserted in outer 4-face.

Fig. 4. Illustration of Lemma 7.

3.2   Constructing Almost Realizing Vertex Placements

Let T be a plane triangulation with an area assignment $\mathcal{A}$. We call a vertex placement D of T *almost $\mathcal{A}$-realizing* if there exist two faces f_a and f_b such that D is a real solution of the equation system $\mathrm{AEQ}(T, \mathcal{A}, \tilde{F})$ with $\tilde{F} := F \setminus \{f_a, f_b\}$. In particular, we insist that the orientation and area of each face, except for f_a and f_b be correct, i.e., the area equations are fulfilled. Note that an almost realizing vertex placement does not necessarily correspond to a crossing-free drawing.

Observation. An almost $\mathcal{A}$-realizing vertex placement D fulfilling the area equations of all faces except for f_a and f_b, certifies the realizability of $\mathcal{A}$ if additionally the area equation of f_a is satisfied.

We construct almost realizing vertex placements with the following lemma.

Lemma 7. *Let $a, b \geq 0$ and let $q_{\mathrm{F}}, q_{\mathrm{M}}, q_{\mathrm{L}}$ be three vertices with a non-collinear placement in the plane. Then there exists a unique placement for vertex v such the ccw triangles $q_{\mathrm{F}} q_{\mathrm{M}} v$ and $q_{\mathrm{M}} q_{\mathrm{L}} v$ fulfill the area equations for a and b, respectively.*

Proof. Consider Fig. 4. To realize the areas, v must be placed on a specific line ℓ_a and ℓ_b, respectively. Note that ℓ_a is parallel to the segment $q_{\mathrm{F}}, q_{\mathrm{M}}$ and ℓ_b is parallel to the segment $q_{\mathrm{M}}, q_{\mathrm{L}}$. Consequently, ℓ_a and ℓ_b are not parallel and their intersection point yields the unique position for vertex v. The coordinates of v are specified by the two equations $\mathrm{Det}(q_{\mathrm{F}}, q_{\mathrm{M}}, v) \overset{!}{=} a$ and $\mathrm{Det}(q_{\mathrm{M}}, q_{\mathrm{L}}, v) \overset{!}{=} b$.

Note that if ℓ_a and ℓ_b are parallel and do not coincide, then there is no position for v realizing the area equations of the two triangles. Based on Lemma 7, we obtain our key lemma.

Lemma 8. *Let T be a plane 4-connected triangulation with a p-order $\mathcal{P}$ specified by $(v_1, v_2, \ldots, v_n)$. Let f_a, f_b be the faces incident to $e_\mathcal{P}$ and $f_0 := v_1 v_2 v_3$. Then there exists a constant $c > 0$ such that for a dense subset $\mathbb{A}_D$ of $\mathbb{A}^c$, every $\mathcal{A} \in \mathbb{A}_D$ has a finite set $\mathcal{B}(\mathcal{A}) \subset \mathbb{R}$, rational functions $x_i(\cdot, \mathcal{A})$, $y_i(\cdot, \mathcal{A})$, $\mathfrak{f}(\cdot, \mathcal{A})$ and a triangle $\triangle$, such that for all $x_4 \in \mathbb{R} \setminus \mathcal{B}(\mathcal{A})$, there exists a vertex placement $D(x_4)$ with the following properties:*

(i) f_0 coincides with the triangle $\triangle$,
(ii) $D(x_4)$ is almost realizing, i.e., a real solution of $\mathrm{AEQ}(T, \mathcal{A}, F \setminus \{f_a, f_b\})$,

(iii) every vertex v_i is placed at the point $\big(x_i(x_4, \mathcal{A}), y_i(x_4, \mathcal{A})\big)$, and
(iv) the area of face f_a in $D(x_4)$ is given by $\mathfrak{f}(x_4, \mathcal{A})$.

The idea of the proof is to use Lemma 7 in order to construct $D(x_4)$ inductively. Therefore, given a vertex placement $v_1, \dots, v_{i-1}$, we have to ensure that the vertices of $\mathrm{pred}(v_i)$ are not collinear. To do so, we consider algebraically independent area assignments. We say an area assignment $\mathcal{A}$ of T is *algebraically independent* if the set $\{\mathcal{A}(f) | f \in F'\}$ is algebraically independent over $\mathbb{Q}$. In fact, the subset of algebraically independent area assignments $\mathbb{A}_I$ of $\mathbb{A}^c$ is dense when c is transcendental.

We call the function $\mathfrak{f}$, constructed in the proof of Lemma 8, the *last face function* of T and interpret it as a function in x_4 whose coefficients depend on $\mathcal{A}$.

3.3   Almost Surjectivity and Area-Universality

In the following, we show that almost surjectivity of the last face function implies area-universality. Let A and B be sets. A function $f\colon A \to B$ is *almost surjective* if f attains all but finitely many values of B, i.e., $B \setminus f(A)$ is finite.

Theorem 1. *Let T be a 4-connected plane triangulation with a p-order $\mathcal{P}$ and let $\mathbb{A}_D, \mathbb{A}^c, \mathfrak{f}$ be obtained by Lemma 8. If the last face function $\mathfrak{f}$ is almost surjective for all area assignments in $\mathbb{A}_D$, then T is area-universal.*

Proof. By Lemma 3, it suffices to show that every $\mathcal{A} \in \mathbb{A}_D$ is realizable. Let f_0 be the triangle formed by v_1, v_2, v_3 and $\mathbb{A}^+ := \mathbb{A}^{\leq 2c}|_{f_0 \to \mathcal{A}(f_0)}$. By Proposition 1, $\mathcal{A}$ is realizable if every open neighborhood of $\mathcal{A}$ in $\mathbb{A}^+$ contains a realizable area assignment. Let f_a and f_b denote the faces incident to $e_{\mathcal{P}}$ and $a := \mathcal{A}(f_a)$. Lemma 8 guarantees the existence of a finite set $\mathcal{B}$ such that for all $x_4 \in \mathbb{R} \setminus \mathcal{B}$, there exists an almost $\mathcal{A}$-realizing vertex placement $D(x_4)$. Since $\mathcal{B}$ is finite and $\mathfrak{f}$ is almost surjective, for every ε with $0 < \varepsilon < c$, there exists $\tilde{x} \in \mathbb{R} \setminus \mathcal{B}$ such that $a \leq \mathfrak{f}(\tilde{x}) \leq a + \varepsilon$, i.e., the area of face f_a in $D(\tilde{x})$ is between a and $a + \varepsilon$. (If f_a and f_b are both inner faces, then the face f_b has an area between $b - \varepsilon$ and b, where $b := \mathcal{A}(f_b)$. Otherwise, if f_a or f_b is the outer face, then the total area changes and face f_b has area between b and $b + \varepsilon$.) Consequently, for some $\mathcal{A}'$ in the ε-neighborhood of $\mathcal{A}$ in $\mathbb{A}^+$, $D(\tilde{x})$ is a real solution of $\mathrm{AEQ}(T, \mathcal{A}', F \setminus \{f_b\})$ and Proposition 3 ensures that $\mathcal{A}'$ is realizable. By Proposition 1, $\mathcal{A}$ is realizable. Thus, T is area-universal. $\qquad\square$

To prove area-universality, we use the following sufficient condition for almost surjectivity. We say two real polynomials p and q are *crr-free* if they do not have common real roots. For a rational function $f := \frac{p}{q}$, we define the *max-degree* of f as $\max\{|p|, |q|\}$, where $|p|$ denotes the degree of p. Moreover, we say f is crr-free if p and q are. The following property follows from the fact that polynomials of odd degree are surjective.

Lemma 9. *Let $p, q\colon \mathbb{R} \to \mathbb{R}$ be polynomials and let Q be the set of the real roots of q. If the polynomials p and q are crr-free and have odd max-degree, then the function $f\colon \mathbb{R} \backslash Q \to \mathbb{R}$, $f(x) = \frac{p(x)}{q(x)}$ is almost surjective.*

For the final result, we make use of several convenient properties of algebraically independent area assignments. For $\mathcal{A}$, let $\mathfrak{f}_{\mathcal{A}}$ denote the last face function and $d_1(\mathfrak{f}_{\mathcal{A}})$ and $d_2(\mathfrak{f}_{\mathcal{A}})$ the degree of the numerator and denominator polynomial of $\mathfrak{f}_{\mathcal{A}}$ in x_4, respectively. Since $\mathfrak{f}_{\mathcal{A}}$ is a function in x_4 whose coefficients depend on $\mathcal{A}$, algebraic independence directly yields the following property.

Claim 1. *For two algebraically independent area assignments $\mathcal{A}, \mathcal{A}' \in \mathbb{A}_I$ of a 4-connected triangulation with a $p-$order $\mathcal{P}$, the degrees of the last face functions $\mathfrak{f}_{\mathcal{A}}$ and $\mathfrak{f}_{\mathcal{A}'}$ with respect to $\mathcal{P}$ coincide, i.e., $d_i(\mathfrak{f}_{\mathcal{A}}) = d_i(\mathfrak{f}_{\mathcal{A}'})$ for $i \in [2]$.*

In fact, the degrees do not only coincide for all algebraically independent area assignments, but also for different embeddings of the plane graph. For a plane triangulation T, let T^* denote the corresponding planar graph and $[T]$ the set (of equivalence classes) of all plane graphs of T^*.

Claim 2. *Let T be a plane 4-connected triangulation with a p-order $\mathcal{P}$. Then for every plane graph $T' \in [T]$, and algebraically independent area assignments $\mathcal{A}$ of T and $\mathcal{A}'$ of T', the last face functions $\mathfrak{f}_{\mathcal{A}}$ and $\mathfrak{f}'_{\mathcal{A}'}$ with respect to $\mathcal{P}$ have the same degrees, i.e., $d_i(\mathfrak{f}_{\mathcal{A}}) = d_i(\mathfrak{f}'_{\mathcal{A}'})$ for $i \in [2]$.*

This implies our final result:

Corollary 1. *Let T be a plane triangulation with a p-order $\mathcal{P}$. If the last face function $\mathfrak{f}$ of T is crr-free and has odd max-degree for one algebraically independent area assignment, then every plane graph in $[T]$ is area-universal.*

4 Applications

We now use Theorem 1 and Corollary 1 to prove area-universality of some classes of triangulations. The considered graphs rely on an operation that we call *diamond addition*. Consider the left image of Fig. 5. Let G be a plane graph and let e be an inner edge incident to two triangular faces that consist of e and the vertices u_1 and u_2, respectively. Applying a *diamond addition of order k* on e results in the graph G' which is obtained from G by subdividing edge e with k vertices, $v_1, \ldots, v_k$, and inserting the edges $v_i u_j$ for all pairs $i \in [k]$ and $j \in [2]$. Figure 5 illustrates a diamond addition on e of order 3.

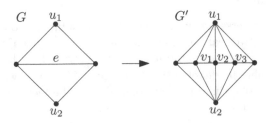

Fig. 5. Obtaining G' from G by a diamond addition of order 3 on edge e.

4.1   Accordion Graphs

An accordion graph can be obtained from the plane octahedron graph $\mathcal{G}$ by a diamond addition: Choose one edge of the central triangle of $\mathcal{G}$ as the *special edge*. The *accordion* graph $\mathcal{K}_\ell$ is the plane graph obtained by a diamond addition of order ℓ on the special edge of $\mathcal{G}$. Consequently, $\mathcal{K}_\ell$ has $\ell+6$ vertices. We speak of an *even accordion* if ℓ is even and of an *odd accordion* if ℓ is odd. Figure 1 illustrates the accordion graphs $\mathcal{K}_i$ for $i \leq 3$. Note that $\mathcal{K}_0$ is $\mathcal{G}$ itself and $\mathcal{K}_1$ is the unique 4-connected plane triangulation on seven vertices. Due to its symmetry, it holds that $[\mathcal{K}_\ell] = \{\mathcal{K}_\ell\}$.

Theorem 2. *The accordion graph $\mathcal{K}_\ell$ is area-universal if and only if ℓ is odd.*

Proof (Sketch). Performing a diamond addition of order ℓ on some plane graph changes the degree of exactly two vertices by ℓ while all other vertex degrees remain the same. Consequently, if ℓ is even, all vertices of $\mathcal{K}_\ell$ have even degree, and hence, $\mathcal{K}_\ell$ as an Eulerian triangulation is not area-universal as shown by the author in [15, Theorem 1].

It remains to prove the area-universality of odd accordion graphs with the help of Theorem 1. Consider an arbitrary but fixed algebraically independent area assignment $\mathcal{A}$. We use the p-order depicted in Fig. 6 to construct an almost realizing vertex placement. We place the vertices v_1 at $(0,0)$, v_2 at $(1,0)$, v_3 at $(1, \Sigma\mathcal{A})$, and v_4 at (x_4, a) with $a := \mathcal{A}(v_1 v_2 v_4)$. Consider also Fig. 6.

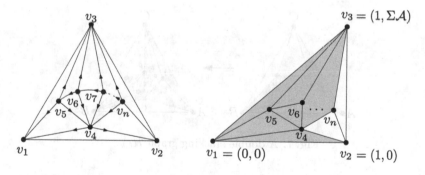

Fig. 6. A p-order of an accordion graph (left) and an almost realizing vertex placement (right), where the shaded faces are realized.

We use Lemma 8 to construct an almost realizing vertex placement. Note that for all vertices v_i with $i > 5$, the three predecessors of v_i are $p_\text{F} = v_3$, $p_\text{M} = v_{i-1}$ and $p_\text{L} = v_4$. One can show that the vertex coordinates of v_i can be expressed as $x_i = \mathcal{N}_i^x / \mathcal{D}_i$ and $y_i = \mathcal{N}_i^y / \mathcal{D}_i$, where $\mathcal{N}_i^x, \mathcal{N}_i^y, \mathcal{D}_i$ are polynomials in x_4. Moreover, the polynomials fulfill the following crucial properties.

Lemma 10. *For all $i \geq 5$, it holds that $|\mathcal{D}_5| = 1$ and*

$$|\mathcal{N}_{i+1}^x| = |\mathcal{D}_{i+1}| = |\mathcal{N}_{i+1}^y| + 1 = |\mathcal{D}_i| + 1.$$

Consequently, $|\mathcal{N}_i^x| = |\mathcal{D}_i|$ is odd if and only if i is odd. In particular, for odd ℓ, $|\mathcal{N}_n^x| = |\mathcal{D}_n|$ is odd since the number of vertices $n = \ell + 6$ is odd.

Lemma 11. *For all $i \geq 5$ and $\circ \in \{x, y\}$, it holds that $\mathcal{N}_i^\circ$ and $\mathcal{D}_i$ are crr-free.*

Consequently, the area of the ccw triangle $v_2 v_3 v_n$ in $D(x_4)$ is given by the crr-free last face function

$$\mathfrak{f}(x) := \mathrm{Det}(v_2, v_3, v_n) = \Sigma \mathcal{A}(1 - x_n) = \Sigma \mathcal{A}\left(1 - \frac{\mathcal{N}_n^x}{\mathcal{D}_n}\right).$$

Since $|\mathcal{N}_n^x|$ and $|\mathcal{D}_n|$ are odd, the max-degree of $\mathfrak{f}$ is odd. Thus, Lemma 9 ensures that $\mathfrak{f}$ is almost surjective. By Theorem 1, $\mathcal{K}_\ell$ is area-universal for odd ℓ.

This result can be generalized to double stacking graphs.

4.2 Double Stacking Graphs

Denote the vertices of the plane octahedron $\mathcal{G}$ by ABC and uvw as depicted in Fig. 7. The *double stacking graph* $\mathcal{H}_{\ell,k}$ is the plane graph obtained from $\mathcal{G}$ by applying a diamond addition of order $\ell - 1$ on Au and a diamond addition of order $k - 1$ on vw. Note that $\mathcal{H}_{\ell,k}$ has $(\ell + k + 4)$ vertices. Moreover, $\mathcal{H}_{\ell,1}$ is isomorphic to $\mathcal{K}_{\ell-1}$; in particular, $\mathcal{H}_{1,1}$ equals $\mathcal{G}$. Note that $[\mathcal{H}_{\ell,k}]$ usually contains several (equivalence classes of) plane graphs.

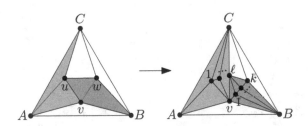

Fig. 7. A double stacking graph $\mathcal{H}_{\ell,k}$.

Theorem 3. *A plane graph in $[\mathcal{H}_{\ell,k}]$ is area-universal if and only if $\ell \cdot k$ is even.*

If $\ell \cdot k$ is odd, every plane graph in $[\mathcal{H}_{\ell,k}]$ is Eulerian and hence not area-universal by [15, Theorem 1]. If $\ell \cdot k$ is even, we consider an algebraically independent area assignment of $\mathcal{H}_{\ell,k}$, show that its last face function is crr-free and has odd max-degree. Then we apply Corollary 1.

Theorem 3 implies that

Corollary 2. *For every $n \geq 7$, there exists a 4-connected triangulation on n vertices that is area-universal.*

5   Discussion and Open Problems

For triangulations with p-orders, we introduced a sufficient criterion to prove area-universality of all embeddings of a planar graph which relies on checking properties of one area assignments of one plane graph. We used the criterion to present two families of area-universal triangulations. Since area-universality is maintained by taking subgraphs, area-universal triangulations are of special interest. For instance, the area-universal double stacking graphs are used in [10,17] to show that all plane quadrangulations with at most 13 vertices are area-universal. The analysis of accordion graphs showns that area-universal and non-area-universal graphs can be structural very similar. The class of accordion graphs gives a hint why understanding area-universality seems to be a difficult problem. In conclusion, we pose the following open questions:

- Is area-universality a property of plane or planar graphs?
- What is the complexity of deciding the area-universality of triangulations?
- Can area-universal graphs be characterized by *local* properties?

Acknowledgements. I thank Udo Hoffmann and Sven Jäger for helpful comments.

References

1. Alam, M.J., Biedl, T.C., Felsner, S., Kaufmann, M., Kobourov, S.G., Ueckerdt, T.: Computing cartograms with optimal complexity. Discrete Comput. Geom. **50**(3), 784–810 (2013). https://doi.org/10.1007/s00454-013-9521-1
2. Angelini, P., et al.: Windrose planarity: embedding graphs with direction-constrained edges. In: Proceedings of the Twenty-Seventh Annual ACM-SIAM Symposium on Discrete Algorithms, pp. 985–996 (2016). https://doi.org/10.1137/1.9781611974331
3. de Berg, M., Mumford, E., Speckmann, B.: On rectilinear duals for vertex-weighted plane graphs. Discrete Math. **309**(7), 1794–1812 (2009). https://doi.org/10.1016/j.disc.2007.12.087
4. Bernáthová, A.: Kreslení grafů s podmínkami na velikosti stěn, Bachelor thesis. Charles University Prague (2009). https://is.cuni.cz/webapps/zzp/detail/63479/20281489/
5. Biedl, T.C., Velázquez, L.E.R.: Orthogonal cartograms with few corners per face. In: Algorithms and Data Structures Symposium (WADS), pp. 98–109 (2011). https://doi.org/10.1007/978-3-642-22300-6_9
6. Biedl, T.C., Velázquez, L.E.R.: Drawing planar 3-trees with given face areas. Comput. Geom. **46**(3), 276–285 (2013). https://doi.org/10.1016/j.comgeo.2012.09.004
7. Dobbins, M.G., Kleist, L., Miltzow, T., Rzążewski, P.: Is area universality ∀∃ℝ-complete? In: Brandstädt, A., KöhlerKlaus, E., Meer, K. (eds.) Graph-Theoretic Concepts in Computer Science, WG 2018. LNCS, vol. 11159. Springer, Heidelberg (2018). https://doi.org/10.1007/978-3-030-00256-5. Accepted
8. Eppstein, D., Mumford, E., Speckmann, B., Verbeek, K.: Area-universal and constrained rectangular layouts. SIAM J. Comput. **41**(3), 537–564 (2012). https://doi.org/10.1137/110834032

9. Evans, W., et al.: Table cartogram. Comput. Geom. **68**, 174–185 (2017). https://doi.org/10.1016/j.comgeo.2017.06.010. Special Issue in Memory of Ferran Hurtado
10. Evans, W., Felsner, S., Kleist, L., Kobourov, S.G.: On area-universal quadrangulations. In preparation (2018)
11. Fáry, I.: On straight line representations of planar graphs. Acta Scientiarum Mathematicarum **11**, 229–233 (1948)
12. Felsner, S.: Exploiting air-pressure to map floorplans on point sets. J. Graph. Algorithms Appl. **18**(2), 233–252 (2014). https://doi.org/10.7155/jgaa.00320
13. Heinrich, H.: Ansätze zur Entscheidung von Flächenuniversalität. Master thesis, Technische Universität Berlin (2018)
14. Kawaguchi, A., Nagamochi, H.: Orthogonal drawings for plane graphs with specified face areas. In: Theory and Applications of Model of Computation (TAMC), pp. 584–594 (2007). https://doi.org/10.1007/978-3-540-72504-6_53
15. Kleist, L.: Drawing planar graphs with prescribed face areas. J. Comput. Geom. (JoCG) **9**(1), 290–311 (2018). https://doi.org/10.20382/jocg.v9i1a9
16. Kleist, L.: On the area-universality of triangulations. preprint, ArXiv:1808.10864v2 (2018). http://arxiv.org/abs/1808.10864v2
17. Kleist, L.: Planar graphs and face areas - area-universality. PhD thesis, Technische Universität Berlin (2018)
18. Nusrat, S., Kobourov, S.: The state of the art in cartograms. In: Computer Graphics Forum, vol. 35, pp. 619–642. Wiley Online Library (2016). https://doi.org/10.1111/cgf.12932
19. Ringel, G.: Equiareal graphs. In: Bodendiek, R. (ed.) Contemporary Methods in Graph Theory, pp. 503–505. BI Wissenschaftsverlag Mannheim (1990). In honour of Prof. Dr. K. Wagner
20. Stein, S.K.: Convex maps. Proc. Am. Math. Soc **2**(3), 464 (1951). https://doi.org/10.1090/S0002-9939-1951-0041425-5
21. Thomassen, C.: Plane cubic graphs with prescribed face areas. Comb. Probab. Comput. **1**(4), 371–381 (1992). https://doi.org/10.1017/S0963548300000407
22. Wagner, K.: Bemerkungen zum Vierfarbenproblem. Jahresbericht der Deutschen Mathematiker-Vereinigung **46**, 26–32 (1936). https://gdz.sub.uni-goettingen.de/id/PPN37721857X_0046
23. Wimer, S., Koren, I., Cederbaum, I.: Floorplans, planar graphs, and layouts. IEEE Trans. Circuits Syst. **35**, 267–278 (1988). https://doi.org/10.1109/31.1739

Monotone Drawings of k-Inner Planar Graphs

Anargyros Oikonomou and Antonios Symvonis[⊠]

School of Applied Mathematical and Physical Sciences,
National Technical University of Athens, Athens, Greece
ar.economou@outlook.com, symvonis@math.ntua.gr

Abstract. A *k-inner planar graph* is a planar graph that has a plane drawing with at most k *internal vertices*, i.e., vertices that do not lie on the boundary of the outer face of its drawing. An outerplanar graph is a 0-inner planar graph. In this paper, we show how to construct a monotone drawing of a k-inner planar graph on a $2(k+1)n \times 2(k+1)n$ grid. In the special case of an outerplanar graph, we can produce a planar monotone drawing on a $n \times n$ grid, improving the results in [2,11].

1 Introduction

A *straight-line drawing* Γ of a graph G is a mapping of each vertex to a distinct point on the plane and of each edge to a straight-line segment between the vertices. A path $P = \{p_0, p_1, \ldots, p_n\}$ is *monotone* if there exists a line l such that the projections of the vertices of P on l appear on l in the same order as on P. A straight-line drawing Γ of a graph G is *monotone*, if a *monotone* path connects every pair of vertices. We say that an angle θ is *convex* if $0 < \theta \leq \pi$. A tree T is called *ordered* if the order of edges incident to any vertex is fixed. We call a drawing Γ of an ordered tree T rooted at r *near-convex monotone* if it is monotone and any pair of consecutive edges incident to a vertex, with the exception of a single pair of consecutive edges incident to r, form a convex angle.

Monotone graph drawing has been lately a very active research area and several interesting results have appeared since its introduction by Angelini et al. [1]. In the case of trees, Angelini et al. [1] provided two algorithms that used ideas from number theory (Stern-Brocot trees [3,15] [6, Sect. 4.5]) to produce monotone tree drawings. Their BFS-based algorithm used an $O(n^{1.6}) \times O(n^{1.6})$ grid while their DFS-based algorithm used an $O(n) \times O(n^2)$ grid. Later, Kindermann et al. [12] provided an algorithm based on Farey sequence (see [6, Sect. 4.5]) that used an $O(n^{1.5}) \times O(n^{1.5})$ grid. He and He in a series of papers [7–9] gave algorithms, also based on Farey sequences, that eventually reduced the required grid size to $O(n) \times O(n)$. Their monotone tree drawing uses a $12n \times 12n$ grid which is asymptotically optimal as there exist trees which require at least $\frac{n}{9} \times \frac{n}{9}$ area [8]. In a recent paper, Oikonomou and Symvonis [14] followed a different approach from number theory and gave an algorithm based on a simple weighting method and some simple facts from geometry, that draws a tree on an $n \times n$ grid.

© Springer Nature Switzerland AG 2018
T. Biedl and A. Kerren (Eds.): GD 2018, LNCS 11282, pp. 347–353, 2018.
https://doi.org/10.1007/978-3-030-04414-5_24

Hossain and Rahman [11] showed that by modifying the embedding of a connected planar graph we can produce a planar monotone drawing of that graph on an $O(n) \times O(n^2)$ grid. To achieve that, they reinvented the notion of an *orderly spanning tree* introduced by Chiang et al. [4] and referred to it as *good spanning tree*. Finally, He and He [10] proved that Felsner's algorithm [5] builds convex monotone drawings of 3-connected planar graphs on $O(n) \times O(n)$ grids in linear time.

A k-*inner planar graph* is a planar graph that has a plane drawing with at most k *inner vertices*, i.e., vertices that do not lie on the boundary of the outer face of its drawing. An outerplanar graph is a 0-inner planar graph. In this paper, we show how to construct a monotone drawing of a k-inner planar graph on a $2(k+1)n \times 2(k+1)n$ grid. This yields monotone drawings for outerplanar graphs on a $2n \times 2n$ grid, improving the results in [2,11]. Due to space limitation, some proofs appear in the arXiv version of the paper [13].

2   Preliminaries

Let T be a tree rooted at a vertex r. Denote by T_u the sub-tree of T rooted at vertex u. By $|T_u|$ we denote the number of vertices of T_u. In the rest of the paper, we assume that all tree edges are directed away from the root.

Our algorithm for obtaining a monotone plane drawing of a k-inner planar graph in based on our ability: (i) to produce for any rooted tree a compact monotone drawing that satisfies specific properties and (ii) to identify for any planar graph a *good spanning tree*.

Theorem 1 (Oikonomou, Symvonis [14]). *Given an n-vertex tree T rooted at vertex r, we can produce in $O(n)$ time a monotone drawing of T where: (i) the root r is drawn at $(0,0)$, (ii) the drawing in non-strictly slope-disjoint[1], (iii) the drawing is contained in the first quadrant, and (iv) it fits in an $n \times n$ grid.*

By utilizing and slightly modifying the algorithm that supports Theorem. 1, we can obtain a specific monotone tree drawing that we later use in our algorithm for monotone k-inner planar graphs.

Theorem 2. *Given an n-vertex tree T rooted at vertex r, we can produce in $O(n)$ time a monotone drawing of T where: (i) the root r is drawn at $(0,0)$, (ii) the drawing in non-strictly slope-disjoint, (iii) the drawing is near-convex, (iv) the drawing is contained in the second octant (defined by two half-lines with a common origin and slope $\frac{\pi}{4}$ and $\frac{\pi}{2}$, resp.), and (v) it fits in a $2n \times 2n$ grid.*

The following definition of a *good spanning tree* is due to Hossain and Rahman [11]. Let G be a connected embedded plane graph and let r be a vertex of G lying on the boundary of its outer face. Let T be an ordered spanning tree of G rooted in r that respects the embedding of G. Let $P(r,v) = \langle u_1(= r), u_2, ..., u_k(= v) \rangle$ be the unique path in T from the root r to a vertex $v \neq r$.

---

[1] See [14] for the definition of *non-strictly slope-disjoint* drawings of trees.

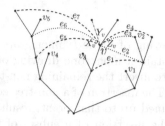
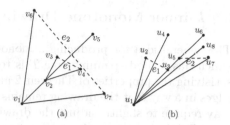

Fig. 1. A good spanning tree.   **Fig. 2.** Dependencies between leader edges.

The path $P(r, v)$ divides the children of u_i $(1 \leq i < k)$, except u_{i+1}, into two groups; the left group L and the right group R. A child x of u_i is in group L and, more specifically in a subgroup of L denoted by u_i^L, if the edge (u_i, x) appears before edge (u_i, u_{i+1}) in clockwise order of the edges incident to u_i when the ordering starts from edge (u_i, u_{i-1}). Similarly, a child x of u_i is in group R and more specifically in a subgroup of R denoted by u_i^R if the edge (u_i, x) appears after edge (u_i, u_{i+1}) in clockwise order of the edges incident to u_i when the ordering starts from edge (u_i, u_{i-1}). A tree T is called a good spanning tree of G if every vertex v $(v \neq r)$ of G satisfies the following two conditions with respect to $P(r, v)$ (see Fig. 1 for an example of a good spanning tree (solid edges)):

C1: G does not have a non-tree edge $(v, u_i), i < k$; and

C2: The edges of G incident to vertex v, excluding (u_{k-1}, v), can be partitioned into three disjoint (and possibly empty) sets X_v, Y_v and Z_v satisfying the following conditions:

   a Each of X_v and Z_v is a set of consecutive non-tree edges and Y_v is a set of consecutive tree edges.

   b Edges of X_v, Y_v and Z_v appear clockwise in this order after edge (u_{k-1}, v).

   c For each edge $(v, v') \in X_v$, $v' \in T_w$ for some $w \in u_i^L$, $i < k$, and for each edge $(v, v') \in Z_v$, $v' \in T_w$ for some $w \in u_i^R$, $i < k$.

Theorem 3 ([4,11]). *Let G be a connected planar graph of n vertices. Then G has a planar embedding G_ϕ that contains a good spanning tree. Furthermore, G_ϕ and a good spanning tree T of G_ϕ can be found in $O(n)$ time.*

Consider an embedded plane graph G for which a good spanning tree T exists. We say that a non-tree edge e of G covers vertex u if u lies in the inner face delimited by the simple cycle formed by tree-edges of T and e. Vertices on the cycle are not covered by edge e.

Lemma 1. *If a planar graph G has a k-inner embedding, then G has a k'-inner embedding which contains a good spanning tree T, where $k' \leq k$. Moreover, each non-tree edge covers at most k of T's leaves.*

3   k-inner Monotone Drawings

The general idea for producing a monotone drawing of plane k-inner graph G which has a good spanning tree T is to first obtain a monotone drawing of T satisfying the properties of Theorem 2 and then to insert the remaining non-tree edges in a way that the drawing remains planar. The insertion of a non-tree edge may require to slightly adjust the drawing obtained up to that point, resulting in a slightly larger drawing. As it turns out, the insertion of a subset of the non-tree edges may violate the planarity of the drawing and, moreover, if these edges are considered in a proper order, the increase on the size of the drawing can be kept small (up to a factor of k for each dimension).

Consider a plane graph G that has a good spanning tree T. For any non-tree edge e of G, we denote by $C(e)$ the set of leaf-vertices of T covered by e. A non-tree edge e is called a *leader edge* if $C(e) \neq \emptyset$ and there doesn't exist another edge e' such that $C(e) = C(e')$ with e' lying in the inner of the cycle induced by the edges of T and e. In Fig. 1, leader edges are drawn as dashed edges. We also have that $C(e_1) = C(e_2) = C(e_3) = \{v_1\}$, $C(e_4) = \{v_1, v_2\}$, $C(e_5) = C(e_6) = \{v_3, v_4\}$, and $C(e_7) = \{v_3, v_4, v_5\}$.

Lemma 2. *Let G be a k-inner plane graph that has a good spanning tree. Then, there exist at most k leader edges in G.*

Proof (Sketch). Firstly observe that a boundary vertex of G that is a leaf-vertex in T cannot be covered by any edge of G. That is, the set $C(e)$, for any non-tree edge e, contains only inner-vertices of G, and thus, $|C(e)| \leq k$. The proof then easily follows by observing that for any two distinct leader edges e_1, e_2, exactly one of the following statements holds: (i) $C(e_1) \subset C(e_2)$, (ii) $C(e_2) \subset C(e_1)$, (iii) $C(e_1) \cap C(e_2) = \emptyset$.   □

Lemma 3. *Let G be a plane graph that has a good spanning tree T and let Γ_T be a monotone drawing of T that is near-convex and non-strictly slope-disjoint. Let Γ_L be the drawing produced if we add all leader edges of G to Γ_T and Γ be the drawing produced if we add all non-tree edges of G to Γ_T. Then, Γ_L is planar if and only if Γ is planar.*

Lemma 3 indicates that we only have to adjust the original drawing of T so that after the addition of all leader edges it is still planar. Then, the remaining non-tree edges can be drawn without violating planarity. Note that, for the proof of Lemma 3, it is crucial that the original drawing Γ_T of T is near-convex.

Consider the near-convex drawing of a good spanning tree T rooted at r that is non-strictly slope-disjoint. Assume that T is drawn in the first quadrant with its root r at $(0,0)$. Let $u \neq r$ be a vertex of T and p^u be its parent. Vertices u and p^u are drawn at grid points (u_x, u_y) and (p^u_x, p^u_y), resp. We define the *reference vector* of u with respect to its parent in T, denoted by $\overrightarrow{u}$, to be vector $\overrightarrow{u} = (u_x - p^u_x, u_y - p^u_y)$. We emphasize that the reference vector $\overrightarrow{u}$ of a tree vertex u is defined wrt the original drawing of T and does not change even if the drawing of T is modified by our drawing algorithm.

The *elongation of edge* (p^u, u) *by a factor of* $\lambda, \lambda \in \mathbb{N}^+$, (also referred to as a λ-*elongation*) translates the drawing of subtree T_u along the direction of edge (p^u, u) so that the new length of (p^u, u) increases by λ times the length of the reference vector $\overrightarrow{u}$ of u. Since the elongation factor is a natural number, the new drawing is still a grid drawing. Let $\overrightarrow{u} = (u_x^d, u_y^d)$. After a λ-elongation of (p^u, u), vertex u is drawn at point $u' = (u_x + \lambda u_x^d, \ u_y + \lambda u_y^d)$. A 0-elongation leaves the drawing unchanged. Note that, by appropriately selecting the elongation factor λ for an upward tree edge (p^u, u), we can reposition u so that it is placed above any given point $z = (z_x, z_y)$.

If we insert the leader edges in the drawing of the good spanning tree in an arbitrary order, we may have to adjust the drawing more than one time for each inserted edge. This is due to dependencies between leader edges. Figure 2 describes the two types of possible dependencies. In the case of Fig. 2(a), the leader edge must by inserted first since $C(e_1) = \{v_5\} \subset \{v_3, v_4, v_5\} = C(e_2)$. Inserting leader e_1 so that it is not intersected by any tree edge can be achieved by elongating edges (v_2, v_3) and (v_2, v_4) by appropriate factors so that vertices v_3 and v_4 are both placed at grid points above (i.e., with larger y coordinate) vertex v_5.

In the case of Fig. 2(b), we have that $C(e_1) = \{u_4\}$, $C(e_2) = \{u_8\}$, and $C(e_1) \cap C(e_2) = \emptyset$. However, leader e_1 must be inserted first since one of its endpoints (u_3) is an ancestor of an endpoint (u_5) of e_2. Again, inserting leader e_1 so that it is not intersected by any tree edge can be achieved by elongating edges (u_1, u_2) and (u_1, u_3) by appropriate factors so that vertices u_2 and u_3 are both placed at grid points above vertex u_4.

Lemma 4. *Let G be a plane graph that has a good spanning tree T and let Γ_T be a drawing of T that satisfies the properties of Theorem 2. Then, there exists an ordering of the leader edges, such that if they are inserted into Γ_T (with the appropriate elongations) in that order they need to be examined exactly once.*

Our method for producing monotone drawings of k-inner planar graphs is summarized in Algorithm 1. A proof that the produced drawing is actually a monotone plane drawing follows from the facts that (i) there is always a good spanning tree with at most k inner vertices (Lemma 1), (ii) there is a monotone tree drawing satisfying the properties of Theorem 2, (iii) the operation of edge elongation on the vertices of a near-convex monotone non-strictly slope-disjoint tree drawing maintains these properties, (iv) the ability to always insert the leader edges into the drawing without violating planarity (through elongation), and (v) the ability to insert the remaining non-tree edges (Lemma 3).

Let $\overrightarrow{u} = (u_x^d, u_y^d)$ and $\overrightarrow{v} = (v_x^d, v_y^d)$ be the reference vectors of u and v, resp. When we process leader edge $e = (u, v)$ (lines 9–15 of Algorithm 1), factors $\lambda_u = \max\left(0, \left\lceil \frac{w_y - u_y}{u_y^d} \right\rceil\right)$ and $\lambda_v = \max\left(0, \left\lceil \frac{w_y - v_y}{v_y^d} \right\rceil\right)$ are used for the elongation of edges (p^u, u) and (p^v, v), resp. The use of these elongation factors ensures that both u and v are placed above vertex w, and thus the insertion of edge e leaves the drawing planar. When the leader edges are processed in the order dictated by their dependencies (line 6 of Algorithm 1), we can show that:

Algorithm 1. Monotone drawing of k-inner planar graphs

1: **procedure** k-INNERPLANARMONOTONEDRAWING(G)
2: Input: An n-vertex planar graph G and an embedding of G with k inner vertices.
3: Output: A monotone planar drawing of G on a $2(k+1)n \times 2(k+1)n$ grid.
4:
5:    Let G_ϕ be an embedding of G that has a good spanning tree T.
6:    $L \leftarrow$ List of the *leader* edges of G, ordered wrt their dependencies (Lemma 4).
7:    $\Gamma_T \leftarrow$ Monotone drawing of T satisfying all properties of Theorem 2.
8:    $\Gamma \leftarrow \Gamma_T$
9:    **while** L is not empty **do**
10:        Let $e = (u, v)$ be the next edge in L. Remove e from L.
11:        Let w be the vertex of $C(e)$ (drawn in Γ at (w_x, w_y)) with the largest
12:            y-coordinate.
13:        Let p^u and p^v be the parents of vertices u and v in T, resp.
14:        *Elongate* Line edges (p^u, u) and (p^v, v) by appropriate factors so that u and
15:            v are placed in Γ above vertex w.
16:        Insert edge e into drawing Γ.
17:    Insert all remaining non-tree edges into drawing Γ.
18:    Output Γ.

Lemma 5. *Let Γ_T be the drawing of the good spanning tree T satisfying the properties of Theorem 2 and let $\Gamma_{(u,v)}$ be the drawing of G immediately after the insertion of leader edge $e = (u, v)$ by Algorithm 1. Let u and v be drawn in Γ_T at points (u_x, u_y) and (v_x, v_y), resp. Then, in $\Gamma_{(u,v)}$ the drawings of T_u and T_v are contained in $(2n - u_x) \times (2n - u_y)$ and $(2n - v_x) \times (2n - v_y)$ grids, resp.*

Based on Lemma 5, we can easily show the main result of our paper.

Theorem 4. *Let G be an n-vertex k-inner planar graph. Algorithm 1 produces a planar monotone drawing of G on a $2(k+1)n \times 2(k+1)n$ grid.*

A corollary of Theorem 4 is that for an n-vertex outerplanar graph G Algorithm 1 produces a planar monotone drawing of G on a $2n \times 2n$ grid. However, we can further reduce the grid size down to $n \times n$.

Theorem 5. *Let G be an n-vertex outerplanar graph. Then, there exists an $n \times n$ planar monotone grid drawing of G.*

Proof. Simply observe that since an outerplanar graph has no leader edges, the drawing produced by Algorithm 1 is identical to that of the original drawing of the good spanning tree T. In Algorithm 1 we used a $2n \times 2n$ drawing of T in the second octant in order to simplify the elongation operation. Since outerplanar graphs have no leader edges, they require no elongations and we can use instead the (first quadrant) $n \times n$ monotone tree drawing of [14], appropriately modified so that it yields a near-convex drawing. $\square$

4   Conclusion

We defined the class of k-inner planar graphs which bridges the gap between outerplanar and planar graphs. For an n-vertex k-inner planar graph G, we provided an algorithm that produces a $2(k+1)n \times 2(k+1)n$ monotone grid drawing of G. Building algorithms for k-inner graphs that incorporate k into their time complexity or into the quality of their solution is an interesting open problem.

References

1. Angelini, P., Colasante, E., Di Battista, G., Frati, F., Patrignani, M.: Monotone drawings of graphs. J. Graph Algorithms Appl. **16**(1), 5–35 (2012). https://doi.org/10.7155/jgaa.00249

2. Angelini, P., et al.: Monotone drawings of graphs with fixed embedding. Algorithmica **71**(2), 233–257 (2015). https://doi.org/10.1007/s00453-013-9790-3

3. Brocot, A.: Calcul des rouages par approximation, nouvelle methode. Revue Chronometrique **6**, 186–194 (1860)

4. Chiang, Y., Lin, C., Lu, H.: Orderly spanning trees with applications. SIAM J. Comput. **34**(4), 924–945 (2005). https://doi.org/10.1137/S0097539702411381

5. Felsner, S.: Convex drawings of planar graphs and the order dimension of 3-polytopes. Order **18**(1), 19–37 (2001). https://doi.org/10.1023/A:1010604726900

6. Graham, R.L., Knuth, D.E., Patashnik, O.: Concrete Mathematics: A Foundation for Computer Science, 2nd edn. Addison-Wesley Longman Publishing Co., Inc., Boston (1994)

7. He, D., He, X.: Nearly optimal monotone drawing of trees. Theor. Comput. Sci. **654**, 26–32 (2016). https://doi.org/10.1016/j.tcs.2016.01.009

8. He, D., He, X.: Optimal monotone drawings of trees. SIAM J. Discret. Math. **31**(3), 1867–1877 (2017). https://doi.org/10.1137/16M1080045

9. He, X., He, D.: Compact monotone drawing of trees. In: Xu, D., Du, D., Du, D. (eds.) COCOON 2015. LNCS, vol. 9198, pp. 457–468. Springer, Cham (2015). https://doi.org/10.1007/978-3-319-21398-9_36

10. He, X., He, D.: Monotone drawings of 3-connected plane graphs. In: Bansal, N., Finocchi, I. (eds.) ESA 2015. LNCS, vol. 9294, pp. 729–741. Springer, Heidelberg (2015). https://doi.org/10.1007/978-3-662-48350-3_61

11. Hossain, M.I., Rahman, M.S.: Good spanning trees in graph drawing. Theor. Comput. Sci. **607**, 149–165 (2015). https://doi.org/10.1016/j.tcs.2015.09.004

12. Kindermann, P., Schulz, A., Spoerhase, J., Wolff, A.: On monotone drawings of trees. In: Duncan, C., Symvonis, A. (eds.) GD 2014. LNCS, vol. 8871, pp. 488–500. Springer, Heidelberg (2014). https://doi.org/10.1007/978-3-662-45803-7_41

13. Oikonomou, A., Symvonis, A.: Monotone drawings of k-inner planar graphs. CoRR abs/1808.06892v1 (2017). http://arxiv.org/abs/1808.06892v2

14. Oikonomou, A., Symvonis, A.: Simple compact monotone tree drawings. In: Frati, F., Ma, K.L. (eds.) GD 2017. LNCS, vol. 10692, pp. 326–333. Springer, Cham (2018). https://doi.org/10.1007/978-3-319-73915-1_26

15. Stern, M.: Ueber eine zahlentheoretische funktion. Journal fur die reine und angewandte Mathematik **55**, 193–220 (1858)

On L-Shaped Point Set Embeddings of Trees: First Non-embeddable Examples

Torsten Mütze[ID] and Manfred Scheucher[(✉)][ID]

Institut für Mathematik, Technische Universität Berlin, Berlin, Germany
{muetze,scheucher}@math.tu-berlin.de

Abstract. An *L-shaped* embedding of a tree in a point set is a planar drawing of the tree where the vertices are mapped to distinct points of the set and every edge is drawn as a sequence of two axis-aligned line segments. Let $f_d(n)$ denote the minimum number N of points such that every n-vertex tree with maximum degree $d \in \{3, 4\}$ admits an L-shaped embedding in every point set of size N, where no two points have the same abscissa or ordinate. The best known upper bounds for this problem are $f_3(n) = O(n^{1.22})$ and $f_4(n) = O(n^{1.55})$, respectively. However, no lower bound besides the trivial bound $f_d(n) \geq n$ is known to this date. In this paper, we present the first examples of n-vertex trees for $n \in \{13, 14, 16, 17, 18, 19, 20\}$ that require strictly more points than vertices to admit an L-shaped embedding, proving that $f_4(n) \geq n + 1$ for those n. Moreover, using computer assistance, we show that every tree on $n \leq 11$ vertices admits an L-shaped embedding in every set of n points, proving that $f_d(n) = n$, $d \in \{3, 4\}$, for those n.

Keywords: L-shaped embedding · Point set · Tree · SAT

1 Introduction

An *L-shaped* embedding of a tree in a point set is a planar drawing of the tree where the vertices are mapped to distinct points of the set and every edge is drawn as a sequence of two axis-aligned line segments; see Fig. 1. Here and throughout this paper, all point sets are such that no two points have the same abscissa or ordinate. The investigation of L-shaped embeddings was initiated in [5–7]. In particular, Di Giacomo et al. [5] showed that $O(n^2)$ points are always sufficient to embed any n-vertex tree, and they asked for a tree that does not admit an L-shaped embedding. Note that an L-shaped embedding of a tree is possible only if the maximum degree of the tree is at most 4. Moreover, if the maximum degree is 2, then the tree is a path and can be embedded greedily on any point set of the same size. Formally, let $f_d(n)$ denote the minimum number N of points such that every n-vertex tree with maximum degree $d \in \{3, 4\}$ admits an L-shaped embedding in every point set of size N.

Partially supported by the DFG Grant FE 340/12-1. We gratefully acknowledge the computing time granted by TBK Automatisierung und Messtechnik GmbH. We also thank the anonymous reviewers for valuable comments.

© Springer Nature Switzerland AG 2018
T. Biedl and A. Kerren (Eds.): GD 2018, LNCS 11282, pp. 354–360, 2018.
https://doi.org/10.1007/978-3-030-04414-5_25

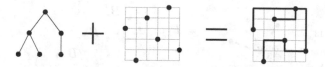

Fig. 1. An L-shaped embedding of a tree in a given point set.

The second author's master's thesis [11] proposed a method to recursively construct an L-shaped embedding of any n-vertex tree in any point set of size $O(n^{1.58})$ (see also [1]). Biedl et al. [3] gave a more precise analysis of this method, proving that $f_3(n) = O(n^{1.22})$ and $f_4(n) = O(n^{1.55})$ points are enough. No lower bound besides the trivial bound $f_d(n) \geq n$ is known to this date. However, the authors of the aforementioned paper also considered a more restrictive setting, where the cyclic order of the edges around each vertex in the embedding is prescribed. For this setting they presented a 14-vertex tree which does not admit an L-shaped embedding in a particular point set of size 14, and they raised the problem to find an infinite family of such non-embeddable trees.[1] All of our results in this paper are for the unrestricted setting, that is, there are no constraints on the cyclic order of the edges around each vertex.

Besides the problem of finding L-shaped embeddings of arbitrary trees in arbitrary point sets, various special classes of trees and point sets have also been studied. For instance, perfect binary and perfect ternary n-vertex trees can be embedded in any point set of size $O(n^{1.142})$ or $O(n^{1.465})$, respectively [3]. Moreover, trees with pathwidth k can be embedded in any set of $2^k n$ points [11, Chap. 3.3.2] (see also [1]). When point sets are chosen uniformly at random (i.e., the y-coordinates are a random permutation), it is known that $O(n \log n (\log \log n)^2)$ and $O(n^{1.332})$ points are enough to embed any tree with maximum degree 3 or degree 4, respectively, with probability at least $1/2$ [11, Chap. 4] (see also [1]).

2   Our Results

To search for n-vertex trees that do not admit an L-shaped embedding in certain point sets of size n, we formulated a SAT instance to test a given pair of tree and point set for embeddability; see Sect. 4. The solver found an embedding of all pairs of trees and point sets up to size $n \leq 11$. Moreover, we found a 13-vertex tree that does not admit an embedding in a particular point set.

Theorem 1 (Computer-assisted). *Every tree on $n \leq 11$ vertices admits an L-shaped embedding in every set of n points, hence $f_d(n) = n$ for $n \leq 11$ and $d \in \{3, 4\}$.*

Theorem 2. *The tree T_{13} in Fig. 2 does not admit an L-shaped embedding in the point set P_{13} shown in the figure, hence $f_4(13) \geq 14$.*

---

[1] Specifically, their counterexample is the 14-vertex caterpillar with 6 vertices on the spine and a pending edge on each side of the four inner vertices of the spine. The point set is a $(4, 6, 4)$-staircase in our terminology (see Definition 1).

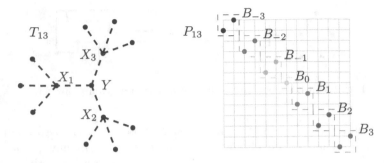

Fig. 2. The tree T_{13} (left) that does not admit an L-shaped embedding in the staircase point set P_{13} (right). Each block of P_{13} is depicted with a different color. (Color figure online)

Even though the 13-vertex tree T_{13} was found using the help of a SAT solver, a human-verifiable proof of Theorem 2 is not hard to obtain; see Sect. 3. Besides the pair (T_{13}, P_{13}), we also found pairs of trees and point sets that do not admit an embedding for larger values of n; see the full version [9]. Overall, we found trees with $n \in \{13, 14, 16, 17, 18, 19, 20\}$ vertices, showing that $f_4(n) \geq n + 1$ for those values of n. For $n = 15$, however, our computer search did not yield any non-embeddable example. We remark that all known non-embeddable trees are lobsters (i.e., trees with pathwidth 2) and they contain T_{13} as a subtree.

As it turns out, the point sets that appear to be difficult for embedding have a regular staircase shape; see Fig. 2.

Definition 1 (Staircase point set). *For any partition $n = a_1 + \ldots + a_k$ with $k, a_1, \ldots, a_k \in \mathbb{N}$, the $(a_1, \ldots, a_k)$-staircase is the point set consisting of a sequence of k blocks, ordered from top-left to bottom-right, and the i-th block contains a sequence of a_i points with increasing x- and y-coordinate.*

3   Proof of Theorem 2

Consider the tree T_{13} and the point set P_{13} depicted in Fig. 2. We label the degree 3 vertex of T_{13} by Y and the three degree 4 vertices of T_{13} as X_1, X_2, X_3, respectively. Moreover, we label the blocks in the $(2, 2, 2, 1, 2, 2, 2)$-staircase point set P_{13} from left to right by $B_{-3}, B_{-2}, \ldots, B_3$. Note that T_{13} is symmetric, as the removal of Y leaves three isomorphic trees. Moreover, P_{13} has reflection symmetries along both diagonals of the grid.

For the sake of contradiction, we assume that an L-shaped embedding of T_{13} to P_{13} exists. We first derive three lemmas that capture to which blocks the vertices X_1, X_2, X_3, Y can be mapped in such an embedding, and we then complete the proof by distinguishing two main cases.

Lemma 1. *Neither of the four vertices X_1, X_2, X_3, Y is mapped to B_{-3} (black) or to B_3 (purple).*

Proof. All points in B_{-3} and B_3 lie on the bounding box of the point set, so if one of the X_i is mapped onto such a point, then one of the four edges incident with X_i would leave the bounding box, which is impossible. Moreover, Y cannot be mapped onto one of these two blocks, as otherwise one of the X_i, which are the only neighbors of Y in T_{13}, would be mapped onto the other point of that same block. □

Lemma 2. *Each of the X_i is mapped to a distinct block (distinct color).*

Proof. Assume that X_i and X_j are mapped to the same block. By symmetry, we may assume that X_j is right above X_i, and that Y is right below of X_i and X_j; see Fig. 3(a). Note that the edge YX_i enters X_i from below and the edge YX_j enters X_j from the right. As X_i and X_j both have degree 4, and their block only contains two points, the edge leaving X_i to the right and the edge leaving X_j to the bottom must cross, a contradiction. □

Lemma 3. *Not all three points X_1, X_2, X_3 lie on the same side (above, below, left or right) of Y.*

Proof. It suffices to prove one of the statements, then the others follow by symmetry. Suppose for the sake of contradiction that X_1, X_2, X_3 all lie above Y. As one edge leaving Y has to go right, one of the X_i, say X_3, is mapped to the same block, and Y is left below of X_3 in that block; see Fig. 3(b). Moreover, the edge YX_3 enters X_3 at the bottom. As X_3 has degree 4, and each block contains at most two points, the edge that leaves Y on the top towards X_1 or X_2 crosses the edge that leaves X_3 to the left, a contradiction. □

By Lemmas 1 and 3, Y is mapped to one of the blocks B_{-1} (orange), B_0 (yellow), or B_1 (green). By Lemma 2 we may assume that X_1, X_2, X_3 appear in distinct blocks in exactly this order from left to right and also from top to bottom, and none of them is in B_{-3} (black) or B_3 (purple). Moreover, from Lemma 3 we conclude that X_1 and X_3 are in other blocks than Y, so at most Y and X_2 are in the same block. We now distinguish two cases.

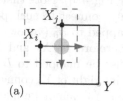

(a)

Case 1: Y and X_2 are mapped to the same block. By symmetry, we may assume that they are mapped to B_1 (green) and that X_2 lies right above Y. Then the vertex X_3 must be mapped to the block B_2 (blue); see Fig. 4(a). If the edge YX_3 would leave Y to the right, then the edge leaving X_2 at the bottom would cross the edge YX_3. It follows that the edge YX_3 leaves Y at the bottom and enters X_3 from the left. Note that the edge that leaves X_2 to the right can only connect to a leaf L that is mapped to $B_2 \cup B_3$, and L must

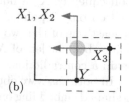

(b)

Fig. 3. Illustration of the proofs of (a) Lemma 2 and (b) Lemma 3. Crossing edges are highlighted red. (Color figure online)

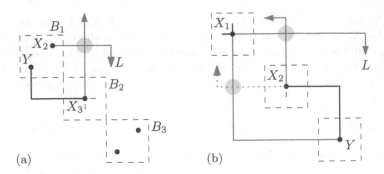

Fig. 4. Illustration of the proof of Theorem 2 in (a) Case 1 and (b) Case 2. (Color figure online)

be mapped to the right of X_3, as otherwise the edges X_2L and YX_3 would cross. The edges leaving X_3 at the bottom and right can only connect to points from $B_2 \cup B_3$, so together with X_3 and L we already have four vertices that are mapped to $B_2 \cup B_3$. Consequently, the edge leaving X_3 at the top must connect to a point outside of $B_1 \cup B_2 \cup B_3$, and therefore this edge crosses the edge X_2L, a contradiction.

Case 2: Y and X_2 are mapped to distinct blocks, so all four points X_1, X_2, X_3, Y are in different blocks. By symmetry, we assume that X_1 and X_2 both lie above and left of Y, and X_3 lies below and right of Y. Moreover, we assume that the edge YX_1 enters X_1 from below and that the edge YX_2 enters X_2 from the right; see Fig. 4(b). Note that X_2 cannot connect to any points right of Y, and X_1 can only connect to such points by the edge leaving it to the right. As Y is either mapped to B_0 (yellow) or B_1 (green), there are at most 7 points left above of Y. Therefore, as X_1 and X_2 together with their leaves form a set of 8 points, Y must be mapped to B_1 (green), and exactly one leaf L of X_1 is mapped to a point right of Y, connected to X_1 by the edge that leaves X_1 to the right. Note that X_2 cannot be mapped to B_0 (yellow), as then the edge leaving X_2 at the bottom could not connect to any point without either crossing YX_1 or YX_2. Consequently, X_2 is mapped to B_{-1} (orange). However, as B_{-1} and B_0 together contain only 3 points, and X_2 together with its leaves form a set of 4 vertices, at least one of the two edges that leave X_2 to the left or top must connect to a point above or left of X_1, and this edge will cross either the edge YX_1 or X_1L, again a contradiction.

In both cases we obtain a contradiction to the assumption that T_{13} admits an L-shaped embedding on the point set P_{13}. This completes the proof of Theorem 2.

4   The SAT Model

To test whether a given tree with vertex set $\{1, \ldots, n\}$ admits an L-shaped embedding on a given point set $\{P_1, \ldots, P_n\}$, we formulated a Boolean satisfiability problem that has a solution if and only if the tree admits an embedding in the point set. Our SAT model has variables $x_{i,j}$ to indicate whether the vertex i is mapped to the point P_j, and for every edge ab in the tree a variable $y_{a,b}$ to indicate whether the edge is connected horizontally to a (otherwise it is connected vertically to b). The following constraints are necessary and sufficient to guarantee the existence of an L-shaped embedding:

(i) Injective mapping from vertices to points: Each vertex is mapped to a point, and no two vertices are mapped to the same point.

(ii) L-shaped edges: For each edge ab of the tree, a is either connected horizontally or vertically to b.

(iii) No overlapping edge segments: For each pair of adjacent edges ab and ac, if b and c are mapped to the right of a, then a cannot be connected horizontally to both b and c. An analogous statement holds if b and c are both mapped to the left, above, or below a.

(iv) No crossing edge segments: For each pair of edges ab and cd, the vertices a, b, c, d must not be be mapped so that segments cross. More specifically, for each four points p, q, r, s (to which a, b, c, d may map), there are at most four cases that have to be forbidden in the mapping, depending on the relative position of p, q, r, s.

The resulting CNF formula thus has $\Theta(n^2)$ variables and $\Theta(n^4)$ clauses.

Our Python program that creates a SAT instance for a given pair of tree and staircase point set is available online [10]. The resulting instances can be solved, e.g., using the solver PicoSAT [4].

5 Discussion

Theorems 1 and 2 leave open whether all 12-vertex trees embed in all point sets of the same size. In our experiments we were only able to test all 12-vertex trees on certain symmetric point sets of that size. Hence, we would not be surprised if T_{13} is indeed a minimal non-embeddable example. We currently do not know of any infinite family of trees which do not always admit an L-shaped embedding. Moreover, since all examples that we know are lobsters (trees with pathwidth 2), it would be interesting to know whether caterpillars (trees with pathwidth 1) always admit an L-shaped embedding. So far we only know of trees with maximum degree 4 which do not always admit an L-shaped embedding — the question for trees with maximum degree 3 remains open [5–7]. Kano and Suzuki [7] even conjectured that $f_3(n) = n$.

A more general class of embeddings are *orthogeodesic* embeddings, where the edges are drawn with minimal ℓ_1-length and consist of segments along the grid induced by the point set [2,5,8,11]. The best known bounds are due to Bárány et al. [2] who showed that every n-vertex tree with maximum degree 4 admits an orthogeodesic embedding on every point set of size $\lfloor 11n/8 \rfloor$. Unfortunately, our example T_{13} allows an orthogeodesic embedding on P_{13} (see the full version [9]), so the question whether n points are always sufficient to guarantee an orthogeodesic embedding of any n-vertex tree [2,5], is still open.

References

1. Aichholzer, O., Hackl, T., Scheucher, M.: Planar L-shaped point set embeddings of trees. In: Proceedings of 32nd European Workshop on Computational Geometry (EuroCG 2016), pp. 51–54 (2016). http://www.eurocg2016.usi.ch/sites/default/files/paper_26.pdf
2. Bárány, I., Buchin, K., Hoffmann, M., Liebenau, A.: An improved bound for ortho-geodesic point set embeddings of trees. In: Proceedings of 31st European Workshop on Computational Geometry (EuroCG 2015), pp. 47–50 (2016). http://www.eurocg2016.usi.ch/sites/default/files/paper_44.pdf
3. Biedl, T., Chan, T.M., Derka, M., Jain, K., Lubiw, A.: Improved bounds for drawing trees on fixed points with L-shaped edges. In: Frati, F., Ma, K.-L. (eds.) GD 2017. LNCS, vol. 10692, pp. 305–317. Springer, Cham (2018). https://doi.org/10.1007/978-3-319-73915-1_24
4. Biere, A.: PicoSAT essentials. J. Satisf. Boolean Model. Comput. (JSAT) **4**, 75–97 (2008). http://satassociation.org/jsat/index.php/jsat/article/view/45
5. Di Giacomo, E., Frati, F., Fulek, R., Grilli, L., Krug, M.: Orthogeodesic point-set embedding of trees. Comput. Geom. **46**(8), 929–944 (2013). https://doi.org/10.1016/j.comgeo.2013.04.003
6. Fink, M., Haunert, J.-H., Mchedlidze, T., Spoerhase, J., Wolff, A.: Drawing graphs with vertices at specified positions and crossings at large angles. In: van Kreveld, M., Speckmann, B. (eds.) GD 2011. LNCS, vol. 7034, pp. 441–442. Springer, Heidelberg (2012). https://doi.org/10.1007/978-3-642-25878-7_43
7. Kano, M., Suzuki, K.: Geometric graphs in the plane lattice. In: Márquez, A., Ramos, P., Urrutia, J. (eds.) EGC 2011. LNCS, vol. 7579, pp. 274–281. Springer, Heidelberg (2012). https://doi.org/10.1007/978-3-642-34191-5_26
8. Katz, B., Krug, M., Rutter, I., Wolff, A.: Manhattan-geodesic embedding of planar graphs. In: Eppstein, D., Gansner, E.R. (eds.) GD 2009. LNCS, vol. 5849, pp. 207–218. Springer, Heidelberg (2010). https://doi.org/10.1007/978-3-642-11805-0_21
9. Mütze, T., Scheucher, M.: On L-shaped point set embeddings of trees: first non-embeddable examples (2018). http://arXiv.org/abs/1807.11043
10. Scheucher, M.: Python program to generate the SAT model. http://page.math.tu-berlin.de/~scheucher/suppl/LShaped/test_T13.py
11. Scheucher, M.: Orthogeodesic point set embeddings of outerplanar graphs. Master's thesis, Graz University of Technology (2015). http://page.math.tu-berlin.de/~scheucher/publ/masters_thesis_2015.pdf

How to Fit a Tree in a Box

Hugo A. Akitaya[1], Maarten Löffler[2], and Irene Parada[3(✉)]

[1] Tufts University, Medford, MA, USA
hugo.alves_akitaya@tufts.edu
[2] Utrecht University, Utrecht, The Netherlands
m.loffler@uu.nl
[3] Graz University of Technology, Graz, Austria
iparada@ist.tugraz.at

Abstract. We study compact straight-line embeddings of trees. We show that perfect binary trees can be embedded optimally: a tree with n nodes can be drawn on a $\sqrt{n}$ by $\sqrt{n}$ grid. We also show that testing whether a given binary tree has an upward embedding with a given combinatorial embedding in a given grid is NP-hard.

1 Introduction

Let $T = (V, E)$ be combinatorial tree; that is, a connected graph without cycles. A *straight-line embedding* of T onto a grid is an injective map $f : V \to \mathbb{Z}^2$. An embedding is *planar* if for every pair of edges $(v_1, v_2), (w_1, w_2) \in E$ the line segments $f(v_1)f(v_2)$ and $f(w_1)f(w_2)$ do not intersect except at common endpoints. The *size* or *dimensions* of an embedding (or, with slight abuse of terminology, the size of the grid) is the width and height of the portion of $\mathbb{Z}^2$ used by f; that is,

$$\dim_f(T) = \left(\max_{v \in V} x_{f(v)} - \min_{v \in V} x_{f(v)} + 1, \max_{v \in V} y_{f(v)} - \min_{v \in V} y_{f(v)} + 1 \right).$$

We are interested in finding embeddings with as small a size as possible.

A *rooted* tree is a tree T with a special vertex $r \in V$ marked as root. Because a tree has no cycles, a rooted tree has an induced partial order on its vertices: for two vertices $v, w \in V$, we say $v \prec w$ if and only if v lies on the path from r to w. An embedding is *upward* if, for all $v, w \in V$ with $v \prec w$, we have $y_{f(v)} > y_{f(w)}$. An embedding is *weakly upward* if, for all $v, w \in V$ with $v \prec w$, we have $y_{f(v)} \geq y_{f(w)}$.

Related Work. Drawing graphs with small area has a long and rich history [7]. By now, we are starting to have some understanding of when graphs admit drawings with *linear area* (a graph with n nodes can be embedded on a $w \times h$ grid with $wh \in O(n)$), and when superlinear area is required. Chan [5] shows that every tree admits a drawing with $n2^{O(\sqrt{\log \log n \log \log \log n})}$ area, improving the long-standing $O(n \log n)$ bound one obtains by a simple divide-and-conquer layout algorithm.

© Springer Nature Switzerland AG 2018
T. Biedl and A. Kerren (Eds.): GD 2018, LNCS 11282, pp. 361–367, 2018.
https://doi.org/10.1007/978-3-030-04414-5_26

However, not much is known about the exact minimum area requirements for graphs that do admit linear-area drawings. It is clear that not every tree admits a perfect drawing on a grid with exactly n points: for instance, when the graph is a star, some grid points are "blocked" and cannot be used. The star graph *can* be drawn on a linear-area grid: Euler already showed that the fraction of points visible from the center of a square grid tends to $\frac{6}{\pi^2}$ more than 300 years ago [8]. For graphs of bounded degree, there is hope that we can do better. Clearly, every path admits a perfect drawing. Garg and Rusu [9,10] show that trees of degree $d = O(n^\delta)$ with $\delta < 1/2$, and in particular of degree 3, have linear-area drawings onto a square grid, and even onto grids of different aspect ratio; their main concern is studying the relation between the aspect ratio and the area, but they do not give concrete bounds on the constant factor.

We conjecture that every degree 3 tree admits a *perfect* drawing onto a square grid, and we prove here that this is the case for perfect binary trees.

When drawing rooted trees, a natural restriction is to require drawings to be *upward*. In this case, clearly, perfect drawings are impossible unless the tree is a path, but we may still investigate almost-perfect drawings that leave only few grid points unused. Chan [5] shows that for strictly upward drawings, we cannot do better than $\Theta(n \log n)$ area. He does give an improved bound for *weakly* upward drawings.

Biedl and Mondal [4] proved NP-hardness for strictly upward unordered straight-line high-degree trees. Later, Biedl [3] gave an algorithm to find for every ternary tree T a strictly upward order-preserving straight-line grid drawing of optimum width.

Contribution. We have the following results.

- It is NP-hard to test whether binary trees with fixed combinatorial embedding admit upward drawings on a given grid.
- Perfect binary trees with n vertices admit drawings on a $\sqrt{n} \times \sqrt{n}$ grid.

2   Optimal Embeddings of Perfect Binary Trees

We consider the following setting. Given a $\sqrt{n} \times \sqrt{n}$ grid and a tree with n vertices, can we draw it with straight non-crossing edges? Clearly this is not always possible, for instance if the tree is a star.

Conjecture 1. If the tree has max degree 3, it is always possible.

In particular, if $n = 2^{k+1}$, a perfect binary tree of odd height k with additional parent of the root (to make the number of vertices exactly n) can be drawn on the $\sqrt{n} \times \sqrt{n}$ grid. We use a recursive strategy to show it. Similar approaches recursively embedding trees have been previously used to show asymptotic bounds (but disregarding smaller order terms); in particular to prove that perfect binary trees and Fibonacci trees can be upward drawn in linear area [6] and to bound the area of complete ternary and 7-ary trees on the 8-grid [2].

Theorem 1. *The perfect binary tree on* $n = 2^{k+1} - 1$ *vertices with* k *odd can be embedded in the* $\sqrt{n} \times \sqrt{n}$ *square grid.*

Proof. We will recursively argue that perfect binary trees can be embedded in square grids in two ways. Let T_k be the perfect binary tree on $n = 2^{k+1} - 1$ vertices. We will recursively define two straight-line crossing-free drawings, F_k and G_k, of T_k. The vertices in these drawings are placed in the grid points $\{(x, y) \in \mathbb{Z}^2 : 1 \leq x \leq 2^{(k+1)/2}, \ 1 \leq y \leq 2^{(k+1)/2}\}$.

We first list the required properties of F_k and G_k, also illustrated in Fig. 1a:

(i) both F_k and G_k map the root of T_k to the point $(2^{(k-1)/2} + 1, 2^{(k-1)/2})$;

(ii) both F_k and G_k do not place any edges in the vertical strip between $x = 2^{(k-1)/2}$ and $x = 2^{(k-1)/2} + 1$, except for the edges incident to the root of T_k;

(iii) F_k leaves the point $(2^{(k-1)/2}, 1)$ unused; and

(iv) G_k leaves the point $(1, 1)$ unused.

Observe that $F_1 = G_1$ is trivial to draw: both are drawings of a path of length 2, drawn by connecting the point $(1, 2)$ to the point $(2, 1)$ to the point $(2, 2)$.

What remains is to argue that we can recursively draw F_k and G_k using drawings of F_{k-2} and G_{k-2}. The argument is illustrated in Fig. 1; the detailed proof can be found in the full version [1].   □

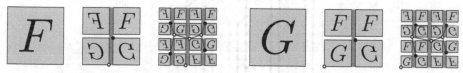

(a) Tiles F and G, and their recursive definition.

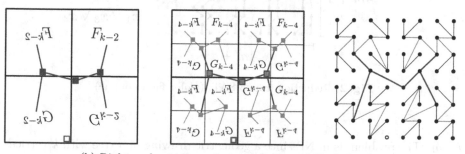

(b) Right and center: recursive definition of F_k. Left: F_5.

Fig. 1. Recursive embedding of perfect binary trees.

3   Upward Embedding of Trees in a Given Grid is NP-Hard

Recall that an embedding of a rooted tree is *upward* if the y-coordinate of a node is strictly greater than the y-coordinate of its children. A *combinatorial embedding* is given by a circular order of incident edges around each vertex. In this section we show that deciding if a rooted binary tree with a fixed combinatorial embedding can be drawn upward and without crossings in a given square grid is NP-complete.

Theorem 2. *Deciding whether an upward planar straight-line drawing of a fixed combinatorial embedding of a rooted binary tree on a grid of given size $(w \times h)$ exists is NP-complete.*

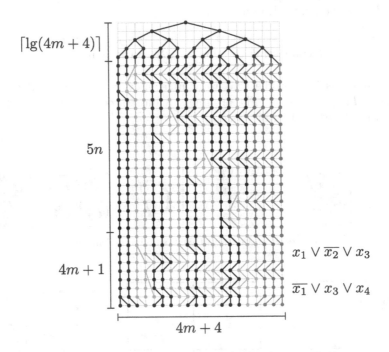

Fig. 2. Reduction from 3-SAT. (Color figure online)

Proof. The problem is in NP since a geometric drawing of a tree with k vertices in the grid can be expressed in $O(k)$ size by assigning vertices to grid points. Checking whether the drawing is an embedding can trivially be done in $O(k^2)$ time by checking pairwise edge crossings. Checking whether the drawing preserves the given rotation system takes $O(k^2)$ time and checking whether it is upward can be done in $O(k)$ time.

We prove NP-hardness by a reduction from 3SAT which is an NP-complete problem [11]. An instance of 3SAT is given by a set $\{x_1, \ldots, x_n\}$ of n variables and a set $\{c_1, \ldots, c_m\}$ of m clauses. Each variable can assume one of two values in $\{\text{true}, \text{false}\}$. Each clause is defined by 3 literals, i.e., positive or negative copies of a variable. A clause is *satisfied* if at least one of its literals is true. The problem 3SAT asks for an assignment from the variables to $\{\text{true}, \text{false}\}$ that satisfies all clauses. We give an arbitrary order for the variables and say that x_i appears *before* x_j if $i < j$. The first (resp., second, resp., third) literal of a clause is the literal (among the 3 literal that define the clause) of the variable that appears first (resp., second, resp., third) in the order assigned to variables. Given an instance of 3SAT we build a rooted tree with $O(m^2 + mn)$ vertices and set $w = 4m + 4$ and $h = \lceil \lg(4m + 4) \rceil + 5n + 4m + 1$.

Overview. Refer to Fig. 2. This paragraph gives a brief informal overview of the reduction. The following paragraphs will give a full proof. The reduction is divided into 3 parts. The top part (spanning the top $\lceil \lg(4m+4) \rceil$ rows in Fig. 2) is a perfect binary tree with $2^{\lceil \lg(4m+4) \rceil - 1}$ leaves. The middle part (spanning the next $5n$ rows in Fig. 2) is where the variables are assigned a boolean value. The bottom part (spanning the last $4m + 1$ rows in Fig. 2) enforces that every clause of the original instance of 3SAT is satisfied. Each variable is represented by a red subtree with two long paths that have to span all but one row below the least common ancestor. The left (resp., right) path represents a positive (resp., negative) literal of the variable. The construction forces one of the paths to be drawn one unit above the other and that encodes the boolean assignment. If the left path does not span the last row, then the variable is set to true. The variable is set to false otherwise. In Fig. 2, x_2 and x_3 (resp., x_1 and x_4) are set to true (resp., false). The blue subtrees encode the clauses by allowing the rest of the construction to occupy specific extra grid positions. The incidence of a variable in a clause is encoded by an extra leaf child in one of the paths that represent the incident literal corresponding to the variable. If none of the incident literals of a clause are set to true, the drawing would require the use of an extra row or column. Otherwise, the extra leaves can be accommodated exactly by the space provided by the blue subtrees.

Construction. There are exactly $4m+4$ subtrees attached to the perfect binary tree on the top of the construction. The fixed combinatorial embedding prescribes a left-to-right order of such subtrees. For each variable x_i, do the following. Set the $4(i-1) + 1$-th subtree to be a path p of length $5n + 4m$; attach another path of length $5n + 4m - 5(i-1) - 4$ to the right of the $5(i-1) + 4$-th vertex of p. Attach a right child to the second to last vertex of p. Set the $4(i-1) + 2$-th subtree to be a path of length $5(i-1) + 1$. Set the $4(i-1) + 3$-th subtree to be a path of length $5(i-1)$. At the end of the path, attach two paths p_t and p_f of length $5(n - i + 1) + 4m - 2$ each as left and right subtrees respectively. Attach a right (resp., left) child to the first vertex of p_t (resp., p_f). We now describe the position of the vertices that encode the incidence of a variable in a clause. We call such vertices *literal leaves*. If x_i (resp., $\overline{x_i}$) is the first or second literal of c_j, then add a right child $\ell_{i,j}$ to the $5(n - i + 1) + 4(m - 1)$-th vertex of p_t

(resp., p_f). If x_i (resp., $\overline{x_i}$) is the third literal of c_j, then add a left child $\ell_{i,j}$ to the $5(n-i+1)+4(m-1)$-th vertex of p_t (resp., p_f). The $4(i-1)+3$-th subtree is shown in red in Fig. 2. Set the $4(i-1)+4$-th subtree to be a path of length $5(i-1)$. Finally, we describe the four last subtrees (shown in blue in Fig. 2). Set the $4m+1$-th, $4m+2$-th, and $4m+3$-th subtrees to be paths of length $5n+4m$ each. For every clause c_j, attach a right leaf child to the $5n+4j$-th vertex of the $4m+3$-th subtree. Set the last subtree to be a path of length $5n$. For each variable x_i, attach a right leaf child to the $5i-4$-th vertex of the path. This finalizes the construction.

Correctness. We argue that the construction is correct, by showing that every satisfiable 3SAT instance can indeed be embedded in a $w \times h$ grid, and that every drawing that fits in a $w \times h$ grid must correspond to a satisfiable 3SAT instance. The details can be found in the full version [1].   □

4   Conclusions

We studied tree drawings in small areas.

For arbitrary drawings, we gave a construction for embedding a perfect binary tree on a square grid. The main remaining open question here is whether every low-degree tree with wh vertices (or fewer) can be embedded on an $w \times h$ grid. We conjecture that this is the case when $w = h$. Another intriguing question is whether, for general trees, testing if they can be embedded on a given grid is computationally tractable.

For upward drawings, we showed that even for bounded-degree trees, testing whether a given tree can be embedded in a $w \times h$ rectangle is already NP-hard, if the combinatorial embedding of the tree is fixed. It would be interesting to know whether the same is true when one can freely choose the combinatorial embedding. Another question is whether the problem is also NP-hard for *weakly* upward drawings, where adjacent vertices may be embedded using the same y-coordinate.

Acknowledgements. This research was initiated at the 33rd Bellairs Winter Workshop on Computational Geometry in 2018. We would like to thank all participants of the workshop for fruitful discussions on the topic. H.A.A. was supported by NSF awards CCF-1422311 and CCF-1423615, and the Science Without Borders scholarship program. M.L. was partially supported by the Netherlands Organisation for Scientific Research (NWO) through grant number 614.001.504. I.P. was supported by the Austrian Science Fund (FWF) grant W1230.

References

1. Akitaya, H.A., Löffler, M., Parada, I.: How to fit a tree in a box. CoRR (2018). http://arxiv.org/abs/1808.10572v1
2. Bachmaier, C., Matzeder, M.: Drawing unordered trees on k-grids. In: Rahman, M.S., Nakano, S. (eds.) WALCOM 2012. LNCS, vol. 7157, pp. 198–210. Springer, Heidelberg (2012). https://doi.org/10.1007/978-3-642-28076-4_20

3. Biedl, T.: Ideal drawings of rooted trees with approximately optimal width. J. Graph Algorithms Appl. **21**(4), 631–648 (2017). https://doi.org/10.7155/jgaa.00432

4. Biedl, T., Mondal, D.: On upward drawings of trees on a given grid. In: Frati, F., Ma, K.-L. (eds.) GD 2017. LNCS, vol. 10692, pp. 318–325. Springer, Cham (2018). https://doi.org/10.1007/978-3-319-73915-1_25

5. Chan, T.M.: Tree drawings revisited. In: Speckmann, B., Tóth, C.D. (eds.) 34th International Symposium on Computational Geometry (SoCG 2018). Leibniz International Proceedings in Informatics (LIPIcs), vol. 99, pp. 23:1–23:15. Schloss Dagstuhl-Leibniz-Zentrum fuer Informatik, Dagstuhl (2018). https://doi.org/10.4230/LIPIcs.SoCG.2018.23

6. Crescenzi, P., Battista, G.D., Piperno, A.: A note on optimal area algorithms for upward drawings of binary trees. Comput. Geom. **2**(4), 187–200 (1992). https://doi.org/10.1016/0925-7721(92)90021-J

7. Díaz, J., Petit, J., Serna, M.: A survey of graph layout problems. ACM Comput. Surv. **34**(3), 313–356 (2002). https://doi.org/10.1145/568522.568523

8. Dunham, W.: Euler : The Master of Us All. Dolciani Mathematical Expositions (Book 22). Mathematical Association of America, Washington (1999)

9. Garg, A., Rusu, A.: Straight-line drawings of general trees with linear area and arbitrary aspect ratio. In: Kumar, V., Gavrilova, M.L., Tan, C.J.K., L'Ecuyer, P. (eds.) ICCSA 2003. LNCS, vol. 2669, pp. 876–885. Springer, Heidelberg (2003). https://doi.org/10.1007/3-540-44842-X_89

10. Garg, A., Rusu, A.: Straight-line drawings of binary trees with linear area and arbitrary aspect ratio. J. Graph Algorithms Appl. **8**(2), 135–160 (2004). https://doi.org/10.7155/jgaa.00086

11. Karp, R.M.: Reducibility among combinatorial problems. In: Complexity of Computer Computations: Proceedings of a symposium on the Complexity of Computer Computations, pp. 85–103 (1972). https://doi.org/10.2307/2271828

Best Paper Track 1

Pole Dancing: 3D Morphs for Tree Drawings

Elena Arseneva[1]([⊠]), Prosenjit Bose[2], Pilar Cano[2,3], Anthony D'Angelo[2], Vida Dujmović[4], Fabrizio Frati[5], Stefan Langerman[6], and Alessandra Tappini[7]

[1] St. Petersburg State University (SPbU), Saint Petersburg, Russia
ea.arseneva@gmail.com
[2] Carleton University, Ottawa, Canada
jit@scs.carleton.ca, anthonydangelo@cmail.carleton.ca
[3] Universitat Politècnica de Catalunya, Barcelona, Spain
m.pilar.cano@upc.edu
[4] University of Ottawa, Ottawa, Canada
vida@cs.mcgill.ca
[5] Roma Tre University, Rome, Italy
frati@dia.uniroma3.it
[6] Université libre de Bruxelles (ULB), Brussels, Belgium
stefan.langerman@ulb.ac.be
[7] Università degli Studi di Perugia, Perugia, Italy
alessandra.tappini@studenti.unipg.it

Abstract. We study the question whether a crossing-free 3D morph between two straight-line drawings of an n-vertex tree can be constructed consisting of a small number of linear morphing steps. We look both at the case in which the two given drawings are two-dimensional and at the one in which they are three-dimensional. In the former setting we prove that a crossing-free 3D morph always exists with $O(\log n)$ steps, while for the latter $\Theta(n)$ steps are always sufficient and sometimes necessary.

We here refer to pole dancing as a fitness and competitive sport. The authors hope that many of our readers try this activity themselves, and will in return introduce many pole dancers to Graph Drawing, thereby alleviating the gender imbalance in both communities. The authors do not condone any pole activity used for sexual exploitation or abuse of women or men.

E. A. was partially supported by F.R.S.-FNRS and SNF grant P2TIP2-168563 under the SNF Early PostDoc Mobility program. P.C. was supported by CONACyT, projects MINECO MTM2015-63791-R and Gen. Cat. 2017SGR1640. P.B, A.D and V.D. were supported by NSERC. F.F. was partially supported by MIUR Project "MODE" under PRIN 20157EFM5C and by H2020-MSCA-RISE project 734922, "CONNECT". S.L. is Directeur de Recherches du F.R.S.-FNRS. A.T. was partially supported by the project "Algoritmi e sistemi di analisi visuale di reti complesse e di grandi dimensioni" - Ric. di Base 2018, Dip. Ingegneria, Univ. Perugia.

T. Biedl and A. Kerren (Eds.): GD 2018, LNCS 11282, pp. 371–384, 2018.
https://doi.org/10.1007/978-3-030-04414-5_27

1   Introduction

A *morph* between two drawings of the same graph is a continuous transformation from one drawing to the other. Thus, any time instant of the morph defines a different drawing of the graph. Ideally, the morph should preserve the properties of the initial and final drawings throughout. As the most notable example, a morph between two planar graph drawings should guarantee that every intermediate drawing is also planar; if this happens, then the morph is called *planar*.

Planar morphs have been studied for decades and find nowadays applications in animation, modeling, and computer graphics; see, e.g., [11,12]. A planar morph between any two topologically-equivalent[1] planar straight-line[2] drawings of the same planar graph always exists; this was proved for maximal planar graphs by Cairns [8] back in 1944, and then for all planar graphs by Thomassen [16] almost forty years later. Note that a planar morph between two planar graph drawings that are not topologically equivalent does not exist.

It has lately been well investigated whether a planar morph between any two topologically-equivalent planar straight-line drawings of the same planar graph always exists such that the vertex trajectories have low complexity. This is usually formalized as follows. Let Γ and Γ' be two topologically-equivalent planar straight-line drawings of the same planar graph G. Then a morph $\mathcal{M}$ is a sequence $\langle \Gamma_1, \Gamma_2, \ldots, \Gamma_k \rangle$ of planar straight-line drawings of G such that $\Gamma_1 = \Gamma$, $\Gamma_k = \Gamma'$, and $\langle \Gamma_i, \Gamma_{i+1} \rangle$ is a planar linear morph, for each $i = 1, \ldots, k - 1$. A *linear morph* $\langle \Gamma_i, \Gamma_{i+1} \rangle$ is such that each vertex moves along a straight-line segment at uniform speed; that is, assuming that the morph happens between time $t = 0$ and time $t = 1$, the position of a vertex v at any time $t \in [0, 1]$ is $(1 - t)\Gamma_i(v) + t\Gamma_{i+1}(v)$. The complexity of a morph $\mathcal{M}$ is then measured by the number of its *steps*, i.e., by the number of linear morphs it consists of.

A recent sequence of papers [3–6] culminated in a proof [2] that a planar morph between any two topologically-equivalent planar straight-line drawings of the same n-vertex planar graph can always be constructed consisting of $\Theta(n)$ steps. This bound is asymptotically optimal in the worst case, even for paths.

The question we study in this paper is whether morphs with sub-linear complexity can be constructed if a third dimension is allowed to be used. That is: Given two topologically-equivalent planar straight-line drawings Γ and Γ' of the same n-vertex planar graph G does a morph $\mathcal{M} = \langle \Gamma = \Gamma_1, \Gamma_2, \ldots, \Gamma_k = \Gamma' \rangle$ exist such that: (i) for $i = 1, \ldots, k$, the drawing Γ_i is a crossing-free straight-line 3D drawing of G, i.e., a straight-line drawing of G in $\mathbb{R}^3$ such that no two edges cross; (ii) for $i = 1, \ldots, k - 1$, the step $\langle \Gamma_i, \Gamma_{i+1} \rangle$ is a crossing-free linear morph, i.e., no two edges cross throughout the transformation; and (iii) $k = o(n)$? A morph $\mathcal{M}$ satisfying properties (i) and (ii) is a *crossing-free 3D morph*.

---

[1] Two planar drawings of a connected graph are *topologically equivalent* if they define the same clockwise order of the edges around each vertex and the same outer face.

[2] A *straight-line drawing* Γ of a graph G maps vertices to points in a Euclidean space and edges to open straight-line segments between the images of their end-vertices. We denote by $\Gamma(v)$ (by $\Gamma(G')$) the image of a vertex v (of a subgraph G' of G, resp.).

Our main result is a positive answer to the above question for trees. Namely, we prove that, for any two planar straight-line drawings Γ and Γ' of an n-vertex tree T, there is a crossing-free 3D morph with $O(\log n)$ steps between Γ and Γ'. More precisely the number of steps in the morph is linear in the *pathwidth* of T. Notably, our morphing algorithm works even if Γ and Γ' are not topologically equivalent, hence the use of a third dimension overcomes another important limitation of planar two-dimensional morphs. Our algorithm morphs both Γ and Γ' to an intermediate suitably-defined *canonical 3D drawing*; in order to do that, a root-to-leaf path H of T is moved to a vertical line and then the subtrees of T rooted at the children of the vertices in H are moved around that vertical line, thus resembling a pole dance, from which the title of the paper comes.

We also look at whether our result can be generalized to morphs of crossing-free straight-line 3D drawings of trees. That is, the drawings Γ and Γ' now live in $\mathbb{R}^3$, and the question is again whether a crossing-free 3D morph between Γ and Γ' exists with $o(n)$ steps. We prove that this is not the case: Two crossing-free straight-line 3D drawings of a path might require $\Omega(n)$ steps to be morphed one into the other. The matching upper bound can always be achieved: For any two crossing-free straight-line 3D drawings Γ and Γ' of the same n-vertex tree T there is a crossing-free 3D morph between Γ and Γ' with $O(n)$ steps.

The rest of the paper is organized as follows. In Sect. 2 we deal with crossing-free 3D morphs of 3D tree drawings. In Sect. 3 we show how to construct 2-step crossing-free 3D morphs between planar straight-line drawings of a path. In Sect. 4 we present our main result about crossing-free 3D morphs of planar tree drawings. Finally, in Sect. 5 we conclude and present some open problems.

Because of space limitations, some proofs are omitted or just sketched; they can be found in the full version of the paper.

2 Morphs of 3D Drawings of Trees

In this section we give a tight $\Theta(n)$ bound on the number of steps in a crossing-free 3D morph between two crossing-free straight-line 3D tree drawings.

Theorem 1. *For any two crossing-free straight-line 3D drawings Γ, Γ' of an n-vertex tree T, there exists a crossing-free 3D morph from Γ to Γ' that consists of $O(n)$ steps.*

Proof (sketch). The proof is by induction on n. The base case, in which $n = 1$, is trivial. If $n > 1$, then we remove a leaf v and its incident edge uv from T, Γ, and Γ'. This results in an $(n-1)$-vertex tree T' and two drawings Δ and Δ' of it. By induction, there is a crossing-free 3D morph between Δ and Δ'. We introduce v in such a morph so that it is arbitrarily close to u throughout the transformation; this significantly helps to avoid crossings in the morph. The number of steps is the one of the recursively constructed morph plus one initial step to bring v close to u, plus two final steps to bring v to its final position. $\square$

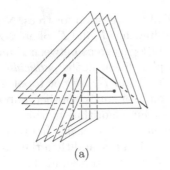 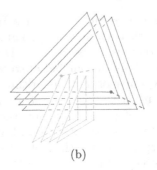

(a)   (b)

Fig. 1. Illustration for the proof of Theorem 2: (a) The drawing Γ of P, with $n = 26$; (b) the link obtained from Γ; the invisible edges are dashed.

Theorem 2. *There exist two crossing-free straight-line 3D drawings* Γ, Γ' *of an n-vertex path* P *such that any crossing-free 3D morph from* Γ *to* Γ' *consists of* $\Omega(n)$ *steps.*

Before proving Theorem 2, we review some definitions and facts from knot theory; refer, e.g., to the book by Adams [1]. A *knot* is an embedding of a circle S^1 in $\mathbb{R}^3$. A *link* is a collection of knots which do not intersect, but which may be linked together. For links of two knots, the (absolute value of the) *linking number* is an invariant that classifies links with respect to ambient isotopies. Intuitively, the linking number is the number of times that each knot winds around the other. The linking number is known to be invariant with respect to different projections of the same link [1]. Given a projection of the link, the linking number can be determined by orienting the two knots of the link, and for every crossing between the two knots in the projection adding $+1$ or -1 if rotating the understrand respectively clockwise or counterclockwise lines it up with the overstrand (taking into account the direction).

Proof (Theorem 2). The drawing Γ of P is defined as follows. Embed the first $\lfloor n/2 \rfloor$ edges of P in 3D as a spiral of monotonically decreasing height. Embed the rest of P as a same type of spiral affinely transformed so that it goes around one of the sides of the former spiral. See Fig. 1a. The drawing Γ' places the vertices of P in order along the unit parabola in the plane $y = 0$.

Cut the edge joining the two spirals (the bold edge in Fig. 1a). Removing an edge makes morphing easier so any lower bound would still apply. Now close the two open curves using two *invisible* edges to obtain a *link* of two knots; see Fig. 1b. It is easy to verify that the (absolute value of the) linking number of this link is $\Omega(n^2)$: indeed, determining it by the above procedure for the projection given by Fig. 1 results in the linking number being equal to the number of crossings between the two links in this projection. In the drawing Γ', each of the two halves of P (and their invisible edges) are separated by a plane and so their linking number is 0.

In a valid linear morph, the edges of P cannot cross each other, but they can cross invisible edges. However, during a linear morph between two straight-line

3D drawings of a graph G any two non-adjacent edges of G intersect $O(1)$ times. Thus each invisible edge can only be crossed $O(n)$ times during a linear morph. A single crossing can only change the linking number by 1. Therefore the linking number can only decrease by $O(n)$ in a single linear morph.   □

3   Morphing Two Planar Drawings of a Path in 3D

In this section we show how to morph two planar straight-line drawings Γ and Γ' of an n-vertex path $P := (v_0, \ldots v_{n-1})$ into each other in two steps.

The *canonical 3D drawing* of P, denoted by $\mathcal{C}(P)$, is the crossing-free straight-line 3D drawing of P that maps each vertex v_i to the point $(0, 0, i) \in \mathbb{R}^3$, as shown in Fig. 2. We now prove the following.

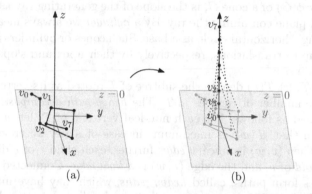

Fig. 2. (a) A straight-line planar drawing Γ of an n-vertex path P and (b) a morph from Γ to $\mathcal{C}(P)$. The vertex trajectories are represented by dotted lines.

Theorem 3. *For any two planar straight-line drawings Γ and Γ' of an n-vertex path P, there exists a crossing-free 3D morph $\mathcal{M} = \langle \Gamma, \mathcal{C}(P), \Gamma' \rangle$ with 2 steps.*

Proof. It suffices to prove that the linear morph $\langle \Gamma, \mathcal{C}(P) \rangle$ is crossing-free, since the morph $\langle \mathcal{C}(P), \Gamma' \rangle$ is just the morph $\langle \Gamma', \mathcal{C}(P) \rangle$ played backwards.

Since $\langle \Gamma, \mathcal{C}(P) \rangle$ is linear, the speed at which the vertices of P move is uniform (though it might be different for different vertices). Thus the speed at which their projections on the z-axis move is uniform as well. Since v_i moves uniformly from $(x_i, y_i, 0)$ to $(0, 0, i)$, at any time during the motion (except at the time $t = 0$) we have $z(v_0) < z(v_1) < \ldots < z(v_{n-1})$. Therefore, in any intermediate drawing any edge (v_i, v_{i+1}) is separated from any other edge by the horizontal plane through one of its end-points. Hence no crossing happens during $\langle \Gamma, \mathcal{C}(P) \rangle$.   □

4   Morphing Two Planar Drawings of a Tree in 3D

Let T be a tree with n vertices, arbitrarily rooted at any vertex. In this section we show that any two planar straight-line drawings of T can be morphed into one another by a crossing-free 3D morph with $O(\log n)$ steps (Theorem 4). Similarly to Sect. 3, we first define a canonical 3D drawing $\mathcal{C}(T)$ of T (see Sect. 4.1), and then show how to construct a crossing-free 3D morph from any planar straight-line drawing of T to $\mathcal{C}(T)$. We describe the morphing procedure in Sect. 4.2; then in Sect. 4.3 we present a procedure $Space()$ that carries out the computations required by the morphing procedure; finally, in Sect. 4.4 we analyze the correctness and efficiency of both procedures.

Before proceeding, we introduce some necessary definitions and notation. By a *cone* we mean a straight circular cone induced by a ray rotated around a fixed vertical line (the *axis*) while keeping its origin fixed at a point (the *apex*) on this line. The *slope* $\phi(C)$ of a cone C, is the slope of the generating ray as determined in the vertical plane containing the ray. By a *cylinder* we always mean a straight cylinder having a horizontal circle as a base. Such cones or cylinders are uniquely determined, up to translations, respectively by their apex and slope or by their height and radius.

For a tree T, let $T(v)$ denote the subtree of T rooted at its vertex v. Also let $|T|$ denote the number of vertices in T. The *heavy-path decomposition* [15] of a tree T is defined as follows. For each non-leaf vertex v of T, let w be the child of v in T such that $|T(w)|$ is maximum (in case of a tie, we choose the child arbitrarily). Then (v, w) is a *heavy edge*; further, each child z of v different from w is a *light child* of v, and the edge (v, w) is a *light edge*. Connected components of heavy edges form paths, called *heavy paths*, which may have many incident light edges. Each path has a vertex, called the *head*, that is the closest vertex to the root of T. See Fig. 3 for an example. A *path tree* of T is a tree whose vertices correspond to heavy paths in T. The parent of a heavy path P in the path tree is the heavy path that contains the parent of the head of P. The root of the path tree is the heavy path containing the root of T. It is well-known [15] that the height of the path tree is $O(\log n)$. We denote by $H(T)$ the root of the path tree of T; let $v_0, \ldots, v_{k-1}$ be the ordered sequence of the vertices of $H(T)$, where v_0 is the root of T. For $i = 0, \ldots, k - 1$, we let $v_i^0, \ldots, v_i^{t_i}$ be the light children of v_i in any order. Let $L(T) = u_0, u_1, \ldots, u_{l-1}$ be the sequence of the light children of $H(T)$ ordered so that: (i) any light child of a vertex v_j precedes any light child of a vertex v_i, if $i < j$; and (ii) the light child v_i^{j+1} of a vertex v_i precedes the light child v_i^j of v_i. When there is no ambiguity we refer to $H(T)$ and $L(T)$ simply as H and L, respectively.

4.1   Canonical 3D Drawing of a Tree

We define the *canonical 3D drawing* $\mathcal{C}(T)$ of a tree T as a straight-line 3D drawing of T that maps each vertex v of T to its *canonical position* $\mathcal{C}(v)$ defined as follows (see Fig. 3b). Note that our canonical drawing is equivalent to the *"standard"* straight-line upward drawing of a tree [7, 9, 10].

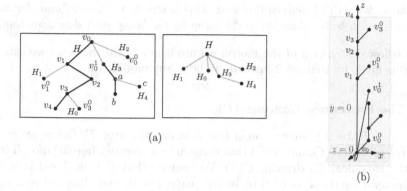

(a)

(b)

Fig. 3. (a) A tree T; (left) its heavy edges (bold lines) forming the heavy paths $H = H(T), H_0, \ldots, H_4$, and (right) the path tree of T; (b) $\mathcal{C}(T)$ for the tree T in (a).

First, we set $\mathcal{C}(v_0) = (0, 0, 0)$ for the root v_0 of T. Second, for each $i = 1, \ldots, k-1$, we set $\mathcal{C}(v_i) = (0, 0, z_{i-1} + |T(v_{i-1})| - |T(v_i)|)$, where z_{i-1} is the z-coordinate of $\mathcal{C}(v_{i-1})$. Third, for each $i = 1, \ldots, k-1$ and for each light child v_i^j of v_i, we determine $\mathcal{C}(v_i^j)$ as follows. If $j = 0$, we set $\mathcal{C}(v_i^j) = (1, 0, 1 + z_i)$, where z_i is the z-coordinate of $\mathcal{C}(v_i)$; otherwise, we set $\mathcal{C}(v_i^j) = (1, 0, z_i^{j-1} + |T(v_i^{j-1})|)$, where z_i^{j-1} is the z-coordinate of $\mathcal{C}(v_i^{j-1})$. Finally, in order to determine the canonical positions of the vertices in $T(v_i^j) \setminus \{v_i^j\}$, we recursively construct the canonical 3D drawing $\mathcal{C}(T(v_i^j))$ of $T(v_i^j)$, and translate all the vertices by the same vector so that v_i^j is sent to $\mathcal{C}(v_i^j)$.

Remark 1. *Notice that the canonical position $\mathcal{C}(v)$ of any vertex v of T is $(\mathrm{dpt}(v), 0, \mathrm{dfs}(v))$. Here $\mathrm{dpt}(v)$ is the depth, in the path tree of T, of the node that corresponds to the heavy path of T that contains v; and $\mathrm{dfs}(v)$ is the position of v in a depth-first search on T in which the children of any vertex are visited as follows: first visit the light children in reverse order with respect to L, and then visit the child incident to the heavy edge.*

The following lemma is a direct consequence of the construction of $\mathcal{C}(T)$.

Lemma 1. *The canonical 3D drawing $\mathcal{C}(T)$ of T lies on a rectangular grid in the plane $y = 0$, where the grid has height n and width equal to the height $h = O(\log n)$ of the path tree of T. Moreover, $\mathcal{C}(T)$ is on or above the line $z = x$.*

Remark 2. *In the above definition of the canonical 3D drawing $\mathcal{C}(T)$, instead of the heavy-path decomposition of T, we can use the decomposition based on the Strahler number of T, see [7] where the Strahler number is used under the name rooted pathwidth of T. With this change, the width of $\mathcal{C}(T)$ will be equal to the Strahler number of T, which is the instance-optimal width of an upward drawing of a tree [7]. Moreover, since the Strahler number is linear in the pathwidth of T,* *so is the width of* $\mathcal{C}(T)$ *defined this way. This is clearly not worse, and, for some instances, much better than the width given by the heavy-path decomposition.*

In the below description of the morph we use heavy paths, however we can use the paths given by Remark 2 instead, without any modification.

4.2   The Procedure *Canonize*(Γ)

Let $\Gamma = \Gamma(T)$ be a planar straight-line drawing of a tree T. Below we give a recursive procedure *Canonize*(Γ) that constructs a crossing-free 3D morph from Γ to the canonical 3D drawing $\mathcal{C}(T)$. We assume that Γ is enclosed in a disk of diameter 1 centered at $(0,0,0)$ in the plane $z = 0$, and that the root v_0 of T is placed at $(0,0,0)$ in Γ. This is not a loss of generality, up to a suitable modification of the reference system.

Step 1 (set the pole). The first step of the procedure *Canonize*(Γ) aims to construct a linear morph $\langle \Gamma, \Gamma_1 \rangle$, where Γ_1 is such that the heavy path $H = (v_0, \ldots, v_{k-1})$ of T lies on the vertical line through $\Gamma(v_0)$ and the subtrees of T rooted at the light children of each vertex v_i lie on the horizontal plane through v_i. More precisely, the vertices of T are placed in Γ_1 as follows. For $i = 0, \ldots, k-1$, place v_i at the point $\mathcal{C}(v_i)$. Every vertex that belongs to a subtree rooted at a light child of v_i is placed at a point such that its trajectory in the morph defines the same vector as the trajectory of v_i.[3] Below we refer to $\Gamma_1(H)$ as the *pole*. The pole will remain still throughout the rest of the morph.

Step 2 (lift). The aim of the second step of the procedure *Canonize*(Γ) is to construct a linear morph $\langle \Gamma_1, \Gamma_2 \rangle$, where Γ_2 is such that the drawings of any two subtrees $T(u_i)$ and $T(u_j)$ rooted at different light children u_i and u_j of vertices in H are vertically and horizontally separated. The separation between $\Gamma_2(T(u_i))$ and $\Gamma_2(T(u_j))$ is set to be large enough so that the recursively computed morphs *Canonize*($\Gamma_2(T(u_i))$) and *Canonize*($\Gamma_2(T(u_j))$) do not interfere with each other.

We describe how to construct Γ_2. As anticipated, $\Gamma_2(v_i) = \Gamma_1(v_i)$, for each vertex v_i in H. In order to determine the placement of the vertices not in H we use l cones $C_{u_0}^{in}, \ldots, C_{u_{l-1}}^{in}$ and l cones $C_{u_0}^{out}, \ldots, C_{u_{l-1}}^{out}$, namely one cone $C_{u_t}^{in}$ and one cone $C_{u_t}^{out}$ per vertex u_t in L. We also use, for each u_t, a cylinder $Space(\Gamma_2(T(u_t)))$ that bounds the volume used by *Canonize*($\Gamma_2(T(u_t))$). We defer the computation of these cones and cylinders to Sect. 4.3, and for now assume that they are already available. For each $t = 0, \ldots, l-1$ and for each $j = 0, \ldots, t-1$, assume that $\Gamma_2(T(u_j))$ has been computed already – this is indeed the case when $t = 0$. Let $\mathcal{P}_t$ be the horizontal plane $z = |T| - 1 + \sum_{j=0}^{t-1} h(u_j)$, where $h(u_j)$ is the height of the cylinder $Space(\Gamma_2(T(u_j)))$. The drawing Γ_2 maps the subtree $T(u_t)$ to the plane $\mathcal{P}_t$, just outside the cone $C_{u_t}^{in}$ and just inside the

---

[3] Since the morph $\langle \Gamma, \Gamma_1 \rangle$ is linear, the trajectory of any vertex v is simply the line segment connecting the positions of v in Γ and in Γ_1. To define a vector, we orient the segment towards the position of v in Γ_1.

cone $C_{u_t}^{\text{out}}$. See Fig. 4. We proceed with the formal definition of Γ_2. Let v be any vertex of $T(u_t)$ and let (v_x, v_y, v_z) be the coordinates of $\Gamma_1(v)$. Then $\Gamma_2(v)$ is the point $(v_x \frac{r_t}{r}, v_y \frac{r_t}{r}, z_t)$. Here z_t is the height of the plane $\mathcal{P}_t$, r_t is the radius of the section of $C_{u_t}^{\text{in}}$ by the plane $\mathcal{P}_t$, and r is the distance from $\Gamma_1(v_i)$ to its closest point of the drawing $\Gamma_1(T(u_t))$, where v_i is the parent of u_t. See Fig. 4. Note that the latter closest point can be a point on an edge.

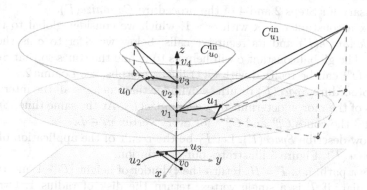

Fig. 4. The vertices v_0, v_1, v_2, v_3, v_4 are in the heavy path H of T. The lower gray disk has its center at v_1 and has radius equal to the distance from $\Gamma_1(v_i)$ to its closest point in $\Gamma_1(T(u_1))$. Blue arrows show the mapping of vertices in subtrees $T(u_0)$ and $T(u_1)$.

Step 3 (recurse). For each $u_t \in L$, we make a recursive call $Canonize(\Gamma_2(T(u_t)))$. The resulting morphs are combined into a unique morph $\langle \Gamma_2, \ldots, \Gamma_3 \rangle$, whose number of steps is equal to the maximum number of steps in any of the recursively computed morphs. Indeed, the first step of $\langle \Gamma_2, \ldots, \Gamma_3 \rangle$ consists of the first steps of all the recursively computed morphs that have at least one step; the second step of $\langle \Gamma_2, \ldots, \Gamma_3 \rangle$ consists of the second steps of all the t recursively computed morphs that have at least two steps; and so on.

Step 4 (rotate, rotate, rotate). The next morph transforms Γ_3 into a drawing Γ_4 such that each vertex $u_t \in L$ is mapped to the intersection of the cone $C_{u_t}^{\text{in}}$, the planes $y = 0$, $\mathcal{P}_t$, and the half-space $x > 0$. Note that going from Γ_3 to Γ_4 in one linear crossing-free 3D morph is not always possible. Refer to Lemma 2 for the implementation of the morph from Γ_3 to Γ_4 in $O(1)$ steps. After Step 4 the whole drawing lies on the plane $y = 0$.

Step 5 (go down). This step consists of a single linear morph $\langle \Gamma_4, \Gamma_5 \rangle$, where Γ_5 is defined as follows. For every vertex v_i in H, $\Gamma_5(v_i) = \Gamma_4(v_i)$; further, for every vertex $u_t \in L$, all the vertices of $T(u_t)$ have the same x- and y-coordinates in Γ_5 as in Γ_4, however their z-coordinate is decreased by the same amount so that u_t lies on the horizontal plane through $\mathcal{C}(u_t)$.

Step 6 (go left). The final part of our morphing procedure consists of a single linear morph $\langle \Gamma_5, \Gamma_6 \rangle$, where Γ_6 is the canonical 3D drawing $\mathcal{C}(T)$ of T. Note that this linear morph only moves the vertices horizontally.

4.3 The Procedure $Space(\Gamma)$

In this section we give a procedure to compute the cylinders and the cones which are necessary for Steps 2 and 4 of the procedure $Canonize(\Gamma)$.

We fix a constant $c \in \mathbb{R}$ with $c > 1$, which we consider global to the procedure $Canonize(\Gamma)$ and its recursive calls; below we refer to c as the *global constant*. The global constant c will help us to define the cones so that Step 4 of $Canonize(\Gamma)$ can be realized with $O(1)$ linear morphs, see Lemma 2.

The procedure $Space(\Gamma)$ returns a cylinder that encloses all the intermediate drawings of the morph determined by $Canonize(\Gamma)$. At the same time, $Space(\Gamma)$ determines the cones $C_{u_t}^{in}$ and $C_{u_t}^{out}$ for every vertex $u_t \in L$.

We now describe $Space(\Gamma)$. Let Γ_1 be the result of the application of Step 1 of $Canonize(\Gamma)$. Figure 5 illustrates our description.

If T is a path, i.e., $T = H$, return the cylinder of height $|T| - 1$ and radius 1. In particular, if T is a single vertex, return the disk of radius 1. Otherwise, construct the cylinder and the cones in the following fashion:

- Set the current cone C to be an infinite cone of slope 1. The apex of C is determined as follows: starting with the apex being at the highest point of the pole, slide C vertically downwards until it touches the drawing $\Gamma_1(T(u_0))$. That is, the apex of C is at the lowest possible position on the pole such that the whole drawing $\Gamma_1(T(u_0))$ is outside of C. See Fig. 5a.
- Set the current height h to be $|T| - 1$.
- Iterate through the light children of H in the order as they appear in L. For every u_t in L:
  - Set $C_{u_t}^{in}$ to be the current cone C.
  - Add the height of $Space(\Gamma_2(T(u_t)))$ to the current height h.
  - Let C' be the cone with the same apex as C and with a slope defined so that the drawing $\Gamma_1(T(u_t))$ is in-between C and C', and C is *well-separated* from C' with the global constant c. That is, $\phi(C') = \min(\phi(C)/Sp(u_t, \Gamma_1), \phi(C)/c)$, where $Sp(u_t, \Gamma_1)$ is the spread of the drawing $\Gamma_1(T(u_t))$ with respect to the parent v_i of u_t in H. Namely $Sp(u_t, \Gamma_1)$ is the ratio between the outer and the inner radius of the minimum annulus centered at v_i and enclosing the drawing $\Gamma_1(T(u_t))$. See Fig. 5a.
  - Let $\mathcal{S}_t$ be the cylinder $Space(\Gamma_2(T(u_t)))$ translated so that the center of its lower base is at the point $\Gamma_2(u_t)$.
  - Decrease $\phi(C')$ so that C' encloses the entire cylinder $\mathcal{S}_t$.
  - Set $C_{u_t}^{out}$ to be the cone C'.
  - If u_t is not the last element of L (i.e., $t < l - 1$), then let $u_t = v_i^j$ and define an auxiliary cone $\tilde{C}$ as follows. The apex of $\tilde{C}$ is at $\Gamma_1(v_x)$ where v_x is the parent of u_{t+1}; note that $v_x = v_i$ iff $j > 0$. The slope of $\tilde{C}$ is the

maximum slope that satisfies the following requirement: (i) the slope of $\tilde{C}$ is at most the slope of C'. In addition, only for the case when $v_x = v_i$, we require: (ii) in the closed half space $z \leq h$, the portion of $\tilde{C}$ encloses the portion of C'. See Fig. 5b. Update the cone C to be the lowest vertical translate of $\tilde{C}$ so that $\Gamma_1(T(u_{t+1}))$ is still outside the cone.

- Return the cylinder of height h (the current height), and radius equal to the radius of the section of the current cone C cut by the plane $z = h$.

4.4   Correctness of the Morphing Procedure

In this section, we analyze the correctness and the efficiency of the procedure $Canonize(\Gamma)$ (see Theorem 4) and we give the details of Step 4 (see Lemma 2).

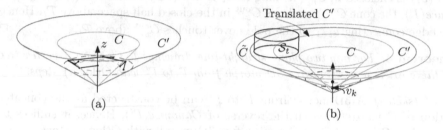

Fig. 5. Illustration for $Space(\Gamma)$: (a) construction of C and C'; (b) construction of $\tilde{C}$.

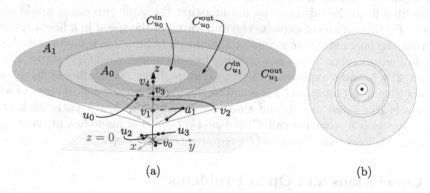

Fig. 6. (a) Annuli for the subtrees rooted at u_0 and u_1; (b) top view of the annuli.

Lemma 2. *Step 4 of the procedure $Canonize(\Gamma)$ can be realized as a crossing-free 3D morph whose number of steps is bounded from above by a constant that depends on the global constant c.*

Proof. Let A_t be the annulus formed by the section of $C_{u_t}^{in}$ and $C_{u_t}^{out}$ cut by the plane $\mathcal{P}_t$. See Fig. 6. The morph performed in Step 4 consists of a sequence of linear morphs; in each of these morphs all the vertices of $T(u_t)$ are translated by the same vector. This is done so that u_t stays in A_t during the whole Step 4. Thus, the trajectory of u_t during Step 4 defines a polygon inscribed in A_t. Since the ratio between the outer and the inner radius of A_t is at least the global constant c, we can inscribe a regular $O(1)$-gon in A_t, and the trajectory of u_t can be defined so that it follows this $O(1)$-gon plus at most one extra line segment.

We now prove that since each $u_t \in L$ stays in A_t, all the steps of the above morph are crossing-free. Recall that at any moment during the morph, the drawing of $T(u_t)$ is a translation of the canonical 3D drawing $\mathcal{C}(T(u_t))$. By Lemma 1, the space below the line of slope 1 passing through u_t in plane $y = 0$ does not contain any point of $\mathcal{C}(T(u_t))$. Since the slope of $C_{u_t}^{out}$ is at most 1, the drawing of $T(u_t)$ is enclosed in $C_{u_t}^{out}$ as long as u_t is in A_t. By conditions (i) and (ii) of $Space(\Gamma)$, the cone $C_{u_{t+1}}^{in}$ encloses $C_{u_t}^{out}$ in the closed half-space above $\mathcal{P}_t$. Hence the edge connecting u_{t+1} to the pole never touches $C_{u_t}^{out}$ above $\mathcal{P}_t$.   □

Theorem 4. *For any two plane straight-line drawings Γ, Γ' of an n-vertex tree T, there exists a crossing-free 3D morph from Γ to Γ' with $O(\log n)$ steps.*

Proof (sketch). A 3D morph from Γ to Γ' can be constructed as the concatenation of $Canonize(\Gamma)$ with the reverse of $Canonize(\Gamma')$. Hence, it suffices to prove that $Canonize(\Gamma)$ is a crossing-free 3D morph with $O(\log n)$ steps.

It is easy to see that Steps 1, 5, and 6 of $Canonize(\Gamma)$ are crossing-free linear morphs. The proof that Step 2 is a crossing-free linear morph is more involved. In particular, for any two light children u_s and u_t with $s < t$ of the same vertex v_i of H, the occurrence of a crossing between the edge $v_i u_s$ and an edge of $T(u_t)$ during Step 2 can be ruled out by arguing that the same two edges would also cross in Γ_1; this argument exploits the uniformity of the speed in a linear morph and that the horizontal component of the morph of Step 2 is a uniform scaling. Lemma 2 ensures that Step 4 is a crossing-free 3D morph with $O(1)$ steps. Thus, Steps 1, 2, 4, 5, and 6 require a total of $O(1)$ steps. Since the number of morphing steps of Step 3 of $Canonize(\Gamma)$ is equal to the maximum number of steps of any recursively computed morph and since, by definition of heavy path, each tree $T(u_t)$ for which a recursive call $Canonize(\Gamma_2(T(u_t)))$ is made has at most $n/2$ vertices, it follows that $Canonize(\Gamma)$ requires $O(\log n)$ steps.   □

5   Conclusions and Open Problems

In this paper we studied crossing-free 3D morphs of tree drawings. We proved that, for any two planar straight-line drawings of the same n-vertex tree, there is a crossing-free 3D morph between them which consists of $O(\log n)$ steps.

This result gives rise to two natural questions. First, is it possible to bring our logarithmic upper bound down to constant? In this paper we gave a positive answer to this question for paths. In fact our algorithm to morph planar straight-line tree drawings has a number of steps which is linear in the pathwidth of the tree (see Remark 2), thus for example it is constant for caterpillars. Second, does a crossing-free 3D morph exist with $o(n)$ steps for any two planar straight-line drawings of the same n-vertex planar graph? The question is interesting to us even for subclasses of planar graphs, like outerplanar graphs and planar 3-trees.

We also proved that any two crossing-free straight-line 3D drawings of an n-vertex tree can be morphed into each other in $O(n)$ steps; such a bound is asymptotically optimal in the worst case. An easy extension of our results to graphs containing cycles seems unlikely. Indeed, the existence of a deterministic algorithm to construct a crossing-free 3D morph with a polynomial number of steps between two crossing-free straight-line 3D drawings of a cycle would imply that the unknot recognition problem is polynomial-time solvable. The *unknot recognition* problem asks whether a given knot is equivalent to a circle in the plane under an ambient isotopy. This problem has been the subject of investigation for decades; it is known to be in NP [13] and in co-NP [14], however determining whether it is in P has been an elusive goal so far.

Acknowledgments. We thank Therese Biedl for pointing out Remark 2. The research for this paper started during the Intensive Research Program in Discrete, Combinatorial and Computational Geometry, which took place in Barcelona, April-June 2018. We thank Vera Sacristán and Rodrigo Silveira for a wonderful organization and all the participants for interesting discussions.

References

1. Adams, C.C.: The Knot Book: An Elementary Introduction to the Mathematical Theory of Knots. American Mathematical Society, Providence (2004)
2. Alamdari, S., et al.: How to morph planar graph drawings. SIAM J. Comput. **46**(2), 824–852 (2017)
3. Alamdari, S., et al.: Morphing planar graph drawings with a polynomial number of steps. In: Khanna, S. (ed.) 24th Annual ACM-SIAM Symposium on Discrete Algorithms (SODA 2013), pp. 1656–1667. SIAM (2013)
4. Angelini, P., Da Lozzo, G., Di Battista, G., Frati, F., Patrignani, M., Roselli, V.: Morphing planar graph drawings optimally. In: Esparza, J., Fraigniaud, P., Husfeldt, T., Koutsoupias, E. (eds.) ICALP 2014. LNCS, vol. 8572, pp. 126–137. Springer, Heidelberg (2014). https://doi.org/10.1007/978-3-662-43948-7_11
5. Angelini, P., Frati, F., Patrignani, M., Roselli, V.: Morphing planar graph drawings efficiently. In: Wismath, S., Wolff, A. (eds.) GD 2013. LNCS, vol. 8242, pp. 49–60. Springer, Cham (2013). https://doi.org/10.1007/978-3-319-03841-4_5
6. Barrera-Cruz, F., Haxell, P., Lubiw, A.: Morphing planar graph drawings with unidirectional moves. In: Mexican Conference on Discrete Mathematics and Computational Geometry, pp. 57–65 (2013). http://arxiv.org/abs/1411.6185
7. Biedl, T.: Optimum-width upward drawings of trees. arXiv preprint arXiv:1506.02096 (2015)
8. Cairns, S.S.: Deformations of plane rectilinear complexes. Am. Math. Mon. **51**(5), 247–252 (1944)
9. Chan, T.M.: Tree drawings revisited. arXiv preprint arXiv:1803.07185 (2018)
10. Crescenzi, P., Di Battista, G., Piperno, A.: A note on optimal area algorithms for upward drawings of binary trees. Comput. Geom. **2**(4), 187–200 (1992)

11. Floater, M.S., Gotsman, C.: How to morph tilings injectively. J Comput. Appl. Math. **101**(1–2), 117–129 (1999)
12. Gotsman, C., Surazhsky, V.: Guaranteed intersection-free polygon morphing. Comput. Graph. **25**(1), 67–75 (2001)
13. Hass, J., Lagarias, J.C., Pippenger, N.: The computational complexity of knot and link problems. J. ACM **46**(2), 185–211 (1999)
14. Lackenby, M.: The efficient certification of knottedness and Thurston norm. CoRR, abs/1604.00290 (2016)
15. Sleator, D.D., Tarjan, R.E.: A data structure for dynamic trees. In: Proceedings of the Thirteenth Annual ACM Symposium on Theory of Computing, pp. 114–122 (1981)
16. Thomassen, C.: Deformations of plane graphs. J. Comb. Theory, Ser. B **34**(3), 244–257 (1983)

Partially Fixed Drawings

The Complexity of Drawing a Graph in a Polygonal Region

Anna Lubiw[1]($\boxtimes$), Tillmann Miltzow[2], and Debajyoti Mondal[3]

[1] Cheriton School of Computer Science, University of Waterloo, Waterloo, Canada
alubiw@uwaterloo.ca
[2] Université libre de Bruxelles (ULB), Brussels, Belgium
t.miltzow@gmail.com
[3] Department of Computer Science, University of Saskatchewan, Saskatoon, Canada
dmondal@usask.ca

Abstract. We prove that the following problem is complete for the existential theory of the reals: Given a planar graph and a polygonal region, with some vertices of the graph assigned to points on the boundary of the region, place the remaining vertices to create a planar straight-line drawing of the graph inside the region. A special case is the problem of extending a partial planar graph drawing, which was proved NP-hard by Patrignani. Our result is one of the first showing that a problem of drawing planar graphs with straight-line edges is hard for the existential theory of the reals. The complexity of the problem is open for a simply connected region.

We also show that, even for integer input coordinates, it is possible that drawing a graph in a polygonal region requires some vertices to be placed at irrational coordinates. By contrast, the coordinates are known to be bounded in the special case of a convex region, or for drawing a path in any polygonal region.

1 Introduction

There are many examples of structural results on graphs leading to beautiful and efficient geometric representations. Two highlights are: Tutte's polynomial-time algorithm [31] to draw any 3-connected planar graph with convex faces inside any fixed convex drawing of its outer face; and Schnyder's tree realizer result [28] that provides a drawing of any n-vertex planar graph on an $n \times n$ grid.

On the other hand, there are geometric representations that are intractable, either in terms of the required coordinates or in terms of computation time. As an example of the former, a representation of a planar graph as touching disks (Koebe's theorem) is not always possible with rational numbers, nor even with roots of low-degree polynomials [5]. As an example of the latter, Patrignani

This work was partially supported by NSERC and by ERC Consolidator Grant 615640-ForEFront. A video explaining this work can be found at https://youtu.be/JbmWLnY1hGk.

© Springer Nature Switzerland AG 2018
T. Biedl and A. Kerren (Eds.): GD 2018, LNCS 11282, pp. 387–401, 2018.
https://doi.org/10.1007/978-3-030-04414-5_28

considered a generalization of Tutte's theorem and proved that it is NP-hard to decide whether a graph has a straight-line planar drawing when part of the drawing is fixed [25]. He was unable to show that the problem lies in NP because of coordinate issues.

This, and many other geometric problems, most naturally lie not in NP, but in a larger class, $\exists\mathbb{R}$, defined by formulas in existentially quantified real (rather than Boolean) variables. Showing that a geometric representation problem is complete for $\exists\mathbb{R}$ is a stronger intractability result, often implying lower bounds on coordinate sizes. For example, McDiarmid and Müller [20] showed that deciding if a graph can be represented as intersecting disks is $\exists\mathbb{R}$-complete. The relaxation from touching disks (Koebe's theorem) to intersecting disks implies that disk centers and radii can be restricted to integers, but McDiarmid and Müller show that an exponential number of bits may be required.

In this paper we prove that an extension of Tutte's problem is $\exists\mathbb{R}$-complete. We call it the "GRAPH IN POLYGON" problem. See Fig. 1. The input is a graph G and a closed polygonal region R (not necessarily simply connected), with some vertices of G assigned fixed positions on the boundary of R. The question is whether G has a straight-line planar drawing inside R respecting the fixed vertices. We regard the region R as a closed region which means that boundary points of R may be used in the drawing. A straight-line planar drawing (see Fig. 2(a, b)) means that vertices are represented as distinct points, and every edge is represented as a straight-line segment joining its endpoints, and no two of the closed line segments intersect except at a common vertex. (In particular, no vertex point may lie inside an edge segment, and no two segments may cross.)

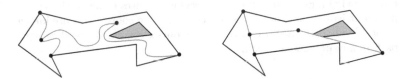

Fig. 1. The GRAPH IN POLYGON problem. Left: a polygonal region with one hole and a graph to be embedded inside the region. The three vertices on the boundary are fixed; the others are free. Right: a straight-line embedding of the graph in the region. Note that we allow an edge of the drawing (in red) to include points of the region boundary. (Color figure online)

Furthermore, we give a simple instance of GRAPH IN POLYGON with integer coordinates where a vertex of G may need irrational coordinates in any solution, thus defeating the naive approach to placing the problem in NP.

The GRAPH IN POLYGON problem is a very natural one that arises in practical applications such as dynamic and incremental graph drawing. Questions of the coordinates (or grid size) required for straight-line planar drawings of graphs are fundamental and well-studied [33]. It is surprising that a problem as simple and natural as GRAPH IN POLYGON is so hard and requires irrational coordinates.

We state our results below, but first we give some background on existential theory of the reals and on relevant graph drawing results. In particular, we explain that our problem is a generalization of the problem of extending a partial drawing of a planar graph to a straight-line drawing of the whole graph, called PARTIAL DRAWING EXTENSIBILITY. See Fig. 2(c, d).

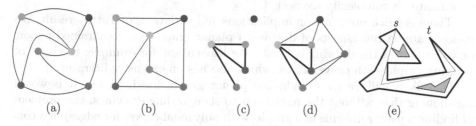

Fig. 2. (a) A planar graph G. (b) a straight-line drawing of G. (c) a partial drawing Γ of G. (d) extension of Γ to a straight-line drawing of G. (e) A minimum-link s-t path in a polygonal region.

Existential Theory of the Reals. In the study of geometric problems, the complexity class $\exists\mathbb{R}$ plays a crucial role, connecting purely geometric problems and real algebra. Whereas NP is defined in terms of Boolean formulas in existentially quantified Boolean variables, $\exists\mathbb{R}$ deals with first-order formulas in existentially quantified real variables.

Consider a first-order formula over the reals that contains only existential quantifiers, $\exists x_1, x_2, \ldots, x_n : \Phi(x_1, x_2, \ldots, x_n)$, where $x_1, x_2, \ldots, x_n$ are real-valued variables and Φ is a quantifier-free formula involving equalities and inequalities of real polynomials. The EXISTENTIAL THEORY OF THE REALS (ETR) problem takes such a formula as an input and asks whether it is satisfiable. The complexity class $\exists\mathbb{R}$ consists of all problems that reduce in polynomial time to ETR. Many problems in combinatorial geometry and geometric graph representation naturally lie in this class, and furthermore, many have been shown to be $\exists\mathbb{R}$-complete, e.g., stretchability of a pseudoline arrangement [19,23,27], recognition of segment intersection graphs [17] and disk intersection graphs [20], computing the rectilinear crossing number of a graph [6], etc. For surveys on $\exists\mathbb{R}$, see [8,19,26]. A recent proof that the ART GALLERY PROBLEM is $\exists\mathbb{R}$-complete [2] provides the framework we follow in our proof.

Planar Graph Drawing. The field of Graph Drawing investigates many ways of representing graphs geometrically [24], but we focus on the most basic representation of planar graphs, with points for vertices and straight-line segments for edges, such that segments intersect only at a common endpoint. By Fáry's theorem [12], every planar graph admits such a straight-line planar drawing.

In Tutte's famous paper, "How to Draw a Graph," he gave a polynomial time algorithm to find a straight-line planar drawing of a graph by first augmenting to a 3-connected graph. Given a combinatorial planar embedding (a specification of the faces) of a 3-connected graph and given a convex polygon drawing of the outer face of the graph, his algorithm produces a planar straight-line drawing respecting both by reducing the problem to solving a linear system involving barycentric coordinates for each internal vertex. Tutte proved that the linear system has a unique solution and that the solution yields a drawing with convex faces. The linear system can be solved in polynomial time. For a discussion of coordinate bit complexity see Sect. 4.

There is a rich literature on implications and variations of Tutte's result. We concentrate on the aspects of drawing a planar graph in a constrained region, or when part of the drawing is fixed. (We leave aside, for example, the issue of drawing graphs with convex faces, which also has an extensive literature.)

Our focus will be on straight-line planar graph drawings, but it is worth mentioning that without the restriction to straight-line drawings, the problem of finding a planar drawing of a graph (with polygonal curves for edges) in a constrained region is equivalent to the problem of extending a partial planar drawing (with polygonal curves for edges), and there is a polynomial time algorithm for the decision version of the problem [3]. Furthermore, there is an algorithm to construct such a drawing in which each edge is represented by a polygonal curve with linearly many segments [10].

For the rest of this paper we assume straight-line planar drawings, which makes the problems harder. The problem of drawing a graph in a constrained region is formalized as GRAPH IN POLYGON, defined above, and more precisely in Subsect. 1.1. The problem of finding a planar straight-line drawing of a graph after part of the drawing has been fixed is called PARTIAL DRAWING EXTENSIBILITY in the literature—its complexity was formulated as an open question in [7].

The relationship between the two problems is that GRAPH IN POLYGON generalizes PARTIAL DRAWING EXTENSIBILITY, as we now argue. Given an instance of partial drawing extensibility, for graph G with fixed subgraph H, we construct an instance of GRAPH IN POLYGON by making a point hole for each vertex of H and assigning the vertex to the point. Then an edge of H can only be drawn as a line segment joining its endpoints, so we have effectively fixed H. To complete the bounded region R, we enclose the point holes in a large box. Clearly, we now have an instance of GRAPH IN POLYGON, and that instance has a solution if and only if G has a planar straight-line drawing that extends the drawing of H. There is no easy reduction in the other direction because GRAPH IN POLYGON involves a closed polygonal region, so an edge may be drawn as a segment that touches, or lies on, the boundary of the region, and it is not clear how to model this as PARTIAL DRAWING EXTENSIBILITY. However, the version of GRAPH IN POLYGON for an open region is equivalent to PARTIAL DRAWING EXTENSIBILITY.

We now summarize results on PARTIAL DRAWING EXTENSIBILITY, beginning with positive results. Besides Tutte's result that a convex drawing of the outer face can always be extended, there is a similar result for a star-shaped drawing of the outer face [14]. There is also a polynomial-time algorithm to decide the case when a convex drawing of a subgraph is fixed [21]. Urhausen [32] examined the case when a star-shaped drawing of one cycle in the graph is given, and proved that there always exists an extension with at most one bend per edge. Gortler et al. [13] gave an algorithm, extending Tutte's algorithm, that succeeds in some (not well-characterized) cases for a simple non-convex drawing of the outer face. The PARTIAL DRAWING EXTENSIBILITY problem was shown to be NP-hard by Patrignani [25]. This implies that GRAPH IN POLYGON is NP-hard. However, there are two natural questions about partial drawing extensibility that remain open: (a) does the problem belong to the class NP (discussed in detail by Patrignani [25]), and (b) does the problem remain NP-hard when a combinatorial embedding of the graph is given and must be respected in the drawing. Our results shed light on these questions for the more general GRAPH IN POLYGON problem: the problem cannot be shown to lie in NP by means of giving the vertex coordinates, and the problem is still ∃R-hard when a combinatorial embedding of the graph is given.

Besides Tutte's result, there is another special case of GRAPH IN POLYGON that is well-solved, namely when the graph is just a path with its two endpoints s and t fixed on the boundary of the region. See Fig. 2(c). This problem is equivalent to the MINIMUM LINK PATH problem—to find a path from s to t inside the region with a minimum number of segments. This is because a path of k edges can be drawn inside the region if and only if the minimum link distance between s and t is less than or equal to k. Minimum link paths in a polygonal region can be found in polynomial time [22], and in linear time for a simple polygon [29]. The complexity of the coordinates is well-understood (see Sect. 4).

1.1 Our Contributions

Our problem is defined as follows.

Graph in Polygon
Input: A planar graph G and a polygonal region R with some vertices of G assigned to fixed positions on the boundary of R.
Question: Does G admit a planar straight-line drawing inside R respecting the fixed vertices?

The graph may be given abstractly, or via a *combinatorial embedding* which specifies the cyclic order of edges around each vertex, thus determining the faces of the embedding. When a combinatorial embedding is specified then the final drawing must respect that embedding.

Note that we regard R as a closed region. Thus, points on the boundary of R may be used in the drawing of G. In particular, an edge of G may be drawn as a segment that touches, or lies on, the boundary of R. See Fig. 1. Note that we still require the drawing of G to be "simple" in the conventional sense that no two edge segments may intersect except at a common endpoint.

Our first result is that solutions to GRAPH IN POLYGON may involve irrational points. This will in fact follow from the proof of our main hardness result, but it is worth seeing a simple example.

Theorem 1. *There is an instance of* GRAPH IN POLYGON *with all coordinates given by integers, in which some vertices need irrational coordinates.*

Note that the theorem does not rule out membership of the problem in NP, since it may be possible to demonstrate that a graph can be drawn in a region without giving explicit vertex coordinates. We prove Theorem 1 by adapting an example from Abrahamsen, Adamaszek and Miltzow [1] that proves a similar irrationality result for the ART GALLERY PROBLEM. Further discussion of bit complexity for special cases of the problem can be found in Sect. 4.

Our main result is the following, which holds whether the graph is given abstractly or via a combinatorial embedding.

Theorem 2. GRAPH IN POLYGON *is* $\exists\mathbb{R}$*-complete.*

We prove Theorem 2 using a reduction from a problem called ETR-INV which was introduced and proved $\exists\mathbb{R}$-complete by Abrahamsen, Adamaszek and Miltzow [2].

Definition 1 (ETR-INV). *In the problem ETR-INV, we are given a set of real variables* $\{x_1, \ldots, x_n\}$, *and a set of equations of the form* $x = 1$, $x + y = z$, $x \cdot y = 1$, *for* $x, y, z \in \{x_1, \ldots, x_n\}$. *The goal is to decide whether the system of equations has a solution when each variable is restricted to the range* $[1/2, 2]$.

Reducing from ETR-INV, rather than from ETR, has several crucial advantages. First, we can assume that all variables are in the range $[1/2, 2]$. Second, we do not have to implement a gadget that simulates multiplication, but only inversion, i.e., $x \cdot y = 1$. For our purpose of reducing to GRAPH IN POLYGON, we will find it useful to further modify ETR-INV to avoid equality and to ensure planarity of the variable-constraint incidence graph, as follows:

Definition 2 (Planar-ETR-INV*). *In the problem Planar-ETR-INV*, we are given a set of real variables* $\{x_1, \ldots, x_n\}$, *and a set of equations and inequalities of the form* $x = 1$, $x + y \leq z$, $x + y \geq z$, $x \cdot y \leq 1$, $x \cdot y \geq 1$, *for* $x, y, z \in \{x_1, \ldots, x_n\}$. *Furthermore, we require planarity of the* variable-constraint inci-*dence graph, which is the bipartite graph that has a vertex for every variable and every constraint and an edge when a variable appears in a constraint. The goal is to decide whether the system of equations has a solution when each variable is restricted to lie in* $[1/2, 4]$.

As a technical contribution, we prove the following.

Theorem 3. *Planar-ETR-INV* is* $\exists\mathbb{R}$*-complete.*

The proof, which is in the long version [18], builds on the work of Dobbins, Kleist, Miltzow and Rzążewski [11] who showed that ETR-INV is $\exists\mathbb{R}$-complete even when the variable-constraint incidence graph is planar. We cannot use their result directly, but follow similar steps in our proof.

2 Irrational Coordinates

Theorem 1. *There is an instance of* GRAPH IN POLYGON *with all coordinates given by integers, in which some vertices need irrational coordinates.*

In fact, the result follows from our proof of Theorem 2, but it is interesting to have a simple explicit example, which is given in Fig. 4. This example is adapted from a result of Abrahamsen et al. [1]. Details can be found in the long version [18], but we outline the idea here. Abrahamsen et al. studied the ART GALLERY PROBLEM, where given a polygon P and a number k, and we want to find a set of at most k guards (points) that together see the entire polygon. We say a guard g sees a point p if the entire line-segment gp is contained inside the polygon P. Abrahamsen et al. gave a simple polygon with integer coordinates such that there exists only one way to guard it optimally, with three guards. Those guards have irrational coordinates. See Fig. 3 for a sketch of their polygon. A key ingredient of their construction is to create notches in the polygon boundary that force there to be a guard on each of the three so-called *guard segments*. The coordinates of the polygon then force the guards to be at irrational points.

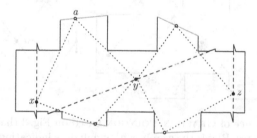

Fig. 3. A sketch of the polygon from Abrahamsen et al. the three guards (black dots) must lie on the *guard segments* (dashed lines).

We adapt their example by using *variable segments* (shown in green) instead of guard segments, and vertices instead of guards. By placing notches in the polygon boundary with fixed vertices of the graph in the notches, we can force there to be a vertex on each variable segment. We create two cycles that replicate the guarding constraints, and use a hole in order to keep our graph drawing planar. From their proof we show that x', y' and z' must be at irrational coordinates.

3 ∃ℝ-completeness

Theorem 2. GRAPH IN POLYGON *is* ∃ℝ-*complete.*

Proof. First note that GRAPH IN POLYGON lies in $\exists\mathbb{R}$ since we can express it as an ETR formula. To prove that the problem is $\exists\mathbb{R}$-hard we give a reduction from Planar-ETR-INV*. Let I be an instance of Planar-ETR-INV*. We will build an instance J of GRAPH IN POLYGON such that J admits an affirmative answer if and only if I is satisfiable. The idea is to construct gadgets to represent variables, and gadgets to enforce the addition and inversion inequalities, $x+y \leq z, x+y \geq z, x\cdot y \leq 1, x\cdot y \geq 1$. We also need gadgets to copy and replicate variables—"wires" and "splitters" as conventionally used in reductions. Thereafter, we have to describe how to combine those gadgets to obtain an instance J of Planar-ETR-INV*.

Encoding Variables. We will encode the value of a variable in $[1/2, 4]$ as the position of a vertex that is constrained to lie on a line segment of length 3.5, which we call a *variable-segment*. One end of a variable-segment encodes the value $\frac{1}{2}$, the other end encodes the value 4, and linear interpolation fills in the values between. Figure 5 shows one side of the construction that forces a vertex to lie on a variable-segment. The other side is similar.

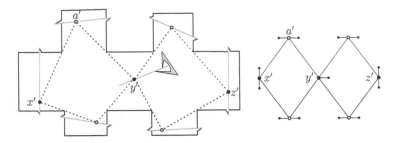

Fig. 4. Left: an instance of GRAPH IN POLYGON based on Fig. 3 that requires vertices at irrational coordinates. Right: the graph, with small dots indicating the fixed vertices.

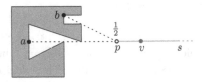

Fig. 5. Variable v is represented as a point on variable-segment s (shown in green). The construction of one end of s is illustrated. In the graph, vertex v is adjacent to fixed vertices a and b on the boundary of a hole of the region (shaded). Adjacency with a forces v to lie on the line of s. Adjacency with b forces v to lie at, or to the right of, point p which is associated with the value 1/2. Note that p is not a vertex. (Color figure online)

By slight abuse of notation, we will identify a variable and the vertex representing it, if there is no ambiguity. For the description of the remaining gadgets, our figures will show variable-segments (in green) without showing the polygonal holes that determine them.

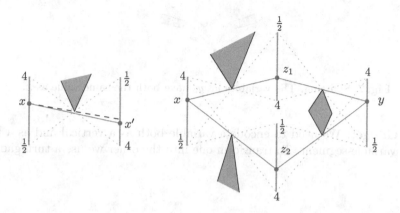

Fig. 6. Copying. Left: a gadget to enforce $x \leq x'$. Right: the full gadget enforcing $x = y$.

Copy Gadget. Given a variable-segment for a variable x, we will need to transmit its value along a "wire" to other locations in the plane. We do this using a copy gadget in which we construct a variable-segment for a new variable y and enforce $x = y$. We show how to construct a gadget that ensures $x \leq x'$ for a new variable x', and then combine four such gadgets, enforcing $x \leq z_1, z_1 \leq y, x \geq z_2, z_2 \geq y$. This implies $x = y$.

The gadget enforcing $x \leq x'$ is shown at the left of Fig. 6. It consists of two parallel variable-segments. In general, these two segments need not be horizontally aligned. In the graph we connect the corresponding vertices by an edge. The left and the right variables are encoded in opposite ways, i.e., x increases as the vertex moves up and x' increases as the corresponding vertex goes down. We place a hole of the polygonal region (shaded in the figure) with a vertex at the intersection point of the lines joining the top of one variable-segment to the bottom of the other. The hole must be large enough that the edge from x to x' can only be drawn to one side of the hole. An argument about similar triangles, or the "intercept theorem", also known as Thales' theorem, implies $x \leq x'$.

We combine four of these gadgets to construct our copy gadget, as illustrated on the right of Fig. 6.

Splitter Gadget. Since a single variable may appear in several constraints, we may need to split a wire into two wires, each holding the correct value of the same variable. Figure 7 shows a gadget to split the variable x to variables y_1 and y_2. The gadget consists of two copy gadgets sharing the variable-segment for x. We can construct the two copy gadgets to avoid any intersections between them.

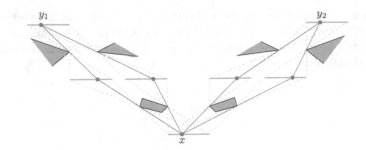

Fig. 7. Splitting. The variables y_1, y_2 have both the same value as x.

Turn Gadget. We need to encode a variable both as a vertical and as a horizontal variable-segment. To transform one into the other we use a turn gadget.

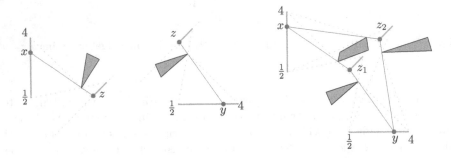

Fig. 8. Turning. Left: gadget to encode $x \leq f(z)$. Middle: symmetric gadget to encode $y > f(z)$. Right: four gadgets of the previous type combined to enforce $x = y$, for x and y on a vertical and horizontal variable-segment, respectively.

The key idea is to construct two diagonal variable-segments for variables z_1 and z_2, and then transfer the value of the vertical variable-segment to the horizontal variable-segment using z_1, z_2. This is in fact very similar to the copy gadget, except that the intermediate variable-segments are placed on a line of slope 1. We do not know if it is possible to enforce the constraint $x \leq z$ directly. However, it is sufficient to enforce $x \leq f(z)$ for some function f. See the left side of Fig. 8. Interestingly, we don't even know the function f. However, we do know that f is monotone and we can construct another gadget enforcing $y \geq f(z)$, for the same function f, by making another copy of the first gadget reflected through the line of the variable-segment for z.

Combining four such gadgets, as on the right of Fig. 8, yields the following inequalities: $x \leq f_1(z_1), f_1(z_1) \leq y, y \leq f_2(z_2), f_2(z_2) \leq x$. This implies $x = y$.

Addition Gadget. The gadget to enforce $x + y \geq z$ is depicted in Fig. 9. Important for correctness is that the gaps between the dotted auxiliary lines have

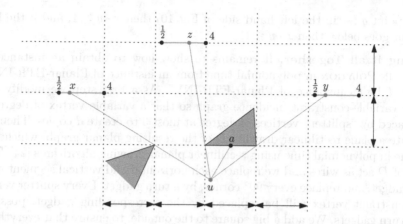

Fig. 9. Addition. The three vertices x, y, z can only be connected to u if $x + y \geq z$ holds.

equal lengths. This is essentially the same gadget that was used by Abrahamsen et al. [2, Lemma 31]. An alternative proof can be found in the long version [18].

Lemma 1 ([2]). *The gadget in Fig. 9 enforces $x + y \geq z$.*

The gadget that enforces $x + y \leq z$ is just a mirror copy of the previous gadget.

Inversion Gadget. The inversion gadgets to enforce $x \cdot y \leq 1$ and $x \cdot y \geq 1$ are depicted in Fig. 10. We use a horizontal variable-segment for x and a vertical variable-segment for y and align them as shown in the figure, 1.5 units apart both horizontally and vertically. We make a triangular hole with its apex at point q as shown in the figure. The graph has an edge between x and y.

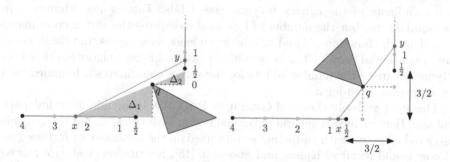

Fig. 10. Inversion. Left: gadget enforcing $x \cdot y \geq 1$. Right: gadget enforcing $x \cdot y \leq 1$.

For correctness, observe that if x and y are positioned so that the line segment joining them goes through point q, then, because triangles Δ_1 and Δ_2 (as shown in the figure) are similar, we have $\frac{x}{1} = \frac{1}{y}$, i.e. $x \cdot y = 1$. If the line segment goes above point q (as in the left hand side of Fig. 10) then $x \cdot y \geq 1$, and if the line segment goes below then $x \cdot y \leq 1$.

Putting it all Together. It remains to show how to obtain an instance of GRAPH IN POLYGON in polynomial time from an instance of Planar-ETR-INV*.

Let I be an instance of Planar-ETR-INV*. As a first step we modify the planar variable-constraint incidence graph so that a variable vertex of degree d is replaced by "splitter" vertices of degree at most 3 to create d copies. Then we compute a plane rectilinear drawing D of the resulting planar graph, which can be done in polynomial time using rectilinear planar drawing algorithms [24]. The edges of D act as wires and we replace each horizontal and vertical segment by a copy gadget, and replace every 90° corner, by a turn gadget. Every splitter vertex and constraint vertex will be replaced by the corresponding gadget, possibly using turn gadgets. We add a big square to the outside, to ensure that everything is inside one polygon. See the long version [18] for an illustration.

It is easy to see that this can be done in polynomial time, as every gadget has a constant size description and can be described with rational numbers, although, we did not do it explicitly. In order to see that we can also use integers, note that we can scale everything with the least common multiple of all the denominators of all numbers appearing. This can also be done in polynomial time.   □

4   Vertex Coordinates

Since we have shown that GRAPH IN POLYGON may require irrational coordinates for vertices in general, it is interesting to examine bounds on coordinates for special cases. In this section we discuss the bit complexity of vertex coordinates needed for two well-solved special cases of GRAPH IN POLYGON.

Tutte's algorithm [30] finds a straight-line planar drawing of a graph inside a fixed convex drawing of its outer face. Suppose the graph has n vertices and each coordinate of the convex polygon uses t bits. Tutte's algorithm runs in polynomial time, but the number of bits used to express the vertex coordinates is a polynomial function of t and n. The dependence on n means that the drawing uses "exponential area." Chambers et al. [9] gave a different algorithm that uses polynomial area—the number of bits for the vertex coordinates is bounded by a polynomial in t and $\log n$.

The other well-solved case of GRAPH IN POLYGON is the minimum link path problem. Here we have a general polygonal region with holes, R, but the graph is restricted to a path with endpoints s and t fixed on the boundary of R. Based on a lower bound result of Kahan and Snoeyink [15], Kostitsyna et al. [16] proved a tight bound of $\Theta(n \log n)$ bits for the coordinates of the bends on a minimum link path. Note that the dependence on n means that this bound is exponentially larger than the bound for drawing a graph inside a convex polygon. Problem 3 below asks about the complexity of drawing a tree in a polygonal region.

5 Conclusion and Open Questions

Our result that GRAPH IN POLYGON is ∃ℝ-complete is one of the first ∃ℝ-hardness results about drawing planar graphs with straight-line edges—along with a recent result about drawings with prescribed face areas [11]. We conclude with some open questions:

1. Our proofs of Theorems 1 and 2 used the fact that the polygonal region may have holes and may have collinear vertices. Is GRAPH IN POLYGON polynomial-time solvable for a simple polygon (a polygonal region without holes) whose vertices lie in general position (without collinearities)?

2. Our proofs also used the assumption that the polygonal region is closed. For an open region, the problem GRAPH IN POLYGON is equivalent to the problem PARTIAL DRAWING EXTENSIBILITY. Is this problem ∃ℝ-hard? There are two versions, depending on whether the graph is given abstractly or via a combinatorial embedding. In the first case the problem is known to be NP-hard [25], but in the second case even that is not known.

3. What is the complexity of GRAPH IN POLYGON when the graph is a tree? Can vertex coordinates still be bounded as for the minimum link path problem? When the tree is a caterpillar, the problem might be related to the minimum link watchman tour problem, which is known to be NP-hard [4].

Acknowledgment. We would like to thank Günter Rote, who discussed with the second author the turn gadget in the context of the ART GALLERY PROBLEM.

References

1. Abrahamsen, M., Adamaszek, A., Miltzow, T.: Irrational guards are sometimes needed. In: Proceedings of the 33rd International Symposium on Computational Geometry (SoCG), pp. 3:1–3:15. LIPIcs (2017)
2. Abrahamsen, M., Adamaszek, A., Miltzow, T.: The art gallery problem is ∃ℝ-complete. In: Proceedings of the 50th Annual ACM SIGACT Symposium on Theory of Computing. STOC 2018, 25–29 June 2018, Los Angeles, CA, USA, pp. 65–73 (2018). https://doi.acm.org/10.1145/3188745.3188868
3. Angelini, P., et al.: Testing planarity of partially embedded graphs. ACM Trans. Algorithms **11**(4), 32:1–32:42 (2015)
4. Arkin, E.M., Mitchell, J.S.B., Piatko, C.D.: Minimum-link watchman tours. Inf. Process. Lett. **86**(4), 203–207 (2003)
5. Bannister, M.J., Devanny, W.E., Eppstein, D., Goodrich, M.T.: The Galois complexity of graph drawing: why numerical solutions are ubiquitous for force-directed, spectral, and circle packing drawings. J. Graph Algorithms Appl. **19**(2), 619–656 (2015)
6. Bienstock, D.: Some provably hard crossing number problems. Discrete Comput. Geom. **6**, 443–459 (1991)
7. Brandenburg, F., Eppstein, D., Goodrich, M.T., Kobourov, S., Liotta, G., Mutzel, P.: Selected open problems in graph drawing. In: Liotta, G. (ed.) GD 2003. LNCS, vol. 2912, pp. 515–539. Springer, Heidelberg (2004). https://doi.org/10.1007/978-3-540-24595-7_55

8. Cardinal, J.: Computational geometry column 62. ACM SIGACT News **46**(4), 69–78 (2015)
9. Chambers, E.W., Eppstein, D., Goodrich, M.T., Löffler, M.: Drawing graphs in the plane with a prescribed outer face and polynomial area. J. Graph Algorithms Appl. **16**(2), 243–259 (2012)
10. Chan, T.M., Frati, F., Gutwenger, C., Lubiw, A., Mutzel, P., Schaefer, M.: Drawing partially embedded and simultaneously planar graphs. J. Graph Algorithms Appl. **19**(2), 681–706 (2015)
11. Dobbins, M.G., Kleist, L., Miltzow, T., Rzążewski, P.: Proceedings of the 44th International Workshop on Graph-Theoretic Concepts in Computer Science (WG). In: Brandstädt, A., Köhler, E., Meer, K. (eds.) ∀∃R-completeness and area-universality. LNCS, vol. 11159, pp. 164–175. Springer, Cham (2018). https://doi.org/10.1007/978-3-030-00256-5_14. http://arxiv.org/abs/1712.05142
12. Fáry, I.: On straight line representations of planar graphs. Acta Sci. Math. Szeged **11**, 229–233 (1948)
13. Gortler, S.J., Gotsman, C., Thurston, D.: Discrete one-forms on meshes and applications to 3D mesh parameterization. Comput. Aided Geom. Des. **23**(2), 83–112 (2006)
14. Hong, S., Nagamochi, H.: Convex drawings of graphs with non-convex boundary constraints. Discrete Appl. Math. **156**(12), 2368–2380 (2008)
15. Kahan, S., Snoeyink, J.: On the bit complexity of minimum link paths: superquadratic algorithms for problems solvable in linear time. Comput. Geom. **12**(1–2), 33–44 (1999)
16. Kostitsyna, I., Löffler, M., Polishchuk, V., Staals, F.: On the complexity of minimum-link path problems. J. Comput. Geom. **8**(2), 80–108 (2017)
17. Kratochvíl, J., Matoušek, J.: Intersection graphs of segments. J. Comb. Theory Ser. B **62**(2), 289–315 (1994)
18. Lubiw, A., Miltzow, T., Mondal, D.: The complexity of drawing a graph in a polygonal region. arXiv preprint arXiv:1802.06699 (2018)
19. Matoušek, J.: Intersection graphs of segments and ∃R. CoRR abs/1406.2636 (2014). http://arxiv.org/abs/1406.2636
20. McDiarmid, C., Müller, T.: Integer realizations of disk and segment graphs. J. Comb. Theory Ser. B **103**(1), 114–143 (2013)
21. Mchedlidze, T., Nöllenburg, M., Rutter, I.: Extending convex partial drawings of graphs. Algorithmica **76**(1), 47–67 (2016)
22. Mitchell, J.S., Rote, G., Woeginger, G.: Minimum-link paths among obstacles in the plane. Algorithmica **8**(1–6), 431–459 (1992)
23. Mnev, N.E.: The universality theorems on the classification problem of configuration varieties and convex polytopes varieties. In: Viro, O.Y., Vershik, A.M. (eds.) Topology and Geometry—Rohlin Seminar. LNM, vol. 1346, pp. 527–543. Springer, Heidelberg (1988). https://doi.org/10.1007/BFb0082792
24. Nishizeki, T., Rahman, M.S.: Planar Graph Drawing. Lecture Notes Series on Computing. World Scientific, Singapore (2004)
25. Patrignani, M.: On extending a partial straight-line drawing. Int. J. Found. Comput. Sci. **17**(05), 1061–1069 (2006)
26. Schaefer, M.: Complexity of some geometric and topological problems. In: Eppstein, D., Gansner, E.R. (eds.) GD 2009. LNCS, vol. 5849, pp. 334–344. Springer, Heidelberg (2010). https://doi.org/10.1007/978-3-642-11805-0_32
27. Schaefer, M., Štefankovič, D.: Fixed points, nash equilibria, and the existential theory of the reals. Theory Comput. Syst. **60**(2), 172–193 (2017)

28. Schnyder, W.: Embedding planar graphs on the grid. In: Proceedings of the First Annual ACM-SIAM Symposium on Discrete Algorithms (SODA), pp. 138–148. Society for Industrial and Applied Mathematics (1990)
29. Suri, S.: A linear time algorithm with minimum link paths inside a simple polygon. Comput. Vis. Graph. Image Process. **35**(1), 99–110 (1986)
30. Tutte, W.T.: Convex representations of graphs. Proc. Lond. Math. Soc. **10**(3), 304–320 (1960)
31. Tutte, W.T.: How to draw a graph. Proc. Lond. Math. Soc. **13**(3), 743–769 (1963)
32. Urhausen, J.: Extending drawings with fixed inner faces with and without bends. Master's thesis. Karlsruhe Institute of Technology (2017)
33. Vismara, L.: Planar straight-line drawing algorithms. In: Tamassia, R. (ed.) Handbook of Graph Drawing and Visualization, pp. 697–736. CRC Press (2014). Chap. 23

Inserting an Edge into a Geometric Embedding

Marcel Radermacher[1][(✉)] and Ignaz Rutter[2]

[1] Department of Computer Science, Karlsruhe Institute of Technology, Karlsruhe, Germany
radermacher@kit.edu
[2] Department of Computer Science and Mathematics, University of Passau, Passau, Germany
rutter@fim.uni-passau.de

Abstract. The algorithm to insert an edge e in linear time into a planar graph G with a minimal number of crossings on e [10], is a helpful tool for designing heuristics that minimize edge crossings in drawings of general graphs. Unfortunately, some graphs do not have a geometric embedding Γ such that $\Gamma + e$ has the same number of crossings as the embedding $G + e$. This motivates the study of the computational complexity of the following problem: Given a combinatorially embedded graph G, compute a geometric embedding Γ that has the same combinatorial embedding as G and that minimizes the crossings of $\Gamma + e$. We give polynomial-time algorithms for special cases and prove that the general problem is fixed-parameter tractable in the number of crossings. Moreover, we show how to approximate the number of crossings by a factor $(\Delta - 2)$, where Δ is the maximum vertex degree of G.

1 Introduction

Crossing minimization is an important task for the construction of readable drawings. The problem of minimizing the number of crossings in a given graph is a well-known $\mathcal{NP}$-complete problem [8]. A very successful heuristic for minimizing the number of crossings in a topological drawing of a graph G is to start with a spanning planar subgraph H of G and to iteratively *insert* the remaining edges into a drawing of H. The edge insertion problem for a planar graph G and two vertices $s, t \in V(G)$ asks to find a drawing $\Gamma + st$ of $G + st$ with the minimum number of crossings such that the induced drawing Γ of G is planar. The problem comes with several variants depending on whether the drawing Γ can be chosen arbitrarily or is fixed [9,10]. In the planar topological case both problems can be solved in linear time. More general problems such as inserting several edges simultaneously [2] or inserting a vertex together with all its incident edges [1] have also been studied.

Work was partially supported by grant WA 654/21-1 of the German Research Foundation (DFG).

T. Biedl and A. Kerren (Eds.): GD 2018, LNCS 11282, pp. 402–415, 2018.
https://doi.org/10.1007/978-3-030-04414-5_29

All these approaches have in common that they focus on topological drawings where edges are represented as arbitrary curves between their endpoints. By contrast, we focus on geometric embeddings, i.e., planar straight-line drawings, and the corresponding rectilinear crossing number. In this scenario we are only aware of a few heuristics that compute straight-line drawings of general graphs [12,13]. Clearly, if a geometric embedding Γ of the input graph G is provided as part of the input, there is no choice left; we can simply insert the straight-line segment from s to t into the drawing and count the number of crossings it produces. If, however, only the combinatorial embedding is specified, but one may still choose the outer face and choose the vertex positions so that this results in a straight-line drawing with the given combinatorial embedding, then the problem becomes interesting and non-trivial. We call this problem *geometric edge insertion*.

Contribution and Outline. We show several results on the complexity of geometric edge insertion with a fixed combinatorial embedding. Namely, we give a linear-time algorithm for the case that the maximum degree Δ of G is at most 5 (Sect. 3). For the general case, we give a $(\Delta - 2)$-approximation that runs in linear time. Moreover, we give an efficient algorithm for testing in special cases whether there exists a way to insert the edge st so that it does not produce more crossings than when we allow to draw it as an arbitrary curve (Sect. 4). Finally, we give a randomized FPT algorithm that tests in $O(4^k n)$ time whether an edge can be inserted with at most k crossings (Sect. 5).

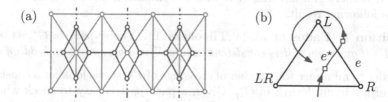

Fig. 1. (a) The extended dual (red + blue) of the primal graph (grey) and the red vertices corresponding to s and t. (b) Labeling induced by the blue path. (Color figure online)

2 Preliminaries

Let $G = (V, E)$ be a planar graph with a given combinatorial embedding where only the choice of the outer face is free. Additionally, let s and t be two distinct vertices with $st \notin E$. Denote by $G + st$ the graph G together with the edge st. We want to insert the *edge st into the embedded graph G*. That is, we seek a straight-line drawing Γ of G (with the given embedding) such that st can be inserted into Γ with a minimum number of crossings. In Γ, the edge st starts at s, traverses a set of faces and ends in t. Topologically, this corresponds to a path

$p(\Gamma)$ from s to t in the *extended dual* $G^\star_{st}$ of G, i.e., in the dual graph $G^\star$ plus s and t connected to all vertices of their dual faces; see Fig. 1a. The number of crossings in $\Gamma + st$ corresponds to the length of the path minus two. However, not all st-paths in $G^\star_{st}$ are of the form $p(\Gamma)$ for a straight-line drawing Γ of G.

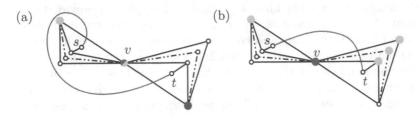

Fig. 2. Ratio between length of the shortest st path and the length of a shortest consistent st-path. The solid black edges induce a graph of maximum degree 6. Red vertices have label L, blue vertices have label R. (a) The shortest path from s to t in $G^\star_{st}$ is not consistent. (Color figure online)

A labeling of G is a mapping $l : V \to \{L, R\}$ that labels vertices as either left or right. Consider an edge uv of G that is crossed by a path p such that u and v are to the left and to the right of p, respectively. The edge uv is *compatible* with a labeling l if $l(u) = L$ and $l(v) = R$. A path p of $G^\star_{st}$ and a labeling l of G are *compatible* if l is compatible with each edge that is crossed by p. A path p is *consistent* if there is a labeling of G that is compatible with p. Eades et al. [4] show the following result.

Proposition 1 (Eades et al. [4],Theorem 1). *An st-path in $G^\star_{st}$ is of the form $p(\Gamma)$ if and only if it is consistent, where Γ is a geometric embedding of G.*

In order to minimize the number of crossings of $\Gamma + st$, we look for a consistent st-path of minimum length in $G^\star_{st}$. Given a path p, it is easy to check whether p is consistent. Figure 2 shows that the ratio between the length of a shortest st-path and the length of a shortest consistent st-path can be arbitrarily large. Thus, our goal is to find short consistent st-paths.

Let $H = (V', E')$ be a directed acyclic graph. A path $p = \langle v_1, v_2, \ldots, v_k \rangle$ is a *directed path* if for each $1 \le i < k$, $v_i v_{i+1} \in E'$. It is *undirected* if for each $1 \le i < k$, either $v_i v_{i+1} \in E'$ or $v_{i+1} v_i \in E'$. We refer to the number $|p|$ of edges of a path as the *length of p*. Two paths p and p' are *edge-disjoint* if they do not share an edge. Two paths p and p' of an embedded graph are *non-crossing* if at each common vertex v, the edges of p and p' incident to v do not alternate in the cyclic order around v in the graph induced by p and p'. We denote by $p[u, v]$ the subpath of a path p from u to v.

3   Bounded Degree

The shortest st-path of the graph in Fig. 2a is not consistent. Note that the maximum vertex degree is 6. In this section, we show that every shortest st-path in graphs of bounded degree 3 is consistent, and that in each planar graph with vertex degree at most 5, there is a shortest st-path that is consistent. Finally, we prove that there is a consistent st-path of length $(\Delta - 2)l$ in a graph with maximum vertex degree Δ and a shortest st-path of length l in $G^\star_{st}$.

Let p be an st-path in $G^\star_{st}$ and let $e^\star$ be an edge of p. An endpoint u of the primal edge e of $e^\star$ is *left of* $e^\star$ if it is locally left of p on e (Fig. 1b). A vertex v of G is *left (right)* of p if v is left (right) of an edge of p. We now consider a labeling extended by two more labels $LR, \perp$. We define the labeling l_p induced by p as follows. Each vertex that is left and right of p gets the label LR. The remaining vertices that are either left or right of p get labels L and R, respectively. Vertices neither left nor right of p get the label $\perp$. Obviously, there is a labeling l of G compatible with p if and only if l_p does not use the label LR. The proof of the following lemma can be found in the full verison on arXiv [14].

Theorem 2. *Let G be a planar embedded graph of degree at most 3. Then every shortest st-path in $G^\star_{st}$ is consistent.*

Theorem 3. *Let G be a planar embedded graph with maximum degree 5. Then there is a shortest st-path in $G^\star_{st}$ that is consistent.*

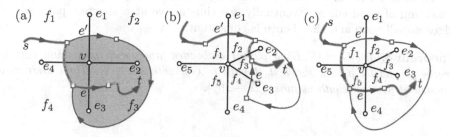

Fig. 3. Inconsistent path around (a) a degree-4 vertex and (b,c) a degree-5 vertex.

Proof. Let p be a shortest st-path in $G^\star_{st}$. We call an edge e of p *good* if the vertices left and right of it do not have label LR in the labeling l_p induced by p.

If p is not consistent, then let e denote the last edge of p that is not good. Then an endpoint v of the primal edge corresponding to e has label LR. Without loss of generality, we may assume that v lies left of e. Since $l_p(v) = LR$, there is an edge e' of p that has v to its right. By the choice of e, it follows that e' lies before e on p. We now distinguish cases based on the degree of v.

If $\deg(v) \leq 3$, then we find that p enters or leaves a face twice, which contradicts the assumption that it is a shortest st-path.

If $\deg(v) = 4$, we denote the edges around v in clockwise order as $e_1, \dots, e_4$ such that e' crosses e_1. Moreover, we denote the faces incident to v in clockwise order as $f_1, \dots, f_4$ where f_1 is the starting face of e'.

Since no face has two incoming or two outgoing edges of p, it follows that $e' = f_1 f_2$ crosses e_1 and $e = f_4 f_3$ crosses e_3; see Fig. 3a. Let p' be the path obtained from p by replacing the subpath $p[f_1, f_4]$ by the edge $f_1 f_4$ that crosses e_4. Since p is a shortest path, it follows that $f_2 = f_4$. By construction, it is $l_{p'}(v) - L$. Observe that $p'[f_4, t] = p[f_4, t]$ lies inside the region ρ bounded by $p[f_1, f_4]$ and a curve connecting f_1 and f_4 that crosses e_4. The only vertex inside this region whose label changed is v. Therefore, the path $p'[f_1, t]$ consists of good edges, and we have thus increased the length of the suffix of the shortest path that consists of good edges.

Now assume that $\deg(v) = 5$. We denote the edges around v as $e_1, \ldots, e_5$ in clockwise order such that e' crosses e_1. We further denote the faces incident to v in clockwise order as $f_1, \ldots, f_5$ such that e' starts in f_1. Since no face has two incoming or two outgoing edges, it follows that either e crosses e_4 from f_5 to f_4 or e crosses e_3 from f_4 to f_3.

If e crosses e_3, then we consider the path p' obtained from p by replacing the subpath $p[f_2, f_3]$ by the edge that crosses e_3; see Fig. 3b. As above, it follows that $f_2 = f_4$ and v is a cutvertex and that $p'[f_1, t]$ consists of good edges.

If e crosses e_4, then we obtain p' by replacing $p[f_1, f_5]$ by the single edge that crosses e_5; see Fig. 3c. As above, we find that $f_2 = f_5$ and v is a cutvertex and that $p'[f_1, t]$ consists of good edges.

Thus, in all cases, we increase the length of the suffix of the shortest path consisting of good edges. Eventually, we thus arrive at a shortest path whose edges are all good and that hence is consistent.   □

Theorem 4. *Let $G = (V, E)$ a planar embedded graph with maximum vertex-degree Δ and let p be a shortest st-path in $G^\star_{st}$ with $s, t \in V$. Then there is a consistent path of length at most $(\Delta - 2)|p|$.*

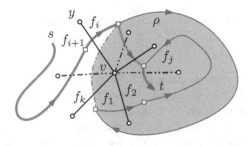

Fig. 4. Inconsistent path around a degree k vertex.

Proof. Let p be an st-path in $G^\star_{st}$. Assume that p is not consistent. Then there is a shortest prefix $p_2 = p[s, f_2] = p[s, f_1] \cdot f_1 f_2$ of p that is not consistent; refer to Fig. 4. Let v be a vertex incident to the primal edge of $f_1 f_2$ with $l_{p_2}(v) = LR$.

Without loss of generality let $f_1, f_2, \ldots, f_k$ be the faces around v in counter-clockwise order, i.e., v lies left of $f_1 f_2$.

Since p_2 is not consistent, there is a second edge of p_2 that crosses a primal edge incident to v. Let e be the last edge of $p[s, f_1]$ that crosses a primal edge incident to v. Since p_2 is the shortest inconsistent prefix of p, v lies right of e, i.e., $e = f_{i+1} f_i$ for some i with $2 < i \le k - 1$. Moreover, let f_j be the first vertex in clockwise order from f_i that lies on the path $p[f_2, t]$. Note that such a vertex f_j exists, since at the latest f_2 satisfies the condition.

Let q be the path $f_i f_{i-1} \cdots f_j$. We obtain a path p' from p by replacing $p[f_i, f_j]$ by q, i.e., $p' = p[s, f_i] \cdot q \cdot p[f_j, t]$. Note that, since f_j is the first vertex in clockwise order on $p[f_2, t]$, p' is a simple path. Since q does not contain the edges $f_k f_1$ and $f_1 f_2$, and $p[f_i, f_j]$ contains at least one edge, the path p' has length at most $|p| + (k - 2) - 1$. We claim that the prefix $p'_j = p'[s, f_j]$ is consistent.

Then, since $p'[f_j, t]$ is a subpath of $p[f_2, t]$ and $p'[s, f_j]$ is consistent, it follows that we have decreased the maximum length of a suffix of the path whose removal results in an inconsistent path. Since this suffix has initially length at most $|p|$, we inductively find a consistent st-path of length at most $(\Delta - 2)|p|$.

It remains to prove that $p'[s, f_j]$ is consistent. Since $p[s, f_2]$ is the shortest inconsistent prefix of p, the prefix $p[s, f_1]$ is consistent. Therefore, v is right of $p[s, f_i] = p'[s, f_i]$. By construction, v is right of q. Thus, we have $l_{p'_j}(v) = R$. The only vertices w of G with $l_{p'_j}(w) = LR$ can be neighbors of v, as otherwise $p[s, f_1]$ would not be consistent.

Consider the region ρ enclosed by the path $p[f_i, f_1]$ and $f_1, f_k, \ldots, f_i$ that contains v; refer to Fig. 4. The prefix $p[s, f_1] = p'[s, f_1]$ lies outside of ρ and the path q lies entirely in ρ. Moreover, in case that vw is crossed by $p'[s, f_i]$, w lies outside of ρ. On the other hand, if q crosses an edge vw, then w lies inside ρ. Thus, in both cases we immediately get that $l_{p'_j}(w) = L$. Therefore, the prefix $p'[s, f_j]$ is consistent.   □

4   Consistent Shortest st-paths

In Sect. 3 we showed that every shortest st-path in the extended dual $G^\star_{st}$ of a graph G with vertex degree at most 3 is consistent. For every graph of maximum degree 5, there is a shortest st-path $G^\star_{st}$ that is consistent. On the other hand, Fig. 2 shows that, starting from degree 6, there are graphs whose shortest st-paths are not consistent. In this section we investigate the problem of deciding whether $G^\star_{st}$ contains a consistent shortest st-path. As a consequence of Proposition 1 this problem is in $\mathcal{NP}$.

In Lemma 5 we show that finding a consistent st-path p in $G^\star_{st}$ is closely related to finding two edge-disjoint paths in G. Especially, we are interested in two edge-disjoint paths where the length of one is minimized. Eilam-Tzoreff [5] proved that this problem is in general $\mathcal{NP}$-complete. In planar graphs the sum of the length of two vertex-disjoint paths can be minimized efficiently [11]. In general directed graphs the problem is $\mathcal{NP}$-hard [7]. Finding two edge-disjoint paths in acyclic directed graphs is $\mathcal{NP}$-complete [6].

The closest relative to our problem is certainly the work of Eilam-Tzoreff. In fact their result can be modified to show that it is $\mathcal{NP}$-hard to decide whether a graph contains two edge-disjoint st-paths such that one of them is a shortest path. We study this problem in the planar setting with the additional restriction that s and t lie on a common face of the subgraph G_{sp} of $G^\star_{st}$ that contains all shortest paths from s to t.

Lemma 5. *An st-path p in $G^\star_{st}$ is consistent if and only if there is an st-path p' in $G^\star_{st}$ that is edge-disjoint from p and that does not cross p.*

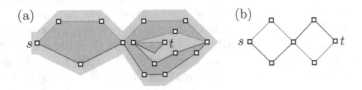

Fig. 5. (a) The green regions are right of p (blue) and the blue left of p. (b) The outer region that is not bounded by maximal subpaths of p and p'. (Color figure online)

Proof. The paths p and p' define a set of regions in the plane. Since p and p' are non-crossing, each region is bounded by one maximal subpath of p and one maximal subpath of p' (Fig. 5). We label each region ρ with either L or R, depending on whether ρ lies left or right of the unique maximal subpath of p on its boundary. We define a labeling l of G by giving each vertex v the label of the region ρ that contains it. We claim that l is compatible with p.

Since p and p' are edge-disjoint, every primal edge connects vertices of the same or adjacent regions. Moreover, by construction, vertices of adjacent regions have different labels. Thus all vertices left of p have label L and all vertices right of p have label R. That is l is compatible with p, i.e., p is consistent.

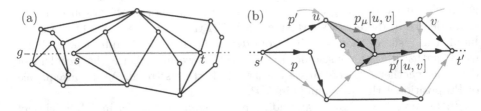

Fig. 6. (a) The line g through the segment st induces a path in $G^\star_{st}$. (b) Modification of the undirected path p' edge-disjoint from p.

Conversely, assume that p is consistent. By Proposition 1 there is a straight-line drawing of G such that the segment st intersects the same edges as p and in the same order (Fig. 6a). Let g be the line that contains the segment st. Each edge of G intersects g at most once. Thus, the complement of st in g defines a path from s to t in $G^\star_{st}$ that is edge-disjoint from p and does not cross p. □

Thus, we now consider the problem of finding a consistent shortest st-path as an edge-disjoint path problem in $G^\star_{st}$. Our proof strategy consists of three steps. Step (1) We first show that the problem is equivalent to finding two edge-disjoint paths p and q in a directed graph $\overrightarrow{G}_{st}$ such that p is directed and q is undirected. Step (2) We modify $\overrightarrow{G}_{st}$ such that p is a path in a specific subgraph G_{sp} and q lies in the subgraph $\overrightarrow{G}_{sp}$. These two graphs may share an edge set $\hat{E}$ such that each edge in $\hat{E}$ can be an edge of p or of q. Moreover, we find pairs of edges e and e' in $\hat{E}$ such that the path p in G_{sp} (the path q in $\overrightarrow{G}_{sp}$) contains either e or e'. Step (3) Finally, we use these properties to reduce our problem to 2-SAT.

We begin with Step 1. A directed graph $\overrightarrow{G}_{st} = (V' \cup \{s, t\}, E')$ is st-*friendly* if $G^\star_{st}$ contains a consistent shortest st-path if and only if $\overrightarrow{G}_{st}$ contains a directed st-path p and an undirected st-path p' that is edge-disjoint from p and does not cross p. We obtain an st-friendly graph $\overrightarrow{G}_{st} = (\overrightarrow{V}, \overrightarrow{E})$ from $G^\star_{st}$ as follows. Denote by G_{sp} the directed acyclic graph that contains all shortest paths from s to t in $G^\star_{st} = (V, E)$. If an edge $uv \in E$ is an edge of G_{sp}, we add it to $\overrightarrow{G}_{st}$. For all remaining edges uv, we add a subdivision vertex x to $\overrightarrow{G}_{st}$ and add the directed edges xu, xv to $\overrightarrow{G}_{st}$ in this direction. We claim that $\overrightarrow{G}_{st}$ is st-friendly.

Let p be a consistent shortest st-path in $G^\star_{st}$. By Lemma 5 there is a path p' in $G^\star_{st}$ that is edge-disjoint from p and does not cross p. By construction p corresponds to a directed path in G_{sp} and p corresponds to an undirected path in $\overrightarrow{G}_{st}$. Conversely, due to the directions of the edges xv, xu, every directed st-path q in $\overrightarrow{G}_{st}$ is a directed path in G_{sp}, and therefore it is a shortest st-path in $G^\star_{st}$. If there is an undirected path q' that is edge-disjoint from q and does not cross q, we obtain a path p' from q' by contracting edges incident to split vertices x. Hence, $\overrightarrow{G}_{st}$ is st-friendly.

We consider the following special case, where s and t lie on a common face o of the subgraph G_{sp} of $\overrightarrow{G}_{st}$. Without loss of generality, let o be the outer face of G_{sp} and let t lie on the outer face of $\overrightarrow{G}_{st}$. We denote by p_μ and p_λ the upper and lower st-path of G_{sp} on the boundary of o. A vertex v of G_{sp} is an *interior vertex* if v does not lie on o. An edge uv of G_{sp} is an *interior edge* if u and v are interior vertices. An edge e of G_{sp} is a *chord* if both its endpoints lie on o but e is not an edge on the boundary of o.

Lemma 6. *For a directed st-path p and an undirected st-path p', that are edge-disjoint and non-crossing, there is an undirected st-path p'' that is edge-disjoint from p, does not cross p, and that does not use interior vertices of G_{sp}.*

Proof. Since p and p' are non-crossing, there are two distinct vertices u, v on p_λ or on p_μ, say p_μ, such that the inner vertices of $p'[u, v]$ lie in the interior of G_{sp}; refer to Fig. 6b. Moreover, since p' and p are non-crossing, the region enclosed by $p'[u,v]$ and $p_\mu[u,v]$ does not contain a vertex of p in its interior. Therefore, we obtain p'' by iteratively replacing pieces in the form of $p'[u,v]$ by $p_\mu[u,v]$. □

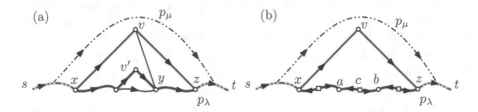

Fig. 7. (a) The red directed path can be circumvented with the blue directed path via vertex v. (b) The red path consists of avoidable edges. (Color figure online)

This finishes Step 1, and we continue with Step 2. In the following, we iteratively simplify the structure of G_{sp} while preserving st-friendliness of $\overrightarrow{G}_{st}$. Due to Lemma 6, the graph G_{sp}/e, obtained from contracting an edge e of G_{sp}, is st-friendly, if e is an interior edge. This may generate a separating triangle xyz. Let v be a vertex in the interior of xyz and let p be a directed st-path that contains v. Then, p contains at least two vertices of x, y, z. Hence, p can be rerouted using an edge of xyz. Thus, the graph after removing all vertices in the interior of xyz is st-friendly. After contracting all interior edges of G_{sp}, each neighbor of an interior vertex of G_{sp} lies either on p_λ or on p_μ. The remaining edges are edges on $p_\lambda \cup p_\mu$ and chords.

Consider three vertices x, y, z that lie in this order on p_λ (p_μ) and two interior vertices v and v', with $xv, v'y, vz \in \overrightarrow{E}$; refer to Fig. 7a. Note that v and v' can coincide. Then, every directed st-path p that contains y also contains x and z. Hence, p can be rerouted through the edges xv, vz and as a consequence of Lemma 6, the graph $G_{sp} - v'y$ is st-friendly. Analogously, if G_{sp} contains the edge yv', $G_{sp} - yv'$ remains st-friendly. We call such edges *circumventable*.

We refer to edges of a subpath $p_\lambda[x,z]$ ($p_\mu[x,z]$) as *avoidable* if there exists an interior vertex v with $xv, vz \in \overrightarrow{E}$ (Fig. 7b). If there exists a directed path p that uses an avoidable edge ab it can be rerouted by replacing the corresponding path $p_\lambda[x,z]$ with the edges xv, vz. Thus, we can split the edge ab with a vertex c and we direct the resulting edges from c towards a and b, respectively, and remove the edge ab from $\overrightarrow{G}_{st}$. Finally, we iteratively contract edges incident to vertices with in- and out-degree 1, and we iteratively remove vertices of degree at most 1, except for s and t. Since all interior edges of G_{sp} are contracted, circumventable interior edges are removed and avoidable edges are replaced, each 2-edge connected component of G_{sp} is an outerplanar graph whose weak dual (excluding the outer face) is a path; compare Fig. 8a. Each face f of G_{sp}, with $f \neq o$, contains at least one edge e_λ of p_λ and one edge e_μ on p_μ. Moreover, every directed st-path contains either e_λ or e_μ. We refer to the edge sets $E_{f,\lambda} = E(f) \cap E(p_\lambda)$ and $E_{f,\mu} = E(f) \cap E(p_\mu)$ as *interior partners*.

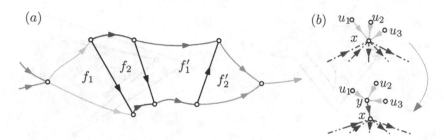

Fig. 8. (a) Interior partners decoded by color of 2-edge connected component of G_{sp}. (b) Split a vertex x on the boundary of $H_{st}^{\star}$.

Property 7. Choosing a directed st-path in G_{sp} is equivalent to choosing for each face f of G_{sp} one of the interior partners $E_{f,\mu}$ or $E_{f,\lambda}$ such that the following condition holds. Let f_1, f_2 be two adjacent faces that are separated by a chord e that ends at p_λ (p_μ) such that f_1 is right of e (left of e), then the choice of $E_{f_2,\mu}$ ($E_{f_2,\lambda}$) implies the choice of $E_{f_1,\mu}$ ($E_{f_1,\lambda}$).

In the following, we modify the *exterior of* $\overrightarrow{G}_{st}$, i.e., $\overline{G_{sp}} = \overrightarrow{G}_{st} - E(G_{sp})$, with the aim to obtain an analog property for the choice of the undirected path. We refer to edges of $\overline{G_{sp}}$ as *exterior edges*. A vertex in $V(\overline{G_{sp}}) \setminus V(G_{sp})$ is an *exterior vertex*.

Since the undirected path is not allowed to cross the directed path, we split each cut vertex x into an upper copy x_μ and a lower copy x_λ. We reconnect edges of p_λ and p_μ incident to x to x_λ and x_μ, respectively. Exterior edges incident to x that are embedded to the right of p_λ are reconnected to x_λ. Likewise, edges embedded to the left of p_μ are reconnected to x_μ. Note that this operation duplicates bridges of G_{sp}. Thus, we forbid the undirected path to traverse these duplicates. Observe that after this operation the outer face o of G_{sp} is bounded by a simple cycle.

Let x be a vertex on o that is incident to an exterior edge. In this case, we insert a vertex y to $\overrightarrow{G}_{st}$ and we remove each exterior edge ux from $\overrightarrow{G}_{st}$ and insert as a replacement edges yx and yu; see Fig. 8b. We refer to the edge yx as a *barrier*. Since the barrier yx is directed from y to x, the modification preserves the st-friendliness of $\overrightarrow{G}_{st}$. We now exhaustively contract exterior edges that are not barrier edges, and remove vertices in the interior of separating triangles.

Recall that s and t lie on a common face o of the subgraph G_{sp} of $\overrightarrow{G}_{st}$ and t lies on the outer face of G_{sp}. Let v be an exterior vertex such that its neighbor x comes before its neighbor y on $p_i, i = \lambda, \mu$, refer to Fig. 9a. Let z be a vertex between x and y on p_i that is connected to a vertex v' such that the edge $v'z$ (zv') lies in the interior of the region bounded by yvx and $p_i[x, y]$. Consider a directed st-path p in G_{sp} and an undirected st-path p' in $\overrightarrow{G}_{st}$ that is edge-disjoint from p, that does not cross p and that contains v'. Due to Lemma 6 we can assume, that p' does not contain an interior vertex of G_{sp}. Thus, it contains x and y. We obtain a new path p'' by replacing the subpath $p'[x, y]$ by vx, vy. Since vx, vy

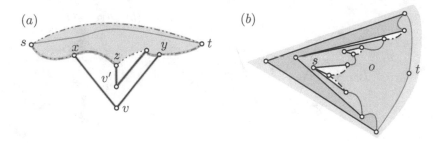

Fig. 9. (a) If the undirected path contains z, it can be rerouted to use vertex v. (b) The color coding of the faces indicate the exterior partners.

are exterior edges, p'' and p are edge-disjoint and non-crossing. Thus, the graph $\overrightarrow{G}_{st} - v'z$ ($\overrightarrow{G}_{st} - zv'$) is st-friendly. After removing all such edges, for any two neighbors x and y of an exterior vertex v, the paths $o[x, y]$ and $o[y, x]$ each contains either s and t. Hence, the region bounded by yvx and $o[x, y]$ contains a second exterior vertex v' if and only if $o[x, y]$ contains either s or t.

Hence, the dual of $\overline{G_{sp}}$, with the dual vertex of o removed, is a caterpillar C, refer to Fig. 9b. In case that s or t is incident to an exterior vertex v, we can assume that the undirected path p' contains the edge sv (vt). Thus, for simplicity, we now assume that neither s nor t is connected to an exterior vertex. Let a and b be the vertices in C whose primal faces are incident to s and t, respectively. Then every undirected st-path in $\overline{G_{sp}}$ from s to t traverses the primal faces of the simple path q from a to b in C. Let f be a primal face of a vertex on q. Since we inserted the barrier edges to $\overrightarrow{G}_{st}$, every face contains at least one edge e_λ of p_λ and one edge e_μ of p_μ. Therefore, every undirected st-path in $\overline{G_{sp}}$ either contains e_λ or e_μ. We refer to the sets $E_{f,\lambda} = E(f) \cap E(p_\lambda)$ and $E_{f,\mu} = E(f) \cap E(p_\mu)$ as *exterior partners*.

Property 8. Choosing an undirected st-path in $\overline{G_{sp}}$ is equivalent to choosing for each face $f \neq o$ of $\overline{G_{sp}}$ one of the exterior partners $E_{f,\lambda}$ or $E_{f,\mu}$.

This finishes Step 2, and we proceed to Step 3. The problem of finding a directed st-path p and an undirected st-path p' in $\overrightarrow{G}_{st}$ reduces to a 2-SAT instance as follows. For each exterior and interior partner we introduce variables x_f and x_g, respectively, where f and g correspond to the faces of the partners. If x_f is true, p' contains the edge of $E_{f,\lambda}$, otherwise it contains $E_{f,\mu}$. The conditions on the choice of p in Property 7 can be formulated as implications. Let $E_{f,\mu}$ an $E_{f,\lambda}$ be exterior partners and let $E_{g,\mu}$ and $E_{g,\lambda}$ be interior partners. In case that $E_{f,\lambda} \cap E_{g,\lambda} \neq \emptyset$, either p can contain edges of $E_{g,\lambda}$ or p' can contains edges of $E_{f,\lambda}$ but not both. Thus, x_f and x_g are not allowed to be true at the same time, i.e., $x_f = \overline{x_g}$. Hence, we have the following Theorem.

Theorem 9. *If s and t lie on a common face of G_{sp}, it is decidable in polynomial time whether $\overrightarrow{G}_{st}$ has a directed st-path and an undirected st-path that are edge-disjoint and non-crossing.*

Corollary 10. *If s and t lie on a common face of G_{sp}, it is decidable in polynomial time whether $G_{st}^\star$ contains a consistent shortest st-path.*

5 Parametrized Complexity of Short Consistent st-Paths

In this section we show that edge insertion can be solved in FPT time with respect to the minimum number of crossings of a straight-line drawing of $G + st$ where G is drawn without crossings and has the specified embedding. Let l be an arbitrary labeling of G. Observe that l defines a directed subgraph of $G_{st}^\star$ by removing each edge whose dual edge has endpoints with the same label and by directing all other edges e such that the endpoint of its primal edge left of e has label L and its other endpoint has label R. We denote this graph by $G_{st}^\star(l)$. Obviously, a shortest st-path in $G_{st}^\star(l)$ is compatible with l, and thus a corresponding drawing exists. Clearly, given the labeling l a shortest st-path in $G_{st}^\star(l)$ can be computed in linear time by a BFS.

Now assume that the length of a shortest consistent path in $G_{st}^\star$ is k. We propose a randomized FPT algorithm with running time $O(4^k n)$ for finding a shortest consistent path in $G_{st}^\star$, based on the color-coding technique [3].

The algorithm works as follows. First, we pick a random labeling of G by labeling each vertex independently with L or R with probability $1/2$. We then compute a shortest path in $G_{st}^\star(l)$. We repeat this process 4^k times and report the shortest path found in all iterations.

Clearly the running time is $O(4^k n)$. Moreover, each reported path is consistent, and therefore the algorithm outputs only consistent paths. It remains to show that the algorithm finds a path of length k with constant probability.

Consider a single iteration of the procedure. If the random labeling l is compatible with p, then the algorithm finds a path of length k. Therefore the probability that our algorithm finds a consistent path of length k is at least as high as the probability that p is compatible with the random labeling l. Let $V_L, V_R \subseteq V$ denote the vertices of V that are left and right of p, respectively. Clearly it is $|V_L|, |V_R| \le k$. A random labeling l is consistent with p if it labels all vertices in V_L with L and all vertices in V_R with R. Since vertices are labeled independently with probability $1/2$, it follows that $\Pr[p$ is consistent with $l] = (1/2)^{|V_L|} \cdot (1/2)^{|V_R|} \ge (1/2)^{2k} = (1/4)^k$.

Therefore, the probability that no path of length k is found in 4^k iterations is at most $(1 - (1/4)^k)^{4^k}$, which is monotonically increasing and tends to $1/e \approx 0.368$. Thus the algorithm succeeds with a probability of $1 - 1/e \approx 0.632$. The success probability can be increased arbitrarily to $1 - \delta$, $\delta > 0$ by repeating the algorithm $\log(1/\delta)$ times. The probability that each iteration fails is then bounded from above by $(1/e)^{\log 1/\delta} = 1/e^{\log 1/\delta} = \delta$. E.g., to reach a success probability of 99%, it suffices to do $\log 100 \le 5$ repetitions. The algorithm can be derandomized with standard techniques [3].

Theorem 11. *There is a randomized algorithm $\mathcal{A}$ that computes a consistent path of length k if one exists with a success probability of $1 - \delta$. The running time of $\mathcal{A}$ is $O(\log(\delta^{-1})4^k n)$.*

6   Conclusion

We have shown that the problem of finding a short consistent st-paths in $G^\star_{st}$ is tractable in special cases and fixed-parameter tractable in general. Whether $G^\star_{st}$ has a short consistent st-path is equivalent to the question of whether $G^\star_{st}$ has two edge-disjoint and non-crossing st-paths, where the length of one path is minimized. Surprisingly, this is related to yet another purely graph theoretic problem: does a directed graph G have two edge-disjoint paths where one is directed and the other is only undirected? By the result of Eilam-Tzoreff [5] the former problem is in general $\mathcal{NP}$-hard. For planar graphs the computational complexity of these problems remains an intriguing open question.

In this paper, we only considered planar graphs with a fixed combinatorial embedding. Allowing for arbitrary embeddings opens new perspectives on the problem and is interesting future work.

References

1. Chimani, M., Gutwenger, C., Mutzel, P., Wolf, C.: Inserting a vertex into a planar graph. In: Proceedings of the 20th Annual ACM-SIAM Symposium on Discrete Algorithms, SODA 2009, pp. 375–383 (2009)
2. Chimani, M., Hlinený, P.: Inserting multiple edges into a planar graph. In: Fekete,S., Lubiw, A. (eds.) Proceedings of the 32nd Annual Symposium on Computational Geometry, SoCG 2016. Leibniz International Proceedings in Informatics (LIPIcs), vol. 51, pp. 30:1–30:15. Schloss DagstuhlLeibniz-Zentrum fuer Informatik (2016). https://doi.org/10.4230/LIPIcs.SoCG.2016.30
3. Cygan, M., et al.: Parameterized Algorithms. Springer, Heidelberg (2015). https://doi.org/10.1007/978-3-319-21275-3
4. Eades, P., Hong, S.H., Liotta, G., Katoh, N., Poon, S.H.: Straight-line drawability of a planar graph plus an edge. In: Dehne, F., Sack, J.R., Stege, U. (eds.) Proceedings of the 14th International Symposium on Algorithms and Data Structures, WADS 2015, pp. 301–313 (2015). https://doi.org/10.1007/978-3-319-21840-3_25
5. Eilam-Tzoreff, T.: The disjoint shortest paths problem. Discrete Appl. Math. **85**(2), 113–138 (1998). https://doi.org/10.1016/S0166-218X(97)00121-2
6. Even, S., Itai, A., Shamir, A.: On the complexity of time table and multi-commodity flow problems. In: 16th Annual Symposium on Foundations of Computer Science, SFCS 1975, pp. 184–193, October 1975. https://doi.org/10.1109/SFCS.1975.21
7. Fortune, S., Hopcroft, J., Wyllie, J.: The directed subgraph homeomorphism problem. Theory Comput. Syst. **10**(2), 111–121 (1980). https://doi.org/10.1016/0304-3975(80)90009-2
8. Garey, M.R., Johnson, D.S.: Crossing number is NP-complete. SIAM J. Algebraic Discrete Methods **4**(3), 312–316 (1983)
9. Gutwenger, C., Klein, K., Mutzel, P.: Planarity testing and optimal edge insertion with embedding constraints. J. Graph Algorithms Appl. **12**(1), 73–95 (2008). https://doi.org/10.7155/jgaa.00160
10. Gutwenger, C., Mutzel, P., Weiskircher, R.: Inserting an edge into a planar graph. Algorithmica **41**(4), 289–308 (2005). https://doi.org/10.1007/s00453-004-1128-8

11. Kobayashi, Y., Sommer, C.: On shortest disjoint paths in planar graphs. Discrete Optim. **7**(4), 234–245 (2010). https://doi.org/10.1016/j.disopt.2010.05.002
12. Kobourov, S.G.: Force-directed drawing algorithms. In: Tamassia, R. (ed.) Handbook of Graph Drawing and Visualization, pp. 383–408. Chapman and Hall/CRC, Boca Raton (2013)
13. Radermacher, M., Reichard, K., Rutter, I., Wagner, D.: A geometric heuristic for rectilinear crossing minimization. In: Pagh, R., Venkatasubramanian, S. (eds.) Proceedings of the 20th Workshop on Algorithm Engineering and Experiments, ALENEX 2018, pp. 129–138 (2018). https://doi.org/10.1137/1.9781611975055.12
14. Radermacher, M., Rutter, I.: Inserting an edge into a geometric embedding. ArXiv e-prints (2018). https://arxiv.org/abs/1807.11711v1

β-Stars or On Extending a Drawing of a Connected Subgraph

Tamara Mchedlidze[1] and Jérôme Urhausen[2(✉)]

[1] Karlsruhe Institute of Technology, Karlsruhe, Germany
mched@iti.uka.de
[2] Utrecht University, Utrecht, The Netherlands
j.e.urhausen@uu.nl

Abstract. We consider the problem of extending the drawing of a subgraph of a given plane graph to a drawing of the entire graph using straight-line and polyline edges. We define the notion of star complexity of a polygon and show that a drawing Γ_H of an induced connected subgraph H can be extended with at most $\min\{h/2, \beta + \log_2(h) + 1\}$ bends per edge, where β is the largest star complexity of a face of Γ_H and h is the size of the largest face of H. This result significantly improves the previously known upper bound of $72|V(H)|$ [5] for the case where H is connected. We also show that our bound is worst case optimal up to a small additive constant. Additionally, we provide an indication of complexity of the problem of testing whether a star-shaped inner face can be extended to a straight-line drawing of the graph; this is in contrast to the fact that the same problem is solvable in linear time for the case of star-shaped outer face [9] and convex inner face [12].

1 Introduction

In this paper we study the problem of extending a given partial drawing of a graph. In particular, given a plane graph $G = (V, E)$, i.e. a planar graph with a fixed combinatorial embedding and a fixed outer face, a subgraph H of G and a planar straight-line drawing Γ_H of H, we ask whether Γ_H can be extended to a planar straight-line drawing of G (see Fig. 1). We study both the decision question and the relaxed variation of using bends for the drawing extension.

It is known that a drawing extension always exists even if $H = (V, \emptyset)$, where each edge is represented by a polyline with at most $120n$ bends, here $n = |V|$ [14]. This bound was improved to $3n + 2$ by Badent et al. [1]. These upper bounds are asymptotically optimal as there are instances that require $\Omega(n)$ bends on $\Omega(n)$ edges [1]. In terms of the size of the pre-drawn graph H, Chan et al. [5] showed that a drawing extension with $72|V(H)|$ bends per edge is possible for a general subgraph H.

In order to pinpoint the source of multiple necessary bends for the drawing extension we define the notion of a β-star (resp. β-outer-star), a polygon where β bends are necessary and sufficient to reach the kernel of the polygon (resp.

© Springer Nature Switzerland AG 2018
T. Biedl and A. Kerren (Eds.): GD 2018, LNCS 11282, pp. 416–429, 2018.
https://doi.org/10.1007/978-3-030-04414-5_30

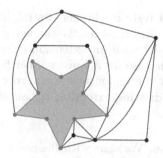

Fig. 1. An embedding of a plane graph alongside a fixed drawing of an inner face (blue) as a star-shaped polygon (gray). (Color figure online)

infinity). We study the upper bounds on the number of bends in a drawing extension as a function of β. We show that a drawing Γ_H of an induced connected subgraph H can be extended with at most $\min\{h/2, \beta + \log_2(h) + 1\}$ bends per edge if each face of H is represented in Γ_H as a β-(outer)-star and h is the size of the largest face of H (Theorem 7). We show that this bound is worst case optimal up to a small additive constant. We observe that in case both G and H are trees a closer to optimal bound of $1 + 2\lceil |V(H)|/2 \rceil$ bends per edge had been provided by Di Giacomo et al. [7].

In case a planar embedding is not provided as a part of the input, it is NP-hard to test whether a straight-line drawing extension exists [15]. The problem is not known to belong to the class NP, as a possible solution may have coordinates which can not be represented with a polynomial number of bits [15]. Very recently, Lubiw et al. have studied a related problem of drawing a graph inside a (not-necessarily simply connected) closed polygon [11]. They showed that this problem can not be shown to lie in NP by the mean of providing vertex coordinates, as these are sometimes irrational numbers. They have also shown that the problem is hard for the existential theory of reals ($\exists\mathbb{R}$-hard) even if a planar embedding of the graph is provided as a part of the input. This problem would be equivalent to partial graph drawing extendability, if the polygon would be open, however this situation has not been investigated. Bekos et al. [2,3] have studied the problem of extending a given partial drawing of bipartite graphs, where one side of the bipartition is pre-drawn. They have shown that this problem lies in NP if each free vertex is required to lie in the convex hull of its pre-drawn neighbors. Regarding drawing extensions with bends, it is NP-hard to test whether a drawing extension with at most k bend per edge exists [2,8].

Despite all the hardness results, it is long known that a straight-line drawing extension always exists if H is the outer face and Γ_H is a convex polygon [4, 16]; and H is a chordless outer face and Γ_H is a star-shaped polygon [9]. An existence of a straight-line drawing extension can be checked by the mean of necessary and sufficient conditions in case where H is an inner face and Γ_H is a convex polygon [12]. As an extension of this work, and with the general goal to better understand the boundary between the easy and the difficult cases, we investigated the question of testing whether a straight-line drawing extension exists for an inner face H drawn as a star-shaped polygon Γ_H. We observe that one can not test whether such an extension exists by just checking each vertex individually, as in the case for a convex inner face, and show that there exists an instance such that the region where a vertex of $V(G) \setminus V(H)$ can lie to allow for a straight-line drawing extension is bounded by a curve of degree $2^{\Omega(|H|)}$ (Theorem 8).

Contribution and Outline. We start with the necessary definitions in Sect. 2. In Sect. 3, we show that a star-shaped drawing of an inner face can be extended with at most 1 bend per edge. Section 4 is devoted to the study of generalizations of stars. In Sect. 4.1, we start with a generalization of star-shaped polygons to β-star and β-outer-star polygons (β is referred to as star complexity), and show that the number of bends per edge necessary for a drawing extension of an inner face H with a star complexity β is not bounded in terms of β (Theorems 2 and 3). Motivated by the proof of Sect. 3 we define the notion of planar-β-star and planar-β-outer-star (this β is referred to as planar star complexity) and show that the planar star complexity determines the number of bends per edge in a drawing extension (Theorems 4 and 5). In Sect. 4.2, we study the planar star complexity of an arbitrary simple polygon and the relationship between the star complexity and the planar star complexity of a polygon. In particular, we show that every β-star with n vertices is a planar-$\beta + \delta$-star where $\delta \leq \log_2(n)$ (Theorem 6). In Sect. 5, we state the implications of Sect. 4 to the drawing extension of (induced) connected subgraphs. In particular, we prove that a drawing Γ_H of an induced connected subgraph H can be extended with at most $\min\{h/2, \beta + \log_2(h) + 1\}$ bends per edge if the star complexity of Γ_H is β and h is the size of the largest face of H (Theorem 7). Last but not least, in Sect. 6 we provide an indication of complexity of the problem of testing whether a star-shaped inner face H admits a straight-line drawing extension. In particular, we prove that there exists an instance such that the region where a vertex of $V(G) \setminus V(H)$ can lie to allow a straight-line drawing extension is bounded by a curve of degree $2^{\Omega(|H|)}$ (Theorem 8). All omitted proofs can be found in the full version [13].

2   Preliminaries

Basic Geometric Terms. The segment (resp. line) induced by two points a and b is designated by $s(a, b)$ (resp. $l(a, b)$). We denote a curve between a and b by $c(a, b)$. We refer to the ray along $l(a, b)$ starting at a and (not) containing b as $r(a, b)$ ($q(a, b)$). For a polyline c, $\#c$ designates the number of bends on c.

Let P be a polygon. Two points a, b *see* each other if the open segment $s(a, b)$ does not intersect the boundary of P. A simple polygon P is *convex* if each pair of points inside P see each other. A simple polygon P is *star-shaped* or a *star* if there is a non-empty set of points K called the *kernel* inside the polygon such that any point of the kernel can see any vertex of the polygon. By assuming that the vertices of P are in general position, we have that a kernel of P contains an open ball of positive radius.

Graphs and Drawings of Graphs. A *drawing* Γ of a graph is a function that assigns to each vertex a unique point in the plane and to each edge $\{a, b\}$ a curve connecting the points assigned to a and b. A drawing is *straight-line* (resp. *k-bend*) if each edge is drawn as a segment (resp. a polyline with at most k bends). A graph is *planar* if it has a *planar* drawing, i.e. a drawing without edge crossings. A planar drawing Γ subdivides the plane into connected regions called *faces*; the unbounded region is the *outer* and the other regions are the *inner* faces. The cyclic ordering of the edges around each vertex of Γ together with the description of the outer face of Γ characterize a class of drawings with the same combinatorial properties, which is called an *embedding* of G. A planar graph G with a planar embedding is called *plane graph*. A *plane subgraph* H of G is a subgraph of G together with a planar embedding that is the restriction of the embedding of G to H. A plane graph G is (*internally*) *triangulated* if each (inner) face of G is a triangle. For a given cycle, a chord is an edge between two non-consecutive vertices of the cycle.

Let G be a plane graph and let H be a plane subgraph of G. Let Γ_H be a planar straight-line drawing of H. We say that the instance (G, Γ_H) admits a *k-bend* (resp. *straight-line*) *extension* if drawing Γ_H can be completed to a planar k-bend (resp. straight-line) drawing Γ_G of the plane graph G. We refer to k as the *curve complexity* of the drawing Γ_G.

For a given graph $G = (V, E)$, let $N(v) = \{w \in V \mid \{v, w\} \in E\}$ be the *neighbors* of $v \in V$. For a plane graph G and a face F, let $N_F(v) = N(v) \cap F = (w_1, w_2, \ldots, w_\ell)$ be the sequence of neighbors of v that belong to F. For v outside F, let the list $N_F(v)$ be ordered clockwise around F with w_1 chosen such that the area delimited by the cycle C composed of edges $\{v, w_1\}$, $\{v, w_\ell\}$ and the clockwise path H from w_1 to w_ℓ in F does not contain F (see Fig. 2a). A vertex $z \in V \setminus V(F)$ lying in the cycle C is said to be *enclosed* by vertex v.

Let F be a face of G and Γ_F its planar drawing. The *feasibility area* of a vertex $v \in V \setminus V(F)$ is the set of all possible positions of v, such that the implied straight-line drawing of $F \cup \{v\}$ can be extended to a planar straight-line drawing of $V(F) \cup \{v\} \cup Q_v$, where Q_v is the set of all vertices enclosed by v.

3   Star-Shaped Polygons

Let G be a plane graph with n vertices, F be a chordless face of G with h vertices and Γ_F a star-shaped drawing of F. In this section we prove that the instance (G, Γ_F) admits a 1-bend-extension. While the proof itself is rather straightforward, we still present it here as it motivates a specific way to generalize star-shaped polygons by considering planarity issues.

In our construction we place vertices $V \setminus V(F)$ one by one with the property that a vertex is placed only after all vertices enclosed by it have already been placed. This property is achieved by a canonical ordering [10] that lists vertices starting from the face F. The following lemma can be proven along the same lines as the existence of a usual canonical ordering [10]. We say $G \setminus F$ is triangulated if each face of G is triangulated with the exception of the face F.

Lemma 1. *Let $G = (V, E)$ be a plane graph, $|V| = n$, and let F be an inner face with h vertices of G, such that $G \setminus F$ is triangulated. There is an ordering $\mathcal{J} = (v_1, \ldots, v_{n-h})$ of the vertices of $V \setminus V(F)$, such that for each j, $1 \leq j \leq n-h$, the following holds: (1) the graph G_j induced by the vertices $\{v_1, \ldots, v_j\} \cup F$ is biconnected, (2) $G_j \setminus F$ is internally triangulated, (3) v_{j+1} lies in the outer face of G_j, (4) vertices $N(v_{j+1}) \cap V(G_j)$ belong to the outer face of G_j.*

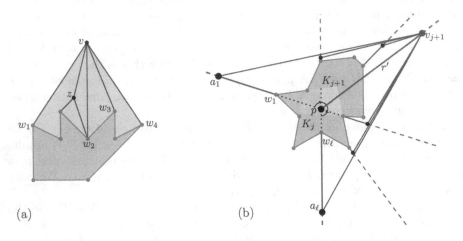

(a)   (b)

Fig. 2. (a) Yellow area contains vertices enclosed by v. (b) Proof of Theorem 1. (Color figure online)

Theorem 1. *Each instance $(G = (V, E), \Gamma_F)$ where Γ_F is a star-shaped drawing of a chordless inner face F allows a 1-bend-extension.*

Proof. We start with triangulating G by placing a vertex in each non-triangular face and connecting it to the vertices of the face. We delete the added vertices and edges after the triangulated graph has been drawn. We refer to the new graph as G as well. Let $\mathcal{J} = (v_1, \ldots, v_{n-h})$ be an ordering of the vertices $V \setminus V(F)$ as defined by Lemma 1. For $1 \leq j \leq n - h = |V(G)| - |V(F)|$, let G_j be the graph as defined by Lemma 1 and let F_j be the outer face of G_j. Additionally we set $G_0 = F_0 = F$.

We prove the theorem by induction. Assume that for a $0 \leq j \leq n-h$ we have a drawing of G_j, such that F_j forms a star-shaped polygon P_j with kernel K_j. This is true for $j = 0$. Let v_{j+1} be the next vertex according to $\mathcal{J}$ and let p be a point of the kernel of the already drawn star-shaped polygon P_j. For each $w \in N_{F_j}(v_{j+1})$ consider the ray $q(p, w)$. Due to P_j being star-shaped and due to property (4) of Lemma 1, they all lie outside of P_j. Since G_j is biconnected, v_{j+1} has at least two neighbors, i.e. $\ell = |N_{F_j}(v_{j+1})| \geq 2$.

Now we consider the ray r' that is the bisector of the clockwise angle formed by the rays $q(p, w_1)$ and $q(p, w_\ell)$, see Fig. 2b. If we place v_{j+1} sufficiently far away from p on r', v_{j+1} sees $q(p, w_1)$ and $q(p, w_\ell)$, i.e. $\exists a_1 \in q(p, w_1), a_\ell \in q(p, w_\ell)$, with $s(v_{j+1}, a_1) \cap P_j = \emptyset = s(v_{j+1}, a_\ell) \cap P_j$. This is due to the fact that the angles between $q(p, w_1)$ and r' and between $q(p, w_\ell)$ and r' are strictly smaller than π.

Since v_{j+1} is between $q(p, w_1)$ and $q(p, w_\ell)$, v_{j+1} also sees a point a_i on the ray $q(p, w_i)$, $i = 2, \ldots, \ell - 1$. For each $i \in \{1, \ldots, \ell\}$ we draw the edge $\{w_i, v_{j+1}\}$ using the segments $s(w_i, a_i)$ and $s(a_i, v_{j+1})$. Observe that the points $a_1, \ldots, a_\ell$ should be chosen so that they appear around v_{j+1} in a counterclockwise order.

The lines $l(p, w_1)$ and $l(p, w_\ell)$ separate the plane into four quadrants. The new kernel K_{j+1} of the polygon P_{j+1} is the intersection of the old kernel K_j and the quadrant containing v_{j+1}. Since the kernel K_j was an open set, p could not have been on the boundary of K_j, therefore K_{j+1} is a non-empty open set. □

We observe that, according to the proof of Theorem 1, the class of the polygons that allows a 1-bend-extension is wider than stars. In particular, these are the polygons from the vertices of which we can shoot rays to infinity which neither intersect mutually nor intersect the polygon itself. We call such polygons *planar outer-stars*. This gives the following:

Corollary 1. *Each instance* (G, Γ_F) *where* F *is a chordless inner face and* Γ_F *is a planar outer-star, allows a 1-bend-extension.*

4 Generalization of Stars

In this section we generalize the notion of stars and planar outer-star polygons and investigate the lower and upper bounds for the number of bends per edge in the drawing extensions.

4.1 β-Stars

A simple polygon P is a *β-star* if there is an open set of points K called the *kernel* inside P with the following property: for each point $p \in K$ and for each vertex v of P there is a polyline $c(v)$ connecting v and p with at most β bends such that $c(v)$ touches P only at v. The smallest such β is referred to as *star complexity* of the polygon P. This set of curves is referred to as *curve-set* $\mathcal{C}$ of P and p is the *center* of $\mathcal{C}$. In the literature this kernel is also known as the link center of the polygon and it can be calculated in $O(n \log n)$ time [6]. The straight-forward extension of this definition to act "outside" the polygons is as follows: a simple polygon P is a *β-outer-star* if for each vertex v of P there is an infinite polyline $c(v)$ outside of P starting at v with at most β bends. The smallest such β is referred to as *outer star complexity* of the polygon P. Again, $\mathcal{C} = \{c(v) \mid v \in P\}$ is called *curve-set*. The *center* of this set is a point at infinity. One can think about β-outer-star as of β-star with the kernel in infinity.

While β-star and β-outer-star are straight-forward ways to extend the notion of a star inside and outside, and these definitions capture an inherent complexity of the polygon, we can show that restricting the fixed inner face to be a 1-star is not sufficient to ensure a c-bend-extension for any constant c (Theorem 2). Even more, restricting the fixed inner face to a β-outer-star still does not imply the existence of a $c+\beta$-bend-extension for any constant c (Theorem 3).

Theorem 2. *There exist instances (G, Γ_F) where F is an inner face with h vertices and Γ_F is a 1-star such that any drawing extension of (G, Γ_F) contains an edge with at least $\lfloor \frac{h-3}{2} \rfloor$ bends.*

Theorem 3. *There exist instances (G, Γ_F) where F is an inner face with h vertices and Γ_F is a β-outer-star such that any drawing extension of (G, Γ_F) has an edge with at least $\beta + \log_2(\frac{h+5}{6}) + 1$ bends.*

The above lower bounds and the fact that a planar outer-star admits an extension with one bend per edge guided us to extend definitions of β-star and β-outer star to include planarity. A simple polygon P is a *planar-β-star* if there is an open set of points K called the kernel inside P with the following property: for a fixed point $p \in K$ and for each vertex v of P there is an oriented polyline $c(v)$ inside P from v to p with at most β bends such that for any v and v', $c(v)$ and $c(v')$ share the single point p.

A simple polygon P is a *planar-β-outer-star* if for each vertex v of P there is an oriented infinite polyline $c(v)$ outside of P starting at v with at most β bends such that for any v and v', $c(v)$ and $c(v')$ neither cross nor touch each other. The smallest such β is referred to as *planar (outer) star complexity* of the polygon P. The set of curves are referred to as *planar curve-set* centered at the fixed point p. Due to these definitions the following two theorems can be proven.

Theorem 4. *Each instance (G, Γ_F) where F is a chordless outer face and Γ_F is a planar-β-star allows a β-bend-extension.*

Theorem 5. *Each instance (G, Γ_F) where F is a chordless inner face and Γ_F is a planar-β-outer-star allows a $\beta + 1$-bend-extension.*

4.2 Planar Star Complexity of Polygons

While planar (outer) star complexity nicely bounds the required number of bends per edge in a drawing extension, it does not represent a simple and inherent polygon characteristic. Thus, in the following we first provide an upper bound on the planar (outer) star complexity of a polygon in terms of the size of the polygon (Lemma 2). Then, after preliminary results, we provide an upper bound of a planar (outer) star complexity in terms of (outer) star complexity (Theorem 6).

Lemma 2. *A simple polygon with h vertices is a planar $\frac{h-2}{2}$-star and a planar $\frac{h-2}{2}$-outer-star.*

Proof. For the interior, we set a kernel K to be an intersection of the interior of P with an ε-ball around a vertex u of P. Let p be a point in K. Notice, that by just following the boundary of the polygon it is possible to reach p from any vertex $v \neq u$ with a polyline $c(v)$ with at most $\frac{h-2}{2}$ bends. A set of such curves $\{c(v_i)|v_i \in P\}$, drawn in an appropriate order in order to avoid mutual intersections, represents a planar curve-set of P.

For the exterior, observe that by following the boundary of the polygon from any vertex u of P it is possible to reach a vertex belonging to the convex hull of P with a polyline $c(u)$ with at most $\frac{h-4}{2}$ bends because the convex hull contains at least three vertices. A set of such curves, drawn in appropriate order in order to avoid mutual crossing, augmented by infinite rays, result in a planar curve-set of P with curve complexity at most $\frac{h-2}{2}$.   □

Observe that, in general, the planar star complexity of a polygon may be much lower than $\frac{h-2}{2}$. Thus, in the following we aim to bound the planar star complexity in terms of the star complexity. We rely on the following definitions: let $\mathcal{C}$ be a planar curve-set of a planar-β(-outer)-star. For a curve $c(v)$ from $\mathcal{C}$ and a point p on $c(v)$, we denote by $c_v(p)$ the part of the curve split at p, not containing v and by $\#c_v(p)$ the number of bends on $c_v(p)$. Furthermore, $c(v, p)$ designates the part of the curve $c(v)$ between v and p. An intersection between the curves $c(v)$ and $c(w)$ of $\mathcal{C}$ at a point p is called *avoidable* if one of the curves has more bends after the intersection than the other, i.e. if $\#c_v(p) \neq \#c_w(p)$. The term "avoidable" stems from the fact that if $\#c_v(p) > \#c_w(p)$, we can modify $c(v)$ by rerouting it along $c(w)$ starting just before the point p and this way eliminate the intersection without increasing the number of bends per curve. Concerning said avoidable intersections the following holds:

Lemma 3. *For a given β(-outer)-star P there is a curve-set of P with at most β bends each without avoidable intersections.*

In order to resolve all remaining intersections we consider pairs of curves a and b intersecting at a point p, such that p is the first intersection for both a and b. In that case we call p *initial* intersection. However, we first have to show that if there are intersections, then there is always at least one initial intersection. We formalize this in the following definition and Lemma 4. A sequence of vertices $(w_1, \ldots, w_m)$ of P, with respective curves $(c(w_1), \ldots, c(w_m))$ is called *cyclic ordering*, if for each $1 \leq j \leq m$, the first curve that $c(w_j)$ intersects is the curve $c(w_{(j \bmod m)+1})$. We can prove the following:

Lemma 4. *For a given polygon P with a curve-set $\{c(v) \mid v \in V(P)\}$ without avoidable intersections there is no cyclic ordering.*

Using Lemmas 3 and 4 we prove a relation between β-stars and planar-β-stars.

Theorem 6. *Every β-star (resp. outer-star) with n vertices is a planar-$(\beta+\delta)$-star (resp. outer-star), where $\delta \leq \log_2(h)$.*

Proof. Let P be a β(-outer)-star with h vertices. By Lemma 3, P has a curve-set with at most β bends per curve without avoidable intersections. Let p be an initial intersection of two curves, which exists by Lemma 4. We resolve the intersection p by adding a bend to one of the curves and rerouting it along and sufficiently close to the other to ensure that they have the same intersections with other curves. We call such curves that follow each other after a resolved intersection a *group*. We then repeat resolving intersections of groups until there are no more intersections. As a final part of the proof we show that during this process for each curve at most $\log_2(h)$ bends have been added.

For a curve c, let $\#^a c$ be the number of bends that were added to c during this algorithm. During the execution of the algorithm we maintain a set of groups $\mathbb{G}$. Each group $\mathcal{G}r_i \in \mathbb{G}$ is a set of curves. For each group $\mathcal{G}r_i$ let $\#^a \mathcal{G}r_i$ be the maximum number of additional bends over all curves in $\mathcal{G}r_i$, i.e. $\#^a \mathcal{G}r_i = \max_{c \in \mathcal{G}r_i}(\#^a c)$. In the beginning each curve is in its own group, that means we start with $\mathbb{G} = \{\{c(v)\} \mid v \in V(P)\}$ and for each $\mathcal{G}r_i \in \mathbb{G}$, $\#^a \mathcal{G}r_i = 0$.

The following step is repeated until there are no more intersections. Let p be an initial intersection of two groups $\mathcal{G}r_i$ and $\mathcal{G}r_j$. We reroute the curves of one of $\mathcal{G}r_i$ and $\mathcal{G}r_j$. If we choose to reroute $\mathcal{G}r_j$, then we add a bend to each curve of $\mathcal{G}r_j$ and then the curves of $\mathcal{G}r_j$ follow along the curves of $\mathcal{G}r_i$, thus increasing $\#^a c$ by one for each $c \in \mathcal{G}r_j$. Resolving the intersection p creates a new group $\mathcal{G}r_k = \mathcal{G}r_i \cup \mathcal{G}r_j$. In order to keep $\#^a \mathcal{G}r_k$ bounded we apply the following strategy: if $\#^a \mathcal{G}r_i \neq \#^a \mathcal{G}r_j$, then we reroute the group with less additional bends and get $\#^a \mathcal{G}r_k = \max\{\#^a \mathcal{G}r_i, \#^a \mathcal{G}r_j\}$. Otherwise, $\#^a \mathcal{G}r_i = \#^a \mathcal{G}r_j$ and we arbitrarily choose one of the groups, so $\#^a \mathcal{G}r_k = \#^a \mathcal{G}r_j + 1$. With each resolved intersection two groups are merged into one, thus the overall number of groups reduces by one. As a result, after at most $h - 1$ resolved crossings between groups this iteration stops.

After the above procedure no two curves intersect, thus P is a planar-$\beta + \delta$-star (resp. outer-star) with $\delta = \max_{\mathcal{G}r \in \mathbb{G}}(\#^a \mathcal{G}r)$. In the following we prove by induction over the group size that for each group $\mathcal{G}r$ it holds that $\#^a \mathcal{G}r \leq \log_2(|\mathcal{G}r|)$. For the induction base we observe that if $|\mathcal{G}r| = 1$ we have $\#^a \mathcal{G}r = 0 = \log_2(|\mathcal{G}r|)$. As an induction hypothesis, assume that for a $k \geq 1$ and each group $\mathcal{G}r$ with $|\mathcal{G}r| \leq k$, it holds that $\#^a \mathcal{G}r \leq \log_2(|\mathcal{G}r|)$. Let $\mathcal{G}r_l$ be a group with $|\mathcal{G}r_l| = k + 1$, which is the result of merging two groups $\mathcal{G}r_i$ and $\mathcal{G}r_j$. Since $|\mathcal{G}r_i|, |\mathcal{G}r_j| < |\mathcal{G}r_l|$, the induction hypothesis holds for both $\mathcal{G}r_i$ and $\mathcal{G}r_j$. If $\#^a \mathcal{G}r_i \neq \#^a \mathcal{G}r_j$, we have $\#^a \mathcal{G}r_l = \max\{\#^a \mathcal{G}r_i, \#^a \mathcal{G}r_j\} \leq \log_2(\max\{|\mathcal{G}r_i|, |\mathcal{G}r_j|\}) < \log_2(|\mathcal{G}r_l|)$. Otherwise, if $\#^a \mathcal{G}r_i = \#^a \mathcal{G}r_j$, lets assume w.l.o.g. $|\mathcal{G}r_i| \geq |\mathcal{G}r_j|$, and therefore $|\mathcal{G}r_l| \geq 2|\mathcal{G}r_j|$. We have $\#^a \mathcal{G}r_l = \#^a \mathcal{G}r_j + 1 \leq \log_2(|\mathcal{G}r_j|) + 1 \leq \log_2(|\mathcal{G}r_l|/2) + \log_2(2) = \log_2(|\mathcal{G}r_l|)$.

Since for each v of P the curve $c(v)$ appears in exactly one group, we have that the maximum size of a group is h. It follows that P is a planar-$\beta + \delta$-star (resp. outer-star) with $\delta = \max_{\mathcal{G}r \in \mathbb{G}}(\#^a \mathcal{G}r) \leq \log_2(h)$. □

5 Drawing Extensions of Connected Subgraphs

In this section we apply the results from the previous section to provide a tight upper bound on the number of bends in a drawing extension of a connected subgraph.

Theorem 7. *Each instance (G, Γ_H) where H is an induced connected subgraph of G allows a $\min\{h/2, \beta + \log_2(h) + 1\}$-bend-extension, where h is the maximum face size of H and β is the maximum (outer) star complexity of a face in Γ_H. This bound is tight up to an additive constant.*

Above theorem implies an upper bound on the number of bends in case of a non-induced subgraph by simply subdividing the induced edges by dummy vertices and removing them after construction. The tightness of the bound follows from the fact that the lower bound proofs (Theorems 2 and 3) can easily be adapted to work for chords.

Corollary 2. *Each instance (G, Γ_H) where H is a connected subgraph of G, allows a $\min\{h + 1, 2\beta + 2\log_2(h) + 3\}$-bend-extension, where h is the maximum face size of H and β is the maximum star complexity of a face in Γ_H. This bound is tight up to an additive constant.*

6   Extending Stars with Straight Lines

Let $G = (V, E)$ be a plane graph and F a chordless face, fixed on the plane as a star-shaped polygon Γ_F. In this section we study the question whether (G, Γ_F) admits a straight-line extension. Note that for F being the outer face of G, Hong and Nagamochi [9] showed that (G, Γ_F) always admits a straight-line extension. In the following F is an inner face.

If F is an inner face fixed as a convex polygon Γ_F, Mchedlidze et al. [12] showed that it can easily be tested if an instance (G, Γ_F) admits a straight-line extension. In their case a necessary and sufficient condition for an extension to exist is that for each vertex individually there is a valid position outside Γ_F. For stars a comparable result is not possible. Even if each vertex could be drawn individually this does not mean that the whole instance admits a straight-line extension. Even more, testing whether pairs of vertices can be drawn together would not be sufficient as the construction in Fig. 3 suggests.

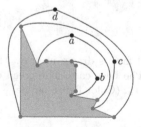

Fig. 3. The drawing cannot be extended to a straight-line drawing of the entire graph, even though this is not revealed when testing individual parts. (Color figure online)

In case of Γ_F being a convex inner face [12], the feasibility area of a vertex adjacent to the fixed face is just a wedge, formed by the intersection of two half planes induced by two edges of Γ_F. In this section we show that the situation for the star shaped inner face is dramatically different, thus there exists an instance for which the feasibility area of a vertex is partially bounded by a curve of exponential complexity.

Theorem 8. *There is an instance (G, Γ_F) where Γ_F is a star-shaped inner face, such that the feasibility area of some vertex $v \in G$ is partially bounded by a curve whose implicit representation is a polynomial of degree $2^{\Omega(|V|)}$.*

Sketch of Proof. A curve is *i-exponentially-complex* if it has a parametric representation of the form $\left\{ \left(\frac{r(t)}{u(t)}, \frac{s(t)}{u(t)} \right) \mid t \in \mathcal{I} \right\}$ where r, s and u are polynomials of degree 2^i and $\mathcal{I}$ is an interval. In the following we describe an instance (G, Γ_F), for which a feasibility area of a vertex v is bounded by an $2^{\Omega(|V|)}$-exponentially-complex curve. By slightly perturbing the positions of the vertices of Γ_F to achieve points in general position we have that the implicit representation of this curve is a polynomial of degree at least $2^{\Omega(|V|)}$.

Let $k \geq 1$ be a fixed integer. Figure 4a displays the plane graph $G_k = (V, E)$ and the drawing of its inner face as a star-shaped polygon. The vertices v_i and w_i still need to be drawn. For $0 \leq i \leq k$, the feasibility area of v_i is denoted by A_i and the boundary of A_i is referred to as B_i. We show that B_k contains a 2^k-exponentially-complex curve. The proof is by induction on $0 \leq i \leq k$.

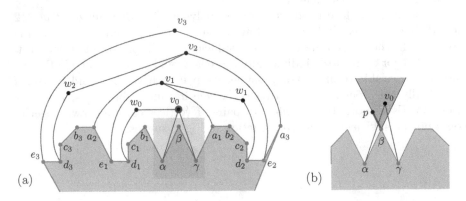

Fig. 4. (a) Graph G_3, the fixed face is drawn in gray. The vertices in the green area are part of the base case. (b) The base case. The green curve $\mathcal{C}_0$ is on the boundary B_0 of the feasibility area A_0 of v_0. (Color figure online)

As the base of the induction we consider the boundary B_0 of vertex v_0 as shown in Fig. 4b. The feasibility area of v_0 is the upper quadrant formed by the lines $l(\alpha, \beta)$ and $l(\gamma, \beta)$. Let $p = (-1/2, 1)$ be a point on the left boundary of A_0. Let $\mathcal{C}_0$ be the segment $s(\beta, p)$ not containing the point p. It holds that $\mathcal{C}_0 = \{(\frac{-t}{t+1}, \frac{2t}{t+1}) \mid t \in \mathcal{I} = [0, 1)\}$. The curve $\mathcal{C}_0$ is 0-exponentially-complex. An implicit equation of $l(\beta, p)$ is $y + 2x = 0$.

In the following we assume that the feasibility area A_{i-1} of v_{i-1} is partially bounded by an $i-1$-exponentially-complex curve satisfying additional invariants and prove that the feasibility area A_i of v_i is partially bounded by an i-exponentially-complex curve that also satisfies these invariants. The invariants are given in three groups, the *universal* invariants, holding after each inductive step, the *even* and the *odd* invariants holding after each even and odd step $i \geq 0$, respectively. Below are the universal and even invariants, with the odd invariants being symmetric.

Universal Invariants

$\mathcal{UI}$.1: A_i is partially bounded by an i-exponentially-complex curve $\mathcal{C}_i = \{v_i(t) = (v_i^x(t), v_i^y(t)) \mid t \in \mathcal{I}\}$, where $\mathcal{I} = [0, \mathcal{I}_{\max})$ and $\mathcal{I}_{\max} > 0$,

$\mathcal{UI}$.2: $v_i^y(t)$ is strictly increasing for $t \in [0, \mathcal{I}_{\max})$.

Even Invariants

$\mathcal{EI}$.1: $b_{i+1}^x < v_i^x(0) < a_{i+1}^x$ and $v_i^y(0) = 0$,

$\mathcal{EI}$.2: A_i is on the right of $\mathcal{C}_i$,

$\mathcal{EI}$.3: Ray $q(a_{i+1}, v_i(0))$ intersects no point of A_i to the left of $v_i(0)$.

We observe that universal and even invariants hold for the base case $i = 0$.

Let $\mathcal{C}_{i-1} = \left\{ v_{i-1}(t) = \left(\frac{r(t)}{u(t)}, \frac{s(t)}{u(t)} \right) \mid t \in \mathcal{I} \right\}$. The position $w_{i-1}(t)$ of vertex w_{i-1} is described as the intersection of the rays $q(v_{i-1}(t), b_i)$ and $q(d_i, c_i)$. The position $v_i(t)$ of v_i is described as $q(a_i, v_{i-1}(t)) \cap q(e_i, w_{i-1}(t))$.

Using this we calculate the curve $\mathcal{C}_i$, i.e. we calculate the position of v_i as a function of t. This can be done by calculating the equation of the line $l(v_{i-1}(t)), b_i)$, the position of the vertex $w_{i-1}(t)$ and then the equations if the lines $l(a_i, v_{i-1}(t))$ and $l(e_i, w_{i-1}(t))$. The intersection of the latter lines is $v_i(t)$ and we obtain $\mathcal{C}_i = \{ (\frac{r_i(t)}{u_i(t)}, \frac{s_i(t)}{u_i(t)}) \mid t \in \mathcal{I} \}$, where each of $r_i(t)$, $s_i(t)$, $u_i(t)$ is quadratic in $r(t)$, $s(t)$ and $u(t)$. By induction hypothesis, $\mathcal{C}_{i-1}$ is an $i-1$-exponentially complex curve, i.e. $u(t)$, $s(t)$, $r(t)$ contain terms $t^{2^{i-1}}$. So the curve $\mathcal{C}_i$ is i-exponentially complex, provided that the coefficients of highest degree do no cancel themselves out, which can be avoided by slightly perturbing the position of vertex e_i. This proves Invariant $\mathcal{UI}$.1. A proof that the remaining invariants hold after the induction step concludes the proof of the theorem.   □

7   Conclusion

We have shown that a drawing Γ_H of an induced connected subgraph H can be extended with at most $\min\{h/2, \beta + \log_2(h) + 1\}$ bends per edge if the star complexity of Γ_H is β and h is the size of the largest face of H and that this bound is tight up to a small additive constant. In the event of a disconnected subgraph H the known upper bound is $72|V(H)|$. It is tempting to investigate whether the constant 72 can be lowered and to provide a matching lower bound.

We have proven that there is an instance (G, Γ_F) where Γ_F is a star-shaped inner face, such that the feasibility area of some vertex $v \in G$ is partially bounded by an exponential degree curve. This is an indication that for a given instance (G, Γ_F) it is difficult to test whether (G, Γ) admits a straight-line extension. It would be interesting to establish the computational complexity of this problem. We were not able to show the NP-hardness of the problem. Due to its similarity with visibility and stretchability problems we conjecture that the problem is as hard as the existential theory of reals.

Acknowledgment. The authors thank Martin Nöllenburg and Ignaz Rutter for the discussions of this problem back in 2012. Jérôme Urhausen was supported by the Netherlands Organisation for Scientific Research under project 612.001.651.

References

1. Badent, M., Giacomo, E.D., Liotta, G.: Drawing colored graphs on colored points. In: Amato, N.M., Lee, D.T., Pietracaprina, A., Tamassia, R. (eds.) Theoretical Computer Science, vol. 408, pp. 129–142 (2008). https://doi.org/10.1016/j.tcs.2008.08.004. http://www.sciencedirect.com/science/article/pii/S0304397508005562
2. Bekos, M.A., et al.: Planar drawings of fixed-mobile bigraphs. In: Frati, F., Ma, K.-L. (eds.) GD 2017. LNCS, vol. 10692, pp. 426–439. Springer, Cham (2018). https://doi.org/10.1007/978-3-319-73915-1_33
3. Bekos, M.A., et al.: Planar drawings of fixed-mobile bigraphs (2017). http://arxiv.org/abs/1708.09238
4. Chambers, E.W., Eppstein, D., Goodrich, M.T., Löffler, M.: Drawing graphs in the plane with a prescribed outer face and polynomial area. **16**, 243–259 (2012). https://doi.org/10.7155/jgaa.00257
5. Chan, T.M., Frati, F., Gutwenger, C., Lubiw, A., Mutzel, P., Schaefer, M.: Drawing partially embedded and simultaneously planar graphs. In: Duncan, C., Symvonis, A. (eds.) J. Graph Algorithms Appl. **19**, 681–706 (2015). https://doi.org/10.1007/978-3-662-45803-7_3
6. Djidjev, H.N., Lingas, A., Sack, J.R.: An O(n log n) algorithm for computing the link center of a simple polygon. Discrete Comput. Geom. **8**(2), 131–152 (1992). https://doi.org/10.1007/BF02293040
7. Giacomo, E.D., Didimo, W., Liotta, G., Meijer, H., Wismath, S.K.: Point-set embeddings of trees with given partial drawings. Comput. Geom. **42**(6), 664–676 (2009). https://doi.org/10.1016/j.comgeo.2009.01.001. http://www.sciencedirect.com/science/article/pii/S0925772109000108
8. Goaoc, X., Kratochvíl, J., Okamoto, Y., Shin, C.S., Spillner, A., Wolff, A.: Untangling a planar graph. Discrete Comput. Geom. **42**(4), 542–569 (2009). https://doi.org/10.1007/s00454-008-9130-6
9. Hong, S.H., Nagamochi, H.: Convex drawings of graphs with non-convex boundary constraints. Discrete Appl. Math. **156**(12), 2368–2380 (2008). https://doi.org/10.1016/j.dam.2007.10.012
10. Kant, G.: Drawing planar graphs using the canonical ordering. Algorithmica **16**(1), 4–32 (1996). https://doi.org/10.1007/BF02086606

11. Lubiw, A., Miltzow, T., Mondal, D.: The complexity of drawing a graph in a polygonal region. In: Biedl, T., Kerren, A. (eds.) Graph Drawing and Network Visualization (2018, to appear)
12. Mchedlidze, T., Nöllenburg, M., Rutter, I.: Extending convex partial drawings of graphs. Algorithmica **76**(1), 47–67 (2016). https://doi.org/10.1007/s00453-015-0018-6
13. Mchedlidze, T., Urhausen, J.: β-stars or on extending a drawing of a connected subgraph (2018). https://arxiv.org/abs/1808.10366
14. Pach, J., Wenger, R.: Embedding planar graphs at fixed vertex locations. Graphs Comb. **17**(4), 717–728 (2001). https://doi.org/10.1007/PL00007258
15. Patrignani, M.: On extending a partial straight-line drawing. Int. J. Found. Comput. Sci. **17**(05), 1061–1069 (2006)
16. Tutte, W.T.: How to draw a graph. Proc. Lond. Math. Soc. **s3-13**(1), 743–767 (1963). https://doi.org/10.1112/plms/s3-13.1.743. http://plms.oxfordjournals.org/cgi/reprint/s3-13/1/743

Experiments

Perception of Symmetries in Drawings of Graphs

Felice De Luca[1]([⊠]), Stephen Kobourov[1], and Helen Purchase[2]

[1] University of Arizona, Tucson, USA
felicedeluca@email.arizona.edu, kobourov@cs.arizona.edu
[2] University of Glasgow, Glasgow, UK
helen.purchase@glasgow.ac.uk

Abstract. Symmetry is an important factor in human perception in general, as well as in the visualization of graphs in particular. There are three main types of symmetry: reflective, translational, and rotational. We report the results of a human subjects experiment to determine what types of symmetries are more salient in drawings of graphs. We found statistically significant evidence that vertical reflective symmetry is the most dominant (when selecting among vertical reflective, horizontal reflective, and translational). We also found statistically significant evidence that rotational symmetry is affected by the number of radial axes (the more, the better), with a notable exception at four axes.

1 Introduction

Many objects in nature, from plants and animals to crystals and snowflakes, have symmetric patterns. Humans and other animals have a nearly perfect reflective symmetry along a single axis; sea stars and snowflakes have repetitive patterns along two or more radial axes; leaves and flowers often have translational patterns of symmetry; see Fig. 1.

The perception of symmetry is one of the key concepts in Gestalt theory which studies how humans perceive different types of objects. Symmetry has also been considered an important feature of well-drawn graphs, on the basis that depicting symmetries will reveal a graph's structure and properties [8]. A natural question that arises is: which types of symmetry are easier to perceive and how does this affect drawings of graphs? In this study we investigate this question, focusing on the reflective (also called "mirror"), translational and rotational (also called "radial") types of symmetries:

- **Vertical:** A pattern is reflected across a vertical axis (*reflective symmetry*)
- **Horizontal:** A pattern is reflected across an horizontal axis (*reflective symmetry with a 90° rotation*)
- **Translational:** A pattern is repeated and shifted in the space
- **Rotational:** A pattern is repeated across radial axes with a given angle

© Springer Nature Switzerland AG 2018
T. Biedl and A. Kerren (Eds.): GD 2018, LNCS 11282, pp. 433–446, 2018.
https://doi.org/10.1007/978-3-030-04414-5_31

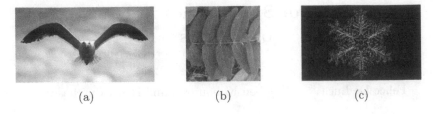

Fig. 1. Symmetry in nature: (a) reflective, (b) translational, (c) rotational.

We used synthetically generated graph drawings in a human subjects experiment to answer several questions related to the perception of symmetry. Specifically, we created graph layouts that exhibit different types of symmetries and asked our participants to select the more symmetric ones. We found statistically significant evidence that vertical reflective symmetry is the most dominant (when selecting among vertical, horizontal and translational). We also found statistically significant evidence that rotational symmetry is affected by the number of radial axes (the more, the better) with a notable exception at four axes.

2 Related Work

Gestalt theorists studied how objects are perceived, their view being that the perception of a whole object cannot be reduced to the sum of the perception of its parts [17]. They focused on describing the patterns we see in visual stimuli [2, 30], and in how we distinguish background from foreground [23], devising a set of laws describing fundamental perceptual phenomena. Well-known Gestalt principles include proximity (things that are close together are perceived as being in a group), similarity (things that look "similar" are perceived as being in a group), closure (closed shapes are preferred to open shapes), and symmetry, which we discuss in more detail next.

Giannouli [9] defines symmetries on the plane as "transformations that preserve equal geometric distance," and identifies the three standard types: translational, rotational and reflective symmetries. Each of these have variations: the distance and direction of the transposition for translational symmetry, the number of times around a circle the object is repeated (known as its "order") for rotational symmetry, and the angle of the axis (with horizontal and vertical being the most common) for reflective symmetry.

The human perception of symmetry has been studied in different contexts. For example, there is evidence that people remember figures as more symmetric and closer than they really are [28] and that symmetry aids in the recall of abstract patterns [12, 26].

Several studies compare the effectiveness of different types of symmetry. An early experiment by Corbalis and Roldan [7] compared vertical reflective symmetry with translation symmetry (limited to only horizontal translation). They compared these two conditions in two forms – where the two components were touching each other (i.e., creating a single holistic visual object) or where there was a horizontal distance between the two components (i.e., they were perceived as two separate objects). They concluded that vertical symmetry was more salient in the holistic case, but translation more salient when the two components are disjoint. Other early work [3] concluded that vertical symmetry is easier to detect than translational symmetry. Royer [24] prioritizes horizontal and vertical reflective symmetry over diagonal reflection, with "centric" (loosely comparable to rotational, despite Royer's stimuli being square in form) performing worse. The work by Palmer and Hemenway [18] confirms the ordering: vertical reflective, horizontal reflective, diagonal reflective.

Cattaneo *et al.* [6] investigate the neurological basis for the perception of vertical and horizontal reflective symmetry and conclude that there is a "partial" difference between the regions of the brain used in detecting these two symmetries. Giannouli's [9] review of research on the visual perception of symmetry finds that vertical reflective symmetry is more readily perceived than any other type. Similar findings are reported in an earlier review by Wageman [29], who ranks reflective symmetries as follows: vertical produces better recognition performance (faster or more accurate) than horizontal, which in turn performs better than diagonal.

Jennings and Kingdom [10] conducted experiments to compare the perception of different orders of rotational symmetry (3, 5 and 7), together with a vertical reflection condition. They conclude that it is easier to detect rotational symmetry as the order increases (as measured by response time), and that, in comparison with rotational symmetry, the vertical reflective symmetry condition performs better than 3rd order, the same as 5th order, and worse than 7th order.

Note that none of this work involved graphs or drawings of graph.

While some researchers have considered the application of Gestalt principles to graph drawing, such work is rather fragmented. Wong and Sun [27] create key criteria for the depiction of UML class diagrams based on the Gestalt theories, and then evaluate three UML diagram tools based on these criteria. Bennet *et al.* [1] review the literature on graph drawing aesthetics with reference to Norman's [16] stages of perception (visceral, behavioral and reflective), including the Gestalt theories in the visceral stage. They conclude that more work needs to be done to validate the common graph drawing aesthetic criteria with respect to perceptual theories. Nesbitt and Freidrich [15] discuss some of the Gestalt theories in relation to graphs that evolve over time, although symmetry is not explicitly considered. Lemon *et al.*'s experiments [13] show that the principles of similarity, proximity and continuity affect the comprehension of complex software diagrams. Rusu *et al.* [25] focus on the principle of continuity, and proposed a method for reducing visual clutter created by edge crossings by creating gaps in the edges. This is embodied in the partial edge drawing algorithms of Bruckdorfer *et al.* [4] and Burch *et al.* [5]. Marriott *et al.* [14] conduct an experiment looking at what features of small graph drawings made them most memorable. Their experimental conditions explicitly relate to the Gestalt principles of symmetry, continuity, orientation and proximity and their finding indicate that drawings that exhibit symmetry and continuity are amongst those most readily recalled.

Eades advocates the use of algorithms that aim to draw graphs with "as much symmetry as possible" [8]. Early experiments investigating the relative importance of different graph drawing aesthetics find support for the depiction of symmetry in terms of performance on graph-reading tasks [20,21]. Two computational methods have been proposed for measuring the extent of symmetry in a graph drawing, a non-trivial task in and of itself. The method proposed by Purchase [19] considers only reflective symmetry. It generates potential axes of symmetry between all pairs of vertices, and determines the existence of symmetric sub-graphs (edges reflected around the axis, with a tolerance) for each axis. Klapaukh's method [11] uses an edge-based metric that includes rotational and translation symmetries in addition to vertical ones. Welch and Kobourov [31] studied which of these two algorithms best correlates with the human perception of symmetry, with results that suggest that a graph drawing with vertical symmetry is considered more symmetric than the identical drawing presented at a slightly different orientation, and that the greater the extent of symmetry in a drawing, the faster the participants' response.

While there has been extensive experimental research in the perception literature comparing the different types of symmetry in a variety of artificial stimuli, no comparable work has been performed to investigate the perception of symmetry in graph drawings. We therefore extend the work done so far by conducting experiments that specifically considers which of the three types of symmetry are more salient in drawings of graphs, including several additional variations.

3 Research Questions

We investigate the perception of graph drawings that exhibit three types of symmetry: reflective (vertical and horizontal), translational and rotational. Rather than attempting to draw existing graphs that embody such symmetries (a difficult task), we create symmetric graph drawings by duplicating graph-substructures. Specifically, we draw a small graph, make a duplicate, place the duplicate(s) appropriately (according to the type of symmetry), and join the components together to create a graph drawing that exhibits the desired symmetry.

Since rotational symmetry is visually very different from reflective and translational symmetry, we address two separate research questions:

1. What is the relative ranking of reflective and translational symmetries for drawings of graphs?
2. What is the impact of the number of axes (order) for rotational symmetry?

3.1 The Symmetric Graph Drawings

Our experiment considers several different types of symmetry: horizontal (H), horizontal with rotation (Hr), vertical (V), vertical with rotation (Vr), translational (T), translational with rotation (Tr), rotational with fixed components

(RC) and rotational with fixed vertices (RV). For a baseline, we also have a non-symmetric version (NS); see Fig. 2. We consider two variants of rotational symmetry (RC and RV) in order to take into account the effect of the number of rotational axes and the effect of different graph sizes.

Ideally, all the stimuli should represent exactly the same graph, but since this would be impossible (especially for the rotational drawings), we attempt to impose some consistency by using the same "base graph" from which the larger graphs are derived.

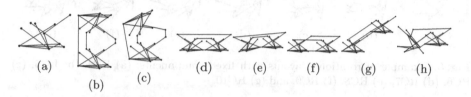

(a)   (b)   (c)   (d)   (e)   (f)   (g)   (h)

Fig. 2. Example layouts in the dataset: (a) base graph, and its (b) H, (c) Hr, (d) V, (e) Vr, (f) T, (g) Tr, and (h) NS.

We create the horizontal, vertical and translational drawings (HVT, for short) as follows. Let $G_b = (V_b, E_b)$ be a base graph drawn with a random layout such that each vertex $v_b \in V_b$ has positive coordinates; see Fig. 2a. Let $G_c = (V_c, E_c)$ be a copy of G_b with the same layout. Then $G_s = (V_b \cup V_c, E_b \cup E_c \cup E)$ is created from the two graphs together with edge set E connecting the vertices in V_b to their copied version in V_c. We fix $|E| = 3$ and the vertices that are chosen for the connection are chosen at random. We use G_s to create the layouts of the graphs in the HVT set by changing the coordinates of the vertices $v_c \in V_s$ as follows:

- **H:** If $v_b = (x, y)$ then $v_c = (x, -y)$; see Fig. 2b.
- **Hr:** H version with a rotation with angle in $[0, 45]$; see Fig. 2c.
- **V:** If $v_b = (x, y)$ then $v_c = (-x, y)$; see Fig. 2d.
- **Vr:** V version with a rotation with angle in $[0, 45]$; see Fig. 2e.
- **T:** If $v = (x, y)$ then $v_c = (x - \delta, y)$ where δ is a shifting factor such that the bounding boxes of V_b and V_s do not overlap; see Fig. 2f.
- **Tr:** T version with a rotation with angle in $[0, 45]$; see Fig. 2g.
- **NS:** Non symmetric (random) placement of the vertices in V_s; see Fig. 2h.

We have two types of rotational drawings: maintaining the base graph component (the "fixed component" version), and limiting the maximum number of vertices (the "fixed vertices" version). We use both methods in order to control for the possible confounding factor of different graph sizes in the first variant.

The rotational fixed component (RC for short) versions are symmetric layouts that repeat the base graph drawing around each axis. The base graph $G_b = (V_b, E_b)$ is a component of the layout of the symmetric graph G_s and we create the different graphs as follows:

- **RC{X}:** The drawing of each component is replicated on each of the $X = [4, \ldots, 10]$ axes of symmetry. By choosing two random vertices from a component, we connect each pair of rotationally consecutive components with edges to the corresponding vertices; see Fig. 3.

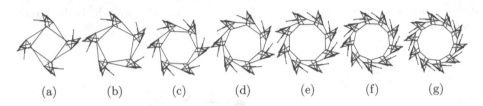

Fig. 3. Example of rotational layouts with fixed components: (a) RC4, (b) RC5, (c) RC6, (d) RC7, (e) RC8, (f) RC9, and (g) RC10.

The rotational fixed vertices (RV for short) versions are symmetric layouts with a limited maximum number of total vertices. They are created as follows:

Table 1. RV sizes

	V	Avg E
RV4	40	52
RV5	50	65
RV6	48	60.3
RV7	49	60.9
RV8	48	59.6
RV9	45	55.8
RV10	50	62

- **RV{X}:** The base graph is reduced in size by removing as many vertices (at random) as needed so that when it is replicated on each of the $X = [4, \ldots, 10]$ axes of symmetry, the total number of vertices does not exceed 50. As before, we connect each pair of rotationally consecutive components with two edges; see Fig. 4. Inevitably, the graphs in this set are not exactly of the same size, but they are within 20% of each other; Table 1 shows the number of vertices and average number of edges.

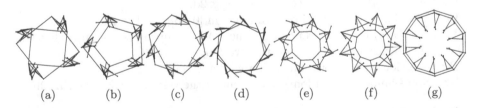

Fig. 4. Example of rotational layout with fixed number of vertices: (a) RV4, (b) RV5, (c) RV6, (d) RV7, (e) RV8, (f) RV9, and (g) RV10.

We use small and sparse base graphs ($|V| = 10$ and $|E| = 11$) as these would be copied and interconnected when creating the various experimental sets. Each base graph is drawn by placing the vertices at random, to avoid accidental symmetries within the base graph and in order to focus on the symmetries created by the base graph replication. We use two (RC and RV) or three (HVT) edges to connect the replicated components for the same reasons.

By construction, all pairs of HVT graphs used as stimuli are the same, while pairs of RC and RV graphs are structurally different. We used the yFiles library [32] to generate and draw our stimuli.

4 Experimental Methodology

We conducted three separate experiments: Reflective and Translational (RT), Rotational Fixed Component (RFC) and Rotational Fixed Vertices (RFV). We used the same methodology but different sets of stimuli. The participants' task was to look at a series of pairs of graph drawings, and, for each pair, indicate which one of them they thought was "more symmetric." This task was designed to address the following experimental questions:

- **Q1:** Which type of symmetry among H, V, T, Hr, Vr, Tr is most recognizable as symmetry?
- **Q2:** How many rotations is most recognizable as rotational symmetry, using the fixed-component generation method?
- **Q3:** How many rotations is most recognizable as rotational symmetry, using the fixed-vertices generation method?

For Q1, we expected vertical symmetry (V) will be most prominent, with horizontal (H) being noticed more than translational (T), and that non-rotated versions will be easier to detect than the corresponding rotated versions [3, 24, 29].

For Q2 and Q3 we expected that the higher the degree, the greater the extent of symmetry recognition [10]. We were unsure about the effect of the additional visual clutter inherent in the RC drawings (as the size of graph increases with the degree) but anticipated that both should follow the same trend.

4.1 Stimuli

Each experiment uses five base graphs. Specifically, we generated 20 random simple graphs, called base graphs (each with 10 vertices and 11 edges), and drew them using a random layout. We randomly chose five of these base graphs as generators for the stimuli in each experiment.

Reflective and Translational (RT): There are 7 conditions in this experiment: horizontal (H), vertical (V), translational (T), horizontal rotated (Hr), vertical rotated (Vr), translational rotated (Tr), and non-symmetric (NS). In this experiment, all drawings are of the same graph; they are all based on the same base component. With 5 different versions of the base graph, we have $5 * 7 = 35$ stimuli for this experiment.

Rotational Fixed Component (RFC): There are 7 conditions in this experiment: rotational orders of 4−10 (designated as RC4, ..., RC10). Since the number of incidences of the base component drawing is increased for each order, and two edges are added every time a new base component is added, the size of the graph increases with every order. With 5 different base graphs, we have $5 * 7 = 35$ stimuli for this experiment.

Rotational Fixed Vertices (RFV): There are 7 conditions in this experiment: rotational orders of 4−10 (designated as RV4, ..., RV10). Here, the number of vertices in each graph varies from 40 to 50, as described in Table 1. With 5 different versions of the base graph, we have $5 * 7 = 35$ stimuli for this experiment.

5   Experimental Process

We use a "two-alternative forced choice" methodology, where a pair of stimuli are presented and participants must choose one of them. Specifically, the participants are asked to select the drawing that they considered "more symmetric." In each experiment, we show all possible pairs twice (switching between left and right for the second presentation); with 7 stimuli we get $2 * (7 * 6)/2 = 42$ pairs and this is done for each of the 5 different versions of the base graph for a total of $5 * 42 = 210$ trials. As a within-participants' experiment, all participants see all 210 pairs. To mitigate against the learning effect, the experimental stimuli are preceded by 10 practice trials for which the data is not collected (using a sixth version of the base graph), and each participant is presented with the 210 trials in a different random order. We collected data on the choice made by the participant for each pair, and the time taken to make the choice.

The experiment is conducted online with the online system randomly assigning one of the three experiments, and randomly selecting 5 base graphs from the 20 available for that participant-experiment combination. By asking each participant to do only one of the three experiments and by choosing only five versions of the base graph, we anticipated that the experiment would not be too lengthy (therefore minimizing the chance of participants not finishing the experiment). We expected the experiment to take approximately 10 min.

Participants are required to give consent at the start, and a set of instructions followed before the practice trials began. A self-timed break is offered every 20 trials. At the end of the experiment, participants are asked to give demographic data: gender, age, educational background, familiarity with networks and with symmetries. As a reward for taking part of the study, statistics about the participant's answers in relation to the answers of other participants are presented. These statistics show an example of the layout used in the task, the number of selections from the participants and from the specific participant grouped by each version, the number of clicks for any pair of versions, the number of left and right clicks, and the average answer time for the current and all participants.

Participants are not given details about the problem we are considering or the task that they would perform and we are intentionally vague: we just ask them to *"choose the layout that looks more symmetric"* between the proposed pair of layouts. We do not provide information about the concept of symmetry or our interpretations thereof, so as to not accidentally influence the participants.

5.1 Pilot Study

We use a simple online system made of four main parts: an introduction page, the main experiment, a demographics page (gender, age, graph expertise, symmetry expertise), and the statistics page. We conducted a pilot study to test our setup. In its early stages our system had an introductory screen with a brief description about the concept of *"graphs"* and *"graph drawings"* but several participants suggested to use *"network"* instead, as it is more common and easier to understand. We also added information about the expected duration of the experiment. Initially we showed pairs of graphs starting from 10 base graphs for a total of 420 pairs, but the pilot study showed that this resulted in a test that was too long. We decided to reduce the number of base graphs to 5 and to introduce a break every 20 pairs. The demographics page was augmented to allow for feedback from the participants. The end of the study was modified from a "Thank you" page to a page that provides statistics about the participants results in comparison to the average previous participants. Finally, following feedback from the pilot study, we changed the background color from white to light-gray and added a black border to highlight the selected image.

6 Data Analysis

We used Reddit [22] and personal communications to crowdsource our study. We collected data from a total of 97 participants.

6.1 Experimental Conduct

We removed data from all participants who did not complete the full experiment ($n = 39$). It's disappointing that so many people did not finish experiment (which was intended to take less than ten minutes); informal feedback from participants are that they found the task boring and, in some cases, difficult. One participant in the RT experiment gave an exceptionally high number of votes to the non-symmetric drawing; we removed this participant from our analysis. Two response times were particularly higher than others (17 s and 100 s) - these data points were replaced by the mean of the other response times for the respective participants. One participant gave almost exactly the same votes to all conditions in the RFV experiment: we removed this participant's data. This left us with 19 participants for RT, 19 for RFC and 18 for RFV. The demographic information of the 56 participants (13 female and 43 male) is summarized in Fig. 5.

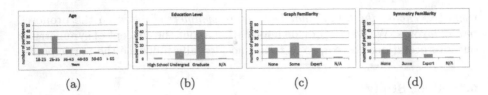

Fig. 5. Participants demographic information (a) age, (b) education level, (c) graph familiarity, and (d) symmetry familiarity.

6.2 Analysis Process

We (conveniently) have seven conditions for each of the three experiments, so the form of the data is the same. Since we include only those participants who completed the entire experiment, each participant "voted" 210 times, each vote being associated with one of the seven conditions. We also have an average response time associated with each vote. For both response times and votes, we use ANOVA and adjusted planned comparison pair-wise tests between the pairs of conditions of interests to determine which conditions are (a) favored over the others, and (b) responded to most quickly. We use a significance level of 0.05 throughout, adjusted as appropriate for the number of planned comparisons made. We do not compare all pairs of conditions, only those of interest to our research question as doing this reduces the extent of required adjustments. The mean vote and mean response time (over all participants) charts are depicted in Fig. 6, while the exact values are shown in Table 2.

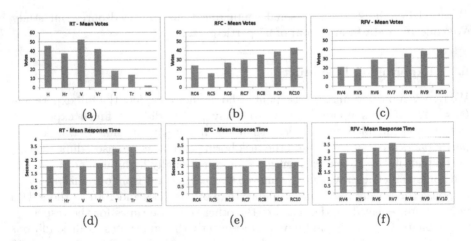

Fig. 6. Mean votes charts for (a) RT (b) RFC, (c) RFV and mean response time charts for (d) RT (e) RFC, and (f) RFV.

Voting Responses

Reflective and Translational (RT): A repeated-measures ANOVA reveals a significant difference between the average votes ($F = 240.5$, $df = 6$, $p < 0.001$). Five pairwise comparisons (at adjusted $p = 0.01$), reveal the following results;

- H obtained significantly more votes than Hr ($p < 0.001$)
- V obtained significantly more votes than Vr ($p < 0.001$)
- T obtained significantly more votes than Tr ($p < 0.001$)
- V obtained significantly more votes than H ($p < 0.001$)
- H obtained significantly more votes than T ($p < 0.001$)

Rotational Fixed Component (RFC): A repeated-measures ANOVA reveals a significant difference between the average votes ($F = 12.2$, $df = 6$, $p < 0.001$). Two pairwise comparisons (at adjusted $p = 0.025$), reveal the following results;

- RC6 obtained significantly more votes than RC5 ($p < 0.001$)
- No significant difference is between the votes for RC4 and RC5 ($p = 0.050$)

Rotational Fixed Vertices (RFV): A repeated-measures ANOVA reveals a significant difference between the average votes ($F = 10.9$, $df = 6$, $p < 0.001$). Two pairwise comparisons (at adjusted $p = 0.025$), reveal the following results;

- RV6 obtained significantly more votes than RV5 ($p = 0.021$)
- No significant difference is between the votes for RV4 and RV5 ($p = 0.63$)

Table 2. Mean vote and mean response time for each task.

	Number of participants	H	Hr	V	Vr	T	Tr	NS
Mean vote	19	45.63	36.89	52.16	41.74	18	13.63	1.95
Mean response time (s)	19	2.05	2.53	2.03	2.25	3.27	3.41	1.93
		RC4	RC5	RC6	RC7	RC8	RC9	RC10
Mean vote	19	23.63	15.11	26.42	29.05	35.21	38.32	42.26
Mean response time (s)	19	2.30	2.21	1.98	1.96	2.36	2.18	2.24
		RV4	RV5	RV6	RV7	RV8	RV9	RV10
Mean vote	18	20.78	18.67	28.78	29.44	35.22	37.72	39.39
Mean response time (s)	18	2.89	3.15	3.25	3.58	2.94	2.65	2.93

Response Time

Reflective and Translational (RT): A repeated-measures ANOVA reveals a significant difference between the average response time ($F = 4.84$, $df = 6, p < 0.001$). Five pairwise comparisons (at adjusted $p = 0.01$), reveal;

- The response time for H is significantly less than that for T ($p < 0.001$)
- No significant differences between the response times for the following pairings: H/Hr; V/Vr; T/Tr; V/H.

Rotational Fixed Component (RFC): A repeated-measures ANOVA reveals no significant difference between the average response time ($F = 1.29$, $df = 6$, $p = 0.267$).

Rotational Fixed Vertices (RFV): A repeated-measures ANOVA reveals no significant difference between the average response time ($F = 1.535$, $df = 6$, $p = 0.174$).

7   Discussion

With respect to question **Q1**, we found statistically significant effects confirming that vertical symmetry is more recognizable as symmetry, followed by horizontal and translational. In all three variants adding even slight rotation has a significant effect. Vertical symmetry is more recognizable than vertical with rotation, horizontal is more recognizable than horizontal with rotation, and translational is more recognizable than translational with rotation. Mean response time also follows this trend, although we have only one statistically significant finding that horizontal is faster than translational. An immediate implication of this is that vertical symmetry is the best perceived because it is frequently seen by people and that can be exploited in future graph layout algorithms.

The RFC and RFV experiments helped us answer questions **Q2** and **Q3**. Our experimental results provide evidence of a greater symmetry recognition for high number of rotation axes. We believe that this is due to the fact that layouts with high number of rotational axes tend to become more and more circular, and the circle is a very symmetric shape. We also find that the increased size of the graphs in the RFC experiments does not seem to affect the better perception for high number of rotational axes as the results for RFC and RFV are similar.

Of particular interest is one exception: RC4 is considered more symmetric than the RC5, which goes against the general trend of better perception for high number of rotation axes. We discuss possible explanations below.

8   Conclusions and Future Work

The conclusions from our experiments are limited by the specifics of our study; e.g., we only consider graphs of similar sizes, starting with a base graph of a fixed size, and connecting copies thereof in symmetric sub-structures using two or three edges. Despite such limitations, our experiment does provide potentially useful information about the relative effects of different types of symmetries in drawings of graphs. The results from our study suggest that humans recognize vertical reflective symmetry over all other types of symmetry, followed by horizontal and translational symmetries and that rotational symmetry is affected by the number of radial axes. These findings can help guide algorithms that identify features to be displayed using these types of symmetries. Vertical symmetry can be used to call attention to isomorphic subgraphs and cycles can be highlighted by laying them out as regular n-gons that have high rotational symmetry.

The results of our experiment, in particular for the RFC and RFV tasks, show that a rotationally symmetric layout with 4 axes is considered more symmetric than that with 5 axes, which goes against the general tendency of recognizability towards higher number of axes. This leads to an interesting question: what is it about these rotationally symmetric layouts that gives such different results? Is it because of something in particular about the layout with 5 axes or the one with 4? Is it because rotational 4 is not perceived as a rotational symmetry but as a combination of horizontal and vertical symmetries? If so, is the rotational symmetry with many axes perceived better than the reflective and translational ones? Answers to such questions can help guide the design of algorithms that visualize symmetries. It would also be interesting to repeat this study using different base graphs that were not drawn with random layout.

Acknowledgements. We thank Eric Welch for help with the experiment, Hang Chen for help with statistics, and yFiles for providing their software for research and education. This work is supported by NSF grants CCF-1423411 and CCF-1712119.

References

1. Bennett, C., Ryall, J., Spalteholz, L., Gooch, A.: The aesthetics of graph visualization. Comput. Aesthet. **2007**, 57–64 (2007)
2. Bruce, V., Green, P.R., Georgeson, M.A.: Visual Perception: Physiology, Psychology, & Ecology. Psychology Press, Hove (2003)
3. Bruce, V.G., Morgan, M.J.: Violations of symmetry and repetition in visual patterns. Perception **4**(3), 239–249 (1975)
4. Bruckdorfer, T., Kaufmann, M.: Mad at edge crossings? break the edges!. In: Kranakis, E., Krizanc, D., Luccio, F. (eds.) FUN 2012. LNCS, vol. 7288, pp. 40–50. Springer, Heidelberg (2012). https://doi.org/10.1007/978-3-642-30347-0_7
5. Burch, M., Vehlow, C., Konevtsova, N., Weiskopf, D.: Evaluating partially drawn links for directed graph edges. In: van Kreveld, M., Speckmann, B. (eds.) GD 2011. LNCS, vol. 7034, pp. 226–237. Springer, Heidelberg (2012). https://doi.org/10.1007/978-3-642-25878-7_22
6. Cattaneo, Z., Bona, S., Silvanto, J.: Not all visual symmetry is equal: partially distinct neural bases for vertical and horizontal symmetry. Neuropsychologia **104**, 126–132 (2017)
7. Corballis, M.C., Roldan, C.E.: On the perception of symmetrical and repeated patterns. Percept. Psychophys. **16**(1), 136–142 (1974)
8. Eades, P., Hong, S.H.: Symmetric graph drawing. In: Tamassia, R. (ed.) Handbook on Graph Drawing and Visualization, pp. 87–113. Chapman and Hall/CRC, Boca Raton (2013)
9. Giannouli, V.: Visual symmetry perception. Encephalos **50**, 31–42 (2013)
10. Jennings, B.J., Kingdom, F.A.A.: Searching for radial symmetry. i-Perception **8**(4), 2041669517725758 (2017)
11. Klapaukh, R.: An empirical evaluation of force-directed graph layout. Ph.D. thesis, Victoria University of Wellington (2014)
12. Lai, H.C., Chien, S.H.L., Kuo, W.Y.: Visual short-term memory for abstract patterns: effects of symmetry, element connectedness, and probe quadrant. J. Vis. **9**(8), 593 (2009)

13. Lemon, K., Allen, E.B., Carver, J.C., Bradshaw, G.L.: An empirical study of the effects of gestalt principles on diagram understandability. In: 2007 First International Symposium on Empirical Software Engineering and Measurement. ESEM 2007, pp. 156–165. IEEE (2007)
14. Marriott, K., Purchase, H., Wybrow, M., Goncu, C.: Memorability of visual features in network diagrams. IEEE Trans. Visual. Comput. Graph. 18(12), 2477–2485 (2012)
15. Nesbitt, K.V., Friedrich, C.: Applying gestalt principles to animated visualizations of network data. In: 2002 6th International Conference on Information Visualisation, pp. 737–743. IEEE (2002)
16. Norman, D.A.: Emotion Design: Why We Love (or Hate) Everyday Things (2004)
17. Palmer, S.E.: Theoretical approaches to vision. In: Vision Science: Photons to Phenomenology, pp. 45–92 (1999)
18. Palmer, S.E., Hemenway, K.: Orientation and symmetry: effects of multiple, rotational, and near symmetries. J. Exp. Psychol. Hum. Percept. Perform. 4(4), 691–702 (1978)
19. Purchase, H.: Metrics for graph drawing aesthetics. J. Vis. Lang. Comput. 13(5), 501–516 (2002)
20. Purchase, H.: Which aesthetic has the greatest effect on human understanding? In: DiBattista, G. (ed.) GD 1997. LNCS, vol. 1353, pp. 248–261. Springer, Heidelberg (1997). https://doi.org/10.1007/3-540-63938-1_67
21. Purchase, H.C., Cohen, R.F., James, M.: Validating graph drawing aesthetics. In: Brandenburg, F.J. (ed.) GD 1995. LNCS, vol. 1027, pp. 435–446. Springer, Heidelberg (1996). https://doi.org/10.1007/BFb0021827
22. Reddit: The front page of the internet. https://www.reddit.com
23. Rock, I., Palmer, S.: The legacy of gestalt psychology. Sci. Am. 263(6), 84–91 (1990)
24. Royer, F.L.: Detection of symmetry. J. Exp. Psychol. Hum. Percept. Perform. 7(6), 1186–1210 (1981)
25. Rusu, A., Fabian, A.J., Jianu, R., Rusu, A.: Using the gestalt principle of closure to alleviate the edge crossing problem in graph drawings. In: 2011 15th International Conference on Information Visualisation, pp. 488–493. IEEE (2011)
26. Schnore, M., Partington, J.: Immediate memory for visual patterns: symmetry and amount of information. Psychon. Sci. 8(10), 421–422 (1967)
27. Sun, D., Wong, K.: On evaluating the layout of UML class diagrams for program comprehension. In: 2005 Proceedings of 13th International Workshop on Program Comprehension. IWPC 2005, pp. 317–326. IEEE (2005)
28. Tversky, B.: Visuospatial reasoning. In: The Cambridge Handbook of Thinking and Reasoning, pp. 209–240 (2005)
29. Wagemans, J.: Detection of visual symmetries. Spat. Vis. 9(1), 9–32 (1995)
30. Ware, C.: Information Visualization: Perception for Design. Elsevier, Amsterdam (2004)
31. Welch, E., Kobourov, : S.: Measuring symmetry in drawings of graphs, vol. 36, pp. 341–351. Wiley (2017)
32. yWorks: The diagramming company. https://www.yworks.com

Network Alignment by Discrete Ollivier-Ricci Flow

Chien-Chun Ni[1] (ID), Yu-Yao Lin[2] (ID), Jie Gao[3](✉) (ID), and Xianfeng Gu[3] (ID)

[1] Yahoo! Research, Sunnyvale, CA 94089, USA
cni02@oath.com
[2] Intel Inc., Hillsboro, OR 97124, USA
yu-yao.lin@intel.com
[3] Stony Brook University, Stony Brook, NY 11794, USA
{jgao,gu}@cs.stonybrook.edu

Abstract. In this paper, we consider the problem of approximately aligning/matching two graphs. Given two graphs $G_1 = (V_1, E_1)$ and $G_2 = (V_2, E_2)$, the objective is to map nodes $u, v \in G_1$ to nodes $u', v' \in G_2$ such that when u, v have an edge in G_1, very likely their corresponding nodes u', v' in G_2 are connected as well. This problem with subgraph isomorphism as a special case has extra challenges when we consider matching complex networks exhibiting the small world phenomena. In this work, we propose to use 'Ricci flow metric', to define the distance between two nodes in a network. This is then used to define similarity of a pair of nodes in two networks respectively, which is the crucial step of network alignment. Specifically, the Ricci curvature of an edge describes intuitively how well the local neighborhood is connected. The graph Ricci flow uniformizes discrete Ricci curvature and induces a Ricci flow metric that is insensitive to node/edge insertions and deletions. With the new metric, we can map a node in G_1 to a node in G_2 whose distance vector to only a few preselected landmarks is the most similar. The robustness of the graph metric makes it outperform other methods when tested on various complex graph models and real world network data sets (Emails, Internet, and protein interaction networks) (The source code of computing Ricci curvature and Ricci flow metric are available: https://github.com/saibalmars/GraphRicciCurvature).

1 Introduction

Given two graphs G_1 and G_2 with approximately the same graph topology, we want to find the correspondence of their nodes – node v in G_1 is mapped to a node $f(v)$ in G_2 such that whenever u, v is connected in G_1, $f(u), f(v)$ are likely to be connected in G_2 as well. This is called the *network alignment problem* and has been ·heavily studied [10] with numerous applications including database schema matching [17], protein interaction alignment [9,13,42,56], ontology matching [55], pattern recognition [45] and social networks [20,68].

Network alignment is a hard problem. A special case is the classical problem of *graph isomorphism*, in which we test whether or not two graphs have

© Springer Nature Switzerland AG 2018
T. Biedl and A. Kerren (Eds.): GD 2018, LNCS 11282, pp. 447–462, 2018.
https://doi.org/10.1007/978-3-030-04414-5_32

exactly the same topology under a proper correspondence. The most recent breakthrough by Babai [3] provides an algorithm with quasi-polynomial running time $O(\exp((\log n)^{O(1)}))$, where n is the number of nodes. It still remains open whether the problem is NP-complete or not. For special graphs, polynomial time algorithms have been developed (e.g. trees [2], planar graphs [25], graphs of bounded valence [40]). For general graphs what has been commonly used include heuristic algorithms such as spectral methods [1,51,56], random walks [60], optimal cost matching method [29] and software packages such as NAUTY, VF, and VF2, see [11,44]. The subgraph isomorphism problem, i.e., testing whether one graph is the subgraph of the other, is NP-complete and has been heavily studied as well. See the survey [10,14,66].

For practical settings, the problem is often formulated as finding a matching in a complete bipartite graph $H = (V_1, V_2, E)$, in which V_1 are nodes in G_1, V_2 are nodes in G_2, and edges E carry weights that indicate similarity of the two nodes. Here similarity can be either attribute similarity or structural similarity (quantifying their positions in the network). The two graphs are aligned by taking a matching of high similarity in graph H. Many algorithms use this approach and they differ by how to define node similarity: L-Graal [43], Natalie [12], NetAlignMP++ [6], NSD [32], and IsoRank [56]. In this paper, our main contribution is to provide a new method to compute *node structural similarity* using the idea of graph curvature and curvature flow.

Our Setting. We mainly focus on complex networks that appear in the real world. Consider the Internet backbone graphs captured at two different points in time, we wish to match the nodes of the two graphs in order to understand how the network has evolved over time. Or, consider two social network topologies on the same group of users. It is likely that when two users are connected by a social tie in one network (say LinkedIn), they are also connected in the other network (say Facebook). A good alignment of two networks can be useful for many applications such as feature prediction [16], link prediction [64], anomaly detection [48], and possibly de-anonymization [19,52].

We work with the assumption that a small number k (say 2 or 3) of nodes, called landmarks, are identified. Often a small number of landmarks could be discovered by using external information or properties of the networks. For example, in networks with power law degree distributions, there are often nodes with really high degree that can be identified easily. Given the landmarks, we can find a coordinate for each node u as $[d(u, \ell_1), d(u, \ell_2), \cdots, d(u, \ell_k)]$, where $d(u, \ell_i)$ captures the similarity of u with the ith landmark. This coordinate vector captures the structural position of a node in the network and can be used in the network alignment problem – two nodes $u_1 \in V_1$ and $u_2 \in V_2$ can be aligned if their respective coordinates are similar. This approach is motivated by localization in the Euclidean setting, in which three landmarks are used to define the barycentric coordinates of any node in the plane. In the complex network setting, this approach faces two major challenges that need to be addressed.

First, a proper metric that measures the distance from any node to the landmarks in the network is needed. The easiest metric is probably the hop

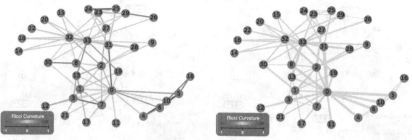

(a) Ricci Curvature: The Original   (b) Ricci Curvature: After Ricci Flow

Fig. 1. An example of Ricci curvature on the karate club graph before and after Ricci flow. The colors represent the Ricci curvature while the thickness represents the edge weight. Ricci flow deformed the edge weights until Ricci curvatures converged (-0.0027 in this case). (Color figure online)

count – the number of edges on the shortest path connecting two nodes in the network. While the hop count is often used for communication networks, for complex networks it is less helpful as these networks often have a small diameter. To look for a better measure, edges shall be properly weighed which is highly application dependent. For the Internet setting [36], measures such as Round-Trip Time (RTT) are used. In a social network setting, it is natural to weight each edge by its tie strength. But these weights are not easy to obtain. Many graph analysis methods use measures that capture local graph structures by common neighborhood, statistics generated by random walks [21,53], etc.

Second, the approach of using distances to landmarks to locate a node in the network is truly geometric in nature [18,30]. But a fundamental question is to decide what underlying metric space shall be used. Low-dimensional Euclidean spaces are often used [23,34] and spectral embedding or Tutte embedding are popular. But it is unclear what is the dimensionality of a general complex network and robustness of these embeddings is not fully understood.

Our Approach. We address the above problems by using the tool of graph curvature and curvature flow. In the continuous setting of a two dimensional surface, the curvature of a point captures how much it deviates from being flat at the point. The tip of a bump has positive curvature and a saddle point has negative curvature. Curvature flow is a process that deforms the surface (changing the metric) and eventually makes the curvature to be uniform everywhere. In our setting, we look at *discrete* curvatures and curvature flow defined on a *graph* and argue that by curvature one can encode and summarize graph structures. As will be explained later in more details and rigor, edges that are in a densely connected 'community' are positively curved while edges that connect two dense communities are negatively curved. Further, we can define curvature flow (which is an adaptation of surface Ricci flow to the graph setting) such that weights are given to the edges to make all edges have the same curvature. This process in some sense 'uniformizes' the network and can be imagined as

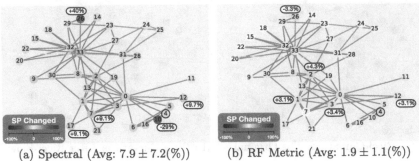

(a) Spectral (Avg: $7.9 \pm 7.2(\%)$)   (b) RF Metric (Avg: $1.9 \pm 1.1(\%)$)

Fig. 2. A comparison of stretch ratios under different metrics started from node 4 labeled in yellow. Here the stretch ratio defined by the ratio of shortest path changed when nodes or edges are removed from graph. For graph G and $G \setminus \{\overline{v_2 v_7}, \overline{v_{26} v_{29}}\}$, the stretch ratios of shortest paths from node 4 to all other nodes are labeled in color. Here the top five stretches paths are labeled by its ratios next to the nodes. With just two edges removed, the stretch ratios of spring and spectral metric changed drastically while RF metric remains stable. (Color figure online)

embedding the network in some intrinsic geometric space. These weights are again intuitive – edges in a dense community are short while edges that connect two far away communities are long, see Fig. 1. We call the shortest path length under such weights as the Ricci flow metric. The Ricci flow metric is designed to improve robustness under insertions and deletions of edges/nodes – when an edge is removed, the shortest path distance between the two endpoints in the original graph can possibly change a lot and Ricci flow reduces such imbalance. This is useful for matching two graphs that are different topologically.

Our curvature based coordinates depend only on the network topology and thus are purely structural. But they can also be combined with node attributes (when available) to produce an alignment score. In our evaluation we have focused on methods that use structural similarities and embeddings such as hop counts, metrics induced by spectral embedding [41] or spring/Tutte embedding [61], and topology based network alignment algorithm IsoRank [56] and NSD [32]. We evaluated our network alignment algorithms on a variety of real world data sets (Internet AS graphs [57], Email networks [22,33], protein networks [16,33]) and generated model networks (random regular graphs, Erdös-Rényi graphs [15], preferential attachment model [5] and Kleinberg's small world model [31]). Experimental evaluations show that the Ricci flow metric greatly outperforms other alternatives. In particular, most of the embedding methods perform poorly on random regular graphs due to identical node degree. Methods using hop counts suffer from the problem of not being descriptive, especially when there are only a small number of landmarks. There might be too many nodes with same coordinate under the landmark based coordinates. The spectral embedding and the spring embedding are less robust under edge insertion and deletion. IsoRank compares local similarity in a pair of graphs. In a large complex network, it is possible to have nodes that look similar in terms of local structures but globally should not be matched. This may explain why IsoRank does not scale well.

In the following, we first present background knowledge on curvature and Ricci flow [24,59]. We then present graph curvature and curvature flow, how they are used in network alignment, and empirical evaluations.

2   Background

Curvature and Ricci Flow on Surfaces. Curvature is a measure of the amount by which a geometric object deviates from being flat/straight and has multiple definitions depending on the context. Ricci flow was introduced by Richard Hamilton for Riemannian manifolds in 1982 [24]. A surface Ricci flow is the process to deform the Riemannian metric of the surface, proportional to the Gaussian curvature, such that the curvature evolves like heat diffusion and becomes uniform at the limit. Intuitively, this behaves as flattening a piece of crumpled paper. Surface Ricci flow is a key tool in the proof of the Poincaré conjecture on 3-manifolds, and has numerous applications in image and shape analysis. In engineering fields, surface Ricci flow has been broadly applied on a triangulated surface setting for tackling many important problems, such as parameterization in graphics [27] and deformable surface registration in vision [67].

Discrete Ricci Curvature. Curvatures for general graphs have only been studied over the past few years [4,7,8,37,39,50,58]. The definitions of curvatures that are easier to generalize to a discrete graph setting are sectional curvature and Ricci curvature. Consider a point x on a surface M and a tangent vector v at x whose endpoint is y. Take another tangent vector w_x at x and imagine transporting w_x along vector v to be a tangent vector w_y at y. Denote the endpoints of w_x, w_y as x', y'. If the surface is flat, then x, y, x', y' would constitute a parallelogram. Otherwise, the distance between x', y' differs from $|v|$. The difference can be used to define sectional curvature [50].

Sectional curvature depends on two tangent vectors v, w. Averaging sectional curvature over all directions w gives the Ricci curvature which only depends on v. Intuitively, if we think of a direction w at x as a point on a small ball S_x centered at x, on average Ricci curvature controls whether the distance between a point of S_x and the corresponding point of S_y is smaller or larger than the distance $d(x, y)$. To allow for a general study, Ollivier [50] defined a Ricci curvature by using a probability measure m_x to represent the ball S_x. Later Ollivier Ricci curvature has been applied in different fields, for distinguishing between cancer-related genes and normal genes [54], for understanding phylogenetic trees [65], and for detecting network features such as backbone and congestions [28,46, 47,62,63]. There was very limited amount of theory on discrete Ollivier-Ricci curvature on graphs, and nearly none on graph Ricci flow. In the future work section of [49] Ollivier suggested that people should study the discrete Ricci flow in a metric space (X, d) by evolving the distance $d(x, y)$ on X according to the Ricci curvature $\kappa(x, y)$ between two points $x, y \in X$:

$$\frac{d}{dt} d(x, y) = -\kappa(x, y) d(x, y).$$

We are the first to study Ollivier-Ricci flow in the graph setting. We also suggest variants of the Ollivier-Ricci curvature that admits much faster computation and empirically evaluate properties of the Ricci flow metrics and applications in network alignment.

3   Theory and Algorithms

Ricci Curvature on Graphs. For an undirected graph $G = (V, E)$, the Ollivier-Ricci curvature of an edge $\overline{xy}$ is defined as follows. Let π_x denote the neighborhood of a node $x \in V$ and $\mathrm{Deg}(x)$ is the degree of x. For a parameter $\alpha \in [0, 1]$, define a probability measure m_x^α:

$$m_x^\alpha(x_i) = \begin{cases} \alpha & \text{if } x_i = x \\ (1 - \alpha) / \mathrm{Deg}(x) & \text{if } x_i \in \pi_x \\ 0 & \text{otherwise,} \end{cases}$$

Suppose $w(x, y)$ is the weight of edge $\overline{xy}$ and $d(x, y)$ the shortest path length between x and y in the weighted graph. The *optimal transportation distance* (OTD) between m_x^α and m_y^α is defined as the best way of transporting the mass distribution m_x^α to the mass distribution m_y^α:

$$W(m_x^\alpha, m_y^\alpha) = \inf_M \sum_{x_i, y_j \in V} d(x_i, y_j) M(x_i, y_j)$$

where $M(x_i, y_j)$ is the amount of mass moved from x_i to y_j along the shortest path (of length $d(x_i, y_j)$) and we would like to take the best possible assignment (transport plan) that minimizes the total transport distance. The discrete Ollivier-Ricci curvature [50] is defined as follows

$$\kappa^w(x, y) = 1 - \frac{W(m_x^\alpha, m_y^\alpha)}{d(x, y)}.$$

In this paper, $\kappa^w(x, y)$ is called the OTD-Ricci curvature.

The OTD-Ricci curvature describes the connectivity in the local neighborhood of $\overline{xy}$ [47]. If $\overline{xy}$ is a bridge so the nodes in π_x have to travel through the edge $\overline{xy}$ to get to nodes in π_y, $W(m_x^\alpha, m_y^\alpha) > d(x, y)$ and the curvature of $\overline{xy}$ is negative. Similarly, if the neighbors of x and the neighbors of y are well connected (such that $W(m_x^\alpha, m_y^\alpha) < d(x, y)$), the curvature on $\overline{xy}$ is positive.

The optimal transportation distance can be solved by linear programming (LP) to find the best values for $M(x_i, y_j)$:

Minimize: $\sum_{i,j} d(x_i, y_j) M(x_i, y_j)$

s.t. $\sum_j M(x_i, y_j) = m_x^\alpha(x_i), \forall i$ and $\sum_i M(x_i, y_j) = m_y^\alpha(y_j), \forall j$

The computation of the linear program on large networks may be time-consuming. To address the computational challenges, we define a variant of the OTD-Ricci curvature by using a specific transportation plan instead of the optimal transportation plan. We take the **average transportation distance** (ATD) $A(m_x^\alpha, m_y^\alpha)$, in which we transfer an equal amount of mass from each neighbor x_i of x to each neighbor y_j of y and transfers the mass of x to y. One could easily verify that this is a valid transportation plan. Thus, the discrete Ricci curvature by the average transportation distance (ATD) is defined as:

$$\kappa^a(x, y) = 1 - \frac{A(m_x^\alpha, m_y^\alpha)}{d(x, y)},$$

Since we remove the LP step in the computation, the computational complexity of the ATD-Ricci curvature is drastically improved. As will be presented later, our experimental results show that computing discrete Ricci flow using the ATD-Ricci curvature maintains and even enhances the robustness of graph alignment. In this paper, we fix $\alpha = 0.5$ and simplify the notation of discrete Ricci curvature by $\kappa(x, y)$ in the discussion of discrete Ricci flow.

Discrete Ricci Flow. Ricci flow is a process that deforms the metric while the Ricci curvature evolves to be uniform everywhere. For any pair of adjacent nodes x and y on a graph $G = (V, E)$, we adjust the edge weight of $\overline{xy}$, $w(x, y)$, by the curvature $\kappa(x, y)$:

$$w_{i+1}(x, y) = w_i(x, y) - \epsilon \cdot \kappa_i(x, y) \cdot w_i(x, y), \quad \forall \overline{xy} \in E,$$

where $\kappa_i(x, y)$ is computed using the current edge weight $w_i(x, y)$. The step size is controlled by $\epsilon > 0$ and we take $\epsilon = 1$ in our experiment.

After each iteration we rescale the edge weights so the total edge weight in the graph remains the same, since only relative distances between nodes matter in a graph metric. A pseudo-code is presented in the appendix.

Ricci Flow Metric. When graph Ricci flow converges, each edge $\overline{xy}$ is given a weight $w(x, y)$. We denote the shortest path metric with such weights to be the *Ricci Flow Metric*, denoted as $d(x, y)$. To understand the metric, notice that when the Ricci flow converges the following is true. Here we take step size $\epsilon = 1$:

$$(w(x, y) - \kappa(x, y) \cdot w(x, y)) \cdot N \approx w(x, y)$$

where

$$N = \frac{|E|}{\sum_{\overline{xy} \in E} (w(x, y) - \kappa(x, y) w(x, y))}.$$

Denote the transportation distance between the two probability measures m_x^α and m_y^α to be $T(x, y)$, we have

$$\kappa(x, y) \approx 1 - \frac{T(x, y)}{d(x, y)}, \quad \frac{T(x, y)}{d(x, y)} \approx \frac{1}{N}.$$

To understand what this means, recall that $T(x, y)$ represents the distances from x's neighborhood to y's neighborhood. Before we run Ricci flow, this value for different edges can vary a lot – in the neighborhood of positively curved edges there are many 'shortcuts' making $T(x, y)$ to be significantly shorter than $d(x, y)$, while in the neighborhood of negatively curved edges $T(x, y)$ is longer than $d(x, y)$. The purpose of Ricci flow is to re-adjust the edge weights to reduce such imbalance. Suppose we remove an edge $\overline{xy}$ with negative curvature before Ricci flow and break the shortest path from x' (a neighbor of x) to y' (a neighbor of y). The alternative path from x' to y' in the neighborhood of $\overline{xy}$ tends to become much longer, as $T(x, y)/d(x, y)$ is large. Thus, the change to the shortest path metric is significant. However, after the Ricci flow, with the new edge weights, the alternative path from x' to y' in the neighborhood of $\overline{xy}$ may still get longer but not as long. Thus, the change in the shortest path length is less significant. The Ricci flow metric is more robust when edges and nodes are randomly removed. To capture this property, we define the uniformity of graph metric using the variation of $T(x, y)/d(x, y)$ for different edges $\overline{xy}$.

Definition 1 (Metric Uniformity). *Given a graph $G = (V, E)$ with edge weight $w(x, y)$ on $\overline{xy} \in E$, the metric uniformity is defined by the interquartile range (IQR), i.e., of $T(x, y)/d(x, y)$ over all edges, the difference between 75th and 25th percentiles.*

The metric uniformity measures the diversity of $T(x, y)/d(x, y)$ over all edges around the median. Lower metric uniformity indicates that $T(x, y)/d(x, y)$ are less dispersed about the median. Therefore, the corresponding metric is more robust upon node/edge insertions and deletions.

4 Network Alignment by Ricci Flow Metric

Given two graphs G_1 and G_2, suppose there are k landmarks $L = \{\ell_i | i = 1, 2, \cdots, k\}$ in both G_1 and G_2 with known correspondence, k is a small constant such as 3 or 4. These landmarks may be known beforehand by external knowledge. We use these landmarks to find the correspondence of nodes in G_1 and G_2. Specifically, we represent each node $v \in G_j$ by its relative positions to the landmarks $v_L = [d_j(v, \ell_1), d_j(v, \ell_2), \cdots, d_j(v, \ell_k)]$, where $d_j(v, \ell_i)$ denotes the shortest distance from v to ℓ_i in graph G_j using the Ricci flow metric. We define the cost of matching $u \in G_1$ with $v \in G_2$ by the 2-norm of the difference between u_L and v_L,

$$C_{uv} = ||u_L - v_L||_2.$$

The smaller C_{uv} is, the more similar u, v are. The alignment problem can be formulated as finding a low-cost matching in the complete bipartite graph $H = (V_1 \cup V_2, E)$ where the edge in E connecting $u \in V_1$ and $v \in V_2$ has weight C_{uv}.

Matching Algorithms. To find a low cost matching in the bipartite graph or a similarity matrix, we can apply the following two algorithms.

- The Hungarian min-cost matching algorithm [26] finds a matching which minimizes $\sum_{u \in G_1, v \in G_2} C_{uv}$ in $O(|V|^3)$ time.
- A greedy matching method iteratively locates the minimal C_{uv}, records the node pair (u, v), and removes all elements involving either u or v, until the nodes of either G_1 or G_2 are all paired.

We note that in prior network alignment algorithms, such as IsoRank and NSD, greedy matching was unanimously selected due to its efficiency. We tested both algorithms with the results on Hungarian algorithms presented in the appendix.

Matching Accuracy. To evaluate the accuracy of the matching results, one idea is to count how many nodes are correctly matched with respect to their IDs in the ground truth. But this measure may be too strict. For example, if G_1 and G_2 are both complete graphs of the same size, any matching of nodes in G_1 and nodes in G_2 should be considered to be correct. It is well known that social networks have community structures and many different levels of node equivalence and symmetry. There are different definitions for two nodes u, v in a graph G to be equivalent:

- *Structural equivalence*: u and v have exactly the same set of neighbors;
- *Automorphic equivalence*: if we relabel the nodes in an automorphic transformation (i.e., the graph after relabeling is isomorphic to the original graph),the labels of u, v are exchanged;
- *Regular equivalence* [38]: u, v are connected if they are equally related to equivalent others. That is, regular equivalence sets are composed of nodes who have similar relations to members of other regular equivalence sets.

Regular equivalence is the least restrictive of the three definitions of equivalence, but probably the most important for the sociologists as it captures the sociological concept of a "role". Previous quantitative measures of regular equivalence [35] unfortunately, are not very effective in identifying the global symmetric structures. For example, the two terminal nodes of a path graph cannot be classified to the same regular equivalence class.

Connected Equivalence. Motivated by regular equivalence, we would like to consider two nodes to be *connected equivalence* if they have similar connections to other nodes. Specifically, we compute the length of shortest paths (using the Ricci flow metric) from u and v to all other nodes, except u, v themselves, in a fixed order in G_1 or G_2, and denote the results by two vectors u_L and v_L. If the two vectors are similar (i.e., $\|u_L - v_L\|_2 < \epsilon$ for some small $\epsilon > 0$), then we say that u, v are equivalent and the matching of u to v is correct.

5   Evaluation

In this section, we demonstrate the performance of Ricci flow metric and its power on network alignment. Since Ricci flow metric only used topology feature, we compare RF-OTD, RF-ATD with two topology based network alignment algorithms IsoRank [56] and NSD [32] (other network alignment algorithms require node label/attributes), and three different embedding metrics: spectral embedding, spring embedding, and hop count. We evaluated the performance of these algorithms in both model graphs and real world data. Our main observations are as follows:

1. Ricci flow metrics (both RF-OTD and RF-ATD) are much more robust against edge/node insertion and deletions compared to others on noisy graph alignment problem.
2. Ricci flow metrics greatly outperform previous topological based network alignment algorithms (IsoRank, NSD) as well as other metrics and similarity measures on noisy graph alignment problem. In many experiments our algorithm achieves more than 90% matching accuracy while other methods achieve accuracy below 30%.
3. While greedy matching performs well in model graphs, min-cost matching (i.e., Hungarian algorithm) is more suitable for real world graph.

Experiment Setup. Given a graph G_1, we remove n nodes or edges (less than 1%) uniform randomly to create G_2 as a noisy graph. We then perform the alignment of G_1 and G_2. Notice that in this case, we can also regard G_2 as a graph with random noises added to G_1. Therefore, we only present the results of node deletion cases.

The graph alignment problem is solved in two steps. First we construct a similarity matrix that records the cost of matching node i in G_1 with node j in G_2 for all possible i, j; then we perform Hungarian algorithm or greedy matching to match nodes in G_1 with nodes in G_2. We evaluate the algorithm performance by using connected equivalence.

For IsoRank and NSD, the similarity matrix is the output of the algorithm; for landmark based method, the similarity matrix is defined by ℓ_2 norm of their landmark distance vectors. Here the distance metric is defined by RF-OTD and RF-ATD[1], distances induced by spectral embedding (dim $= 2$), spring embedding, and hop count respectively. The performance of each metric is evaluated by varying the number of removed nodes/edges n and the number of landmarks k. To eliminate the dependency of landmark selection, the matching accuracy is averaged by 10 experiments on different sets of landmarks. These landmarks are chosen such that every newly added landmark is furthest away from the landmarks chosen so far.

Metric Uniformity. We first analyze the metric uniformity over all metrics on a random regular graph with 1000 nodes and 6000 edges. The box plot result

---

[1] With 50 Ricci flow iterations, $\epsilon = 1$, $\alpha = 0.5$.

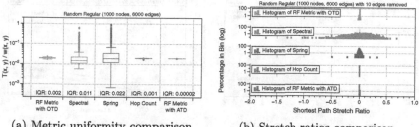

(a) Metric uniformity comparison   (b) Stretch ratios comparison

Fig. 3. (a) Shows metric uniformity comparison of all metrics. The box plot is for $T(x,y)/w(x,y)$ with respect to each metric. We define the IQR (InterQuartile Range) of the box plot to be the metric uniformity. The smaller the metric uniformity is, the more robust the metric is under random edge removals. (b) Demonstrates a comparison of the stretch ratios of shortest path length over all methods.

is illustrated in Fig. 3(a). RF-ATD yields the best metric uniformity with the smallest IQR of $2e-5$. RF-OTD also performs well with IQR as 0.002. Spectral and spring embedding behave poorly. These performances are directly related to the accuracy in graph alignment. The metric with low metric uniformity is more stable when nodes/edges are missing, see Fig. 2. Notice that hop count metric generates a small IQR $= 0.001$. This is because there are often multiple shortest paths (in terms of hop count) connecting two nodes in the network so hop count is actually a fairly robust measure. But using it for graph alignment is still limited by its lack of descriptive power.

For further analysis that metric uniformity indeed captures the robustness of shortest path metric, we compute the stretch of the shortest path length between a pair of nodes in a random regular graph when 10 edges are randomly removed, shown in Fig. 3(b). Consider two nodes u, v in both G_1 and G_2, we denote the length of the shortest path from u to v by $d_{G_i}^m(u,v)$ under a metric m, where $i = 1, 2$. We define the stretch ratio of G_1 and G_2 as $s(u,v) = (d_{G_1}^m(u,v) - $

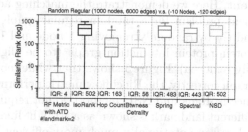

Fig. 4. The comparison of the similarity rank of all methods. The similarity matrix represents the pairwise similarity between a random regular graph with 1000 nodes and 6000 edges and the previous graph with 10 random nodes and 120 corresponding edges removed.

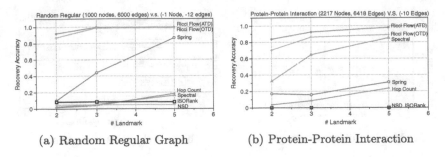

(a) Random Regular Graph   (b) Protein-Protein Interaction

Fig. 5. A comparison of noisy graph alignment results by Hungarian algorithm and connected equivalence on a random regular graph (a) and protein-protein interaction graph (b).

$d^m_{G_2}(u,v))/d^m_{G_1}(u,v)$. The stretch ratio captures the changes of the shortest path length. A larger stretch ratio means the shortest path length changes more. In Fig. 3(b), we collect the stretch ratios from one random node to all of the other nodes based on different metrics, and plot the distribution of these stretch ratios as a histogram. It turns out that hop count, Ricci flow metric with OTD and with ATD result in smaller stretch ratios, while spectral embedding is the most vulnerable one with edge deletions.

Similarity Matrix. Here we test the similarity matrix of every pairwise node similarity of two graphs G_1, G_2. G_1 is a random regular graph with 1000 nodes and 6000 edges and G_2 removes 10 nodes randomly from G_1. We check the performance of the similarity matrix as follows. For every node $n_1 \in G_1$ and the corresponding node n_2 in G_2 in the ground truth, we check the rank of n_2 in the sorted list ranked by similarity values with n_1. A good similarity matrix should rank n_2 as the most similar one. A lower ranking indicates better performance of the similarity matrix. We show the results in Fig. 4 with 2 landmarks. Thanks to the metric uniformity, our method yields the best performance with average similarity rank of 2 while other methods are at least 10 times higher.

Network Alignment. Here we demonstrate the matching accuracy results on noisy graph alignment problem for a random regular graph (1000 nodes and degree 12 as G_1, and G_2 by randomly removing 1 node and 12 edges from G_1) and a real word protein-protein interaction graph (to be aligned with a graph of 10 edges randomly removed). Figure 5 shows that while most of the methods failed to align the graph correctly, the Ricci flow metric performs well with only 2 landmarks. Here since IsoRank and NSD method do not require landmarks, the performance over different landmark is shown as a straight line. The result shows that RF-ATD, which is computationally much more efficient, performs equally well as RF-OTD. Notice that the methods with poor metric uniformity as shown in Fig. 3 also result in poor performance here. This supports the importance of metric uniformity in network alignment. More evaluation on different model networks and real networks also support this claim. The performance of alignment accuracy also effected by the portion of nodes and edges added/removed, the noisy graph alignment problem become harder with more noise.

6 Conclusion

In this paper, we have presented a framework to endow a graph with a novel metric through the notion of discrete Ricci flow with an application to network alignment. From the experimental results, we found that (1) the graph Ricci curvature converges through the discrete Ricci flow and; (2) Ricci flow metric on a graph is fairly stable when edges are inserted or removed. Providing theoretic proofs of these observations are currently open and will be the next direction of our future work.

Acknowledgement. The authors would like to thanks the funding agencies NSF DMS-1737812, CNS-1618391, CCF-1535900, DMS-1418255, and AFOSR FA9550-14-1-0193.

References

1. Aflalo, Y., Bronstein, A., Kimmel, R.: On convex relaxation of graph isomorphism. Proc. Nat. Acad. Sci. **112**(10), 2942–2947 (2015)
2. Aho, A.V., Hopcroft, J.E.: The Design and Analysis of Computer Algorithms, 1st edn. Addison-Wesley Longman Publishing Co. Inc., Boston (1974)
3. Babai, L.: Graph isomorphism in quasipolynomial time. CoRR abs/1512.03547 (2015). http://arxiv.org/abs/1512.03547
4. Bakry, D., Émery, M.: Diffusions hypercontractives. In: Azéma, J., Yor, M. (eds.) Séminaire de Probabilités XIX 1983/84. LNM, vol. 1123, pp. 177–206. Springer, Heidelberg (1985). https://doi.org/10.1007/BFb0075847
5. Barabasi, A., Albert, R.: Emergence of scaling in random networks. Science **286**, 509–512 (1999)
6. Bayati, M., Gleich, D.F., Saberi, A., Wang, Y.: Message-Passing algorithms for sparse network alignment. ACM Trans. Knowl. Discov. Data **7**(1), 3:1–3:31 (2013)
7. Bonciocat, A.I.: A rough curvature-dimension condition for metric measure spaces. Central Eur. J. Math. **12**(2), 362–380 (2014)
8. Chung, F.R.K., Yau, S.T.: Logarithmic Harnack inequalities. Math. Res. Lett **3**(6), 793–812 (1996)
9. Clark, C., Kalita, J.: A comparison of algorithms for the pairwise alignment of biological networks. Bioinformatics **30**(16), 2351–2359 (2014)
10. Conte, D., Foggia, P., Sansone, C.: Thirty years of graph matching in pattern recognition. Int. J. Pattern Recogn. Artif. Intell. **18**(03), 265–298 (2004)
11. Cordella, L.P., Foggia, P., Sansone, C., Vento, M.: A (sub)graph isomorphism algorithm for matching large graphs. IEEE Trans. Pattern Anal. Mach. Intell. **26**(10), 1367–1372 (2004)
12. El-Kebir, M., Heringa, J., Klau, G.W.: Natalie 2.0: sparse global network alignment as a special case of quadratic assignment. Algorithms **8**(4), 1035–1051 (2015)
13. Elmsallati, A., Clark, C., Kalita, J.: Global alignment of protein-protein interaction networks: a survey. IEEE/ACM Trans. Comput. Biol. Bioinform. **13**(4), 689–705 (2016)

14. Emmert-Streib, F., Dehmer, M., Shi, Y.: Fifty years of graph matching, network alignment and network comparison. Inf. Sci. **346–347**, 180–197 (2016)
15. Erdos, P., Renyi, A.: On random graphs. Publicationes Math. **6**, 290–297 (1959)
16. Ewing, R.M., et al.: Large-scale mapping of human protein-protein interactions by mass spectrometry. Mol. Syst. Biol. **3**, 89 (2007)
17. Fan, W., Wang, X., Wu, Y.: Incremental graph pattern matching. ACM Trans. Database Syst. (TODS) **38**(3), 18 (2013)
18. Fang, Q., Gao, J., Guibas, L., de Silva, V., Zhang, L.: GLIDER: gradient landmark-based distributed routing for sensor networks. In: Proceedings of the 24th Conference of the IEEE Communication Society (INFOCOM), vol. 1, pp. 339–350, March 2005
19. Fu, H., Zhang, A., Xie, X.: Effective social graph deanonymization based on graph structure and descriptive information. ACM Trans. Intell. Syst. Technol. (TIST) **6**(4), 49 (2015)
20. Goga, O., Loiseau, P., Sommer, R., Teixeira, R., Gummadi, K.P.: On the reliability of profile matching across large online social networks. In: KDD 2015, pp. 1799–1808. ACM, New York (2015)
21. Grover, A., Leskovec, J.: node2vec. In: Proceedings of the 22nd ACM SIGKDD International Conference on Knowledge Discovery and Data Mining - KDD 2016 (2016)
22. Guimerà, R., Danon, L., Díaz-Guilera, A., Giralt, F., Arenas, A.: Self-similar community structure in a network of human interactions. Phys. Rev. E **68**(6), 065103 (2003)
23. Ham, J., Lee, D.D., Saul, L.K.: Semisupervised alignment of manifolds. In: AISTATS, pp. 120–127 (2005)
24. Hamilton, R.S.: Three manifolds with positive Ricci curvature. J. Differ. Geom. **17**, 255–306 (1982)
25. Hopcroft, J.E., Wong, J.K.: Linear time algorithm for isomorphism of planar graphs (preliminary report). In: STOC 1974, pp. 172–184. ACM, New York (1974). https://doi.org/10.1145/800119.803896
26. Hopcroft, J.E., Karp, R.M.: An $n^{5/2}$ algorithm for maximum matchings in bipartite graphs. SIAM J. Comput. **2**(4), 225–231 (1973)
27. Jin, M., Kim, J., Luo, F., Gu, X.: Discrete surface Ricci flow. IEEE TVCG **14**(5), 1030–1043 (2008)
28. Kennedy, W.S., Narayan, O., Saniee, I.: On the hyperbolicity of large-scale networks. arXiv preprint arXiv:1307.0031 (2013)
29. Khan, A., Wu, Y., Aggarwal, C.C., Yan, X.: NeMa: fast graph search with label similarity. Proc. VLDB Endowment **6**(3), 181–192 (2013)
30. Kleinberg, J., Slivkins, A., Wexler, T.: Triangulation and embedding using small sets of beacons. In: Proceedings of the 45th IEEE Symposium on Foundations of Computer Science, pp. 444–453 (2004)
31. Kleinberg, J.M.: The small-world phenomenon - an algorithmic perspective. In: STOC (2000)
32. Kollias, G., Mohammadi, S., Grama, A.: Network similarity decomposition (NSD): a fast and scalable approach to network alignment. IEEE Trans. Knowl. Data Eng. **24**(12), 2232–2243 (2012)
33. Kunegis, J.: KONECT. In: Proceedings of the 22nd International Conference on World Wide Web - WWW 2013, Companion, pp. 1343–1350. ACM Press, New York (2013)
34. Lafon, S., Keller, Y., Coifman, R.R.: Data fusion and multicue data matching by diffusion maps. IEEE Trans. Pattern Anal. Mach. Intell. **28**(11), 1784–1797 (2006)

35. Leicht, E.A., Holme, P., Newman, M.E.: Vertex similarity in networks. Phys. Rev. E **73**(2), 026120 (2006)
36. Lim, H., Hou, J.C., Choi, C.H.: Constructing internet coordinate system based on delay measurement. IEEE/ACM Trans. Netw. **13**(3), 513–525 (2005)
37. Lin, Y., Lu, L., Yau, S.T.: Ricci curvature of graphs. Tohoku Math. J. **63**(4), 605–627 (2011)
38. Lorrain, F., White, H.C.: Structural equivalence of individuals in social networks. J. Math. Sociol. **1**(1), 49–80 (1971)
39. Lott, J., Villani, C.: Ricci curvature for metric-measure spaces via optimal transport. Ann. Math. **169**(3), 903–991 (2009)
40. Luks, E.M.: Isomorphism of graphs of bounded valence can be tested in polynomial time. J. Comput. Syst. Sci. **25**(1), 42–65 (1982)
41. Luxburg, U.: A tutorial on spectral clustering. Stat. Comput. **17**(4), 395–416 (2007). https://doi.org/10.1007/s11222-007-9033-z
42. Malod-Dognin, N., Ban, K., Pržulj, N.: Unified alignment of protein-protein interaction networks. Sci. Rep. **7**(1), 953 (2017)
43. Malod-Dognin, N., Pržulj, N.: L-GRAAL: lagrangian graphlet-based network aligner. Bioinformatics **31**(13), 2182–2189 (2015)
44. McKay, B.D., Piperno, A.: Practical graph isomorphism, ii. J. Symbolic Comput. **60**, 94–112 (2014)
45. Melnik, S., Garcia-Molina, H., Rahm, E.: Similarity flooding - a versatile graph matching algorithm and its application to schema matching. In: ICDE (2002)
46. Narayan, O., Saniee, I.: Large-scale curvature of networks. Phys. Rev. E **84**(6), 066108 (2011)
47. Ni, C.C., Lin, Y.Y., Gao, J., Gu, X., Saucan, E.: Ricci curvature of the Internet topology. In: 2015 IEEE Conference on Computer Communications (INFOCOM), pp. 2758–2766. IEEE (2015)
48. Noble, C.C., Cook, D.J.: Graph-based anomaly detection. In: Proceedings of the ninth ACM SIGKDD International Conference on Knowledge Discovery and Data Mining - KDD 2003, pp. 631–636. University of Texas at Arlington, Arlington (2003)
49. Ollivier, Y.: A survey of Ricci curvature for metric spaces and Markov chains. Probab. Approach Geom. **57**, 343–381 (2010)
50. Ollivier, Y.: Ricci curvature of Markov chains on metric spaces. J. Funct. Anal. **256**(3), 810–864 (2009)
51. Patro, R., Kingsford, C.: Global network alignment using multiscale spectral signatures. Bioinformatics **28**(23), 3105–3114 (2012)
52. Peng, W., Li, F., Zou, X., Wu, J.: A two-stage deanonymization attack against anonymized social networks. IEEE Trans. Comput. **63**(2), 290–303 (2014)
53. Perozzi, B., Al-Rfou, R., Skiena, S.: Deepwalk: online learning of social representations. In: Proceedings of the 20th ACM SIGKDD International Conference on Knowledge Discovery and Data Mining - KDD 2014, New York, pp. 701–710 (2014)
54. Sandhu, R., et al.: Graph curvature for differentiating cancer networks. Sci. Rep. **5**, 12323 (2015)
55. Shvaiko, P., Euzenat, J.: Ontology matching: state of the art and future challenges. IEEE Trans. Knowl. Data Eng. **25**(1), 158–176 (2013)
56. Singh, R., Xu, J., Berger, B.: Global alignment of multiple protein interaction networks with application to functional orthology detection. Proc. Nat. Acad. Sci. **105**(35), 12763–12768 (2008)
57. Spring, N., Mahajan, R., Wetherall, D.: Measuring ISP topologies with rocketfuel. SIGCOMM Comput. Commun. Rev. **32**(4), 133–145 (2002)

58. Sturm, K.T.: On the geometry of metric measure spaces. Acta Mathematica **196**(1), 65–131 (2006)
59. Tao, T.: Ricci flow. Technical report, Department of Mathematics, UCLA (2008)
60. Tong, H., Faloutsos, C., Gallagher, B., Eliassi-Rad, T.: Fast best-effort pattern matching in large attributed graphs. In: Proceedings of the 13th ACM SIGKDD International Conference on Knowledge Discovery and Data Mining - KDD 2007, pp. 737–746. Carnegie Mellon University, Pittsburgh (2007)
61. Tutte, W.T.: How to draw a graph. Proc. London Math. Soc. **13**, 743–767 (1963)
62. Wang, C., Jonckheere, E., Banirazi, R.: Wireless network capacity versus Ollivier-Ricci curvature under Heat-Diffusion (HD) protocol. In: Proceedings of the American Control Conference, pp. 3536–3541. University of Southern California, Los Angeles, IEEE, January 2014
63. Wang, C., Jonckheere, E., Banirazi, R.: Interference constrained network control based on curvature. In: Proceedings of the American Control Conference, pp. 6036–6041. University of Southern California, Los Angeles, IEEE, July 2016
64. Wang, D., Pedreschi, D., Song, C., Giannotti, F., Barabasi, A.L.: Human mobility, social ties, and link prediction. ACM, New York, August 2011
65. Whidden, C., Matsen IV, F.A.: Ricci-Ollivier curvature of the rooted phylogenetic subtree-prune-regraft graph. In: ANALCO, pp. 106–120 (2016)
66. Yan, J., Yin, X.C., Lin, W., Deng, C., Zha, H., Yang, X.: A short survey of recent advances in graph matching. In: ICMR 2016, pp. 167–174. ACM, New York (2016). https://doi.org/10.1145/2911996.2912035
67. Zeng, W., Samaras, D., Gu, X.D.: Ricci flow for 3D shape analysis. IEEE Trans. Pattern Anal. Mach. Intell. (IEEE TPAMI) **32**(4), 662–677 (2010)
68. Zhang, J., Yu, P.S.: Multiple anonymized social networks alignment. In: 2015 IEEE International Conference on Data Mining, pp. 599–608, November 2015

Same Stats, Different Graphs
(Graph Statistics and Why We Need Graph Drawings)

Hang Chen[1], Utkarsh Soni[2], Yafeng Lu[2], Ross Maciejewski[2],
and Stephen Kobourov[1(✉)]

[1] University of Arizona, Tucson, AZ, USA
kobourov@cs.arizona.edu
[2] Arizona State University, Tempe, AZ, USA

Abstract. Data analysts commonly utilize statistics to summarize large datasets. While it is often sufficient to explore only the summary statistics of a dataset (e.g., min/mean/max), Anscombe's Quartet demonstrates how such statistics can be misleading. We consider a similar problem in the context of graph mining. To study the relationships between different graph properties and statistics, we examine all low-order (≤ 10) non-isomorphic graphs and provide a simple visual analytics system to explore correlations across multiple graph properties. However, for graphs with more than ten nodes, generating the entire space of graphs becomes quickly intractable. We use different random graph generation methods to further look into the distribution of graph statistics for higher order graphs and investigate the impact of various sampling methodologies. We also describe a method for generating many graphs that are identical over a number of graph properties and statistics yet are clearly different and identifiably distinct.

Keywords: Graph mining · Graph properties · Graph generators

1 Introduction

Statistics are often used to summarize a large dataset. In a way, one hopes to find the "most important" statistics that capture one's data. For example, when comparing two countries, we often specify the population size, GDP, employment rate, etc. The idea is that if two countries have a "similar" statistical profile, they are similar (e.g., France and Germany have a more similar demographic profile than France and USA). However, Anscombe's quartet [3] convincingly illustrates that datasets with the same

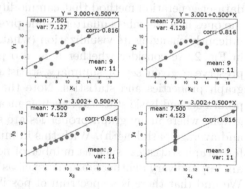

Fig. 1. Anscombe's quartet: all four datasets have the same mean and st. deviation in x and y and (x, y)-correlation.

T. Biedl and A. Kerren (Eds.): GD 2018, LNCS 11282, pp. 463–477, 2018.
https://doi.org/10.1007/978-3-030-04414-5_33

values over a limited number of statistical properties can be fundamentally different – a great argument for the need to visualize the underlying data; see Fig. 1.

Similarly, in the graph analytics community, a variety of statistics are being used to summarize graphs, such as graph density, average path length, global clustering coefficient, etc. However, summarizing a graph with a fixed set of graph statistics leads to the problem illustrated by Anscombe. It is easy to construct several graphs that have the same basic statistics (e.g., number of vertices, number of edges, number of triangles, girth, clustering coefficient) while the underlying graphs are clearly different and identifiably distinct; see Fig. 2. From a graph theoretical point of view, these graphs are very different: they differ in connectivity, planarity, symmetry, and other structural properties.

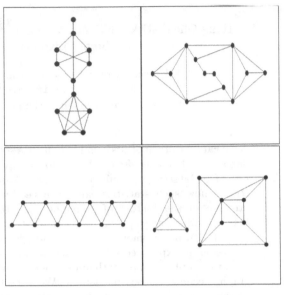

Fig. 2. These four graphs share the same 5 common statistics: $|V| = 12$, $|E| = 21$, number of triangles $|\triangle| = 10$, girth = 3 and global clustering coefficient $GCC = 0.5$. However, structurally the graphs are very different: some are planar others are not, some show regular patterns and are symmetric others are not, and finally, one of the graphs is disconnected, another is 1-connected and the rest are 2-connected.

Recently, Matejka and Fitzmaurice [31] proposed a dataset generation method that can modify a given 2-dimensional point set (like the ones in Anscombe's quartet) while preserving its summary statistics but significantly changing its visualization (what they call "graph"). Given the graphs in Fig. 2, we consider whether it is also possible to modify a given graph and preserve a given set of summary statistics while significantly changing other graph properties and statistics. Note that the problem is much easier for 2D point sets and basic statistics, such as mean, deviation and correlation, than for graphs where many graph properties are structurally correlated (e.g., diameter and average path length). With this in mind, we first consider how can we fix a few graph statistics (such as number of nodes, number of edges, number of triangles) and vary another statistic (such as clustering coefficient or connectivity). We find that there is a spectrum of possibilities. Sometimes the "unrestricted" statistic can vary dramatically, sometimes not, and the outcome depends on two issues: (1) the inherent correlation between some statistics (e.g., density and number of triangles), and (2) the bias in graph generators.

We begin by studying the correlation between graph summary statistics across the set of all non-isomorphic graphs with up to 10 vertices. The statistical properties derived for all graphs for a fixed number of vertices provide further information about certain "restrictions." In other words, the range of one statistic may be restricted if another statistical property is fixed. However, we cannot explore the entire space of graph statistics and correlations. As the number of vertices grows, the number of different non-isomorphic graphs grows super-exponentially. For $|V| = 1, 2 \ldots 9$ the numbers are $1, 2, 4, 11, 34, 156, 1044, 12346, 274668$, but already for $|V| = 16$ we have 6×10^{22} non-isomoprhic graphs.

To go beyond ten vertices we use graph generators based on models, such as Erdös-Rényi and Watts-Strogatz. However, different graph generators have different biases and these can significantly impact the results. We study the extent to which sampling using random generators can represent the whole graph set for an arbitrary number of vertices with respect to their coverage of the graph statistics. One way to evaluate the performance of random generators is based on the ground-truth graph sets that are available: all non-isomorphic graphs for $|V| \leq 10$ vertices. If we randomly generate a small set of graphs (also for $|V| \leq 10$ vertices) using a given graph generator, we can explore how well the sample and generator cover the space of graph statistics. In this way, we can begin exploring the issues of "same stats, different graphs" for larger graphs.

Data and tools are available at http://vader.lab.asu.edu/sameStatDiff Graph/. Specifically, we have a basic visual analytics system and basic exploration tools for the space of all low-order (≤ 10) non-isomorphic graphs and sampled higher order graphs. We also include a generator for "same stats, different graphs," i.e., multiple graphs that are identical over a number of graph statistics, yet are clearly different.

2 Related Work

We briefly review the graph mining literature, paying special attention to commonly collected graph statistics. We also consider different graph generators.

Graph Statistics: Graph mining is applied in different domains from bioinformatics and chemistry, to software engineering and social science. Essential to graph mining is the efficient calculation of various graph properties and statistics that can provide useful insight about the structural properties of a graph. A review of recent graph mining systems identified some of the most frequently extracted statistics. We list those, along with their definitions, in Table 1. These properties range from basic, e.g., vertex count and edge count, to complex, e.g., clustering coefficients and average path length. Many of them can be used to derive further properties and statistics. For example, graph density can be determined directly as the ratio of the number of edges $|E|$ to the maximum number of edges possible $|V| \times (|V| - 1)/2$, and real-world networks are often found to have a low graph density [33]. Node connectivity and edge connectivity measures may

Table 1. The set of graph statistics considered in this paper.

Name	Formula	Reference
Average clustering coefficient	$ACC(G) = \frac{1}{n}\sum_{i=1}^{n} c(u_i), u_i \in V, n = |V|$	[10, 11, 25, 27, 34]
	$c(v) = \frac{|\{(u,w)|u,w\in \Gamma(v),(u,w)\in E\}|}{|\Gamma(v)|(|\Gamma(v)|-1)/2}, v, u, w \in V$	
Global clustering coefficient	$GCC(G) = \frac{3\times|triangles|}{|connected\ triples|\ in\ the\ graph}$	[10, 25]
Square clustering	$SCC(G) = \frac{\sum_{u=1}^{k_v}\sum_{w=u+1}^{k_v} q_v(u,w)}{\sum_{u=1}^{k_v}\sum_{w=u+1}^{k_v}[a_v(u,w)+q_v(u,w)]}$	[28]
Average path length	$APL = ave\{\frac{n-1}{\sum_{v\in V} d(u,v), u\neq v}\}$	[10, 11, 27, 34]
Degree assortativity	$r = \frac{\sum_{xy} xy(e_{xy}-a_x b_y)}{\sigma_a \sigma_b}$	[34, 36]
Diameter	$diam(G) = max\{dist(v,w), v, w \in V\}$	[11, 25, 32, 34]
Density	$den = \frac{2|E|}{|V|(|V|-1)}$	
Ratio of triangles	$Rt = \frac{|triangles|}{|V|(|V|-1)/2}$	
Node connectivity	Cv: the minimum number of nodes to remove to disconnect the graph	[17]
Edge connectivity	Ce: the minimum number of edges to remove to disconnect the graph	[17]

be used to describe the resilience of a network [9,29], and graph diameter [24] captures the maximum among all pairs of shortest paths [2,8].

Other graph statistics measure how tightly nodes are grouped in a graph. For example, clustering coefficients have been used to describe many real-world networks, and can be measured locally and globally. Nodes in a highly connected clique tend to have a high local clustering coefficient, and a graph with clear clustering patterns will have a high global clustering coefficient [18,19,26,37]. Studies have shown that the global clustering coefficient has been found to be nearly always larger in real-world graphs than in Erdös-Rényi graphs with the same number of vertices and edges [10,37,42], and a small-world network should have a relatively large average clustering coefficient [13,15,44]. The average path length (APL) is also of interest; small-world networks have APL that is logarithmic in the number of vertices, while real-world networks have small (often constant) APL [13,15,37,42–44].

Degree distribution is one frequently used property describing the graph degree statistics. Many real-world networks, including communication, citation, biological and social networks, have been found to follow a power-law shaped degree distribution [6,10,37]. Other real world networks have been found to follow an exponential degree distribution [22,40,45]. Degree assortativity is of particular interest in the study of social networks and is calculated based on the Pearson correlation between the vertex degrees of connected pairs [35]. A random graph generated by Erdös-Rényi model has an expected assortative coefficient of 0. Newman [35] extensively studied assortativity in real-world networks and found that social networks are often assortative (positive assortativity), i.e., vertices with a similar degree preferentially connect together, whereas techno-logical and biological networks tend to be disassortative (negative assortativity) implying that vertices with a smaller degree tend to connect to high degree ver-tices. Assortativity has been shown to affect clustering [30], resilience [35], and epidemic-spread [7] in networks.

Graph Generators: Basic graph statistics have been used to describe various classes of graphs (e.g., geometric, small-world, scale-free) and a variety of algo-rithms have been developed to automatically generate graphs that mimic these various properties. Charkabati et al. [11] divide graph models and generators into four broad categories:

1. Random Graph Models: The graphs are generated by a random process.
2. Preferential Attachment Models: In these models, the "rich get richer," as the network grows, leading to power law effects.
3. Optimization-Based Models: Here, power laws are shown to evolve when risks are minimized using limited resources.
4. Geographical Models: These models consider the effects of geography (i.e., the positions of the nodes) on the topology of the network. This is relevant for modeling router or power grid networks.

The Erdös-Rényi (ER) network model is a simple graph generation model [10] that creates graphs either by choosing a network randomly with equal probability from a set of all possible networks of size $|V|$ with $|E|$ edges [20] or by creating each possible edge of a network with $|V|$ vertices with a given probability p [16]. The latter process gives a binomial degree distribution that can be approximated with a Poisson distribution. Note that fixing the number of nodes and using $p = 1/2$ results in a good sampling of the space of isomorphic graphs. However, this model (and others discussed below) does not sample well the space of non-isomorphic graphs, which are the subject of our study.

Watts and Strogatz [44] addressed the low clustering coefficient limitation of the ER model in their model (WS) which can be used to generate small-world graphs. The WS model can generate disconnected graphs, but the variation suggested by Newman and Watts [38] ensures connectivity. Models have also been proposed for generating synthetic scale-free networks with a varying scal-ing exponent(γ). The first scale-free directed network model was given by de Solla Price [39]. Barabási and Albert (BA) [5] described another popular net-work model for generating undirected networks. It is a network growth model in which each added vertex has a fixed number of edges $|E|$, and the probability of each edge connecting to an existing vertex v is proportional to the degree of v. Dorogovtsev et al. [14] and Albert and Barabási [1] also developed a variation of the BA model with a tunable scaling exponent.

Bach et al. [4] introduce an interactive system to create random graphs that match user-specified statistics based on a genetic algorithm. The statistics con-sidered are $|V|, |E|$, average vertex degree, number of components, diameter, ACC, density, and the number of clusters (as defined by Newman and Gir-van [21]). The goal is to generate graphs that get as close as possible to a set of target statistics; however, there are no guarantees that the target values can be obtained. Somewhat differently, we are interested in creating graphs that match several target statistics exactly, but differ drastically in other parameters.

3   Preliminary Experiments and Findings

In a recent study of the ability to perceive different graph properties such as edge density and clustering coefficient in different types of graph layouts (e.g., force-directed, circular), we generated a large number of graphs with 100 vertices. Specifically, we generated graphs that vary in a controlled way in edge density and graphs that vary in a controlled way in the average clustering coefficient [41]. A post-hoc analysis of this data (http://vader.lab.asu.edu/GraphAnalytics/), reveals some interesting patterns among the statistics described in Table 1.

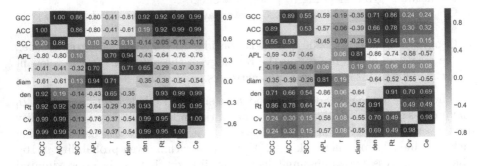

Fig. 3. Graph property correlation matrix plots for the edge density dataset (left) and the ground truth set of all non-isomorphic graphs on $|V| = 9$ vertices (right). (Color figure online)

The edge density dataset has 4,950 graphs and We compute all ten statistics from Table 1 and compute Pearson correlation coefficients; see Fig. 3. We observed high positive (blue) correlations and negative (yellow) correlations for many property pairs. For example, the average clustering coefficient is highly correlated with the global clustering coefficient, the number of triangles, and graph connectivity.

Note, however, these graphs were created for a very specific purpose and cover only limited space of all graphs with $|V| = 100$. The type of generators we used, and the way we used them (some statistical properties were controlled), could bias the results and influence the correlations. The fact that these correlations exist when some properties are fixed indicates that we can keep certain graph statistics fixed while manipulating others. This motivated us to conduct the following experiments:

1. Generate all non-isomorphic lower order graphs ($|V| \leq 10$) and analyze the relationships between statistical properties. We consider this type of data as ground truth due to its completeness.

2. Use different graph generators and compare how well they represent the space of non-isomorphic graphs and how well they cover the range of possible values in the ground truth data.

An analysis of the set of 274,668 non-isomorphic graphs on $|V| = 9$ vertices shows that the correlations are quite different than those in graphs from our edge density experiment; see Fig. 3.

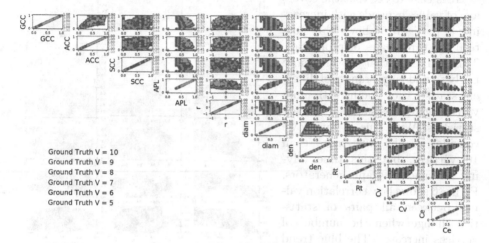

Fig. 4. Correlations between graph statistics in the ground truth for $|V| = 5, 6, 7, 8, 9$. Note that for $|V| = 9$ there are already 274,668 points. Points are plotted to overlap, with the largest sets plotted first (i.e., $|V| = 9, \ldots |V| = 5$) to enable us to identify the range of statistics that can be covered with a given number of vertices.

4   Analysis of Graph Statistics for Low-Order Graphs

We start the experiment by looking at pairwise relationship of graph statistics of low-order graphs, where all non-isomorphic graphs can be enumerated. If two statistics, say s_1 and s_2, are highly correlated, then fixing s_1 is likely to restrict the range of possible values for s_2. On the other hand, if s_1 and s_2 are independent, fixing s_1 might not impact the range of values for s_2, yielding same stats (s_1) different graphs (s_2). With this in mind, we first study the correlations between the statistics under consideration.

We compute all statistics for all non-isomorphic graphs on $|V| = 4, 5, \ldots, 10$ vertices (we exclude graphs with fewer vertices as many of the statistics are not well defined and there are only a handful of graphs). We then consider the pairwise correlations between the different statistics and how this changes as the graph order increases; see Fig. 4. To compare the coverage of statistics with different $|V|$, we scale the statistic values into the same range. By definition, clustering coefficients (ACC, GCC, SCC) are in the $[0, 1]$ range and degree assortativity is in the $[-1, 1]$ range. We keep their values and ranges without scaling. Edge density, number of triangles, diameter and connectivity measures (C_v and C_e), are normalized into $[0, 1]$ (dividing by the corresponding maximum value). The last statistic, APL, is also normalized into $[0, 1]$, subject to some complications: we compute the exact average path length to divide by in our ground truth datasets, but not when we use the generators, where we use the maximal path length encountered instead (which may not be the same as the maximum).

It is easy to see that the coverage of values expands with increasing $|V|$. Figure 5 shows this pattern for three pairs of properties. This indicates that we are more likely to find larger ranges of different statistics for graphs with more vertices given the same set of fixed statistics. With this in mind, we consider graphs with more than 10 vertices, but this time relying on random graph generators. Figure 6 shows how correlation values between all pairs of statistics change when the number of vertices increases. The blue trend lines for the ground truth data show the correlation values calculated using the set of all possible graphs for a given number of nodes. The orange trend lines show the correlation values calculated from graphs generated with the ER model. Specifically, the ER data is created as follows: for each value of $|V| = 5, 6, \ldots, 15$ we generate $100,000$ graphs with p selected uniformly at random in the $[0, 1]$ range.

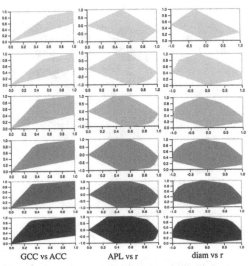

GCC vs ACC   APL vs r   diam vs r

Fig. 5. The convex hull of graph coverage across several statistical properties. Each row (starting from the top) represents all graphs for a fixed number of vertices ($|V| = 5 \ldots |V| = 10$). Columns are pairs of graph properties.

For most of the cells in the matrix shown in Fig. 6, the correlation values seem to converge as $|V|$ becomes larger than 8. (both in the ground truth and the ER-model generated graph sets). Moreover, for most of the cells, the pattern of the change in correlation values appears to be the same for both sets. Analyzing the trend lines of the ER-model, we observe four patterns of change in the correlation values: convergence to a constant value, monotonic decrease, monotonic increase, and non-monotonic change. These patterns are highlighted in Fig. 6 by enclosing boxes of different colors. There are exceptions that do not fit these patterns, e.g., (S_c, r) and in two cases, (r, C_v) and (r, C_e), the trend lines show different patterns.

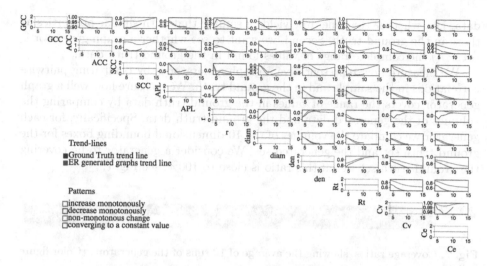

Fig. 6. Trends in the correlations with increasing $|V|$: the x-axis shows the number of vertices and the y-axis shows the correlation value for the pair of graph statistics.

5   Graph Statistics and Graph Generators

While we can explore statistical coverage and correlations in low-order graphs, it is difficult to generate all non-isomorphic graphs with more than 10 vertices due to the super-exponential increase in the number of different graphs (e.g., for $|V| = 16$ we there are 6×10^{22} different graphs). However, these higher order graphs are common in many domains. As such, we want to further explore this issue of "same stats, different graphs" for larger graphs. As such, we turn to graph generators to help us explore the same-stats-different-graphs problem.

We select four different random generators that cover the four categories [11] of graph generation: the ER random graph model, the WS small-world model, the BA preferential attachment model, and the geometric random graph model.

Coverage for Ground-Truth Graph Set: We use implementations of all four generators (ER, WS, BA, geometric) from NetworkX [23], with three variants of ER ($p = 0.5$, p selected uniformly at random from the $[0, 1]$ range, and p selected to match edge density in the ground truth). More details about the graph generators and how well they perform for our tasks are provided in the full version of the paper [12]. For each generator, we generate 1%, 0.1% and 0.01% of the total number of graphs in ground-truth graph set. We use low sampling rates as for high order graphs the ground truth set is huge and any sampling strategy will have just a fraction of the total. Our goal here is to explore whether a small sample of graphs could be representative of the ground truth set of non-isomorphic graphs and cover the space well.

We evaluate the different graph generators in two different ways. First we want to see whether a graph generator is *representative* of the ground truth data, i.e., whether the generator yields a sample that with similar properties as those in the ground truth. Second, we want to see whether a graph generator is *covering* the complete range of values found in the ground truth data.

We measure how representative a graph generator is by comparing pairwise correlations in the sample and in the ground truth. We measure how well a graph generator covers the range of values in the ground truth data by comparing the volumes of the generated data and the ground truth data. Specifically, for each generator we compare the volumes of the 10-dimensional bounding boxes for the ground truth set and the generated set. We consider a generator to be covering the ground truth set well if this ratio is close to 100%; see Fig. 7.

WS model	BA model	ER model p = 0.5	ER model p~Uniform	ER model p~Population	Geometric model
22.04%	0.10%	0.98%	90.83%	12.37%	83.87%

Fig. 7. Coverage ratios, showing the average of 10 runs of the generators. (Color figure online)

Both of these measures can be visualized by plotting each of the graphs in ground-truth graph set as dots in the 2D matrix of correlations and then drawing the generated graph set on top of the first plot to see how well the generator set covers the ground-truth graph set. We color the ground-truth graph set in blue and the generated data in red. Because the ground-truth graph set includes all possible graphs for a fixed $|V|$, there is at least one blue point under each red point. Detailed illustrations can be found in the full version of the paper [12] but here we include one example of the most representative model: ER with $p = 0.5$; see Fig. 8. From this figure it is easy to see that nearly all pairwise correlations are very similar in the ground truth and in the generated data. Note, however, that from the same figure we can also see that this generator does not cover the range of possible values in the ground truth data well (e.g., in the columns corresponding to APL, r, diameter and density, the leftmost and rightmost points in the plots are blue).

6   Finding Different Graphs with the Same Statistics

While our exploration of graph statistics, correlation, and generation revealed some challenges, it is still possible to explore the fundamental question of whether we can identify graphs that are similar across some statistics while being drastically different across others. To find graphs that are identical over a number of graph statistics and yet are different, we use the ground truth data for small non-isomorphic graphs. For larger graphs, we use the graph generators together with some filters.

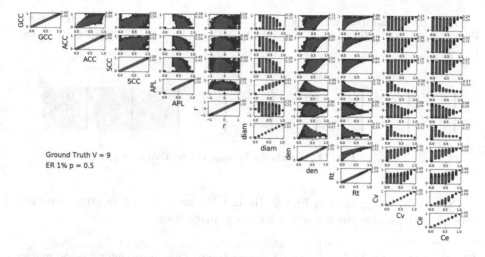

Fig. 8. Ground truth (blue) and ER with $p = 0.5$ (red). (Color figure online)

Finding Graphs in the Ground Truth: For $|V| \leq 10$, we directly use all possible non-isomorphic graphs as our dataset. In fact, we can fix different combinations of 5 statistics and still get multiple distinct graphs. We visualize this with figures that encapsulate the variability of one statistic in 10 slots, covering the ranges $[0.0, 0.1], [0.1, 0.2], \ldots [0.9, 1]$ and in each slot we show a graph (if it exists) drawn by a spring layout; see Fig. 9.

For the first experiment, we fix $|V| = 9$, $APL \subset (1.42, 1.47)$, $den \in (0.52, 0.57)$, $GCC \in (0.5, 0.6)$, $R_t \in (0.15, 0.25)$. Since all our statistics are normalized to $[0, 1]$ and assortativity is in $[-1, 1]$, each of the ten slots has a range of 0.2. We find graphs for seven of the ten possible slots; see Fig. 9. This figure also illustrates the output of our "same stats, different graphs" generator: fix several statistics and generate graphs that vary in another statistic.

Fig. 9. Variability in assortativity.

Similarly, for the second experiment, we fix $|V| = 9$, $APL \in (1.47, 1.69)$, $diam = 3$, $Cv = 2$, $Ce = 2$, and $r \in (-0.22, -0.29)$ to obtain GCC in the range $(0, 0.8)$; see Fig. 10.

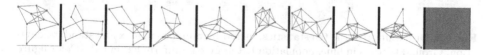

Fig. 10. Variability in GCC.

As a final example, we fix $|V| = 9$, $SCC \in (0.75, 0.85)$, $ACC \in (0.75, 0.8)$, $r \in (-0.3, -0.2)$, $R_t \in (0.35, 0.45)$ and find graphs with C_e from 0 to 5; see Fig. 11.

Fig. 11. Variability in edge connectivity.

Note that the graphs in Figs. 9, 10 and 11. are different in structure even though they possess similar values for many properties.

Finding Graphs Using Graph Generators: This approach relies on generating many graphs and filtering graphs based on several fixed statistics. For the two most important statistics of a graph, $|V|$ and $|E|$, we generate all graphs with a fixed $|V|$ and choose $|E|$ as follows:

1. uniform: select $|E|$ uniformly from its range. This is equivalent to forcing the edge density in the generated set to follow a uniform distribution;
2. population: select $|E|$ by forcing the edge density in the generated set to match the distribution in the ground truth (population) graph set.

Using both edge selection strategies for all four generators, we compare the statistics distribution to the ground truth for $|V| = 9$. Figure 12 illustrates how different statistics are distributed given uniform edge sampling and population-based edge sampling for the ER model. It shows that although the population-based sampling approach generates a distribution that is more similar to the

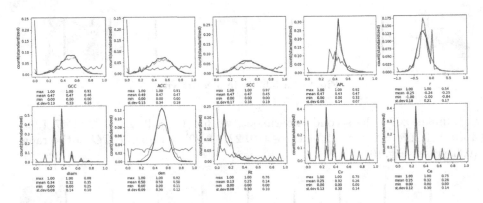

Fig. 12. Distribution of the ten statistics, including min/mean/max and standard deviation. Ground truth is in blue, population ER in green, uniform ER in red. (Color figure online)

ground truth, it has a narrower coverage (larger min and smaller max) than the uniform sampling. The WS and BA models also do not provide good coverage of the various statistics.

7   Discussion and Future Work

Random graph generators have been designed to model different types of graphs, but by design such algorithms sample the space of isomorphic graphs. For the purpose of studying graph properties and structure, we need generators that represent and cover the space of non-isomorphic graphs.

We considered how to explore the space of graphs and graph statistics that make it possible to have multiple graphs that are identical in a number of graph statistics, yet are clearly different. To "see" the difference, it often suffices to look at the drawings of the graphs. However, as graphs get larger, some graph drawing algorithms may not allow us to distinguish differences in statistics between two graphs purely from their drawings. We recently studied how the perception statistics, such as density and ACC, is affected by different graph drawing algorithms [41]. The results confirm the intuition that some drawing algorithms are more appropriate than others in aiding viewers to perceive differences between underlying graph statistics. Further work in this direction might help ensure that differences between graphs are captured in the different drawings.

References

1. Albert, R., Barabási, A.L.: Statistical mechanics of complex networks. Rev. Mod. Phys. **74**(1), 47 (2002)
2. Albert, R., Jeong, H., Barabási, A.L.: Internet: diameter of the world-wide web. Nature **401**(6749), 130 (1999)
3. Anscombe, F.J.: Graphs in statistical analysis. Am. Stat. **27**(1), 17–21 (1973). http://www.jstor.org/stable/2682899
4. Bach, B., Spritzer, A., Lutton, E., Fekete, J.-D.: Interactive random graph generation with evolutionary algorithms. In: Didimo, W., Patrignani, M. (eds.) GD 2012. LNCS, vol. 7704, pp. 541–552. Springer, Heidelberg (2013). https://doi.org/10.1007/978-3-642-36763-2_48
5. Barabási, A.L., Albert, R.: Emergence of scaling in random networks. Science **286**(5439), 509–512 (1999)
6. Boccaletti, S., Latora, V., Moreno, Y., Chavez, M., Hwang, D.U.: Complex networks: structure and dynamics. Phys. Rep. **424**(4–5), 175–308 (2006)
7. Boguná, M., Pastor-Satorras, R.: Epidemic spreading in correlated complex networks. Phys. Rev. E **66**(4), 047104 (2002)
8. Broder, A., et al.: Graph structure in the web. Comput. Netw. **33**(1–6), 309–320 (2000)
9. Cartwright, D., Harary, F.: Structural balance: a generalization of Heider's theory. Psychol. Rev. **63**(5), 277 (1956)
10. Chakrabarti, D., Faloutsos, C.: Graph mining: laws, generators, and algorithms. ACM Comput. Surv. (CSUR) **38**(1), 2 (2006)

11. Chakrabarti, D., Faloutsos, C.: Graph patterns and the R-MAT generator. In: Mining Graph Data, pp. 65–95 (2007)
12. Chen, H., Soni, U., Lu, Y., Maciejewski, R., Kobourov, S.: Same stats, different graphs (graph statistics and why we need graph drawings). ArXiv e-prints arXiv:1808.09913, August 2018
13. Davis, G.F., Yoo, M., Baker, W.E.: The small world of the American corporate elite, 1982–2001. Strateg. Org. 1(3), 301–326 (2003)
14. Dorogovtsev, S.N., Mendes, J.F.F., Samukhin, A.N.: Structure of growing networks with preferential linking. Phys. Rev. Lett. 85(21), 4633 (2000)
15. Ebel, H., Mielsch, L.I., Bornholdt, S.: Scale-free topology of e-mail networks. Phys. Rev. E 66(3), 035103 (2002)
16. Erdös, P., Rényi, A.: On random graphs. Publicationes mathematicae 6, 290–297 (1959)
17. Even, S., Tarjan, R.E.: Network flow and testing graph connectivity. SIAM J. Comput. 4(4), 507–518 (1975)
18. Feld, S.L.: The focused organization of social ties. Am. J. Sociol. 86(5), 1015–1035 (1981)
19. Frank, O., Harary, F.: Cluster inference by using transitivity indices in empirical graphs. J. Am. Stat. Assoc. 77(380), 835–840 (1982)
20. Gilbert, E.N.: Random graphs. Ann. Math. Stat. 30(4), 1141–1144 (1959)
21. Girvan, M., Newman, M.E.: Community structure in social and biological networks. Proc. Natl. Acad. Sci. 99(12), 7821–7826 (2002)
22. Guimera, R., Danon, L., Diaz-Guilera, A., Giralt, F., Arenas, A.: Self-similar community structure in a network of human interactions. Phys. Rev. E 68(6), 065103 (2003)
23. Hagberg, A., Swart, P., S Chult, D.: Exploring network structure, dynamics, and function using networkx. Technical report, Los Alamos National Lab. (LANL), Los Alamos, NM, United States (2008)
24. Hanneman, R.A., Riddle, M.: Introduction to social network methods (2005)
25. Kairam, S., MacLean, D., Savva, M., Heer, J.: GraphPrism: compact visualization of network structure. In: Proceedings of the International Working Conference on Advanced Visual Interfaces, pp. 498–505. ACM (2012)
26. Karlberg, M.: Testing transitivity in graphs. Soc. Netw. 19(4), 325–343 (1997)
27. Li, G., Semerci, M., Yener, B., Zaki, M.J.: Graph classification via topological and label attributes. In: Proceedings of the 9th International Workshop on Mining and Learning with Graphs (MLG), San Diego, USA, vol. 2 (2011)
28. Lind, P.G., Gonzalez, M.C., Herrmann, H.J.: Cycles and clustering in bipartite networks. Phys. Rev. E 72(5), 056127 (2005)
29. Loguinov, D., Kumar, A., Rai, V., Ganesh, S.: Graph-theoretic analysis of structured peer-to-peer systems: routing distances and fault resilience. In: Proceedings of the 2003 Conference on Applications, Technologies, Architectures, and Protocols for Computer Communications, pp. 395–406. ACM (2003)
30. Maslov, S., Sneppen, K., Zaliznyak, A.: Detection of topological patterns in complex networks: correlation profile of the internet. Physica A: Stat. Mech. Appl. 333, 529–540 (2004)
31. Matejka, J., Fitzmaurice, G.: Same stats, different graphs: generating datasets with varied appearance and identical statistics through simulated annealing. In: Proceedings of the 2017 CHI Conference on Human Factors in Computing Systems, pp. 1290–1294. ACM (2017)

32. McGlohon, M., Akoglu, L., Faloutsos, C.: Statistical properties of social networks. In: Aggarwal, C. (ed.) Social Network Data Analytics, pp. 17–42. Springer, Boston (2011). https://doi.org/10.1007/978-1-4419-8462-3_2

33. Melancon, G.: Just how dense are dense graphs in the real world?: a methodological note. In: Proceedings of the 2006 AVI Workshop on BEyond Time and Errors: Novel Evaluation Methods for Information Visualization, pp. 1–7. ACM (2006)

34. Mislove, A., Marcon, M., Gummadi, K.P., Druschel, P., Bhattacharjee, B.: Measurement and analysis of online social networks. In: Proceedings of the 7th ACM SIGCOMM Conference on Internet Measurement, pp. 29–42. ACM (2007)

35. Newman, M.E.: Assortative mixing in networks. Phys. Rev. Lett. **89**(20), 208701 (2002)

36. Newman, M.E.: Mixing patterns in networks. Phys. Rev. E **67**(2), 026126 (2003)

37. Newman, M.E.: The structure and function of complex networks. SIAM Rev. **45**(2), 167–256 (2003)

38. Newman, M.E., Watts, D.J.: Scaling and percolation in the small-world network model. Phys. Rev. E **60**(6), 7332 (1999)

39. de Solla Price, D.: A general theory of bibliometric and other cumulative advantage processes. J. Assoc Inf. Sci. Technol. **27**(5), 292–306 (1976)

40. Sen, P., Dasgupta, S., Chatterjee, A., Sreeram, P., Mukherjee, G., Manna, S.: Small-world properties of the indian railway network. Phys. Rev. E **67**(3), 036106 (2003)

41. Soni, U., Lu, Y., Hansen, B., Purchase, H., Kobourov, S., Maciejewski, R.: The perception of graph properties in graph layouts. In: 20th IEEE Eurographics Conference on Visualization (EuroVis) (2018)

42. Uzzi, B., Spiro, J.: Collaboration and creativity: the small world problem. Am. J. Sociol. **111**(2), 447–504 (2005)

43. Van Noort, V., Snel, B., Huynen, M.A.: The yeast coexpression network has a small-world, scale-free architecture and can be explained by a simple model. EMBO Rep. **5**(3), 280–284 (2004)

44. Watts, D.J., Strogatz, S.H.: Collective dynamics of 'small-world' networks. Nature **393**(6684), 440 (1998)

45. Wei-Bing, D., Long, G., Wei, L., Xu, C.: Worldwide marine transportation network: efficiency and container throughput. Chin. Phys. Lett. **26**(11), 118901 (2009)

Orthogonal Drawings

Bend-Minimum Orthogonal Drawings in Quadratic Time

Walter Didimo[1]([✉]), Giuseppe Liotta[1], and Maurizio Patrignani[2]

[1] Università degli Studi di Perugia, Perugia, Italy
{walter.didimo,giuseppe.liotta}@unipg.it
[2] Università Roma Tre, Rome, Italy
maurizio.patrignani@uniroma3.it

Abstract. Let G be a planar 3-graph (i.e., a planar graph with vertex degree at most three) with n vertices. We present the first $O(n^2)$-time algorithm that computes a planar orthogonal drawing of G with the minimum number of bends in the variable embedding setting. If either a distinguished edge or a distinguished vertex of G is constrained to be on the external face, a bend-minimum orthogonal drawing of G that respects this constraint can be computed in $O(n)$ time. Different from previous approaches, our algorithm does not use minimum cost flow models and computes drawings where every edge has at most two bends.

1 Introduction

A pioneering paper by Storer [22] asks whether a crossing-free orthogonal drawing with the minimum number of bends can be computed in polynomial time. The question posed by Storer is in the fixed embedding setting, i.e., the input is a plane 4-graph (an embedded planar graph with vertex degree at most four) and the wanted output is an embedding-preserving orthogonal drawing with the minimum number of bends. Tamassia [23] answers Storer's question in the affirmative by describing an $O(n^2 \log n)$-time algorithm. The key idea of Tamassia's result is the equivalence between the bend minimization problem and the problem of computing a min-cost flow on a suitable network. To date, the most efficient known solution of the bend-minimization problem for orthogonal drawings in the fixed embedding setting is due to Cornelsen and Karrenbauer [6], who show a novel technique to compute a min-cost flow on an uncapacitated network and apply this technique to Tamassia's model achieving $O(n^{\frac{3}{2}})$-time complexity.

A different level of complexity for the bend minimization problem is encountered in the variable embedding setting, that is when the algorithm is asked to find a bend-minimum solution over all planar embeddings of the graph. For example, the orthogonal drawing of Fig. 1(c) has a different planar embedding

Research supported in part by the project: "Algoritmi e sistemi di analisi visuale di reti complesse e di grandi dimensioni - Ricerca di Base 2018, Dipartimento di Ingegneria dell'Università degli Studi di Perugia" and by MIUR project "MODE – MOrphing graph Drawings Efficiently", prot. 20157EFM5C_001.

T. Biedl and A. Kerren (Eds.): GD 2018, LNCS 11282, pp. 481–494, 2018.
https://doi.org/10.1007/978-3-030-04414-5_34

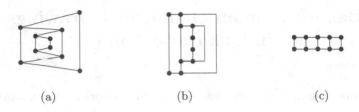

Fig. 1. (a) A planar embedded 3-graph G. (b) An embedding-preserving bend-minimum orthogonal drawing of G. (c) A bend-minimum orthogonal drawing of G.

than the graph of Fig. 1(a) and it has no bends, while the drawing of Fig. 1(b) preserves the embedding but it is suboptimal in terms of bends.

Garg and Tamassia [13] prove that the bend-minimization problem for orthogonal drawings is NP-complete for planar 4-graphs, while Di Battista et al. [8] show that it can be solved in $O(n^5 \log n)$ time for planar 3-graphs. Generalizations of the problem in the variable embedding setting where edges have some flexibility (i.e., they can bend a few times without cost for the optimization function) have also been the subject of recent studies by Bläsius et al. [2].

Improving the $O(n^5 \log n)$ time complexity of the algorithm by Di Battista et al. [8] has been an elusive open problem for more than a decade (see, e.g., [3]), until a paper by Chang and Yen [4] has shown how to compute a bend-minimum orthogonal drawing of a planar 3-graph in the variable embedding setting in $\tilde{O}(n^{\frac{17}{7}})$ time, which can be read as $O(n^{\frac{17}{7}} \log^k n)$ time for a positive constant k.

Similar to [8], the approach in [4] uses an SPQR-tree to explore all planar embeddings of a planar 3-graph and combines partial solutions associated with the nodes of this tree to compute a bend-minimum drawing. Both in [8] and in [4], the computationally most expensive task is computing min-cost flows on suitable variants of Tamassia's network. However, Chang and Yen elegantly prove that a simplified flow network where all edges have unit capacity can be adopted to execute this task. This, combined with a recent result [5] about min-cost flows on unit-capacity networks, yields the improved time complexity.

Contribution and Outline. This paper provides new algorithms to compute bend-minimum orthogonal drawings of planar 3-graphs, which improve the time complexity of the state-of-the-art solution. We prove the following.

Theorem 1. *Let G be an n-vertex planar 3-graph. A bend-minimum orthogonal drawing of G can be computed in $O(n^2)$ time. If either a distinguished edge or a distinguished vertex of G is constrained to be on the external face, a bend-minimum orthogonal drawing of G that respects the given constraint can be computed in $O(n)$ time. Furthermore, the computed drawings have at most two bends per edge, which is worst-case optimal.*

As in [8] and in [4], the algorithmic approach of Theorem 1 computes a bend-minimum orthogonal drawing by visiting an SPQR-tree of the input graph. However, it does not need to compute min-cost flows at any steps of the visit, which is the fundamental difference with the previous techniques. This makes it possible to design the first quadratic-time algorithm to compute bend-minimum orthogonal drawings of planar 3-graphs in the variable embedding setting.

The second part of the statement of Theorem 1 extends previous studies by Nishizeki and Zhou [26], who give a first example of a linear-time algorithm in the variable embedding setting for planar 3-graphs that are partial two-trees. The bend-minimum drawings of Theorem 1 have at most two bends per edge, which is a desirable property for an orthogonal representation. We recall that every planar 4-graph (except the octahedron) has an orthogonal drawing with at most two bends per edge [1,17], but minimizing the number of bends may require some edges with a $\Omega(n)$ bends [8,24]. It is also proven that every planar 3-graph (except K_4) has an orthogonal drawing with at most one bend per edge [16], but the drawings of the algorithm in [16] are not bend-minimum. Finally, a non-flow based algorithm having some similarities with ours is given in [12]; it neither computes bend-minimum drawings nor guarantees at most two bends per edge.

The paper is organized as follows. Preliminary definitions and results are in Sect. 2. In Sect. 3 we prove key properties of bend-minimum orthogonal drawings of planar 3-graphs used in our approach. Sect. 4 describes our drawing algorithms. Open problems are in Sect. 5. All full proofs and more figures can be found in [11].

2 Preliminaries

We assume familiarity with basic definitions on graph connectivity and planarity (see Appendix A of [11]). If G is a graph, $V(G)$ and $E(G)$ denote the sets of vertices and edges of G. We consider *simple* graphs, i.e., graphs with neither self-loops nor multiple edges. The *degree* of a vertex $v \in V(G)$, denoted as $\deg(v)$, is the number of its neighbors. $\Delta(G)$ denotes the maximum degree of a vertex of G; if $\Delta(G) \leq h$ ($h \geq 1$), G is an *h-graph*. A graph G is *rectilinear planar* if it admits a planar drawing where each edge is either a horizontal or a vertical segment (i.e., it has no bend). Rectilinear planarity testing is NP-complete for planar 4-graphs [13], but it is polynomially solvable for planar 3-graphs [4,8] and linear-time solvable for subdivisions of planar triconnected cubic graphs [18]. By extending a result of Thomassen [25] on those 3-graphs that have a rectilinear drawing with all rectangular faces, Rahman et al. [21] characterize rectilinear plane 3-graphs. For a plane graph G, let $C_o(G)$ be its external cycle ($C_o(G)$ is simple if G is biconnected). Also, if C is a simple cycle of G, $G(C)$ is the plane subgraph of G that consists of C and of the vertices and edges inside C. An edge $e = (u, v) \notin E(G(C))$ is a *leg* of C if exactly one of the vertices u and v belongs to C; such a vertex is a *leg-vertex* of C. If C has exactly k legs and no edge embedded outside C joins two of its vertices, C is a *k-legged cycle* of G.

Theorem 2. [21] *Let G be a biconnected plane 3-graph. G admits an orthogonal drawing without bends if and only if: (i) $C_o(G)$ contains at least four vertices of degree 2; (ii) each 2-legged cycle contains at least two vertices of degree 2; (iii) each 3-legged cycle contains at least one vertex of degree 2.*

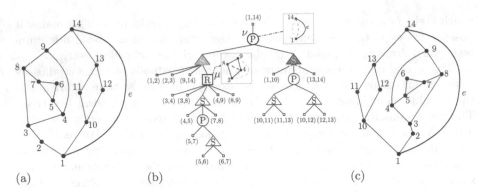

Fig. 2. (a) A plane 3-graph G. (b) The SPQR-tree of G with respect to e; the skeletons of a P-node ν and of an R-node μ are shown. (c) A different embedding of G obtained by changing the embedding of skel(ν) and of skel(μ).

As in [21], we call *bad* any 2-legged and any 3-legged cycle that does not satisfy Condition (*ii*) and (*iii*) of Theorem 2, respectively.

SPQR-Trees of Planar 3-Graphs. Let G be a biconnected graph. An *SPQR-tree* T of G represents the decomposition of G into its triconnected components and can be computed in linear time [7,14,15]. Each triconnected component corresponds to a node μ of T; the triconnected component itself is called the *skeleton* of μ and denoted as skel(μ). A node μ of T can be of one of the following types: (*i*) *R-node*, if skel(μ) is a triconnected graph; (*ii*) *S-node*, if skel(μ) is a simple cycle of length at least three; (*iii*) *P-node*, if skel(μ) is a bundle of at least three parallel edges; (*iv*) *Q-nodes*, if it is a leaf of T; in this case the node represents a single edge of the graph and its skeleton consists of two parallel edges. Note that, neither two S- nor two P-nodes are adjacent in T. A *virtual edge* in skel(μ) corresponds to a tree node ν adjacent to μ in T. If T is rooted at one of its Q-nodes ρ, every skeleton (except the one of ρ) contains exactly one virtual edge that has a counterpart in the skeleton of its parent: This virtual edge is the *reference edge* of skel(μ) and of μ, and its endpoints are the *poles* of skel(μ) and of μ. The edge of G corresponding to the root ρ of T is the *reference edge* of G, and T is the SPQR-tree of G *with respect to e*. For every node $\mu \neq \rho$ of T, the subtree T_μ rooted at μ induces a subgraph G_μ of G called the *pertinent graph* of μ, which is described by T_μ in the decomposition: The edges of G_μ correspond to the Q-nodes (leaves) of T_μ. Graph G_μ is also called a *component* of G with respect to the reference edge e, namely G_μ is a P-, an R-, or an S-component depending on whether μ is a P-, an R-, or an S-component, respectively.

The SPQR-tree T rooted at a Q-node ρ implicitly describes all planar embeddings of G with the reference edge of G on the external face. All such embeddings are obtained by combining the different planar embeddings of the skeletons of P- and R-nodes: For a P-node μ, the different embeddings of skel(μ) are the different permutations of its non-reference edges. If μ is an R-node, skel(μ) has two possible planar embeddings, obtained by flipping skel(μ) minus its reference

edge at its poles. See Fig. 2 for an illustration. The child node of ρ and its pertinent graph are called the *root child* of T and the *root child component* of G,
respectively. An *inner node* of T is neither the root nor the root child of T. The
pertinent graph of an inner node is an *inner component* of G. The next lemma
gives basic properties of T when $\Delta(G) \leq 3$.

Lemma 1. *Let G be a biconnected planar 3-graph and let T be the SPQR-tree
of G with respect to a reference edge e. The following properties hold:*

T1 *Each P-node μ has exactly two children, one being an S-node and the other
being an S- or a Q-node; if μ is the root child, both its children are S-nodes.*
T2 *Each child of an R-node is either an S-node or a Q-node.*
T3 *For each inner S-node μ, the edges of $\text{skel}(\mu)$ incident to the poles of μ are
(real) edges of G. Also, there cannot be two incident virtual edges in $\text{skel}(\mu)$.*

3 Properties of Bend-Minimum Orthogonal Representations of Planar 3-Graphs

We prove relevant properties of bend-minimum orthogonal drawings of planar 3-
graphs that are independent of vertex and bend coordinates, but only depend on
the vertex angles and edge bends. To this aim, we recall the concept of *orthogonal
representation* [23] and define some types of "shapes" that we use to construct
bend-minimum orthogonal representations.

Orthogonal Representations. Let G be a plane 3-graph. If $v \in V(G)$ and if
e_1 and e_2 are two (possibly coincident) edges incident to v that are consecutive
in the clockwise order around v, we say that $a = \langle e_1, v, e_2 \rangle$ is an *angle at v of G*
or simply an *angle of G*. Let Γ and Γ' be two embedding-preserving orthogonal
drawings of G. We say that Γ and Γ' are *equivalent* if: (i) For any angle a of G,
the geometric angle corresponding to a is the same in Γ and Γ', and (ii) for any
edge $e = (u, v)$ of G, the sequence of left and right bends along e moving from u
to v is the same in Γ and in Γ'. An *orthogonal representation* H of G is a class
of equivalent orthogonal drawings of G; H can be described by the embedding
of G together with the geometric value of each angle of G (90, 180, 270°)[1] and
with the sequence of left and right bends along each edge. Figure 3(a) shows a
bend-minimum orthogonal representation of the graph in Fig. 2(a).

Let p be a path between two vertices u and v in H. The *turn number* of p is
the absolute value of the difference between the number of right and the number
of left turns encountered along p moving from u to v (or vice versa). The turn
number of p is denoted by $t(p)$. A turn along p is caused either by a bend on an
edge of p or by an angle of 90/270 degrees at a vertex of p. For example, $t(p) = 2$
for the path $p = \langle 3, 4, 5, 6, 7 \rangle$ in the orthogonal representation of Fig. 3(a). We
remark that if H is a bend-minimum orthogonal representation, the bends along
an edge, going from an end-vertex to the other, are all left or all right turns [23].

---
[1] Angles of 360 degrees only occur at 1-degree vertices; we can avoid to specify them.

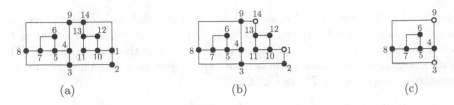

Fig. 3. (a) A bend-minimum orthogonal representation H with four bends of the graph in Fig. 2(a). (b) The component H_ν, which is L-shaped; the two poles of the component are the white vertices. (c) The component H_μ, which is D-shaped.eps

Shapes of Orthogonal Representations. Let G be a biconnected planar 3-graph, T be the SPQR-tree of G with respect to a reference edge $e \in E(G)$, and H be an orthogonal representation of G with e on the external face. For a node μ of T, denote by H_μ the restriction of H to a component G_μ. We also call H_μ a *component* of H. In particular, H_μ is a P-, an R-, or an S-component depending on whether μ is a P-, an R-, or an S-component, respectively. If μ is the root child of T, then H_μ is the *root child component* of H. Denote by u and v the two poles of μ and let p_l and p_r be the two paths from u to v on the external boundary of H_μ, one walking clockwise and the other walking counterclockwise. These paths are the *contour paths* of H_μ. If μ is an S-node, p_l and p_r share some edges (they coincide if H_μ is just a sequence of edges). If μ is either a P- or an R-node, p_l and p_r are edge disjoint; in this case, we define the following *shapes* for H_μ, depending on $t(p_l)$ and $t(p_r)$ and where the poles are external corners:

- H_μ is *C-shaped*, or ▢-*shaped*, if $t(p_l) = 4$ and $t(p_r) = 2$, or vice versa;
- H_μ is *D-shaped*, or ▢-*shaped*, if $t(p_l) = 0$ and $t(p_r) = 2$, or vice versa;
- H_μ is *L-shaped*, or ▢-*shaped*, if $t(p_l) = 3$ and $t(p_r) = 1$, or vice versa;
- H_μ is *X-shaped*, or ▢-*shaped*, if $t(p_l) = t(p_r) = 1$.

For example, H_ν in Fig. 3(b) is ▢-shaped, while H_μ in Fig. 3(c) is ▢-shaped. Concerning S-components, the following lemma rephrases a result in [8, Lemma 4.1], and it is also an easy consequence of Property T3 in Lemma 1.

Lemma 2. *Let H_μ be an inner S-component with poles u and v and let p_1 and p_2 be any two paths connecting u and v in H_μ. Then $t(p_1) = t(p_2)$.*

Based on Lemma 2, we describe the shape of an inner S-component H_μ in terms of the turn number of any path p between its two poles: We say that H_μ is k-*spiral* and has *spirality* k if $t(p) = k$. The notion of spirality of an orthogonal component was introduced in [8]. Differently from [8], we restrict the definition of spirality to inner S-components and we always consider absolute values, instead of both positive and negative values depending on whether the left turns are more or fewer than the right turns. For instance, in the representation of Fig. 3(a) the two series with poles $\{1, 14\}$ (the two filled S-nodes in Fig. 2(b)) have spirality three and one, respectively; the series with poles $\{4, 8\}$ (child of the R-node) has spirality zero, while the series with poles $\{5, 7\}$ has spirality two.

We now give a key result that claims the existence of a bend-minimum orthogonal representation with specific properties for any biconnected planar 3-graph.

This result will be used to design our drawing algorithm. Given an orthogonal representation H, we denote by $\overline{H}$ the orthogonal representation obtained from H by replacing each bend with a dummy vertex: $\overline{H}$ is the *rectilinear image* of H; a dummy vertex in $\overline{H}$ is a *bend vertex*. Also, if w is a degree-2 vertex with neighbors u and v, *smoothing* w is the reverse operation of an edge subdivision, i.e., it replaces the two edges (u, w) and (w, v) with the single edge (u, v).

Lemma 3. *A biconnected planar 3-graph G with a distinguished edge e has a bend-minimum orthogonal representation H with e on the external face such that:*

O1 *Every edge of H has at most two bends, which is worst-case optimal.*
O2 *Every inner P-component or R-component of H is either ▢- or ▢-shaped.*
O3 *Every inner S-component of H has spirality at most four.*

Proof (sketch). We prove in three steps the existence of a bend-minimum orthogonal representation H that satisfies O1-O3. We start by a bend-minimum orthogonal representation of G with e on the external face, and in the first step we prove that it either satisfies O1 or it can be locally modified, without changing its planar embedding, so to satisfy O1. In the second step, we prove that from the orthogonal representation obtained in the first step we can derive a new orthogonal representation (still with same embedding) that satisfies O2 in addition to O1. Finally, we prove that this last representation also satisfies O3.

Step 1: Property O1. Suppose that H is a bend-minimum orthogonal representation of G with e on the external face and having an edge g (possibly $g = e$) with at least three bends. Let $\overline{H}$ be the rectilinear image of H, and let $\overline{G}$ be the plane graph underlying $\overline{H}$. Since $\overline{H}$ has no bend, $\overline{G}$ satisfies Conditions $(i) - (iii)$ of Theorem 2. Let v_1, v_2, v_3 be three bend vertices in $\overline{H}$ that correspond to three bends of g in H. Assume first that g is an internal edge of G and let $\overline{G}'$ be the plane graph obtained from $\overline{G}$ by smoothing v_1. We claim that $\overline{G}'$ still satisfies Conditions $(i) - (iii)$ of Theorem 2. Indeed, if this is not the case, there must be a bad cycle in $\overline{G}'$ that contains both v_2 and v_3. This is a contradiction, because no bad cycle can contain two vertices of degree two. Hence, there exists an (embedding-preserving) representation $\overline{H}'$ of $\overline{G}'$ without bends, which is the rectilinear image of an orthogonal representation of G with fewer bends than H, a contradiction. Assume now that g is on the external cycle $C_o(G)$ of G. If $C_o(\overline{G})$ contains more than four vertices of degree two, we can smooth v_1 and apply the same argument as above to contradict the optimality of H (note that, such a smoothing does not violate Condition (i) of Theorem 2). Suppose vice versa that $C_o(\overline{G})$ contains exactly four vertices of degree two (three of them being v_1, v_2, and v_3). In this case, just smoothing v_1 violates Condition (i) of Theorem 2. However, we can smooth v_1 and subdivide an edge of $C_o(\overline{G}) \cap C_o(G)$ (such an edge exists since $C_o(G)$ has at least three edges and, by hypothesis and a simple counting argument, at least one of its edges has no bend in H). The resulting plane graph $\overline{G}''$ still satisfies the three conditions of Theorem 2 and admits a representation $\overline{H}''$ without bends; the representation of which $\overline{H}''$ is the rectilinear image is a bend-minimum orthogonal representation of G with at most two bends per edge. To see that two bends per edge is worst-case optimal, just consider a bend-minimum representation of the complete graph K_4.

Step 2: Property O2. Let H be a bend-minimum orthogonal representation of G that satisfies O1 and let $\overline{H}$ be its rectilinear image. The plane underlying graph $\overline{G}$ of $\overline{H}$ satisfies the three conditions of Theorem 2. Rhaman, Nishizeki, and Naznin [21, Lemma 3] prove that, in this case, $\overline{G}$ has an embedding-preserving orthogonal representation $\overline{H'}$ without bends in which every 2-legged cycle C is either ⬗-shaped or ⬖-shaped, where the two poles of the shape are the two leg-vertices of C. On the other hand, if G_μ is an inner P- or R-component, the external cycle $C_o(G_\mu)$ is a 2-legged cycle of G, where the two leg-vertices of $C_o(G_\mu)$ are the poles of G_μ. Hence, the representation H' of G whose rectilinear image is $\overline{H'}$ satisfies O2, as H'_μ is either ⬗-shaped or ⬖-shaped. Also, the bends of H' are the same as in H, because the bend vertices of $\overline{H}$ coincide with those of $\overline{H'}$. Hence, H' still satisfies O1 and has the minimum number of bends.

Step 3: Property O3. Suppose now that H is a bend-minimum orthogonal representation of G (with e on the external face) that satisfies both O1 and O2. More precisely, assume that $H = H'$ is the orthogonal representation obtained in the previous step, where its rectilinear image $\overline{H}$ is computed by the algorithm of Rhaman et al. [21]. By a careful analysis of how this algorithm works, we prove that each series gets spirality at most four in H (see Appendix B of [11]).

4 Drawing Algorithm

Let G be a biconnected 3-planar graph with a distinguished edge e and let T be the SPQR-tree of G with respect to e. Section 4.1 gives a linear-time algorithm to compute bend-minimum orthogonal representations of the inner components of T. Section 4.2 handles the root child of T to complete a bend-minimum representation with e on the external face and it proves Theorem 1. Lemma 3 allows us to restrict our algorithm to search for a representation satisfying Properties O1-O3.

4.1 Computing Orthogonal Representations for Inner Components

Let T be the SPQR-tree of G with respect to reference edge e and let μ be an inner node of T. A key ingredient of our algorithm is the concept of 'equivalent' orthogonal representations of G_μ. Intuitively, two representations of G_μ are equivalent if one can replace the other in any orthogonal representation of G. Similar equivalence concepts have been used for orthogonal drawings [8,10]. As we shall prove (see Theorem 3), for planar 3-graphs a simpler definition of equivalent representations suffices. If μ is a P- or an R-node, two representations H_μ and H'_μ are *equivalent* if they are both ⬖-shaped or both ⬗-shaped. If μ is an inner S-node, H_μ and H'_μ are *equivalent* if they have the same spirality.

Lemma 4. *If H_μ and H'_μ are two equivalent orthogonal representations of G_μ, the two contour paths of H_μ have the same turn number as those of H'_μ.*

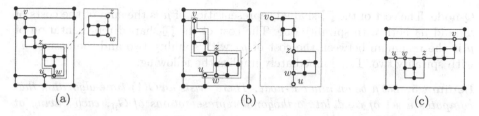

Fig. 4. (a) An orthogonal representation H; a D-shaped R-component with poles $\{w, z\}$ and an equivalent representation of it are in the blue frames. (b) A representation obtained from H by replacing the R-component with the equivalent one; a 1-spiral S-component with poles $\{u, v\}$ and an equivalent one are shown in the red frames. (c) The representation obtained by replacing the S-component with the equivalent one.

Suppose that H_μ is an inner component of H with poles u and v, and let p_l and p_r be the contour paths of H_μ. *Replacing H_μ in H with an equivalent representation H'_μ* means to insert H'_μ in H in place of H_μ, in such a way that: (i) if H_μ and H'_μ are ▨-shaped, the contour path p' of H'_μ for which $t(p') = t(p_l)$ is traversed clockwise from u to v on the external boundary of H'_μ (as for p_l on the external boundary of H_μ); (ii) in all cases, the external angles of H'_μ at u and v are the same as in H_μ. This operation may require to mirror H'_μ (see Fig. 4). The next theorem uses arguments similar to [8].

Theorem 3. *Let H be an orthogonal representation of a planar 3-graph G and H_μ be the restriction of H to G_μ, where μ is an inner component of the SPQR-tree T of G with respect to a reference edge e. Replacing H_μ in H with an equivalent representation H'_μ yields a planar orthogonal representation H' of G.*

We are now ready to describe our drawing algorithm. It is based on a dynamic programming technique that visits bottom-up the SPQR-tree T with respect to the reference edge e of G. Based on Lemma 3 and Theorem 3, the algorithm stores for each visited node μ of T a *set of candidate orthogonal representations* of G_μ, together with their cost in terms of bends. For a Q-node, the set of candidate orthogonal representations consists of three representations, with 0, 1, and 2 bends, respectively. This suffices by Property O1. For a P- or an R-node, the set of candidate representations consists of a bend-minimum ▨-shaped and a bend-minimum ▢-shaped representation. This suffices by Property O2. For an S-node, the set of candidate representations consists of a bend-minimum representation for each value of spirality $0 \leq k \leq 4$. This suffices by Property O3. In the following we explain how to compute the set of candidate representations for a node μ that is a P-, an S-, or an R-node (computing the set of a Q-node is trivial). To achieve overall linear-time complexity, the candidate representations stored at μ are described incrementally, linking the desired representation in the set of the children of μ for each virtual edge of skel(μ).

Candidate Representations for a P-node. By property T1 of Lemma 1, μ has two children μ_1 and μ_2, where μ_1 is an S-node and μ_2 is an S-node or a

Q-node. The cost of the ⌐⌐-shaped representation of μ is the sum of the costs of μ_1 and μ_2 both with spirality one. The cost of the ⌐-shaped representation of μ is the minimum between the cost of μ_1 with spirality two and the cost of μ_2 with spirality two. This immediately implies the following.

Lemma 5. *Let μ be an inner P-node. There exists an $O(1)$-time algorithm that computes a set of candidate orthogonal representations of G_μ, each having at most two bends per edge.*

Candidate Representations for an S-node. By property T3 of Lemma 1, skel(μ) without its reference edge is a sequence of edges such that the first edge and the last edge are real (they correspond to Q-nodes) and at most one virtual edge, corresponding to either a P- or an R-node, appears between two real edges. Let c_0 be the sum of the costs of the cheapest (in terms of bends) orthogonal representations of all P-nodes and R-nodes corresponding to the virtual edges of skel(μ). By Property O2, each of these representations is either ⌐- or ⌐⌐-shaped. Let n_Q be the number of edges of skel(μ) that correspond to Q-nodes and let n_D be the number of edges of skel(μ) that correspond to P- and R-nodes whose cheapest representation is ⌐⌐-shaped. Obviously, any bend-minimum orthogonal representation of G_μ satisfying O2 has cost at least c_0. We have the following.

Lemma 6. *An inner S-component admits a bend-minimum orthogonal representation respecting Properties O1-O3 and with cost c_0 if its spirality $k \leq n_Q + n_D - 1$ and with cost $c_0 + k - n_Q - n_D + 1$ if $k > n_Q + n_D - 1$.*

Note that the possible presence in skel(μ) of virtual edges corresponding to P- and R-nodes whose cheapest representation is ⌐⌐-shaped does not increase the spirality reachable at cost c_0 by the S-node. Lemma 6 also provides an alternative proof of a known result ([8, Lemma 5.2]), stating that for a planar 3-graph the number of bends of a bend-minimum k-spiral representation of an inner S-component does not decrease when k increases. Moreover, since for an inner S-component $n_Q \geq 2$, a consequence of Lemma 6 is Corollary 1. It implies that every bend-minimum k-spiral representation of an inner S-component does not require additional bends with respect to the bend-minimum representations of their subcomponents when $k \in \{0, 1\}$.

Corollary 1. *For each $k \in \{0, 1\}$, every inner S-component admits a bend-minimum orthogonal representation of cost c_0 with spirality k.*

Lemma 7. *Let μ be an inner S-node and n_μ be the number of vertices of skel(μ). There exists an $O(n_\mu)$-time algorithm that computes a set of candidate orthogonal representations of G_μ, each having at most two bends per edge.*

Candidate Representations for an R-node. If μ is an R-node, its children are S- or Q-nodes (Property T2 of Lemma 1). To compute a bend-minimum orthogonal representation of G_μ that satisfies Properties O1-O3, we devise a variant of the linear-time algorithm by Rahman, Nakano, and Nishizeki [19] that exploits the properties of inner S-components.

Lemma 8. *Let μ be an inner R-node and n_μ be the number of vertices of* skel(μ). *There exists an $O(n_\mu)$-time algorithm that computes a set of candidate orthogonal representations of G_μ, each having at most two bends per edge.*

Proof (sketch). Let $\{u, v\}$ be the poles of μ. Our algorithm works in two steps. First, it computes an ▢-shaped orthogonal representation $\tilde{H}_\mu^{\square}$ and a ▢-shaped orthogonal representation $\tilde{H}_\mu^{\square}$ of $\tilde{G}_\mu = $ skel$(\mu) \setminus (u, v)$, with a variant of the recursive algorithm in [19]. Then, it computes a bend-minimum ▢-shaped representation $H_\mu^{\square}$ and a bend-minimum ▢-shaped representation $H_\mu^{\square}$ of G_μ, by replacing each virtual edge e_S in each of $\tilde{H}_\mu^{\square}$ and $\tilde{H}_\mu^{\square}$ with the representation in the set of the corresponding S-node whose spirality equals the number of bends of e_S. Every time the algorithm needs to insert a degree-2 vertex along an edge of a bad cycle, it adds this vertex on a virtual edge, if such an edge exists. By Corollary 1, this vertex does not cause an additional bend in the final representation when the virtual edge is replaced by the corresponding S-component.

4.2   Handling the Root Child Component

Let T be the SPQR-tree of G with respect to edge $e = (u, v)$ and let μ be the root child of T. Assuming to have already computed the set of candidate representations for the children of μ, we compute an orthogonal representation H_μ of G_μ and a bend-minimum orthogonal representation H of G (with e on the external face) depending on the type of μ.

Algorithm P-root-child. Let μ be a P-node with children μ_1 and μ_2. By Property T1 of Lemma 1, both μ_1 and μ_2 are S-nodes. Let k_1 (k_2) be the maximum spirality of a representation H_{μ_1} (H_{μ_2}) at the same cost $c_{0,1}$ ($c_{0,2}$) as a 0-spiral representation. W.l.o.g., let $k_1 \geq k_2$. We have three cases:
Case 1: $k_1 \geq 4$. Compute a ▣-shaped H_μ by merging a 4-spiral and a 2-spiral representation of μ_1 and μ_2, respectively; add e with 0 bends to get H. Case 2: $k_1 = 3$. Compute an ▚-shaped H_μ by merging a 3-spiral and a 1-spiral representation of μ_1 and μ_2, respectively; add e with 1 bend to get H. Case 3: $k_1 = 2$ or $k_2 = k_1 = 1$. Compute a ▢-shaped H_μ by merging a 2-spiral and a 0-spiral representation of μ_1 and μ_2, respectively; add e with 2 bends to get H.

Lemma 9. *P-root-child computes a bend-minimum orthogonal representation of G with e on the external face and at most two bends per edge in $O(1)$ time.*

Algorithm S-root-child. Let μ be an S-node. if G_μ starts and ends with one edge, we compute the candidate orthogonal representations of G_μ as if it were an inner S-node, and we obtain H by adding e with zero bends to the 2-spiral representation of G_μ. Else, if G_μ only starts or ends with one edge, we add e to the other end of G_μ, compute the candidate representations of $G_\mu \cup \{e\}$ as if it were an inner S-node, and obtain G by adopting the representation of $G_\mu \cup \{e\}$ with spirality 3 and by identifying the first and last vertex. Finally, if skel$(\mu) \setminus \{e\}$ starts and ends with an R- or a P-node, we add two copies e', e'' of e

at the beginning and at the end of G_μ, compute the candidate representations of $G_\mu \cup \{e', e''\}$ as if it were an inner S-node, and obtain H from the representation of $G_\mu \cup \{e', e''\}$ with spirality 4, by identifying the first and last vertex of $G_\mu \cup \{e', e''\}$ and by smoothing the resulting vertex.

Lemma 10. *S-root-child computes a bend-minimum orthogonal representation of G with e on the external face and at most two bends per edge in $O(n_\mu)$ time, where n_μ is the number of vertices of* skel(μ).

Algorithm R-root-child. Let μ be an R-node and let ϕ_1 and ϕ_2 be the two planar embeddings of skel(μ) obtained by choosing as external face one of those incident to e. For each ϕ_i, compute an orthogonal representation H_i of G by: (i) finding a representation $\tilde{H}_i$ of skel(μ) (included e) with the variant of [19] given in the proof of Lemma 8, but this time assuming that all the four designated corners of the external face in the initial step must be found; (ii) replacing each virtual edge that bends $k \geq 0$ times in $\tilde{H}_i$ with a minimum-bend k-spiral representation of its corresponding S-component. H is the cheapest of H_1 and H_2. Since the variant of [19] applied to skel(μ) still causes at most two bends per edge, with the same arguments as in Lemma 8 we have:

Lemma 11. *R-root-child computes a bend-minimum orthogonal representation of G with e on the external face and at most two bends per edge in $O(n_\mu)$ time, where n_μ be the number of vertices of* skel(μ).

Proof of Theorem 1. If G is biconnected, Lemmas 5, 7, 8, 9–11 yield an $O(n)$-time algorithm that computes a bend-minimum orthogonal representation of G with a distinguished edges e on the external face and at most two bends per edge. Call BendMin-RefEdge this algorithm. An extension of BendMin-RefEdge to a simply-connected graph G, which still runs in $O(n)$ time, is easily derivable by exploiting the block-cut-vertex tree of G (see Appendix C of [11]). Running BendMin-RefEdge for every possible reference edge, we find in $O(n^2)$ time a bend-minimum orthogonal representation of G over all its planar embeddings. If v is a distinguished vertex of G, running BendMin-RefEdge for every edge incident to v, we find in $O(n)$ time a bend-minimum orthogonal representation of G with v on the external face (recall that $\deg(v) \leq 3$). Finally, an orthogonal drawing of G is computed in $O(n)$ time from an orthogonal representation of G [7].

5   Open Problems

We suggest two research directions related to our results: (i) Is there an $O(n)$-time algorithm to compute a bend-minimum orthogonal drawing of a 3-connected planar cubic graph, for every possible choice of the external face? (ii) It is still unknown whether an $O(n)$-time algorithm for the bend-minimization problem in the fixed embedding setting exists [9]. This problem could be tackled with non-flow based approaches. A positive result in this direction is given in [20] for plane 3-graphs.

References

1. Biedl, T.C., Kant, G.: A better heuristic for orthogonal graph drawings. Comput. Geom. **9**(3), 159–180 (1998). https://doi.org/10.1016/S0925-7721(97)00026-6
2. Bläsius, T., Rutter, I., Wagner, D.: Optimal orthogonal graph drawing with convex bend costs. ACM Trans. Algorithms **12**(3), 33:1–33:32 (2016). https://doi.org/10.1145/2838736
3. Brandenburg, F., Eppstein, D., Goodrich, M.T., Kobourov, S., Liotta, G., Mutzel, P.: Selected open problems in graph drawing. In: Liotta, G. (ed.) GD 2003. LNCS, vol. 2912, pp. 515–539. Springer, Heidelberg (2004). https://doi.org/10.1007/978-3-540-24595-7_55
4. Chang, Y., Yen, H.: On bend-minimized orthogonal drawings of planar 3-graphs. In: Aronov, B., Katz, M.J. (eds.) 33rd International Symposium on Computational Geometry, SoCG 2017, 4–7 July 2017, Brisbane, Australia. LIPIcs, vol. 77, pp. 29:1–29:15. Schloss Dagstuhl - Leibniz-Zentrum fuer Informatik (2017). https://doi.org/10.4230/LIPIcs.SoCG.2017.29, http://www.dagstuhl.de/dagpub/978-3-95977-038-5
5. Cohen, M.B., Madry, A., Tsipras, D., Vladu, A.: Matrix scaling and balancing via box constrained newton's method and interior point methods. In: Umans, C. (ed.) 58th IEEE Annual Symposium on Foundations of Computer Science, FOCS 2017, Berkeley, CA, USA, October 15–17, 2017. pp. 902–913. IEEE Computer Society (2017). https://doi.org/10.1109/FOCS.2017.88, http://ieeexplore.ieee.org/xpl/mostRecentIssue.jsp?punumber=8100284
6. Cornelsen, S., Karrenbauer, A.: Accelerated bend minimization. J. Graph Algorithms Appl. **16**(3), 635–650 (2012). https://doi.org/10.7155/jgaa.00265
7. Di Battista, G., Eades, P., Tamassia, R., Tollis, I.G.: Graph Drawing: Algorithms for the Visualization of Graphs. Prentice-Hall, Englewood Cliffs (1999)
8. Di Battista, G., Liotta, G., Vargiu, F.: Spirality and optimal orthogonal drawings. SIAM J. Comput. **27**(6), 1764–1811 (1998). https://doi.org/10.1137/S0097539794262847
9. Di Giacomo, E., Liotta, G., Tamassia, R.: Graph drawing. In: Goodman, J., O'Rourke, J., Toth, C. (eds.) Handbook of Discrete and Computational Geometry, 3rd edn, pp. 1451–1477. Chapman and Hall/CRC, Boca Raton (2017)
10. Didimo, W., Liotta, G., Patrignani, M.: On the Complexity of HV-rectilinear Planarity Testing. In: Duncan, C., Symvonis, A. (eds.) GD 2014. LNCS, vol. 8871, pp. 343–354. Springer, Heidelberg (2014). https://doi.org/10.1007/978-3-662-45803-7_29
11. Didimo, W., Liotta, G., Patrignani, M.: Bend-minimum orthogonal drawings inquadratic time. CoRR 1804.05813v3 (2018). http://arxiv.org/abs/1804.05813v3
12. Garg, A., Liotta, G.: Almost bend-optimal planar orthogonal drawings of biconnected degree-3 planar graphs in quadratic time. In: Kratochvíyl, J. (ed.) GD 1999. LNCS, vol. 1731, pp. 38–48. Springer, Heidelberg (1999). https://doi.org/10.1007/3-540-46648-7_4
13. Garg, A., Tamassia, R.: On the computational complexity of upward and rectilinear planarity testing. SIAM J. Comput. **31**(2), 601–625 (2001). https://doi.org/10.1137/S0097539794277123
14. Gutwenger, C., Mutzel, P.: A linear time implementation of SPQR-trees. In:Marks, J. (ed.) Graph Drawing, 8th International Symposium, GD 2000,Colonial Williamsburg, VA, USA, September 20-23, 2000, Proceedings. LectureNotes in Computer Science, vol. 1984, pp. 77–90. Springer (2000).https://doi.org/10.1007/3-540-44541-2_8

15. Hopcroft, J.E., Tarjan, R.E.: Dividing a graph into triconnected components. SIAM J. Comput. **2**(3), 135–158 (1973). https://doi.org/10.1137/0202012
16. Kant, G.: Drawing planar graphs using the canonical ordering. Algorithmica 16(1), 4–32 (1996). https://doi.org/10.1007/BF02086606
17. Liu, Y., Morgana, A., Simeone, B.: A linear algorithm for 2-bend embeddings of planar graphs in the two-dimensional grid. Discrete Appl. Math. **81**(1–3), 69–91 (1998). https://doi.org/10.1016/S0166-218X(97)00076-0
18. Rahman, M.S., Egi, N., Nishizeki, T.: No-bend orthogonal drawings of subdivisions of planar triconnected cubic graphs. IEICE. Transactions **88-D(1)**, 23–30 (2005)
19. Rahman, M.S., Nakano, S., Nishizeki, T.: A linear algorithm for bend-optimal orthogonal drawings of triconnected cubic plane graphs. J. Graph Algorithms Appl. **3**(4), 31–62 (1999). http://www.cs.brown.edu/publications/jgaa/accepted/99/SaidurNakanoNishizeki99.3.4.pdf
20. Rahman, M.S., Nishizeki, T.: Bend-minimum orthogonal drawings of plane 3-graphs. In: Goos, G., Hartmanis, J., van Leeuwen, J., Kučera, L. (eds.) WG 2002. LNCS, vol. 2573, pp. 367–378. Springer, Heidelberg (2002). https://doi.org/10.1007/3-540-36379-3_32
21. Rahman, M.S., Nishizeki, T., Naznin, M.: Orthogonal drawings of plane graphs without bends. J. Graph Algorithms Appl. **7**(4), 335–362 (2003). http://jgaa.info/accepted/2003/Rahman+2003.7.4.pdf
22. Storer, J.A.: The node cost measure for embedding graphs on the planar grid (extended abstract). In: Miller, R.E., Ginsburg, S., Burkhard, W.A., Lipton, R.J. (eds.) Proceedings of the 12th Annual ACM Symposium on Theory of Computing, 28–30 April 1980, Los Angeles, California, USA, pp. 201–210. ACM (1980). https://doi.org/10.1145/800141.804667
23. Tamassia, R.: On embedding a graph in the grid with the minimum number of bends. SIAM J. Comput. 16(3), 421–444 (1987). https://doi.org/10.1137/0216030
24. Tamassia, R., Tollis, I.G., Vitter, J.S.: Lower bounds for planar orthogonal drawings of graphs. Inf. Process. Lett. **39**(1), 35–40 (1991). https://doi.org/10.1016/0020-0190(91)90059-Q
25. Thomassen, C.: Plane representations of graphs. In: Bondy, J., Murty, U. (eds.) Progress in Graph Theory, pp. 43–69 (1987)
26. Zhou, X., Nishizeki, T.: Orthogonal drawings of series-parallel graphs with minimum bends. SIAM J. Discrete Math. **22**(4), 1570–1604 (2008). https://doi.org/10.1137/060667621

Greedy Rectilinear Drawings

Patrizio Angelini[1], Michael A. Bekos[1], Walter Didimo[2], Luca Grilli[2],
Philipp Kindermann[3]([envelope]) [ORCID], Tamara Mchedlidze[4], Roman Prutkin[4],
Antonios Symvonis[5], and Alessandra Tappini[2]

[1] Institut für Informatik, Universität Tübingen, Tübingen, Germany
{angelini,bekos}@informatik.uni-tuebingen.de
[2] Università degli Studi di Perugia, Perugia, Italy
{walter.didimo,luca.grilli}@unipg.it,
alessandra.tappini@studenti.unipg.it
[3] David R. Cheriton School of Computer Science, University of Waterloo,
Waterloo, Canada
philipp.kindermann@uwaterloo.ca
[4] Institute of Theoretical Informatics, Karlsruhe Institute of Technology,
Karlsruhe, Germany
mched@iti.uka.de, roman.prutkin@kit.edu
[5] School of Applied Mathematical and Physical Sciences, NTUA, Athens, Greece
symvonis@math.ntua.gr

Abstract. A drawing of a graph is *greedy* if for each ordered pair of vertices u and v, there is a path from u to v such that the Euclidean distance to v decreases monotonically at every vertex of the path. The existence of greedy drawings has been widely studied under different topological and geometric constraints, such as planarity, face convexity, and drawing succinctness. We introduce *greedy rectilinear drawings*, in which each edge is either a horizontal or a vertical segment. These drawings have several properties that improve human readability and support network routing. We address the problem of testing whether a planar *rectilinear representation*, i.e., a plane graph with specified vertex angles, admits vertex coordinates that define a greedy drawing. We provide a characterization, a linear-time testing algorithm, and a full generative scheme for *universal* greedy rectilinear representations, i.e., those for which every drawing is greedy. For general greedy rectilinear representations, we give a combinatorial characterization and, based on it, a polynomial-time testing and drawing algorithm for a meaningful subset of instances.

1 Introduction

In a *greedy drawing* of a graph in the plane every vertex is mapped to a distinct point and, for each ordered pair of vertices u and v, there is a *distance-decreasing*

This work started at the Bertinoro Workshop on Graph Drawing 2017, Italy. Research was partially supported by DFG grant Ka812/17-1 and by the project "Algoritmi e sistemi di analisi visuale di reti complesse e di grandi dimensioni" - Ricerca di Base 2018, Dipartimento di Ingegneria dell'Università degli Studi di Perugia.

T. Biedl and A. Kerren (Eds.): GD 2018, LNCS 11282, pp. 495–508, 2018.
https://doi.org/10.1007/978-3-030-04414-5_35

path from u to v, i.e., a path such that the Euclidean distance to v decreases monotonically at every vertex of the path. Greedy drawings have been originally proposed to support *greedy routing schemes* for ad hoc wireless networks [20–22]. In such schemes, a node that has to send a packet to a destination v just forwards the packet to one of its neighbors that is closer to v than itself. In their seminal work, Papadimitriou and Ratajczak [20,21] showed that 3-connected planar graphs form the largest class of graphs for which every instance may admit a greedy drawing, and they formulated two conjectures. *Weak conjecture*: Every 3-connected planar graph admits a greedy drawing. *Strong conjecture*: Every 3-connected planar graph admits a *convex* greedy drawing, i.e., a planar greedy drawing with convex faces. Concerning the weak conjecture, Dhandapani [8] provided an existential proof of greedy drawings for maximal planar graphs. Later on, Leighton and Moitra [16] and Angelini et al. [4] independently settled the weak conjecture positively, by also describing constructive algorithms. Da Lozzo et al. [7] strengthened these results, showing that in fact every 3-connected planar graph admits a *planar* greedy drawing. However, the strong conjecture is still open. For graphs that are not 3-connected, Nöllenburg and Prutkin [18] characterized the trees that admit a greedy drawing.

Greedy drawings have also been investigated in terms of *succinctness*, an important property that helps to make greedy routing schemes work in practice. A drawing is succinct if the vertex coordinates are represented by a polylogarithmic number of bits. Since there exist greedy-drawable graphs in the Euclidean sense that do not admit a succinct greedy drawing [3], several papers also studied succinct greedy drawings in spaces different from the Euclidean one or according to a metric different from the Euclidean distance [11–14,17,24].

We finally mention another model, called *self-approaching drawing*, that reinforces the properties of greedy drawings [1,19]. A straight-line drawing is self-approaching if for any ordered pair of vertices u and v, there exists a path P from u to v such that, for any point p on P, the distance from u to p always decreases while continuously moving along P in the drawing. Clearly, every self-approaching drawing is greedy, but not vice versa. In particular, the *dilation* of self-approaching drawings is bounded by a constant [15], while for greedy drawings it may be unbounded [1]. The dilation (or "stretch-factor") of a straight-line drawing is the maximum value of the ratio between the length of the shortest path between two vertices in the drawing and their Euclidean distance.

Motivation and Contribution. Our work is motivated by the rich literature on greedy drawings that satisfy some interesting topological or geometric requirements, such as planarity [7] and face convexity [13,14,20,24]. We study greedy drawings in the popular *orthogonal drawing* convention [9,10,23]: Vertices are mapped to points and edges are sequences of horizontal and vertical segments (each vertex has degree at most 4). More precisely, we introduce planar *greedy rectilinear drawings*, that is, crossing-free greedy drawings where each edge is either a horizontal or a vertical segment. We address the following question: "Let H be a *planar rectilinear representation*, i.e., a plane graph with given values ($90°$, $180°$, $270°$) for the geometric angles around each vertex; is it

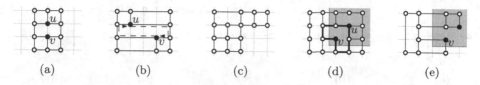

Fig. 1. (a) A rectilinear drawing that is not greedy; (b) A greedy rectilinear drawing of the same representation (the distance-decreasing paths between u and v are dashed). (c) Drawing of a universal greedy rectilinear representation. (d)–(e) H is not greedy realizable if an internal face is not a rectangle or the external face is not orthoconvex.

possible to assign coordinates to the vertices of H so that the resulting drawing is greedy rectilinear?". Figure 1a shows a rectilinear drawing that is not greedy; however, the corresponding rectilinear representation has a greedy drawing, as shown in Fig. 1b. Our question fits into the effective *topology-shape-metrics* approach [5,23], which first computes a planar embedding of the graph, then finds an embedding-preserving orthogonal representation, and finally assigns coordinates to vertices and bends to complete the drawing; we address this last step, but our representations have no bend. Our contribution is as follows.

Section 2 discusses basic properties of greedy rectilinear drawings. We prove that the faces are convex and the dilation is bounded by a small constant, and we show convex (non-rectilinear) greedy drawings in which every distance-decreasing path between two nodes is arbitrarily longer than the Euclidean distance.

Section 3 focuses on planar *universal greedy* rectilinear representations for which *every* drawing is greedy (see Fig. 1c). We give a linear-time recognition algorithm that, in the positive case, computes a greedy drawing of minimum area on an integer grid. We also describe a generative scheme for constructing any possible universal greedy rectilinear representation starting from a rectangle.

Section 4 extends the study to general rectilinear greedy representations. We give a non-geometric characterization of this class, which leads to a linear-time testing algorithm for a meaningful subset of instances. If the condition of the characterization is satisfied, a greedy drawing of minimum area within that condition can be computed in quadratic time. However, we show that in general greedy rectilinear representations may require exponential area.

We assume familiarity with basic concepts of graph drawing and planarity [9]; for space reasons, terminology and some proofs are moved to the full version [2].

2   Basic Properties of Greedy Rectilinear Representations

We denote by $x(v)$ and $y(v)$ the x- and the y-coordinate of a vertex v in a drawing Γ of a graph. For two vertices u and v of Γ, $d(u, v)$ is the Euclidean distance between u and v and a path from u to v in Γ is a *u-v-path*. The degree of v is denoted as $\deg(v)$. If v has neighbors $u_1, u_2, \ldots, u_h$, the *cell* of v in Γ, denoted as cell(v), is the (possibly unbounded) region of all points of the plane that are

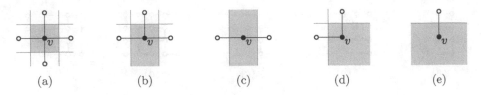

Fig. 2. Different types of cells (shaded regions) of a vertex v in a rectilinear drawing of a graph: (a) $\deg(v) = 4$; (b) $\deg(v) = 3$; (c)–(d) $\deg(v) = 2$, (e) $\deg(v) = 1$.

closer to v than to any u_i. Figure 2 shows all types of cells of a vertex v in a rectilinear drawing, depending on $\deg(v)$ and on the angles at v; if $\deg(v) \le 3$, cell(v) is unbounded. The following geometric characterization is proven in [21].

Theorem 1 (Papadimitriou and Ratajczak [21]). *A drawing of a graph is greedy if and only if for every vertex v, cell(v) contains no vertex other than v.*

If a rectilinear representation H admits a greedy rectilinear drawing, H is *greedy realizable* or, equivalently, it is a *greedy rectilinear representation*. W.l.o.g., we shall assume that H comes with a fixed "rotation", i.e., for any edge (u, v), it is fixed whether u is to the left, to the right, above, or below v in every rectilinear drawing Γ of H. A *flat vertex* of H (or of Γ) is a vertex with a flat angle (180°). A flat angle formed by two horizontal segments is *north-oriented* (*south-oriented*) if it is above (below) the two segments. A flat angle between two vertical segments is either *east-oriented* or *west-oriented*.

We restrict our study to biconnected graphs, as otherwise the set of greedy rectilinear drawings may be very limited (see the full version [2]). Lemma 1 (proved in the full version [2]) allows us to further restrict to *convex* rectilinear representations, i.e., those having rectangular internal faces and an orthoconvex polygon as external boundary. Indeed, if H is not convex, there exist two vertices u, v such that $u \in$ cell(v) in any drawing of H (see also Figs. 1d–e).

Lemma 1. *H is greedy realizable only if it is convex.*

For a rectilinear drawing of a convex rectilinear representation H, let $R(u, v)$ denote the minimum bounding box (rectangle or segment) including u and v. The next property immediately follows from the convexity of H.

Property 1. Let f be a face of H and w be any vertex of H with an angle of 90° inside f. Denote by u and v the two neighbors of w along the boundary of f. In any rectilinear drawing of H, there is no vertex properly inside $R(u, v)$.

We exploit Property 1 to prove that rectilinear greedy drawings have bounded dilation, where the paths that determine the dilation are distance-decreasing. An analogous statement does not hold for general convex greedy drawings; an example of this fact is in the full version [2], together with the proof of Theorem 2.

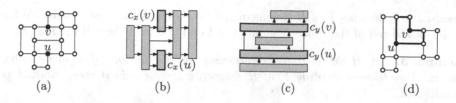

(a)   (b)   (c)   (d)

Fig. 3. (a) A greedy realizable convex rectilinear representation H that is not universal greedy due to conflict $\{u,v\}$; (b)–(c) The DAGs D_x and D_y for H. (d) A non-convex representation such that $u \in \text{cell}(v)$ for any drawing, even though $u \prec_x v$ and $u \prec_y v$.

Theorem 2. *In a rectilinear greedy drawing on an integer grid, for every two vertices s, t there is a distance-decreasing s-t-path of length at most $3\sqrt{2} \cdot d(s,t)$.*

Conflicts in Rectilinear Representations. We now define two directed acyclic graphs (DAGs) D_x and D_y associated with H, already used for orthogonal compaction [6,23]. They are fundamental tools for the rest of the paper. D_x is obtained from H by orienting the horizontal edges from left to right and by contracting each maximal path of vertical edges into a node. D_y is defined symmetrically on the maximal paths of horizontal edges; see Fig. 3. D_x and D_y may have multiple edges and they are *st-digraphs* (they have a single source and a single sink), since the external face of H is orthoconvex. For any vertex u of H, we denote by $c_x(u)$ ($c_y(u)$) the node of D_x (D_y) corresponding to the maximal vertical (horizontal) path containing u in H. If $c_x(u) \neq c_x(v)$, the notation $u \prec_x v$ ($u \not\prec_x v$) denotes the existence (absence) of a directed path from $c_x(u)$ to $c_x(v)$ in D_x. The notation $u \sim_x v$ means that either $u \prec_x v$ or $v \prec_x u$ holds, while $u \not\sim_x v$ means that none of them holds. The notations $u \sim_y v$, $u \not\sim_y v$, $u \sim_y v$, and $u \not\sim_y v$ are symmetric for D_y. Clearly, $\prec_x$ and $\prec_y$ are transitive relations. The next lemma (proved in the full version [2]) states that there is a directed path between any two vertices of H in at least one of D_x and D_y.

Lemma 2. *For any two vertices u and v of a convex rectilinear representation H, at least one of the following holds: (i) $u \sim_x v$ or (ii) $u \sim_y v$.*

Let u and v be two vertices of H such that $c_x(u) \neq c_x(v)$ and $c_y(u) \neq c_y(v)$. By Lemma 2, at least one of $u \sim_x v$ and $u \sim_y v$ holds, say the latter. If $u \sim_x v$ also holds, the relative positions (left/right/top/bottom) of u and v are fixed (they are the same in any drawing of H); in this case, we prove that none of the two vertices lies in the cell of the other in any drawing of H (Lemma 3). Conversely, this is not guaranteed if $u \not\prec_x v$ (Theorem 3), and we say that u and v are in a *conflict*, denoted by $\{u,v\}$. In this case, suppose that $u \prec_y v$ (the case $v \prec_y u$ is symmetric) and consider the topmost (flat) vertex u' of the vertical path corresponding to $c_x(u)$ and the bottommost (flat) vertex v' of the vertical path corresponding to $c_x(v)$. We say that u' and v' are *responsible* for the conflict $\{u,v\}$. A conflict $\{u,v\}$ is an x-conflict if $u \not\prec_x v$ and a y-conflict if $u \not\prec_y v$. In Fig. 3a, $\{u,v\}$ is an x-conflict, with $u' = u$ and $v' = v$. A conflict is *resolved* in a drawing Γ of H if none of the two vertices that are responsible for it lies in the cell of the other. The proof of Lemma 3 is in the full version [2].

Lemma 3. *Let H be a convex rectilinear representation of a biconnected graph. A rectilinear drawing Γ of H is greedy if and only if every conflict is resolved in Γ.*

3   Universal Greedy Rectilinear Representations

A convex rectilinear representation H is *conflict-free* if it has no conflict. The following concise characterization holds for universal rectilinear representations.

Theorem 3. *Let H be a convex rectilinear representation of a biconnected plane graph. H is universal greedy if and only if it is conflict-free.*

Proof. By Lemma 3, if H is conflict-free, every rectilinear drawing of H is greedy (note that a rectilinear representation may be conflict-free without being convex, which would imply that it is not universal greedy; see Fig. 3d).

Suppose that H is universal greedy but not conflict-free. Let Γ be any rectilinear drawing of H. Consider two vertices u and v that are responsible for a conflict in H; assume w.l.o.g. that $u \not\prec_x v$. We can further assume that there is no vertex w such that $x(u) < x(w) < x(v)$ in Γ. Indeed, if such a vertex w exists (which implies $w \not\prec_x u$ and $v \not\prec_x w$), at least one of $u \not\prec_x w$ and $w \not\prec_x v$ holds, as otherwise $u \prec_x v$. Hence, we could have selected either u and w or w and v instead of u and v. If $x(u) = x(v)$, Γ is not greedy, because $u \in \text{cell}(v)$ and $v \in \text{cell}(u)$. If $x(u) < x(v)$, we can transform Γ into a drawing Γ' by moving u and all the vertices in its vertical path to the right until $x(u) = x(v)$. Since u and v are consecutive along the x-axis in Γ and since H is convex, Γ' is still planar but not greedy, which contradicts the fact that H is universal greedy. □

Theorem 4. *Let H be a rectilinear representation of an n-vertex biconnected plane graph. There exists an $O(n)$-time algorithm to test if H is universal greedy.*

Proof. The algorithm first checks in linear time if H is convex. If not, the instance is rejected. Otherwise, it checks whether both D_x and D_y contain a (directed) Hamiltonian path, which can be done in linear time in the size of D_x and D_y, which is $O(n)$. Namely, since each of D_x and D_y is an st-digraph, computing a longest path from s to t is done in $O(n)$ time from a topological sorting. We claim that H is universal greedy if and only if this test succeeds. By Theorem 3, to prove this claim, it is enough to show that a DAG D contains a Hamiltonian path if and only if for any two vertices u and v of D, there is a directed path either from u to v or from v to u. If D has a Hamiltonian path π, a directed path between any two vertices of D is a subpath of π. Conversely, if there is a directed path between any two vertices of D, then a linear extension of a topological sorting of the vertices corresponds to a Hamiltonian path. □

Since conflict-free rectilinear representations are a subclass of the *turn-regular* orthogonal representations [6], for which a minimum-area drawing can be found in linear time, we can also state the following.

Corollary 1. *Let H be a universal greedy rectilinear representation. There is a linear-time algorithm to compute a (greedy) drawing of H with minimum area.*

Generative Scheme. Let H be a biconnected universal greedy rectilinear representation. Each of the following operations on H produces a new biconnected universal greedy rectilinear representation, which gives a generative scheme for universal greedy rectilinear representations. The proof is in the full version [2].
– *k-reflex vertex addition.* Attach to the external face of H a path of $1 \leq k \leq 4$ reflex vertices (corners) that forms a new rectangular internal face, provided that the resulting representation is convex.
– *flat vertex addition.* Subdivide an external edge (u, v) of H with a flat vertex of degree two, provided that the strip of the plane between the two lines orthogonal to (u, v) and passing through u and v, respectively, has no vertices in its interior.

Theorem 5. *Let H be a universal greedy rectilinear representation of a biconnected planar graph. H can be obtained by a suitable sequence of k-reflex vertex and flat vertex additions, starting from a rectangle.*

4   General Greedy Rectilinear Representations

We now consider convex rectilinear representations H of biconnected plane graphs that may contain conflicts, and investigate conditions under which they are greedy realizable. We present a characterization (Theorem 6), which yields a polynomial-time testing algorithm for a meaningful subclass of instances, namely when D_x and D_y are series-parallel (Theorem 9). Proofs are in the full version [2].

Let D be one of the two DAGs D_x and D_y associated with H. Since D is an *st*-digraph, it has an *st*-ordering $\mathcal{S} = v_1, \ldots, v_m$. For two indices i, j, with $1 \leq i < j \leq m$, $D\langle i, j \rangle$ denotes the subgraph of D induced by $v_i, \ldots, v_j$. We say that $\mathcal{S}$ is *good* if: (**S.1**) For any two indices i, j, with $1 \leq i < j \leq m$, $D\langle i, j \rangle$ has at most two connected components, and (**S.2**) if $D\langle i, j \rangle$ has exactly two components, then all nodes of one component precede those of the other in $\mathcal{S}$.

Further, we say that a drawing of H *respects* an *st*-ordering $\mathcal{S}_x$ of D_x ($\mathcal{S}_y$ of D_y) if for any two vertices $u, w \in H$, we have that u lies to the left of w (below w) in the drawing if and only if $c_x(u)$ precedes $c_x(w)$ in $\mathcal{S}_x$ ($c_y(u)$ precedes $c_y(w)$ in $\mathcal{S}_y$). Finally, when we refer to the x-coordinate (y-coordinate) of a node v_i of D_x (of D_y), we mean the one of all the vertices $w \in H$ with $c_x(w) = v_i$ (with $c_y(w) = v_i$), as these vertices belong to the same vertical (horizontal) path.

Theorem 6. *A convex rectilinear representation H of a biconnected plane graph is greedy realizable if and only if both DAGs D_x and D_y admit good st-orderings.*

We start by proving the necessity of Theorem 6.

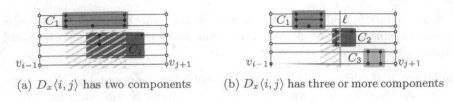

(a) $D_x\langle i,j\rangle$ has two components   (b) $D_x\langle i,j\rangle$ has three or more components

Fig. 4. Illustration for the proof of Lemma 4

Lemma 4. *If D_x or D_y admits no good st-ordering, H is not greedy realizable.*

Proof Sketch. If H admits a greedy drawing Γ respecting an st-ordering $\mathcal{S}_x$ of D_x that is not good, there exist i, j, with $1 \leq i < j \leq m$, such that $D_x\langle i,j\rangle$ has at least two connected components C_1 and C_2. First, note that the vertices of C_1 and C_2 are vertically separated by a horizontal line in Γ (say, C_1 lies above C_2; see Fig. 4), as otherwise there would be at least a pair of vertical paths whose corresponding nodes in D_x are joined by a directed path.

Since every internal face of H is rectangular, the vertices of the bottom (top) boundary of C_1 (C_2) are part of a horizontal path, spanning all x-coordinates between the ones of v_i and v_j. Thus, all vertices on the bottom (top) boundary of C_1 (C_2) are south-oriented (north-oriented) flat vertices, and the union of their cells is a connected region spanning all the x-coordinates between their leftmost and rightmost vertices; see Fig. 4a. So, if a vertex of C_1 appears between two vertices of C_2 in $\mathcal{S}_x$, then it lies inside the cell of a vertex of C_2 in Γ, and vice versa. Thus, the vertices of each component are consecutive in $\mathcal{S}_x$. Since $\mathcal{S}_x$ is not good, there is at least another component C_3 in $D_x\langle i,j\rangle$; see Fig. 4b. Consider the vertical line ℓ that is horizontally equidistant to v_i and v_j in Γ. Any two components of $D_x\langle i,j\rangle$ must be separated by ℓ in order for the cells of the vertices of these components to be empty, which is not possible for three components. $\square$

To prove the sufficiency of Theorem 6, we assign the x- and y-coordinates in two steps, which can be performed independently due to the following lemma.

Lemma 5. *Let Γ_1 and Γ_2 be two drawings of H such that all x-conflicts are resolved in Γ_1 and all y-conflicts are resolved in Γ_2. Then, the drawing Γ_3 of H in which the x-coordinate of each vertex is the same as in Γ_1 and the y-coordinate of each vertex is the same as in Γ_2 is greedy.*

We describe the assignment of the x-coordinates based on the good st-ordering $\mathcal{S}_x = v_1, \ldots, v_m$ of D_x. The assignment of the y-coordinates based on the good st-ordering of D_y works symmetrically. We first prove in Lemma 6 that, to guarantee that every x-conflict is resolved, it suffices to resolve a specific subset of them, which is called *minimal*. Namely, an x-conflict $\{u, v\}$ *dominates* an x-conflict $\{w, z\}$, with $c_x(u) = v_i$, $c_x(v) = v_j$, $c_x(w) = v_k$, and $c_x(z) = v_\ell$, if $k \leq i < j \leq \ell$. A *minimal* x-conflict is not dominated by any x-conflict.

By Lemmas 3 and 6, we conclude that a greedy rectilinear drawing can be obtained by resolving all the minimal conflicts. We finally give a constructive

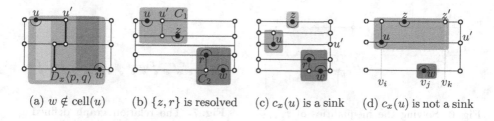

(a) $w \notin cell(u)$ (b) $\{z, r\}$ is resolved   (c) $c_x(u)$ is a sink   (d) $c_x(u)$ is not a sink

Fig. 5. Illustration for the proof of Lemma 6

proof that this can always be done since $\mathcal{S}_x$ is good. In particular, we encode that a minimal x-conflict is resolved with a single inequality on the horizontal distances between the vertices in the x-conflict. Then, in Lemma 7, we prove that for a minimal x-conflict $\{u, v\}$ the nodes $c_x(u)$ and $c_x(v)$ of D_x are consecutive in $\mathcal{S}_x$. We use this property to show that the system of inequalities describing the conditions for the minimal x-conflicts to be resolved always admits a solution.

Lemma 6. *Let Γ be a rectilinear drawing of H respecting $\mathcal{S}_x$. If every minimal x-conflict dominating an x-conflict $\{u, w\}$ is resolved in Γ, $\{u, w\}$ is resolved.*

Proof Sketch. We may assume that u and w are responsible for $\{u, w\}$. Let $v_i = c_x(u)$ and $v_j = c_x(w)$, with $i < j$. Since $\mathcal{S}_x$ is good, graph $D_x\langle i, j \rangle$ has at most two connected components C_1 and C_2. Assume that $v_i \in C_1$.

Suppose first that $v_j \in C_1$. Let u' be the right neighbor of u in H; see Fig. 5a. Since $v_i, v_j \in C_1$, node $c_x(u')$ precedes $c_x(w)$ in $\mathcal{S}_x$ and $c_x(u') \in C_1$. So, u' lies to the left of w in any drawing of H respecting $\mathcal{S}_x$. Hence, the mid-point of edge (u, u'), which defines the right boundary of cell(u), lies to the left of w, and thus $w \notin cell(u)$. Symmetrically, $u \notin cell(w)$.

Suppose now that $v_j \in C_2$. By symmetry, we assume that C_1 lies above and to the left of C_2; see Fig. 5b. Let z (r) be the bottommost (topmost) vertex of the vertical path corresponding to the last node $c_x(z)$ of C_1 (first node $c_x(r)$ of C_2) in $\mathcal{S}_x$, i.e., z and r are responsible for a minimal (and thus resolved) x-conflict.

We show that $w \notin cell(u)$ (by symmetry, $u \notin cell(w)$). If $u' \in C_1$, the previous case applies. Otherwise, u lies on the right boundary of C_1. If $c_x(u)$ is a sink of C_1, then C_1 contains only $c_x(u)$; see Fig. 5c. Thus, either $c_x(u)$ is not a sink of C_1, or $c_x(u) = c_x(z)$. In the latter case, $r \notin cell(u)$, since the minimal x-conflict $\{z, r\}$ is resolved, which implies $w \notin cell(u)$. Hence, $c_x(u) \neq c_x(z)$ and $c_x(u)$ is not a sink of C_1; see Fig. 5d. Since $u' \notin C_1$ and u is south-oriented, u lies below z. Let z' be the right neighbor of z, with $c_x(z') = v_k$. If z' is to the left of u', $D_x\langle i, k\rangle$ has two connected components; one contains v_i and v_k, the other contains v_j, as (u, u') cannot be crossed. This contradicts (**S.2**), as $i < j < k$. So, z' is to the right of u'. Since z is to the right of u, the right boundary of cell(z) is to the right of the one of cell(u). Since $r \notin cell(z)$, also $r \notin cell(u)$, and thus $w \notin cell(u)$. □

Lemma 7. *For any two vertices u and w of H such that $\{u, w\}$ is a minimal x-conflict, we have that $c_x(u)$ and $c_x(w)$ are consecutive in a good st-ordering $\mathcal{S}_x$.*

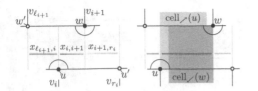 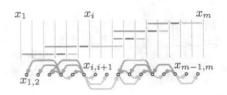

Fig. 6. Solving the inequalities of $x_{i,i+1}$ implies $u \notin \mathrm{cell}(w)$ and $w \notin \mathrm{cell}(u)$.

Fig. 7. The relation graph defined by the left and right inequalities

Proof. Suppose that there is a vertex $z \in H$ such that $c_x(z) = v_j$ lies between $c_x(u) = v_i$ and $c_x(w) = v_k$ in $\mathcal{S}_x$, i.e., $i < j < k$. If $c_x(u)$ and $c_x(w)$ belong to the same connected component C of $D_x\langle i, k \rangle$, then by definition, v_i and v_k are a source and a sink of C, respectively. Since $\{u, w\}$ is an x-conflict, we have $u \nprec_x w$; hence, there is another source $c_x(s)$ in C, for some vertex $s \in H$, such that $s \prec_x w$. Since $c_x(u)$ and $c_x(s)$ are different sources of C, we have $u \nprec_x s$, and thus $\{u, s\}$ is an x-conflict dominating the minimal x-conflict $\{u, w\}$; a contradiction. If $c_x(u)$ and $c_x(w)$ belong to different components, then $c_x(z)$ does not belong to the same component as one of them, say $c_x(u)$. Thus, $u \nprec_x z$, i.e., $\{u, z\}$ is an x-conflict dominating $\{u, w\}$; a contradiction. □

We now present our algorithm to assign x-coordinates so that all minimal x-conflicts are resolved. We extend some definitions from vertices of H to nodes of D_x. Namely, we say $v_i \prec_x v_j$ if there is a directed path in D_x from v_i and v_j. Also, we say that there is a (minimal) x-conflict $\{v_i, v_j\}$ in D_x if there is a (minimal) x-conflict $\{u, w\}$ in H such that $c_x(u) = v_i$ and $c_x(w) = v_j$.

For $0 < i, j \le m$, let $x_{i,j} := x_j - x_i$ be the x-*distance* between v_i and v_j. To prove that a good st-ordering $\mathcal{S}_x$ allows for a greedy realization, we set up a system of inequalities describing the geometric requirements for the x-distance of consecutive nodes in $\mathcal{S}_x$ in a greedy drawing, and then prove that this system always admits a solution since $\mathcal{S}_x$ is good. First note that, for every $0 < i < m$ such that there is no minimal x-conflict $\{v_i, v_{i+1}\}$, we only require the x-distance to be positive, so we define the *trivial inequality* $x_{i,i+1} > 0$.

For every $0 < i < m$ such that there is a minimal x-conflict $\{v_i, v_{i+1}\}$, we define two inequalities that describe the necessary conditions for the x-conflict to be resolved. Let u and w, with $c_x(u) = v_i$ and $c_x(w) = v_{i+1}$, be responsible for $\{v_i, v_{i+1}\}$. We assume that $u \prec_y w$; the other case is symmetric.

By assumption, v_i lies to the bottom left of v_{i+1}, so we only have to consider the part $\mathrm{cell}_{\nearrow}(w)$ of $\mathrm{cell}(w)$ to the bottom left of w (dark region in Fig. 6). Let (w', w) be the bottommost incoming edge of v_{i+1} with $c_x(w') = v_{\ell_{i+1}}$. Then, the left boundary of $\mathrm{cell}_{\nearrow}(w)$ is delimited by the vertical line through the midpoint of (w', w). Thus, we require $x_{i,i+1} > x_{\ell_{i+1},i+1}/2 \Leftrightarrow x_{i,i+1} > x_{\ell_{i+1},i}$. Symmetrically, we only consider the part $\mathrm{cell}_{\nearrow}(u)$ of $\mathrm{cell}(u)$ to the top right of u (light region in Fig. 6), which is bounded by the vertical line through the midpoint of the topmost outgoing edge (u, u') of v_i with $c_x(u') = v_{r_i}$. Thus, we require $x_{i,i+1} > x_{i,r_i}/2 \Leftrightarrow x_{i,i+1} > x_{i+1,r_i}$. Since $v_{\ell_{i+1}}$ and v_i (and v_{i+1} and v_{r_i})

are not necessarily consecutive in the st-ordering, we express the x-distance $x_{\ell_{i+1},i}$ (and x_{i,r_i}) as the sum of the x-distances between the consecutive nodes between them in the st-ordering. This gives the *left* and the *right inequality*.

$$x_{i,i+1} > \sum_{j=\ell_{i+1}}^{i-1} x_{j,j+1} \quad \text{(left inequality)} \quad x_{i,i+1} > \sum_{j=i+1}^{r_i-1} x_{j,j+1} \quad \text{(right inequality)}$$

Note that for every variable $x_{i,i+1}$ there exists either a trivial inequality or a left and right inequality. Consider the following triangulated matrices, where $c_{i,j} = -1$ if $i > j \geq \ell_{i+1}$ or $i < j \leq r_i$, where $c_{i,j} = 1$ if $i = j$, and $c_{i,j} = 0$ otherwise.

$$A = \begin{pmatrix} c_{1,1} & & 0 \\ \vdots & \ddots & \\ c_{m-1,1} & \cdots & c_{m-1,m-1} \end{pmatrix} \quad B = \begin{pmatrix} c_{1,1} & \cdots & c_{1,m-1} \\ & \ddots & \vdots \\ 0 & & c_{m-1,m-1} \end{pmatrix} \quad x = \begin{pmatrix} x_{1,2} \\ \vdots \\ x_{m-1,m} \end{pmatrix}$$

We express the left and trivial (right and trivial) inequalities as $Ax > 0$ (as $Bx > 0$). Any vector $x > 0$ determines a unique rectilinear drawing: we assign to each vertex the y-coordinate defined by S_y, we assign to v_1 the x-coordinate $x_1 = 0$ and to every other v_i the x-coordinate $x_i = x_{i-1} + x_{i-1,i}$. Since $x > 0$, the x-coordinates preserve the good st-ordering and resolve all x-conflicts.

Lemma 8. *A vector* $x = (x_{1,2}, \ldots, x_{m-1,m})^\top > 0$ *solves both* $Ax > 0$ *and* $Bx > 0$ *if and only if it determines a drawing where all x-conflicts are resolved.*

Note that we can always solve $Ax > 0$ and $Bx > 0$ independently by solving the linear equation systems $Ax = 1$ and $Bx = 1$ via forward substitution, since A and B are triangular. We prove that there is always a vector x solving $Ax > 0$ and $Bx > 0$ simultaneously. Let $C = A + B - I_{m-1}$ be the matrix defined by the values of $c_{i,j}$. Any solution to the linear inequality system $Cx > 0$ is also a solution to both $Ax > 0$ and $Bx > 0$. We show that C can be triangulated. For this, we define the *relation graph* corresponding to the adjacency matrix $I_{m-1} - C$ that contains a vertex u_i for each interval $x_{i,i+1}$, $1 \leq i < m$, and a directed edge from a vertex u_i to a vertex u_j if and only if $c_{i,j} = -1$; see Fig. 7.

Lemma 9. *The relation graph of a good st-ordering is acyclic.*

Proof Sketch. We show that a shortest cycle $\mathcal{C}$ in the relation graph has length 2, finding a "shortcut" for every longer cycle. Then, we analyze the relative order of the y-coordinates of the responsible vertices for the two minimal x-conflicts in H corresponding to $\mathcal{C}$, and find a contradiction for every combination. □

Lemma 10. *The matrix C is triangularizable.*

Proof. By Lemma 9, the relation graph described by the matrix $I_{m-1} - C$ is acyclic. Hence, there is a permutation matrix P (corresponding to a topological sort) such that $P(I_{m-1} - C)P^{-1}$ is triangulated with only 0's on the diagonal. Thus, $PI_{m-1}P^{-1} - PCP^{-1} = I_{m-1} - PCP^{-1}$ is triangulated with only 0's on the diagonal, so PCP^{-1} is triangulated with only 1's on the diagonal. □

Since C is triangularizable by Lemma 10, the system of linear equations $Cx = 1$ always has a solution, which solves $Ax > 0$ and $Bx > 0$ simultaneously. This concludes the sufficiency proof for Theorem 6.

Note that the given construction ensures that all the coordinates are integer; however, the area of the drawing is in general not minimum. A rectilinear greedy drawing with minimum area respecting the given st-orderings can be constructed in polynomial time by solving a linear program that minimizes $\sum_{i=1}^{m-1} x_{i,i+1}$ under the constraints $Ax \geq 1$ and $Bx \geq 1$. We analyze the integrality of the solution and the running time in the full version [2].

Theorem 7. *Let H be a convex rectilinear representation of a biconnected plane graph and let S_x and S_y be good st-orderings of D_x and D_y. We can compute a greedy drawing of H that respects S_x and S_y with minimum area in $O(n^2)$ time.*

Although minimum, the area of the drawings yielded by our algorithm may be non-polynomial in some cases; Theorem 8 states that there exist convex rectilinear representations (see Fig. 8) whose DAGs admit good st-orderings, but there is no combination of them resulting in a succinct greedy drawing, since the solutions of the corresponding system of inequalities are always exponential in the input size. On the contrary, every universal greedy rectilinear representation of an n-vertex graph is succinct, since by Corollary 1 it has a (greedy) drawing of minimum area on an integer grid of size $O(n^2)$ [6,23].

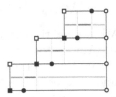

Fig. 8. Exponential area lower bound

Theorem 8. *There exist rectilinear representations whose every greedy rectilinear drawing has exponential area, even if D_x and D_y are series-parallel.*

When D_x and D_y are series-parallel, the conditions can be tested efficiently.

Theorem 9. *Let H be a convex rectilinear representation of a biconnected plane graph. If D_x and D_y are series-parallel, we can test in $O(n)$ time if H is greedy realizable. If the test succeeds, a greedy drawing of H is computed in $O(n^2)$ time.*

Proof Sketch. To find a good st-ordering for D_x, we recursively apply the following procedure. Let D_x be composed of a set of subgraphs $D_1, \ldots, D_k$, forming a parallel or a series composition. If $D_1, \ldots, D_k$ form a parallel composition, then either $k = 2$, or $k = 3$ and D_3 is a single edge; otherwise, the graph violates Condition (**S.1**). If we remove the sink and source from G, then, by Condition (**S.2**), we have a good st-ordering if and only if all nodes of D_1 precede all nodes of D_2, there is exactly one sink in D_1, and there is exactly one source in D_2. If $D_1, \ldots, D_k$ form a series composition, we construct good st-orderings of $D_1, \ldots, D_k$ recursively and merge them in a good st-ordering of D_x. $\qquad\square$

5   Open Problems

We introduced rectilinear greedy drawings, i.e., planar greedy drawings in the orthogonal drawing style with no bends. Our work reveals several interesting open problems. (1) What if we allow bends along the edges? (2) Can we always test in polynomial time whether a planar DAG admits a good st-ordering? (3) Given a biconnected plane graph G, what is the complexity of deciding whether G admits a (universal) greedy rectilinear representation? This question pertains the intermediate step of the topology-shape-metrics approach [23].

References

1. Alamdari, S., Chan, T.M., Grant, E., Lubiw, A., Pathak, V.: Self-approaching graphs. In: Didimo, W., Patrignani, M. (eds.) GD 2012. LNCS, vol. 7704, pp. 260–271. Springer, Heidelberg (2013). https://doi.org/10.1007/978-3-642-36763-2_23
2. Angelini, P., et al.: Greedy rectilinear drawings. Arxiv report 1808.09063 (2018). http://arxiv.org/abs/1808.09063
3. Angelini, P., Di Battista, G., Frati, F.: Succinct greedy drawings do not always exist. Networks **59**(3), 267–274 (2012). https://doi.org/10.1002/net.21449
4. Angelini, P., Frati, F., Grilli, L.: An algorithm to construct greedy drawings of triangulations. J. Graph Algorithms Appl. **14**(1), 19–51 (2010). https://doi.org/10.7155/jgaa.00197
5. Batini, C., Nardelli, E., Tamassia, R.: A layout algorithm for data flow diagrams. IEEE Trans. Software Eng. **12**(4), 538–546 (1986). https://doi.org/10.1109/TSE.1986.6312901
6. Bridgeman, S.S., Di Battista, G., Didimo, W., Liotta, G., Tamassia, R., Vismara, L.: Turn-regularity and optimal area drawings of orthogonal representations. Comput. Geom. **16**(1), 53–93 (2000). https://doi.org/10.1016/S0925-7721(99)00054-1
7. Da Lozzo, G., D'Angelo, A., Frati, F.: On planar greedy drawings of 3-connected planar graphs. In: Aronov, B., Katz, M.J. (eds.) Proceedings 33rd International Symposium on Computational Geometry (SoCG 2017). LIPIcs, vol. 77, pp. 33:1–33:16 (2017). https://doi.org/10.4230/LIPIcs.SoCG.2017.33
8. Dhandapani, R.: Greedy drawings of triangulations. Discrete Comput. Geom. **43**(2), 375–392 (2010). https://doi.org/10.1007/s00454-009-9235-6
9. Di Battista, G., Eades, P., Tamassia, R., Tollis, I.G.: Graph Drawing: Algorithms for the Visualization of Graphs. Prentice-Hall, Englewood Cliffs (1999)
10. Duncan, C.A., Goodrich, M.T.: Planar orthogonal and polyline drawing algorithms. In: Tamassia, R. (ed.) Handbook on Graph Drawing and Visualization. Chapman and Hall/CRC (2013). https://cs.brown.edu/~rt/gdhandbook/
11. Eppstein, D., Goodrich, M.T.: Succinct greedy geometric routing using hyperbolic geometry. IEEE Trans. Comput. **60**(11), 1571–1580 (2011). https://doi.org/10.1109/TC.2010.257
12. Goodrich, M.T., Strash, D.: Succinct greedy geometric routing in the euclidean plane. In: Dong, Y., Du, D.-Z., Ibarra, O. (eds.) ISAAC 2009. LNCS, vol. 5878, pp. 781–791. Springer, Heidelberg (2009). https://doi.org/10.1007/978-3-642-10631-6_79
13. He, X., Zhang, H.: On succinct convex greedy drawing of 3-connected plane graphs. In: Randall, D. (ed.) Proceedings 22nd Annual ACM-SIAM Symposium on Discrete Algorithms (SODA 2011), pp. 1477–1486. SIAM (2011). https://doi.org/10.1137/1.9781611973082.115

14. He, X., Zhang, H.: On succinct greedy drawings of plane triangulations and 3-connected plane graphs. Algorithmica **68**(2), 531–544 (2014). https://doi.org/10.1007/s00453-012-9682-y
15. Icking, C., Klein, R., Langetepe, E.: Self-approaching curves. Math. Proc. Camb. Phil. Soc. **125**, 441–443 (1999). https://doi.org/10.1017/S0305004198003016
16. Leighton, T., Moitra, A.: Some results on greedy embeddings in metric spaces. Discrete Comput. Geom. **44**(3), 686–705 (2010). https://doi.org/10.1007/s00454-009-9227-6
17. Leone, P., Samarasinghe, K.: Geographic routing on virtual raw anchor coordinate systems. Theor. Comput. Sci. **621**, 1–13 (2016). https://doi.org/10.1016/j.tcs.2015.12.029
18. Nöllenburg, M., Prutkin, R.: Euclidean greedy drawings of trees. In: Bodlaender, H.L., Italiano, G.F. (eds.) ESA 2013. LNCS, vol. 8125, pp. 767–778. Springer, Heidelberg (2013). https://doi.org/10.1007/978-3-642-40450-4_65
19. Nöllenburg, M., Prutkin, R., Rutter, I.: On self-approaching and increasing-chord drawings of 3-connected planar graphs. J. Comput. Geom. **7**(1), 47–69 (2016). https://doi.org/10.20382/jocg.v7i1a3
20. Papadimitriou, C.H., Ratajczak, D.: On a conjecture related to geometric routing. In: Nikoletseas, S.E., Rolim, J.D.P. (eds.) ALGOSENSORS 2004. LNCS, vol. 3121, pp. 9–17. Springer, Heidelberg (2004). https://doi.org/10.1007/978-3-540-27820-7_3
21. Papadimitriou, C.H., Ratajczak, D.: On a conjecture related to geometric routing. Theor. Comput. Sci. **344**(1), 3–14 (2005). https://doi.org/10.1016/j.tcs.2005.06.022
22. Rao, A., Papadimitriou, C.H., Shenker, S., Stoica, I.: Geographic routing without location information. In: Johnson, D.B., Joseph, A.D., Vaidya, N.H. (eds.) Proceedings of 9th Annual International Conference on Mobile Computing and Networking (MOBICOM 2003), pp. 96–108. ACM (2003). https://doi.org/10.1145/938985.938996
23. Tamassia, R.: On embedding a graph in the grid with the minimum number of bends. SIAM J. Comput. **16**(3), 421–444 (1987). https://doi.org/10.1137/0216030
24. Wang, J., He, X.: Succinct strictly convex greedy drawing of 3-connected plane graphs. Theor. Comput. Sci. **532**, 80–90 (2014). https://doi.org/10.1016/j.tcs.2013.05.024

Orthogonal and Smooth Orthogonal Layouts of 1-Planar Graphs with Low Edge Complexity

Evmorfia Argyriou[1], Sabine Cornelsen[2], Henry Förster[3(✉)],
Michael Kaufmann[3], Martin Nöllenburg[4], Yoshio Okamoto[5],
Chrysanthi Raftopoulou[6], and Alexander Wolff[7]

[1] yWorks GmbH, Tübingen, Germany
evmorfia.argyriou@yworks.com
[2] University of Konstanz, Konstanz, Germany
sabine.cornelsen@uni-konstanz.de
[3] University of Tübingen, Tübingen, Germany
{foersth,mk}@informatik.uni-tuebingen.de
[4] TU Wien, Vienna, Austria
noellenburg@ac.tuwien.ac.at
[5] RIKEN Center for Advanced Intelligence Project,
University of Electro-Communications, Chōfu, Japan
okamotoy@uec.ac.jp
[6] National Technical University of Athens, Athens, Greece
crisraft@mail.ntua.gr
[7] University of Würzburg, Würzburg, Germany

Abstract. While *orthogonal* drawings have a long history, *smooth orthogonal* drawings have been introduced only recently. So far, only planar drawings or drawings with an arbitrary number of crossings per edge have been studied. Recently, a lot of research effort in graph drawing has been directed towards the study of beyond-planar graphs such as *1-planar* graphs, which admit a drawing where each edge is crossed at most once. In this paper, we consider graphs with a fixed embedding. For 1-planar graphs, we present algorithms that yield orthogonal drawings with optimal curve complexity and smooth orthogonal drawings with small curve complexity. For the subclass of outer-1-planar graphs, which can be drawn such that all vertices lie on the outer face, we achieve optimal curve complexity for both, orthogonal and smooth orthogonal drawings.

1 Introduction

Orthogonal drawings date back to the 1980's, with Valiant's [24], Leiserson's [17] and Leighton's [16] work on VLSI layouts and floor-planning applications and

This work started at Dagstuhl seminar 16452 "Beyond-Planar Graphs: Algorithmics and Combinatorics". We thank the organizers and the other participants. The full version of this article is available at arxiv.org/abs/1808.10536.

© Springer Nature Switzerland AG 2018
T. Biedl and A. Kerren (Eds.): GD 2018, LNCS 11282, pp. 509–523, 2018.
https://doi.org/10.1007/978-3-030-04414-5_36

have been extensively studied over the years. The quality of an orthogonal drawing can be judged based on several aesthetic criteria such as the required area, the total edge length, the total number of bends, or the maximum number of bends per edge. While schematic drawings such as orthogonal layouts are very popular for technical applications (such as UML diagrams) still to date, from a cognitive point of view, schematic drawings in other applications like subway maps seem to have disadvantages over subway maps drawn with smooth Bézier curves, for example, in the context of path finding [19]. In order to "smoothen" orthogonal drawings and to improve their readability, Bekos et al. [6] introduced *smooth orthogonal drawings* that combine the clarity of orthogonal layouts with the artistic style of Lombardi drawings [11] by replacing sequences of "hard" bends in the orthogonal drawing of the edges by (potentially shorter) sequences of "smooth" inflection points connecting circular arcs. Formally, our drawings map vertices to points in $\mathbb{R}^2$ and edges to curves of one of the following two types.

Orthogonal Layout: Each edge is drawn as a sequence of vertical and horizontal line segments. Two consecutive segments of an edge meet in a bend.

Smooth Orthogonal Layout [6]**:** Each edge is drawn as a sequence of vertical and horizontal line segments as well as circular arcs: quarter arcs, semicircles, and three-quarter arcs. Consecutive segments must have a common tangent.

The maximum vertex degree is usually restricted to four since every vertex has four available *ports* (North, South, East, West), where the edges enter and leave a vertex with horizontal or vertical tangents. In addition, the usual model insists that no two edges incident to the same vertex can use the same port. Throughout this paper, we restrict ourselves to graphs of maximum degree four.

The *curve complexity* of a drawing is the maximum number of segments used for an edge. An OC_k-*layout* is an orthogonal layout with curve complexity k, that is, an orthogonal layout with at most $k-1$ bends per edge. An SC_k-*layout* is a smooth orthogonal layout with curve complexity k. For results, see Table 1.

The well-known algorithm of Biedl and Kant [7] draws any connected graph of maximum degree 4 orthogonally on a grid of size $n \times n$ with at most $2n + 2$ bends, bending each edge at most twice (and, hence, yielding OC$_3$-layouts). For the output of their algorithm applied to K_5, see Fig. 1a. Note that their approach introduces crossings to the produced drawing. For planar graphs, they describe how to obtain planar orthogonal drawings with at most two bends per edge, except possibly for one edge on the outer face.

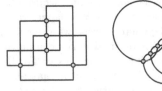

(a) OC$_3$-layout   (b) SC$_1$-layout

Fig. 1. Two 2-planar drawings of K_5.

So far, smooth orthogonal drawings have been studied nearly exclusively for planar graphs. Bekos et al. [5] showed how to compute an SC$_1$-layout for any maximum degree 4 graph, but their algorithm does not consider the embedding of the given graph. For a drawing of K_5 computed by their algorithm, see Fig. 1b.

Also, in the produced drawings, the number of crossings that an edge may have is not bounded. Bekos et al. also showed that, if one does not restrict vertex degrees, many planar graphs do not admit (planar) SC_1-layouts under the *Kandinsky model*, where the number of edges using the same port is unbounded. They proved, however, that all planar graphs of maximum degree 3 admit an SC_1-layout (under the usual port constraint). For the same class of graphs, Alam et al. [1] showed how to get a polynomial drawing area ($O(n^2) \times O(n)$) when increasing the curve complexity to SC_2. Further, they showed that every planar graph of maximum degree 4 admits an SC_2-layout, but not every such graph admits an SC_1-layout where the vertices lie on a polynomial-sized grid. They also proved that every biconnected outerplane graph of maximum degree 4 admits an SC_1-layout (respecting the given embedding).

In this paper, we study orthogonal and smooth orthogonal layouts of non-planar graphs, in particular, 1-planar graphs. Recall that k-planar graphs are those graphs that admit a drawing in the plane where each edge has at most k crossings. Our goal is to extend the well-established aesthetic criterion 'curve complexity' of (smooth) orthogonal drawings from planar to 1-planar graphs.

1-planar graphs, introduced by Ringel [18], probably form the most-studied class of the *beyond-planar* graphs, which extend the notion of planarity. There are recent surveys on both 1-planar graphs [15] and beyond-planar graphs [10]. Mostly, straight-line drawings have been studied for 1-planar graphs. While every planar graph has a planar straight-line drawing (due to Fáry's theorem), this is not true for 1-planar graphs [12,23]. For the 3-connected case, the statement holds except for at most one edge on the outer face [2]. Given a drawing of a 1-planar graph, one can decide in linear time whether it can be "straightened" [14].

An important subclass of 1-planar graphs are *outer-1-planar* graphs. These are the graphs that have a 1-planar drawing where every vertex lies on the outer (unbounded) face. They are planar graphs, can be recognized in linear time [4, 13], and can be drawn with straight-line edges and right-angle crossings [9].

We are specifically interested in 1-plane and outer-1-plane graphs, which are 1-planar and outer-1-planar graphs together with an embedding. Such an embedding determines the order of the edges around each vertex, but also which edges cross and in which order. By the *layout of a 1-plane graph* we mean that the layout respects the given embedding, without stating this again. In contrast, the *layout of a 1-planar graph* can have any 1-planar embedding.

Our Contribution. Previous results and our contribution on (smooth) orthogonal layouts are listed in Table 1. We present new layout algorithms for 1-planar graphs in the orthogonal model (Sect. 3) and in the smooth orthogonal model (Sect. 4), achieving low curve complexity and preserving 1-planarity. We study 1-plane graphs as well as the special case of outer-1-plane graphs, where all vertices lie on the outer face. We conclude with some open problems; see Sect. 5.

In particular, we show that all 1-plane graphs admit OC_4-layouts (Theorem 2) and SC_3-layouts (Theorem 5). We also prove that all biconnected outer-1-plane graphs admit OC_3-layouts (Theorem 4) and SC_2-layouts (Theorem 7). Three out of these four results are worst-case optimal: There exist biconnected

1-plane graphs that do not admit an OC_3-layout (Theorem 1) and biconnected outer-1-plane graphs that do not admit OC_2-layouts (Theorem 3) and SC_1-layouts (Theorem 6).

Table 1. Comparison of our results to previous work. The model K(andinsky)-SC_1 does not restrict the number of edges per port to one. (*) except for the octahedron (OC_4). "Super-poly" means that the drawings are not known to be of polynomial size.

Graph class	Max. deg.	Curve complexity	Drawing area	Reference
Orthogonal drawings				
General	4	OC_3	$n \times n$	[7]
Planar	4	OC_3 (*)	$n \times n$	[7]
1-plane	4	$\not\subseteq OC_3$		Theorem 1
	4	OC_4	$O(n) \times O(n)$	Theorem 2
Biconnected outer-1-plane	4	$\not\subseteq OC_2$		Theorem 3
	4	OC_3	$O(n) \times O(n)$	Theorem 4
Smooth orthogonal drawings				
Planar	4	SC_2	super-poly	[1]
Planar, poly-area	4	$\not\supseteq SC_1$	—	[1]
Planar, OC_2	4	$\not\subseteq SC_1$	—	[1]
Planar	3	SC_2	$\lfloor n^2/4 \rfloor \times \lfloor n/2 \rfloor$	[1]
Planar	3	SC_1	super-poly	[5]
Biconnected outerplane	4	SC_1	super-poly	[1]
General (non-planar)	4	SC_1	$2n \times 2n$	[5]
Planar	∞	$\not\subseteq$ K-SC_1		[6]
	∞	K-SC_2	$O(n) \times O(n)$	[5]
Biconnected 1-plane	4	SC_3	$O(n) \times O(n^2)$	Theorem 5
Biconnected outer-1-plane	4	$\not\subseteq SC_1$		Theorem 6
	4	SC_2	super-poly	Theorem 7

2   1-Planar Bar Visibility Representation

As an intermediate step towards orthogonal drawings, we introduce *1-planar bar visibility representations*: Each vertex is represented as a horizontal segment – called bar – and each edge is represented as either a vertical segment or a polyline composed of a vertical segment and a horizontal segment between the bars of its adjacent vertices. Edges must not intersect other bars. If an edge has a horizontal segment, we call it *red*. The horizontal segment of a red edge must be on top of its vertical segment and crosses exactly one vertical segment of another edge – which is called *blue*. The vertical segment of a red edge must not be crossed; see Fig. 2. We consider every edge as a pair of two *half-edges*, one for each of its two endpoints. Red edges are split at their bend – the *construction bend*, such that each half-edge consists of either a vertical or a horizontal segment. Observe that horizontal half-edges are always red. We show that every 1-planar graph has a 1-planar bar visibility representation, following the approach of Brandenburg [8]:

For a 1-planar embedding, we define a *kite* to be a K_4 induced by the end vertices of two crossing edges with the property that each of the four triangles induced by the crossing point and one end vertex of each of the two crossing edges is a face. A crossing is *caged* if its end vertices induce a kite. Let now G be a 1-planar graph. As a preprocessing step, G is augmented to a not necessarily simple graph G', with the property that any crossing is caged and no planar edge can be added to G' without creating a new crossing or a double edge [2].

After the preprocessing step, all crossing edges are removed and a bar visibility representation for the produced plane graph G_p is computed [20,22]. To this end an *st*-ordering of a biconnected supergraph of G_p is computed, i.e., an ordering $s = v_0, v_1, \ldots, v_{n-2}, v_{n-1} = t$ of the vertices such that each vertex except s and t is adjacent to both, a vertex with a greater and a lower index. The *st*-number is the index of a vertex. The y-coordinate of each bar is chosen to be the *st*-number of the respective vertex.

Faces of size four that correspond to the kites of G have three possible configurations: left/right wing or diamond configuration. Figure 2 shows the configurations and how to insert the crossing edges in order to obtain a 1-planar bar visibility representation of G'. Removing the caging edges results in a 1-planar bar visibility representation of G.

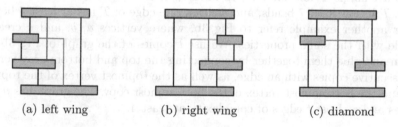

(a) left wing   (b) right wing   (c) diamond

Fig. 2. Different configurations for kites in a 1-planar bar visibility representation (Color figure online).

An edge is a *left, right, top* or *bottom edge* for a bar if it is attached to the respective side of that bar. Note that only red edges of G can be left or right edges for exactly one of their endpoints (and top edge for their other endpoint). If a bar has no bottom (top) edges, it is a *bottom* (*top*) bar, respectively. Otherwise it is a *middle bar*. For a bottom (top) bar, consider the x-coordinates of the touching points of its edges. We define its *leftmost* and *rightmost edge* to be the edge with the smallest and largest x-coordinate, respectively. If such a bar has a left or right edge then, by the previous definition, this is its leftmost or rightmost edge, respectively. Note that by the construction of the bar visibility representation, each bar has at most one left and at most one right red edge.

3   Orthogonal 1-Planar Drawings

In this section, we examine orthogonal 1-planar drawings. In particular, we give a counterexample showing that not every biconnected 1-plane graph of maximum degree 4 admits an OC_3-layout. On the other hand, we prove that every 1-plane graph of maximum degree 4 admits an OC_4-layout that preserves the given embedding. For biconnected outer-1-plane graphs we achieve optimal curve complexity 3.

3.1   Orthogonal Drawings for General 1-Planar Graphs

Theorem 1. *Not every biconnected 1-plane graph of maximum degree 4 admits an OC_3-layout. Moreover, there is a family of graphs requiring a linear number of edges of complexity at least 4 in any OC_4-layout respecting the embedding.*

Proof. Consider the 1-planar embedding of a K_5 as shown in Fig. 3a. The outer face is a triangle T and all vertices have their free ports in the interior of T. Hence, T has at least 7 bends, and at least one edge of T has at least 3 bends.

For another example refer to Fig. 3b, where vertices a, b, and c create a triangle with the same properties. We use t copies of the graph of Fig. 3b in a column and glue them together by connecting the top and bottom gray vertices of consecutive copies with an edge, as well as the topmost vertex of the topmost copy and the bottommost vertex of the bottommost copy. The graph has $n = 9t$ vertices and at least t edges of complexity at least 4.   □

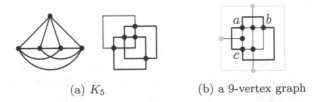

(a) K_5 (b) a 9-vertex graph

Fig. 3. Biconnected 1-plane graphs without OC_3-layout

In order to achieve an OC_4-layout for 1-plane graphs, we will use a general property of orthogonal drawingsof planar graphs: Consider two consecutive bends on an edge e with an incident face f. We say that the pair of bends forms a *U-shape* if they are both convex or both concave in f and an *S-shape*, otherwise. It follows from the flow model of Tamassia [21] that if a planar graph has an orthogonal drawing with an S-shape then it also has an orthogonal drawing with the identical sequence of bends on all edges except for the two bends of the S-shape that are removed. Thus, by planarization, any pair of S-shape bends can be removed as long as the two bends are not separated by crossings.

Theorem 2. *Every n-vertex 1-plane graph of maximum degree 4 admits an OC_4-layout on a grid of size $O(n) \times O(n)$.*

Proof. Let G be a 1-planar graph of maximum degree 4 and consider a 1-planar bar visibility representation of G. If G is not connected, we draw each connected component separately, therefore we assume that G is connected.

Each vertex is placed on its bar. Figures 4 and 5 indicate how to route the adjacent half-edges. Recall that the S-shape bend pairs can be eliminated. Thus, a horizontal half-edge gets at most one extra bend and a vertical half-edge gets at most two extra bends; see Fig. 5. We call a half-edge *extreme* if it was horizontal and got one bend or vertical and got two bends that create a U-shape.

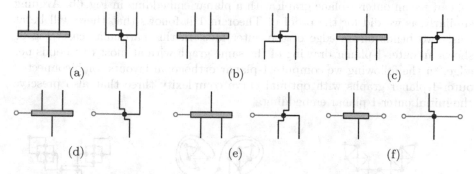

Fig. 4. Replacing a middle bar with a vertex in the presence of (a)–(c) zero, (d)–(e) one, and (f) two horizontal half-edges

It suffices to show that the edges can be routed such that no edge is composed of two extreme half-edges. Even for red edges where we have the construction bend, we either get one extra bend from the horizontal (extreme) half-edge or two extra bends from the vertical (extreme) half-edge. Observe that an edge is extreme if and only if it is the rightmost or leftmost edge of a bottom or top bar, respectively, and it is attached to the bottom or top of the vertex, respectively. For each bottom or top bar we have the free choice to set either its rightmost or leftmost half-edge to become extreme. Consider the following bipartite graph H. The vertices of H are the top and bottom bars, as well as their leftmost and rightmost edges. A bar-vertex and an edge-vertex are adjacent in H if and only if the bar and the edge are incident. Observe that each bar-vertex has degree two and each edge-vertex has degree at most two, thus H is a union of disjoint paths

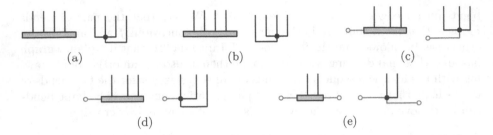

Fig. 5. Replacing a bottom bar of degree 4 with a vertex.

and cycles and there is a matching of H in which each bar-vertex is matched. This matching defines the extreme half-edges. It assigns exactly one half-edge to every bottom or top-bar and matches at most one half-edge of each edge. □

3.2   Orthogonal Drawings of Outer-1-Plane Graphs

Since outer-1-planar graphs are planar graphs [4], a planar orthogonal layout could be computed with curve complexity at most three. For example, in Fig. 6a we can see an outer-1-plane graph with a planar embedding in Fig. 6b. Arguing similarly as we did for the proof of Theorem 1 it follows that there will be at least two bends on an edge of the outer face. In this particular case, Fig. 6c shows an outer-1-planar drawing of the same graph with at most two bends per edge. In the following we compute 1-planar orthogonal layouts for biconnected outer-1-planar graphs with optimal curve complexity three that also preserve the initial outer-1-planar embedding.

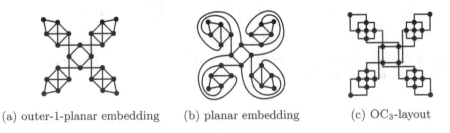

(a) outer-1-planar embedding   (b) planar embedding   (c) OC$_3$-layout

Fig. 6. An outer-1-plane graph.

Theorem 3. *Not every biconnected outer-1-plane graph of maximum degree 4 admits an OC$_2$-layout.*

Proof. K_4 is a biconnected outer-1-plane graph. Actually, it has a unique OC$_2$-layout as shown in Fig. 7a. When connecting two copies of K_4 by two intersecting edges as in Fig. 7b, it is not possible to draw the resulting graph such that the connector edges intersect and have curve complexity two. □

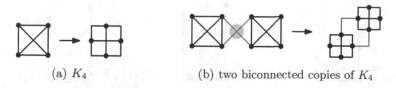

(a) K_4 (b) two biconnected copies of K_4

Fig. 7. Constructing a biconnected outer-1-plane graph that does not admit an OC_2-layout with the same embedding.

Theorem 4. *Every biconnected outer-1-plane graph of maximum degree 4 admits an OC_3-layout in an $O(n) \times O(n)$ grid, where n is the number of vertices.*

Proof (sketch). Let G be an outer-1-planar graph of maximum degree 4. Observe that all crossings can be caged without changing the embedding: A maximal outer-1-planar graph always admits a straight-line outer-1-planar drawing in which all faces are convex [9,12]. We would directly obtain the required curve complexity if there were no top or bottom bars of degree 4. Instead, our proof is based on a 1-planar bar visibility representation of G produced by a specific st-ordering. Let s and t be two vertices on the outer face. Define S_l and S_r to be the sequences of vertices on the left path and on the right path from s to t along the outer face of G, respectively. We choose s, S_l, S_r, t as our st-ordering. Observe that this is also an st-ordering of the caged and planarized graph G_p.

We process middle bars as in the algorithm of Theorem 2. For the top and bottom bars of degree 4 we choose differently which half-edge will be attached to the north or south port, respectively. Let v be a vertex such that $b(v)$ is a top or bottom bar of degree 4. Let $e_l = (v, v_l)$ and $e_r = (v, v_r)$ be its leftmost and rightmost edge, respectively. Assume that $v \in S_l \cup \{s\}$ and $b(v)$ is a bottom bar. If $v_l \in S_l$, we choose edge e_l to be attached to the south port of v, otherwise we choose edge e_r. If $b(v)$ is a top bar of degree 4 we choose its leftmost edge e_l to be attached to the north port of v. Symmetrically, if $v \in S_r \cup \{t\}$ and $b(v)$ is a top bar, we choose e_r for the north port of v if $v_r \in S_r$, otherwise we choose e_l. If $b(v)$ is a bottom bar we choose its rightmost edge e_r for the south port of v.

The above choice has the following property (see the full version [3] for a detailed proof): Any edge with three or four bends contains two consecutive bends that create an S-shape. The two bends are always connected with a vertical segment. If this is an uncrossed edge of G, the S-shape can be eliminated. For crossing edges, we prove that only one edge per crossing may have more than two bends. If the vertical segment connecting the two bends of the S-shape is crossed, we apply the flow technique of Tamassia [21] around the crossing point and reduce the number of bends (for details see the full version [3]). □

4   Smooth Orthogonal 1-Planar Drawings

In this section we examine smooth orthogonal 1-planar drawings. In particular, we show that every 1-plane graph of maximum degree 4 admits an SC_3-layout

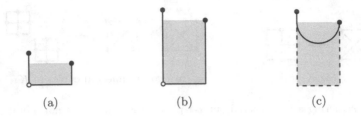

(a)   (b)   (c)

Fig. 8. Smoothing process of U-shapes created by top (bottom) bars.

that preserves the given embedding. For biconnected outer-1-plane graphs, we achieve SC_2, which is optimal for this graph class.

4.1   Smooth Orthogonal Drawings for General 1-Planar Graphs

Theorem 5. *Every 1-plane graph of maximum degree 4 admits an SC_3-layout in $O(n) \times O(n^2)$ area.*

Proof. We compute an SC_3-layout based on an OC_4-layout computed by the algorithm of Theorem 2. Observe that in the OC_4-layouts calculated by our approach, the area bounded U-shaped half-edges created at top and bottom bars is vertex-free (see gray area in Fig. 8a), and, each vertex is located on a separate level. We replace one bend of each U-shaped half-edge by a dummy vertex; see Fig. 8a. By doing so, we split each U-shaped half-edge into a vertical edge and an L-shaped half-edge. In the following, we treat the L-shaped half-edge as if the bend was on an L-shaped half-edge incident to the dummy vertex. We process $V = \{v_1, v_2, \ldots, v_n\}$ in the ascending vertical order of vertices (including dummy vertices). For v_i, let $\Delta_i^\uparrow$ be the largest horizontal distance between v_i and any bend on incident L-shaped half-edges leading to neighbors with larger index. Let $\Delta_i^\downarrow$ be the corresponding value for bends at incident L-shaped half-edges and construction bends of red edges incident to edges leading to neighbors with smaller index. We increase the y-coordinate of all v_j with $j \geq i$ by $\Delta_i^\downarrow$ units and then the y-coordinate of all v_k with $k > i$ by $\Delta_i^\uparrow$ units. Bends on L-shaped half-edges and construction bends of red edges leading to neighbors with smaller index will be moved together with the corresponding vertex. Note that the region enclosed by U-shapes created at top and bottom bars remains empty; see Fig. 8b. After the stretching, we remove the additional dummy vertices.

Each U-shaped half-edge will be replaced by a semi-circle which fits into the corresponding stretched empty region. We place the semi-circle directly incident to the endpoint which created the U-shape; see Fig. 8c. Then we replace each intersected S-shaped half-edge formed by a construction bend of a red edge by two consecutive quarter arcs incident to the top endpoint of the edge. Recall that if a red edge has an S-shape from its top vertex, it has no bend from its bottom vertex. Further we replace each remaining bend by a quarter arc starting at the corresponding endpoint. Arcs at the two endpoints will be connected by a vertical segment. The correctness follows from the fact that the regions stretched to make space for drawing arcs were empty in the initial drawing.

The area of the resulting drawing is $O(n) \times O(n^2)$ as the input drawing had $O(n) \times O(n)$ area and for every vertex the stretching operation increases the height by at most the length of the longest horizontal segment (i.e. $O(n)$).   □

4.2   Smooth Orthogonal Drawings for Outer-1-Plane Graphs

We focus on smooth layouts of outer-1-plane graphs. We demonstrate that curve complexity one is not always possible, but curve complexity two can be achieved for biconnected outer-1-plane graphs. We start with the following observation. The complete graph on four vertices with free ports towards its outer face has a unique SC_1-layout, shown in Fig. 9a. Removing one edge and restricting all ports towards its outer face, there exist two SC_1-layouts, see Figs. 9b and c.

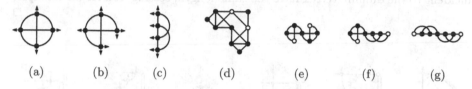

| (a) | (b) | (c) | (d) | (e) | (f) | (g) |

Fig. 9. (a) SC_1-layouts for K_4 and (b)–(c) for $K_4 - e$ with restricted ports. (d) A biconnected outer-1-plane graph that does not have an SC_1-layout. (e)-(g) SC_1-layouts of a subgraph of (d).

Theorem 6. *Not every biconnected outer-1-plane graph of maximum degree 4 has an SC_1-layout.*

Proof. Take the graph in Fig. 9d. It has two subgraphs isomorphic to $K_4 - e$ (with restricted ports) that share a vertex. Combining two drawings for both copies gives rise to the three drawings in Figs. 9e–g in which the edge between the two highlighted vertices cannot be added with curve complexity one.   □

To achieve SC_2-layouts for biconnected outer-1-plane graphs (see Fig. 11 for an example), we modify the algorithm of Alam et al. [1] for outerplane graphs; see the full version [3] for details.

Theorem 7. *Every biconnected outer-1-plane graph of maximum degree 4 has an SC_2-layout. The drawing area may be super-polynomial.*

Proof (sketch). The algorithm of Alam et al. [1] processes the faces of the graph along the *weak-dual*, i.e., the dual graph omitting the outer face and rooted at some inner face. For the next face, one of its edges (the *reference edge*) is already drawn and imposes the drawing of the face. Figures 10a–f show the different cases.

We define an auxiliary graph G': Let G be a biconnected outer-1-plane graph, and let G_p be the planarized graph of G, where crossing points are replaced with dummy vertices. Three types of dummy vertices exist in G_p: *dummy-cuts* (cut vertices), *in-dummies* (only incident to inner faces), and *out-dummies*. G' contains all in-dummy and out-dummy vertices of G_p, while dummy-cuts are replaced by a caging cycle. The face inside a caging cycle is called a *cut-face*. All other faces are called *normal*. Faces are processed along a traversal of the weak dual of G'. As G' may not be outerplanar, its weak dual does not have to be acyclic. It contains cycles of length four around in-dummies (see Fig. 10m). The auxiliary graph G' also contains *virtual edges* that are red. These are edges added for caging dummy-cuts and edges added to complete the process of faces around an in-dummy. Figures 10g–j show how to process normal faces not appearing in Alam et al. [1]. When processing a cut-face, we draw the crossing edges instead of the caging cycles; see Figs. 10k–l for two out of ten cases. Finally, in order to draw the fourth face around an in-dummy, we ensure that the edge-segments incident to the dummy vertex have the same length; see Fig. 10n for an example.

$\qquad\qquad\qquad\qquad\qquad\qquad\qquad\qquad\qquad\qquad\qquad\qquad\qquad\qquad$ □

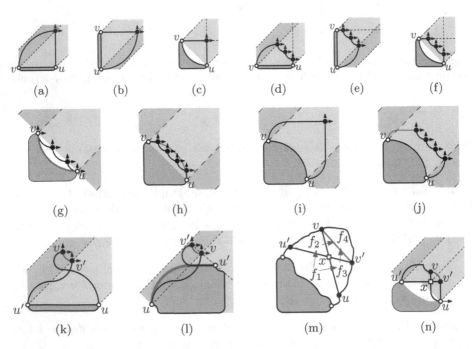

Fig. 10. Constructing an SC$_2$-drawing of biconnected outer 1-planar graphs (Color figure online).

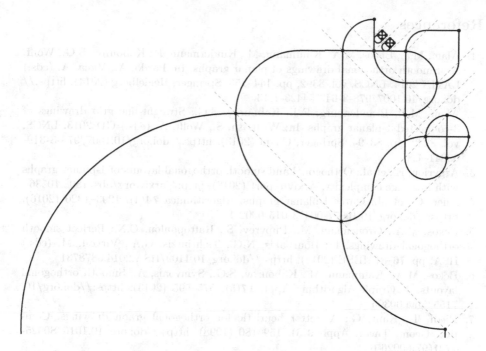

Fig. 11. SC$_2$-layout of an outer-1-plane graph produced by our algorithm, which is based on the algorithm of Alam et al. [1]. The largest 3/4-arc is only partially drawn.

5 A List of Open Problems

- Can we improve our curve complexity bounds if we restrict ourselves to more strongly connected classes of graphs (of maximum degree 4)?
- Candidate subclasses of outer-1-plane graphs for SC$_1$-layouts are for example outer-IC-plane graphs where crossings are independent. A possible variant would be to allow degenerate layouts where pairs of edges can touch but not cross.
- Is there a 1-plane graph that does not admit an SC$_2$-layout?
- Do biconnected outer-1-plane graphs admit an SC$_2$-layout with polynomial drawing area?
- Do similar results also hold for 2-planar graphs and more generally beyond-planar graphs?

References

1. Alam, M.J., Bekos, M.A., Kaufmann, M., Kindermann, P., Kobourov, S.G., Wolff, A.: Smooth orthogonal drawings of planar graphs. In: Pardo, A., Viola, A. (eds.) LATIN 2014. LNCS, vol. 8392, pp. 144–155. Springer, Heidelberg (2014). https://doi.org/10.1007/978-3-642-54423-1_13
2. Alam, M.J., Brandenburg, F.J., Kobourov, S.G.: Straight-line grid drawings of 3-connected 1-planar graphs. In: Wismath, S., Wolff, A. (eds.) GD 2013. LNCS, vol. 8242, pp. 83–94. Springer, Cham (2013). https://doi.org/10.1007/978-3-319-03841-4_8
3. Argyriou, E., et al.: Orthogonal and smooth orthogonal layouts of 1-planar graphs with low edge complexity. Arxiv report (2018). http://arxiv.org/abs/1808.10536
4. Auer, C., et al.: Outer 1-planar graphs. Algorithmica **74**(4), 1293–1320 (2016). https://doi.org/10.1007/s00453-015-0002-1
5. Bekos, M.A., Gronemann, M., Pupyrev, S., Raftopoulou, C.N.: Perfect smooth orthogonal drawings. In: Bourbakis, N.G., Tsihrintzis, G.A., Virvou, M. (eds.) IISA, pp. 76–81. IEEE (2014). https://doi.org/10.1109/IISA.2014.6878731
6. Bekos, M.A., Kaufmann, M., Kobourov, S.G., Symvonis, A.: Smooth orthogonal layouts. J. Graph Algorithms Appl. **17**(5), 575–595 (2013). https://doi.org/10.7155/jgaa.00305
7. Biedl, T., Kant, G.: A better heuristic for orthogonal graph drawings. Comput. Geom. Theor. Appl. **9**(3), 159–180 (1998). https://doi.org/10.1016/S0925-7721(97)00026-6
8. Brandenburg, F.J.: 1-visibility representations of 1-planar graphs. J. Graph Algorithms Appl. **18**(3), 421–438 (2014). https://doi.org/10.7155/jgaa.00330
9. Dehkordi, H.R., Eades, P.: Every outer-1-plane graph has a right angle crossing drawing. Int. J. Comput. Geom. Appl. **22**(6), 543–558 (2012). https://doi.org/10.1142/S021819591250015X
10. Didimo, W., Liotta, G., Montecchiani, F.: A survey on graph drawing beyond planarity. Arxiv report (2018). http://arxiv.org/abs/1804.07257
11. Duncan, C.A., Eppstein, D., Goodrich, M.T., Kobourov, S.G., Nöllenburg, M.: Lombardi drawings of graphs. J. Graph Algorithms Appl. **16**(1), 85–108 (2012). https://doi.org/10.7155/jgaa.00251
12. Eggleton, R.B.: Rectilinear drawings of graphs. Utilitas Math. **29**, 146–172 (1986)
13. Hong, S.H., Eades, P., Katoh, N., Liotta, G., Schweitzer, P., Suzuki, Y.: A linear-time algorithm for testing outer-1-planarity. Algorithmica **72**(4), 1033–1054 (2015). https://doi.org/10.1007/s00453-014-9890-8
14. Hong, S.-H., Eades, P., Liotta, G., Poon, S.-H.: Fáry's theorem for 1-planar graphs. In: Gudmundsson, J., Mestre, J., Viglas, T. (eds.) COCOON 2012. LNCS, vol. 7434, pp. 335–346. Springer, Heidelberg (2012). https://doi.org/10.1007/978-3-642-32241-9_29
15. Kobourov, S.G., Liotta, G., Montecchiani, F.: An annotated bibliography on 1-planarity. Comput. Sci. Rev. **25**, 49–67 (2017). https://doi.org/10.1016/j.cosrev.2017.06.002
16. Leighton, F.T.: New lower bound techniques for VLSI. Math. Syst. Theor. **17**(1), 47–70 (1984). https://doi.org/10.1007/BF01744433
17. Leiserson, C.E.: Area-efficient graph layouts. In: Foundations of Computer Science (FOCS 1980), pp. 270–281. IEEE (1980). https://doi.org/10.1109/SFCS.1980.13
18. Ringel, G.: Ein Sechsfarbenproblem auf der Kugel. Abh. Math. Semin. Univ. Hamburg **29**(1–2), 107–117 (1965)

19. Roberts, M.J., Newton, E.J., Lagattolla, F.D., Hughes, S., Hasler, M.C.: Objective versus subjective measures of Paris metro map usability: investigating traditional octolinear versus all-curves schematics. Int. J. Hum. Comput. Stud. **71**(3), 363–386 (2013). https://doi.org/10.1016/j.ijhcs.2012.09.004

20. Rosenstiehl, P., Tarjan, R.E.: Rectilinear planar layouts and bipolar orientations of planar graphs. Discrete Comput. Geom. **1**, 343–353 (1986). https://doi.org/10.1007/BF02187706

21. Tamassia, R.: On embedding a graph in the grid with the minimum number of bends. SIAM J. Comput. **16**(3), 421–444 (1987). https://doi.org/10.1137/0216030

22. Tamassia, R., Tollis, I.G.: A unified approach a visibility representation of planar graphs. Discrete Comput. Geom. **1**, 321–341 (1986). https://doi.org/10.1007/BF02187705

23. Thomassen, C.: Rectilinear drawings of graphs. J. Graph Theor. **12**(3), 335–341 (1988). https://doi.org/10.1002/jgt.3190120306

24. Valiant, L.G.: Universality considerations in VLSI circuits. IEEE Trans. Comput. **30**(2), 135–140 (1981). https://doi.org/10.1109/TC.1981.6312176

Ortho-Polygon Visibility Representations of 3-Connected 1-Plane Graphs

Giuseppe Liotta, Fabrizio Montecchiani[✉], and Alessandra Tappini

Università degli Studi di Perugia, Perugia, Italy
{giuseppe.liotta,fabrizio.montecchiani}@unipg.it,
alessandra.tappini@studenti.unipg.it

Abstract. An ortho-polygon visibility representation Γ of a 1-plane graph G (OPVR of G) is an embedding preserving drawing that maps each vertex of G to a distinct orthogonal polygon and each edge of G to a vertical or horizontal visibility between its end-vertices. The representation Γ has vertex complexity k if every polygon of Γ has at most k reflex corners. It is known that 3-connected 1-plane graphs admit an OPVR with vertex complexity at most twelve, while vertex complexity at least two may be required in some cases. In this paper, we reduce this gap by showing that vertex complexity five is always sufficient, while vertex complexity four may be required in some cases. These results are based on the study of the combinatorial properties of the B-, T-, and W-configurations in 3-connected 1-plane graphs. An implication of the upper bound is the existence of a $\tilde{O}(n^{\frac{10}{7}})$-time drawing algorithm that computes an OPVR of an n-vertex 3-connected 1-plane graph on an integer grid of size $O(n) \times O(n)$ and with vertex complexity at most five.

1 Introduction

Let G be a graph embedded in the plane. An *ortho-polygon visibility representation* of G (*OPVR of G*) is an embedding preserving drawing that maps every vertex of G to a distinct orthogonal polygon and every edge of G to a vertical or horizontal visibility between its end-vertices (it is assumed the ϵ-visibility model, where the visibilities can be replaced by strips of non-zero width, see also [8]). The *vertex complexity* of an OPVR of G is the minimum k such that every polygon has at most k reflex corners. For example, Fig. 1(b) shows an OPVR Γ of the graph G of Fig. 1(a). All vertices of Fig. 1(b) are rectangles except vertex u, and thus the vertex complexity of Γ is one.

The notion of ortho-polygon visibility representation generalizes the classical concept of *rectangle visibility representation*, that is, in fact, an OPVR with vertex complexity zero (see, e.g., [2,6,14,18,19]). In this context, Biedl et al. [2] characterize the 1-plane graphs that admit a rectangle visibility representation in terms of forbidden subgraphs, called B-, T-, and W-configurations (see Fig. 2

Research supported in part by: "Algoritmi e sistemi di analisi visuale di reti complesse e di grandi dimensioni" - Ricerca di Base 2018, Dip. Ingegneria - Univ. Perugia.

T. Biedl and A. Kerren (Eds.): GD 2018, LNCS 11282, pp. 524–537, 2018.
https://doi.org/10.1007/978-3-030-04414-5_37

for examples and Sect. 2 for definitions). We recall that 1-*plane graphs* are graphs embedded in the plane such that every edge is crossed by at most one other edge, and that the 1-*planar graphs* are those graphs that admit such an embedding; these graphs are a classical subject of investigation in the constantly growing research field called graph drawing beyond-planarity (refer to [1,9,15]).

Partly motivated by the result of Biedl et al. [2], Di Giacomo et al. [8] study the vertex complexity of ortho-polygon visibility representations of 1-plane graphs. They prove that an OPVR of a 1-plane graph may require $\Omega(n)$ vertex complexity. However, if the graph is 3-connected, then vertex complexity twelve is always sufficient, while vertex complexity two is sometimes necessary.

The idea behind the approach of Di Giacomo et al. [8] to prove a constant upper bound can be shortly described as follows. Let G be a 3-connected 1-plane graph. For each crossing in G, one of the two edges that form the crossing is suitably chosen and removed from G. The removed edges are such that each vertex of G is incident to at most six of them. After this edge removal, the obtained graph is planar, and hence it admits a bar-visibility representation Γ (vertices are represented as horizontal bars and edges as vertical visibilities) [7]. An OPVR of G is now computed by turning the bars of Γ into orthogonal polygons and by inserting horizontal visibilities for the (at most six per vertex) removed edges. The paper shows how to compute a transformation of the bars that adds at most two reflex corners per removed edge, which implies a vertex complexity of at most twelve. Reducing the gap between the upper bound of twelve and the lower bound of two is left as an open problem in [8], and it is the question that motivates our research. We prove the following theorem.

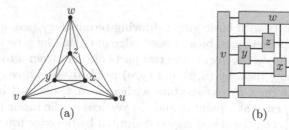

(a)   (b)

Fig. 1. (a) A 1-plane graph G; (b) An OPVR of G with vertex complexity one.

Theorem 1. *Let G be a 3-connected 1-plane graph with n vertices. There exists an $\tilde{O}(n^{\frac{10}{7}})$-time algorithm that computes an ortho-polygon visibility representation of G with vertex complexity at most five on an integer grid of size $O(n) \times O(n)$. Also, there exists an infinite family of 3-connected 1-plane graphs such that any ortho-polygon visibility representation of a graph in the family has vertex complexity at least four.*

Concerning the upper bound stated in Theorem 1, the main difference between our approach and the one in [8] is that we do not aim at removing all crossings so to make G planar. Instead, we define a *subset* F of the B-, T-, and W-configurations of G such that F has two fundamental properties: (i) Removing the elements of F removes *all* B-, T-, and W-configurations from G; and (ii) Each vertex of G can be associated with at most five elements of F. We remove F from G and compute a rectangle visibility representation by using the algorithm of Biedl et al. [2]. We then carefully reinsert the removed configurations by "bending" each rectangle with at most five reflex corners. We remark that the study of the combinatorial properties of the B-, T-, and W-configurations in 3-connected 1-plane graphs is a contribution of independent interest that fits in the rich literature about the properties of 1-plane graphs (see, e.g. [15]).

Finally, we recall that some authors recently studied OPVRs with fixed vertex complexity. Evans et al. [11] consider OPVRs of directed acyclic graphs where vertices are L-shapes (i.e., with vertex complexity one). OPVRs with L-shapes are also studied in [16], where it is shown that a particular subclass of 1-planar graphs admits such a representation. Brandenburg [4] studies OPVRs where vertices are T-shapes (i.e., with vertex complexity two) and proves that all 1-planar graphs admit such a representation if the embedding of the input graph can be changed, and hence the final representation may be not 1-planar.

The rest of the paper is organized as follows. Preliminaries are in Sect. 2. The lower bound and the upper bound on the vertex complexity are proved in Sect. 3 and in Sect. 4, respectively. Section 5 contains open problems. For space reasons some proofs have been omitted or sketched, and can be found in [17] (the corresponding statements are marked with [*]).

2   Preliminaries

We assume familiarity with basic graph drawing terminology (see, e.g. [7]). Let G be a 1-plane graph, let (u, v) be a crossed edge of G, and let p be the crossing along (u, v). We call *edge fragments* the two parts of (u, v) from u to p and from p to v, and we denote them by (u, p) and (p, v) respectively. Three edges (u, v), (w, z), (u, z) of G form a *B-configuration* with *poles* u, z, denoted by $b(u, z)$, if (i) (u, v) and (w, z) cross at a point p, and (ii) vertices v, z lie inside the external boundary of $b(u, z)$, i.e., the closed region delimited by the edge fragment (u, p), the edge fragment (p, z), and the edge (u, z); see Fig. 2(a). Four edges (u, v),

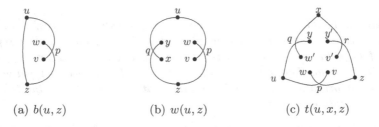

(a) $b(u, z)$ (b) $w(u, z)$ (c) $t(u, x, z)$

Fig. 2. (a) B-configuration; (b) W-configuration; (c) T-configuration.

(w, z), (u, x), (y, z) of G form a *W-configuration* with poles u, z, denoted by $w(u, z)$, if (i) (u, v) and (w, z) cross at a point p, (ii) (u, x), (y, z) cross at a point q, (iii) vertices v, w, x, y lie inside the external boundary of $w(u, z)$, i.e., the closed region delimited by the edge fragments (u, p), (p, z), (z, q), and (q, u); see Fig. 2(b). Finally, six edges (u, v), (w, z), (u, y), (x, w'), (z, y'), and (v', x) of G form a *T-configuration* with poles u, x, z, denoted by $t(u, x, z)$, if (i) (u, v) and (w, z) cross at a point p, (ii) (u, y), (x, w') cross at a point q, (iii) (z, y') and (v', x) cross at a point r, (iv) vertices v, v', w, w', y, y' lie inside the external boundary of $t(u, x, z)$, i.e., the closed region delimited by edge fragments (u, p), (p, z), (u, q), (q, x), (x, r), and (r, z); see Fig. 2(c). For example, the graph G of Fig. 1(a) contains the T-configuration $t(u, v, w)$ and hence any OPVR of G has at least one reflex corner. A 1-plane graph has a rectangle visibility representation (*RVR*) if and only if it contains no B-, no T-, and no W-configurations [2].

3   Lower Bound on the Vertex Complexity

Let $S(i)$ be the *nested triangle graph* with i levels, i.e., a maximal plane graph with $3i$ vertices recursively defined as follows [10]. Graph $S(1)$ is a triangle. Denote by u_1, u_2, and u_3 the vertices on the outer face of $S(i - 1)$. Graph $S(i)$ is obtained by adding three vertices v_1, v_2, v_3 on the outer face of $S(i - 1)$ and edges (u_1, v_1), (u_2, v_2), (u_3, v_3), (u_1, v_2), (u_2, v_3), and (u_3, v_1). Also, we mark as *T-faces* a set of faces of $S(i)$ such that: (1) $S(i)$ has $3i - 2$ T-faces, and (2) no two T-faces share an edge. All other faces of $S(i)$ are marked as *NT-faces*. Figure 3(a) shows an assignment for $S(3)$ that satisfies these two conditions (the T-faces are gray, while the NT-faces are white). Graph $G(3i)$ is the 3-connected 1-plane graph with $3i$ poles obtained from $S(i)$ as follows. For each T-face of $S(i)$, whose boundary contains the three vertices u, x, z, we add in its interior a T-configuration $t(u, x, z)$ and three B-configurations $b(u, x)$, $b(u, z)$ and $b(x, z)$ as shown in Fig. 3(b). The resulting graph is 1-plane and it has $3i$ poles. In particular, we have one B-configuration for each of the $3(3i) - 6$ edges of $S(i)$, and we have $3i - 2$ T-configurations. However, this graph is not 3-connected. To achieve 3-connectivity, for each NT-face of $S(i)$, whose boundary contains the three vertices u, v, w, we first add a vertex c in its interior and we then connect it to one vertex that is not a pole for each of $b(u, v)$, $b(u, w)$, and $b(v, w)$; the added edges are crossed exactly once each by an edge on the boundary of the NT-face, as shown in Fig. 3(c). Finally, we add crossing-free edges until all faces are triangles. One can easily verify that the resulting graph is 3-connected.

Theorem 2. *For every $n_p > 8$ with $n_p \pmod 3 = 0$, there exists a 3-connected 1-plane graph $G(n_p)$ whose OPVRs all have vertex complexity at least four.*

Proof. Consider the graph $G(n_p)$ described above, with $n_p > 8$. Let Γ be any OPVR of G. We first prove that, for each forbidden configuration f of $G(n_p)$, Γ contains at least one reflex corner on one of the poles of f and that this reflex corner lies inside the *interior region* of f, i.e., inside the bounded region of Γ delimited by the external boundary of f. We follow an argument similar to the

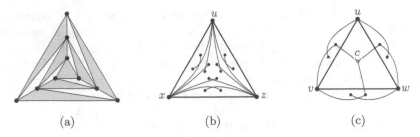

Fig. 3. (a) The graph $S(3)$; the T-faces are gray while the NT-faces are white. (b) Insertion of a T-configuration and three B-configurations in a T-face. (c) Insertion of a vertex in an NT-face to achieve 3-connectivity.

one in [2]. Suppose first that f is a B-configuration $b(u, z)$. Consider a closed walk in clockwise direction along the external boundary of $b(u, z)$ in Γ. The crossing point is a left turn, as well as any attaching point of a visibility to a polygon, while the corners of the polygons are right turns. Since the external boundary is an orthogonal polygon, the number of right turns equals the number of left turns plus four. Let a be the number of attaching points of a visibility to a polygon, let k be the number of crossings, and let r be the number of corners. We have that $r = k + a + 4$. Since $k = 1$ and $a \geq 4$, we have that $r \geq 9$, which implies that at least one of the two polygons representing u and z, say u, has at least five corners. As a consequence, there exists at least one reflex corner along the polygon of u that lies inside the interior region of $b(u, z)$. Similarly, for a T-configuration $t(u, x, z)$, we have that $k = 3$ and $a \geq 6$, which implies that $r \geq 13$, and thus at least one of its poles has a reflex inside the interior region of $t(u, x, z)$. Since $G(n_p)$ contains $4n_p - 8$ forbidden configurations, and since any pair of forbidden configurations of $G(n_p)$ is such that the intersection of their two interior regions is empty, it follows that Γ contains at least $4n_p - 8$ distinct reflex corners distributed among its n_p poles. Let c be the maximum number of reflex corners on a polygon of $G(n_p)$, it follows that $c\,n_p \geq 4n_p - 8$, which implies $c \geq 4 - 8/n_p > 3$ because $n_p > 8$. $\qquad\square$

4   Upper Bound on the Vertex Complexity

In this section, we first show the existence of an assignment between the set of forbidden configurations in G and their poles such that each pole is assigned at most five forbidden configurations (Sect. 4.1). Then we make use of this assignment and of a suitable modification of the algorithm in [2] to obtain an OPVR of G with vertex complexity at most ten (Sect. 4.2). Finally, we apply a post-processing step to reduce the vertex complexity to five (Sect. 4.3).

4.1   Forbidden Configurations in 3-Connected 1-Plane Graphs

Two forbidden configurations of a 3-connected 1-plane graph G are called *independent* if they share no crossing (although they may share poles), while they are called *dependent* otherwise. The next lemma proves some basic properties of the independent forbidden configurations in G.

Lemma 1. [∗] *Let G be a 3-connected 1-plane graph and let $G' \subseteq G$. The following properties hold. **P1:** There are no three independent forbidden configurations of G' that share a pair of poles. **P2:** If G' contains a W-configuration $w(u, z)$, all vertices of G', except u and z, are inside $w(u, z)$. **P3:** If G' contains a W-configuration $w(u, z)$, no other forbidden configuration of G' that is independent of $w(u, z)$ has u, z as poles. **P4:** There are no two B-configurations of G' sharing their two poles. The only exception is when two B-configurations form a W-configuration. **P5:** Two T-configurations of G' that are dependent share exactly one pair of crossing edges.*

Intuitively, two dependent forbidden configurations may be drawn by inserting only one reflex corner on a common pole. By following this intuition, our goal is to find a set of forbidden configurations that "cover" all others and such that they can be drawn by introducing only a small number of reflex corners per vertex. To formalize this idea, we give the following definition. A set F of forbidden configurations of G is *non-redundant* if it contains: (1) all B-configurations of G; (2) all T-configurations of G independent of B-configurations; (3) zero, one, or two copies of the W-configuration in G (there is at most one by **P2**), if the W-configuration

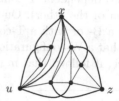

Fig. 4. A non-redundant set contains only $b(u, x)$.

exists and has two, one, or zero, respectively, dependent B-configurations. For example, in the graph of Fig. 4, $t(u, x, z)$ and $b(u, x)$ are dependent, and thus $b(u, x) \in F$ while $t(u, x, z) \notin F$.

A T-configuration t of G is *separating* if G contains a pole v that is not a pole of t and that lies in the interior region of t (i.e., inside the bounded region delimited by the external boundary of t). Let β, τ, and ω be the number of B-/T-/W-configurations in F, respectively. Note that, if G is (a subgraph of) a 3-connected 1-plane graph, F contains at most one W-configuration by **P2**, i.e., $\omega \leq 1$. It follows that, if F contains zero or one copy of the (at most one) W-configuration of G, we have $|F| = \beta + \tau + \omega$, else $|F| = \beta + \tau + \omega + 1$.

Lemma 2. *Let G be (a subgraph of) a 3-connected 1-plane graph, let F be a set of non-redundant forbidden configurations of G, and let P be the set of its poles. If G has no separating T-configurations, then $|F| \leq 4|P| - 8$ if $\omega = 0$ and $|F| \leq 4|P| - 7$ otherwise. Also, if $\omega = 0$, then $|F| = 4|P| - 8$ if and only if $\beta = 3|P| - 6$ and $\tau = |P| - 2$.*

Proof. By Lemma 1, properties **P1–P5** hold for G. We define an auxiliary graph G_A whose edges represent the crossings of the forbidden configurations in F. More precisely, let f be a forbidden configuration. For each crossing k of f there exist two poles of f, denoted by u_k and z_k, such that the edge fragments (u_k, k) and (k, z_k) belong to the external boundary of f. Let G_A be the graph with

$n_A = |P|$ vertices and m_A edges obtained from G as follows; see, e.g., Fig. 5. Remove first all edges of G and then all vertices of G that are not poles of any forbidden configuration. For each forbidden configuration f of F and for each crossing k of f, draw an edge (u_k, z_k) on the external boundary of f by following the two edge fragments (u_k, k) and (k, z_k). Note that G_A is plane and may have parallel edges. By **P1**, each pair of adjacent vertices of G_A is connected by at most two parallel edges, that is, $m_A \leq 2(3n_A - 6) = 6n_A - 12$. Also, G_A contains an edge for each B-configuration, two edges for the W-configuration (if it exists), and three edges for each T-configuration in F. A B-configuration does not share an edge with a T-configuration by construction of F, also, two B-configurations do not share an edge by **P4**, and finally, no two T-configurations share an edge as otherwise one of them would be a separating T-configuration (they would be two dependent T-configurations such that one has a pole inside the interior region of the other). On the other hand, a W-configuration can share an edge with a B- or with a T-configuration. Let $0 \leq s \leq 2$ be the number of edges of G_A that a W-configuration shares with other configurations. From the argument above, it follows that $m_A = \beta + 3\tau + 2\omega - s \leq 6n_A - 12$.

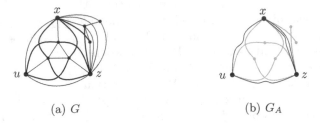

(a) G (b) G_A

Fig. 5. Illustration for the proof of Lemma 2: Construction of the auxiliary graph G_A from G. The vertices of G that are not poles are smaller, and the edges of G that are not part of forbidden configurations are thinner. The edges of G_A are drawn by following the remaining (bold) pairs of crossing edges of G.

If $\omega = 0$ (and hence $s = 0$), then $|F| = \beta + \tau$ and $m_A = \beta + 3\tau \leq 6n_A - 12$. Also, $\beta \leq 3n_A - 6$ by **P4**. For a fixed value of n_A (note that $n_A \geq 2$ if $|F| > 0$), consider the function $f(\beta, \tau) = \beta + \tau$ in the domain defined by the two inequalities $\beta + 3\tau \leq 6n_A - 12$ and $\beta \leq 3n_A - 6$. By studying the function $f(\beta, \tau)$, it is easy to verify that its maximum value in the above domain is $4n_A - 8$, and that this value is obtained if and only if $\beta = 3n_A - 6$ and $\tau = n_A - 2$. If $\omega = 1$ then either $|F| = \beta + \tau + 1$ and $\beta = 3n_A - 6$, or $|F| = \beta + \tau + 2$ and $\beta \leq 3n_A - 7$. In both cases $|F| \leq 4n_A - 7$.   □

Theorem 3. *Let G be a 3-connected 1-plane graph and let F be a set of non-redundant forbidden configurations of G. For each configuration $f \in F$, it is possible to assign f to one of its poles such that every pole is assigned at most five elements of F.*

Proof. Let $H = (F \cup P, E \subseteq F \times P)$ be the bipartite graph with vertex set $F \cup P$ (where P is the set of poles of G), and having an edge (f, u) with $f \in F$ and $u \in P$ if u is a pole of f. A *k-matching from F into P* is a set $M \subseteq E$ such that each vertex in F is incident to exactly one edge in M and each vertex in P is incident to at most k edges in M. For a subset $F' \subseteq F$, we denote by $N(F')$ the set of all vertices in P that are adjacent to a vertex in F'. We prove the existence of a 5-matching of F into P by using Hall's theorem [13], i.e., we show that $\forall F' \subseteq F : |F'| \leq 5|N(F')|$.

Let G' be any subgraph of G that contains all and only the forbidden configurations in F'. By Lemma 1, G' of G satisfies **P1–P5**. The proof is by induction on the number h of separating T-configurations of G'. Let G'_A be the auxiliary graph of G' constructed as in the proof of Lemma 2. In the base case $h = 0$, we have that $|F'| \leq 4|N(F')| - 7$ by Lemma 2. Suppose now that the claim holds for $h - 1 > 0$. Let $t(u, x, z)$ be a separating T-configuration of G' such that it does not contain any other separating T-configuration in its interior. Let G'_{IN} be any subgraph of G' containing $t(u, x, z)$ and all and only the forbidden configurations of F' that are inside $t(u, x, z)$, that is, its auxiliary graph $G'_{A,IN}$ is the subgraph of G'_A having the three edges of $t(u, x, z)$ as outer face. Note that G'_A may contain some of the possible B-configurations $b(u, x)$, $b(u, z)$, and $b(x, z)$, but their corresponding edges of G'_A are not part of $G'_{A,IN}$. We denote by F'_{IN} the set of forbidden configurations of F' in G'_{IN}. Let G'_{OUT} be any subgraph of G' containing all and only the forbidden configurations of F' except those in F'_{IN}, but including $t(u, x, z)$. We denote by F'_{OUT} the set of forbidden configurations of F' in G'_{OUT}. Since G'_{OUT} contains $h - 1$ separating T-configurations, by induction we have that $|F'_{OUT}| \leq 5|N(F'_{OUT})|$. On the other hand, G'_{IN} does not contain separating T-configurations and it is a subgraph of G, thus $|F'_{IN}| \leq 4|N(F'_{IN})| - 7$ by Lemma 2. In particular, G'_{IN} does not contain any W-configuration, since G can have at most one and it must be part of its outer face. Hence, $|F'_{IN}| \leq 4|N(F'_{IN})| - 8$, and in particular $|F'_{IN}| = 4|N(F'_{IN})| - 8$ if and only if its number of B-configurations is such that $\beta_{IN} = 3|F'_{IN}| - 6$ (Lemma 2). However, the (at most) three B-configurations $b(u, x)$, $b(u, z)$, and $b(x, z)$ are not part of G'_{IN} by construction, and therefore $|F'_{IN}| \leq 4|N(F'_{IN})| - 11$. Since $|N(F'_{IN})| + |N(F'_{OUT})| = |N(F')| + 3$ (we have to consider the three vertices u, x, z that are poles in both graphs), and since $|F'_{IN}| + |F'_{OUT}| \geq |F'|$, it follows that $|F'| \leq |F'_{IN}| + |F'_{OUT}| \leq 4|N(F'_{IN})| - 11 + 5|N(F'_{OUT})|$, and thus $|F'| \leq 5|N(F')| + 4 - |N(F'_{IN})|$, which implies that $|F'| \leq 5|N(F')|$ when $|N(F'_{IN})| \geq 4$. Since $t(u, x, z)$ is a separating T-configuration, G'_{IN} contains at least one pole more than u, x, z, thus $|N(F'_{IN})| \geq 4$. □

4.2 Proving Vertex Complexity at Most 10

We briefly recall an algorithm by Biedl et al. [2], called 1P-RVDRAWER, which takes as input a 1-plane graph with no forbidden configurations and that returns an RVR of this graph. First, the planarization G_p of G is computed. The plane graph G_p is then triangulated in such a way that the degree of dummy vertices remains four, i.e., avoiding the addition of edges incident to dummy vertices. The resulting graph G_t does not contain any planarized forbidden configuration (i.e., any subgraph such that by replacing dummy vertices with crossings we obtain a forbidden configuration). Moreover, if G is 3-connected, G_t does not contain parallel edges (and hence is 3-connected as well). As next step, 1P-RVDRAWER decomposes G_t into its 4-connected components, it computes an RVR for each 4-connected component, and finally it patches the drawings by suitably identifying the outer face of each component with the corresponding inner face of its parent component. The algorithm guarantees that each dummy vertex is represented by a rectangle having one visibility on each of its four sides. This property allows to replace that rectangle with a crossing. Also, we observe that for each 4-connected component C of G_t, 1P-RVDRAWER chooses one edge e on the outer face of C called the *surround edge* of C. This edge is chosen so to satisfy the following two conditions: (1) The inner face of C containing e on its boundary consists of the two end-vertices of e plus a third vertex which is not dummy; (2) If the surround edge of the parent component C' of C (if C' exists) is an edge e' of the outer face of C, then $e = e'$. The feasibility of this choice is guaranteed by the absence of planarized forbidden configurations in G_t. The resulting RVR is such that all edges of C incident to an end-vertex of e are represented by horizontal visibilities.

Lemma 3. [*] *Every 3-connected 1-plane graph admits an OPVR with vertex complexity at most ten.*

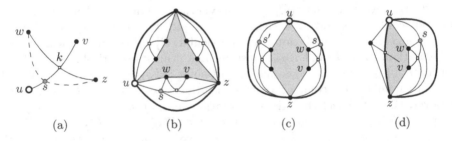

(a)   (b)   (c)   (d)

Fig. 6. Illustration for Lemma 3: (a) Inserting a subdivision vertex (orange) to remove a forbidden configuration; (b–d) The separating triangles (bold) facing the subdivision vertices in (b) T-, (c) W-, and (d) B-configurations. (Color figure online)

Sketch of Proof. Let G^* be a 3-connected 1-plane graph. Let P be the set of poles of G^*, and let F be a set of non-redundant forbidden configurations of G^*. By Theorem 3, there exists a 5-matching of F into P. Let f be a forbidden configuration of F, and let u be the pole of P matched with f. Let (u, v) and (w, z) be two crossing edges of f such that z is another pole of f, and denote by k their crossing. Note that this pair of edges is unique if f is a B-configuration, while if f is a T-configuration there are two such pairs. Also, if f is a W-configuration, by construction of F we have that each of its two crossings is matched with one of its two poles. We subdivide the edge fragment (k, u) with a *subdivision vertex* s and we add the uncrossed edges (z, s) and (s, w); see Fig. 6(a) for an illustration. It is easy to verify that this operation removes f from G^* and it does not introduce any new forbidden configuration. Let G be the 1-plane graph obtained by introducing a subdivision vertex for each forbidden configuration of F; one can show that G does not contain any forbidden configuration and is 3-connected; hence, it admits an RVR γ. In particular, it is possible to compute γ such that every subdivision vertex introduced when going from G^* to G is incident to a face containing a surround edge of G_t. Recall that G_t is a triangulated plane graph and that dummy vertices have degree four (see the description of 1P-RVDRAWER above and [2] for details), which implies that every dummy vertex is inside a 4-cycle of uncrossed edges (called *kite* in [2]). In particular, it can be shown that every forbidden configuration f whose matched pole is denoted by u, whose subdivision vertex is denoted by s (which is incident to u), and whose other pole adjacent to s is denoted by z, is such that there exists a separating triangle Δ_f in G_t having u and z as two of its vertices (for ease of description, we view the outer face of G_t as a separating triangle) and such that (u, z) can be chosen as surround edge.

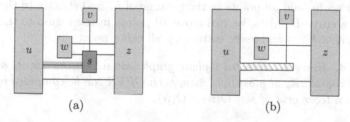

(a)   (b)

Fig. 7. Illustration for Lemma 3: Replacing a subdivision vertex, denoted by s in (a), with a spoke, which is represented in (b) with a dashed fill.

We finally turn γ into an OPVR Γ of G^* with vertex complexity at most ten. Let u be a pole of G^* and let $r(u)$ be the rectangle representing it in γ. Observe that, since u has at most five matched forbidden configurations, u is adjacent to at most five subdivision vertices of G. On the other hand, in order to turn γ into the desired OPVR, we need to replace all u's visibilities towards subdivision vertices with visibilities towards the other endpoints of the subdivided edges. To this aim, we attach on a side of $r(u)$ a rectangle so that $r(u)$ becomes an orthogonal polygon with two reflex corners for each attached rectangle. Let s be a subdivision vertex adjacent to u and let (u, v) be the edge subdivided by s. Also, let (w, z) be the edge that crosses (u, v). From the argument above, (u, z) is a surround edge and all visibilities incident to u and z are horizontal. We can remove $r(s)$ from γ, and attach to $r(u)$ a *spoke*, i.e., a rectangle around the visibility (u, s) (see the shaded blue region in Fig. 7(a)) so that the visibility between $r(s)$ and $r(v)$ is now attached to this spoke, as shown in Fig. 7(b). By repeating this procedure for all poles and for all their subdivision vertices, we obtain the desired OPVR Γ of G. In particular, since each pole u is adjacent to at most five subdivision vertices in G^*, we attached to $r(u)$ at most five spokes, hence we created at most ten reflex corners along the boundary of $r(u)$.   □

4.3   Reducing the Vertex Complexity to 5

An OPVR can be interpreted as a planar orthogonal drawing whose vertices are the corners of the polygons, the crossing points, and the attaching points between visibilities and polygons, and whose edges are (pieces of) sides of the polygons and (pieces of) visibilities. We now recall a well-known method used to modify an orthogonal drawing in order to move a desired set of vertices and edges while keeping stationary all other elements of the drawing. This can be achieved with *zig-zag-bend-elimination-slide curves* [3]. Any such a curve D contains: (1) a horizontal segment s_h that intersects neither edges nor vertices of the drawing; (2) an "upward" vertical half-line h_l that originates at the leftmost endpoint of s_h and whose points are above it; and (3) a "downward" vertical half-line h_r that originates at the rightmost endpoint of s_h and whose points are below it. The two half-lines can intersect edges and vertices of the drawing. The *region to the right of* D is the set of all points that are in the y-range of h_l and strictly to the right or on h_l, and all points in the y-range of h_r and strictly to the right of h_r. If such a curve D exists, we can move all points in the region to the right of D by any given $\delta > 0$ and leave stationary all other points [3].

Theorem 4. *Every 3-connected 1-plane graph with n vertices admits an OPVR with vertex complexity at most five. Also, such OPVR can be computed in $\tilde{O}(n^{\frac{10}{7}})$ time on an integer grid of size $O(n) \times O(n)$.*

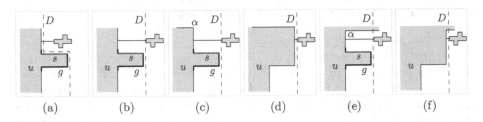

$$\text{(a)} \qquad \text{(b)} \qquad \text{(c)} \qquad \text{(d)} \qquad \text{(e)} \qquad \text{(f)}$$

Fig. 8. Illustration for Theorem 4: Removing reflex corners through zig-zag-bend-elimination curves. The spoke g is bold and its free side is red. (Color figure online)

Proof. Let G be a 3-connected 1-plane graph with n vertices. Let Γ be an OPVR of G with vertex complexity at most 10, which exists by Lemma 3. We show how to reduce the number of reflex corners around a vertex by using the above defined zig-zag-bend-elimination-slide curves. Let $p(u)$ be a polygon representing a vertex u of G in Γ. Note that $p(u)$ is either a rectangle, or a polygon obtained by attaching spokes (rectangles) on the sides of an initial rectangle $r(u)$ in the intermediate RVR (see the proof sketch of Lemma 3). Moreover, each spoke g of $p(u)$ has only one visibility attached to it on a side, while the opposite side of g does not have any visibility, and we say it is *free*. Suppose that $p(u)$ contains $0 < k \leq 5$ spokes and hence $2k$ reflex corners. We apply k zig-zag-bend-elimination-slide curves in order to remove k reflex corners. Since the application of a zig-zag-bend-elimination-slide curve on a spoke may alter the shape of another spoke, in what follows a spoke g is more generally defined as a chain of segments in $p(u)$ such that g contains exactly one reflex corner followed by two inflex corners and by one more reflex corner. The free side of g is hence a segment of g that is between a reflex and an inflex corner and that does not contain any visibility on it. Consider any spoke g of $p(u)$. Without loss of generality, we can assume that the free side s of g is horizontal and is the topmost side of g (up to a mirroring/rotation of the drawing). Let $[x_1, x_2]$ be the x-range of g, and let $\varepsilon > 0$ be a value smaller than the smallest distance between any two points on the boundary of two distinct polygons of Γ. Let α be the first angle encountered when walking along $p(u)$ counter-clockwise, starting from the leftmost point of s. Consider the zig-zag-bend-elimination curve D constructed by using a horizontal segment above s by ε and with x-range $[x_1 + \varepsilon, x_2 + \varepsilon]$, as shown in Fig. 8(a). We move all points in the region to the right of D by $\delta = |x_2 - x_1|$. After this operation, there is no polygon above s (other than $p(u)$), as shown in Fig. 8(b). Hence, we can modify $p(u)$ as shown in Figs. 8(c)–(d) if α is an inflex corner, or as shown in Figs. 8(e)–(f) if α is a reflex corner. In both cases, after this operation, $p(u)$ contains exactly one less reflex corner and one less spoke.

By repeating this argument for all vertices of Γ, we obtain an OPVR Γ' of G with at most five reflex corners per polygon. It remains to show how to compute an OPVR of G with vertex complexity at most five in time $\tilde{O}(n^{\frac{10}{7}})$ time and on an integer grid of size $O(n) \times O(n)$. Di Giacomo et al. described an algorithm that computes an OPVR of G with minimum vertex complexity (which is at most five as shown above) in time $O(n^{\frac{7}{4}} \sqrt{\log n})$ and on an integer grid of $O(n) \times O(n)$ (Theorem 5 in [8]). This algorithm requires the computation of a feasible flow in a flow network with $O(n)$ nodes and edges. For such a flow network, Di Giacomo et al. used the min-cost flow algorithm of Garg and Tamassia [12], whose time complexity is $O(\chi^{\frac{3}{4}} n \sqrt{\log n})$, where χ is the cost of the flow, which is $O(n)$. Instead, we can use a recent result by Cohen et al. [5] as follows. We first replace all arcs of the flow network with capacity $k > 1$ with k arcs having unit capacity (note that $k \in O(1)$). Then the unit-capacity min-cost flow problem can be solved on the resulting flow network in $\tilde{O}(n^{\frac{10}{7}} \log W)$ time, where W is the maximum cost of an arc, which is $O(1)$. Thus, we can compute an OPVR in $\tilde{O}(n^{\frac{10}{7}})$ time on an integer grid of size $O(n) \times O(n)$, as desired.   □

Theorem 4, together with Theorem 2, proves Theorem 1.

5 Open Problems

We conclude by mentioning three open problems that are naturally suggested by the research in this paper. (i) Close the gap between the lower bound and the upper bound stated in Theorem 1. (ii) Can the time complexity of Theorem 1 be improved? (iii) An immediate consequence of Theorem 3 is that a 3-connected 1-plane graph G with $|P|$ poles has a set of non-redundant forbidden configurations whose size is at most $5|P|$. Is this upper bound tight?

References

1. Bekos, M.A., Kaufmann, M., Montecchiani, F.: Guest editors' foreword and overview for the special issue on graph drawing beyond planarity. J. Graph Algorithms Appl. **22**(1), 1–10 (2018)
2. Biedl, T., Liotta, G., Montecchiani, F.: Embedding-preserving rectangle visibility representations of nonplanar graphs. Discrete Comput. Geom. **60**(2), 345–380 (2018)
3. Biedl, T., Lubiw, A., Petrick, M., Spriggs, M.J.: Morphing orthogonal planar graph drawings. ACM Trans. Algorithms **9**(4), 29:1–29:24 (2013)
4. Brandenburg, F.J.: T-shape visibility representations of 1-planar graphs. Comput. Geom. **69**, 16–30 (2018)
5. Cohen, M.B., Madry, A., Sankowski, P., Vladu, A.: Negative-weight shortest paths and unit capacity minimum cost flow in õ ($m^{10/7} \log W$) time. In: ACM-SIAM SODA 2017, pp. 752–771. SIAM (2017)
6. Dean, A.M., Hutchinson, J.P.: Rectangle-visibility representations of bipartite graphs. Discrete Appl. Math. **75**(1), 9–25 (1997)
7. Di Battista, G., Eades, P., Tamassia, R., Tollis, I.G.: Graph Drawing: Algorithms for the Visualization of Graphs. Prentice-Hall, Upper Saddle River (1999)
8. Di Giacomo, E., et al.: Ortho-polygon visibility representations of embedded graphs. Algorithmica **80**(8), 2345–2383 (2018)
9. Didimo, W., Liotta, G., Montecchiani, F.: A survey on graph drawing beyondplanarity. CoRR **abs/1804.07257** (2018). http://arxiv.org/abs/1804.07257
10. Dolev, D., Leighton, F.T., Trickey, H.: Planar embedding of planar graphs. In: Preparata, F.P. (ed.) VLSI Theory, Advances in Compututing Research, vol. 2, pp. 147–161. JAI Press, Greenwich (1985)
11. Evans, W.S., Liotta, G., Montecchiani, F.: Simultaneous visibility representations of plane st-graphs using L-shapes. Theor. Comput. Sci. **645**, 100–111 (2016)
12. Garg, A., Tamassia, R.: A new minimum cost flow algorithm with applications to graph drawing. In: North, S. (ed.) GD 1996. LNCS, vol. 1190, pp. 201–216. Springer, Heidelberg (1997). https://doi.org/10.1007/3-540-62495-3_49
13. Hall, P.: On representatives of subsets. J. London Math. Soc. **s1–10**(1), 26–30 (1935)
14. Hutchinson, J.P., Shermer, T.C., Vince, A.: On representations of some thickness-two graphs. Comput. Geom. **13**(3), 161–171 (1999)
15. Kobourov, S.G., Liotta, G., Montecchiani, F.: An annotated bibliography on 1-planarity. Comput. Sci. Rev. **25**, 49–67 (2017)
16. Liotta, G., Montecchiani, F.: L-visibility drawings of IC-planar graphs. Inf. Process. Lett. **116**(3), 217–222 (2016)

17. Liotta, G., Montecchiani, F., Tappini, A.: Ortho-polygon visibilityrepresentations of 3-connected 1-plane graphs. CoRR **abs/1807.01247** (2018). http://arxiv.org/abs/1807.01247
18. Shermer, T.C.: On rectangle visibility graphs. III. External visibility and complexity. In: CCCG 1996, pp. 234–239. Carleton University Press (1996)
19. Streinu, I., Whitesides, S.: Rectangle visibility graphs: characterization, construction, and compaction. In: Alt, H., Habib, M. (eds.) STACS 2003. LNCS, vol. 2607, pp. 26–37. Springer, Heidelberg (2003). https://doi.org/10.1007/3-540-36494-3_4

Realizability

Realization and Connectivity of the Graphs of Origami Flat Foldings

David Eppstein[✉]

Department of Computer Science, University of California, Irvine, USA
eppstein@uci.edu

Abstract. We investigate the graphs formed from the vertices and creases of an origami pattern that can be folded flat along all of its creases. As we show, this is possible for a tree if and only if the internal vertices of the tree all have even degree greater than two. However, we prove that (for unbounded sheets of paper, with a vertex at infinity representing a shared endpoint of all creased rays) the graph of a folding pattern must be 2-vertex-connected and 4-edge-connected.

1 Introduction

This work concerns the following question: Which graphs can be drawn as the graphs of origami flat folding patterns?

In origami and other forms of paper folding, a *flat folding* is a type of construction in which an initially-flat piece of paper is folded so that the resulting folded shape lies flat in a plane and has a desired shape or visible pattern. This style of folding may be used as the initial base from which a three-dimensional origami figure is modeled, or it may be an end on its own. Flat foldings have been extensively studied in research on the mathematics of paper folding. The folding patterns that can fold flat with only a single vertex have been completely characterized, for standard models of origami [1–8], for *rigid origami* in which the paper must continuously move from its unfolded state to its folded state without bending anywhere except at its given creases [9], and even for single-vertex folding patterns whose paper does not form a single flat sheet [10]. However, the combinatorics of multi-vertex flat folding patterns is much less well understood, and testing whether a multi-vertex pattern folds flat is NP-hard [11].

From the point of view of graph drawing, origami folding patterns can be thought of as planar graphs, drawn with straight line edges in the Euclidean plane, with each edge representing a crease that must be folded. For instance, the familiar bird base, a starting point for the classic three-dimensional origami crane, can be thought of as a graph drawing of a planar graph with 13 vertices (Fig. 1). This naturally raises the question (analogous to similar questions for other types of geometric graphs such as Voronoi diagrams [12]): which graphs can be drawn this way? The NP-completeness of recognizing multi-vertex flat folding patterns does not extend to this question, because the completeness result is for folding patterns that have already been embedded with a given geometry

© Springer Nature Switzerland AG 2018
T. Biedl and A. Kerren (Eds.): GD 2018, LNCS 11282, pp. 541–554, 2018.
https://doi.org/10.1007/978-3-030-04414-5_38

and its proof depends on the specific geometry of the embedding. Here, instead, we ask whether an embedding exists. We do not resolve this question, but we provide partial answers to it in two different directions.

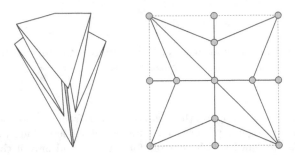

Fig. 1. Origami bird base (as illustrated by Fred the Oyster at https://commons.wikimedia.org/wiki/File:Bird_base.svg) and the corresponding folding pattern, interpreted as a graph drawing. The black lines indicate the final creases of the bird base. Temporary creases made while folding the base but later flattened out are not included. Blue dashed lines indicate the boundary of the sheet of paper; these lines are not considered as edges of the graph because they are not creased. (Color figure online)

First, we investigate the trees that may be drawn as flat folding patterns. For this problem, we make the simplifying assumption that the sheet of paper to be folded is infinite, with internal vertices of the tree at points where multiple creases come together, and with the leaves of the tree corresponding to creases along infinite rays. Cutting the infinite paper of such a drawing along a square that surrounds all the internal vertices would produce a finite representation of the same tree with its leaves on the boundary of the square, like the representation of a non-tree graph in Fig. 1. Similar tree-drawing styles, with infinite rays for the leaves of the trees, have been used in past work on drawings of trees as Voronoi diagrams [12], straight skeletons [13],[1] or with optimized angular resolution [15]. For this model of origami folding and tree realization, we provide a complete characterization: a tree may be drawn in this way if and only if all of its internal vertices have even degree greater than two.

Second, we investigate the connectivity restrictions on the graphs that may be drawn as flat folding patterns. This type of constraint has proven very fruitful in past questions about the geometric realizations of planar graphs, providing complete characterizations of the graphs of convex polyhedra (Steinitz's theorem) [16], drawings with rectangular faces ("rectangular duals") [17–20], orthogonal polyhedra [21], and two-dimensional soap bubble clusters [22].

Trees are not highly connected, and may be drawn as flat folding diagrams, but it turns out that these diagrams remain highly connected through the bound-

---

[1] Straight skeletons have also been used to construct folding patterns [14]. However, this technique adds extra folds to the skeleton, so the realizations of trees as straight skeletons do not yield realizations of the same trees as flat folding patterns.

ary of the drawing. To capture this boundary connectivity, we modify our mathematical model of flat folding. We again assume an infinite sheet of paper, but we treat creases along infinite rays as all having a single shared endpoint at infinity, which forms another vertex of the graph. In this model, the tree foldings of the other model become series-parallel graphs, in which all the leaves of the tree have been merged into a single supervertex.

We prove that, for this model of graphs as folding patterns, the graphs that may be realized are highly restricted, beyond even the graphs of polyhedra and beyond the immediate restriction (from the one-vertex case) that all vertices have even degree. In particular, they are necessarily 2-vertex-connected and 4-edge-connected. More strongly, the vertex at infinity is not an articulation vertex, and any subset of vertices that separates the graph and does not include the vertex at infinity must include at least four other vertices. These connectivity restrictions hold even for a weaker model of *local flat foldability* in which we seek a piecewise linear map from the folding pattern to its folded state in the plane without regard to whether this folding can be embedded without self-intersections into three-dimensional space. Our realizations of trees as flat folding patterns show that the 2-vertex-connectivity and 4-edge-connectivity conditions are both tight: no higher restriction on connectivity is possible.

2   Preliminaries

2.1   Mathematical Model of Folding

Departing from the usual square-paper model of origami in order to avoid complications from its boundary conditions, we model the sheet of paper to be folded as the entire Euclidean plane. We first define a *local flat folding*. This is a highly simplified model of how a piece of paper might be folded that only takes into account local constraints (the paper can only be folded, not stretched, sheared, or crumpled), does not prevent self-intersections, and does not even represent the most basic information about how the folding might occur in three dimensions, such as whether a given fold is a mountain fold or a valley fold.

Definition 1. *We define a continuous function φ from the plane to itself to be a local flat folding if every point p of the plane has one of the following three types:*

- *An* unfolded point *of a local flat folding is a point p such that φ is a local isometry: there is a neighborhood of p that is mapped by φ in a distance-preserving way (necessarily a combination of translation, rotation, or reflection of the plane).*
- *A* crease point *of a local flat folding is a point p that has a neighborhood N that can be covered by two subsets, each containing p and each mapped by φ in a different distance-preserving way. Necessarily, the boundary between these two subsets must be a line containing p. To preserve continuity of the mapping, the two distinct isometric mappings for the two subsets must be*

reflections of each other across the image of this line. The points within N that belong to this fold line are also crease points, and the other points within N are unfolded points.

- *A* vertex point *of a local flat folding is a point p that has a neighborhood N that can be covered by finitely many (and at least three) subsets, each containing p and each mapped by φ in a distance-preserving way so that there are at least three distinct isometric mappings among these subsets. Necessarily, each subset must be a wedge. The points within N that belong to the rays between pairs of wedges are crease points, and the points within N that do not belong to these rays are unfolded points.*

Then, as stated above, a local flat folding is a continuous function ϕ such that all points of the plane are unfolded points, crease points, and vertex points. We add one more restriction: we consider only local flat foldings that have at least one vertex point. We do not require the number of vertex points to be finite.

As a simple example, consider the function $\varphi : (x, y) \mapsto (f(x), f(y))$ where $f(x) = |(x \bmod 2) - 1|$. Here f is a continuous function that maps the intervals $[2i, 2i + 1]$ to $[0, 1]$ in reverse order, and that maps the intervals $[2i + 1, 2i + 2]$ to $[0, 1]$ linearly. φ corresponds to a folding pattern in which we *pleat* the plane along the integer-coordinate vertical lines (that is, we create a sequence of folds that alternates between mountain and valley folds, like an accordion; see [23, p. 31]), and then we pleat it again along the integer-coordinate horizontal lines, so that the whole plane is mapped to the unit square. Its folding pattern has vertex points at points of the plane where both coordinates are integers, crease points at points with one integer coordinate, and unfolded points everywhere else. That is, it is a drawing of the infinite square grid graph.

In general, the graph of a local flat folding is almost a graph drawing, in that its vertex points form a discrete set, connected in pairs by line segments consisting of crease points. For the grid example, it is a graph drawing. However, for other local flat foldings, some of the crease points may belong to semi-infinite rays rather than forming bounded line segments. To make a graph that also includes these rays as edges, we add a special vertex ∞ that is not represented by any geometric point, and we treat this special vertex as an endpoint of each ray of crease points.

Definition 2. *We define the* graph *of a local flat folding φ to be a graph G that has a vertex for each vertex point of φ and (if φ includes any infinite rays of crease points) another special vertex ∞. Two vertex points form adjacent vertices in G when the line segment between them consists only of crease points. A vertex point p and the special vertex ∞ are adjacent when there exists a ray with apex p consisting only (other than at its apex) of crease points. This graph may have multiple adjacencies between ∞ and other vertices (for instance, it will do so in any one-vertex flat folding pattern) but it can have at most one edge between any two vertex points.*

The folding pattern provides a topological planar embedding for the whole graph G, and a geometric straight-line planar embedding for all vertices except ∞. As usual, we call the maximal regions of the plane that are disjoint from the vertices and edges of the embedding (the vertex and crease points of φ) the *faces* of the embedding. These are possibly-unbounded polygonal regions, the connected components of the unfolded points of φ. Because the action of φ on each face of the graph is determined from its action on adjacent faces, the embedding of G completely determines the mapping of φ, up to a congruence transformation of the whole plane.

For our realizations of trees, we will use a slightly different graph, that can be derived from the graph of the folding. (It will not be interesting to study the graph connectivity of this graph, because it will have many degree-one vertices.)

Definition 3. *We define the* truncated graph *of a local flat folding to be the graph formed in either of the following two equivalent ways:*

- *From the graph of the folding, subdivide each edge incident to ∞, and then delete vertex ∞.*
- *Form a graph with a vertex for each vertex point of the folding and another vertex for each ray of crease points of the folding. Connect two vertex points by an edge if the line segment between them consists only of crease points. Add an edge for each ray of crease points, connecting the vertex point at the apex of the ray to the additional vertex for the same ray.*

Truncated graphs of local flat foldings can also be interpreted as the type of graph drawn in Fig. 1 for a folding pattern on a sheet of square paper with the additional property that the creases reaching the boundary form diverging rays. However, the folding pattern in Fig. 1 has creases that instead meet at the boundary, and it is also possible to form converging pairs of rays. Therefore the type of graph shown in the figure, of a folding pattern on a bounded square of paper, is somewhat more general. However, for the purposes for which we use truncated graphs (realization of trees), a less general model is better, as any realization in such a model will also be a realization for the more general model.

It remains to define a mathematical model of foldings as global structures, accounting for how paper can fold in three dimensions and how some parts of the paper can block other parts of paper from passing through them (disallowing self-intersections). It is possible to model precisely the above-below relation of the faces of φ, and the nesting structure of the folding at the creases of φ; see, for instance, [10] for a similar model of lower-dimensional flat-folded structures. However, we will forgo the complexity of such a model in favor of the following simpler topological approach.

Definition 4. *A* global flat folding *is a local flat folding φ with the additional property that, for every $\epsilon > 0$, there exists a topological embedding $\varphi_\epsilon : \mathbb{R}^2 \to \mathbb{R}^3$ (without self-intersections) such that composing φ_ϵ with the coordinatewise vertical projection from $\mathbb{R}^3$ to $\mathbb{R}^2$ produces a mapping that, for every point p, is within distance ϵ of the mapping given by φ.*

Intuitively, a global flat folding is a local flat-folding that, for every $\epsilon > 0$, is ϵ-close to a topological embedding of the plane into three-dimensional space.

2.2 Single-Vertex Restrictions

The geometry of single-vertex folding patterns, such as the one in Fig. 2, is characterized by Maekawa's theorem and Kawasaki's theorem [1–8]. These apply as well to each vertex of a multi-vertex folding pattern.

Theorem 1 (Maekawa's theorem for one-vertex folding patterns without mountain-valley assignments). *Each vertex point of a folding pattern must be incident to an even number of creases.*

This follows easily from the observation that, at each crease, the paper alternates between having its top side up (a region within which φ is an orientation-preserving isometric mapping) and having its bottom side up (a region within which φ is an orientation-reversing isometric mapping).

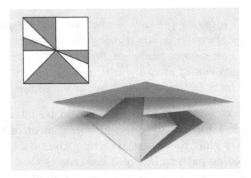

Fig. 2. A single-vertex flat folding and its pattern, demonstrating Maekawa's theorem (the number of folds is even) and Kawasaki's theorem (the face-up orange total angle equals the bottom-up white total angle). Image by the author for Wikipedia, 2011. (Color figure online)

Theorem 2 (Kawasaki's theorem). *At each vertex point of a folding pattern, the alternating sum of wedge angles totals to zero.*

This again follows from the fact that, near the vertex in the flat-folded state of the pattern, each point is covered by equal numbers of upward-facing and downward-facing regions, so the total amount of upward-facing paper must equal the amount of downward-facing paper.

Corollary 1. *Each wedge of a vertex point of a flat folding has angle strictly less than π. Therefore, each face of a flat folding pattern is a (possibly unbounded) convex polygon.*

3   Realization of Trees

Let T be any plane tree. Then by Maekawa's theorem, if T is to be realized as the truncated graph of a local flat folding, its internal vertices must have even degree greater than two. Our purpose in this section is to prove that this condition is necessary as well as sufficient.

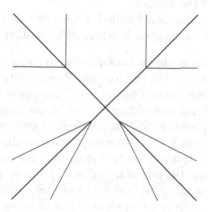

Fig. 3. A tree folding pattern that can be locally flat folded, but not globally flat folded.

We are interested here in global flat foldings, not just local flat foldings, and for this reason some care must be taken. It is not sufficient merely to embed T as a graph in the plane, with its leaf edges drawn as rays, and with each internal vertex meeting the angle sum condition of Kawasaki's theorem. Figure 3 depicts a counterexample. It obeys Kawasaki's theorem, and can be locally flat folded, but not globally flat folded. The four heavier diagonal lines of the figure can be flat folded in only one way up to combinatorial equivalence. Their folding is obtained by first folding along one diagonal line, and then along the other. The four creases of this fold are then modified by subsidiary folds that are each individually possible. But one of the four heavier creases must be nested tightly within another one. The two subsidiary creases of these two nested creases are arranged in such a way that, no matter which crease is nested within the other, the subsidiary crease of one will be blocked by the paper from the other nested crease. (Try it!)

To evade this problem, we seek a stronger type of realization, one in which each crease is "protected" by a wedge surrounding it, within which we can add modifications (such as the subsidiary wedges of Fig. 3) without interfering with other parts of the folding.

Theorem 3. *Let T be any finite tree with all internal vertices having even degree greater than two. Then T can be realized as the truncated graph of a global flat folding.*

Proof. We use induction on the number of internal nodes of T to prove a stronger statement: that T can be realized in such a way that each ray r of T is associated with a wedge W_r, satisfying the following properties:

- Ray r and wedge W_r have the same apex, and r is the median ray of its wedge.
- Each two rays have interior-disjoint wedges. Each edge of T that is not a ray is disjoint from all of the wedges.
- There exists a three-dimensional folded state such that the two halves of each wedge W_r are placed touching each other, with no other paper between them.

The third property above is phrased informally, so let us relate it to our earlier topological definition of a global flat folding. Recall that, in order to formalize the notion of a "three-dimensional folded state" we really have a parameterized family of three-dimensional embeddings. That is, we have both a folding map $\varphi : \mathbb{R}^2 \to \mathbb{R}^2$ and, for each $\epsilon > 0$, a topological embedding $\varphi_\epsilon : \mathbb{R}^2 \to \mathbb{R}^3$ whose vertical projection to $\mathbb{R}^2$ is ϵ-close to φ. We formalize the "no other paper between them" constraint, again up to ϵ-closeness: for each point $p \in \mathbb{R}^2$ at a distance of ϵ or more from the boundary of $\varphi(W_r)$, the preimage of p (according to the vertical projection) in $\varphi_\epsilon(\mathbb{R}^2)$ should have two points from the two sides of W_r consecutive with each other in the vertical ordering of the points.

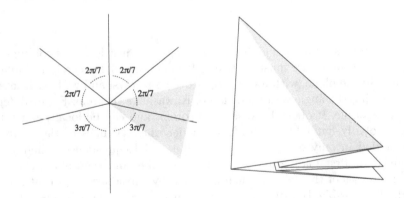

Fig. 4. The base case for realizing a one-internal-vertex tree (here with degree $d = 6$), showing the wedge W_r for one of the rays r both in the folding pattern and in the folded state.

The base case of the induction is a tree T with one internal node v of even degree d greater than four. In this case, we let $\theta = \pi/(d+1)$. We draw T as a set of d rays, all meeting at a common point. We make two of the angles between consecutive rays of T equal to 3θ, and all remaining angles equal to 2θ. For instance, when $d = 7$, we get $\theta = \pi/7$ and six rays separated by angles of $3\pi/7, 3\pi/7, 2\pi/7, 2\pi/7, 2\pi/7, 2\pi/7$. We fold this in three dimensions by placing the two wider wedges on the top and bottom of the folded pattern, and pleating the remaining wedges between them. For this fold, we make each wedge W_r for a ray r of the folding pattern be the wedge centered on that ray with opening angle 2θ. This opening angle is sufficient to make all the wedges interior-disjoint, and it is straightforward to verify that the 3D realization of this fold places no paper between the two halves of any wedge. This case is depicted in Fig. 4.

Otherwise, if T has more than one internal vertex, let v be any internal vertex that has only a single non-leaf neighbor. (For instance, v may be found by choosing any vertex u arbitrarily and letting v be an internal vertex that is maximally far from u.) Let T' be the tree formed from T by removing the leaf neighbors of v, so that v itself becomes a leaf. Then by the induction hypothesis, T' can be realized by a global flat folding, with a ray r that is associated with its leaf v and that is surrounded by a wedge W_r, whose two halves touch each other without being blocked by other paper in the folding. Let θ denote the opening angle of wedge W_r. Suppose also that, in T, v has degree d, and therefore it also has $d-1$ leaf children.

Then we modify the folding that represents T' to form a folding representing T, as follows. We place v at an arbitrarily chosen point along r (for instance, at the point a unit distance away from the apex of ray r). Then, we form $d-1$ creases, along $d-1$ rays with v as apex, to represent the $d-1$ leaf children of r. We choose the angles of these rays so that they are separated from each other and from the two boundary rays of W_r by an angle of θ/d. Finally, we assign each of these rays its own wedge, with v as its apex and with opening angle θ/d. (See Fig. 5.)

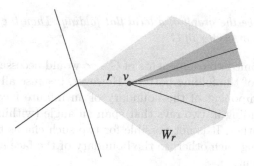

Fig. 5. Adding a vertex v to the folding of T' to create a folding for T. We choose the angles of the new rays incident to v so that they and the two boundary rays of the outer wedge W_r are equally spaced. The wedge surrounding each new ray has opening angle equal to the spacing of the rays. The crease pattern of the figure corresponds to a tree with two degree-four internal nodes.

The 3D folding of the crease pattern for T' can also be modified in the same way to form a 3D folding for the crease pattern for T. At v, the rays and segments representing incident edges of T form d wedges, two of which have opening angle greater than π and the rest of which have opening angle θ/d. As before, we fold this part of the paper so that the two large wedges are outermost and the other wedges are pleated between them. The angles of the creased rays are chosen so that, after this pleat, the creases that are folded to become the closest to the boundary rays of W_r (such as the middle ray of the figure) become parallel to these boundary rays. Because of this, the folded state stays within the region of $\mathbb{R}^3$ previously occupied by the paper for wedge W_r, and the empty space between the two sides of that wedge, so it does not interfere with any other part of the global flat folding. Each of the wedges of opening angle θ/d surrounding the new rays of the folding has its two sides mapped directly above and below each other in the pleating, maintaining the invariant of the induction.   □

We remark that, because the pleating pattern used for this realization does not ever tightly nest one crease inside another, it is possible to find a 3D realization that projects exactly to the two-dimensional local flat folding, rather than approaching it through ϵ-approximations.

4   Connectivity

Although we have seen that truncated graphs of flat foldings may be trees (graphs that are not very highly connected), we now show that the full graph, including the special vertex ∞, is (when finite) always well connected. We assume throughout this section that the full graph has at least one finite vertex; otherwise, as a one-vertex graph, the full graph is trivially k-vertex-connected and k-edge-connected for all k.

Lemma 1. *Let G be the graph of a local flat folding. Then the special vertex ∞ is not an articulation vertex of G.*

Proof. If it were, some two components of $G - \infty$ would necessarily be separated by an infinite face of the folding pattern. However, because all faces are convex each connected component of the boundary of an infinite face forms a convex polygonal chain, ending in two rays that span an angle (within the face) of less than π with each other. It is not possible for two such chains to bound a single face without crossing each other, so the boundary of the face can have only one connected component.   □

Lemma 2. *Let u and v be two vertex points of a local flat folding φ that belong to the same face of φ and let d denote Euclidean distance. Then $d(u,v) = d(\varphi(u), \varphi(v))$.*

Proof. Because the faces of φ are strictly convex, the line segment between u and v must either consist entirely of crease points (on an edge of the graph of the folding) or unfolded points (if u and v are not consecutive on their shared face). In either case this line segment is mapped to an equal-length line segment by φ.   □

Lemma 3. *Let G be the finite graph of a local flat folding. Then removing up to three of the vertex points of the folding from G cannot cause the remaining graph to become disconnected.*

Proof. Suppose for a contradiction that S is a set of at most three vertex points whose removal disconnects G. Since G is a plane graph, there must exist a simple closed curve C in the plane that passes through S and is otherwise disjoint from the vertices and edges of G, with at least one vertex inside the curve and at least one vertex outside the curve. (For folding patterns that include a ray of crease points, we count ∞ as being outside all such curves.) But as we show in the case analysis below, this is not possible:

- If $|S| = 1$, any curve C through the single vertex of S that is otherwise disjoint from G must remain within a single convex face of G, and cannot enclose anything.
- If S consists of two non-adjacent vertices, they can only have one face of G in common. Any curve C through these two vertices that is otherwise disjoint from G must remain within that face, and cannot enclose anything.
- If S consists of two adjacent vertices, then a curve C through the two vertices u and v of S that is otherwise disjoint from G can either stay within one of the two faces incident to edge uv (not enclosing anything) or have one arc in one of these two faces and one arc in the other of the two faces, enclosing edge uv but not enclosing any vertices.
- If S consists of three collinear vertex points, then curve C must visit each of these three points in turn. But the outermost of these two vertex points cannot belong to any convex face of the folding pattern (because this face would also contain the middle point), and cannot be connected by an arc of C.
- If S consists of three non-collinear vertex points u, v, and w, then C can only enclose any vertex points that might lie interior to triangle uvw. However, triangle uvw is mapped by the local flat folding map φ to a congruent triangle, by Lemma 2 and by the fact that there is only one Euclidean triangle (up to congruence) for any triple of distances between its vertices. In order to avoid stretching, every line segment formed by intersecting a line with triangle uvw must be mapped by φ to the corresponding line segment of the image triangle. In particular, there can be no creases within triangle uvw, because whenever a line segment properly crosses a crease of a local flat folding, it is not mapped to a congruent line segment. Therefore, every point inside triangle uvw must be an unfolded point, and C cannot contain a vertex point.

Because there is no way to construct curve C, the hypothesized set S cannot exist. □

The assumption that G is finite is used in the existence of C. If G could be infinite, our tree realization construction could be used to construct a realization of an infinite tree in which ∞ is a degree-one leaf. This does not have the connectivity described by the lemma, but this is not a contradiction because it does not meet the assumptions of the lemma.

Theorem 4. *If G is the finite graph of a local flat folding φ, then G is 2-vertex-connected and 4-edge-connected.*

Proof. G can have no articulation vertex, because neither ∞ nor any vertex point of φ can be an articulation vertex (Lemma 1 and Lemma 3 respectively).

Assume for a contradiction that G could have three edges e_1, e_2, and e_3 whose removal disconnects G. Choose a vertex point v_i as one of the two endpoints of each of these edges (as each edge in G has at least one vertex point as its endpoint). The separation of G caused by the removal of the edges e_i cannot separate any subset of the three vertices v_i from the rest of G, because G has minimum degree four and, in a graph of this degree, any set of up to three vertices is connected to the rest of the graph by at least four incident edges. Therefore, there must be at least one vertex of G on each side of the separation that is not one of the three chosen vertices v_i. However, this implies that these three vertices also separate G, contradicting Lemma 3. This contradiction implies that our assumption is false, and therefore that G is 4-edge-connected.   □

We remark that our realizations of 4-regular trees show that both 2-vertex-connectivity and 4-edge-connectivity are tight: some graphs that can be realized as global flat foldings are neither 3-vertex-connected nor 5-edge-connected.

5   Conclusions

We have shown that trees can be realized as the (truncated) graphs of flat folding patterns, and that despite this the (non-truncated) graphs of flat folding patterns must be highly connected. However we have not succeeded in completely characterizing the graphs of flat folding patterns. We leave the following questions as open for future research:

- Which plane graphs (with specified vertex ∞) are the graphs of global flat foldings?
- What is the computational complexity of recognizing and realizing these graphs?
- Is there any graph-theoretic difference between the graphs of global flat foldings and the graphs of local flat foldings? In particular does the folding-assignment version of Maekawa's theorem, that each vertex must have two more mountain folds than valley folds or vice versa, impose any nontrivial constraints on the graphs of flat foldings?
- In the full version of this paper (arXiv:1808.06013) we describe another class of graphs, the *dual orthotrees*, that can always be realized as the graphs of local flat foldings. Can they always be realized as the graphs of global flat foldings?
- What (if anything) changes when we consider folding patterns on a square sheet of paper (or other bounded shape) rather than on an infinite sheet? In the full version we begin a preliminary investigation of this case, in the special case where we restrict the vertex points to the boundary of the paper. On

circular paper, all outerplanar graphs are possible, but on square paper, not even all trees can be folded; we find an exact characterization of the foldable trees, different from the characterization in Sect. 3. However, similar questions without the restriction to boundary points remain open.

- Previously we studied algorithms for realizing trees as convex subdivisions of the plane while optimizing the angular resolution of the resulting tree drawing [15]. Can we use similar ideas to optimize the angular resolution of a folding pattern realization of a tree?

Acknowledgements. This work was supported in part by NSF grants CCF-1618301 and CCF-1616248.

References

1. Justin, J.: Mathematics of origami, part 9, pp. 28–30. British Origami (1986)
2. Hull, T.: On the mathematics of flat origamis. In: Proceedings of the Twenty-Fifth Southeastern International Conference on Combinatorics, Graph Theory and Computing, Boca Raton, FL, 1994, vol. 100, Congressus Numerantium, pp. 215–224 (1994)
3. Huffman, D.A.: Curvature and creases: a primer on paper. IEEE Trans. Comput. C–25(10), 1010–1019 (1976)
4. Husimi, K., Husimi, M.: The Geometry of Origami. Nihon Hyouronsha, Tokyo (1979)
5. Robertson, S.A.: Isometric folding of Riemannian manifolds. Proc. R. Soc. Edinb. Ser. A 79(3–4), 275–284 (1977)
6. Kawasaki, T.: On the relation between mountain-creases and valley-creases of a flat origami. In: Huzita, H. (ed.) Proceedings of the 1st International Meeting on Origami Science and Technology, Comune di Ferrara and Centro Origami Diffusion, Ferrara, Italy, pp. 229–237 (1989)
7. Murata, S.: The theory of paper sculpture, I. Bull. Junior Coll. Art 4, 61–66 (1966)
8. Murata, S.: The theory of paper sculpture, II. Bull. Junior Coll. Art 5, 29–37 (1966)
9. Abel, Z., et al.: Rigid origami vertices: conditions and forcing sets. J. Comput. Geom. 7(1), 171–184 (2016)
10. Abel, Z., Demaine, E.D., Demaine, M.L., Eppstein, D., Lubiw, A., Uehara, R.: Flat foldings of plane graphs with prescribed angles and edge lengths. J. Comput. Geom. 9(1), 71–91 (2018)
11. Bern, M., Hayes, B.: The complexity of flat origami. In: Proceedings of the 7th ACM-SIAM Symposium on Discrete Algorithms (SODA 1996), Philadelphia, PA, pp. 175–183. Society for Industrial and Applied Mathematics (1996)
12. Liotta, G., Meijer, H.: Voronoi drawings of trees. Comput. Geom. Theor. Appl. 24(3), 147–178 (2003)
13. Aichholzer, O., et al.: What makes a tree a straight skeleton? In: Proceedings of the 24th Canadian Conference on Computational Geometry (CCCG 2012) (2012)
14. Demaine, E.D., Demaine, M.L., Lubiw, A.: Folding and cutting paper. In: Akiyama, J., Kano, M., Urabe, M. (eds.) JCDCG 1998. LNCS, vol. 1763, pp. 104–118. Springer, Heidelberg (2000). https://doi.org/10.1007/978-3-540-46515-7_9

15. Carlson, J., Eppstein, D.: Trees with convex faces and optimal angles. In: Kaufmann, M., Wagner, D. (eds.) GD 2006. LNCS, vol. 4372, pp. 77–88. Springer, Heidelberg (2007). https://doi.org/10.1007/978-3-540-70904-6_9
16. Steinitz, E.: Polyeder und Raumeinteilungen. In: Meyer, W.F., Mohrmann, H. (eds.) Encyclopädie der mathematischen Wissenschaften, Band 3 (Geometries), vol. IIIAB12, pp. 1–139. B. G. Teubner, Leipzig (1922)
17. Koźmiński, K., Kinnen, E.: Rectangular duals of planar graphs. Networks 15(2), 145–157 (1985)
18. Bhasker, J., Sahni, S.: A linear algorithm to find a rectangular dual of a planar triangulated graph. Algorithmica 3(2), 247–278 (1988)
19. He, X.: On finding the rectangular duals of planar triangular graphs. SIAM J. Comput. 22(6), 1218–1226 (1993)
20. Kant, G., He, X.: Two algorithms for finding rectangular duals of planar graphs. In: van Leeuwen, J. (ed.) WG 1993. LNCS, vol. 790, pp. 396–410. Springer, Heidelberg (1994). https://doi.org/10.1007/3-540-57899-4_69
21. Eppstein, D., Mumford, E.: Steinitz theorems for simple orthogonal polyhedra. J. Comput. Geom. 5(1), 179–244 (2014)
22. Eppstein, D.: A Möbius-invariant power diagram and its applications to soap bubbles and planar Lombardi drawing. Discrete Comput. Geom. 52(3), 515–550 (2014)
23. Lang, R.J.: Origami Design Secrets: Mathematical Methods for an Ancient Art, 2nd edn. CRC Press, Boca Raton (2012)

Arrangements of Pseudocircles: On Circularizability

Stefan Felsner(ID) and Manfred Scheucher(✉)(ID)

Institut für Mathematik, Technische Universität Berlin, Berlin, Germany
{felsner,scheucher}@math.tu-berlin.de

Abstract. An arrangement of pseudocircles is a collection of simple closed curves on the sphere or in the plane such that any two of the curves are either disjoint or intersect in exactly two crossing points. We call an arrangement intersecting if every pair of pseudocircles intersects twice. An arrangement is circularizable if there is a combinatorially equivalent arrangement of circles.

In this paper we present the results of the first thorough study of circularizability. We show that there are exactly four non-circularizable arrangements of 5 pseudocircles (one of them was known before). In the set of 2131 digon-free intersecting arrangements of 6 pseudocircles we identify the three non-circularizable examples.

Most of our non-circularizability proofs depend on incidence theorems like Miquel's. In other cases we contradict circularizability by considering a continuous deformation where the circles of an assumed circle representation grow or shrink in a controlled way.

The claims that we have all non-circularizable arrangements with the given properties are based on a program that generated all arrangements up to a certain size. Given the complete lists of arrangements, we used heuristics to find circle representations. Examples where the heuristics failed were examined by hand.

Keywords: Circularizability · Incidence theorems Great-(pseudo)circles

1 Introduction

Arrangements of pseudocircles generalize arrangements of circles in the same vein as arrangements of pseudolines generalize arrangements of lines. The study of arrangements of pseudolines was initiated by Levi [12] in 1918. Since then arrangements of pseudolines were intensively studied. The handbook article on

Partially supported by the DFG Grants FE 340/11-1 and FE 340/12-1. Manfred Scheucher was partially supported by the ERC Advanced Research Grant no. 267165 (DISCONV). We gratefully acknowledge the computing time granted by TBK Automatisierung und Messtechnik GmbH and by the Institute of Software Technology, Graz University of Technology. We also thank the anonymous reviewers for valuable comments.

© Springer Nature Switzerland AG 2018
T. Biedl and A. Kerren (Eds.): GD 2018, LNCS 11282, pp. 555–568, 2018.
https://doi.org/10.1007/978-3-030-04414-5_39

the topic [5] lists more than 100 references. To the best of our knowledge the study of arrangements of pseudocircles was initiated by Grünbaum [8] in the 1970s.

A *pseudocircle* is a simple closed curve in the plane or on the sphere. An *arrangement of pseudocircles* is a collection of pseudocircles with the property that the intersection of any two of the pseudocircles is either empty or consists of two points where the curves cross. Other authors also allow touching pseudocircles, e.g. [1]. A cell of the arrangement with k crossings on its boundary is a *k-cell*. A 2-cell is also called a *digon* (some authors call it a *lens*), and a 3-cell is also called a *triangle*. An arrangement $\mathcal{A}$ of pseudocircles is

simple, if no three pseudocircles of $\mathcal{A}$ intersect in a common point;
connected, if the graph of the arrangement is connected;
intersecting, if any two pseudocircles of $\mathcal{A}$ intersect.

In this paper we assume that arrangements are simple and connected.

Two arrangements $\mathcal{A}$ and $\mathcal{B}$ are *isomorphic* if they induce homeomorphic cell decompositions of the compactified plane, i.e., on the sphere. In particular, the isomorphism class of an arrangement of pseudocircles in the plane is closed under changes of the unbounded cell.

Figure 1 shows the three connected arrangements of three pseudocircles. We call the unique digon-free intersecting arrangement the *Krupp*[1]. The second intersecting arrangement is the *NonKrupp*; this arrangement has digons. The non-intersecting arrangement is the *3-Chain*.

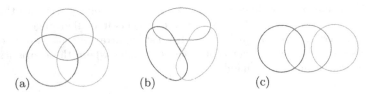

(a)   (b)   (c)

Fig. 1. The 3 connected arrangements of $n = 3$ pseudocircles. (a) *Krupp*, (b) *NonKrupp*, (c) *3-Chain*.

Every triple of great-circles on the sphere induces a Krupp arrangement, hence, we call an arrangement of pseudocircles an *arrangement of great-pseudocircles* if every subarrangement induced by three pseudocircles is a Krupp.

Some authors think of arrangements of great-pseudocircles when they speak about arrangements of pseudocircles, this is e.g. common practice in the theory of oriented matroids. In fact, arrangements of great-pseudocircles serve to represent rank 3 oriented matroids, cf. [2].

Definition 1. *An arrangement of pseudocircles is* circularizable *if there is an isomorphic arrangement of circles.*

Preceeding our work there have been only few results about circularizability of arrangements of pseudocircles. Edelsbrunner and Ramos [4] presented an intersecting arrangement of 6 pseudocircles (with digons) which has no realization with circles, i.e., it is not circularizable. Linhart and Ortner [13] found a non-circularizable non-intersecting arrangement of 5 pseudocircles with digons (see Fig. 2(b)). They also proved that every intersecting arrangement of at most 4 pseudocircles is circularizable. Kang and Müller [9] extended the result by showing that every arrangement of at most 4 pseudocircles is circularizable. They also proved that deciding circularizability for connected arrangements is NP-hard.

2   Overview

In Sect. 3 we present some background on arrangements of pseudocircles and provide tools that will be useful for non-circularizability proofs.

In Sect. 4 we study arrangements of great-pseudocircles – this class of arrangements of pseudocircles is in bijection with projective arrangements of pseudolines. Our main theorem in this section is the Great-Circle Theorem which allows the transfer of knowledge regarding arrangements of pseudolines to arrangements of pseudocircles.

Theorem 1 (Great-Circle Theorem).   *An arrangement of great-pseudocircles is circularizable (i.e., has a circle representation) if and only if it has a great-circle representation.*

In the last two sections we present the full classification of circularizable and non-circularizable arrangements among all connected arrangements of 5 pseudocircles and all digon-free intersecting arrangements of 6 pseudocircles. With the aid of computers we generated the complete lists of connected arrangements of $n \le 6$ pseudocircles and of intersecting arrangements of $n \le 7$ pseudocircles. The respective numbers are shown in Table 1. Given the complete lists of arrangements, we used automatized heuristics to find circle representations. Examples where the heuristics failed had to be examined by hand.

Computational issues and algorithmic ideas are omitted here – we refer the interested reader to the full version of this paper [7]. The encoded lists of arrangements of up to $n = 6$ pseudocircles and circle representations are available on our webpage [6].

The list of circle representations at [6] together with the non-circularizability proofs given in Sect. 5 yields the following theorem.

Theorem 2. *The four isomorphism classes of arrangements $\mathcal{N}_5^1$, $\mathcal{N}_5^2$, $\mathcal{N}_5^3$, and $\mathcal{N}_5^4$ (shown in Fig. 2) are the only non-circularizable ones among the 984 isomorphism classes of connected arrangements of $n = 5$ pseudocircles.*

Note that $\mathcal{N}_5^1$ is the only non-circularizable intersecting arrangement on 5 pseudocircles. Non-circularizability of $\mathcal{N}_5^2$ was previously shown by Linhart and Ortner [13]. We give an alternative proof which also shows the non-circularizability of $\mathcal{N}_5^3$. Jonathan Wild and Christopher Jones, contributed

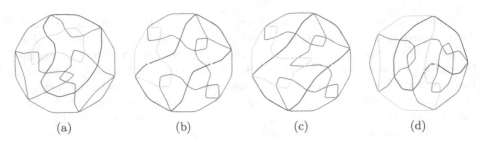

(a)   (b)   (c)   (d)

Fig. 2. The four non-circularizable arrangements on $n = 5$ pseudocircles: (a) $\mathcal{N}_5^1$, (b) $\mathcal{N}_5^2$, (c) $\mathcal{N}_5^3$, and (d) $\mathcal{N}_5^4$.

Table 1. Number of combinatorially different arrangements of n pseudocircles.

n	3	4	5	6	7
connected	3	21	984	609 423	?
+digon-free	1	3	30	4 509	?
intersecting	2	8	278	145 058	447 905 202
+digon-free	1	2	14	2 131	3 012 972
great-p.c.s	1	1	1	4	11

sequences A250001 and A288567 to the On-Line Encyclopedia of Integer Sequences (OEIS). These sequences count certain classes of arrangements of circles and pseudocircles. Wild and Jones also looked at circularizability and independently found Theorem 2 (personal communication).

Concerning arrangements of 6 pseudocircles, we were able to fully classify digon-free intersecting arrangements.

Theorem 3. *The three isomorphism classes of arrangements $\mathcal{N}_6^{\Delta}$, $\mathcal{N}_6^2$, and $\mathcal{N}_6^3$ (shown in Fig. 3) are the only non-circularizable ones among the 2131 isomorphism classes of digon-free intersecting arrangements of $n = 6$ pseudocircles.*

In Sect. 6, we give non-circularizability proofs for $\mathcal{N}_6^{\Delta}$, $\mathcal{N}_6^2$, and $\mathcal{N}_6^3$. In fact, for the non-circularizability of $\mathcal{N}_6^{\Delta}$ and $\mathcal{N}_6^2$, respectively, we have two proofs of different flavors: One proof (see Sect. 6) uses continuous deformations similar to the proof of the Great-Circle Theorem (Theorem 1) and the other proof (omitted in this version) is based on an incidence theorem. The incidence theorem used for $\mathcal{N}_6^{\Delta}$ may be of independent interest.

It may be worth mentioning that, by enumerating and realizing all arrangement of $n \leq 4$ pseudocircles, we have an alternative proof of the Kang and Müller result, that all arrangements of $n \leq 4$ pseudocircles are circularizable [9].

In the full version [7] we have further results, for example, non-circularizability proofs for some intersecting arrangements on $n = 6$ pseudocircles with digons.

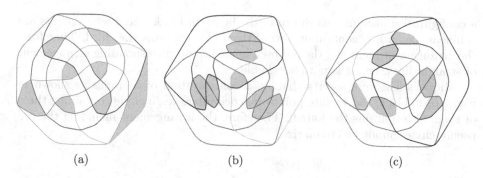

(a)   (b)   (c)

Fig. 3. The three non-circularizable digon-free intersecting arrangements for $n = 6$: (a) $\mathcal{N}_6^{\Delta}$, (b) $\mathcal{N}_6^2$, and (c) $\mathcal{N}_6^3$. Inner triangles are colored gray. Note that in (b) and (c) the outer face is also a triangle.

3   Preliminaries: Basic Properties and Tools

Stereographic projections map circles to circles (if we consider a line to be a circle containing the point at infinity), therefore, circularizability on the sphere and in the plane is the same concept. Arrangements of circles can be mapped to isomorphic arrangements of circles via Möbius transformations. In this context, the sphere is identified with the extended complex plane $\mathbb{C} \cup \{\infty\}$.

Let $\mathcal{C}$ be an arrangement of circles represented on the sphere. Each circle of $\mathcal{C}$ spans a plane in 3-space, hence, we obtain an arrangement $\mathcal{E}(\mathcal{C})$ of planes in $\mathbb{R}^3$. In fact, with a sphere S we get a bijection between (not necessarily connected) circle arrangements on S and arrangements of planes with the property that each plane of the arrangement intersects S.

Consider two circles C_1, C_2 of a circle arrangement $\mathcal{C}$ on S and the corresponding planes E_1, E_2 of $\mathcal{E}(\mathcal{C})$. The intersection of E_1 and E_2 is either empty (i.e., E_1 and E_2 are parallel) or a line ℓ. The line ℓ intersects S if and only if C_1 and C_2 intersect, in fact, $\ell \cap S = C_1 \cap C_2$.

With three pairwise intersecting circles C_1, C_2, C_3 we obtain three planes E_1, E_2, E_3 intersecting in a vertex v of $\mathcal{E}(\mathcal{C})$. It is notable that v is in the interior of S if and only if the three circles form a Krupp in $\mathcal{C}$.

Lemma 1. *Let $\mathcal{C}$ be an arrangement of circles represented on the sphere. Three circles C_1, C_2, C_3 of $\mathcal{C}$ form a Krupp if and only if the three corresponding planes E_1, E_2, E_3 intersect in a single point in the interior of S.*

Digons are also nicely characterized: A pair C_1, C_2 of circles forms a digon of $\mathcal{C}$ if and only if the segment of ℓ in the interior of S contains no vertex of $\mathcal{E}(\mathcal{C})$.

3.1   Incidence Theorems

The smallest non-stretchable arrangements of pseudolines are closely related to the incidence theorems of Pappos and Desargues. A construction already described by Levi [12] is depicted in Fig. 4(a). Pappos's Theorem states that, in a configuration of 8 lines as shown in the figure in black, the 3 white points are collinear, i.e., a line containing two of them also contains the third. Therefore, the arrangement including the red pseudoline has no corresponding arrangement of straight lines, i.e., it is not stretchable.

Miquel's Theorem asserts that, in a configuration of 5 circles as shown in Fig. 4(b) in black, the 4 white points are cocircular, i.e., a circle containing three of them also contains the fourth. Therefore, the arrangement including the red pseudocircle cannot be circularized.

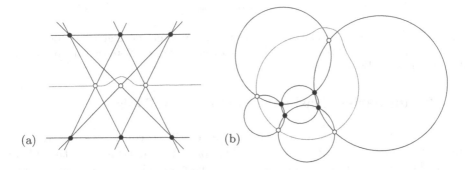

(a)   (b)

Fig. 4. (a) A non-stretchable arrangement of pseudolines from Pappos's Theorem. (b) A non-circularizable arrangement of pseudocircles from Miquel's Theorem. (Color figure online)

Next we state an incidence theorem that will be used in later proofs of non-circularizability. In the course of the paper we will meet further incidence theorems, e.g. Lemmas 4 and 5. (For a proof, see the full version [7].)

Lemma 2 (First Four-Circles Incidence Lemma). *Let $\mathcal{C}$ be an arrangement of four circles C_1, C_2, C_3, C_4 such that none of them is contained in the interior of another one, and such that (C_1, C_2), (C_2, C_3), (C_3, C_4), and (C_4, C_1) are touching. Then there is a circle C^* passing through these four touching points in the given cyclic order.*

3.2   Flips and Deformations of Pseudocircles

Let $\mathcal{C}$ be an arrangement of circles. Imagine that the circles of $\mathcal{C}$ start moving independently, i.e., the position of their centers and their radii depend on a time parameter t in a continuous way. This yields a family $\mathcal{C}(t)$ of arrangements with $\mathcal{C}(0) = \mathcal{C}$. Let us assume that the set T of all t for which $\mathcal{C}(t)$ is not simple or contains touching circles is discrete and for each $t \in T$ the arrangement $\mathcal{C}(t)$ contains either a single point where 3 circles intersect or a single touching. If $t_1 < t_2$ are consecutive in T, then all arrangements $\mathcal{C}(t)$ with $t \in (t_1, t_2)$ are isomorphic. Selecting one representative from each such class, we get a list $\mathcal{C}_0, \mathcal{C}_1, \ldots$ of simple arrangements such that two consecutive (non-isomorphic) arrangements $\mathcal{C}_i, \mathcal{C}_{i+1}$ are either related by a triangle flip or by a digon flip.

We will make use of controlled changes in circle arrangements, in particular, we grow or shrink specified circles of an arrangement to produce touchings or points where 3 circles intersect. The following lemma will be of use frequently. (For a proof, see the full version [7].)

Lemma 3 (Digon Collapse Lemma). *Let C be an intersecting arrangement of $n \geq 3$ circles in the plane and let C be a circles from C. If C has no incident triangle in its interior, then we can shrink C into its interior such that the combinatorics of the arrangement remain the same except that two digons collapse to touchings. Moreover, the two corresponding circles touch C from the outside.*

In the following we will sometimes use the dual version of the lemma, whose statement is obtained from the Digon Collapse Lemma by changing interior to exterior and outside to inside. The validity of the dual lemma is seen by applying a Möbius transformation which exchanges interior and exterior of C.

Triangle flips and digon flips are also central to the work of Snoeyink and Hershberger [17]. They have shown that an arrangement C of pseudocircles can be swept with a sweepfront γ starting at any pseudocircle $C \in C$, i.e., $\gamma_0 = C$. The sweep consists of two stages, one for sweeping the interior of C, the other for sweeping the exterior. At any fixed time t the sweepfront γ_t is a closed curve such that $C \cup \{\gamma_t\}$ is an arrangement of pseudocircles. Moreover, this arrangement is simple except for a discrete set T of times where sweep events happen. The sweep events are triangle flips or digon flips involving γ_t.

4   Arrangements of Great-Pseudocircles

Central projections map between arrangements of great-circles on a sphere S and arrangements of lines on a plane. Changes of the plane preserve the isomorphism class of the projective arrangement of lines. In fact, arrangements of lines in the projective plane are in one-to-one correspondence to arrangements of great-circles.

In this section we generalize this concept to arrangements of pseudolines and show that there is a one-to-one correspondence to arrangements of great-pseudocircles. As already mentioned, this correspondence is not new (see e.g. [2]).

An Euclidean arrangement of n pseudolines can be represented by x-monotone pseudolines, see e.g [5]. As illustrated in Fig. 5, an x-monotone representation can be glued with a mirrored copy of itself to form an arrangement of n pseudocircles. The resulting arrangement is intersecting and has no NonKrupp subarrangement, hence, it is an arrangement of great-pseudocircles.

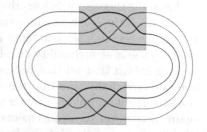

Fig. 5. Obtaining an arrangement of great-pseudocircles from an Euclidean arrangement L of pseudolines and its mirrored copy. The gray boxes highlight the arrangement L and its mirrored copy.

For a pseudocircle C of an arrangement of n great-pseudocircles the cyclic order of crossings on C is *antipodal*, i.e., the infinite sequence corresponding to the cyclic order crossings of C with the other pseudocircles is periodic of order $n - 1$. If we consider projections of projective arrangements of n pseudolines, then this order does not depend on the choice of the projection. In fact, projective arrangements of n pseudolines are in bijection with arrangements of n great-pseudocircles.

Projective arrangements of pseudolines are also known as projective abstract order types or oriented matroids of rank 3. The precise numbers of such arrangements are known for $n \leq 11$, see [10,11]. Hence the numbers of great-pseudocircle arrangements given in Table 1 are not new.

4.1 The Great-Circle Theorem and Its Applications

Let $\mathcal{A}$ be an arrangement of great-pseudocircles and let $\mathcal{L}$ be the corresponding projective arrangement of pseudolines. Central projections show that, if $\mathcal{L}$ is realizable with straight lines, then $\mathcal{A}$ is realizable with great-circles, and conversely. In fact, due to Theorem 1, it is sufficient that $\mathcal{A}$ is circularizable to conclude that $\mathcal{A}$ is realizable with great-circles and $\mathcal{L}$ is realizable with straight lines.

Proof (of Theorem 1). Consider an arrangement of circles $\mathcal{C}$ on the unit sphere S that realizes an arrangement of great-pseudocircles. Let $\mathcal{E}(\mathcal{C})$ be the arrangement of planes spanned by the circles of $\mathcal{C}$. Since $\mathcal{C}$ realizes an arrangement of great-pseudocircles, every triple of circles forms a Krupp, hence, the point of intersection of any three planes of $\mathcal{E}(\mathcal{C})$ is in the interior of S.

Imagine the planes of $\mathcal{E}(\mathcal{C})$ moving towards the origin. To be precise, for time $t \geq 1$ let $\mathcal{E}_t := \{1/t \cdot E : E \in \mathcal{E}(\mathcal{C})\}$. Since all intersection points of the initial arrangement $\mathcal{E}_1 = \mathcal{E}(\mathcal{C})$ are in the interior of the sphere S, the circle arrangement obtained by intersecting the moving planes $\mathcal{E}_t$ with the sphere S remains the same (isomorphic). Moreover, every circle in this arrangement converges to a great-circle as $t \to +\infty$, and the statement follows. $\square$

From the theorem it follows that an arrangement of pseudolines is stretchable if and only if the corresponding arrangement of great-pseudocircles is circularizable. Since deciding stretchability is known to be $\exists\mathbb{R}$-complete (see e.g. [14,15]), the hardness of stretchability directly carries over to hardness of circularizability.

It is known that all (not necessarily simple) arrangements of $n \leq 8$ pseudolines are stretchable and that the simple non-Pappos arrangement is the unique non-stretchable simple projective arrangements of 9 pseudoline, see e.g. [5]. This again carries over to arrangements of great-pseudocircles. Bokowski and Sturmfels [3] have shown that infinite families of minimal non-stretchable arrangements of pseudolines exist, i.e., non-stretchable arrangements where every proper subarrangement is stretchable. Again, this carries over to arrangement of pseudocircles.

5 Arrangements of 5 Pseudocircles

On the webpage [6] we have the data for circle realizations of 980 out of the 984 connected arrangements of 5 pseudocircles. The remaining four arrangements in this class are the four arrangements of Theorem 2. Since all arrangements with $n \leq 4$ pseudocircles have circle representations, there are no disconnected non-circularizable examples with $n \leq 5$. Hence, the four arrangements $\mathcal{N}_5^1$, $\mathcal{N}_5^2$, $\mathcal{N}_5^3$, and $\mathcal{N}_5^4$ are the only non-circularizable arrangements with $n \leq 5$.

For the non-circularizability proof of $\mathcal{N}_5^1$ we need the following additional incidence lemma. (A proof is given in the full version [7].)

Lemma 4 (Second Four-Circles Incidence Lemma). *Let C be an arrangement of four circles C_1, C_2, C_3, C_4 such that every pair of them is touching or forms a digon, and every circle is involved in at least two touchings. Then there is a circle C^* passing through the digon or touching point of each of the following pairs of circles (C_1, C_2), (C_2, C_3), (C_3, C_4), and (C_4, C_1) in this cyclic order.*

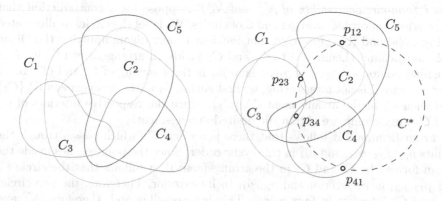

Fig. 6. An illustration of the non-circularizability proof of $\mathcal{N}_5^1$. The auxiliary circle C^* is drawn dashed.

Proof (non-circularizability of $\mathcal{N}_5^1$). Suppose for a contradiction that there is an isomorphic arrangement C of circles. We apply the Digon Collapse Lemma (Lemma 3) to shrink C_2, C_3, and C_4 into their respective interiors. We also use the dual of the Digon Collapse Lemma for C_1. In the resulting subarrangement C' formed by these four transformed circles C_1', C_2', C_3', C_4', each of the four circles is involved in at least two touchings. Moreover, since the intersection of C_i' and C_j' in C' is contained in the intersection of C_i and C_j in C, each of the four points p_{12}, p_{23}, p_{34}, and p_{41} lies in the original digons of C which respectively are touching points or points from the digons of (C_1', C_2'), (C_2', C_3'), (C_3', C_4'), and (C_4', C_1'). It follows that the circle C_5 has p_{12} and p_{34} in its interior but p_{23} and p_{41} in its exterior. Figure 6 gives an illustration.

By applying Lemma 4 to $\mathcal{C}'$ we obtain a circle C^* which passes through the points p_{12}, p_{23}, p_{34}, and p_{41} (in this order). Now the two circles C_5 and C^* intersect in four points. This is impossible, and hence $\mathcal{N}_5^1$ is not circularizable. □

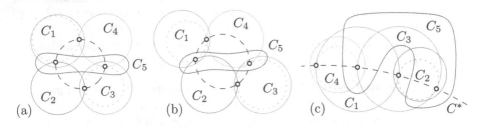

Fig. 7. An illustration of the non-circularizability proofs of (a) $\mathcal{N}_5^2$, (b) $\mathcal{N}_5^3$, and (c) $\mathcal{N}_5^4$. The auxiliary circle C^* is drawn dashed.

Proof (non-circularizability of $\mathcal{N}_5^2$ and $\mathcal{N}_5^3$). Suppose for a contradiction that there is an isomorphic arrangement $\mathcal{C}$ of circles. We label the circles as illustrated in Figs. 7(a) and (b). Since C_5 is not incident to any digon, applying the Digon Collapse Lemma (Lemma 3) to C_1 and C_3 yields an arrangement $\mathcal{C}'$ with four touching points $p_{12}, p_{23}, p_{34}, p_{41}$, where p_{ij} is the touching of C_i' and C_j'. To be precise, since $\mathcal{C}$ is not intersecting, we first shrink C_1 in the arrangement $\mathcal{C} \setminus \{C_3\}$ and then C_3 in the arrangement $\mathcal{C} \setminus \{C_1\}$. Since the respective interiors of C_1 and C_3 are disjoint, we obtain the desired arrangement.

From Lemma 2 it follows that there is a circle C^* which passes trough the points p_{12}, p_{23}, p_{34}, and p_{41} in this cyclic order. Since the point p_{ij} lies inside the digon formed by C_i and C_j in the arrangement $\mathcal{C}$, it follows that the circle C_5 has p_{12}, p_{34} in its interior and p_{23}, p_{41} in its exterior. Therefore, the two circles C_5 and C^* intersect in four points. This is impossible and, therefore, $\mathcal{N}_5^2$ and $\mathcal{N}_5^3$ are not circularizable. □

To prove the non-circularizability of $\mathcal{N}_5^4$, we make use of the following incidence lemma. (A proof and an illustration of this lemma can be found in the full version [7].)

Lemma 5 (Third Four-Circles Incidence Lemma). *Let $\mathcal{C}$ be an arrangement of four circles C_1, C_2, C_3, C_4 such that (C_1, C_2), (C_2, C_3), (C_3, C_4), and (C_4, C_1) are touching, moreover, C_4 is in the interior of C_1 and the exterior of C_3, and C_2 is in the interior of C_3 and the exterior of C_1. Then there is a circle C^* passing through the four touching points in the given cyclic order.*

Proof (non-circularizability of $\mathcal{N}_5^4$). Suppose there is an isomorphic arrangement of circles $\mathcal{C}$. Referring to the labeling as shown in Fig. 7(c) we shrink the circles C_2 and C_4 such that the pairs $(C_1, C_2), (C_2, C_3), (C_3, C_4), (C_4, C_1)$ (which form digons in $\mathcal{C}$) touch. With these touchings the four circles C_1, C_2, C_3, C_4

form the configuration of Lemma 5. Hence there is a circle C^* containing the four touching points in the given cyclic order. Now the two circles C^* and C_5 intersect in four points. This is impossible and, therefore, $\mathcal{N}_5^4$ is not circularizable. □

6   Arrangements of 6 Pseudocircles

On the webpage [6] we have the data of circle realizations of all 2131 intersecting digon-free arrangements of 6 pseudocircles except for the three arrangements mentioned in Theorem 3.

The arrangement $\mathcal{N}_6^\Delta$ (shown in Fig. 3(a)) is an intersecting digon-free arrangement. Since each of the eight triangles of $\mathcal{N}_6^\Delta$ is a NonKrupp, the non-circularizability of $\mathcal{N}_6^\Delta$ is an immediate consequence of the following theorem:

Theorem 4. *Let $\mathcal{A}$ be a connected digon-free arrangement of pseudocircles. If every triple of pseudocircles which forms a triangle is NonKrupp, then $\mathcal{A}$ is not circularizable.*

Proof. Assume for a contradiction that there exists an isomorphic arrangement of circles $\mathcal{C}$ on the unit sphere S. Let $\mathcal{E}(\mathcal{C})$ be the arrangements of planes spanned by the circles of $\mathcal{C}$. Imagine the planes of $\mathcal{E}(\mathcal{C})$ moving away from the origin. To be precise, for time $t \geq 1$ let $\mathcal{E}_t := \{t \cdot E : E \in \mathcal{E}(\mathcal{C})\}$. Consider the arrangement induced by intersecting the moving planes $\mathcal{E}_t$ with the sphere S. Since $\mathcal{C}$ has NonKrupp triangles, it is not a great-circle arrangement and some planes of $\mathcal{E}(\mathcal{C})$ do not contain the origin. All planes from $\mathcal{E}(\mathcal{C})$, which do not contain the origin, will eventually lose the intersection with S, hence some event has to happen.

When the isomorphism class of the intersection of $\mathcal{E}_t$ with S changes, we see a triangle flip, or a digon flip, or some isolated circle disappears. Since initially there is no digon and no isolated circle, the first event is a triangle flip. Triangles of $\mathcal{C}$ correspond to NonKrupp subarrangements, hence, the intersection point of their planes is outside of S (Lemma 1). This shows that a triangle flip event is also impossible. This contradiction implies that $\mathcal{A}$ is non-circularizable. □

The arrangement $\mathcal{N}_6^2$ (shown in Fig. 3(b)) is intersecting, not an arrangement of great-pseudocircles, and each triangle in $\mathcal{N}_6^2$ is Krupp. The following theorem is of the same flavor as Theorem 4 and directly implies the non-circularizability of $\mathcal{N}_6^2$.

Theorem 5. *Let $\mathcal{A}$ be an intersecting arrangement of pseudocircles which is not an arrangement of great-pseudocircles. If every triple of pseudocircles which forms a triangle is Krupp, then $\mathcal{A}$ is not circularizable.*

We outline the proof: Suppose a realization of $\mathcal{A}$ exists on the sphere. Continuously move the planes spanned by the circles towards the origin. The induced arrangement will eventually become isomorphic to an arrangement of great-circles. Now consider the first event that occurs. As the planes move towards the origin, there is no digon collapse. Since $\mathcal{A}$ is intersecting, no digon is created, and, since all triangles are Krupp, the corresponding intersection points of their planes is already inside S. Therefore, no event can occur – a contradiction.

The arrangement $\mathcal{N}_6^3$ is shown in Figs. 3(c) and 8(b). To prove its non-circularizability, we again use an incidence lemma.

The following lemma is mentioned by Richter-Gebert as a relative of Pappos's Theorem, cf. [16, p. 26]. Figure 8(a) gives an illustration.

Lemma 6. *Let ℓ_1, ℓ_2, ℓ_3 be lines, C_1, C_2, C_3 be circles, and $p_1, p_2, p_3, q_1, q_2, q_3$ be points, such that for $\{i, j, k\} = \{1, 2, 3\}$ point p_i is incident to line ℓ_i, circle C_j, and circle C_k, while point q_i is incident to circle C_i, line ℓ_j, and line ℓ_k. Then $C_1, C_2,$ and C_3 have a common point of intersection.*

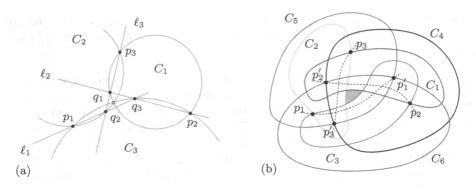

Fig. 8. (a) An illustration for Lemma 6. (b) The non-circularizable arrangement $\mathcal{N}_6^2$ with 3 dashed pseudolines illustrating the proof.

Proof (non-circularizability of $\mathcal{N}_6^3$). Suppose that $\mathcal{N}_6^3$ has a representation $\mathcal{C}$ as a circle arrangement in the plane. We refer to circles and intersection points via the label of the corresponding object in Fig. 8(b). As in the figure, we assume without loss of generality that the triangular cell spanned by C_4, C_5, and C_6 is the outer cell of the arrangement.

Consider the region $R := R_{24} \cup R_{35}$ where R_{ij} denotes the intersection of the respective interiors of C_i and C_j. The two straight-line segments $p_1 p_1'$ and $p_3 p_3'$ are fully contained in R_{35} and R_{24}, respectively, and have alternating end points along the boundary of R, hence they cross inside the region $R_{24} \cap R_{35}$.

From rotational symmetry we obtain that the three straight-line segments $p_1 p_1'$, $p_2 p_2'$, and $p_3 p_3'$ intersect pairwise.

For $i = 1, 2, 3$, let ℓ_i denote the line spanned by p_i and p_i', let q_i denote the intersection-point of ℓ_{i+1} and ℓ_{i+2}, and let C_i' denote the circle spanned by q_i, p_{i+1}, p_{i+2} (indices modulo 3). Note that ℓ_i contains p_i, q_{i+1}, q_{i+2}. These are precisely the conditions for the incidences of points, lines, and circles in Lemma 6. Hence, the three circles C_1', C_2', and C_3' intersect in a common point.

Let T be the triangle with corners p_1, p_2, p_3. Since p_2 and p_3 are on C_1, and q_1 lies inside of C_1, we find that the intersection of the interior of C_1' with T is a subset of the intersection of the interior of C_1 with T. The respective containments also hold for C_2' and C_2 and for C_3' and C_3. Moreover, since C_1', C_2', and C_3' intersect in a common point, the union of the interiors of C_1', C_2', and C_3' contains T. Hence, the union of interiors of the C_1, C_2, and C_3 also contains T. This shows that in $\mathcal{C}$ there is no face corresponding to the gray triangle; see Fig. 8(b). This contradicts the assumption that $\mathcal{C}$ is a realization of $\mathcal{N}_6^3$, whence the arrangement is not circularizable. $\qquad\square$

References

1. Agarwal, P.K., Nevo, E., Pach, J., Pinchasi, R., Sharir, M., Smorodinsky, S.: Lenses in arrangements of pseudo-circles and their applications. J. ACM **51**(2), 139–186 (2004). https://doi.org/10.1145/972639.972641
2. Björner, A., Las Vergnas, M., White, N., Sturmfels, B., Ziegler, G.M.: Oriented Matroids, Encyclopedia of Mathematics and its Applications, 2nd edn. Cambridge University Press, Cambridge (1999). https://doi.org/10.1017/CBO9780511586507
3. Bokowski, J., Sturmfels, B.: An infinite family of minor-minimal nonrealizable 3-chirotopes. Math. Z. **200**(4), 583–589 (1989). https://doi.org/10.1007/BF01160956
4. Edelsbrunner, H., Ramos, E.A.: Inclusion-exclusion complexes for pseudodisk collections. Discrete Comput. Geom. **17**, 287–306 (1997). https://doi.org/10.1007/PL00009295
5. Felsner, S., Goodman, J.E.: Pseudoline arrangements. In: Toth, C.D., O'Rourke, J., Goodman, J.E. (eds.) Handbook of Discrete and Computational Geometry, 3rd edn. CRC Press, Boca Raton (2018). https://doi.org/10.1201/9781315119601
6. Felsner, S., Scheucher, M.: Webpage: Homepage of Pseudocircles. http://www3.math.tu-berlin.de/pseudocircles
7. Felsner, S., Scheucher, M.: Arrangements of Pseudocircles: On Circularizability (2017). http://arXiv.org/abs/1712.02149
8. Grünbaum, B.: Arrangements and Spreads, CBMS Regional Conference Series in Mathematics, vol. 10. AMS (1972). https://doi.org/10.1090/cbms/010. (Reprinted 1980)
9. Kang, R.J., Müller, T.: Arrangements of pseudocircles and circles. Discrete Comput. Geom. **51**, 896–925 (2014). https://doi.org/10.1007/s00454-014-9583-8
10. Knuth, D.E.: Axioms and hulls. In: Knuth, D.E. (ed.) Axioms and Hulls. LNCS, vol. 606, pp. 1–98. Springer, Heidelberg (1992). https://doi.org/10.1007/3-540-55611-7_1
11. Krasser, H.: Order Types of Point Sets in the Plane. Ph.D. thesis, Institute for Theoretical Computer Science, Graz University of Technology, Austria (2003)
12. Levi, F.: Die Teilung der projektiven Ebene durch Gerade oder Pseudogerade. Berichte über die Verhandlungen der Sächsischen Akademie der Wissenschaften zu Leipzig, Mathematisch-Physische Klasse **78**, 256–267 (1926)
13. Linhart, J., Ortner, R.: An arrangement of pseudocircles not realizable with circles. Beiträge zur Algebra und Geometrie **46**(2), 351–356 (2005). https://doi.org/10.1007/s00454-014-9583-8
14. Matoušek, J.: Intersection graphs of segments and $\exists\mathbb{R}$. http://arXiv.org/abs/1406.2636 (2014)

15. Mněv, N.E.: The universality theorems on the classification problem of configuration varieties and convex polytopes varieties. In: Viro, O.Y., Vershik, A.M. (eds.) Topology and Geometry — Rohlin Seminar. LNM, vol. 1346, pp. 527–543. Springer, Heidelberg (1988). https://doi.org/10.1007/BFb0082792
16. Richter-Gebert, J.: Perspectives on Projective Geometry – A Guided Tour through Real and Complex Geometry. Springer, Heidelberg (2011). https://doi.org/10.1007/978-3-642-17286-1
17. Snoeynik, J., Hershberger, J.: Sweeping arrangements of curves. In: Goodman, L.S., Pollack, J.R., Steiger, S. (eds.) Discrete & Computational Geometry. DIMACS Series in Discrete Mathematics and Theoretical Computer Science, vol. 6, pp. 309–349. AMS (1991). https://doi.org/10.1145/73833.73872

The Weighted Barycenter Drawing Recognition Problem

Peter Eades[1], Patrick Healy[2(⊠)], and Nikola S. Nikolov[2]

[1] University of Sydney, Sydney, Australia
peter.d.eades@gmail.com
[2] University of Limerick, Limerick, Republic of Ireland
{patrick.healy,nikola.nikolov}@ul.ie

Abstract. We consider the question of whether a given graph drawing Γ of a triconnected planar graph G is a weighted barycenter drawing. We answer the question with an elegant arithmetic characterisation using the faces of Γ. This leads to positive answers when the graph is a Halin graph, and to a polynomial time recognition algorithm when the graph is cubic.

1 Introduction

The *barycenter algorithm* of Tutte [15,16] is one of the earliest and most elegant of all graph drawing methods. It takes as input a graph $G = (V, E)$, a subgraph $F_0 = (V_0, E_0)$ of G, and a position γ_a for each $a \in V_0$. The algorithm simply places each vertex $v \in V - V_0$ at the barycenter of the positions of its neighbours. The algorithm can be seen as the grandfather of force-directed graph drawing algorithms, and can be implemented easily by solving a system of linear equations. If G is a planar triconnected graph, F_0 is the outside face of G, and the positions γ_a for $a \in V_0$ are chosen so that F_0 forms a convex polygon, then the drawing output by the barycenter algorithm is planar and each face is convex.

The barycenter algorithm can be generalised to planar graphs with positive edge weights, placing each vertex i of $V - V_0$ at the weighted barycenter of the neighbours of i. This generalisation preserves the property that the output is planar and convex [8]. Further, weighted barycenter methods have been used in a variety of theoretical and practical contexts [4,5,11,13]. Examples of weighted barycenter drawings (the same graph with different weights) are in Fig. 1.

In this paper we investigate the following question: given a straight-line planar drawing Γ of a triconnected planar graph G, can we compute weights for the edges of G so that Γ is the weighted barycenter drawing of G? We answer the question with an elegant arithmetic characterisation, using the faces of Γ. This yields positive answers when the graph is a Halin graph, and leads to a polynomial time algorithm when the graph is cubic.

Our motivation in examining this question partly lies in the elegance of the mathematics, but it was also posed to us by Veronika Irvine (see [2,10]), who needed the characterisation to to create and classify "grounds" for bobbin lace

© Springer Nature Switzerland AG 2018
T. Biedl and A. Kerren (Eds.): GD 2018, LNCS 11282, pp. 569–575, 2018.
https://doi.org/10.1007/978-3-030-04414-5_40

drawings; this paper is a first step in this direction. Further, we note that our result relates to the problem of morphing from one planar graph drawing to another (see [1,9]). Previous work has characterised drawings that arise from the Schnyder algorithm (see [3]) in this context. Finally, we note that this paper is the first attempt to characterise drawings that are obtained from force-directed methods.

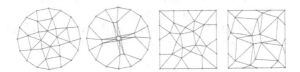

Fig. 1. Weighted barycenter drawings of the same graph embedding with different weights.

2   Preliminaries: The Weighted Barycenter Algorithm

Suppose that $G = (V, E)$ denotes a triconnected planar graph and w is a *weight function* that assigns a non-negative real weight w_{ij} to each edge $(i, j) \in E$. We assume that the weights are positive unless otherwise stated. We denote $|V|$ by n and $|E|$ by m. In this paper we discuss planar straight-line drawings of such graphs; such a drawing Γ is specified by a position γ_i for each vertex $i \in V$. We say that Γ is *convex* if every face is a convex polygon.

Throughout this paper, F_0 denotes the outer face of a plane graph G. Denote the number of vertices on F_0 by f_0. In a convex drawing, the edges of F_0 form a simple convex polygon P_0. Some terminology is convenient: we say that an edge or vertex on F_0 is *external*; a vertex that is not external is *internal*; a face F (respectively edge, e) is *internal* if F (resp. e) is incident to an internal vertex, and *strictly internal* if every vertex incident to F (resp. e) is internal.

The *weighted barycenter algorithm* takes as input a triconnected planar graph $G = (V, E)$ with a weight function w, together with F_0 and P_0, and produces a straight-line drawing Γ of G with F_0 drawn as P_0. Specifically, it assigns a position γ_i to each internal vertex i such that γ_i is the weighted barycenter of its neighbours in G. That is:

$$\gamma_i = \frac{1}{\sum_{j \in N(i)} w_{ij}} \sum_{j \in N(i)} w_{ij} \gamma_j \tag{1}$$

for each internal vertex i. Here $N(i)$ denotes the set of neighbours of i. If $\gamma_i = (x_i, y_i)$ then (1) consists of $2(n - f_0)$ linear equations in the $2(n - f_0)$ unknowns x_i, y_i. The Eq. (1) are called the *(weighted) barycenter equations* for G. Noting that the matrix involved is a submatrix of the Laplacian of G, one can show that the equations have a unique solution that can be found by traditional (see for Example [14]) or specialised (see for Example [12]) methods.

The weighted barycenter algorithm, which can be viewed as a force directed method, was defined by Tutte [15,16] and extended by Floater [8]; the classic theorem says that the output is planar and convex:

Theorem 1. *(Tutte [15,16], Floater [8]) The drawing output by the weighted barycenter algorithm is planar, and each face is convex.*

3 The Weighted Barycenter Recognition Problem

This paper discusses the problem of finding weights w_{ij} so that a given drawing is the weighted barycenter drawing with these weights. More precisely, we say that a drawing Γ is a *weighted barycenter drawing* if there is a positive weight w_{ij} for each internal edge (i,j) such that for each internal vertex i, Eq. (1) hold.

The Weighted Barycenter Recognition Problem
Input: A straight-line planar drawing Γ of a triconnected plane graph $G = (V, E)$, such that the vertices on the convex hull of $\{\gamma_i : i \in V\}$ form a face of G.
Question: Is Γ a weighted barycenter drawing?

Thus we are given the location $\gamma_i = (x_i, y_i)$ of each vertex, and we must compute a positive weight w_{ij} for each edge so that the barycenter Eq. (1) hold for each internal vertex.

Theorem 1 implies that if Γ is a weighted barycenter drawing, then each face of the drawing is convex; however, the converse is false, even for triangulations (see appendix on arXiv [6]).

4 Linear Equations for the Weighted Barycenter Recognition Problem

In this section we show that the weighted barycenter recognition problem can be expressed in terms of linear equations. The equations use *asymmetric* weights z_{ij} for each edge (i,j); that is, z_{ij} is not necessarily the same as z_{ji}. To model this asymmetry we replace each undirected edge (i,j) of G with two directed edges (i,j) and (j,i); this gives a directed graph $\overrightarrow{G} = (V, \overrightarrow{E})$. For each vertex i, let $N^+(i)$ denote the set of *out-neighbours* of i; that is, $N^+(i) = \{j \in V : (i,j) \in \overrightarrow{E}\}$.

Since each face is convex, each internal vertex is inside the convex hull of its neighbours. Thus each internal vertex position is a convex linear combination of the vertex positions of its neighbours. That is, for each internal vertex i there are non-negative weights z_{ij} such that

$$\sum_{j \in N^+(i)} z_{ij} = 1 \quad \text{and} \quad \gamma_i = \sum_{j \in N^+(i)} z_{ij}\gamma_j. \tag{2}$$

The values of z_{ij} satisfying (2) can be determined in linear time. For a specific vertex i, the z_{ij} for $j \in N^+(i)$ can be viewed as a kind of *barycentric coordinates* for i. In the case that $|N^+(i)| = 3$, these coordinates are unique.

Although Eqs. (1) and (2) seem similar, they are not the same: one is directed, the other is undirected. In general $z_{ij} \neq z_{ji}$ for directed edges (i, j) and (j, i), while the weights w_{ij} satisfy $w_{ij} = w_{ji}$. However we can choose a "scale factor" $s_i > 0$ for each vertex i, and scale Eq. (2) by s_i. That is, for each internal vertex i,

$$\gamma_i = \frac{1}{\sum_{j \in N^+(i)} s_i z_{ij}} \sum_{j \in N^+(i)} s_i z_{ij} \gamma_j. \tag{3}$$

The effect of this scaling is that we replace z_{ij} by $s_i z_{ij}$ for each edge (i, j).

We would like to choose a scale factor $s_i > 0$ for each internal vertex i such that for each strictly internal edge $(i, j) \in E$, $s_i z_{ij} = s_j z_{ji}$; that is, we want to find a real positive s_i for each internal vertex i such that

$$s_i z_{ij} - s_j z_{ji} = 0 \tag{4}$$

for each strictly internal edge (i, j).

It can be shown easily that the existence of any nontrivial solution to (4) implies the existence of a positive solution (see appendix in arXiv version).

We note that any solution of (4) for strictly internal edges gives weights w_{ij} such that the barycenter Eq. (1) hold. We choose $w_{ij} = s_i z_{ij}$ for each (directed) edge (i, j) that is incident to an internal vertex i. Equation (4) ensure that $w_{ij} = w_{ji}$ for each strictly internal edge. For edges which are internal but not strictly internal, we can simply choose $w_{ij} = s_i z_{ij}$ for any value of s_i, since z_{ji} is undefined.

Thus if Eq. (4) have a nontrivial solution, then the drawing is a weighted barycenter drawing.

The Main Theorem. We characterise the solutions of Eq. (4) with an arithmetic condition on the faces of Γ. This considers the product of the weights z_{ij} around directed cycles in G: if the product around each strictly internal face in the clockwise direction is the same as the product in the counter-clockwise direction, then Eq. (4) have a nontrivial solution.

Theorem 2. *Equation (4) have a nontrivial solution if and only if for each strictly internal face $C = (v_0, v_1, \ldots, v_{k-1}, v_k = v_0)$ in G, we have*

$$\prod_{i=0}^{k-1} z_{v_i, v_{i+1}} = \prod_{i=1}^{k} z_{v_i, v_{i-1}}. \tag{5}$$

Proof. For convenience we denote $\frac{z_{ji}}{z_{ij}}$ by ζ_{ij} for each directed edge (i, j); note that $\zeta_{ij} = 1/\zeta_{ji}$. Equation (4) can be re-stated as

$$s_i - \zeta_{ij} s_j = 0 \tag{6}$$

for each strictly internal edge (i, j), and the Eq. (5) for cycle C can be re-stated as

$$\prod_{i=0}^{k-1} \zeta_{v_i, v_{i+1}} = 1. \tag{7}$$

First suppose that Eq. (6) have nontrivial solutions s_i for all internal vertices i, and $C = (v_0, v_1, \ldots, v_{k-1}, v_k = v_0)$ is a strictly internal face in G. Now applying (6) around C clockwise beginning at v_0, we can have:

$$s_{v_0} = \zeta_{v_0, v_1} s_{v_1} = \zeta_{v_0, v_1} \zeta_{v_1, v_2} s_{v_2} = \zeta_{v_0, v_1} \zeta_{v_1, v_2} \zeta_{v_2, v_3} s_{v_3} = \ldots$$

We can deduce that

$$s_{v_0} = \left(\prod_{i=0}^{j-1} \zeta_{v_i, v_{i+1}}\right) s_{v_j} = \left(\prod_{i=0}^{k-1} \zeta_{v_i, v_{i+1}}\right) s_{v_k} = \left(\prod_{i=0}^{k-1} \zeta_{v_i, v_{i+1}}\right) s_{v_0}$$

and this yields Eq. (7).

Now suppose that Eq. (7) holds for every strictly internal facial cycle of G. We first show that Eq. (7) holds for *every* strictly internal cycle. Suppose that (7) holds for two cycles C_1 and C_2 that share a single edge, (u, v), and let C_3 be the sum of C_1 and C_2 (that is, $C_3 = (C_1 \cup C_2) - \{(u, v)\}$). Now traversing C_3 in clockwise order gives the clockwise edges of C_1 (omitting (u, v)) followed by the clockwise edges of C_2 (omitting (v, u)). But from Eq. (7), the product of the edge weights ζ_{ij} in the clockwise order around C_1 is one, and the product of the edge weights $\zeta_{i'j'}$ in the clockwise order around C_2 is one. Thus the product of the edge weights ζ_{ij} in clockwise order around C_3 is $\frac{1}{\zeta_{uv}\zeta_{vu}} = 1$. That is, (7) holds for C_3. Since the facial cycles form a cycle basis, it follows that (7) holds for every cycle.

Now choose a reference vertex r, and consider a depth first search tree T rooted at r. Denote the set of directed edges on the directed path in T from i to j by E_{ij}. Let $s_r = 1$, and for each internal vertex $i \neq r$, let

$$s_i = \prod_{(u,v) \in E_{ri}} \zeta_{uv}. \tag{8}$$

Clearly Eq. (6) holds for every edge of T. Now consider a back-edge (i, j) for T (that is, a strictly internal edge of G that is not in T), and let k denote the least common ancestor of i and j in T. Then from (8) we can deduce that

$$\frac{s_i}{s_j} = \frac{\prod_{(u,v) \in E_{ri}} \zeta_{uv}}{\prod_{(u',v') \in E_{rj}} \zeta_{u'v'}} = \frac{\prod_{(u,v) \in E_{ki}} \zeta_{uv}}{\prod_{(u',v') \in E_{kj}} \zeta_{u'v'}}. \tag{9}$$

Now let C be the cycle in Γ that consists of the reverse of the directed path in T from k to j, followed by the directed path in T from k to i, followed by the edge (i, j). Since Eq. (7) holds for C, we have:

$$1 = \left(\prod_{(v',u') \in E_{jk}} \zeta_{v'u'}\right) \left(\prod_{(u,v) \in E_{ki}} \zeta_{uv}\right) \zeta_{ij} = \left(\frac{\prod_{(u,v) \in E_{ki}} \zeta_{uv}}{\prod_{(u'v') \in E_{kj}} \zeta_{u'v'}}\right) \zeta_{ij} \tag{10}$$

Combining Eqs. (9) and (10) we have $s_i = \zeta_{ij}s_j$ and so Eq. (6) holds for each back edge (i, j). We can conclude that (6) holds for all strictly internal edges. □

5   Applications

We list some implications of Theorem 2 for cubic, Halin [7] and planar graphs with degree larger than three. Proofs of the corollaries below are straightforward.

Corollary 1. *A drawing Γ of a cubic graph is a weighted barycenter drawing if and only if Eq. (4) have rank smaller than $n - f_0$.*   □

Corollary 2. *For cubic graphs, there is a linear time algorithm for the weighted barycenter recognition problem.*   □

For cubic graphs, the weights z_{ij} are unique, and thus Eq. (4) give a complete characterisation of weighted barycenter drawings. One can use Theorem 2 to test whether a solution of Eq. (4) exists, checking Eq. (5) in linear time.

Corollary 3. *Suppose that Γ is a convex drawing of a Halin graph such that the internal edges form a tree. Then Γ is a weighted barycenter drawing.*   □

Graphs with Degree Larger Than Three. For a vertex i of degree $d_i > 3$, solutions for Eq. (2) are not unique. Nevertheless, these equations are linear, and we have 3 equations in d_i variables. Thus, for each vertex i, the solution $z_{ij}, j \in N(i)$, form a linear space of dimension at most $d_i - 3$. In this general case, we have:

Corollary 4. *A drawing Γ of a graph G is a weighted barycenter drawing if and only if there are solutions z_{ij} to Eq. (2) such that the cycle Eq. (5) holds for every internal face.*   □

Although Corollary 4 is quite elegant, it does not lead to an immediately practical algorithm because the Eq. (5) are not linear.

6   Conclusion

Force-directed algorithms are very common in practice, and drawings obtained from force-directed methods are instantly recognisable to most researchers in Graph Drawing. However, this paper represents the first attempt to give algorithms to recognise the output of a particular force-directed method, namely the weighted barycenter method. It would be interesting to know if the results of other force-directed methods can be automatically recognised.

Acknowledgements. We wish to thank Veronika Irvine for motivating discussions.

References

1. Barrera-Cruz, F., Haxell, P., Lubiw, A.: Morphing schnyder drawings of planar triangulations. In: Duncan, C., Symvonis, A. (eds.) GD 2014. LNCS, vol. 8871, pp. 294–305. Springer, Heidelberg (2014). https://doi.org/10.1007/978-3-662-45803-7_25
2. Biedl, T., Irvine, V.: Drawing bobbin lace graphs, or, fundamental cycles for a subclass of periodic graphs. In: Frati, F., Ma, K.-L. (eds.) GD 2017. LNCS, vol. 10692, pp. 140–152. Springer, Cham (2018). https://doi.org/10.1007/978-3-319-73915-1_12
3. Bonichon, N., Gavoille, C., Hanusse, N., Ilcinkas, D.: Connections between theta-graphs, delaunay triangulations, and orthogonal surfaces. In: Thilikos, D.M. (ed.) WG 2010. LNCS, vol. 6410, pp. 266–278. Springer, Heidelberg (2010). https://doi.org/10.1007/978-3-642-16926-7_25
4. de Fraysseix, H., de Mendez, P.O.: Stretching of jordan arc contact systems. In: Liotta, G. (ed.) GD 2003. LNCS, vol. 2912, pp. 71–85. Springer, Heidelberg (2004). https://doi.org/10.1007/978-3-540-24595-7_7
5. de Verdière, É.C., Pocchiola, M., Vegter, G.: Tutte's barycenter method applied to isotopies. Comput. Geom. 26(1), 81–97 (2003)
6. Eades, P., Healy, P., Nikolov, N.S.: The Weighted Barycenter Drawing Recognition Problem. ArXiv e-prints, September 2018
7. Eppstein, D.: Simple recognition of halin graphs and their generalizations. J. Graph Algorithms Appl. 20(2), 323–346 (2016)
8. Floater, M.S.: Parametrization and smooth approximation of surface triangulations. Comput. Aided Geom. Des. 14(3), 231–250 (1997)
9. Floater, M.S., Gotsman, C.: How to morph tilings injectively. J. Comput. Appl. Math. 101, 117–129 (1999)
10. Irvine, V.: Tesselace (2018). https://tesselace.com/gallery/
11. Dúnlaing, C.Ó: Nodally 3-connected planar graphs and convex combination mappings. CoRR, abs/0708.0964 (2007)
12. Spielman, D.A., Teng, S.: Spectral sparsification of graphs. SIAM J. Comput. 40(4), 981–1025 (2011)
13. Thomassen, C.: Deformations of plane graphs. J. Comb. Theor. Ser. B 34, 244–257 (1983)
14. Trefethen, L.N., Bau III, D.: Numerical Linear Algebra. SIAM (1997)
15. Tutte, W.T.: Convex representations of graphs. Proc. Lond. Math. Soc. 10, 304–320 (1963)
16. Tutte, W.T.: How to draw a graph. Proc. Lond. Math. Soc. 13, 743–767 (1963)

Miscellaneous

Algorithms and Bounds for Drawing Directed Graphs

Giacomo Ortali[1(✉)] and Ioannis G. Tollis[2,3]

[1] University of Perugia, Perugia, Italy
giacomo.ortali@gmail.com
[2] Computer Science Department, University of Crete, Heraklion, Crete, Greece
tollis@csd.uoc.gr
[3] Tom Sawyer Software, Inc., Berkeley, CA 94707, USA

Abstract. In this paper we present a new approach to visualize directed graphs and their hierarchies that completely departs from the classical four-phase framework of Sugiyama and computes readable hierarchical visualizations that contain the complete reachability information of a graph. Additionally, our approach has the advantage that only the necessary edges are drawn in the drawing, thus reducing the visual complexity of the resulting drawing. Furthermore, most problems involved in our framework require only polynomial time. Our framework offers a suite of solutions depending upon the requirements, and it consists of only two steps: (a) the cycle removal step (if the graph contains cycles) and (b) the channel decomposition and hierarchical drawing step. Our framework does not introduce any dummy vertices and it keeps the vertices of a channel *vertically aligned*. The time complexity of the main drawing algorithms of our framework is $O(kn)$, where k is the number of channels, typically much smaller than n (the number of vertices).

1 Introduction

The visualization of directed (often acyclic) graphs is very important for many applications in several areas of research and business. This is the case because such graphs often represent hierarchical relationships between objects in a structure (the graph). In their seminal paper of 1981, Sugiyama, Tagawa, and Toda [21] proposed a four-phase framework for producing hierarchical drawings of directed graphs. This framework is known in the literature as the "Sugiyama" framework, or algorithm. Most problems involved in the optimization of various phases of the Sugiyama framework are NP-hard. In this paper we present a new approach to visualize directed graphs and their hierarchies that completely departs from the classical four-phase framework of Sugiyama and computes readable hierarchical visualizations that contain the complete reachability information of a graph. Additionally, our approach has the advantage that only the

G. Ortali—This author's research was performed in part while he was visiting the University of Crete.

T. Biedl and A. Kerren (Eds.): GD 2018, LNCS 11282, pp. 579–592, 2018.
https://doi.org/10.1007/978-3-030-04414-5_41

necessary edges are drawn in the drawing, thus reducing the visual complexity of the resulting drawing. Furthermore, most problems involved in our framework require polynomial time.

Let $G = (V, E)$ be a directed graph with n vertices and m edges. The Sugiyama Framework for producing hierarchical drawings of directed graphs consists of four main phases [21]: (a) Cycle Removal, (b) Layer Assignment, (c) Crossing Reduction, and (d) Horizontal Coordinate Assignment. The reader can find the details of each phase and several proposed algorithms to solve various of their problems and subproblems in Chap. 9 of the Graph Drawing book of [2]. Other books have also devoted significant portions of their Hierarchical Drawing Algorithms chapters to the description of this framework [13, 14].

The Sugiyama framework has also been extensively used in practice, as manifested by the fact that various systems have chosen it to implement hierarchical drawing techniques. Several systems such as *AGD* [17], *da Vinci* [5], *GraphViz* [7], *Graphlet* [8], *dot* [6], and others implement this framework in order to hierarchically draw directed graphs. Even commercial software such as the Tom Sawyer Software TS Perspectives [20] and yWorks [23] essentially use this framework in order to offer automatic hierarchical visualizations of directed graphs. More recent information regarding the Sugiyama framework and newer details about various algorithms that solve its problems and subproblems can be found in [14].

Even tough this framework is very popular, it has several limitations: as discussed above, most problems and subproblems that are used to optimize the results of each phase have turned out to be NP-hard. Several of the heuristics employed to solve these problems give results that are not bounded by any approximation. Additionally, the required manipulations of the graph often increase substantially the complexity of the graph itself (such as the number of dummy vertices in phase b can be as high as $O(nm)$). The overall time complexity of this framework (depending upon implementation) can be as high as $O((nm)^2)$, or even higher if one chooses algorithms that require exponential time. Finally, the main limitation of this framework is the fact that the heuristic solutions and decisions that are made during previous phases (e.g., crossing reduction) will influence severely the results obtained in later phases. Nevertheless, previous decisions cannot be changed in order to obtain better results.

In this paper we propose a new framework that departs completely from the typical Sugiyama framework and its four phases. Our framework is based on the idea of partitioning the vertices of a graph G into *channels*, that we call *channel decomposition* of G. Namely, after we partition the vertices of G into channels, we compute a new graph Q which is closely related to G and has the same reachability properties as G. The new graph consists of the vertices of G, *channels edges* that connect vertices that are in the same channel, and *cross edges* that connect vertices that belong to different channels. Our framework draws either (a) graph G without the transitive "channel edges" or (b) a condensed form of the transitive closure of G. Our idea is to compute a hierarchical drawing of Q and, since Q has the same reachability properties as G, this drawing contains most edges of G and gives us all the reachability information of G. The "missing" incident edges of a vertex can be drawn interactively on demand by placing the mouse on top of the vertex and its incident edges will appear at once in red color.

Our framework offers a suite of solutions depending upon the requirements of the user, and it consists of only two steps: (a) the cycle removal step (if the graph contains cycles) and (b) the channel decomposition and hierarchical drawing step. Our framework does not introduce any dummy vertices, keeps the vertices of a channel *vertically aligned* and it offers answers to reachability queries between vertices by traversing at most one cross edge. Let k be the number of channels and m' be the number of cross edges in Q. We show that $m' = O(nk)$. The number of bends we introduce is at most $O(m')$ and the required area is at most $O(nk)$. The number of crossings between cross edges and channels can be minimized in $O(k!k^2)$ time, which is reasonable for small k. If k is large, we present linear-time heuristics that find a small number of such crossings. The total time complexity of the algorithms of our framework is $O(kn)$ plus the time required to compute the channel decomposition of G, which depends upon the type of channel decomposition required.

Our paper is organized as follows: the next section presents necessary preliminaries including a brief description of the phases of the Sugiyama framework, the time complexity of the phases, and a description of "bad" choices. In Sect. 3 we present the concept of path decomposition of a DAG and of path graph (i.e., when the channels are required to be paths of G) and we present the new algorithm for hierarchical drawing which is based on any (computed) path decomposition of a DAG. Section 4 presents the concepts of channel decomposition of a DAG and of channel graph (where channels are not paths) and the new algorithm for hierarchical drawing which is based on any (computed) channel decomposition of a DAG. In Sect. 5 we present the properties of the drawings obtained by our framework, we offer comparisons with the drawings obtained by traditional techniques, and present our conclusions. Due to space limitations, we present the techniques on minimizing the number of crossings between cross edges and channels in [16].

2 Sugiyama Framework

Let $G = (V, E)$ be a directed graph with n vertices and m edges. A *Hierarchical drawing* of G requires that all edges are drawn in the same direction upward (downward, rightward, or leftward) monotonically. If G contains cycles this is clearly not possible, since in a drawing of the graph some edges have to be oriented backwards. The Sugiyama framework contains the Cycle Removal Phase in which a (small) subset of edges is selected and the direction of these edges is reversed. Since it is important to maintain the character of the input graph, the number of the selected edges has to be minimum. This is a well known NP-hard problem, called the *Feedback Arc Set* problem. A well known approximation algorithm, called *Greedy-Cycle-Removal*, runs in linear time and produces sets that contain at most $m/2 - n/6$ edges. If the graph is sparse, the result is further reduced to $m/3$ edges [2].

Since the input graph G may contain cycles our framework also needs to remove or absorb them. One approach is to use a cycle removal algorithm (similar to Sugiyama's first step) but instead of reversing the edges, we propose to remove them, since reversing them could lead to an altered transitivity of the original graph. This is done because the reversed edge will be a transitive edge in the new graph and hence it may affect the drawing. By the way, this is another disadvantage of Sugiyama's framework. Since the removal and/or reversal of such edges will create a graph that will have a "different character" than the original graph we propose another possibility that will work well if the input graphs do not contain long cycles. It is easy to (a) find the *Strongly Connected Components (SCC)* of the graph in linear time, (b) cluster and collapse each SCC into a supernode, and then the resulting graph G' will be acyclic. Even if both techniques are acceptable, we believe that the second one might be able to better preserve the character of the input graph. On the other hand, this technique would not be useful if most vertices of a graph are included in a very long cycle. From now on, we assume that the given graph is acyclic after using either of the techniques described above.

In the Layer Assignment Phase of the Sugiyama framework the vertices are assigned to a layer and the layering is made *proper*, see [2,14,21]. In other words, long edges that span several layers are broken down into many smaller edges by introducing dummy vertices, so that every edge that starts at a layer terminates at the very next layer. Clearly, in a graph that has a longest path of length $O(n)$ and $O(m)$ transitive edges, the number of dummy vertices can be as high as $O(nm)$. This fact will impact the running time (and space) of the subsequent phases, with heaviest impact on the next phase, Crossing Reduction Phase.

The Crossing Reduction Phase is perhaps the most difficult and most time-consuming phase. It deals with various difficult problems that have attracted a lot of attention both by mathematicians and computer scientists. It is outside the scope of this paper to describe the various techniques for crossing reduction, however, the reader may see [2,14] for further details. The most popular technique for crossing reduction is the *Layer-by-Layer Sweep* [2,14]. This technique solves multiple problems of the well known *Two-Layer-Crossing Problem* by considering the layers in pairs going up (or down). Of course, a solution for a specific two layer crossing problem "fixes" the relative order of the vertices (real and dummy) for the next two layer crossing problem, and so on. Therefore, "bad" choices may propagate. Please notice that each two layer crossing problem is NP-complete [4]. The heuristics employed here tend to reduce crossings by various techniques, but notice that the number of crossings may be as high as $O(M'^2)$, where M' is the number of edges between the vertices of two adjacent layers.

Finally, in the last phase the exact x-coordinates of the vertices are computed by quadratic-programming techniques [2,14], which require considerable computational resources. The dummy vertices are replaced by bends. This implies that the number of bends is about equal to the number of dummy vertices (except when the edge segments are completely aligned).

3 Path Constrained Hierarchical Drawing

Let $G = (V, E)$ be a DAG. In this paper we define a *path decomposition* of G as a set of vertex-disjoint paths $S_p = \{P_1, ..., P_k\}$ such that $V(P_1), ..., V(P_k)$ is a partition of $V(G)$. A path $P_h \in S_p$ is called a *decomposition path*. The vertices in a decomposition path are clearly ordered in the path, and we denote by v_i^j the fact that v is the jth vertex of path P_i. The *path decomposition graph*, or simply path graph, of G associated with path decomposition S_p is a graph $H = (V, A)$ such that $e = (u, v) \in A$ if and only if $e \in E$ and (a) u, v are consecutive in a path of S_p (called *path edges*) or (b) u and v belong to different paths (called *cross edges*). In other words, an edge of H is a *path edge* if it connects two consecutive vertices of the same decomposition path, else it is a *cross edge*. Notice that the edges belonging to G but not to H are transitive edges between vertices of the same path of G.

A *path constrained hierarchical drawing* (PCH drawing) Γ of G given S_p is a hierarchical drawing of H such that two vertices are drawn on the same vertical line (i.e., same x-coordinate) if and only if they belong to the same decomposition path. In this section we propose an algorithm that computes PCH drawings assigning to each vertex the x-coordinate of the path it belongs to and for y-coordinate we will use its rank in a topological sorting. We will prove that this assignment lets us obtain good results in terms of both area and number of bends.

Next we present Algorithm PCH-Draw that computes a PCH drawing Γ of G such that every edge of G bends at most once. We denote by $X(P_h)$ the x-coordinate of Path P_h and by $X(v), Y(v)$ the x-coordinate and the y-coordinate of any vertex v. Let P_v be the path of S_p containing v. By definition of PCH drawing we have that $X(v) = X(P_v)$. Suppose that the vertices of G are topologically ordered and let $T(v)$ be the position of v in a topological order of V. PCH-Draw associates to every path, and consequently to every vertex of the path, an x-coordinate that is an even number and to every vertex a y-coordinate that corresponds to its topological order, i.e., $Y(v) = T(v)$ (Steps 1–4). The algorithm draws every edge $e = (u, v)$ as a straight line if the drawn edge doesn't intersect a vertex w different from u and v in Γ (Steps 5–7). Otherwise it draws edge e with one bend b_e such that: its x-coordinate $X(b_e)$ is equal to $X(u) + 1$ if $X(u) < X(v)$, or $X(u) - 1$ if $X(u) > X(v)$. The y-coordinate of bend b_e $Y(b_e)$ is equal to $Y(v) - 1$ (Steps 8–14).

Algorithm PCH-Draw($G = (V, E)$, $S_p = \{P_1, P_2, ..., P_k\}$, $H = (V, A)$)
1.   **For** $i = 1$ to k do
2. $X(P_i) = 2i$
3.   **For** any $v \in V$
4. $(X(v), Y(v)) = (X(P_v), T(v))$
5.   **For** any $e = (u, v) \in A$
6.   **If** the straight line drawing of e does not intersect a vertex different from u, v:
7.   Draw e as a straight line

8.   **Else:**
9.   **If** $X(u) < X(v)$:
10. $X(b_e) = X(u) + 1$
11.   **Else:**
12. $X(b_e) = X(u) - 1$
13. $Y(b_e) = Y(v) - 1$
14.   Draw e with one bend at point $(X(b_e), Y(b_e))$

In Fig. 1 we show an example of a drawing computed by Algorithm PCH-Draw. In (a) we show the drawing of a graph G as computed by Tom Sawyer Perspectives (a tool of Tom Sawyer Software) which follows the Sugiyama Framework. In (b) we show the drawing Γ of H computed by Algorithm PCH-Draw. The path decomposition that we used to compute the drawing is $S_p = \{P_1, P_2, P_3\}$, where: $P_1 = \{0, 1, 4, 7, 12, 13, 15, 16, 17, 20, 22, 24, 25, 26, 29, 30\}$; $P_2 = \{2, 5, 9, 11, 23, 27\}$; $P_3 = \{3, 6, 8, 10, 14, 18, 19, 21, 28\}$. Edge $e = (21, 25)$ is the only one bending. In grey we show edge e drawn as straight line, intersecting vertex 23.

Any drawing Γ computed by Algorithm PCH-Draw has several interesting properties. First, the area of Γ is typically less than $O(n^2)$. By construction, Γ has height $n - 1$ and width of $2k - 1$. Hence $Area(\Gamma) = O(kn)$. Given S_p and the topological order of the vertices of G, every vertex need $O(1)$ time to be placed. Every edge $e = (u, v)$ needs $O(k)$ time to be placed, since before drawing it we need to check if its straight line drawing would intersect a vertex different from u, v (Step 6). Since the drawing of e must be monotonous, it can intersect at most one edge per path, so we just need to check if in correspondence of every path placed between the path of u and v in Γ the drawing of e intersects some vertex. Hence we have:

Theorem 1. *Algorithm PCH-Draw computes a drawing Γ of a DAG G in $O(n + mk)$ time. Furthermore, $Area(\Gamma) = O(kn)$.*

The proofs of Lemma 1 and Lemma 2 are in [16]:

Lemma 1. *A cross edge $e = (u, v)$ does not intersect a vertex different from u and v in Γ.*

Lemma 2. *Let $e = (u, v)$ and $e' = (u', v')$ be two cross edges drawn with a bend in Γ. Their bends are placed in the same point if and only if u and u' are in the same decomposition path and $v = v'$.*

In the case described by the above lemma, two edges have overlapping segments (b_e, v) and $(b_{e'}, v)$. We consider this feature acceptable, or even desirable for two edges that have the same endpoint. This typical merging of edges has been used in the past, see for example [1, 11, 18]. However, in case that this feature is not desirable, we propose two alternative solutions that avoid this overlap. The price to pay is larger area, or less edges drawn:

1. Larger area option: We can shift horizontally by one unit the position of bend $b_{e'}$ and all the vertices v and bends b such that $X(v) > X(b_e)$ and

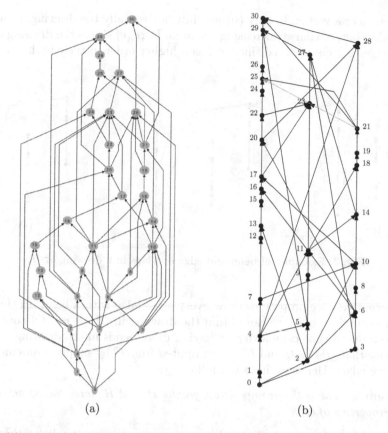

(a)   (b)

Fig. 1. (a)Drawing of a dag G computed by Tom Sawyer Perspectives (a tool of Tom Sawyer Software) (b) PCH drawing of H computed by Algorithm PCH-draw.

$X(b) > X(b_e)$. In this case we have no overlaps, but the area of Γ can be as large as $O(knm)$.

2. Less edges option: We can define the path decomposition graph differently by removing some transitive cross edges from H. For every vertex v we remove the edge (u, v) if there exists an edge (u', v) such that u' and u are in the same decomposition path P and u precedes u' in the order of P. It is easy to prove that H' is a subgraph of H and that $A - A'$ contains only transitive edges of G. By definition of H', given a decomposition path P, for any vertex v there exists at most one cross edge $e = (u, v)$ such that $u \in P$. According to Lemma 2, there are no bends overlapping. The area of a drawing Γ computed using H' is $Area(\Gamma') = O(kn)$. However, we pay for the absence of overlapping bends by the exclusion from the drawing of some transitive cross edges of G.

In Fig. 2 we show an example of the edge overlap described above in a drawing of H. Part (a) shows a simple drawing where two edges, $e_1 = (u, v)$ and $e_2 = (u', v)$, overlap. In grey the drawings of e_1, e_2 as straight lines, please notice that both of them intersect a vertex. In part (b) we shift horizontally the drawing, removing the overlap but, of course, increasing the area. Part (c) shows the drawing of H', where edge (u, v) is removed since u' has a higher order in their path.

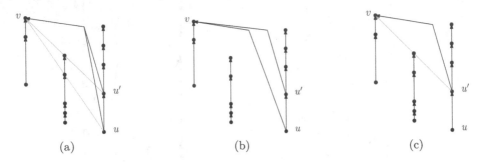

(a)   (b)   (c)

Fig. 2. Examples of bend and edge overlaps in a drawing of H.

Alternatively we propose to draw every cross edge with a bend. In this case we can avoid Step 6, so we can obtain the drawing in $O(n + m)$. Of course, we pay the reduced time complexity by having more bends in the drawing.

Notice that graphs H and H' are computed from G by simply removing some transitive edges. Hence we have the following:

Theorem 2. *The path decomposition graphs H and H' have the same reachability properties of G.*

Theorem 2 is rather simple, but it is very important, since it tells us that visualizing a hierarchical drawing of H or H' we do read and understand correctly any reachability relation between the vertices of G.

A path decomposition S_p of a DAG G with a small number of paths lets us compute a readable PCH drawing of G, since the number of decomposition paths influences the area of the drawing and its number of bends. Indeed, since a cross edge can intersect at most one vertex of every decomposition path, the number of decomposition paths influences the number of bends of the drawing. Furthermore, the number of paths k clearly influences the time to find the minimum number of crossings between cross edges and paths, as it is described in [16]. Several algorithms solve the problem of finding a path decomposition of minimum size [9,12,15,19]. The algorithm of [12] is the fastest one for sparse and medium DAGs. In the next section we introduce a relaxed definition of path, the channel, and a way to obtain hierarchical drawings based on a channel decomposition. Notice that, since paths are constrained versions of channels, we expect the minimum size of a channel decomposition to be lower than or in the worst case equal to the minimum size of a path decomposition. Therefore, we now turn our attention to the concept of a channel decomposition.

4   Channel Constrained Hierarchical Drawing

Let $G = (V, E)$ be a DAG. A *channel* C is an ordered set of vertices such that any vertex $u \in C$ has a path to each of its successors in C. In other words, given any two vertices $v, w \in C$, v precedes w in the order of channel C if and only if w is reachable from v in G. A channel can be seen as a generalization of a path, since a path is always a channel, but a channel may not be a path. A *channel decomposition* $S_c = \{C_1, ..., C_k\}$ is a partition of the vertex set V of the graph into channels. If vertex v belongs to channel C_i we write v_i^j if v is the jth vertex of channel C_i. The channel decomposition graph H'' and a *channel constrained hierarchical drawing* (CCH drawing) of G are defined in a similar fashion as we defined the path decomposition graph H and the PCH drawing of G in the previous section. Notice that, since the channel is a generalization of a path, the concepts of channel decomposition graph and CCH drawing are a generalization of the concepts of path decomposition graph and PCH drawing. We can define Algorithm CCH-Draw in a similar fashion as Algorithm PCH-draw, and its pseudocode is similar to the pseudocode of Algorithm PCH-draw. The only difference is that Algorithm CCH-Draw takes as input a channel decomposition instead of a path decomposition and that its output is a CCH drawing instead of a PCH drawing. Algorithm CCH-Draw is clearly a generalization of Algorithm PCH-Draw. Because of space limitations we do not discuss the complete details of Algorithm CCH-Draw here.

Now we introduce a "special" transitive closure, called compressed transitive closure, which is based on the concept of channel decomposition. This transitive closure is obtained from an ordinary transitive closure by removing some of its transitive edges. Next, we will define a graph Q, based on the compressed transitive closure, that will let us obtain more readable drawings.

Compressed Transitive Closure (CTC): Let L_v be a list of vertices associated with a vertex $v \in V$ such that: L_v contains at most one vertex of any decomposition channel; a vertex w is reached from v in G if and only if list L_v contains a vertex w' such that: w and w' are in the same decomposition channel and w' precedes w in the order of that decomposition channel.

The *compressed transitive closure* (CTC) of G is the set of all the lists L_v. In [10] it is shown how to compute the CTC of a graph in $O(mk)$ time. Next we show how we can store the CTC in $O(nk)$ space and that it contains the complete reachability information of G.

We define the *compressed transitive closure graph (CTC graph)* $Q = (V, I)$ such that $(u, v) \in I$ if and only if u is the highest vertex in the order of its channel such that $v \in L_u$. Notice that an edge of Q may not exist in the original graph G, as is the case in the ordinary transitive closure graph G_c of G. Furthermore, an edge of G may not be included in Q, while G_c contains all the edges of G. Please notice that Q has the same reachability properties (i.e., the same transitive closure) as G, since it is computed directly from the CTC of G. We denote by *channel edge* an edge of Q connecting two vertices of the same channel, else it is a *cross edge*, similar to the definition of the previous section.

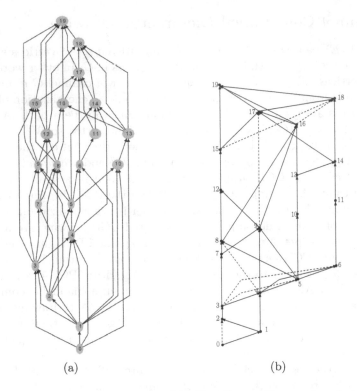

(a)   (b)

Fig. 3. (a)Drawing of a dag G computed by Tom Sawyer Perspectives (a tool of Tom Sawyer Software) (b) CCH drawing of Q computed by Algorithm CCH-draw.

Let u_i^j be a vertex. The list L_u contains by definition the vertex v_i^{j+1}, since it is the lowest vertex in the channel C_i reachable from u. Hence we have the following property:

Lemma 3. $(u, v) \in I$ *for any* u_i^j, v_i^{j+1}.

Lemma 3 implies that the channel decomposition S_c of G is a path decomposition of Q, so a CCH drawing of Q is essentially also a PCH drawing and hence we can compute it using Algorithm CCH-Draw or Algorithm PCH-Draw since in this case the two algorithms produce the same result.

In Fig. 3 an example of a CCH drawing of G computed by Algorithm CCH-Draw using Q as an input is shown: Part (a) shows the original graph G drawn as computed by Tom Sawyer Perspectives that uses the Sugiyama Framework. A channel decomposition of this graph is $S_c = \{C_1, C_2, C_3, C_4\}$, where: $C_1 = \{0, 2, 3, 7, 8, 12, 15, 16, 19\}$; $C_2 = \{1, 4, 9, 17\}$; $C_3 = \{5, 10, 13, 18\}$; $C_4 = \{6, 11, 14\}$. In part (b) we show the drawing of Q as computed by Algorithm CCH-Draw. The dotted edges are edges that do not exist in G. Some channel edges are dotted, since a channel may not be a path of G.

There is one list L_v for every vertex v and every list contains $O(k)$ elements. Since every element of a list L_v corresponds to (at most) one edge of Q we have that Q contains $O(nk)$ edges. Hence we have the following:

Theorem 3. *The number of edges of Q is $O(nk)$.*

The above theorem implies that the number of edges of Q is linear if k is a constant. Also, it requires only $O(nk)$ space to be stored. As we did in the previous section, we denote by *cross edge* an edge of Q connecting two vertices belonging to two different channels. We denote by *mono channel path* (mc-path) a path of Q such that all the edges of it are in the same channel, while we denote by *double channel path* (dc-path) a path of Q composed by two mc-paths and a cross edge.

Theorem 4. *Let v and u be any pair of vertices such that v is reachable from u. Then there exists either an mc-path or a dc-path from u to v in Q.*

Proof. Suppose that u and v are in the same channel C_i. In this case there exists an mc-path from u to v as a consequence of Lemma 3. Suppose that u and v are in two different channels C_i and C_j. If u reaches v, by definition of Q, there must be a vertex $u' \in C_i$ that is a successor of u in C_i and a vertex $v' \in C_j$ which is predecessor of v in C_j, such that $(u', v') \in I$. Let p_1 be the mc-path from u to u' and p_2 be the mc-path from v' to v. The path $p = p_1 + (u', v') + p_2$ is a dc-path from u to v.

We claim that such a CCH drawing of Q is a very useful instrument to visualize the reachability properties of G. Indeed, if we want to check if a vertex reaches another vertex in Q (and consequently in G) we just need to check if there exists an mc-path or a dc-path connecting them (Lemma 4). Morcover, finding an mc-path or a dc-path in Γ is very easy, since every mc-path is drawn on a vertical line and every dc-path is drawn as two different vertical lines connected by a cross edge. Moreover since Q has an almost-linear number of edges ($O(nk)$) by Theorem 3 it makes Q easier to visualize and so it gives us a clear way to visualize the reachability properties of G. The price we have to pay is that we do not visualize many edges of the original graph G. These edges can be visualized (in red) on demand by moving the mouse over a given query vertex.

A channel decomposition with a small number of channels lets us compute a readable CCH drawing of G. The width b of a DAG G is the maximum cardinality of a subset of V of pairwise incomparable vertices of G, i.e., there is no path between any two vertices in the subset. In [3] it is proved that the minimum value of the cardinality of S_c, is b and in [10] an algorithm is given to compute S_c with $k = b$ in $O(n^3)$ time. The time complexity is improved to $O(bn^2)$ in [22]. Clearly, since paths are a restricted type of channels, the minimum size of S_c is less than or equal to the minimum size of S_p.

5   Comparisons and Conclusions

We discussed the results of our algorithms in terms of bends and area. The framework we present in this paper produces results that are far superior to the results produced by the Sugiyama framework with respect to the number of crossings, number of bends, area of the drawing and visual clarity of the existing paths and reachability. Namely, because the hierarchical drawings produced by the Sugiyama framework have (a) many crossings (a bound is not possible to be computed), (b) the total number of bends is large and it depends heavily on the number of dummy vertices introduced, (c) the area is large because the width of the drawing is negatively influenced by the number of dummy vertices, (d) the number of bends per edge is also influenced by the number of dummy vertices on it (although the last phase tries to straighten the edges by aligning its segments, at the expense of the area, of course), (e) most problems and subproblems of each phase are NP-hard, and many of the heuristic are very time consuming, and (f) the reachability information in the graph is not easy to detect from the drawing.

Our framework produces hierarchical drawings that are far superior of the ones produced by the Sugiyama framework in all measures discussed above. Namely, our drawings have (a) a minimum number of channel crossings as an upper bound (see [16]), (b) the total number of bends is low since we introduce at most one bend for some (not all) cross edges, (c) the area is precisely bounded by a rectangle of height $n - 1$ and width $O(k)$, where k is typically a small fraction of n, (d) the reachability and path information is easily visible in our drawings since any path is deduced by following at most one cross edge (which might have at most one bend), (e) the vertices in each channel are vertically aligned and there is a path from each vertex in the channel to all the vertices that are at higher y-coordinates, (f) all our algorithms run in polynomial time (with the exception of the minimization of the number of channel crossings, which requires $O(k!k^2)$ time), and finally, (g) the flexibility of our framework allows a user to decide to have their specified paths as channels, thus allowing for user paths to be drawn aligned.

The only drawback of the drawings produced by our framework is the fact that it does not draw all the edges of the graph, which might be important for some applications. This might be considered as an advantage by some other users since it offers drawings that are not cluttered by the edges. In any case, we offer the remedy to visualize all the edges incident to a vertex interactively when the mouse is placed on top of a vertex.

We believe that the above comparison is convincing of the power of the new framework. Hence we do not offer experimental results here. However, in the future we plan to contact user studies in order to verify that the users we will benefit from the aforementioned properties by showing higher understanding and ease of use of the new drawing framework. We plan to work on allowing to include user specified channels (or paths), and still find the minimum number of channels in a channel decomposition. It would be interesting to find specific topological orderings and/or sophisticated layer assignment that will reduce the height, the number of crossings and the number of bends of the computed drawing. Finally, it would be desirable to avoid the exponential in k (i.e., $k!$) factor in the time complexity of finding the best order of the channels.

References

1. Bannister, M.J., Brown, D.A., Eppstein, D.: Confluent orthogonal drawings of syntax diagrams. In: Di Giacomo, E., Lubiw, A. (eds.) GD 2015. LNCS, vol. 9411, pp. 260–271. Springer, Cham (2015). https://doi.org/10.1007/978-3-319-27261-0_22
2. Di Battista, G., Eades, P., Tamassia, R., Tollis, I.G.: Graph Drawing: Algorithms for the Visualization of Graphs. Prentice-Hall, Upper Saddle River (1999)
3. Dilworth, R.: A decomposition theorem for partially ordered sets. Ann. Math. **51**(1), 161–166 (1950). https://doi.org/10.2307/1969503
4. Eades, P., Wormald, N.C.: Edge crossings in drawings of bipartite graphs. Algorithmica **11**(4), 379–403 (1994). https://doi.org/10.1007/BF01187020
5. Fröhlich, M., Werner, M.: Demonstration of the interactive graph visualization system *da Vinci*. In: Tamassia, R., Tollis, I.G. (eds.) GD 1994. LNCS, vol. 894, pp. 266–269. Springer, Heidelberg (1995). https://doi.org/10.1007/3-540-58950-3_379
6. Gansner, E., Koutsofios, E., North, S.: Drawing graphs with dot. Technical report (2006). http://www.graphviz.org/Documentation/dotguide.pdf
7. Gansner, E.R., North, S.C.: An open graph visualization system and its applications to software engineering. Softw. Pract. Exper. **30**(11), 1203–1233 (2000). https://doi.org/10.1002/1097-024X(200009)30:11⟨1203::AIDSPE338⟩3.0.CO;2-N
8. Himsolt, M.: Graphlet: design and implementation of a graph editor. Softw., Pract. Exper. **30**(11), 1303–1324 (2000). https://doi.org/10.1002/1097-024X(200009)30:11⟨1303::AID-SPE341⟩3.0.CO;2-3
9. Hopcroft, J.E., Karp, R.M.: An $n^{5/2}$ algorithm for maximum matchings in bipartite graphs. SIAM J. Comput. **2**(4), 225–231 (1973). https://doi.org/10.1137/0202019
10. Jagadish, H.V.: A compression technique to materialize transitive closure. ACM Trans. Database Syst. **15**(4), 558–598 (1990). https://doi.org/10.1145/99935.99944
11. Kornaropoulos, E.M., Tollis, I.G.: Algorithms and bounds for overloaded orthogonal drawings. J. Graph Algorithms Appl. **20**(2), 217–246 (2016). https://doi.org/10.7155/jgaa.00391
12. Kuosmanen, A., Paavilainen, T., Gagie, T., Chikhi, R., Tomescu, A., Mäkinen, V.: Using minimum path cover to boost dynamic programming on DAGs: co-linear chaining extended. In: Raphael, B.J. (ed.) RECOMB 2018. LNCS, vol. 10812, pp. 105–121. Springer, Cham (2018). https://doi.org/10.1007/978-3-319-89929-9_7
13. Michael Kaufmann, D.W.: Drawing Graphs: Methods and Models. Springer, Heidelberg (2001). https://doi.org/10.1007/3-540-44969-8
14. Nikolov, N.S., Healy, P.: Hierarchical drawing algorithms, in handbook of graph drawing and visualization. In: Tamassia, R. (ed.), pp. 409–453. CRC Press (2014)
15. Orlin, J.B.: Max flows in O(nm) time, or better. In: Symposium on Theory of Computing Conference, STOC 2013, Palo Alto, CA, USA, 1–4 June 2013, pp. 765–774 (2013). https://doi.org/10.1145/2488608.2488705
16. Ortali, G., Tollis, I.G.: Algorithms and bounds for drawing directed graphs. CoRR **abs/1808.10364** (2018). http://arxiv.org/abs/1808.10364
17. Paulisch, F.N., Tichy, W.F.: EDGE: an extendible graph editor. Softw. Pract. Exper. **20**(S1), S1 (1990)
18. Pupyrev, S., Nachmanson, L., Kaufmann, M.: Improving layered graph layouts with edge bundling. In: Brandes, U., Cornelsen, S. (eds.) GD 2010. LNCS, vol. 6502, pp. 329–340. Springer, Heidelberg (2011). https://doi.org/10.1007/978-3-642-18469-7_30
19. Schnorr, C.: An algorithm for transitive closure with linear expected time. SIAM J. Comput. **7**(2), 127–133 (1978). https://doi.org/10.1137/0207011

20. Software, T.S.: www.tomsawyer.com
21. Sugiyama, K., Tagawa, S., Toda, M.: Methods for visual understanding of hierarchical system structures. IEEE Trans. Syst. Man Cybern. **11**(2), 109–125 (1981). https://doi.org/10.1109/TSMC.1981.4308636
22. Yangjun Chen, Y.C.: On the DAG decomposition. Br. J. Math. Comput. Sci. **10**(6), 1–27 (2015). Article no. BJMCS.19380, ISSN: 2231–0851
23. yWorks. www.yworks.com

Optimal Grid Drawings of Complete Multipartite Graphs and an Integer Variant of the Algebraic Connectivity

Ruy Fabila-Monroy[1], Carlos Hidalgo-Toscano[1], Clemens Huemer[2(✉)],
Dolores Lara[1], and Dieter Mitsche[3]

[1] CINVESTAV-IPN, Mexico City, Mexico
ruyfabila@math.cinvestav.edu.mx, cmhidalgo@math.cinvestav.mx,
dlara@cs.cinvestav.mx
[2] Universitat Politècnica de Catalunya, Barcelona, Spain
clemens.huemer@upc.edu
[3] Laboratoire Dieudonné, Univ. Nice, Nice, France
dmitsche@unice.fr

Abstract. How to draw the vertices of a complete multipartite graph G on different points of a bounded d-dimensional integer grid, such that the sum of squared distances between vertices of G is (i) minimized or (ii) maximized? For both problems we provide a characterization of the solutions. For the particular case $d = 1$, our solution for (i) also settles the minimum-2-sum problem for complete bipartite graphs; the minimum-2-sum problem was defined by Juvan and Mohar in 1992. Weighted centroidal Voronoi tessellations are the solution for (ii). Such drawings are related with Laplacian eigenvalues of graphs. This motivates us to study which properties of the algebraic connectivity of graphs carry over to the restricted setting of drawings of graphs with integer coordinates.

1 Introduction

Let r, d be positive integers. Let $n_1 \leq \cdots \leq n_r$ be positive integers such that $\sum n_i = (2M+1)^d$ for some integer M. We consider straight line drawings of the complete r-partite graph $K_{n_1,\ldots,n_r}$ into the d-dimensional integer grid

$$P := \left\{ (x_1, \ldots, x_d) \in \mathbb{Z}^d : -M \leq x_i \leq M \right\}.$$

No two vertices of the graph are drawn on the same grid point. Note that such a drawing corresponds to a coloring of the points of P with r colors, such that color i appears n_i times, for $i = 1, \ldots, r$. The goal is to find the assignment of colors to the points of P such that the sum of squared distances between points of different colors is (i) minimized or (ii) maximized. The motivation for this problem stems from the following relation between drawings of a graph and spectral theory:

© Springer Nature Switzerland AG 2018
T. Biedl and A. Kerren (Eds.): GD 2018, LNCS 11282, pp. 593–605, 2018.
https://doi.org/10.1007/978-3-030-04414-5_42

Let $G = (V, E)$ be a graph with vertex set $V = \{1, \ldots, N\}$, and let $deg(i)$ denote the degree of vertex i. The *Laplacian* matrix of G is the $N \times N$ matrix, $L = L(G)$, whose entries are

$$L_{i,j} = \begin{cases} deg(i), & \text{if } i = j, \\ -1, & \text{if } i \neq j \text{ and } ij \in E, \\ 0, & \text{if } i \neq j \text{ and } ij \notin E. \end{cases}$$

Let $\lambda_1(G) \leq \lambda_2(G) \leq \cdots \leq \lambda_N(G)$ be the eigenvalues of L. The *algebraic connectivity* (also known as the Fiedler value [9]) of G is the value of $\lambda_2(G)$. It is related to many graph invariants (see [9]), and in particular to the size of the separator of a graph, giving rise to partitioning techniques using the associated eigenvector (see [14]). Spielman and Teng [14] proved the following lemma:

Lemma 1 (Embedding Lemma).

$$\lambda_2(G) = \min \frac{\sum_{ij \in E} \|\boldsymbol{v}_i - \boldsymbol{v}_j\|^2}{\sum_{i \in V} \|\boldsymbol{v}_i\|^2},$$

and

$$\lambda_N(G) = \max \frac{\sum_{ij \in E} \|\boldsymbol{v}_i - \boldsymbol{v}_j\|^2}{\sum_{i \in V} \|\boldsymbol{v}_i\|^2},$$

where the minimum, respectively maximum, is taken over all tuples $(\boldsymbol{v}_1, \ldots, \boldsymbol{v}_N)$ of vectors $\boldsymbol{v}_i \in \mathbb{R}^d$ with $\sum_{i=1}^{N} v_i = \boldsymbol{0}$, and not all v_i are zero-vectors $\boldsymbol{0}$.

In fact, Spielman and Teng [14] proved the Embedding Lemma for $\lambda_2(G)$, but the result for $\lambda_N(G)$ follows by very similar arguments; when adapting the proof of [14] we have to replace the last inequality given there by the inequality $\sum_i x_i / \sum_i y_i \leq \max_i \frac{x_i}{y_i}$, for $x_i, y_i > 0$.

Let $\mathbf{v} = (\boldsymbol{v}_1, \ldots, \boldsymbol{v}_N)$ be a tuple of positions defining a drawing of G (vertex i is placed at $\boldsymbol{v}_i$). Let

$$\lambda(\mathbf{v}) := \frac{\sum_{ij \in E} \|\boldsymbol{v}_i - \boldsymbol{v}_j\|^2}{\sum_{i \in V} \|\boldsymbol{v}_i\|^2}. \tag{1}$$

Note that $\|\boldsymbol{v}_i - \boldsymbol{v}_j\|^2$ is equal to squared length of the edge ij in the drawing defined by $\mathbf{v}$. Lemma 1 provides a link between the algebraic connectivity of G and its straight line drawings. Clearly

$$\lambda_2(G) \leq \lambda(\mathbf{v}) \leq \lambda_N(G).$$

We remark that in dimension $d = 1$, optimal drawings $\mathbf{v}$ are eigenvectors of $L(G)$ and $\lambda(\mathbf{v})$ is the well known Rayleigh quotient.

In this paper we study how well we can approximate $\lambda_2(G)$ and $\lambda_N(G)$ with drawings with certain restrictions. First, we restrict ourselves to drawings in which the vertices are placed at points with integer coordinates and no two vertices are placed at the same point. Since $\lambda(\alpha\mathbf{v}) = \lambda(\mathbf{v})$ for $\alpha \in \mathbb{R} \setminus \{0\}$, we have that $\lambda_2(G)$ and $\lambda_N(G)$ can be approximated arbitrarily

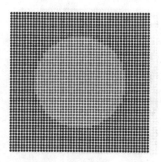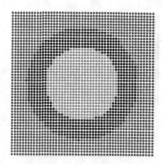

Fig. 1. The best way to minimize the sum of squared distances between points of different colors on a 51×51 integer grid. Left: for $r = 2$ colors with $1/3$ of the points in red and $2/3$ of the points in blue. Right: for $r = 3$ colors, with 1359 red points, 724 blue points, and 518 green points. (Color figure online)

closely with straight line drawings with integer coordinates of sufficiently large absolute value. We therefore bound the absolute value of such drawings and consider only drawings in the bounded d-dimensional integer grid P. Juvan and Mohar [10,11] already studied drawings of graphs with integer coordinates for $d = 1$. More precisely, the authors consider the *minimum-p-sum*-problem: for $0 < p < \infty$, a graph G and a bijective mapping Ψ from V to $\{1, \ldots, N\}$, define $\sigma_p(G, \Psi) = \left(\sum_{uv \in E(G)} |\Psi(u) - \Psi(v)|^p \right)^{1/p}$, and for $p = \infty$, let $\sigma_p(G, \Psi) = \max_{uv \in E(G)} |\Psi(u) - \Psi(v)|$. The quantity $\sigma_p(G) = \min_{\Psi} \sigma_p(G, \Psi)$ (where the minimum is taken over all bijective mappings) is then called the minimum-p-sum of G, and if $p = \infty$, it is also called the bandwidth of G. In [11] relations between the min-p-sum and $\lambda_2(G)$ and $\lambda_N(G)$ are analyzed, and also polynomial-time approximations of the minimum-p-sum based on the drawing suggested by the eigenvector corresponding to $\lambda_2(G)$ are given. In [10] the minimum-p-sums and its relations to $\lambda_2(G)$ and $\lambda_N(G)$ are studied for the cases of random graphs, random regular graphs, and Kneser graphs. For a survey on the history of these problems, see [5] and [6].

The use of eigenvectors in graph drawing has been studied for instance in [12], and we also mention [13] as a recent work on spectral bisection.

In the next two sections we characterize the optimal drawings $\mathbf{v}$ for complete multipartite graphs $K_{n_1, \ldots, n_r}$ which minimize/maximize $\lambda(\mathbf{v})$. The assumption $N = \sum_{i=1}^{r} n_i = (2M+1)^d$ made in the beginning is to ensure that every drawing satisfies the condition $\sum_{i=1}^{N} \mathbf{v}_i = \mathbf{0}$. The Laplacian eigenvalues of $K_{n_1, \ldots, n_r}$ are known to be, see [4],

$$0^1, (N - n_r)^{n_r - 1}, (N - n_{r-1})^{n_{r-1} - 1}, \ldots, (N - n_2)^{n_2 - 1}, (N - n_1)^{n_1 - 1}, N^{r-1}$$

where the superindexes denote the multiplicities of the eigenvalues. Therefore, $N - n_r \leq \lambda(\mathbf{v}) \leq N$.

Two examples of optimal drawings in dimension $d = 2$ which minimize $\lambda(\mathbf{v})$ are given in Fig. 1. Figures 2 and 3 show examples which maximize $\lambda(\mathbf{v})$.

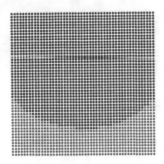 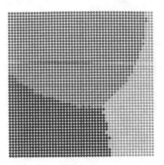

Fig. 2. The best way to maximize the sum of squared distances between points of different colors on a 51×51 integer grid. Left: for $r = 2$ colors with $3/4$ of the points in blue and $1/4$ of the points in red. Right: for $r = 3$ colors, with 1359 red points, 724 blue points, and 518 green points. (Color figure online)

 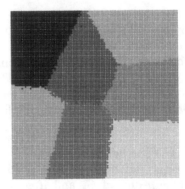

Fig. 3. The best way to maximize the sum of squared distances between points of different colors on a 101×101 integer grid. Left: for $r = 6$ colors, with 1701 purple points and 1700 points of every other color. Right: for $r = 7$ colors, with 1459 yellow points and 1457 points of every other color. (Color figure online)

We mention that we obtained all these drawings with computer simulations, using simulated annealing. The solution for minimizing $\lambda(\mathbf{v})$ shown in Fig. 1 consists of concentric rings and applies to the case when all color classes have different size. While this solution is unique, we will show that if the color classes have the same size, then there are exponentially many drawings that minimize $\lambda(\mathbf{v})$. In that case, the solutions are characterized as those drawings where for each color class all its points sum up to $\mathbf{0}$. As can be observed in Figs. 2 and 3 (and proved in Sect. 3), the solution for maximizing $\lambda(\mathbf{v})$ is given by (weighted) centroidal Voronoi diagrams, which are related to clustering [7]. Let us give the definition of a centroidal Voronoi tessellation, according to [7]. Given an open set $\Omega \subseteq \mathbb{R}^d$, the set $\{V_i\}_{i=1}^r$ is called a *tessellation* of Ω if $V_i \cap V_j = \emptyset$ for $i \neq j$ and $\cup_{i=1}^r \overline{V_i} = \overline{\Omega}$. Given a set of points $\{c_i\}_{i=1}^r$ belonging to $\overline{\Omega}$, the Voronoi region $\hat{V}_i$ corresponding to the point c_i is defined by

$$\hat{V}_i = \{x \in \Omega \mid ||x - c_i|| < ||x - c_j|| \text{ for } j = 1, \ldots, r, j \neq i\}.$$

The points $\{c_i\}_{i=1}^r$ are called *generators* or *sites*. The set $\{\hat{V}_i\}_{i=1}^r$ is a *Voronoi tessellation* or *Voronoi diagram*. A Voronoi diagram is *multiplicatively weighted*, see [3], if each generator c_i has an associated weight $w_i > 0$ and the weighted Voronoi region of c_i is

$$\hat{V}_i = \{x \in \Omega \mid ||x - c_i||w_j < ||x - c_j||w_i \text{ for } j = 1, \ldots, r, j \neq i\}.$$

A Voronoi tessellation is *centroidal* if the generators are the centroids for each Voronoi region. Voronoi diagrams have also been defined for discrete sets P instead of regions Ω [7].

Finally, in Sect. 4 we focus on graph drawings in dimension $d = 1$ and treat the question on what can be said about approximations of eigenvectors with bounded integer vectors. In particular, we study the relation between the algebraic connectivity and an integer version of the algebraic connectivity and the minimum-2-sum. We think analogous relations should also hold for drawings in higher dimension; we leave this for further research.

2 Optimal Drawings for Minimizing $\lambda(\mathbf{v})$

In the following we give bounds on $\lambda(\mathbf{v})$. Note that in Equation (1), the term $\sum_{i \in V} ||\boldsymbol{v}_i||^2$ is the same for all drawings on P. Let $S := \sum_{v \in P} ||v||^2$. We first calculate the value of S which we need later on.

Proposition 1.

$$S = 2d(2M + 1)^{d-1} \frac{M(M + 1)(2M + 1)}{6}.$$

The proof can be found in the full version [8].

Let $A_1, \ldots, A_r$ be the partition classes of $K_{n_1, \ldots, n_r}$ with $|A_i| = n_i$ (for $1 \leq i \leq r$). Let $N = (2M + 1)^d = \sum_{i=1}^r n_i$. Let $\mathbf{v}$ be a fixed straight line drawing of $K_{n_1, \ldots, n_r}$. In what follows we abuse notation and say that a point $v \in P$ is in A_i if a vertex of A_i is mapped to v. We also use A_i to refer to the image of A_i under $\mathbf{v}$.

Let A and B be two finite subsets of $\mathbb{R}^d$. We define

$$A \cdot B := \sum_{\substack{v \in A \\ w \in B}} v \cdot w,$$

where $\cdot$ is the dot product. We will need the following property:

Proposition 2. *Let $A_1, \ldots, A_r$ be $r \geq 2$ finite subsets of $\mathbb{R}^d$ such that*

$$\sum_{i=1}^r \sum_{v \in A_i} v = \mathbf{0}.$$

Then

$$\sum_{i=1}^{r-1} \sum_{j=i+1}^{r} A_i \cdot A_j = -\frac{1}{2} \sum_{i=1}^{r} \left\| \sum_{v \subset \Lambda_i} v \right\|^2 .$$

The proof can be found in the full version [8].

Lemma 2. *Let* $\mathbf{v}$ *be a fixed straight line drawing of* $G = (V, E) = K_{n_1, \ldots, n_r}$. *Then*

$$\lambda(\mathbf{v}) = N + \frac{1}{S} \sum_{i=1}^{r} \left(-n_i \sum_{v \in A_i} \|v\|^2 \right) + \frac{1}{S} \sum_{i=1}^{r} \left\| \sum_{v \in A_i} v \right\|^2 .$$

Proof.

$$\lambda(\mathbf{v}) = \frac{1}{S} \sum_{(v,w) \in E} \|v - w\|^2 = \frac{1}{S} \sum_{(v,w) \in E} \left(\|v\|^2 + \|w\|^2 - 2v \cdot w \right) .$$

Since in the complete multipartite graph each $v \in A_i$ is adjacent to all vertices but the n_i vertices of its class A_i, this further equals

$$\lambda(\mathbf{v}) = \frac{1}{S} \sum_{i=1}^{r} \left((N - n_i) \sum_{v \in A_i} \|v\|^2 \right) - \frac{2}{S} \sum_{i \neq j} A_i \cdot A_j$$

$$= N + \frac{1}{S} \sum_{i=1}^{r} \left(-n_i \sum_{v \in A_i} \|v\|^2 \right) + \frac{1}{S} \sum_{i=1}^{r} \left\| \sum_{v \in A_i} v \right\|^2 .$$

$\square$

The following theorem provides best possible drawings whenever one can draw $K_{n_1, \ldots, n_r}$ on P such that for each class A_i we have $\sum_{v \in A_i} v = \mathbf{0}$. This can be achieved for instance if $|A_i|$ is even for all but one of the classes, and for each point $v \in A_i$ in the drawing, also $-v \in A_i$, and the remaining vertex is drawn at $\mathbf{0}$. If all $|A_i|$ are even, then the theorem also holds under the assumption that no vertex is drawn at $\mathbf{0}$ (recall that $|P|$ is odd). Otherwise, the best drawings are such that $\sum_{i=1}^{r} \left\| \sum_{v \in A_i} v \right\|^2$ is minimized, and the drawing in the second case of the theorem only gives an approximation.

Theorem 1. *Let* $\mathbf{v}$ *be a straight line drawing of* $K_{n_1, \ldots, n_r}$ *that minimizes* $\lambda(\mathbf{v})$. *If* $n_1 = n_2 = \ldots = n_r$, *then* $\mathbf{v}$ *minimizes* $\sum_{i=1}^{r} \left\| \sum_{v \in A_i} v \right\|^2$; *in particular, if* $\sum_{v \in A_i} v = \mathbf{0}$, *for all* $1 \leq i \leq r$, *then* $\lambda(\mathbf{v}) = N - n_r$. *If* $n_1 < n_2 < \ldots < n_r$, *then* $\mathbf{v}$ *has the following structure: For each* $i = 1, \ldots, r - 1$, *the union of the smallest* i *color classes,* $\bigcup_{j=1}^{i} A_j$, *forms a ball centered at* $\mathbf{0}$.

Proof. Consider first the case when all classes A_i have the same number of points $n = n_i$. Take a drawing $\mathbf{v}$. By Lemma 2,

$$\lambda(\mathbf{v}) = N + \frac{1}{S} \sum_{i=1}^{r} \left(-n_i \sum_{v \in A_i} \|v\|^2 \right) + \frac{1}{S} \sum_{i=1}^{r} \left\| \sum_{v \in A_i} v \right\|^2 = N - n + \frac{1}{S} \sum_{i=1}^{r} \left\| \sum_{v \in A_i} v \right\|^2 .$$

Then $\lambda(\mathbf{v})$ is minimized if $\sum_{i=1}^{r} \left\| \sum_{v \in A_i} v \right\|^2$ is minimized. If there are drawings $\mathbf{v}$ such that $\sum_{v \in A_i} v = \mathbf{0}$ for each class, then $\sum_{i=1}^{r} \left\| \sum_{v \in A_i} v \right\|^2 = 0$. Since the algebraic connectivity of $K_{n,n...,n}$ (with $N = r \cdot n$) is $N - n$, such a drawing is best possible. Consider then the case $n_1 < n_2 < \ldots < n_r$. By Lemma 2, $\mathbf{v}$ minimizes $\lambda(\mathbf{v})$ if $\sum_{i=1}^{r} \left\| \sum_{v \in A_i} v \right\|^2$ is minimized and $\sum_{i=1}^{r} \left(n_i \sum_{v \in A_i} \|v\|^2 \right)$ is as large as possible. Both conditions can be guaranteed at the same time. $\sum_{i=1}^{r} \left\| \sum_{v \in A_i} v \right\|^2$ can be kept small (or equal to 0) when drawing each A_i in a symmetric way around the origin. The quantity $\sum_{i=1}^{r} \left(n_i \sum_{v \in A_i} \|v\|^2 \right)$ is maximized when the smallest class A_1 is drawn such that $\sum_{v \in A_1} \|v\|^2$ is as small as possible, which is the case when the vertices of A_1 are drawn as close as possible to the origin; then, in an optimal drawing the vertices of the second smallest class A_2 are drawn as close as possible to the origin on grid points which are not occupied by A_1. In the same way, iteratively, for the i-th smallest class all grid points closest to the origin, that are not yet occupied by smaller classes, are selected. This results in a drawing with concentric rings around the origin. $\qquad \square$

Remark 1. If some of the classes have the same number of elements, then the optimal solutions are given by a combination of the two cases of Theorem 1. That is, several classes with the same number of elements can form one of the concentric rings in the drawing which satisfies $\sum_{v \in A_i} v = \mathbf{0}$ for all color classes.

We show next that the number of optimal drawings of $K_{n_1,\ldots,n_r}$ that minimize $\lambda(\mathbf{v})$ can be exponential if some classes have the same number of elements. For the sake of simplicity of the exposition, we show this only for the case $K_{1,2m,2m}$ and dimension $d = 1$. The argument can be adapted to the general case.

Proposition 3. *Let $d = 1$, and $P = \{-2m, -2m+1, \ldots, 2m-1, 2m\}$. There exists a constant $c > 0$ such that the number $\mathcal{N}$ of straight line drawings $\mathbf{v}$ of $K_{1,2m,2m}$ on P which minimize $\lambda(\mathbf{v})$ satisfies $c16^m/m^5 < \mathcal{N} < 16^m$.*

Proof. Let A_1, A_2, A_3 be the classes of $K_{1,2m,2m}$, with $n_1 = 1$ and $n_2 = n_3 = 2m$. Theorem 1 characterizes the optimal drawings as all drawings that satisfy $\sum_{v \in A_i} v = 0$. Then the only vertex of class A_1 is drawn at position 0 in any optimal drawing. For the upper bound, the number of such drawings is at most $\binom{4m}{2m} < 16^m$, since there are at most $\binom{4m}{2m}$ choices for mapping the vertices of A_2 to $P \backslash \{0\}$, and then the positions of the vertices in A_3 are already determined. Regarding the lower bound, in order to have $\sum_{v \in A_2} v = 0$, we must have $\sum_{v \in A_2, v < 0} -v = \sum_{v \in A_2, v > 0} v$. We may thus consider only drawings with exactly m elements v of A_2 with $v > 0$. There are at most $\sum_{i=1}^{2m} i = 2m^2 + m$ different sums that can be obtained by $\sum_{v \in A_2, v > 0} v$, and the same holds for $\sum_{v \in A_2, v < 0} -v$. Thus, one of these sums, call it s, appears in at least $\frac{\binom{2m}{m}}{2m^2+m}$ of all the drawings of $\{v \in A_2, v > 0\}$, and by symmetry, the same sum s appears also at least $\frac{\binom{2m}{m}}{2m^2+m}$ times when considering $\sum_{v \in A, v < 0} -v$. Any drawing for which at the same time we have $\sum_{v \in A_2, v > 0} v = s$ and $\sum_{v \in A_2, v < 0} -v = s$ is an optimal drawing. There are at least $\left(\frac{\binom{2m}{m}}{2m^2+m}\right)^2 = \Omega\left(\frac{16^m}{m^5}\right)$ such drawings, where we use the asymptotic estimate $\binom{2m}{m} \sim \frac{4^m}{\sqrt{\pi m}}$. Hence the lower bound follows. $\square$

3   Optimal Drawings for Maximizing $\lambda(\mathbf{v})$

We now study drawings of $K_{n_1,\ldots,n_r}$ that maximize $\lambda(\mathbf{v})$. The following solution as a Voronoi diagram has to be considered as an approximation, due to the discrete setting and due to the given bounding box. However, the bigger the numbers n_i, the better the approximation to the boundary curves between adjacent Voronoi regions.

Theorem 2. *Let* $\mathbf{v}$ *be a straight-line drawing of* $K_{n_1,\ldots,n_r}$ *on* P *that maximizes* $\lambda(\mathbf{v})$. *If* $n_1 = n_2 = \ldots = n_r$, *then* $\mathbf{v}$ *defines a centroidal Voronoi diagram. If the* n_i *are not all the same, then* $\mathbf{v}$ *defines a multiplicatively weighted centroidal Voronoi diagram.*

Proof. We make use of the following fact: let Q be an arbitrary set of n points $p_1,\ldots,p_n$ in $\mathbb{R}^d$. Let c be the centroid of Q, $c = \frac{1}{n}\sum_{i=1}^n p_i$. Then, see [2],

$$\sum_{i=1}^{n-1}\sum_{j=i+1}^n \|p_i - p_j\|^2 = n\sum_{i=1}^n \|p_i - c\|^2. \tag{2}$$

In the case of our theorem, let $\mathbf{v}$ be a drawing of $K_{n_1,\ldots,n_r}$ drawn on

$$P = \left\{(x_1,\ldots,x_d) \in \mathbb{Z}^d : -M \leq x_i \leq M\right\}.$$

Denote by $c_{A_1},\ldots,c_{A_r}$ the centroids of the classes $A_1,\ldots,A_r$, respectively. Then, from Equation (1) and $S = \sum_{v\in P} \|v\|^2$ we get

$$\lambda(\mathbf{v})|S| = \sum_{(v,w)\in E} \|v - w\|^2 = \sum_{i<j}\sum_{\substack{v\in A_i \\ w\in A_j}} \|v - w\|^2$$

$$= \sum_{v,w\in P} \|v - w\|^2 - \sum_{i=1}^r \sum_{v,w\in A_i} \|v - w\|^2$$

$$= \sum_{v,w\in P} \|v - w\|^2 - \sum_{i=1}^r n_i \sum_{v\in A_i} \|v - c_{A_i}\|^2,$$

where in the last equation we use (2). The quantity $\sum_{v,w\in P} \|v - w\|^2$ is the same for each drawing of $K_{n_1,\ldots,n_r}$, and $\sum_{i=1}^r n_i \sum_{v\in A_i} \|v - c_{A_i}\|^2$ is minimized if for each class A_i, its vertices are drawn as close as possible to its centroid c_{A_i}. Then the union of the r regions defined by $A_1,\ldots,A_r$ forms a centroidal Voronoi tessellation, see [7]. Note that when the $n_i's$ are different, then this is a multiplicatively weighted Voronoi diagram, see [3]. $\square$

4   An Integer Variant of the Algebraic Connectivity

In this section we consider drawings in dimension $d = 1$ of graphs $G = (V, E)$ with $V = \{1, \ldots, N\}$, that is, drawings $\mathbf{v}$ where the vertices of G are mapped to different points of $P = \{-\lfloor N/2 \rfloor, -\lfloor N/2 \rfloor + 1, \ldots, \lfloor N/2 \rfloor\}$. If N is even, then in order to satisfy the condition $\sum_{i=1}^{N} v_i = 0$ (recall the definition of (1) and the Embedding Lemma), no vertex is mapped to the origin. We denote by

$$\lambda_2^I(G) = \min \lambda(\mathbf{v}),$$

where the minimum is taken over all drawings $\mathbf{v}$ of G on P. Note that when N is odd, then $\lambda_2^I(G)$ is equivalent to the square of the minimum-2-sum, $\sigma_2^2(G)$ (recall the definition of minimum-2-sum in the introduction).

Continuing the investigations by Juvan and Mohar mentioned in the introduction (see [10,11]), we are here interested in properties and bounds for $\lambda_2^I(G)$, similar in spirit to bounds and properties of $\lambda_2(G)$. First, the following relation, analogous to the one for $\lambda_2(G)$ from [9] is obtained easily.

Proposition 4. *If G and H are edge-disjoint graphs with the same set of vertices, then*

$$\lambda_2^I(G) + \lambda_2^I(H) \leq \lambda_2^I(G \cup H).$$

The proof is immediate from the definition of $\lambda_2^I(G)$ and the Embedding Lemma, by splitting the sum of the edge weights for $G \cup H$ into two sums of edge weights, one for G and one for H.

Denote by $G + e$ the graph obtained from the graph G with N vertices by adding an edge e. It is known (see [1]) that $\lambda_2(G) \leq \lambda_2(G + e) \leq \lambda_2(G) + 2$. We have a result in the same spirit for $\lambda_2^I(G)$:

Proposition 5. *Denote by $G + e$ the graph obtained from the graph G with N vertices by adding an edge e. Then*

$$\lambda_2^I(G) + \frac{1}{2\sum_{i=1}^{\lfloor N/2 \rfloor} i^2} \leq \lambda_2^I(G + e) \leq \lambda_2^I(G) + \frac{N^2}{2\sum_{i=1}^{\lfloor N/2 \rfloor} i^2}.$$

Again, the proof is immediate; adding an edge to a drawing of G increases the edge weight by at least 1 and by at most N^2.

Let us then consider the Cartesian product of graphs. Recall that the Cartesian product $G \times H$ is defined as follows: $V(G \times H) = V(G) \times V(H)$, and $(u, u')(v, v') \in E(G \times H)$ iff either $u = v$ and $u'v' \in E(H)$, or $u' = v'$ and $uv \in E(G)$. For the Cartesian product of two graphs G and H, Fiedler [9] proved the relation $\lambda_2(G \times H) = \min\{\lambda_2(G), \lambda_2(H)\}$. The analogous relation does not hold for $\lambda_2^I(G)$; $\lambda_2^I(G \times H)$ can be strictly larger than $\min\{\lambda_2^I(G), \lambda_2^I(H)\}$ as can be seen by the example of $C_3 \times P_2$, the Cartesian product of a triangle with an edge.

In the following, the number of vertices of a graph G is also denoted by $|G|$.

Proposition 6. *Let G and H be two graphs such that $|G|$ and $|H|$ are odd numbers. Then we have*

$$\lambda_2^I(G \times H) \leq \lambda_2^I(G) \left(\frac{|G|^2 - 1}{|G|^2|H|^2 - 1} \right) + \lambda_2^I(H) \left(\frac{|G|^2(|H|^2 - 1)}{|G|^2|H|^2 - 1} \right) \qquad (3)$$

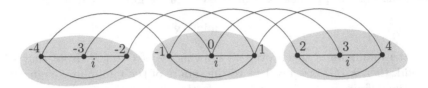

Fig. 4. A drawing of the Cartesian product of cycle C_3 and path P_3. Some straight line edges are drawn as arcs for better visibility.

Proof. We present a drawing of $G \times H$ which attains the claimed bound. First consider an optimal drawing H_{opt} of H which gives $\lambda_2^I(H)$, and then replace each vertex of H by $|G|$ vertices. More precisely, the $|H| \cdot |G|$ vertices of $G \times H$ are drawn on P in such a way that we have $|H|$ consecutive copies G_i of G (each copy G_i occupies an interval of $|G|$ consecutive points of P). Within each G_i the vertices are ordered in the same way such that the drawing of G_i is best possible (minimizing the sum of squared edge lengths); denote this drawing of G_i as G_{opt}. Figure 4 shows such a drawing of $G \times H$ for $G = C_3$ and $H = P_3$. From Proposition 1, with $2M + 1 = |G||H|$ and $d = 1$, we have

$$\sum_{i \in V} ||v_i||^2 = \frac{1}{12} \left(|G|^2|H|^2 - 1 \right) (|G|) (|H|).$$

Between two consecutive copies of a vertex i of H there are exactly $|G| - 1$ points of P. Then an edge $e \in H_{opt}$ with squared length e^2 has squared edge length $(e|G|)^2$ in our drawing of $G \times H$. We get

$$\lambda_2^I(G \times H) \leq \frac{|H| \sum\limits_{e \in G_{opt}} e^2}{\frac{1}{12}\left(|G|^2|H|^2 - 1\right)(|G|)(|H|)} + \frac{|G| \sum\limits_{e \in H_{opt}} (e|G|)^2}{\frac{1}{12}\left(|G|^2|H|^2 - 1\right)(|G|)(|H|)}$$

$$= \frac{|H| \sum\limits_{G_{opt}} e^2}{\frac{1}{12}(|G|)(|H|)(|G|^2 - 1)\left(\frac{|G|^2|H|^2-1}{|G|^2-1}\right)} + \frac{|G|^3 \sum\limits_{H_{opt}} e^2}{\frac{1}{12}(|G|)(|H|)(|H|^2 - 1)\left(\frac{|G|^2|H|^2-1}{|H|^2-1}\right)}$$

$$= \lambda_2^I(G) \left(\frac{|G|^2 - 1}{|G|^2|H|^2 - 1} \right) + \lambda_2^I(H) \left(\frac{|G|^2(|H|^2 - 1)}{|G|^2|H|^2 - 1} \right).$$

$\square$

Corollary 1. *Let G and H be two graphs such that $|G| = |H|$ is an odd number. Then we have*

$$\lambda_2^I(G \times H) \leq \frac{\lambda_2^I(G) + \lambda_2^I(H)}{2}.$$

Proof. This follows from the proof of Proposition 6, by interchanging the role of G and H in the drawing, and then by summing the two inequalities.   □

Corollary 2. *If $\lambda_2^I(G) = \lambda_2(G)$ and $|G|$ is odd, then*

$$\lambda_2^I(G \times G) = \lambda_2^I(G).$$

Proof. On the one hand, $\lambda_2^I(G \times G) \geq \lambda_2(G \times G) = \lambda_2(G) = \lambda_2^I(G)$. On the other hand, by Corollary 1, $\lambda_2^I(G \times G) \leq \lambda_2^I(G)$.   □

The assumption of $|G|$ and $|H|$ being odd numbers in Proposition 6 simplifies the calculations. We believe that a similar bound holds when $|G|$ or $|H|$ are even. Indeed, the drawing for $G \times H$ explained in the proof of Proposition 6 can be optimal when $|G|$ and $|H|$ are even. We illustrate this with the hypercube and mention that its eigenvalues and eigenvectors are well known.

Proposition 7. *For Q_N, the hypercube on N vertices, $\lambda_2^I(Q_N) = \lambda_2(Q_N) = 2$.*

Proof. To see this, note that $Q_N = Q_{N/2} \times P_2$. In this case an optimal drawing of Q_N can be obtained from two copies of an optimal drawing for $Q_{N/2}$ using ideas of the drawing of Proposition 6: indeed, one can take an optimal drawing of $Q_{N/2}$ once shifted towards $\{1, \ldots, N/2\}$ (corresponding to vertices of the hypercube having 0 in the first dimension), and once shifted towards $\{-N/2, \ldots, -1\}$ (corresponding to vertices of the hypercube having 1 in the first dimension), and then connecting them by a matching. This drawing is similar to the one described in Proposition 6; in fact, the only difference is that no vertex is mapped to the origin.   □

Proposition 8. *There are graphs G with $\lambda_2^I(G \times G) < \lambda_2^I(G)$.*

Proof. Let G be the graph consisting of a triangle, with labels of the vertices $1, 2, 3$, and a path of length 2 attached to vertex 3; label these vertices $4, 5$, in this order. Clearly, the function $f : V(G) \to \{-2, \ldots, 2\}$ given by $f(i) = i - 3$, $1 \leq i \leq 5$, defines an optimal drawing of G, yielding $\lambda_2^I(G) = \frac{8}{10}$. On the other hand, consider the drawing $g : V(G) \times V(G) \to \{-12, \ldots, 12\}$ given as follows: $g(i,j) = -12 + 3(i-1) + (j-1)$ for $1 \leq i, j \leq 3$, $g(i,j) = -3 + 2(i-1) + (j-4)$ for $1 \leq i \leq 3, 4 \leq j \leq 5$, and $g(i,j) = 3 + 5(i-4) + (j-1)$ for $4 \leq i \leq 5, 1 \leq j \leq 5$. The drawing given by g gives an upper bound on $\lambda_2^I(G \times G)$, and hence $\lambda_2^I(G \times G) \leq \frac{775}{1300} < 0.6$.   □

Whereas it is obvious that for graphs G with an odd number N of vertices, the optimal drawings of $\lambda_2^I(G)$ and $\sigma_2^2(G)$ coincide, this is not always the case for N even.

Proposition 9. *There exist graphs G with $|G|$ even, for which the optimal draw-ings of $\lambda_2^I(G)$ and $\sigma_2^2(G)$ are different.*

Proof. Consider the graph G shown in Fig. 5. An optimal drawing for $\sigma_2^2(G)$ is given by ordering the vertices in the order 12354678 or 12345678. Indeed, in any drawing, the five edges incident to vertex 5 together have squared edge length at least $2 \cdot 1^2 + 2 \cdot 2^2 + 3^2$ and the other two edges have squared edge length at least 1. It is easily checked that for $\lambda_2^I(G)$, 18275346 is a better embedding than 12354678 or 12345678.   □

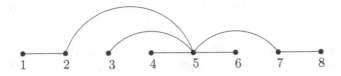

Fig. 5. A graph G which has different drawings for $\lambda_2^I(G)$ and for $\sigma_2^2(G)$.

5   Conclusion

In this paper we gave drawings minimizing as well as maximizing $\lambda(\mathbf{v})$, and we analyzed properties of an integer variant of the algebraic connectivity. It would be interesting to characterize the class of graphs G for which $\lambda_2(G) = \lambda_2^I(G)$.

Acknowledgments. We thank Igsyl Domínguez for valuable discussions on early versions of this work. We also thank the reviewers for their very helpful comments.

 This project has received funding from the European Union's Horizon 2020 research and innovation programme under the Marie Skłodowska-Curie grant agreement No 734922.

Clemens Huemer was supported by projects MINECO MTM2015-63791-R and Gen. Cat. DGR 2017SGR1336. Ruy Fabila-Monroy and Carlos Hidalgo-Toscano were supported by Conacyt grant 253261.

References

1. de Abreu, N.M.M.: Old and new results on the algebraic connectivity of graphs. Linear Algebra Appl. **423**(1), 53–73 (2007). https://doi.org/10.1016/j.laa.2006.08.017
2. Apostol, T.M., Mnatsakanian, M.A.: Sums of squares of distances in m-space. Amer. Math. Monthly **110**, 516–526 (2003). https://doi.org/10.1080/00029890.2003.11919989
3. Aurenhammer, F., Klein, R., Lee, D.-T.: Voronoi Diagrams and Delaunay Trian-gulations. World Scientific, Singapore (2013). https://doi.org/10.1142/8685

4. Bolla, M.: Spectral Clustering and Biclustering. Wiley, Hoboken (2013). https://doi.org/10.1002/9781118650684
5. Chinn, P.Z., Chvátalová, J., Dewdney, A.K., Gibbs, N.E.: The bandwidth problem for graphs and matrices - a survey. J. Graph Theor. **6**, 223–254 (1982). https://doi.org/10.1002/jgt.3190060302
6. Chvátalová, J.: Optimal labeling of a product of two paths. Discrete Math. **11**, 249–253 (1975). https://doi.org/10.1016/0012-365X(75)90039-4
7. Du, Q., Faber, V., Gunzburger, M.: Centroidal voronoi tessellations: applications and algorithms. SIAM Rev. **41**(4), 637–676 (1999). https://doi.org/10.1137/S0036144599352836
8. Fabila-Monroy, R., Hidalgo-Toscano, C., Huemer, C., Lara, D., Mitsche, D.: Optimal Grid Drawings of Complete Multipartite Graphs and an Integer Variant of the Algebraic Connectivity. arXiv:1808.09024 [cs.DM] (2018)
9. Fiedler, M.: Algebraic connectivity of graphs. Czechoslovak Math. J. **23**(2), 298–305 (1973). https://doi.org/10.21136/CMJ
10. Juvan, M., Mohar, B.: Laplace eigenvalues and bandwidth-type invariants of graphs. J. Graph Theor. **17**(3), 393–407 (1993). https://doi.org/10.1002/jgt.3190170313
11. Juvan, M., Mohar, B.: Optimal linear labelings and eigenvalues of graphs. Discrete Appl. Math. **36**, 153–168 (1992). https://doi.org/10.1016/0166-218X(92)90229-4
12. Koren, Y.: Drawing graphs by eigenvectors: theory and practice. Comput. Math. Appl. **49**, 1867–1888 (2005). https://doi.org/10.1016/j.camwa.2004.08.015
13. Rocha, I.: Spectral bisection with two eigenvectors. In: Drmota, M., Kaug, M., Krattenthaler, C., Nešetřil, J. (eds.) EUROCOMB 2017. Electronic Notes in Discrete Mathematics, vol. 61, pp. 1019–1025 (2017). https://doi.org/10.1016/j.endm.2017.07.067
14. Spielman, D.A., Teng, S.-H.: Spectral partitioning works: planar graphs and finite element meshes. Linear Algebra Appl. **421**, 284–305 (2007). https://doi.org/10.1016/j.laa.2006.07.020

Graph Drawing Contest Report

Graph Drawing Contest Report

William Devanny[1], Philipp Kindermann[2], Maarten Löffler[3(✉)],
and Ignaz Rutter[4]

[1] University of California, Irvine, USA
williamdevanny@gmail.com
[2] Universität Würzburg, Würzburg, Germany
philipp.kindermann@uni-wuerzburg.de
[3] Utrecht University, Utrecht, The Netherlands
m.loffler@uu.nl
[4] University of Passau, Passau, Germany
rutter@fim.uni-passau.de

Abstract. This report describes the 25th Annual Graph Drawing Contest, held in conjunction with the 26th International Symposium on Graph Drawing and Network Visualization (GD'18) in Barcelona, Spain. The mission of the Graph Drawing Contest is to monitor and challenge the current state of the art in graph-drawing technology.

1 Introduction

This year, the Graph Drawing Contest was divided into two parts: the *creative topics* and the *live challenge*.

The creative topics had two graphs: the first one was a graph about characters in the Game of Thrones television series, and the second one described adviser-advisee relationships between mathematicians. The data sets were published a year in advance, and contestants submitted their drawings before the conference started. Submissions were evaluated according to aesthetic appeal, domain-specific requirements, and how well the data was visually represented.

The live challenge took place during the conference in a format similar to a typical programming contest. Teams were presented with a collection of challenge graphs and had one hour to submit their highest scoring drawings. This year's topic was to maximize the smallest crossing angle in a straight-line drawing of a graph with vertex locations restricted to a grid.

Overall, we received 44 submissions: 31 submissions for the creative topics and 13 submissions for the live challenge.

2 Creative Topics

The two creative topics for this year were a graph about Game of Thrones and a mathematics genealogy graph. The goal was to visualize each graph with

© Springer Nature Switzerland AG 2018
T. Biedl and A. Kerren (Eds.): GD 2018, LNCS 11282, pp. 609–617, 2018.
https://doi.org/10.1007/978-3-030-04414-5_43

complete artistic freedom, and with the aim of communicating the data in the graph as well as possible.

We received 23 submissions for the first topic, and 8 for the second. For each topic, we selected between 3 and 5 contenders for the prizes, which were printed on large poster boards and presented at the Graph Drawing Symposium. Out of those contenders, we selected the winning submissions. We will now review the top three submissions for each topic (for a complete list of submissions, refer to http://www.graphdrawing.de/contest2018/results.html).

2.1   Game of Thrones

The TV show "Game of Thrones" is based on the book series "A Song of Ice and Fire" by George R. R. Martin and is one of the most popular TV shows in the previous years. For the contest, we extracted the relations between some of the most important characters in the show as of the end of Season 7 from the Game of Thrones Wiki[1]. The graph consists of 84 characters and 216 relations.

Third Place: Velitchko Filipov, Davide Ceneda, Michael Koller, Alessio Arleo, and Silvia Miksch (TU Vienna). The committee likes the overall aesthetics of the drawing, and the clever combination of using both the interior and exterior space for routing edges in this radial layout.

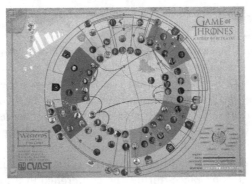

Second Place: Marian Amann, Philipp de Col, and Markus Wallinger (TU Vienna). The committee likes the clarity of this layout, with a good global overview of the graph structure, as well as showing lower-level connections between different individual characters.

---

[1] http://gameofthrones.wikia.com/wiki/Game_of_Thrones_Wiki.

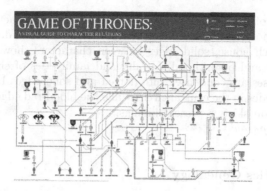

Winner: Evmorfia Argyriou, Michael Baur, Anne Eberle, and Armin Gufler (yWorks). The committee likes the overall clarity of this drawing, and the use of symbols representing houses and individuals. It is a nice idea to use different drawing styles to visualize clusters of family relations, "peaceful" relations, and killings, allowing the viewer to focus on each of these as almost separate subgraphs. The visualization and an explanation of the drawing process is available online: http://yworks.com/got.

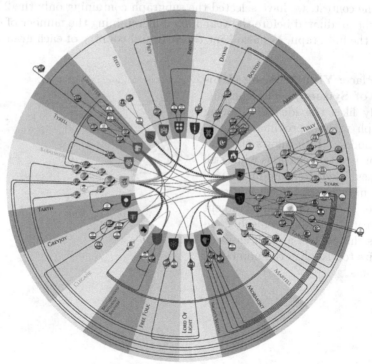

> ❝ In our Game of Thrones visualization we grouped the characters according to their house and placed them on radial layers that depict generations. In each house, solid edges between characters show marriages, love interests and parenthood. Loyalty and family ties of characters of different houses are condensed into dotted and solid edges between the new house nodes in the central circle, while the blood-red edges on radial paths in the outskirts represent killings. ❞
>
> *Evmorfia Argyriou*

2.2 Mathematics Genealogy

The Mathematics Genealogy Project[2] is an initiative of the North Dakota State University to track all advisor-advisee relationships in the broader field of mathematics since the earliest records that are available. The database has 222,360 scientists on record as of today.

In 2016, Cosmin Ionita and Pat Quillen of MathWorks analyzed the graph. The main component had 7323 root vertices and 137,155 leaves, and contained 90% of the vertices. There were 7639 isolated vertices and 1962 components of size two. The graph has some cycles, but is generally very tree-like.

For the contest, we have selected the subgraph containing only the 2277 scientists who graduated before the year 1900, but retaining the number of descendants in the full graph. We also kept the year and country of each graduation.

Third Place: Yixuan Wang (University of Sydney). The committee really likes the idea of drawing this graph with approximately geolocated nodes, which allows the user to interpret vertex locations while still allowing sufficient freedom in the vertex placement to see the actual graph structure. The use of a color gradient for the year of graduation nicely complements this choice, as well as using vertex size to visualize the number of descendants.

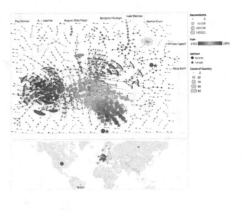

---

2 https://www.genealogy.math.ndsu.nodak.edu.

Mathematical Genealogy

A visualisation of advisor-advisee relations in science preceding the year 1900

Second Place: Gesa Behrends, Maria Hartmann, Johannes Janssen, Artsem Kavaleuski, Andre Mazal Krauss, Duc Do Minh, Alexander Zachrau, Hong Zhu, and Günter Rote (FU Berlin). The committee was impressed by the clarity of this drawing, given the large data size and the authors' decision to include all individual names on a single poster. The use of large empty regions help the viewer to see the global graph structure at a glance, while individual clusters and relationships can still be distinguished.

Winner: Florian Grötschla, Tamar Mirbach, Christian Ortlieb, Tamara Mchedlidze, and Marcel Radermacher (KIT). The committee was impressed by this interactive visualization. The website has some nice functionality; especially the highlighting of advisors and students and the additional information display on hovering over a node should be emphasized. This makes it a great way to explore the data. The drawing can be explored here: https://mathematics-genealogy.de.

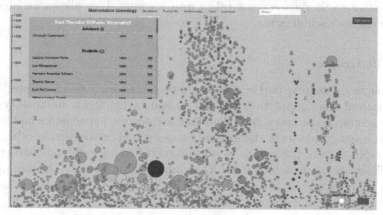

❝ The large size of the genealogy graph, and the fact that it is denser than a tree, defeated our tries to create a readable standalone node-link diagram. Thus, we decided to avoid the display of the edges and instead design an interactive visualization tool where one can focus on smaller portions of the visualization using zooming operation, investigate the relations between mathematicians using an interactive highlight tool and still obtain a big picture of the data, which is enhanced by the coloring of the nodes according to the country of graduation. To keep the node proximity meaningful (connected nodes are close and non-connected far apart) and to express the time of graduation, we employed the Sugiyama framework for the node positions, where horizontal layers correspond to the graduation date. ❞
Florian Grötschla

3   Live Challenge

The live challenge took place during the conference and lasted exactly one hour. During this hour, local participants of the conference could take part in the manual category (in which they could attempt to solve the graphs using a supplied tool[3]), or in the automatic category (in which they could use their own software to solve the graphs). At the same time, remote participants could also take part in the automatic category.

The challenge focused on maximizing the minimum crossing angle in a straight-line embedding of a given graph, with vertex locations restricted to a grid. The results were judged solely with respect to the minimum crossing angle; other aesthetic criteria were not taken into account. This allows an objective way to evaluate each drawing.

3.1   The Graphs

In the manual category, participants were presented with seven graphs. These were arranged from small to large and chosen to highlight different types of graphs and graph structures. In the automatic category, participants had to solve the same seven graphs as in the manual category, and in addition another seven larger graphs. Again, the graphs were constructed to have different structure.

For illustration, we include the third graph in its initial state, the best manual solution we received (by team ToBeDecided), and the best automatic solution we received (by team TübingenColdShower).

---

[3] http://graphdrawing.de/tool.

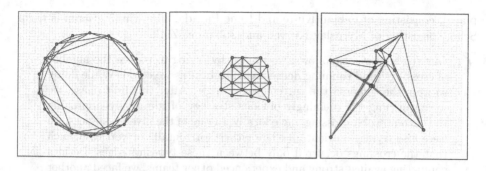

For the complete set of graphs and submissions, refer to the contest website at http://www.graphdrawing.de/contest2018/results.html.

From the resulting drawings, the committee reenforces its conclusion from the previous year; namely that, if the minimum crossing angle is an indicator of the legibility of a graph visualization, it must be so only when combined with other criteria; for instance, penalizing low distances between vertices. The committee also observed that manual (human) drawings of graphs often display a deeper understanding of the underlying graph structure than automatic drawings of the same graph, even when the automatic drawing scores equally high or higher.

3.2   Results: Manual Category

We are happy to present the full list of scores for all teams. The numbers listed are the smallest crossing angle in degrees in each graph; the horizontal bars visualize the corresponding scores.

graph	1	2	3	4	5	6	7
OFMO	47°	90°	74°	53°	39°	31°	23°
MaLaRa	56°	89°	71°	62°	33°	57°	22°
PUK	58°	90°	90°	50°	43°	40°	28°
To Be Decided	33°	90°	90°	40°	36°	35°	21°
Over-Aged	46°	90°	78°	48°	29°	28°	33°
NonAustrians	43°	90°	71°	57°	49°	55°	37°
Scho	59°	90°	90°	63°	39°	71°	21°
Rotes Minions	42°	89°	88°	39°	42°	45°	40°
AFT	45°	90°	90°	47°	32°	27°	23°

The runner-up teams are team PUK (3rd place), consisting of Paul Jungeblut, Jérôme Urhausen, and Peter Stumpf; and team NonAustrianAustrians (2nd place), consisting of Fabian Klute and Irene Parada. The winning team is team Scho, consisting of Myroslav Kryven and Johannes Zink.

> ❝ Our strategy was as follows: we started by trying to reduce the number of crossings (and thinning dense areas) as best as we could. While doing so we rather ignored the current worst angle. After this first phase, we were pursuing the seemingly obvious strategy of iteratively improving the current worst crossing angle locally. In one of the given instances, we were also trying to orient all edges either horizontally or vertically such that the crossings occurred only between these types of edges. Besides competing against strong and experienced other teams, we faced another tough opponent: the time. Since we spent much of it on the first graphs, we had to give up our aforementioned strategy towards the end. On the last graph, we had only spent a few moments to improve the worst angles locally before we submitted it some seconds before the end of the time. In spite of this last bad result, it turned out that we managed to keep our lead barely. Finally, if you are wondering about our team name it is both the Ukrainian word for "what?" and the Franconian-German word for "indeed". ❞
>
> *Johannes Zink*

3.3   Results: Automatic Category

We are happy to present the full list of scores for all teams. The numbers listed are the smallest crossing angle in degrees in each graph; the horizontal bars visualize the corresponding scores.

graph	1	2	3	4	5	6	7	8	9	10	11	12	13	14
Arizona Anglers	49°	86°	87°	37°	30°	41°	25°	4°	28°	3°	1°			
TübingenColdShower	87°	90°	88°	88°	71°	89°	80°	34°	77°	13°	11°	74°	2°	3°
CoffeeVM+	58°	90°	85°	60°	48°	71°	43°	27°	60°	19°	22°	9°	2°	11°
TKMKN		15°	87°									72°		

The runner-up teams are team Arizona Anglers (3rd place), consisting of Reyan Ahmed and Sabin Devkota; and team CoffeeVM+ (2nd place), consisting of Almut Demel, Dominik Dürrschnabel, Lasse Wulf, Tamara Mchedlidze, and Marcel adermacher. The winning team is team TübingenColdShower, consisting of Amadäus Spallek, Christian Geckeler, Henry Förster, and Michalis Bekos!

    ❝ After our last year's failure on the exact same topic, we decided to completely change our approach to become more competitive. It turned out that our new probabilistic hill climbing approach performed way better than our previous force-directed algorithm, and thus the result for the TübingenColdShower team was not of the same kind as the water in the shower earlier the same day. ❞
Christian Geckeler

Acknowledgments. The contest committee would like to thank the organizing committee of the conference for printing the posters and providing a room with hardware for the live challenge; the generous sponsors of the symposium, in particular Springer for contributing prizes; and all the contestants for their participation. Further details including all submitted drawings and challenge graphs can be found at the contest website:

    http://www.graphdrawing.de/contest2018/results.html

Poster Abstracts

Visual Analysis of Temporal Fiscal Networks with TeFNet

Walter Didimo, Luca Grilli, Giuseppe Liotta, Fabrizio Montecchiani,
and Daniele Pagliuca[✉]

Università degli Studi di Perugia, Perugia, Italy
{walter.didimo,luca.grilli,giuseppe.liotta,
fabrizio.montecchiani}@unipg.it, daniele.pagliuca@studenti.unipg.it

The development of software systems for the analysis of economic and financial networks is a fundamental activity to contrast tax evasion, fiscal frauds, and money laundering phenomena (see, e.g., [3, 6, 8]). Of particular interest in this context is the design of visual analytics systems (see, e.g., [1, 4, 5, 7]).

In this poster, we present TEFNET, a new system specifically designed to support public officers in tax evasion discovery and risk analysis. It is an evolution of TAXNET [4]: A visual analytics decision support system for tax evasion discovery. TEFNET inherits all the TAXNET's functionalities while overcoming its main limitation: The lack of a native support for temporal queries and visualizations. TEFNET's main ingredients are: (i) a network data model to represent time-varying relationships between taxpayers, called *temporal fiscal network*; (ii) a visual query language to easily define and search for suspicious (time-dependent) patterns in a temporal fiscal network; (iii) visualization functionalities to interactively explore the subgraphs that match a pattern. Both the visual query language and the graph visualization techniques rely on a suitable *timeline* approach [2, 9], which maps the time dimension to a space dimension.

Temporal Fiscal Networks. A *temporal fiscal network* is a directed graph G, whose nodes represent taxpayers (persons or companies), and whose edges represent oriented relationships between pairs of taxpayers, such as economic transactions, shareholdings and legal acts. Each element (node or edge) of G exists in a specific time interval (the *validity period*) going *from* an initial date *to* an ending date. In addition, an element can have one or more associated attributes, which can be *static* (time-independent), *temporal* (time-dependent), or *periodical*, i.e., time-dependent according to fiscal or business calendar periods.

Visual Query Language. The visual query language of TEFNET allows the user to define time-dependent patterns to be matched in G. A *pattern* p is a pair $\langle G_p, R_p \rangle$, where G_p is a graph that defines the topology of p, and R_p is a set of rules on the nodes and on the edges of G_p. The user can restrict the analysis to data within a desired *time range* and specify a *time slicing unit* (e.g., year or month) to partition the time range into intervals (*slices*) of the same length.

Work in cooperation with the Italian Revenue Agency (IRV). We thank in particular Carlo Palumbo, Mario Landolfi, and Giuseppe De Luca for their support.

Research supported in part by the project: "Algoritmi e sistemi di analisi visuale di reti complesse e di grandi dimensioni" - Ricerca di Base 2018, Dipartimento di Ingegneria dell'Università degli Studi di Perugia.

© Springer Nature Switzerland AG 2018
T. Biedl and A. Kerren (Eds.): GD 2018, LNCS 11282, pp. 621–623, 2018.
https://doi.org/10.1007/978-3-030-04414-5

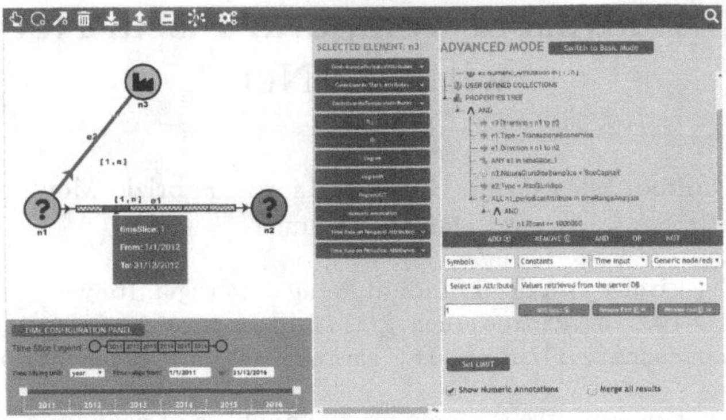

Fig. 1. Visual query language interface of TeFNet for graph pattern definition.

She can also define, in a specific time slice or in the time range, temporal rules on nodes, edges and related attributes. For example, in the pattern of Fig. 1, the timeline edge `e1` is used to express the presence of economic transactions from a node `n1` to a node `n2` in the year 2012. This rule is defined by means of the *quantity operator* `ANY` and it is visually conveyed in G_p by a solid filling of the corresponding slice. Other quantity operators for a slice are `SINGLE` - only one relation in that slice and `NONE` - no relation in that slice. A slice without quantity operators is visually conveyed with a chess filling.

Visual Exploration. In response to a user query, TeFNet returns all the subgraphs that match the specified pattern. The analysis of a result is performed through interactive visual exploration. Edges can be visually displayed in a standard mode or as timeline edges. A timeline edge is visually split into slices as in the query interface. Each slice is filled with a color whose intensity is proportional to the value of some desired function, which may represent the weight of the edge, e.g., the amount of the transaction, or the presence/absence of relations. This makes it possible for the user to easily capture in a unique view the trend of a specific parameter over the time range of analysis. The user can also expand the analysis of a result by introducing other neighbors in the current visualization. For example, Fig. 2 shows a visualization of a subgraph after some exploration steps, starting from a result of the pattern defined in Fig. 1.

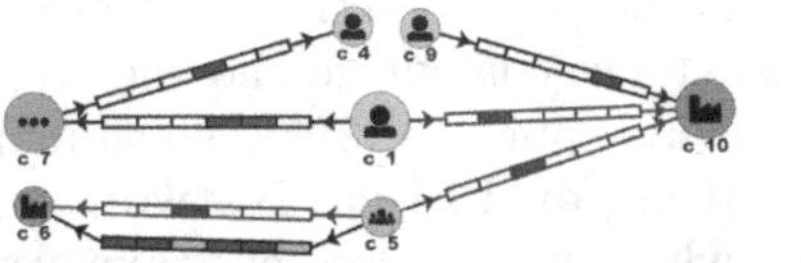

Fig. 2. Visualization of a subgraph after some exploration steps.

We tested TeFNet in a real working environment on a real 3-year fiscal network of approximately 800 K nodes and 1.9M edges. The experimental tasks were performed by expert tax officers, who were asked to find subjects having specific time-varying relations with a given taxpayer. The results show that using TeFNet may significantly improve time and accuracy of the analysis at the IRV.

References

1. Argyriou, E.N., Symvonis, A., Vassiliou, V.: A fraud detection visualization system utilizing radial drawings and heat-maps. In: Laramee, R.S., Kerren, A., Braz, J. (eds.) IVAPP 2014. pp. 153–160. SciTePress (2014). https://doi.org/10.5220/0004735501530160
2. Beck, F., Burch, M., Diehl, S., Weiskopf, D.: A taxonomy and survey of dynamic graph visualization. Comput. Graph. Forum **36**(1), 133–159 (2017). https://doi.org/10.1111/cgf.12791
3. Colladon, A.F., Remondi, E.: Using social network analysis to prevent money laundering. Expert Syst. Appl. **67**, 49–58 (2017). https://doi.org/10.1016/j.eswa.2016.09.029
4. Didimo, W., Giamminonni, L., Liotta, G., Montecchiani, F., Pagliuca, D.: A visual analytics system to support tax evasion discovery. Decis. Support Syst. **110**, 71–83 (2018). https://doi.org/10.1016/j.dss.2018.03.008
5. Didimo, W., Liotta, G., Montecchiani, F.: Network visualization for financial crime detection. J. Vis. Lang. Comput. **25**(4), 433–451 (2014). https://doi.org/10.1016/j.jvlc.2014.01.002
6. González, P.C., Velásquez, J.D.: Characterization and detection of taxpayers with false invoices using data mining techniques. Expert Syst. Appl. **40**(5), 1427–1436 (2013). https://doi.org/10.1016/j.eswa.2012.08.051
7. Huang, M.L., Liang, J., Nguyen, Q.V.: A visualization approach for frauds detection in financial market. In: Banissi, E., Stuart, L.J., Wyeld, T.G., Jern, M., Andrienko, G.L., Memon, N., Alhajj, R., Burkhard, R.A., Grinstein, G.G., Groth, D.P., Ursyn, A., Johansson, J., Forsell, C., Cvek, U., Trutschl, M., Marchese, F.T., Maple, C., Cowell, A.J., Moere, A.V. (eds.) IV 2009. pp. 197–202. IEEE (2009). https://doi.org/10.1109/IV.2009.23https://doi.org/10.1109/IV.2009.23
8. Ngai, E.W.T., Hu, Y., Wong, Y.H., Chen, Y., Sun, X.: The application of data mining techniques in financial fraud detection: A classification framework and an academic review of literature. Decis. Support Syst. **50**(3), 559–569 (2011). https://doi.org/10.1016/j.dss.2010.08.006
9. Schmauder, H., Burch, M., Weiskopf, D.: Visualizing dynamic weighted digraphs with partial links. In: Braz, J., Kerren, A., Linsen, L. (eds.) IVAPP 2015 - Proceedings of the 6th International Conference on Information Visualization Theory and Applications, Berlin, Germany, 11–14 March, 2015. pp. 123–130 (2015). https://doi.org/10.5220/0005303801230130

Multilevel Planarity

Lukas Barth[(✉)], Guido Brückner, Paul Jungeblut, and Marcel Radermacher

Department of Computer Science,
Karlsruhe Institute of Technology, Karlsruhe, Germany
{lukas.barth,brueckner,radermacher}@kit.edu,
paul.jungeblut@student.kit.edu

An upward-planar drawing is a planar drawing where each edge is drawn as a strictly y-monotone curve. While testing upward planarity of a graph is an NP-complete problem in general [11], efficient algorithms are known for single-source graphs and for embedded graphs [5, 6]. One notable specialization of upward planarity is that of level planarity. A level graph is a directed graph $G = (V, E)$ together with a level assignment $\gamma : V \to \mathbb{Z}$ that assigns an integer level to each vertex and satisfies $\gamma(u) < \gamma(v)$ for all $(u, v) \in E$. A drawing of G is level planar if it is upward planar, and for the y-coordinate of each vertex $v \in V$ it holds that $y(v) = \gamma(v)$. Level-planarity testing and embedding is feasible in linear time for single-source graphs and graphs with multiple sources, the latter case being considerably more complex [9, 13]. There exist further level-planarity variants on the cylinder and on the torus [1, 3] and there has been considerable research on further-constrained versions of level planarity [2, 7, 10, 12, 14].

We introduce and study the *multilevel-planarity testing* (MLPT) problem, which is a generalization of upward planarity and level planarity. Let $G = (V, E)$ be a directed graph and let $\ell : V \to \mathcal{P}(\mathbb{Z})$ be a function that assigns a finite set of integers to each vertex. A multilevel-planar drawing of G is an upward planar drawing of G such that the y-coordinate of each vertex $v \in V$ satisfies $y(v) \in \ell(v)$.

We present linear-time algorithms for testing multilevel planarity of embedded graphs with a single source (sT-graphs) and for oriented cycles. To this end, we characterize multilevel-planar sT-graphs as subgraphs of certain planar graphs with a single source and a single sink (st-graphs). Similar characterizations exist for upward planarity and level planarity [9, 15]. The idea behind our characterization is that we can insert edges into any given multilevel-planar drawing of a graph so as to make it an st-graph while maintaining multilevel planarity. This technique is similar to the one found by Bertolazzi et al. [6] for upward planarity, and in fact is built on top of it. For the obtained st-graphs, we may assume without loss of generality that the multilevel assignment ℓ has *normal form*, i.e., for all $(u, v) \in E$ it is $\min \ell(u) < \min \ell(v)$ and $\max \ell(u) < \max \ell(v)$. Then, we can test multilevel planarity by greedily attempting to place the vertices of G in topological order on the lowest possible level. For oriented cycles, we identify sets of vertices of minimal cardinality that have to be placed on the lowest and highest possible levels. Assuming a multilevel-planar drawing exists, the remaining vertices can then be placed greedily as low as possible between them. Both algorithms test multilevel planarity in linear time and generate a multilevel-planar drawing within the same running time.

© Springer Nature Switzerland AG 2018
T. Biedl and A. Kerren (Eds.): GD 2018, LNCS 11282, pp. 624–626, 2018.
https://doi.org/10.1007/978-3-030-04414-5

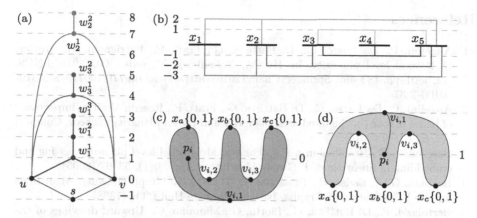

Fig. 1. (a) The sT-graph formed by three task gadgets (red, green and blue subgraphs). (b) A rectilinear embedding of a planar monotone 3-SAT instance. E-shapes above/below the variables are clauses containing only negative/positive literals. (c) The gadget for a positive clause. (d) The gadget for a negative clause.

Complementing these algorithmic results, we show that MLPT is NP-complete even in very restricted cases, namely for sT-graphs without a fixed embedding, for trees and for general embedded ST-graphs (graphs with multiple sources and sinks). This contrasts both the upward planar and level planar setting, where these problems are solvable in polynomial time (see Table 1 for a full comparison).

The first two reductions use the strongly NP-complete single-processor scheduling problem with individual release times, deadlines and processing times. For a set of tasks there exists a non-preemptive one-processor schedule if and only if a crossing free nesting of the task gadgets in the sT-graph shown in Fig. 1(a) exists. Using similar gadgets, this sT-graph can be transformed into a tree. Here the release time and deadline of a task define the interval of possible levels for each of the gadgets vertices. The number of vertices in the gadget is the processing time.

To show NP-completeness for embedded ST-graphs, we give a polynomial reduction from planar monotone 3-SAT [8]. Given a rectilinear embedding of the variables and clauses as in Fig. 1(b), we substitute the E-shaped clauses by the gadgets shown in Fig. 1(c,d). Now there is a multilevel-planar drawing, if and only if there is a truth assignment of the planar monotone 3-SAT instance.

Table 1. Result overview

	Not embedded		Fixed combinatorial embedding		
	Trees	sT-graph	Cycle	sT-graph	ST-graph
Upward planarity	$O(1)$ [4]	$O(n)$ [6]	$O(n)$ [5]	$O(n)$ [5]	P [5]
Multilevel planarity	NPC	NPC	$O(n)$	$O(n)$	NPC
Level planarity	$O(n)$ [13]	$O(n)$ [13]	$O(n)$ [13]	$O(1)$ [13]	?

References

1. Angelini, P., Da Lozzo, G., Di Battista, G., Frati, F., Patrignani, M., Rutter, I.: Beyond level planarity. In: Hu, Y., Nöllenburg, M. (eds.) GD 2016. LNCS, vol. 9801, pp. 482–495. Springer, Cham (2016). https://doi.org/10.1007/978-3-319-50106-2_37

2. Angelini, P., Da Lozzo, G., Di Battista, G., Frati, F., Roselli, V.: The importance of being proper (in clustered-level planarity and t-level planarity). Theor. Comput. Sci. **571**, 1–9 (2015)

3. Bachmaier, C., Brandenburg, F.J., Forster, M.: Radial level planarity testing and embedding in linear time. J. Graph Algorithms Appl. **9**(1), 53–97 (2005)

4. Battista, G.D., Eades, P., Tamassia, R., Tollis, I.G.: Graph Drawing: Algorithms for the Visualization of Graphs. 1st edn. Prentice Hall PTR (1998)

5. Bertolazzi, P., Di Battista, G., Liotta, G., Mannino, C.: Upward drawings of tri-connected digraphs. Algorithmica **12**(6), 476–497 (1994)

6. Bertolazzi, P., Di Battista, G., Mannino, C., Tamassia, R.: Optimal upward planarity testing of single-source digraphs. SIAM J. Comput. **27**(1), 132–169 (1998)

7. Brückner, G., Rutter, I.: Partial and constrained level planarity. In: Klein, P.N. (ed.) Proceedings of the 28th Annual ACM-SIAM Symposium on Discrete Algorithms (SODA 2017), pp. 2000–2011 (2017)

8. De Berg, M., Khosravi, A.: Optimal binary space partitions for segments in the plane. Int. J. Comput. Geom. Appl. **22**(3), 187–205 (2012)

9. Di Battista, G., Tamassia, R.: Algorithms for plane representations of acyclic digraphs. Theor. Comput. Sci. **61**(2), 175–198 (1988)

10. Forster, M., Bachmaier, C.: Clustered level planarity. In: Van Emde Boas, P., Pokorný, J., Bieliková, M., Štuller, J. (eds.) SOFSEM 2004. LNCS, vol. 2932, pp. 218–228. Springer, Heidelberg (2004). https://doi.org/10.1007/978-3-540-24618-3_18

11. Garg, A., Tamassia, R.: On the computational complexity of upward and rectilinear planarity testing. SIAM J. Comput. **31**(2), 601–625 (2002)

12. Harrigan, M., Healy, P.: Practical level planarity testing and layout with embedding constraints. In: Hong, S.-H., Nishizeki, T., Quan, W. (eds.) GD 2007. LNCS, vol. 4875, pp. 62–68. Springer, Heidelberg (2008). https://doi.org/10.1007/978-3-540-77537-9_9

13. Jünger, M., Leipert, S.: Level planar embedding in linear time. In: Kratochvíyl, J. (ed.) GD 1999. LNCS, vol. 1731, pp. 72–81. Springer, Heidelberg (1999). https://doi.org/10.1007/3-540-46648-7_7

14. Klemz, B., Rote, G.: Ordered level planarity, geodesic planarity and bi-monotonicity. In: Frati, F., Ma, K.-L. (eds.) GD 2017. LNCS, vol. 10692, pp. 440–453. Springer, Cham (2018). https://doi.org/10.1007/978-3-319-73915-1_34

15. Leipert, S.: Level Planarity Testing and Embedding in Linear Time. Ph.D. thesis, University of Cologne (1998)

Schematizing Regional Landmarks
for Route Maps

Marcelo Galvão[✉], Jakub Krukar, and Angela Schwering

Institute for Geoinformatics, University of Münster, Münster, Germany
galvao.marcelo@uni-muenster.de

1 Introduction

Regional landmarks play an important role in facilitating wayfinding and orientation in navigation tasks [7]. Information such as "the route goes around the city center" or "there is a right turn after going along the park" represent spatial relations between polygonal objects and a route. Schematic visualizations are used for the cognitively adequate representation of such spatial relations [4]. Some commercial schematic maps make use of regional landmarks (urban areas, parks, lakes, forests). In contrast to topographic maps, they change shape, orientation and scale of the polygons to emphasize their topological function.

Although regional landmarks are used in commercial maps, related work focusing on route [1, 3] or network schematization [5, 6, 10] does not consider such landmarks at all. Publications addressing schematization of regional landmarks or subdivisions [2, 9] do not consider their spatial relation with paths. Since such landmarks are more important as references in route maps for drivers than in transit maps, there is a need for an algorithm that can produce route maps with regional landmarks, highlighting their correct spatial relation with the path.

In this contribution, we describe a new approach for drawing polygonal landmarks over an already schematized route. For the topological correct schematization, the method makes use of affine transformations and an adaptation of Nöllenburg-Wolff's Mixed Integer Programming (MIP) for metro map drawing [6]. The advantage of using MIP over Buchin et al. method for polygon schematization [2] is that it allows results with higher level of abstraction by enforcing the correct topology with hard constraints while aesthetics are optimized in the objective function. We are able to emphasize crossings by constraining angles, and line alongness[1] by manipulating control points. The results resemble regional landmarks drawn by designers in commercial schematic maps.

2 Approach

To test the proposed method we read and planarize OpenStreetMap route data and polygonal geometries in its surrounding area, such as parks, lakes, urban

---

This research was supported by the ERC StRG Grant Agreement No 637645.
[1] Line alongness: the ratio of the region boundary being parallel to a path.

T. Biedl and A. Kerren (Eds.): GD 2018, LNCS 11282, pp. 627–629, 2018.
https://doi.org/10.1007/978-3-030-04414-5

areas. For polygons overlapping the route, extra vertices are added at each crossing dividing the polygon into paths (sections). For polygons disconnected from the route, we select two vertices of the polygon as control points and two corresponding extra vertices are added to the route in order to hold their relative positions. The control points are selected based on their distance to the route (d), and the distance to the beginning/end of the linear referencing of the polygon against the route (l). We want d and l to be simultaneously minimized.

After the planarization process is completed the route is schematized. The schematization rescales the route parts, restricts edges orientation to the octilinear angles, forces fixed angles at decision points, while bends and shape distortion are minimized at the same time. Details on route schematization process are omitted for space reasons but we use similar approach as in [3].

With the newly calculated schematic position of the route vertices, we use affine transformation to transpose the polygons. For polygons crossed by the route, the transformation is made to their resulting paths independently, and the crossings themselves are used as references by the transformation. For polygons disconnected from the route, the pair of control points and their correspondent vertices in the route are used to readjust their position. We allow this adjustment to be looser or tighter depending on the spatial relation we want to highlight.

After the affine transformation is applied to the polygons or their subsections, the transposed geometries are submitted to the schematization process. The schematized geometry of the route is sent to the polygon schematization process to preserve their mutual topology. For the schematization process, we adapt Nöllenburg-Wolff's MIP inequalities. We use the hard constraints for octilinearity and edge spacing [6] to ensure the correct topology with the route. One limitation of the edge spacing constraint, is that it requires the route and the polygons to have the same edge orientation restriction.

For the objective function, we combine three soft constraints that are summed together and can be weighted by independent parameters. To enhance similarity to the original shape of the polygon, we use Nöllenburg-Wolff's function to preserve relative positions. Additionally, we add a new function that preserve location by minimizing distance between old and the new polygon vertices position in the L_1-norm. To enhance abstraction, we use a similar bend cost function as Nöllenburg-Wolff's one, which penalizes bends along the resulting polygon shape. That way similarity and abstraction is balanced by adjusting the parameters.

3   Conclusion and Future Work

Using our application and real data of Münsterland-Germany, we were able to create drawings of schematized regional landmarks that emphasizes particular spatial relations with a route (e.g, line alongness and crossings). Next, we want to formalize the approach for the ten groups of path-polygon topological relations described by Shariff et al. [8], and later extend the method to be applicable with more complex street networks. Finally, we want to develop empirical experiments with participants to test how the spatial relations are interpreted and recalled in navigation tasks as compared to topographic maps.

References

1. Agrawala, M., Stolte, C.: Rendering effective route maps: improving usability through generalization. In: Proceedings of the 28th Annual Conference on Computer Graphics and Interactive Techniques, vol. 1, pp. 241–249. ACM (2001). https://doi.org/10.1145/383259.383286
2. Buchin, K., Meulemans, W., Speckmann, B.: A new method for subdivision simplification with applications to urban-area generalization. In: Proceedings of the 19th ACM SIGSPATIAL International Conference on Advances in Geographic Information Systems, pp. 261–220. ACM (2011). https://doi.org/10.1145/2093973.2094009
3. Delling, D., Gemsa, A., Nöllenburg, M., Pajor, T., Rutter, I.: On d-regular schematization of embedded paths. Comput. Geom. Theory Appl. **47**(3), 381–406 (2014)
4. Klippel, A., Richter, K.F., Barkowsky, T., Freksa, C.: The cognitive reality of schematic maps. In: Meng, L., Reichenbacher, T., Zipf, A. (eds.) Map-based Mobile Services, pp. 55–71. Springer, Heidelberg (2005). https://doi.org/10.1007/3-540-26982-7-5
5. Kopf, J., Agrawala, M., Bargeron, D., Salesin, D., Cohen, M.: Automatic generation of destination maps. ACM Trans. Graph. **29**(6), 1 (2010). https://doi.org/10.1145/1882261.1866184
6. Nöllenburg, M., Wolff, A.: Drawing and labeling high-quality metro maps by mixed-integer programming. IEEE Trans. Vis. Comput. Graph. **17**(5), 626–641 (2011). https://doi.org/10.1109/TVCG.2010.81
7. Schwering, A., Krukar, J., Li, R., Anacta, V.J., Fuest, S.: Wayfinding through orientation. Spat. Cogn. Comput. **17**(4), 273–303 (2017). https://doi.org/10.1080/13875868.2017.1322597
8. Shariff, A.R.B., Egenhofer, M.J., Mark, D.M.: Natural-language spatial relations between linear and areal objects: the topology and metric of English-language terms. Int. J. Geograph. Inf. Sci. **12**(3), 215–246 (1998). https://doi.org/10.1.1.16.918
9. Van Goethem, A., Meulemans, W., Speckmann, B., Wood, J.: Exploring curved schematization of territorial outlines. IEEE Trans. Vis. Comput. Graph. **21**(8), 889–902 (2015). https://doi.org/10.1109/TVCG.2015.2401025
10. Wang, Y.S., Chi, M.T.: Focus plus context metro maps. IEEE Trans. Vis. Comput. Graph. **17**(12), 2528–2535 (2011). https://doi.org/10.1109/TVCG.2011.205

Low-Degree Graphs Beyond Planarity

Patrizio Angelini$^{(\boxtimes)}$, Michael A. Bekos, Michael Kaufmann,
and Thomas Schneck

Institut für Informatik, Universität Tübingen, Tübingen, Germany
{angelini,bekos,mk,schneck}@informatik.uni-tuebingen.de

Abstract. We study beyond-planarity for graphs of low degree. In particular, we aim at establishing tight bounds for values of d such that every graph of degree at most d belongs to a certain beyond-planarity class.

Beyond-planarity is a central topic in graph drawing, studying algorithmic and combinatorial properties of non-planar graphs. The most-studied beyond-planarity classes include: (i) *k-planar graphs*, where each edge crosses at most k edges, (ii) *quasiplanar graphs*, which disallow 3 mutually crossing edges, (iii) *fan-planar* graphs, where an edge only crosses a *fan* (a set of edges incident to a common vertex), (iv) *fan-crossing-free* graphs, where no edge crosses a fan, and (v) *RAC k-bend graphs*, where crossings happen at right angles and edges have at most k bends. For further definitions and state of the art, see [11].

Our goal is to establish upper and lower bounds for values of d such that every graph of degree at most d belongs to a certain beyond-planarity class. Table 1 summarizes the state of the art, including our results.

To prove that for any fixed $k > 0$ there exists an infinite family of bipartite Hamiltonian degree-3 graphs whose members are not k-planar, we employ an argument based on the crossing number of the n-vertex 3-regular graph known in the literature [14] as *cube-connected cycles* CCC_n. This graph is constructed starting from the n-regular *hypercube graph* [12] $Q_n = (V_n, E_n)$, whose 2^n vertices are denoted by distinct n-digit binary numbers; then, two vertices are joined

Table 1. The largest (second column) and smallest (third) value of d such that all (not all) degree-d graphs belong to certain beyond-planarity classes.

Graph class	Feasible	Infeasible
k-planar Hamiltonian bipartite	2	3 (CCC_n, Theorem 1)
fan-planar Hamiltonian bipartite	2	3 (CCC_n, Corollary 1)
quasi-planar	4 [2]	10 (K_{11}, ref. [1])
RAC (0-bend)	2	4 ($K_{4,4}$, ref. [10])
RAC (0-bend) Hamiltonian	3 [5]	4 ($K_{4,4}$, ref. [10])
RAC 1-bend	3 [4]	9 (K_{10}, ref. [3])
RAC 2-bends	6 [4]	148 (K_{149}, ref. [6])
fan-crossing-free	3 [2]	5 ($K_{5,5}$, Theorem 2)

© Springer Nature Switzerland AG 2018
T. Biedl and A. Kerren (Eds.): GD 2018, LNCS 11282, pp. 630–632, 2018.
https://doi.org/10.1007/978-3-030-04414-5

by an edge in E_n if and only if their binary representations differ in a single digit. To obtain CCC_n, each vertex v of Q_n is replaced with a cycle of length n.

Graph CCC_n has $n2^n$ vertices and $3n2^{n-1}$ edges, and its crossing number is known [15] to be larger than $\frac{1}{20}4^n - (9n+1)2^{n-1}$. Hence, there is an edge with at least $\left\lceil \frac{1}{15}\frac{2^n}{n} - 6 - \frac{2}{2n} \right\rceil$ crossings, which shows that, for every $k \geq 1$, graph CCC_n is not k-planar for every n so that $k < \left\lceil \frac{1}{15}\frac{2^n}{n} - 6 - \frac{2}{2n} \right\rceil$. Since, for even values of $n \geq 6$, graph CCC_n is bipartite and Hamiltonian [13], we have the following.

Theorem 1. *For every $k \geq 1$, there exist infinitely many bipartite Hamiltonian 3-regular graphs that are not k-planar.*

Note that Theorem 1 could also be derived from random graph theory [8].

As observed in [2], every degree-4 graph is quasiplanar, since it has thickness 2. Thus, Theorem 1 provides an alternative proof that, for any fixed k, there exist quasiplanar graphs that are not k-planar [7]. Further, since every fan-planar drawing of a 3-regular graph is a 3-planar drawing, we have the following.

Corollary 1. *There exist infinitely many 3-regular bipartite Hamiltonian graphs that are not fan-planar.*

Alam et al. [2] observed that every degree-3 graph that can be decomposed into a matching and a set of cycles is fan-crossing-free and quasiplanar at the same time. This result can be extended to every degree-3 graph as follows[1]. First, contract vertices of degree at most 2 and remove self-loops and bridges, to obtain a 3-regular bridgeless simple graph, which admits the required decomposition by Petersen's theorem; then, reinsert the contracted or removed edges while maintaining the fan-crossing-free and quasi-planarity properties.

We prove that this result cannot be extended to degree-5 graphs, by showing that the 5-regular complete bipartite graph $K_{5,5}$ is not fan-crossing free. We prove this by means of a stronger result, namely a characterization of the complete bipartite fan-crossing-free graphs, analogous to existing characterizations for other beyond-planarity classes [9, 10].

Theorem 2. *The complete bipartite graph $K_{a,b}$, with $a \leq b$, is fan-crossing-free if and only if (i) $a \in \{1,2\}$, or (ii) $a \in \{3,4\}$ and $b \leq 6$. In particular, $K_{5,5}$ is not fan-crossing-free.*

We pose as future goal to further narrow the gaps between the bounds described in Table 1. In particular, the main open question is whether degree-3 graphs are RAC; this long-standing question has been posed already several times and is the one that first triggered our study. Note that the fan-crossing-free and quasiplanarity properties are necessary conditions for a graph to be RAC. In this sense, the extension of the result by Alam et al. [2] to all degree-3 graphs is an important step towards an answer to this question. Another intriguing question that stems from our results is whether degree-4 graphs are fan-crossing-free. Finally, the upper bounds for d concerning quasiplanar, RAC

---

[1] We thank an anonymous reviewer of GD'18 for suggesting this extension.

1-bend, and RAC 2-bend graphs presented in Table 1 descend from the known upper bounds on the maximum edge density of graphs in these classes [1, 3, 6]; it would be interesting to prove the existence of some low-degree graphs not belonging to these classes by exploiting direct arguments.

References

1. Ackerman, E., Tardos, G.: On the maximum number of edges in quasi-planar graphs. J. Comb. Theory, Ser. A **114**(3), 563–571 (2007). https://doi.org/10.1016/j.jcta.2006.08.002
2. Alam, M.J., et al.: Working group B2: beyond-planarity of graphs with bounded degree. In: Hong, S.-H., Kaufmann, M., Kobourov, S.G., Pach, J. (eds.) Beyond-Planar Graphs: Algorithmics and Combinatorics (Dagstuhl Seminar 16452), Dagstuhl Reports, vol. 6, pp. 55–56. Schloss Dagstuhl-Leibniz-Zentrum fuer Informatik (2017)
3. Angelini, P., Bekos, M., Förster, H., Kaufmann, M.: On RAC drawings of graphs with one bend per edge. In: International Symposium on Graph Drawing (GD 2018) (2018)
4. Angelini, P., Cittadini, L., Didimo, W., Frati, F., Di Battista, G., Kaufmann, M., Symvonis, A.: On the perspectives opened by right angle crossing drawings. J. Graph Algorithms Appl. **15**(1), 53–78 (2011)
5. Argyriou, E.N., Bekos, M.A., Kaufmann, M., Symvonis, A.: Geometric RAC simultaneous drawings of graphs. J. Graph Algorithms Appl. **17**(1), 11–34 (2013). https://doi.org/10.7155/jgaa.00282
6. Arikushi, K., Fulek, R., Keszegh, B., Moric, F., Tóth, C.D.: Graphs that admit right angle crossing drawings. Comput. Geom. **45**(4), 169–177 (2012). https://doi.org/10.1016/j.comgeo.2011.11.008
7. Bae, S.W., et al.: Gap-planar graphs. Theor. Comput. Sci. (2018). https://doi.org/10.1016/j.tcs.2018.05.029
8. Bollobás, B.: The isoperimetric number of random regular graphs. Eur. J. Comb. **9**(3), 241–244 (1988). https://doi.org/10.1016/S0195-6698(88)80014-3
9. Czap, J., Hudák, D.: 1-planarity of complete multipartite graphs. Discrete Appl. Math. **160**(4–5), 505–512 (2012). http://dx.doi.org/10.1016/j.dam.2011.11.014
10. Didimo, W., Eades, P., Liotta, G.: A characterization of complete bipartite RAC graphs. Inf. Process. Lett. **110**(16), 687–691 (2010). http://dx.doi.org/10.1016/j.ipl.2010.05.023
11. Didimo, W., Liotta, G., Montecchiani, F.: A survey on graph drawing beyond planarity. CoRR, abs/1804.07257, (2018). http://arxiv.org/abs/1804.07257
12. Harary, F., Hayes, J.P., Wu, H.-J.: A survey of the theory of hypercube graphs. Comput. Math. Appl. **15**(4):277–289 (1988). https://doi.org/10.1016/0898-1221(88)90213-1
13. Hsu, L., Ho, T., Ho, Y., Tsay, C.: Cycles in cube-connected cycles graphs. Discrete Appl. Math. **167**, 163–171 (2014). http://dx.doi.org/10.1016/j.dam.2013.11.021
14. Preparata, F.P., Vuillemin, J.: The cube-connected cycles: a versatile network for parallel computation. Commun. ACM **24**(5), 300–309 (1981). http://dx.doi.org/10.1145/358645.358660
15. Sýkora, O., Vrťo, I.: On crossing numbers of hypercubes and cube connected cycles. BIT **33**, 232–237 (1993)

Bounding the Tripartite-Circle Crossing Number of Complete Tripartite Graphs

Charles Camacho[1], Silvia Fernández-Merchant[2], Marija Jelic[3], Rachel Kirsch[4],
Linda Kleist[5(✉)], Elizabeth Bailey Matson[6], and Jennifer White[7]

[1] Oregon State University, Corvallis, USA
camachoc@math.oregonstate.edu
[2] California State University, Los Angeles, USA
silvia.fernandez@csun.edu
[3] University of Belgrade, Belgrade, Serbia
marijaj@matf.bg.ac.rs
[4] London School of Economics, London, UK
r.kirsch1@lse.ac.uk
[5] Technische Universität Berlin, Berlin, Germany
kleist@math.tu-berlin.de
[6] Alfred University, Alfred, USA
matson@alfred.edu
[7] Saint Vincent College, Latrobe, USA
jennifer.white@stvincent.edu

1 Introduction

The *crossing number* of a graph G, denoted by $\mathrm{cr}(G)$, is the minimum number of edge-crossings over all drawings of G on the plane. To date, even the crossing numbers of complete and complete bipartite graphs are open. For the crossing number of the complete bipartite graph Zarankiewicz [6] showed that

$$\mathrm{cr}(K_{m,n}) \leq \left\lfloor \frac{n}{2} \right\rfloor \left\lfloor \frac{n-1}{2} \right\rfloor \left\lfloor \frac{m}{2} \right\rfloor \left\lfloor \frac{m-1}{2} \right\rfloor,$$

and conjectured that equality holds. Harary and Hill [4] and independently Guy [3] conjectured that the crossing number of the complete graph K_n is

$$\mathrm{cr}(K_n) = \frac{1}{4} \left\lfloor \frac{n}{2} \right\rfloor \left\lfloor \frac{n-1}{2} \right\rfloor \left\lfloor \frac{n-2}{2} \right\rfloor \left\lfloor \frac{n-3}{2} \right\rfloor =: H(n).$$

The construction of Harary and Hill is a so-called *cylindrical drawing*, in which the vertices lie on the circles of a cylinder, and edges of the graph cannot cross the circles. Towards the Zarankiewicz Conjecture, these drawings can be restricted to *bipartite cylindrical drawings*, in which each set of the vertex partition lies on its own circle. A *k-circle drawing* of a graph G is a drawing of G in the plane where the vertices are placed on k disjoint circles and the edges do not cross the circles. The *k-circle crossing number* of a graph G is the minimum number of crossings in a k-circle drawing of G. For the special case when G is a

T. Biedl and A. Kerren (Eds.): GD 2018, LNCS 11282, pp. 633–635, 2018.
https://doi.org/10.1007/978-3-030-04414-5

k-partite graph, we can further require that each of the k vertex classes is placed on one of the k circles. The corresponding crossing number is called the *k-partite-circle crossing number* and is denoted by $cr_{\textcircled{k}}(G)$. Richter and Thomassen [5] showed that $cr_{\textcircled{2}}(K_{n,n}) = n\binom{n}{3}$. Ábrego, Fernández-Merchant, and Sparks [1] generalized this result for $m \leq n$ to

$$cr_{\textcircled{2}}(K_{n,m}) = \binom{n}{2}\binom{m}{2} + \sum_{0 \leq i < j \leq m-1} \left(\left\lfloor \frac{n}{m}j \right\rfloor - \left\lfloor \frac{n}{m}i \right\rfloor \right)\left(\left\lfloor \frac{n}{m}j \right\rfloor - \left\lfloor \frac{n}{m}i \right\rfloor - n \right).$$

2 Our Results

We investigate the tripartite-circle crossing number of the complete tripartite graph. Drawings that minimize the number of crossings are *good*, i.e., no edge crosses itself and any two edges share at most one point. We develop methods to count the number of crossings in good drawings and provide concrete drawings to obtain upper bounds.

Theorem 1. *For any integers m, n, and p,*

$$\sum_{\substack{\{x,y\} \in \binom{\{m,n,p\}}{2} \\ z \in \{m,n,p\} \setminus \{x,y\}}} \left(cr_{\textcircled{2}}(K_{x,y}) + xy \left\lfloor \frac{z}{2} \right\rfloor \left\lfloor \frac{z-1}{2} \right\rfloor \right) \leq cr_{\textcircled{3}}(K_{m,n,p})$$

$$\leq \sum_{\substack{\{x,y\} \in \binom{\{m,n,p\}}{2} \\ z \in \{m,n,p\} \setminus \{x,y\}}} \left(\binom{x}{2}\binom{y}{2} + xy \left\lfloor \frac{z}{2} \right\rfloor \left\lfloor \frac{z-1}{2} \right\rfloor \right).$$

For the balanced case, Fig. 1 illustrates the drawing, and the formulas simplify to

$$3n\binom{n}{3} + 3n^2 \left\lfloor \frac{n}{2} \right\rfloor \left\lfloor \frac{n-1}{2} \right\rfloor \leq cr_{\textcircled{3}}(K_{n,n,n}) \leq 3\binom{n}{2}^2 + 3n^2 \left\lfloor \frac{n}{2} \right\rfloor \left\lfloor \frac{n-1}{2} \right\rfloor.$$

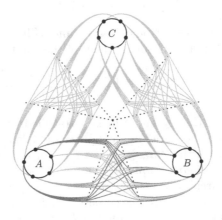

Fig. 1. A tripartite-circle drawing of $K_{n,n,n}$ proving the upper bound.

Connection to the Harary-Hill Conjecture The drawings of K_n presented by Harary and Hill [4] have $H(n)$ crossings and consist of a 2-circle drawing of $K_{n/2,n/2}$ together with all straight line segments joining vertices on the same circle. Moreover, Blažek and Koman [2] presented a 1-circle drawing of K_n with $H(n)$ crossings. Therefore it has been asked whether a 3-circle drawing of $K_{\frac{n}{3},\frac{n}{3},\frac{n}{3}}$ together with all straight line segments joining vertices on the same circle can achieve $H(n)$ crossings. Our result proves that such a drawing does not exist.

References

1. Ábrego, B.M., Fernández-Merchant, S., Sparks, A.: The cylindrical crossing number of the complete bipartite graph (2017, preprint)
2. Blažek, J., Koman, M.: A minimal problem concerning complete plane graphs. Theory Graphs Appl. Czech. Acad. Sci. 113–117 (1964)
3. Guy, R.K.: A combinatorial problem. Nabla (Bull. Malayan Math. Soc.) **7**, 68–72 (1960)
4. Harary, F., Hill, H.: On the number of crossings in a complete graph. Proc. Edinburgh Math. Soc. **13**, 333–338 (1963)
5. Richter, R.B., Thomassen, C.: Relations between crossing numbers of complete and complete bipartite graphs. Am. Math. Mon. **104–2**, 131–137 (1997)
6. Zarankiewicz, K.: On a problem of P. Turan concerning graphs. Fundamenta Mathematicae **41–1**, 137–145 (1955)

On the Edge Density of k-Planar Graphs

Steven Chaplick, Andre Löffler, and Rainer Schmöger[✉]

Lehrstuhl für Informatik I, Universität Würzburg, Würzburg, Germany
rainer.schmoeger@stud-mail.uni-wuerzburg.de,
http://www1.informatik.uni-wuerzburg.de/en/staff

In the 18th century Euler discovered his famous polyhedron formula, which can be used to bound the edge density for planar graphs. Let $G = (V, E)$ be a simple and planar graph with $|V| \geq 3$, then $|E| \leq 3|V| - 6$. Turán co-established Extremal Graph Theory, a branch in which extremal graphs are investigated under the assumption of specified properties. He studied the edge density of graphs which are not necessarily planar but do not contain cliques of fixed size. For several beyond-planar graph classes Turán-type results were discovered:

- k-planar graphs, for which there exists a drawing where no edge is crossed more than k times are studied in [2, 5, 10–12, 15, 16].
- k-quasi-planar graphs with no set of k pairwise crossing edges are investigated in [1, 3–5, 12, 13, 18, 19].
- Fan-planar graphs, where edges can be crossed by one fan, a set of edges sharing one common endpoint [5, 6, 8, 9, 14].

We consider the edge density of (non-) simple k-planar graphs. A *simple* graph does not contain loops or parallel edges. A *non-simple* multigraph has a drawing without homotopic parallel edges and self-loops. Bodendiek et al. first bounded the edge density of 1-planar graphs [10]. Pach and Tóth [16] gave bounds for k-planar graphs with $0 \leq k \leq 4$, namely $|E| \leq (k+3)(|V|-2)$, including Euler's result for planar graphs. Edge density of k-planar graphs strongly relates to the *Crossing Lemma* which provides a lower bound on the crossing number $cr(G)$ for any graph G. They [16] used their bounds to improve it to $cr(G) \geq \frac{1}{33.75} \cdot \frac{|E|^3}{|V|^2}$. Later Pach et al. [15] improved the bound for 3-planar graphs to $|E| \leq 5.5(|V| - 2)$. A charging argument by Ackerman [2] improves the bound for 4-planar graphs to $|E| \leq 6(|V| - 2)$, proving the current best constant $\left(\frac{1}{29}\right)$ for the Crossing Lemma. For $k \geq 5$ only a general bound has been established: considering the number of crossings C, a lower bound by the Crossing Lemma and an upper bound from k-planarity yields (1) and gives $|E| \leq 3.807\sqrt{k}|V|$.

$$\frac{1}{29}\frac{|E|^3}{|V|^2} \leq cr(G) \leq C \leq \frac{|E| \cdot k}{2} \tag{1}$$

Pach et al. [17] recently proved a Crossing Lemma for multigraphs using another constant ($\approx 10^{-7}$), so a similar inequality to (1) leads to the following bound on the edge density of non-simple k-planar graphs. Curiously, this appears to be the first and only upper bound known for arbitrary k.

© Springer Nature Switzerland AG 2018
T. Biedl and A. Kerren (Eds.): GD 2018, LNCS 11282, pp. 636–639, 2018.
https://doi.org/10.1007/978-3-030-04414-5

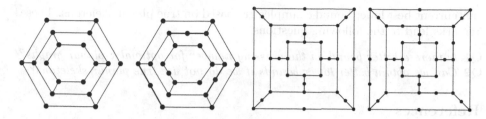

Fig. 1. True planar skeletons for different values of k (left-to-right):
$k = 4$ [15]; $5 \leq k \leq 6$; $7 \leq k \leq 9$; $10 \leq k \leq 12$.

Table 1. Bounds on the number of edges for non-simple and simple graphs; similar bounds can also be obtained for larger values of k.

	Simple	[2, 16]	Non-simple	[Theorem 1]
k	Lower bound	Upper bound	Lower bound	Upper bound
5	$6.00\|V\| - 16.00$	$8.51\|V\|$	$6.20\|V\| - 12.40$	$5000\|V\|$
6	$6.80\|V\| - 23.60$	$9.32\|V\|$	$7.00\|V\| - 14.00$	$5478\|V\|$
7	$7.00\|V\| - 20.00$	$10.07\|V\|$	$7.33\|V\| - 14.67$	$5917\|V\|$
8	$7.33\|V\| - 20.67$	$10.77\|V\|$	$7.67\|V\| - 15.33$	$6325\|V\|$
9	$7.67\|V\| - 21.33$	$11.42\|V\|$	$8.00\|V\| - 16.00$	$6709\|V\|$
10	$7.71\|V\| - 23.43$	$12.04\|V\|$	$8.14\|V\| - 16.29$	$7072\|V\|$
11	$8.00\|V\| - 24.00$	$12.63\|V\|$	$8.43\|V\| - 16.86$	$7417\|V\|$
12	$8.57\|V\| - 25.14$	$13.19\|V\|$	$9.00\|V\| - 18.00$	$7746\|V\|$

Theorem 1. *For $k \geq 1$, a non-simple k-planar graph G has $|E| < 2237\sqrt{k}|V|$.*

We also construct lower bound examples based on the structure of *optimal k-planar graphs* ($k \leq 3$), i.e., k-planar graphs with maximum edge density. Namely, Bekos et al. [7] showed that every optimal non-simple 2-planar (3-planar) graph has a regular *true planar skeleton*: a spanning subgraph consisting of a set of crossing-free edges with only pentagonal (hexagonal) faces. In the original graph, every such face is (almost) a clique, having five (eight) edges inside.

The idea of the true planar skeleton leads us to lower bounds on edge density. Using the patterns in Fig. 1 for $4 \leq k \leq 12$ and adding all possible edges in every face respecting k-planarity produces a family of non-simple k-planar graphs, i.e., establishing lower bounds on the edge density. To use these skeletons for lower bounds on simple k-planar graphs, we have to carefully consider how to avoid multi edges when inserting edges. In particular, by using skeletons such as those shown in Fig. 1, we obtain new bounds for both simple and non-simple k-planar graphs – all discovered bounds can be found in Table 1.

Observe that the general upper bounds on the edge density of k-planar graphs rely on a naive upper bound on the number of crossings; see (1). Additionally, our current best lower bound examples are based on true planar skeletons. These remarks lead to the following questions:

Q1 *Is there a better bound on the crossing number for optimal k-planar graphs?*
Q2 *Can we obtain better lower bounds if we do not use true planar skeletons?*

References

1. Ackerman, E.: On the maximum number of edges in topological graphs with no four pairwise crossing edges. In: Symposium on Computational Geometry, pp. 259–263. ACM (2006)
2. Ackerman, E.: On topological graphs with at most four crossings per edge. CoRR abs/1509.01932 (2015)
3. Ackerman, E., Tardos, G.: On the maximum number of edges in quasi-planar graphs. J. Comb. Theory, Ser. A **114**(3), 563–571 (2007)
4. Agarwal, P.K., Aronov, B., Pach, J., Pollack, R., Sharir, M.: Quasi-planar graphs have a linear number of edges. Combinatorica **17**(1), 1–9 (1997)
5. Angelini, P., Bekos, M.A., Kaufmann, M., Pfister, M., Ueckerdt, T.: Beyond-planarity: Density results for bipartite graphs. CoRR abs/1712.09855 (2017)
6. Bekos, M.A., Cornelsen, S., Grilli, L., Hong, S., Kaufmann, M.: On the recognition of fan-planar and maximal outer-fan-planar graphs. Algorithmica **79**(2), 401–427 (2017)
7. Bekos, M.A., Kaufmann, M., Raftopoulou, C.N.: On optimal 2- and 3-planar graphs. In: Symposium on Computational Geometry, LIPIcs, vol. 77, pp. 16:1–16:16. Schloss Dagstuhl - Leibniz-Zentrum fuer Informatik (2017)
8. Binucci, C., et al.: Algorithms and characterizations for 2-layer fan-planarity: from caterpillar to stegosaurus. J. Graph Algorithms Appl. **21**(1), 81–102 (2017)
9. Binucci, C., et al.: Fan-planarity: properties and complexity. Theor. Comput. Sci. **589**, 76–86 (2015)
10. Bodendiek, R., Schumacher, H., Wagner, K.: Über 1-optimale graphen. Mathematische Nachrichten **117**(1), 323–339 (1984)
11. Brandenburg, F.J., Eppstein, D., Gleißner, A., Goodrich, M.T., Hanauer, K., Reislhuber, J.: On the density of maximal 1-planar graphs. In: Didimo, W., Patrignani, M. (eds.) GD 2012. LNCS, vol. 7704, pp. 327–338. Springer, Heidelberg (2013). https://doi.org/10.1007/978-3-642-36763-2_29
12. Chaplick, S., Kryven, M., Liotta, G., Löffler, A., Wolff, A.: Beyond outerplanarity. In: Frati, F., Ma, K.-L. (eds.) GD 2017. LNCS, vol. 10692, pp. 546–559. Springer, Cham (2018). https://doi.org/10.1007/978-3-319-73915-1_42
13. Fox, J., Pach, J., Suk, A.: The number of edges in k-quasi-planar graphs. SIAM J. Discrete Math. **27**(1), 550–561 (2013)
14. Kaufmann, M., Ueckerdt, T.: The density of fan-planar graphs. CoRR abs/1403.6184 (2014)
15. Pach, J., Radoicic, R., Tardos, G., Tóth, G.: Improving the crossing lemma by finding more crossings in sparse graphs. Discrete Comput. Geom. **36**(4), 527–552 (2006)
16. Pach, J., Tóth, G.: Graphs drawn with few crossings per edge. Combinatorica **17**(3), 427–439 (1997)
17. Pach, J., Tóth, G.: A crossing lemma for multigraphs. In: Symposium on Computational Geometry, LIPIcs, vol. 99, pp. 65:1–65:13. Schloss Dagstuhl - Leibniz-Zentrum fuer Informatik (2018)

18. Suk, A.: *k*-quasi-planar graphs. In: van Kreveld, M., Speckmann, B. (eds.) GD 2011. LNCS, vol. 7034, pp. 266–277. Springer, Heidelberg (2012). https://doi.org/10.1007/978-3-642-25878-7_26
19. Suk, A., Walczak, B.: New bounds on the maximum number of edges in k-quasi-planar graphs. Comput. Geom. **50**, 24–33 (2015)

Confluent* Drawings by Hierarchical Clustering

Jonathan X. Zheng(✉), Samraat Pawar, and Dan F. M. Goodman

Imperial College London, London, UK
jxz12@ic.ac.uk

Recently an edge bundling technique known as confluent* drawing was applied to general graphs by Bach et al. [2] by leveraging power graph decomposition (a form of edge compression that groups similar vertices together, merging edges shared among group members). We explore the technique further by demonstrating the equivalence between confluent drawing and the hierarchical edge bundling of Holten [3], thereby opening the door for existing hierarchical clustering algorithms to be used instead of power graphs to produce confluent drawings for general graphs. We investigate various popular hierarchical clustering methods, and present a qualitative experimental comparison between them. We also introduce a new distance measure for agglomerative clustering that outperforms previous measures, and make recommendations for using the method in practice.

Creating the Routing Graph. The method of bundling we consider consists of two steps: first, we find a suitable auxiliary routing graph; second, we find a layout for this new graph, and then draw the original edges back on top, using invisible routing nodes as spline control points. An example of this can be seen in Fig. 1.

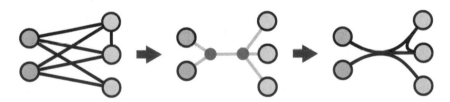

Fig. 1. An example of a simple graph (left), its potential routing graph (middle), and the resultant drawing with edges bundled through routing nodes (right).

This should not be confused with the routing graph used by Pupyrev et al. [4] in their paper on *metro-style* bundling, which generates a routing based on curving around fixed node positions. Our routing is generated using the topology of the graph itself. The benefit of this is that the bundling reflects the actual

---

The original definition of *confluent* by Dickerson et al. [1] forbids edge crossings, while Bach et al. [2] recognize but do not strictly follow this. We continue the use of such terminology here for consistency, but recognize its imprecise usage. A more suitable name for the general edge bundling method of using an auxiliary routing graph should be adopted in the future.

© Springer Nature Switzerland AG 2018
T. Biedl and A. Kerren (Eds.): GD 2018, LNCS 11282, pp. 640–642, 2018.
https://doi.org/10.1007/978-3-030-04414-5

structure of the data, rather than a potentially arbitrary spatial positioning. On the other hand, the popular method of Holten [3] generates its routing using a hierarchy already included in the data, which is easily converted into a graph through a tree representation, where original vertices are the leaves and groups within the hierarchy are branch nodes.

The primary purpose of this poster is to show that the work on confluent drawings by Bach et al. [2] is also based on a routing graph, and therefore should be classified with these previous two techniques, rather than within the realm of confluent drawing. As such, the routing graphs generated here may also be used to produce metro-style bundles [4].

As noted by Bach et al. [2] in their original paper, we also find that power graph decomposition performs poorly on graphs where clusters or cliques are a common motif, producing fractal-like artifacts (see poster for example). To alleviate this issue, we investigate the use of hierarchical clustering to generate the routing instead. This requires the definition of a dissimilarity measure between pairs of vertices, a popular choice being the Jaccard distance measure

$$d_{ij} = 1 - \frac{|N(i) \cap N(j)|}{|N(i) \cup N(j)|} \tag{1}$$

where $N(i)$ is the set of neighbours of vertex i. However, simply using Jaccard distance only captures the dissimilarity of vertices with shared neighbours, and any pair of vertices more than two hops away is automatically given a distance $d_{ij} = 1$. We introduce a method of capturing such longer range dissimilarities, by simply multiplying d_{ij} by the shortest path between them. A visual comparison between the two can be seen in the poster, along with a further example using a popular divisive clustering method.

Drawing the Bundled Graph. The result of a hierarchical clustering algorithm is a dendrogram (a rooted tree used to describe hierarchical relationships) which needs to be converted to a routing graph. One could simply assign unit edge lengths to the branches, but the output of agglomerative methods also includes a merging cost between clusters. We encode this using varying edge lengths, and therefore require a force-directed method that explicitly includes this. In our case we use a multidimensional scaling approach i.e. the popular Kamada-Kawai layout. This was recently improved by Zheng et al. [5] by making use of stochastic gradient descent, and can also be used to easily produce optimal radial layouts (see poster for example).

The original edges are then drawn of top of this layout using b-splines controlled along the shortest path through the routing graph. To improve the drawing aesthetically, we also reduce the bundling strength as in Holten [3], while full strength bundling can be used to reproduce the confluent effect utilised by Bach et al. [2].

References

1. Dickerson, M., Eppstein, D., Goodrich, M.T., Meng, J.Y.: Confluent drawings: visualizing non-planar diagrams in a planar way. J. Graph Algorithms Appl. **9**(1), 31–52 (2005). https://doi.org/10.7155/jgaa.00099
2. Bach, B., Riche, N.H., Hurter, C., Marriott, K., Dwyer, T.: Towards unambiguous edge bundling: investigating confluent drawings for network visualization. IEEE Trans. Vis. Comput. Graph. **23**(1), 541–550 (2017). https://doi.org/10.1109/TVCG.2016.2598958
3. Holten, D.: Hierarchical edge bundles: visualization of adjacency relations in hierarchical data. IEEE Trans. Vis. Comput. Graph. **12**(5), 741–748 (2006). https://doi.org/10.1109/TVCG.2006.147
4. Pupyrev, S., Nachmanson, L., Bereg, S., Holroyd, A.E.: Edge routing with ordered bundles. Comput. Geom. Theory Appl. **52**, 18–33 (2016). https://doi.org/10.1016/j.comgeo.2015.10.005
5. Zheng, X.J., Pawar, S., Goodman, D.F.M.: Graph drawing by stochastic gradient descent. IEEE Trans. Vis. Comput. Graph. (2018, to appear). https://doi.org/10.1109/TVCG.2018.2859997

Examining Weak Line Covers with Two Lines in the Plane

Oksana Firman, Fabian Lipp[ID], Laura Straube[✉], and Alexander Wolff[ID]

Lehrstuhl für Informatik I, Universität Würzburg, Würzburg, Germany
{oksana.firman,fabian.lipp}@uni-wuerzburg.de, laura.straube@gmx.net

Chaplick et al. [2] defined the *l-dimensional affine line cover number* $\rho_d^l(G)$, for $1 \leq l < d$ and an arbitrary graph G, as the minimum number of l-dimensional planes in $\mathbb{R}^d$ such that G admits a crossing-free straight-line drawing whose vertices and edges are contained in the union of these planes. The l-dimensional *weak* affine line cover number $\pi_d^l(G)$ also counts such planes but insists only that the vertices are covered by their union. In particular, the weak line cover number $\pi_2^1(G)$ is the minimum number of lines in the plane that are necessary to cover the vertices of a planar graph G.

Firman et al. [3] asked whether π_2^1 has a sublinear upper bound for the class of planar graphs. In the following we restrict their open problem further and make some progress in characterizing the class of graphs that can be drawn on two lines in the plane (further referred to as *drawable*). In order to verify conjectures (such as Conjecture 1 below), we needed a drawability test. Given that drawability is NP-hard to decide [1], we contented ourselves with exponential-time approaches.

First, we formulated drawability as an integer linear program (ILP). The solution of the ILP yields a drawing on two lines. Without loss of generality, we consider the case that the two lines are perpendicular and view their intersection point as the origin of a Euclidean coordinate system with four quadrants each incident to two half-axes. There are Boolean variables for every combination of a vertex and a half-axis describing whether they are incident. Other variables represent the order of the vertices on a given half-axis. The constraints ensure that every vertex is mapped to exactly one half-axis, that the ordering on each half-axis is transitive, and that the resulting drawing is planar.

Second, we transformed our ILP formulation into a Boolean formula in CNF that can be tested by a SAT solver. The ILP formulation uses only binary variables and was therefore easy to transform. Our hope was that the SAT formulation could be evaluated more efficiently than the ILP formulation. On our test suite with 824 solvable and 304 unsolvable graphs, the SAT solver MiniSat (version 2.2.0) was indeed always faster than the ILP solver IBM ILOG CPLEX Optimization Studio (12.8.0.0). Both in terms of total computation time and only solving time, the difference in speed was an order of magnitude.

Our experiments suggest the following.

Conjecture 1. Every planar graph with maximum degree 3 is drawable.

However, David Eppstein found a 3-regular counterexample with 26 vertices; see Fig. 3. We verified it using our SAT formulation.

© Springer Nature Switzerland AG 2018
T. Biedl and A. Kerren (Eds.): GD 2018, LNCS 11282, pp. 643–645, 2018.
https://doi.org/10.1007/978-3-030-04414-5

We can show that any graph that contains nested triangles in each of its planar embeddings is not drawable. For graphs with maximum degree 3, this case does not arise as they can always be embedded such that there are no nested triangles due to their low connectivity, see Fig. 1. On the other hand, triangulations are not drawable – apart from the tetrahedron and graphs that extend the tetrahedron in a specific way; see Fig. 2. The graph with the dashed edges in Fig. 1 shows that not all 2-outerplanar and not all graphs of maximum degree 4 are drawable. We also managed to extend our nested-triangle condition to nested cycles; this yielded quadrangulations (which are obviously triangle-free) that are not drawable.

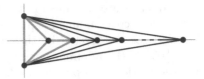

Fig. 1. The triangular prism (solid edges, left) is drawable (right), but its 4-regular supergraph with the dashed edges is non-drawable.

Fig. 2. The tetrahedron (in gray) and a family of drawable triangulations based on it.

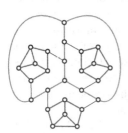
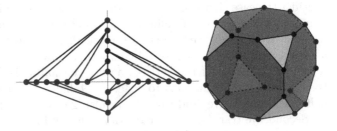

Fig. 3. David Eppstein's 3-regular graph that is not drawable on two lines.

Fig. 4. The truncated hexahedron (a 3-regular graph with 24 vertices) drawn on two lines not using their intersection – and its representation as an Archimedean solid.

Our experiments also showed that all tested graphs of maximum degree 3 except the tetrahedron are not only drawable, but can be drawn on two lines such that no edge contains the intersection of the two lines; see, for example, Fig. 4. This could help in finding a strategy for distributing vertices on two lines. Under this additional condition no triangulation we tested was drawable, including any of those in Fig. 2.

References

1. Biedl, T., et al.: Line and plane cover numbers revisited (2018). Unpublished manuscript
2. Chaplick, S., Fleszar, K., Lipp, F., Ravsky, A., Verbitsky, O., Wolff, A.: Drawing graphs on few lines and few planes. In: Hu, Y., Nöllenburg, M. (eds.) GD 2016. LNCS, vol. 9801, pp. 166–180. Springer, Cham (2016). https://doi.org/10.1007/978-3-319-50106-2_14
3. Firman, O., Ravsky, A., Wolff, A.: On the weak line cover numbers. In: Korman, M., Mulzer, W. (eds.) Proceedings of 34th European Workshop on Computational Geometry (EuroCG 2018), pp. 63:1–63:5 (2018). https://conference.imp.fu-berlin.de/eurocg18/download/paper_63.pdf

1-Gap Planarity of Complete Bipartite Graphs

Christian Bachmaier, Ignaz Rutter, and Peter Stumpf[(✉)]

Faculty of Computer Science and Mathematics,
University of Passau, Passau, Germany
{bachmaier,rutter,stumpf}@fim.uni-passau.de

Abstract. A graph is 1-gap planar if it admits a drawing such that each
crossing can be assigned to one of the two involved edges in such a way
that each edge is assigned at most one crossing. We show that $K_{3,14}$,
$K_{4,10}$ and $K_{6,6}$ are not 1-gap planar.

1   Introduction

A graph is 1-gap planar if it admits a drawing such that each crossing can
be assigned to one of the two involved edges in such a way that each edge is
assigned at most one crossing. The motivation comes from edge casings, where
one creates a small gap in one of the edges involved in each crossing to increase
the readability. In a 1-gap planar drawing each edge receives at most one such
gap. This notion was introduced in GD'17 by Bae et al. [1]. Among others they
showed that a 1-gap planar graph on n vertices has at most $5n - 10$ edges and
this is tight. They further show that the complete graph K_n is 1-gap planar if
and only if $n \leq 8$. An important observation of Bae et al. is that every 1-gap
planar graph G satisfies $\mathrm{cr}(G) \leq |E|$ (since each crossing is assigned to one of the
edges). For complete bipartite graphs, they gave 1-gap planar drawings for $K_{3,12}$,
$K_{4,8}$ and $K_{5,6}$, whereas they exclude $K_{3,15}$, $K_{4,11}$ and $K_{6,7}$ by observing that
their crossing number is strictly greater than their edge number. They leave the
remaining complete bipartite graphs as an open problem. We show the following
theorem.

Theorem 1. *The graphs $K_{3,14}$, $K_{4,10}$ and $K_{6,6}$ are not 1-gap planar.*

This shrinks the open cases to $K_{3,13}$ and $K_{4,9}$. We note that for all the graphs
we exclude, the crossing number equals the edge number [2]. Thus, we know that
in a 1-gap planar drawing of such a graph each edge has at least one crossing.

2   Proof Strategy

Our proof strategy is an extension of the one of Bae et al., who encountered a
similar situation when treating the case of K_9, which has 36 edges and whose
crossing number is 36. For convenience, we briefly sketch their argument. Assume
for the sake of contradiction that Γ is a 1-gap planar drawing of K_9, and consider

© Springer Nature Switzerland AG 2018
T. Biedl and A. Kerren (Eds.): GD 2018, LNCS 11282, pp. 646–648, 2018.
https://doi.org/10.1007/978-3-030-04414-5

the planarization Γ^* of this drawing, where all crossings are replaced by dummy vertices. Observe that Γ has precisely $\mathrm{cr}(K_9) = |E(K_9)| = 36$ crossings [1]. If two vertices of K_9 share a face in Γ^*, we can reroute the edge between them without crossings in this face, thus obtaining a drawing with fewer crossings, which is not possible. Thus, for any two original vertices their incident faces of Γ^* are disjoint. This gives a lower bound of 72 faces. On the other hand, from Euler's formula it follows that Γ^* has only 65 faces; a contradiction.

In contrast, for complete bipartite graphs, vertices may share a face of the planarization if they are independent. Let $G = (R \dot\cup B, E)$ be a complete bipartite graph with $\mathrm{cr}(G) = |E| = |R| \cdot |B|$. The vertices in R and B are *red* and *blue*, respectively. As before, we consider a hypothetical 1-gap planar drawing Γ of G, for which we know that it has $\mathrm{cr}(\Gamma) = \mathrm{cr}(G) = |E|$ crossings, and we denote the planarization by Γ^*. Let F denote the set of faces of Γ^* and let $F_R, F_B \subseteq F$ be the faces that are incident to a red and a blue vertex, respectively. If $F_R \cap F_B \neq \emptyset$, then there is a face in F that is incident to both a red and a blue vertex. We can route the edge between them without crossings and thus reach a contradiction as in the case of K_9. By assumption, Γ^* has $|R| + |B| + |E|$ vertices and $|R| \cdot |B| + 2 \cdot |E|$ edges, and hence $|F| = 2 \cdot |R| \cdot |B| - |R| - |B| + 2$ faces.

Consider the auxiliary bipartite graph $G_R = (R \cup F_R, E_R)$ where a face and a vertex are adjacent if and only if they are incident in Γ^*. The graph $G_B = (B \cup F_B, E_B)$ is defined analogously. Observe that $|E_R| = |E_B| = |R| \cdot |B|$ since each vertex in R has degree $|B|$ and vice versa. We argue that either G_R and G_B are both trees, or one of them, say G_R, is a cycle decorated with leaves in F_R and the other one, G_B, is a forest with two connected components.

In the former case, we obtain $|E_R| = |R| + |F_R| - 1$, which gives $|F_R| = |R| \cdot |B| - |R| + 1$ and likewise $|F_B| = |R| \cdot |B| - |B| + 1$. Hence $|F_B| + |F_R| = 2 \cdot |B| \cdot |R| - |R| - |B| + 2 = |F|$. In the latter case, the number of faces in F_R decreases by 1, but the number of faces in F_B increases by 1. In all cases we find that $|F_R| + |F_B| = |F|$, i.e., each face of Γ^* is either in F_R or in F_B. A contradiction is reached by showing that there exists at least one *white face* of Γ^* that is not incident to any red or blue vertex.

First it follows from the fact that each edge has a gap that there is a cycle C in Γ^* that only contains dummy vertices. This can be seen as follows. We start in any dummy vertex and follow the edge that does not have its gap there to its own gap. Repeating this step eventually produces the desired cycle C. If all red and blue vertices lie inside (outside) C, then C contains a white face in its exterior (interior). Otherwise it separates a component of G_R from a component of G_B. Further analysis yields a contradiction. The details vary depending on whether G is $K_{3,14}$, $K_{4,10}$ or $K_{6,6}$ as well as on the size and structure of the components that are separated by C.

3   Conclusion

We have shown that $K_{3,14}$, $K_{4,10}$ and $K_{6,6}$ are not 1-gap planar. We leave open the cases of $K_{3,13}$ and $K_{4,9}$. It seems difficult to adapt our proof technique to these cases since their crossing numbers are strictly smaller than their edge number, which results in additional freedom for possible 1-gap planar drawings.

References

1. Bae, S.W., et al.: Gap-planar graphs. Theor. Comput. Sci. (2018). https://doi.org/10.1016/j.tcs.2018.05.029
2. Kleitman, D.J.: The crossing number of $K_{5,n}$. J. Comb. Theory **9**(4), 315–323 (1970). https://doi.org/10.1016/S0021-9800(70)80087-4

Extending Drawings of K_n into Arrangements of Pseudocircles

Alan Arroyo[1,3]([⊠]), R. Bruce Richter[1], and Matthew Sunohara[2]

[1] Department of Combinatorics and Optimization,
University of Waterloo, Waterloo, Canada
{amarroyo,brichter}@uwaterloo.ca
[2] Department of Mathematics, University of Toronto, Toronto, Canada
matthew.sunohara@gmail.com
[3] IST Austria, Klosterneuburg, Austria

The Harary-Hill Conjecture states that the crossing number of the complete graph K_n is equal to:

$$H(n) = \frac{1}{4} \left\lfloor \frac{n}{2} \right\rfloor \left\lfloor \frac{n-1}{2} \right\rfloor \left\lfloor \frac{n-2}{2} \right\rfloor \left\lfloor \frac{n-3}{2} \right\rfloor.$$

In general, if $\mathcal{S}$ is the unit sphere in $\mathbb{R}^3$, then a *spherical drawing* of a graph G is one in which the vertices of G are represented as distinct points in $\mathcal{S}$, and every edge is a shortest-arc connecting its corresponding ends. Although the Harary-Hill Conjecture is known to be true for certain classes of drawings of K_n, it is yet unknown that spherical drawings have at least $H(n)$ crossings.

In the proofs of [1,3] showing that rectilinear drawings of K_n have at least $H(n)$ crossings, a crucial point was to relate the number of crossings in a given drawing to the separation properties of the $\binom{n}{2}$ lines extending the edges. Understanding these separation properties, but for the curves extending the edges in spherical drawings, serve as our motivation for studying arrangement of pseudocircles extending the edges of a drawing.

An *arrangement of pseudocircles* is a set of simple closed curves in the sphere in which every two curves intersect at most twice, and every intersection is a crossing. If γ is a simple closed curve, then a *side* of γ is one of the two disks in $\mathcal{S}$ bounded by γ. In spherical drawings, the great circles extending the edge-arcs form an arrangement of pseudocircles.

With the aim of finding a combinatorial extension of spherical drawings analogous to how pseudolinear drawings extend rectilinear drawings, there have been two significant questions under active consideration:

(Q1) Do the edges of every good drawing of K_n in the sphere extend to an arrangement of pseudocircles?

(Q2) If the edges of a drawing of K_n extend to an arrangement of pseudocircles, is there an extending arrangement in which any two pseudocircles intersect exactly twice?

© Springer Nature Switzerland AG 2018
T. Biedl and A. Kerren (Eds.): GD 2018, LNCS 11282, pp. 649–651, 2018.
https://doi.org/10.1007/978-3-030-04414-5

These questions were indeed considered by a working group at the 2015 Crossing Number Workshop in Rio de Janeiro.

In this work, we answer these questions by showing that (1) there is a drawing of K_{10} (Fig. 1c) in which there is no extension of its edges into an arrangement of pseudocircles; and (2) there is a drawing of K_9 in which there is an extension of its edges into an arrangement of pseudocircles, but no such extension exists for which any two curves cross exactly twice.

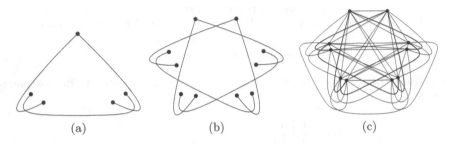

Fig. 1. Construction of a K_{10} with whose edges cannot be extended to an arrangement of pseudocircles.

To construct the examples we consider the *basic gadget* in Fig. 1a. This has the property that if we extend its three edges into an arrangement of pseudocircles, then either the pseudocircle extending the edge with two degree 1 vertices is drawn in the bounded face of the drawing or the two pseudocircles extending the other two edges are drawn in the bounded face. Overlapping two basic gadgets as in Fig. 1b yields a drawing not extendible to an arrangement of pseudocircles; Fig. 1b can be enlarged to the non-extendible drawing of K_{10} in Fig. 1c. A similar construction, but using two disjoint copies of the basic gadget, yields the drawing of K_9 answering (Q2) in the negative form.

Among the five non-isomorphic drawings of K_5 in the sphere, there are two that are non-rectilinear. The class of *convex drawings* of K_n is obtained by forbidding the two non-rectilinear K_5s. Convex drawings (or locally rectilinar drawings) were introduced in [2] in the context of the Harary-Hill Conjecture, where it is shown that there is a possibility that every optimal drawing of K_n is convex.

In this work we show that every h-convex drawing, a special kind of convex drawing, can be extended into an arrangement of pseudocircles. Furthermore, the extension satisfies that if two vertices x and y are on the same side of a curve extending an edge, then the edge xy is drawn on that side of the curve. Moreover, we prove that any drawing of K_n with a pseudocircular extension of such kind is h-convex.

Related to (Q2), we also show that h-convex drawings have a "better" extension in which pseudocircles are pairwise intersecting. This, and the fact that h-convex drawings can be decomposed into two pseudolinear drawings, suggest that h-convex is, possibly, the right definition for pseudospherical drawings of K_n.

References

1. Ábrego, B.M., Fernández-Merchant, S.: A lower bound for the rectilinear crossing number. Graphs Comb. **21**, 293–300 (2005)
2. Arroyo, A., McQuillan, D., Richter, R.B., Salazar, G.: Convex drawings of K_n: topology meets geometry (submitted)
3. Lovász, L., Vesztergombi, K., Wagner, U., Welzl, E.: Convex quadrilaterals and k-sets. Contemp. Math. **342**, 139–148 (2004). Towards a theory of geometric graphs. Amer. Math. Soc., Providence, RI

Topology Based Scalable Graph Kernels

Kin Sum Liu[1($\boxtimes$)], Chien-Chun Ni[2], Yu-Yao Lin[3], and Jie Gao[1]

[1] Stony Brook University, Stony Brook, NY 11794, USA
{kiliu,jgao}@cs.stonybrook.edu
[2] Yahoo Research, Sunnyvale, CA 94089, USA
cni02@oath.com
[3] Intel Inc., Hillsboro, OR 97124, USA
yu-yao.lin@intel.com

1 Introduction

We propose a new graph kernel for graph classification and comparison using Ollivier Ricci curvature. The Ricci curvature of an edge in a graph describes the connectivity in the local neighborhood. An edge in a densely connected neighborhood has positive curvature and an edge serving as a local bridge has negative curvature. We use the edge curvature distribution to form a graph kernel which is then used to compare and cluster graphs. The curvature kernel uses purely the graph topology and thereby works for settings when node attributes are not available. The computation of the curvature for an edge uses only information within two hops from the edge and a random sample of $O(1/\varepsilon^2 \log 1/\varepsilon + 1/\varepsilon^2 \log 1/\delta)$ edges in a large graph can produce a good approximation to the curvature distribution with error bounded by ε with probability at least $1 - \delta$. Thus, one can compute the graph kernel for really large graphs that some other graph kernels cannot handle. This Ricci curvature kernel is extensively tested on graphs generated by different generative models as well as standard benchmark datasets from bioinformatics and Internet AS network topologies.

Graph classification and comparison are widely applied in bioinformatics, vision and social network analysis. One of the most popular approaches in practice is using graph kernels which compute the similarity of two graphs in terms of subgraph structures. Many graph kernels have been developed, which differ by the subgraph structures they focus on, such as random walks [2], shortest paths [1], subtrees [5], and cycles [3]. Graph kernels have been extensively tested on benchmark datasets from bioinformatics to chemistry [1, 4].

In our work, we focus on the setting of unlabeled graphs and propose a new graph kernel based on discrete Ricci curvature which takes only the network topology as an input. Our work is motivated by the use of curvature related kernels in shape matching. Curvature on a smooth surface defines the amount by which a geometric object deviates from being flat or straight. It is a local measure at each point but nevertheless has deep connections to global topology and structures. Despite the success in shape matching, curvature has not been used much for comparing graphs. In this paper, we propose to use curvature distribution of graphs to build new graph kernels. The goal is to demonstrate

© Springer Nature Switzerland AG 2018
T. Biedl and A. Kerren (Eds.): GD 2018, LNCS 11282, pp. 652–654, 2018.
https://doi.org/10.1007/978-3-030-04414-5

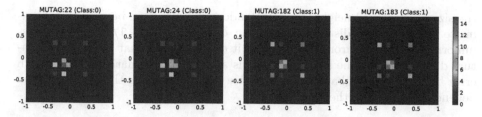

Fig. 1. The 2D Ricci Curvature histogram of MUTAG graphs which describes the curvature distribution for pairs of neighboring edges. A suitable choice of mass distribution on the neighborhood results in curvature always ranging between $[-1, 1]$. Graphs in the same class tend to have similar histogram distributions. Here MUTAG:22 and MUTAG:24 belong to the same class, while MUTAG:182 and MUTAG:183 belong to another class.

that the Ricci curvature distribution and kernels can be efficiently computed and capture interesting graph properties. They add to the family of graph features and kernels, could be combined with other attributes, and used for other classifiers.

2 Discrete Ricci Curvature

For an edge $\overline{uv}$, define a distribution m_u, m_v on the neighborhood of u, v respectively (such as uniform m_u and m_v). Now compute the Earth Mover Distance $W(u, v)$ from m_u to m_v, where the cost of moving mass from a neighbor u_i of u to a neighbor v_j of v is the shortest path distance in the graph. Here the edges are unweighted unless they inherit weights arising from the application domain, such as tie strength or distances. For example, $W(u, v)$ will be upper bounded by 2 for unweighted graph if we allocate 50% mass to the node's neighbor. The Ollivier-Ricci curvature is defined as $w(\overline{uv}) = 1 - W(u, v)/d(u, v)$, where $d(u, v)$ is the length of edge $\overline{uv}$. Intuitively the curvature captures the structural properties of the local neighborhood. If $\overline{uv}$ stays in a well connected, dense neighborhood, the curvature is positive; if $\overline{uv}$ is locally a bridge, its curvature is negative.

3 Ricci Curvature Graph Kernel

We define the Ricci Curvature Kernel as the following. Denote the curvature distribution of all edges in G by $D(G)$ and that of G' by $D(G')$. We use the standard Gaussian RBF kernel: $k(G, G') = \exp(-\|D(G) - D(G')\|_2^2/2\sigma^2)$, where $\|D(G) - D(G')\|_2$ is the ℓ_2 norm of two vectors $D(G)$, $D(G')$. Since the kernel depends on the curvature distribution, the distribution is less robust statistically for small graphs. We could boost up the kernel by considering the curvature distribution for pairs of neighboring edges $\{(w(e), w(e'))\}$. It appears to be more effective in practice. See Fig. 1 for an example. When the graph is really large, computing the curvature distribution might be costly ($O(|G.E| * n^3)$ where n is the size of concerned neighborhood). Random sampling can be used to approximate the curvature distribution. Taking $O(1/\varepsilon^2 \log 1/\varepsilon + 1/\varepsilon^2 \log 1/\delta)$ edges uniformly at random from the graph G, it can be shown that $\hat{D}(G)$, the curvature distribution on the sampled edges, is a good approximation of $D(G)$ with error bound ε with probability $1 - \delta$. Notice that the running time does not even depend on the size of the graph n.

References

1. Borgwardt, K.M., Kriegel, H.P.: Shortest-path kernels on graphs. In: Proceedings of the Fifth IEEE International Conference on Data Mining (ICDM 2005), pp. 74–81. IEEE Computer Society, Washington, DC (2005). http://dx.doi.org/10.1109/ICDM.2005.132
2. Gärtner, T., Flach, P., Wrobel, S.: On graph kernels: hardness results and efficient alternatives. In: Schölkopf, B., Warmuth, M.K. (eds.) COLT-Kernel 2003. LNCS (LNAI), vol. 2777, pp. 129–143. Springer, Heidelberg (2003). https://doi.org/10.1007/978-3-540-45167-9_11
3. Horváth, T., Gärtner, T., Wrobel, S.: Cyclic pattern kernels for predictive graph mining. In: Kim, W., Kohavi, R., Gehrke, J., DuMouchel, W. (eds.) Proceedings of the 10th ACM SIGKDD International Conference on Knowledge Discovery and Data Mining (KDD 2004), 22–25 August 2004, Seattle, WA, USA, pp. 158–167. ACM Press, New York (2004). http://doi.acm.org/10.1145/1014052.1014072
4. Ralaivola, L., Swamidass, S.J., Saigo, H., Baldi, P.: Graph kernels for chemical informatics. Neural Netw. **18**(8), 1093–1110 (2005). http://dx.doi.org/10.1016/j.neunet.2005.07.009
5. Ramon, J., Gärtner, T.: Expressivity versus efficiency of graph kernels. In: Raedt, L.D., Washio, T. (eds.) Proceedings of the First International Workshop on Mining Graphs, Trees and Sequences (MGTS 2003) at the 14th European Conference on Machine Learning and 7th European Conference on Principles and Practice of Knowledge Discovery in Databases (ECML/PKDD 2003), 22 and 23 September 2003, Cavtat-Dubrovnik, Croatia. pp. 65–74 (2003)

New Spectral Sparsification Approach for Drawing Large Graphs

Jingming Hu[(✉)] and Seok-Hee Hong

The School of Information Technologies,
The University of Sydney, Camperdown, Australia
jihu2855@uni.sydney.edu.au, seokhee.hong@sydney.edu.au

1 Introduction

Nowadays many big complex networks are abundant in various application domains, such as the internet, finance, social networks, and systems biology. Examples include web graphs, AS graphs, Facebook networks, Twitter networks, protein-protein interaction networks and biochemical pathways. However, computing good visualization of big complex networks is extremely challenging due to scalability and complexity.

Recent work for visualizing big graphs uses a *proxy graph* approach [3]: the original graph is replaced by a proxy graph, which is much smaller than the original graph. The challenge for the proxy graph approach is to ensure that the proxy graph is a *good representation* of the original graph.

Eades *et al.* [2] presented proxy graphs using the spectral sparsification approach. *Spectral sparsification* is a technique to reduce the number of edges in a graph, while retaining its structural properties, introduced by Spielman *et al.* [5].

More specifically, they present a method for computing proxy graphs, called DSS (Deterministic Spectral Sampling), by selecting *edges* with high *resistance values* [5]. Their experimental results confirmed the promises by Spielman *et al.*: i.e., the spectral sparsification based methods are more effective than *Random Edge sampling* based method.

It was left as an open problem to compare the spectral sparsification based proxy graph approach with other graph sampling based proxy graph methods.

2 Our Results

In this poster, we introduce a new method called Spectral Sparsification Vertex (SSV-I) for computing proxy graphs using the spectral sparsification approach. Roughly speaking, we define resistance values for *vertices*, using the sum of resistance values of incident edges then we select vertices with high resistance values.

Suppose that $G = (V, E)$ is a graph with a vertex set V ($n = |V|$) and an edge set E ($m = |E|$). Let $r(v)$ represents a resistance value of a vertex v, and $r(e)$ represents a resistance value of an edge e. Let $deg(v)$ represents a degree of

© Springer Nature Switzerland AG 2018
T. Biedl and A. Kerren (Eds.): GD 2018, LNCS 11282, pp. 655–657, 2018.
https://doi.org/10.1007/978-3-030-04414-5

a vertex v (i.e., the number edges incident to v), and let E_v represents a set of edges incident to a vertex v.

More specifically, we define *resistance value* for each vertex v as below:

$$r(v) = \sum_{e \in E_v} r(e)$$

We now describe a new method called SSV-I (Spectral Sparsification Vertex) for computing spectral sparsification based proxy graph $G' = (V', E')$ of $G = (V, E)$. Let V' consist of the n' of largest effective resistance. Then G' be the subgraph of G induced by V'.

Our experimental results with both benchmark real-world graphs and synthetic graphs using graph sampling quality metrics, visual comparison with various graph layouts and proxy graph quality metrics [3] show significant improvement by the SSV-I method over the Random Vertex (RV) sampling method.

Our main contribution and findings can be summarised as follows:

1. We introduced a new method called Spectral Sparsification Vertex (SSV-I) for computing proxy graphs using the spectral sparsification approach.
2. Experimental results with sampling metrics confirm that the SSV-I shows significant improvement over RV method. To be precise, around 35% improvement SSV-I over RV method on average in most metrics.
3. We observed that the Backbone layout [4] shows better structure for Benchmark graphs (i.e., real-world data), esp. scale-free graphs, and the Organic layout [1] produces better shape for Black-hole graphs (i.e., synthetic graphs).
4. Visual comparison of proxy graphs computed by SSV-I and RV using Benchmark, GION, and Black-hole data sets using the Backbone and Organic layouts confirms that our new SSV-I method produces proxy graphs with better connectivity structure with similar visual structure to the original graph than RV.
5. Experimental results confirm our hypothesis that the SSV-I method performs better than RV in proxy quality metrics, esp., when the relative density is low.
6. We observed that the Backbone layout performs better than the Organic layout in terms of the improvement in proxy quality metrics computed by SSV-I over RV method.

References

1. yEd - Java Graph Editor. https://www.yworks.com/products/yed
2. Eades, P., Nguyen, Q., Hong, S.-H.: Drawing big graphs using spectral sparsification. In: Frati, F., Ma, K.-L. (eds.) GD 2017. LNCS, vol. 10692, pp. 272–286. Springer, Cham (2018). https://doi.org/10.1007/978-3-319-73915-1_22
3. Nguyen, Q.H., Hong, S.H., Eades, P., Meidiana, A.: Proxy graph: visual quality metrics of big graph sampling. IEEE Trans. Vis. Comput. Graph. 23(6), 1600–1611 (2017)

4. Nocaj, A., Ortmann, M., Brandes, U.: Untangling the hairballs of multi-centered, small-world online social media networks. J. Graph Algorithms Appl. **19**(2), 595–618 (2015)
5. Spielman, D.A., Teng, S.H.: Spectral sparsification of graphs. SIAM J. Comput. **40**(4), 981–1025 (2011)

SPQR Proxy Graphs for Visualization of Large Graphs

Seok-Hee Hong$^{(\boxtimes)}$ and Quan Nguyen

The School of Information Technologies,
University of Sydney, Camperdown, Australia
quan.nguyen@sydney.edu.au

1 Introduction

Recent work for visualizing large graphs uses a *proxy graph* method [3]: the original graph is replaced by a proxy graph, which is much smaller than the original graph. The challenge for the proxy graph approach is to ensure that the proxy graph is a *good representation* of the original graph. However, previous work to compute proxy graphs using the *random sampling* methods often fail to preserve the important global skeletal structure and connectivity of the original graph [4, 5].

For example, Zhang *et al.* presented experimental comparison of different sampling algorithms under various sampling metrics [5]. Wu *et al.* presented user studies to investigate how sampling methods influence graph visualization, in terms of human perception of high degree vertices, clusters and coverage area; it was recommended to use Random Walk (RW) for high degree vertex, Random Jump for clustering, but to avoid Random Vertex (RV) sampling [4]. In particular, Random Vertex and Random Edge sampling often produce a set of disconnected proxy graphs [4].

The *BC (Block Cut-vertex) proxy graph* methods, based on the BC tree decomposition of a connected graph into biconnected components, produced better results than the random sampling based methods. However, the bottleneck was when graphs have giant biconnected components [2]. In particular, the performance gain for real-world graphs was smaller, due to the existence of single dominant component in real-world graphs.

Therefore, it was left as an open problem and future work is to conduct further experiments, by combining with other graph partitioning methods [2].

2 The SPQR Tree

The *SPQR tree* of a biconnected undirected graph G represents the decomposition of G into triconnected components [1], which can be computed in linear time.

We use basic terminology of SPQR trees; for details, see [1]. Each triconnected component consists of *real edges* (i.e., edges in the original graph) and *virtual*

© Springer Nature Switzerland AG 2018
T. Biedl and A. Kerren (Eds.): GD 2018, LNCS 11282, pp. 658–660, 2018.
https://doi.org/10.1007/978-3-030-04414-5

edges. (i.e., edges introduced during the decomposition process, which represents the other triconnected components, sharing the same virtual edges defined by cut-pairs).

Each node ν in the SPQR tree is associated with a graph called the *skeleton* of ν, denoted by $\sigma(\nu)$, which corresponds to a triconnected component. There are four types of nodes ν in the SPQR tree: (i) S-node, where $\sigma(\nu)$ is a simple cycle with at least three vertices; (ii) P-node, where $\sigma(\nu)$ consists of two vertices connected by at least three edges; (iii) Q-node, where $\sigma(\nu)$ consists of two vertices connected by two (real and virtual) edges; and (iv) R-node, where $\sigma(\nu)$ is a simple triconnected graph with at least four vertices.

In this poster, we use the SPR tree, a simplified version of the SPQR tree *without* Q-nodes, since the Q-node consists of two vertices and edges.

3   Our Results

This poster introduces new *SPQR proxy graph* methods, integrating graph sampling methods with the *SPQR tree* [1] to maintain the important global connectivity structure of the original graph.

We present two new families of proxy graph methods SPQR-W and SPQR-E, each contains the five most popular sampling methods, including RV (Random Vertex), RE (Random Edge), IRE (Induced Random Edge), RP (Random Path) and RW (Random Walk), used in previous work [2–5].

More specifically, we first include the *separation pairs* of the original graph to proxy graphs, since separation pairs are structurally important vertices in terms of connectivity. Then, SPQR-W proxy graph methods perform sampling using the original sampling algorithms.

SPQR-E algorithm is a Divide and Conquer algorithm that uses SPQR-W algorithms: it first selects separation pairs, and then performs SPQR-W algorithms for each triconnected component $\nu_i, i = 1, \ldots, k$ of G to compute a proxy graph G_i' of $\sigma(\nu_i)$, the *skeleton* of ν. Finally, it merges $G_i', i = 1, \ldots, k$, into the final proxy graph G' of G.

Note that the skeleton $\sigma(\nu)$ consists of virtual edges and real edges. Since such virtual edges do not exist in the original graph, we only sample real edges of $\sigma(\nu)$.

The main contribution of this poster is summarized as follows:

1. We present two new families of proxy graph methods SPQR-W (SPQR-Whole) and SPQR-E (SPQR-Each). Each family consists of five new methods, integrating the SPQR tree decomposition with the most popular five sampling methods, used in the sampling-based proxy graph method [2, 3].
2. Experimental results using graph sampling quality metrics, proxy quality metrics [3] and visual comparison with real world graphs show that our new SPQR proxy graph methods produce significantly better results than the previous methods [2, 3].

References

1. Di Battista, G., Tamassia, R.: On-line planarity testing. SIAM J. Comput. **25**(5), 956–997 (1996). https://doi.org/10.1137/S0097539794280736
2. Hong, S., Nguyen, Q., Meidiana, A., Li, J., Eades, P.: Bc tree based proxy graphs for visualization of big graphs. In: IEEE PacificVis 2018, pp. 11–20 (2018)
3. Nguyen, Q.H., Hong, S., Eades, P., Meidiana, A.: Proxy graph: visual quality metrics of big graph sampling. IEEE Trans. Vis. Comput. Graph. **23**(6), 1600–1611 (2017). https://doi.org/10.1109/TVCG.2017.2674999
4. Wu, Y., Cao, N., Archambault, D., Shen, Q., Qu, H., Cui, W.: Evaluation of graph sampling: a visualization perspective. IEEE Trans. Vis. Comput. Graph. **23**(1), 401–410 (2017)
5. Zhang, F., Zhang, S., Wong, P.C., Swan II, J.E., Jankun-Kelly, T.: A visual and statistical benchmark for graph sampling methods. In: Exploring Graphs at Scale (EGAS) Workshop, IEEE VIS 2015, October 2015

Taming the Knight's Tour: Minimizing Turns and Crossings

Juan Jose Besa, Timothy Johnson, Nil Mamano$^{(\boxtimes)}$, and Martha C. Osegueda

Department of Computer Science, University of California, Irvine, CA, USA
{jjbesavi,tujohnso,nmamano,mosegued}@uci.edu

1 Introduction

The classic Knight's Tour Problem asks for a sequence of knight moves in an $n \times n$ chess board that allows the knight to visit every square exactly once and return to the starting position. There is a long history of algorithms for producing knight tours (e.g., see [1, 3]). However, most of them produce complex tours. We consider the problem of finding knight's tours minimizing two metrics of complexity: the number of turns and the number of crossings. A *turn* is when two consecutive knight moves in the tour go in different directions (i.e., when the three cells involved are not collinear); a *crossing* is when the line segment connecting the cells of two knight moves intersect. To the best of our knowledge, these metrics are new in this context, but they are often studied in geometric contexts and, in the case of crossings, in graph drawing. (However, people have looked at the related problem of the longest knight path without any crossings [2].)

We use a novel approach to produce a family of knight's tours for $n \times n$ boards (where n is even, since, otherwise, a tour does not exist) with a near-optimal number of turns and crossings (see Results). Our approach also has several other good qualities: **(i)** the knight move at any given cell can be determined in constant time without constructing the tour explicitly; **(ii)** it can be generalized to rectangular boards (as long as both sides do not have odd length, in which case a tour does not exist), and **(iii)** it the tours are easy to visualize and construct without the need of computers or calculations.

Results. Our tours have $10.75n + O(1)$ turns and $13n + O(1)$ crossings, where the constant factors are quite small but vary slightly depending on n mod 8. For instance, if $n \equiv 2$ mod 8, the constants are 40.5 and 91, respectively. Since a knight must turn at any cell next to the edge of the board, *any* knight tour must have at least $4n$ turns.[1] Similarly, by examining the ways to cover the first two rows (or columns) on each side of the board, we can show that there is at least one crossing per column per side. Therefore, the number of crossings must be also be at least $4n$. Therefore (for sufficiently large values of n) our tours are within a factor of ≈ 2.7 and 3.25 of the minimum.

---

[1] This lower bound can be improved to $4.25n$ by observing that cells close to the center of the board must be part of a sequence of moves that must contain a turn that is not in one of the cells along the boundary. We omit this slight improvement for brevity.

© Springer Nature Switzerland AG 2018
T. Biedl and A. Kerren (Eds.): GD 2018, LNCS 11282, pp. 661–663, 2018.
https://doi.org/10.1007/978-3-030-04414-5

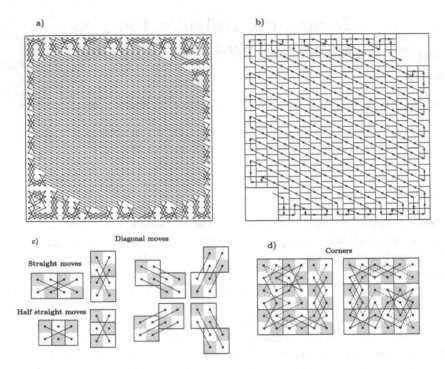

Fig. 1. (a) Knight's tour on a 34×34 chess board with 320 turns and 429 crossings (Other dimensions at [4]). (b) Corresponding sequence of "formation moves". (c) "formation moves" for a group of 2×2 knights. In half straight moves only two of the four knights move maintaining the 2×2 formation (d) corner constructions that pair up four knight paths into two connected paths (solid and dashed).

2   Construction

We begin by covering the board (except two corners) with four knights arranged in a 2×2 formation. By using the "formation moves" depicted in Fig. 1c, they can cover the board while remaining in formation (see Fig. 1b). The main idea is to move in zig-zag along diagonals, because diagonal moves do not create turns or crossings. A little care must be put when selecting the moves along the bottom and top edges of the board to make sure that every cell is visited; nonetheless, it is not hard to do so. The issue of how to transform the four knights into a single knight's tour is resolved by using a special construction at the bottom-left and top-right corners of the board (see Fig. 1d). This construction pairs up the four knights into two connected paths at each of these two corners. This results in either a valid knight's tour or two disjoint cycles. However, note that there are two alternative corners which pair up different knights; therefore, it is always possible to set the corners to make a valid knight tour.

References

1. Ball, W.W.R.: Mathematical Recreations and Essays. Macmillan (1914)
2. Knuth, D.E.: Selected Papers on Fun and Games. In: CSLI Lecture Notes Series (2011)
3. Parberry, I.: An efficient algorithm for the Knight's tour problem. Discrete Appl. Math. **73**(3), 251–260 (1997)
4. https://www.ics.uci.edu/~nmamano/knightstour.html

Author Index

Printed in the United States
By Bookmasters